Human Viruses: Diseases, Treatments and Vaccines

Shamim I. Ahmad
Editor

Human Viruses: Diseases, Treatments and Vaccines

The New Insights

Springer

Editor
Shamim I. Ahmad
Wollaton, Nottingham, UK

ISBN 978-3-030-71164-1 ISBN 978-3-030-71165-8 (eBook)
https://doi.org/10.1007/978-3-030-71165-8

This Springer imprint is published by the registered company Springer Nature Switzerland AG
The registered company address is: Gewerbestrasse 11, 6330 Cham, Switzerland

Shamim I. Ahmad
Editor

Human Viruses: Diseases, Treatments and Vaccines

The New Insights

Springer

Editor
Shamim I. Ahmad
Wollaton, Nottingham, UK

ISBN 978-3-030-71164-1 ISBN 978-3-030-71165-8 (eBook)
https://doi.org/10.1007/978-3-030-71165-8

This Springer imprint is published by the registered company Springer Nature Switzerland AG
The registered company address is: Gewerbestrasse 11, 6330 Cham, Switzerland

The editor dedicates this book to his late father and mother, Abdul Nasir and Anjuman Ara, who played important roles to bring him to this stage of academic achievements with their esteemed love, sound care and sacrifice. Dedication also goes to his wife, Riasat Jan, for her patience and persistent encouragement to produce this book, as well as to their children, Farhin, Mahrin, Tamsin, Alisha and Arsalan, especially the latter two for providing great pleasure with their innocent interruptions, leading to the energy of the editor to be revitalized. Next, his best wishes go to the ailing patients, suffering from various viral diseases especially COVID-19, for their recovery from the ailing diseases and the rest of the life to run smoothly. Finally, his dedication goes to his loving and caring Professors, Dr. Fumio Hanaoka continuing his work in Japan, and Dr. Robert H. Pritchard and Dr. Abe Eisenstark, who passed away now.

Acknowledgements

The editor cordially acknowledges the authors of various chapters to contribute their works, employing their in-depth knowledge, high skills and professional presentations. Without their input, it would not have been possible to bring out this book on this highly topical and important subject. He would also like to acknowledge Melissa Morton, Karthikeyan Durairaj, Ritu Chandwani and Leo Johnson of Springer publication for their hard work, friendly approach, patience and highly professional handling of this project. Final acknowledgement goes to my son-in-law Mir Azam Ali Khan for his valuable assistance with IT that helped to bring this book to the stage of publication. Without this help, it would have been extremely difficult to publish this work.

Viral Pandemics: COVID-19 is Only a Trailer—We Ought Not to Wait for the Whole Film to Run

An Appeal to World Leaders, Health Officers, Medical Organizations, and, Corporations, and those trying to manage the fear of pandemics

Past Pandemics and Epidemics

Before November 2019, most people were unaware what a pandemic is or even what viruses are! But people now, even in the most remote areas of the world, are aware at least of the terms virus and COVID-19. Also, although they cannot see this germ, they have been made aware of the consequences of infection, including high numbers of deaths; and in certain cases, people consider it to be some kind of horror coming from the sky. The sequel to this horror film is unclear—we neither know when it will end nor if and when it will come back. Let us look at how the horror film ran in the past, and what may be expected in any rerun.

Occurrence of pandemics is not new. Indeed, history reports that as far back as 3000 BC, pandemics occurred, but here only those are mentioned which occurred in the last two millennia: *The Black Death*: 1346–1353; *Cocoliztli epidemics*: 1545–1548; *16th Century American Plague; The Great Plague of London*: 1665–1666; *The Great Plague of Marseille*: 1720–1723; *The Russian Plague*: 1770–1772; *Philadelphia Yellow Fever Epidemic of 1793*; *Russian Flu Pandemic, 1889–1890*; *Spanish Flu* of 1918–1920; and *Asian Flu* of 1957–1958.

Pandemics and Epidemics Which Continue Sporadically:

Dengue virus originated 800 years back, Measles virus may have been originated around eleventh to twelfth century, American Polio epidemic in 1916, H1N1 Swine flu pandemic in 2009, the Avian Flu in 2018, West African Ebola pandemic in 1976, Chikungunya virus in 2006, Zika virus epidemic of South East Asia in 2015, Crimean-Congo hemorrhagic fever virus in 2017, AIDS pandemic in 1981, and SARS and MERS in 2003 and 2012, respectively, and now SARS-CoV-2.

Sporadic human viruses lurking around: Please see Appendix

Zoonotic Viruses and Their Reservoirs:
Zoonotic viruses are those organisms which mostly reside in vertebrate animals, birds, bats and insects as their reservoirs and then can jump to humans. The diseases they cause are called zoonoses.

A PubMed search (27 January 2021) for zoonotic viruses showed that there are 8329 research papers on this subject and highlighted the presence of 224 human viruses and their associated traits. Out of a number of reservoirs of human viruses reported, the most common primary viral hosts are humans, ducks, geese, and shorebirds, bats, pigs, dogs, seals, horses and mosquitoes. Other reservoirs regionally found are bamboo bats, Belgian hedgehogs, rhesus monkey, honey and bumble bees, dromedary camel (MERS), palm civets, small and big cats, minks and ferrets.

Current Crises with SARS-CoV-2 and Its Variants:
In early 2021 when the SARS-CoV-2 pandemic was slowly and steadily coming under control, it was recorded that three significant variants of this virus have emerged causing additional issues around the control of the pandemic. These specific variants are more contagious and more virulent than the Wuhan strains. These variants have emerged due to new mutations in their genomes. Questions have been raised as to whether the vaccines currently in use can also confer protection against these variants. It has been seen for some other viral infections that high rates of mutagenesis, giving rise to a large number of variants, make it difficult to produce vaccines which can control all the different strains, e.g. avian viruses of the H1N1 series cause influenza at regular intervals (for details of viral variations, see chapter "SARS-CoV-2 Mutations: An Insight" of this book).

Consequences of the COVID-19 Pandemic:
It was January 2021 when the final material for this book was going to press and the world is still going through the global calamity of COVID-19, which started in December 2019. The whole world is now experiencing very high numbers of infections and deaths. Turmoil includes economic damage due to a large number of factories, industries and shops closing, leading to increased unemployment and reductions in industrial productivity and agriculture. Other human sufferings include increased depression and mental health issues and increased domestic violence. More worryingly, it is not known how long this will last and how many more human lives will perish. The prediction is that even when the current wave of infections is over, the virus may re-appear either in other mutated forms or by finding another suitable environment/host.

Soon after the discovery of COVID-19, strenuous searches started to find effective treatments, and hundreds of antiviral and other putative drugs, including Chinese herbal medicines, were tested. As far as my searches have found, no proven effective therapies for this virus currently exist. Some drugs have been used in clinical trials, but there is no satisfactory data showing that these drugs, either alone or in combination, are effective. Dexamethasone, which reduces mortality by

25% in severe cases, has been in use as an anti-inflammatory drug. Other drugs with limited success may be remdesivir, tocilizumab, chloroquine phosphate, hydroxychloroquine sulphate and monoclonal antibodies. From the start of the pandemic, a number of companies and universities entered into the race to produce a variety of vaccines, and today a number of companies in different countries including the USA, the UK, Germany, France, India, Russia and China have produced and released vaccines with the hope that they will control infections and the resulting disease (for further details, please see chapter "Treatment of COVID-19 by Combination Therapy with 5-fluorouracil, Ribonucleosides and Ribavirin—A Modified Strategy" of this book).

Reasons for Rapid Global Spread of Pathogenic Viruses:
The present SARS-CoV-2, causing COVID-19, started in Wuhan, China. The speculation is that in Wuhan, the virus may have spread from bats or other livestock being sold in market places and infected humans. Subsequently, the virus rapidly spread to other cities in China and to neighbouring countries such as South Korea and Japan. Then fairly rapidly, it spread to almost 160 counties in the world. Spreading of the virus can be imagined to have occurred from infected persons travelling from China to other countries and within the country by travelling from one city to another. Spreading of the virus is exponential, which in its simplest form means one person infecting a second person and the 2 of them to a further 2 making 4 and then 8 making 16 and then 32 and on and on. Imagine when it reaches say 50,000, the next infection is 100,000. It is not difficult to imagine that the infection of one person after 20 rounds of infection can reach over a million people. In fact, the natural infection rate for Sars-CoV-2 is around 3, making the spread even faster.

The main reason for the viral spread is physical person-to-person contact: either within families or via interfamily contacts. It has become apparent that as with other diseases, transmission can occur through contact between an asymptomatic person (unknowingly carrying the infection) and a healthy person.

By March 2020, guidance was issued by governments to people to use face masks, keep a distance of at least 2 meters between individuals, and to wash hands more regularly and thoroughly especially when entering home from outside trips. Unfortunately, the guidance has not been taken, by some, as seriously as required, resulting in a limited downturn in the number of infections. Hence, more stringent actions have been taken including the closure of non-essentials shops, restaurants etc. and the requirement that people stay home as much as possible.

Facts About Human Viruses Affecting Our Life, Unknown or Barely Known to People
Viruses live all around us, in the air, attached to living and non-living things, being viable out of hosts for short or long times, and can enter living bodies. In the body, viruses enter cells and produce their progeny and then release them from the cell to invade other healthy cells. If this process continues especially when the immune system is compromised, it can be lethal. Another important feature of human viruses is that almost all are zoonotic; they can live stably inside their reservoir hosts without causing diseases. But at times, their genomes change by mutation and

when they enter into humans, they can cause catastrophic results. It is also considered that many more viruses have evolved over millions of years, and some may surface as and when they get the opportunity. Recent examples are Avian influenza, Ebola, Dengue, AID, Zika and Chikungunya viruses. Viral impact is highly variable in that it can range from asymptomatic up to lethal.

Vaccines are Used Prophylactically to Prevent Disease Rather Than Treat it:
The fact is vaccines are not for use in the treatment of established diseases. They are given to the general public to safeguard themselves by activating the immune system: vaccine components acting as antigens. Vaccines are prepared in a number of different ways: as attenuated whole virus, inactivated virus or engineered viruses. The third method is prevailing now which involves the use of viral genetic material: specific sequences of DNA or RNA are used, which drive the production of the specific protein which acts as the antigen (such as the recent use of the spike protein of SARS-CoV-2). To protect the sequence from degeneration and provide efficient packaging, the sequence requires a vehicle. A vehicle developed for this purpose is the lipid nanoparticle. This nanoformulation enables the safe delivery of the genetic material, biodistribution, intracellular localization and release. The antigen induces the immune system, which incapacitates the invading live viruses (for more about vaccines, please see chapter "Global Polio Eradication: Progress and Challenges" of this book).

Difficulties in Generating the Antiviral Drugs:
Although a large number of antiviral drugs have either been isolated from natural sources or synthesized in laboratories, only a handful of them have been found to be effective on COVID-19. In contrast to antibiotics which are commonly available to kill bacterial infections, there are only a handful of antiviral agents available which can be used to treat more than one virus. One reason for this failure is that unlike other microbial infections, in which the infective agent can live and grow outside the cells in the body and hence their drugs can find the targets more easily, most viruses live silently inside host cells. This means that to send drugs to the targets, it is important that the drugs be produced which can easily enter the cell without harming it which is not that easy, or only treat the symptoms rather than the cause of infection.

The Appeal

As SARS-CoV-2 may not be the last pandemic, the emergence and re-emergence of pandemics by SARS coronaviruses, Ebola virus, Zika virus and zoonotic avian influenza viruses is highly probable. But when and how and to what extent they will appear and interfere with normal global human life is everybody's guess. According to an estimate, 224 types and traits of human viruses are present globally.

So what is the future of re-emergence of epidemics and pandemics? Are we ready for the next pandemic? A simple answer is "NO" and so should we not learn lessons from the present pandemic and prepare to combat future attacks?

Now is the time to prepare and coordinate local, national and global efforts and determine what should be done in the preparation.

- We should take valuable lessons from the present COVID-19 and use them if and when a new pandemic emerges.
- A centralized organization either independent or a subsection of WHO should be established which can take an oversight of epidemics and pandemics strictly at the global levels.
- More robust and unified policies should be made and introduced by independent or subsection organizations of WHO to combat them at the global level.
- Countries suspecting the start of a pandemic must send an alarm to the central organization (mentioned above), as soon as possible.
- More advanced safety measures must be developed to protect doctors, surgeons, nurses, health workers and caregivers in hospitals, nursing homes and old people homes, and ambulance men and the police on the street.
- We need an infrastructure to provide rapid testing, which exists and can be tweaked for each virus. This will be more valuable for travellers.
- More stringent travel restrictions must be imposed at early stages in potential pandemic events, and travellers should be tested more stringently as soon as possible.
- Especially for the underdeveloped countries, new techniques must be developed and used to produce and distribute vaccines to combat viruses causing epidemics and pandemics.
- More government funded research must be carried out to develop antiviral drugs.
- Safe handling and consumption must be imposed on birds, bats and animals used by viruses as their reservoirs.

I sincerely hope that at least the scientific community all over the world will support my presentation and the proposal to save millions of lives expected to be at risk in future epidemics and pandemics, with various kinds of viruses expected to emerge in future.

February 2021 Shamim I. Ahmad

PS: Excerpts from the speech of Mr. Bill Gates "The Next Outbreaks? We are not ready"—delivered in 2016 and presented on YouTube.

With apology if any part of the presentation improperly understood and presented.

During the 1918 flu pandemic, 3 million people died and it was predicted that any further outbreak may be more devastating. Now, we are living in a much more technologically advanced age and can use all our modern knowledge, facilities and technology to fight emerging diseases.

When our army goes to war, it uses all the technological advancements, equipments and facilities at its disposal to win the war. We should follow a similar approach to fight the pandemic war. For this:

1. Strong health systems must be developed.
2. Medical expertise needs to be pooled.
3. There needs to be a step up in terms of specific diagnostics, medical research and development.
4. Vaccine must be developed and delivered earlier, especially in support of poor countries.
5. Germ-game simulations should be developed to practice scenarios in a similar way to that used in war games.

Editor's salutation: A big salute to Mr. Bill Gates with my hat down in front of him. Also I salute all the doctors, surgeons, nurses, health workers and caregivers both at hospitals, nursing homes and old people homes, ambulance men and the police for their hard work in the most dangerous COVID-19 environment risking their lives, and send condolences to those who have lost their loving and caring relations and friends.

Appendix

Epidemics and Pandemics by Zoonotic Human Viruses

Avian Influenza Viruses

Influenza A virus (IAV): It is zoonotic and infects various living organisms including humans. By mutational changes of its genomes or by its reassembling during co-infection with different IAV strains, it can change its host and properties. The past recorded influenza virus pandemics, linked to IAV subtypes, are H2 (1889-91), H3 (1990), H1 (1918-20), H2 (1957-58) and H3 (1968). Regarding pathogenecity, there are the low pathogenic avian influenza viruses (LPAIVs) which are normally asymptomatic in their natural avian hosts and from this they become highly pathogenic (HPAIVs) with devastating consequences.

H1N1: This strain appeared during 2018–2019 which marks the centennial of the Spanish flu which killed approximately 50 million people. This strain can be considered the parent strain of all the variants of the viruses presented below. Reasons for the variations and pathogenecity presented are the regular sequential induction of mutations in their genome, which can evade the vaccines used for their prevention. Although good achievements have been made for diagnosis, prevention and treatment, the possibility of a severe pandemic to re-emerge remains possible. H1N1, also known as the Swine flu virus, causes upper, and potentially, lower respiratory tract infections in the host it infects.

H2N2: The Asian flu pandemic of 1957 was generated by this virus and circulated among humans from 1957 to 1968 before it was replaced by H3N2 subtype.

H3N2: Then in 1968, there was a 50th A(H3N2) pandemic. Although the pandemic morbidity and mortality were much lower in 1968 than in 1918, influenza A (H3N2) virus infection was the leading cause of seasonal influenza illness and death over the last 50 years. Clinically, influenza is characterized by acute onset fever, chills, runny nose, cough, sore throat, headache and myalgia. Mostly, illness lasts for 3–4 days with a resolution in 7–10 days.

H3N3: Limited information about this virus is that it was isolated from migratory birds in South Korea in November 2016 and in Canadian pigs in 2001 and 2002.

H3N8: This virus is universally present in dogs, cats, horses, donkeys and Bacterian camel in different countries ranging from China (2007, 2014 and 2015) to Malaysia (2015) and Chile (2018). This virus is included in this appendix for the possibility of its spillover in future.

H5 avian and human influenza viruses: This virus was first identified in 2013 as low pathogenic avian influenza virus (LPAI) and in 2017, its derivative LPAIH7N9 became highly pathogenic and existed for two decades.

H5N1: The highly pathogenic virus was first reported in 1997 which had moved from birds to humans and from Eurasia to Asia to going around the world. Since then, the outbreaks of this highly pathogenic subtype have affected a wide variety of mammals in addition to poultry and wild birds.

H6: This is the subtype of IAVs, is commonly detected in wild birds and domestic poultry and can infect humans.

H6N6: In 2010, the H6N6 virus emerged in Southern China, and since then, it has caused sporadic infections among swine. H6N6 swine IAV (SIV) currently poses a moderate risk to public health, but its evolution and spread should be closely monitored.

H7: This variant has its origin from H5 A/goose/Guangdong/1/1996 (GsGd) and Asian H7N9 viruses and has caused several hundred human infections with high mortality rates in the Netherlands, Italy, Canada, the United States and the United Kingdom. Human illness ranges from conjunctivitis to mild upper respiratory illness and pneumonia. While these viruses have not spread beyond infected individuals, if they evolve to new strains, they can transmit efficiently from person-to-person and can initiate a pandemic.

H7N3: This moved from birds to humans in Canada in 2013, and China is undergoing a recent outbreak of a novel H7N9 avian influenza virus (nH7N9) infection. Evolutionary analysis shows that human H7N9 influenza viruses originated from the H9N2, H7N3 and H11N9 avian viruses, and that it is as a novel reassortment influenza virus.

H7N7: This poultry epidemic pathogenic virus moved from birds to humans in the Netherlands in 2003 is now thought to be extinct.

H7N9: This caused high mortality. Since its identification as novel reassortant avian influenza A (H7N9) virus in China in 2013, until 2017, this avian origin as an Asian virus has caused five epidemic waves leading to a total of 1,552 human infections, with a fatality rate of about 40%. In the spring of 2017, it re-immerged with few human infections. It is included in the index in case its variation can cause the next pandemic.

H9N2: This is a subtype of the influenza A virus that originated in China through interspecies transmission that started in poultry. Its receptor specificity and human-to-human transfer was first recorded in 2009 in Finland. This influenza A virus poses a persistent threat to human health, as shown above with H5N1, H7N9 and (see below) the recent surge of H9N2 infections. Recent increase in human infection with the H9N2 virus in China highlights the necessity to closely monitor the interspecies transmission events.

H10N8: This is an avian influenza virus with human infection and mortality; since 2013, it has caused several influenza-related events; the spread of highly pathogenic avian influenza of H10N8 and others, H5N6 zoonotic infections, the ongoing H7N9 infections in China and the continued zoonosis of H5N1 viruses in parts of Asia and the Middle East remains alarming.

H11N9: Although the pathogenecity of this virus is confined to avians such as wild birds and domestic ducks, it is important to note that the N9 gene of H7N9, which has the ability to cause death in humans, originated from an H11N9 influenza strain. In future surveillances, emphasis should be given on reassortments of H11N9 viruses affecting public health.

Rhinoviruses: These, present ubiquitously as respiratory tract pathogens, have been neglected in the past because they were perceived as a virus only capable of causing mild common cold. The virus was first discovered in the 1950s; their epidemiological studies have shown that these viruses are commonly present in adults and children with upper or lower respiratory tract infections. The virus can cause upper respiratory tract illness ("the common cold") and lower respiratory tract illness (pneumonia). Other chronic respiratory diseases include asthma and bronchitis.

Bats and Flying Foxes Viruses

Ebola virus: This virus (EBOV) of the family Filoviridae discovered in 1976 causes hemorrhagic fever that can lead to death in a few days. The first major outbreak in West Africa appeared with patients who suffered from chills, myalgia, fever, diarrhoea, vomiting, and other hemorrhagic complications and failure of multiple organs. The virus spreads by infected persons passing body fluids such as blood, saliva, urine, seminal fluid to uninfected person, and also by contact with contaminated surfaces. This virus has spawned several epidemics during the past 38 years during this period spread from Africa to other continents, Guinea, Liberia, Sierra Leone and Nigeria becoming a pandemic (details can be found in chapters "Ebola Virus: Overview, Genome Analysis and Its Antagonists" and "Global Outbreaks of Ebola Virus Disease and Its Preventive Strategies" of this book).

Nipah and Hendra viruses: Nipah and Hendra viruses were classified in the new genus Henipavirus because of their uncommon features among Paramyxoviridae. These are equine based and were first identified in mid-1990s in Australia and Malaysia, causing epidemics with high mortality rate in affected animals and humans. Then in 1994 and in 2010, the hendra virus infection of horses spread from the viral reservoir in Australian mainland flying foxes, and the transmission occurred to people by exposure from affected horses. Pteropid bats (flying foxes) are the natural host of hendra virus causing severe acute respiratory disease, encephalitis and highly fatal infection in horses and humans.

Marburg viruses (MARV): This is a zoonotic and highly pathogenic and is associated with severe disease and mortality rates as high as 90%. Outbreaks of MARV are sporadic and deadly through hemorrhagic fever. Its genus includes two viruses: MARV and Ravn. The first case of the Marburg virus disease was reported in 1967, but in 2002 the outbreaks spread to Uganda, Angola, Congo, Kenya and in the USA. Bats have been considered as one of the most important reservoirs for this virus.

Middle East respiratory syndrome (MERS) and Severe acute respiratory syndrome (SARS) viruses: Human coronaviruses (HCoVs) in the past although considered to be relatively harmless respiratory pathogens, after the outbreak of SARS and MERS, have received worldwide attention as important pathogens in respiratory tract infection. SARS-CoV and the MERS-CoV outbreaks started in China and Saudi Arabia in 2003 and 2012, respectively. Since then they have caused multiple major public health events that resulted in epidemics. Although both viruses likely originated in bats, their spike protein (S) is slightly different. MERS is mostly known as a lower respiratory tract (LRT) disease involving fever, cough, breathing difficulties and pneumonia that may progress to acute respiratory distress syndrome and multiorgan failure.

SARS-CoV-2/COVID-19: The outbreak in December 2019 has a significant impact on people's health and lives. Like SARS and MERS, this virus can cause viral pneumonia, however, evidence exist that this virus may also affect the central nervous system. Given the outbreak of COVID-19, it seems necessary to perform investigations on the possible neurological complications in patients who suffered from this virus (for further details, see chapter "Treatment of COVID-19 by Combination Therapy with 5-fluorouracil, Ribonucleosides and Ribavirin—A Modified Strategy" of this book).

Nipah virus (NIV): This is an emerging bat-borne pathogen, related to the Hendra virus, first emerged in Malaysia in 1998. Clinically, it can stay asymptomatic but can lead to fatal encephalitis. After its outbreak in Malaysia, subsequently, it moved to other parts of South and South East Asia, Bangladesh and India. It causes severe neurological and respiratory diseases and encephalitis which is highly lethal.

Other Mammalian Viruses

Rabies virus: This is zoonotic and a member of Lyssavirus, infecting humans with a high fatality rate. Mostly, it is transmitted via the saliva of an infected animal especially after dog bites which accounts for >99% of human cases especially children. Pathologically, the virus first infects peripheral motor neurons, and symptoms occur after the virus reaches the central nervous system. Once the clinical disease develops, it is almost certainly fatal. Although rabies has been the subject of large-scale public health interventions chiefly through vaccination efforts, the disease continues to take the lives of about 40,000–70,000 people per year; roughly, 40% of them are children. Most of these deaths occur in poor countries.

Hepatitis A virus: Acute hepatitis due to the hepatitis A virus usually is a short, benign and self-limited disease causing non-chronic hepatitis. Hepatitis A virus is one of the most common infectious virus prevailing worldwide. The virus is transmitted by faecal-orally, resulting in symptoms ranging from asymptomatic infection to fulminant hepatitis, often causing jaundice and mostly remains non-lethal.

Hepatitis B virus: This is a human blood-borne virus which infects hepatocytes and causes significant liver diseases. An estimated 400 million chronic carriers of HBV exist worldwide. This virus is associated with human delta virus (HDV) and 15 to 20 million having serological evidence of exposure to HDV. High rates of endemicity are present in Central and Northern Africa, the Amazon Basin, Eastern Europe and the Mediterranean, the Middle East and parts of Asia.

Hepatitis C virus: This virus, first emerged in 1970, causes hepatitis, liver cirrhosis and hepatocellular carcinoma. It infects around 170–200 million people globally, especially in developing countries. Being a blood-borne virus, its prevalence progressed by parenteral transmission routes associated with medical treatments, immunization, blood transfusion and by injecting drug use. Suspected source of this virus remains non-human primates harbouring viruses and is related to HCV with cross-species transmission of variants. Seven main genotypes have been identified in humans.

Hepatitis D virus and Hepatitis Delta virus: Hepatitis D causes hepatitis disease, was first reported in 1977, and now is estimated that 15–20 million people are infected worldwide. Infection in liver leads to liver cirrhosis and possible liver cancer. To complete the life cycle in hepatocytes, this RNA virus requires hepatitis B surface antigen (HBsAg). In industrialized countries, this virus seems to have diminished since the last two decades. However, it remains a medical issue for injecting drug users, as well as immigrants travelling from infected area to non-infected areas.

Hepatitis E virus: This zoonotic virus infection can lead to acute and chronic hepatitis leading to acute liver failure as well as neurological and renal diseases. Infections occur through transmission or contaminated blood products. At least 20 million people worldwide suffer from HEV infections, with more than 3 million symptomatic cases and about 60,000 fatalities. Hepatitis E is generally self-limiting, with fatality a rate of 0.5–3% in young adults.

Oropouche virus: This zoonotic virus circulates mainly in South and Central America, and is responsible to induce oropouche fever. During the last 60 years, more than 30 epidimecs and over half a million clinical cases are attributed to OROV infection in Brazil, Peru, Panama, and Trinidad and Tobago.

Measles virus: Measles is a highly contagious disease that results from infection with measles virus, and although vaccines are given, it is estimated that about 100,000 deaths occur every year due to measles. The virus is transmitted by the respiratory route and illness begins with fever, cough and conjunctivitis followed by a characteristic rash. Measles virus of genus of Morbillivirus belongs to the family Paramyxoviridae. Persistent infection of this virus can lead to affect the central nervous system including developing encephalitis and damage to certain other inner organs. The infection is self-limiting interacted by virus-specific immune response and leads to lifelong immunity of the sufferers. The viral genomic RNA is single-stranded, non-segmented, and encodes six major structural proteins.

Mumps virus: This ia a member of the paramyxoviridae, enveloped, non-segmented, negative-sense RNA virus mostly infecting children. The disease symptoms are painful inflammation such as parotitis and orchitis. It is also highly neurotropic, affecting the central nervous system in approximately half of the cases. Due to the availability of MMR vaccines, mumps virus infection is now much rarer at the global level.

Parainfluenza viruses and Respiratory syncytial virus (RSV): Human parainfluenza viruses (HPIVs) are single-stranded, enveloped RNA viruses of the Paramyoviridaie family. Their four serotypes cause respiratory illnesses in adults and mostly in children causing otitis, pharyngitis, conjunctivitis, croup, tracheobronchitis and pneumonia. Uncommonly, it can cause acute bronchiolitis, mucosal and sub-mucosal oedema, apnoea, bradycardia, parotitis, and respiratory distress syndrome and rarely disseminated infection. Worldwide, the disease is the leading causes of death of infants of less than 1 year of age, second only to malaria.

Rubella virus: Also known as German measles, its infection and disease are often mild. Body rash with swollen lymph nodes and fever normally start 2 weeks after the infection and may last for 3 days. Although its infection causes mild symptoms or even no symptoms, it can cause serious problems for unborn babies if the mother is infected during pregnancy.

Mosquito-Borne Viruses

Dengue virus: A mosquito-borne flavivirus and a member of arboviruses, this is one of the oldest discovered human viruses. Dengue fever disease was first suspected by outbreaks in 1635 in Martinique and Guadeloupe, and in 1699 in Panama. Viral transmission by *Aedes aegypti* mosquitoes was recorded in the early nineteenth century. This virus most likely is responsible for higher morbidity and mortality than any other Arbovirus illness in humans. Its recorded spread spans 40% of the world's population, and about 2.5 billion people live in Dengue-infested areas. According to WHO, around 400 million people are infected annually by any of the 4 dengue virus serotypes resulting in diverse range of symptoms, from mild undifferentiated fever to life-threatening hemorrhagic fever and shock (for details, please see chapters "Dengue Preventive Strategies Through Entomological Control, Vaccination and Biotechnology", "Dengue Fever Epidemic in Pakistan and Its Control Measures: Where Are We Moving?", "Employing Geographic Information System and Spatiotemporal Analysis of Dengue Outbreaks in a Metropolitan Area in Pakistan", "Identification of Dengue NS2B-NS3 Protease Inhibitors Through High-Throughput Virtual Screening—Impacts on Drug Development Against the Dengue Virus" of this book).

West Nile fever virus: This zoonotic arbovirus has been found in birds, mosquitoes of Culex genus and horses. Isolated in Uganda first in 1937, it spread globally up to North America in 1999 and is responsible for a high rate of morbidity and mortality. Blood transfusion, organ transplantation, and possibly pregnancy and breastfeeding also can transfer this virus from person to person. Clinical symptoms range from mild to asymptomatic pathogenesis to neuroinvasion.

Chikungunya virus: A member of arboviruses (CHIKV), this causes crippling musculoskeletal inflammatory disease in humans characterized by fever, polyarthralgia, myalgia, rash and headache. *Aedes aegypti* is the most common reservoir from where through bite the virus is transmitted to humans. Its long-term progress from infection is yet ill-defined but its prolonged debilitating arthralgia is well identified. Since 2004, CHIKV has spread on a global scale.

Zika virus: This is another mosquito-borne member of arboviruses and is closely related to dengue, yellow fever, West Nile, and Japanese encephalitis viruses. From Africa and Asia, its emergence in Brazil in 2015 peaked the spread throughout the Americas. Its infections are characterized by subclinical or mild influenza-like illness, but severe outcomes such as Guillain-Barre syndrome in adults and microcephaly in babies born to infected mothers have been recorded (for further information, see chapters "Zika Virus Disease: Progress and Prospects" and "The Medicinal Chemistry of Zika Virus" of this book).

Tahyna virus (TAHV): This virus endemic to Europe, Asia and Africa is responsible for California encephalitis. A mosquito-borne California encephalitis, viral sera analysed from wild boar, roe deer and red deer collected in Austria, Hungary and Romania, demonstrate that TAHV transmission to mammals is widespread in Europe, particularly in the wild boar population, but are no more a major health problem in Western Europe.

Inkoo and Chatanga viruses: These viruses of the Bunyavirus group, circulating in Finland, are mosquito-borne California serogroup orthobunyaviruses that have a high sero-prevalence among humans. In Northern Europe, the sero-prevalence against Inkoo virus (INKV) is high, 41% in Sweden and 51% in Finland. Studies indicate that INKV infection is mainly asymptomatic, but can cause mild encephalitis in humans.

Sindbis virus: Sindbis virus (SINV) is another mosquito-borne alphavirus that is locally amplified in a bird-mosquito enzootic cycle and distributed all over the Old World and Australia/Oceania. Its disease can range from mild illness to lethal encephalitis or severe polyarthritis.

Batai virus: Belonging to the Bunyavirus group, this is another mosquito-transmitted Orthobunyavirus. It was first detected in 2009 in Southwest Germany in anopheline and culicine mosquitoes. Little is known about its infectivity in humans as almost no systematic surveillance or infection studies have been carried out to date.

Lednice virus: The Bunyavirus group Lednice virus (LEDV) has been detected in *Culex modestus* mosquitoes in several European countries within the last six decades. Not much information is available about this virus.

Semliki Forest complex virus: Mosquito species, *Aedes camptorhynchus*, *Aedes vigilax and Culex annulirostris* are the prime vectors for this virus. Not much is known about this virus.

Tick-Borne Viruses

Powassan virus (POWV): This is a flavivirus discovered in Powassan, Ontario in 1958, which causes sporadic but severe cases of encephalitis in humans. Its study for infectivity in humans showed that each year in the USA it has steadily increased. The virus is first transmitted by *Ixodes scapularis*, *Ixodes cookei*, and several other Ixodes tick species to small- and medium-sized mammals and then from there to humans. The Powassan virus is a neurovirulent flavivirus consisting of two lineages causing meningoencephalitis.

Crimean-Congo hemorrhagic virus: This is a member of the Bunyaviridae family and Nairovirus genus associated with Hyalomma ticks. Its viral genome consists of 3 RNA segments of 12 kb (L), 6.8 kb (M) and 3 kb (S). Viral induced fever is widespread and infection is reported from many regions of Africa, the Middle East, Southern and Eastern Europe, India and Asia. Transfer of virus by tick bite or body fluids from an infected individual to another is the most common route.

Clinical manifestations include asymptomatic or high fever, headache, malaise, arthralgia, myalgia, nausea, abdominal pain and non-bloody diarrhoea.

Other Less-Important Zoonotic Viruses

Enteric viral co-infections: This has been reported for a diverse group of etiological agents, including rotavirus, norovirus, astrovirus, adenovirus and enteroviruses. These pathogens are causative agents for acute gastroenteritis and diarrheal disease in immunocompromised individuals of all ages. Some of them have been highlighted below.

Human Astrovirus (HAtVs): This is a non-enveloped, positive-sense, single-stranded RNA virus discovered in 1975. It causes gastrointentinal illness (e.g. diarrhoea) in children, elderly and immunocompromised people. Astroviruses were classified into Mamastrovirus and Avastrovirus infecting mammalian and avian groups.

Rotavirus: Group A rotavirus (RVA) is another gastroenteritis disease and worldwide, its yearly prevalence is around 180 million, which remains the leading cause of disease in neonates, children and adults. The virus was first reported in turkey poults in the USA during 1977 and since then, RVs of group A (RVA), D (RVD), F (RVF) and G (RVG) have been identified around the globe. The rotavirus genome is composed of 11 gene segments of double-stranded RNA. This extremely contagious virus is transmitted by the faecal-oral route, and diarrheal infections remain the major cause of morbidity and mortality among young children.

Norovirus: This is a non-enveloped, positive-sense, single-stranded RNA viruses, discovered in 1968, which belongs to the Caliciviridae family. This worldwide prevailing virus remains a cause of acute gastrointestinal illness especially in children, and disease may remain associated with nausea, vomiting, and diarrhoea and significant dehydration. Viral transmission occurs by the faecal-oral route, person-to-person contact, or by the ingestion of contaminated food and water. About 699 million infections and over 200,000 deaths worldwide each year have been recorded.

Oncogenic viruses: As most oncogenic viruses are not associated with pandemics, only their names are given as reference: Epstein-Barr virus, Hepatitis B and C Viruses, HIV-1 and HIV-2 (human immunodeficiency virus 1 and 2), Human papilloma virus, Human T-lymphotropic virus type 1, Human herpesvirus type 8, and Merkel cell polyomavirus.

Editor's Worries About the Changing World

Viral epidemics and pandemics: Most virologists know that there exist more than two hundred human viruses; some are continuously persisting in the human body and others infect humans every now and then causing various diseases and death. Then there are zoonotic viruses present in animal bodies and ready to adapt to infect human. The fourth group of unknown types is present underneath the soil, where they stay for many years and when soil surface will get disturbed, they will emerge and attack humans either directly or through animals. This means the human species will always stay susceptible to viral attacks like epidemics or pandemics. We have to be fully prepared for that in advance as much as possible.

Environmental disturbance and destruction: Environmentalists every now and then have been making us aware over the last four to five decades that disturbances and destructions have been going on in our environment in several different ways: increased pollutions such as the overuse of plastic materials and disposal of unwanted chemicals and garbage, increased amount of carbon-associated exhaust fumes from aeroplanes and cars, deforestation and the reduction of trees, all of which lead to global warming, melting of glaciers, forest fires, increased number of floods, storms, typhoons and tornadoes; also possible is the elimination of small islands in future. With this awareness, I understand that the leaders of the world, although already putting efforts to minimize these damages, they join more hands, take more targeted, serious and rapid actions to eliminate them as much as possible.

Extinction of rare animal and plant species: Another important consequence of environmental damage is the possible elimination of many rare remaining species of animals and plants. Naturalists are keeping a close eye on this issue and including charity organizations who are working for it, but we have to support them as much as possible for the success of their hard work.

Chemical, biological, and nuclear war fares: Although several countries have been suspected to be hoarding chemical and biological weapons, human destruction by these agents are known to be not as destructive as the nuclear weapons. Moreover, it is possible that in future more and more countries will try to develop them, and hence the question is whether this will act as a deterrent or will it be used as a retaliation? It remains an open question. Let us try to stop further production of this weapon of mass destruction.

Continued wealth gap between rich and poor: It is well known that the wealth gap between the rich and the poor is increasing at a faster rate. A few years back, the global list issued was about millionaires, then soon the list of billionaires, and now it is not far that we will see the list of trillionaires. Its direct or indirect effects are increased urbanization, unemployment and increased migration of people from poor to developed countries. Also the increase in automation is resulting in the replacement of human workforce with higher unemployment. What could be done about it is another important question.

Public demands through demonstrations: The current surge of demonstrations, some peaceful some violent, although occurring mostly at local and some at national levels, most are due to the introduction of new unacceptable laws, unhappiness, disappointment, depression and significant reduction in the quality of life. There are cases where demonstrators are brutally treated with sporadic deaths. This issue should stay more effectively to be raised at the United Nation Security Council and dealt with appropriately through international diplomacy.

Appearance of new human diseases and increase in the number of existing ones: No doubt that a number of human conditions globally is on the rise including heart attacks, diabetes (the new type-2 diabetes in children which was rare a few years back), different kinds of cancers and allergic reactions. No doubt some of them are caused by environmental pollution and quality of food consumed. Newly appearing diseases are mostly caused by zoonotic viruses which are generated by certain unusual way of life in certain countries. A combined global effort is required to control and minimize these.

Biotechnology: Biotechnology has now reached a stage that if used irresponsibly, it can have enormous devastation on human life. Although strict rules and regulations have been produced and distributed to the academia and researchers, it may not be ruled out that someone some day can misuse it and then what? It remains a worrying prospect. We should stay vigilant about it.

Cybercrimes: This is something equivalent to virus pandemics, in that they both work at the global level; one infects and kills humans while the other infects computers and make them dysfunctional. This can affect industries like banking, telecom, health care, etc. where hackers not only can hack very huge amounts of money from banks but can also steal confidential/sensitive user information and then threaten and demand ransom from companies. For this reason, computer experts should continue finding certain kinds of deterrent for this problem.

February 2012 Shamim I. Ahmad

Contents

Measles Virus

Corona Virus and SARS-CoV-2

Antiviral Agents

Viral Vaccines

General Virology

Editor and Contributors

About the Editor

Shamim I. Ahmad after obtaining his Master's degree in Botany from Patna University, India, and his Ph.D. in Molecular Genetics from Leicester University, England, in 1972 he joined Nottingham Trent University as a grade-1 lecturer and was subsequently promoted to a Senior Lecturer post. Here, after serving for 35 years, he took an early retirement although yet delivering certain lectures. His research interest culminated in collaboration with the University of Missouri, Columbia, USA; University of Osaka, Japan; University of Copenhagen, Denmark; Institute of Genetics and Industrial Microorganisms, Moscow, Russia; Academic Institute of Photographic Chemistry, Beijing, China; University of Leiden, Holland; Imperial Chemical Industry, Bellingham, UK; CNRS (Centre National de la Recherche Scientifique) Gif-Sur-Yvette, France; and PHLS (Public Health Laboratory Services), Salisbury, England. These exclude a number of universities and organizations where he was invited to give lectures and courses. His active research expanded for over three decades and carried out in different areas of molecular biology/genetics including thymineless death in bacteria, genetic control of nucleotide catabolism, development of anti-AIDs drug, control of microbial infection of burns, phages of thermophilic bacteria, and microbial flora of Chernobyl after the accident at its nuclear power station. His all along research activities remained on DNA damage, repair and mutagenesis and

UV photoreaction of biological and non-biological compounds producing reactive oxygen species: implications on human health. His latest valued publication is “5-fluorouracil in combination with deoxyribonucleosides and deoxyribose as possible therapeutic options for the coronavirus, COVID-19 infection”. He is a strong believer that the phage therapy remains as an alternative to the future treatment of bacterial infections, especially those developing resistant to multiple antibiotics.

In 2003 he received a prestigious “Asian Jewel Award” in the UK for “Excellence in Education”. His long-time ambition to produce medical books started in 2007 and since then, he has published a number of books including Molecular Mechanisms of Fanconi Anemia; Molecular Mechanisms of Xeroderma Pigmentosum; Molecular Mechanisms of Cockayne Syndrome; Molecular Mechanisms of Ataxia Telangiectasia; Diseases of DNA Repair; Neurodegenerative diseases; Diabetes: An Old Disease A New Insight; Obesity: A Practical Guide; Thyroid: Basic Science and Clinical Practice; Ultraviolet Light in Human Health, Diseases and Environment; Diabetes: A Comprehensive Treatise for Patients and Caregivers; Reactive Oxygen Species in Biology and Human Health; Aging: Exploring a Complex Phenomenon; and Handbook of Mitochondrial Dysfunction. Springer Publication, Taylor and Francis, and CRC Press have published most books.

Contributors

Chirom Aarti Research Department of Plant Biology and Biotechnology, Loyola College, Chennai, TN, India

Idris Nasir Abdullahi Department of Medical Laboratory Science, Faculty of Allied Health Sciences, College of Medical Sciences, Ahmadu Bello University, Zaria, Nigeria

Sharafudeen Dahiru Abubakar Department of Medical Laboratory Science, Faculty of Allied Health Sciences, College of Medical Sciences, Ahmadu Bello University, Zaria, Nigeria

Hafeez Aderinsayo Adekola Department of Microbiology, Olabisi Onabanjo University, Ogun State, Nigeria

Nitish Aggarwal Yale New Haven Hospital, New Haven, CT, USA

Pushkar Aggarwal University of Cincinnati College of Medicine, Cincinnati, OH, USA

Abdurrahman El-fulaty Ahmad Department of Medical Laboratory Science, Faculty of Allied Health Sciences, College of Medical Sciences, Ahmadu Bello University, Zaria, Nigeria

Shamim I. Ahmad Former Senior Lecturer, School of Science and Technology, Nottingham Trent University, Nottingham, England

Ali Ahmed School of Pharmacy, Monash University, Bandar Sunway, Malaysia

Leticia Anderson CESMAC University Center, Maceió, Brazil

Abubakar Umar Anka Department of Medical Laboratory Science, Faculty of Allied Health Sciences, College of Medical Sciences, Ahmadu Bello University, Zaria, Nigeria

Pedro Gregório Vieira Aquino Environment Laboratory, Federal University of Agreste From Pernambuco, Garanhuns, Brazil

Jorge L. Arias-Arias Facultad de Microbiología, Centro de Investigación en Enfermedades Tropicales, Universidad de Costa Rica, San José, Costa Rica

Jeremy Baldwin Vaxine Pty Ltd, Warradale, Australia

Ananda Sankar Bandyopadhyay Bill and Melinda Gates Foundation, Seattle, WA, USA

Ênio José Bassi IMUNOREG – Immunoregulation Research Group, Laboratory of Research in Virology and Immunology, Institute of Biological Sciences and Health, Federal University of Alagoas, Campus A.C. Simões, Maceió, Brazil

Amit Bhatt Department of Zoology, Sri Venkateswara College, University of Delhi (South Campus), New Delhi, India

D. Katterine Bonilla-Aldana Semillero de Investigacion en Zoonosis (SIZOO), Biodiversity and Ecosystem Conservation Research Group (BIOECOS), Fundación Universitaria Autónoma de las Américas, Pereira, Risaralda, Colombia;
Latin American Network of COVID-19 Research, LANCOVID, Pereira, Risaralda, Colombia

Seyit Ali Büyüktuna Department of Infectious Disease, School of Medicine, Cumhuriyet University, Sivas, Turkey

Cristina Carolina Laboratorio de Fisiopatología de La Hipófisis, Centro de Investigaciones Básicas Y Aplicadas (CIBA), CIT NOBA, Universidad Nacional del Noroeste de La Pcia. de Bs. As. Consejo Nacional de Investigaciones Científicas Y Técnicas (UNNOBA-CONICET), Junín, Argentina

Hirak Chakraborty School of Chemistry, Sambalpur University, Burla, Odisha, India

Bhudev C. Das Stem Cell and Cancer Research Lab, Amity Institute of Molecular Medicine & Stem Cell Research (AIMMSCR), Amity University Uttar Pradesh, Noida, India

Edeildo Ferreira da Silva-Júnior Institute of Chemistry and Biotechnology, Federal University of Alagoas, Campus A.C. Simões, Maceió, Brazil;
Laboratory of Medicinal Chemistry, Pharmaceutical Sciences Institute, Federal University of Alagoas, Maceió, AL, Brazil

João Xavier de Araújo-Júnior Institute of Chemistry and Biotechnology, Federal University of Alagoas, Campus A.C. Simões, Maceió, Brazil

Elane Conceição dos Santos IMUNOREG – Immunoregulation Research Group, Laboratory of Research in Virology and Immunology, Institute of Biological Sciences and Health, Federal University of Alagoas, Campus A.C. Simões, Maceió, Brazil

Halef Okan Doğan Department of Biochemistry, School of Medicine, Cumhuriyet University, Sivas, Turkey

Daniel Esau Department of Medicine, The University of British Columbia, Vancouver, BC, Canada;
Royal Jubilee Hospital, Victoria, BC, Canada

Kei Fujiwara Department of Gastroenterology and Metabolism, Nagoya City University Graduate School of Medical Sciences, Nagoya, Mizuho, MizuhoAichi, Japan

Shezeen Abdul Gafoor Department of Audiology, All India Institute of Speech and Hearing, Manasagangothri, Naimisham Campus, Mysore, Karnataka, India

Usman Shah Gilani School of Biological Sciences, University of the Punjab, Lahore, Pakistan

Jean-Paul Gonzalez Department of Microbiology & Immunology, Division of Biomedical Graduate Research Organization, Georgetown University, School of Medicine, Washington, DC, USA;
Advisory Board, Centaurus Biotech LLC, Vienna, VA, USA

Todd J. Green Department of Microbiology, School of Medicine, University of Alabama at Birmingham, Birmingham, AL, USA

D. K. Gupta Felix Hospital, Noida, India

Shilpi Gupta Stem Cell and Cancer Research Lab, Amity Institute of Molecular Medicine & Stem Cell Research (AIMMSCR), Amity University Uttar Pradesh, Noida, India;
Molecular Genetics Laboratory, National Institute of Cancer Prevention and Research (ICMR), Noida, India;
Felix Hospital, Noida, India

Stefan Josef Hammerschmidt Institute of Pharmaceutical and Biomedical Science, Johannes Gutenberg-University Mainz, Mainz, Germany

Muhammad Farhan Ul Haque School of Biological Sciences, University of the Punjab, Lahore, Pakistan

Ryan D. Heath Fellow, Division of Gastroenterology and Hepatology, Department of Internal Medicine, University of Missouri, Columbia, MO, USA

Roger Hewson Department of Infection Biology, London School of Hygiene and Tropical Medicine, London, UK

Iqra Hussain School of Biological Sciences, University of the Punjab, Lahore, Pakistan

Sevic Ina Laboratorio de Microambiente Tumoral, Centro de Investigaciones Básicas Y Aplicadas (CIBA), CIT NOBA, Universidad Nacional del Noroeste de La Pcia. de Bs. As. Consejo Nacional de Investigaciones Científicas Y Técnicas (UNNOBA-CONICET), Junín, Argentina

Ahtisham Irshad Department of Geography, GC University Lahore, Lahore, Pakistan

Venkatesan Jayaprakash Department of Pharmaceutical Sciences & Technology, Birla Institute of Technology, Mesra, Ranchi, Jharkhand, India

Satya Ganesh Kakarla ENVIS Resource Partner on Climate Change & Public Health, Applied Biology Division, CSIR-Indian Institute of Chemical Technology (CSIR-IICT), Tarnaka, Hyderabad, Telangana, India;
Academy of Scientific & Innovative Research (AcSIR), Ghaziabad, Uttar Pradesh, India

Suha Abu Khalaf Department of Internal Medicine, University of Missouri-Columbia, Columbia, MO, USA

Gul Majid Khan Department of Pharmacy, Quaid I Azam University Islamabad, Islamabad, Pakistan

Ayesha Khanam Department of Computer Science, Jamia Millia Islamia, New Delhi, India

Ameer Khusro Research Department of Plant Biology and Biotechnology, Loyola College, Chennai, TN, India

Prabhat Kumar Stem Cell and Cancer Research Lab, Amity Institute of Molecular Medicine & Stem Cell Research (AIMMSCR), Amity University Uttar Pradesh, Noida, India

Swatantra Kumar Centre for Advanced Research (CFAR), Faculty of Medicine, King George's Medical University (KGMU), Lucknow, India

Alaniz Laura Laboratorio de Microambiente Tumoral, Centro de Investigaciones Básicas Y Aplicadas (CIBA), CIT NOBA, Universidad Nacional del Noroeste de La Pcia. de Bs. As. Consejo Nacional de Investigaciones Científicas Y Técnicas (UNNOBA-CONICET), Junín, Argentina

Xinyong Liu Department of Medicinal Chemistry, Key Laboratory of Chemical Biology, Ministry of Education, School of Pharmaceutical Sciences, Shandong University, Ji'nan, China;
Institute of Chemistry and Biotechnology, Federal University of Alagoas, Maceió, Brazil

Romano Lucia Laboratorio de Fisiopatología de La Hipófisis, Centro de Investigaciones Básicas Y Aplicadas (CIBA), CIT NOBA, Universidad Nacional del Noroeste de La Pcia. de Bs. As. Consejo Nacional de Investigaciones Científicas Y Técnicas (UNNOBA-CONICET), Junín, Argentina

Grace Ruth Macklin World Health Organization, Genève, Switzerland;
London School of Hygiene and Tropical Medicine, London, UK

Menahil Mahmood School of Biological Sciences, University of the Punjab, Lahore, Pakistan

Shakeel Mahmood Department of Geography, GC University Lahore, Lahore, Pakistan

Madiiha Bibi Mandary Centre for Virus and Vaccine Research, School of Medical and Life Sciences, Sunway University, Kuala Lumpur, Selangor, Malaysia

Bien-Aimé M. Mandja Faculty of Medicine, Microbiology Department, Kinshasa University, Kinshasa, Democratic Republic of Congo;
Chrono-Environment Laboratory, UMR 6249 CNRS, Bourgogne Franche-Comté, University Besançon, Besançon, France

Vimal K. Maurya Centre for Advanced Research (CFAR), Faculty of Medicine, King George's Medical University (KGMU), Lucknow, India

Hannah Maus Institute of Pharmaceutical and Biomedical Science, Johannes Gutenberg-University Mainz, Mainz, Germany

Geetanjali Meher School of Chemistry, Sambalpur University, Burla, Odisha, India

Rajasekhar Mopuri ENVIS Resource Partner on Climate Change & Public Health, Applied Biology Division, CSIR-Indian Institute of Chemical Technology (CSIR-IICT), Tarnaka, Hyderabad, Telangana, India;
Academy of Scientific & Innovative Research (AcSIR), Ghaziabad, Uttar Pradesh, India

Rodrigo Mora-Rodríguez Facultad de Microbiología, Centro de Investigación en Enfermedades Tropicales, Universidad de Costa Rica, San José, Costa Rica

Ankan Mukherjee Das Laboratory of Disease Dynamics and Molecular Epidemiology, Amity Institute of Public Health, Amity University Uttar Pradesh, Noida, India

Sheikh Murtuja Department of Pharmaceutical Sciences & Technology, Birla Institute of Technology, Mesra, Ranchi, Jharkhand, India

Srinivasa Rao Mutheneni ENVIS Resource Partner on Climate Change & Public Health, Applied Biology Division, CSIR-Indian Institute of Chemical Technology (CSIR-IICT), Tarnaka, Hyderabad, Telangana, India;
Academy of Scientific & Innovative Research (AcSIR), Ghaziabad, Uttar Pradesh, India

Igor José dos Santos Nascimento Chemistry and Biotechnology Institute, Federal University of Alagoas, Maceió, AL, Brazil

Rajni Nyodu Centre for Advanced Research (CFAR), Faculty of Medicine, King George's Medical University (KGMU), Lucknow, India

Tomoaki Ogino Department of Medical Microbiology and Immunology, College of Medicine and Life Sciences, University of Toledo, Toledo, OH, USA

Seng-Kai Ong Department of Biological Sciences, School of Medical and Life Sciences, Sunway University, Kuala Lumpur, Selangor, Malaysia

Nikolai Petrovsky Vaxine Pty Ltd, Warradale, Australia;
College of Medicine and Public Health, Flinders University, Adelaide, Australia

Rajendra Phartyal Department of Zoology, Sri Venkateswara College, University of Delhi (South Campus), New Delhi, India

Chit Laa Poh Centre for Virus and Vaccine Research, School of Medical and Life Sciences, Sunway University, Kuala Lumpur, Selangor, Malaysia

Prashanth Prabhu Department of Audiology, All India Institute of Speech and Hearing, Manasagangothri, Naimisham Campus, Mysore, Karnataka, India

Sahar Qazi Department of Computer Science, Jamia Millia Islamia, New Delhi, India

Khalid Raza Department of Computer Science, Jamia Millia Islamia, New Delhi, India

Érica Erlanny da Silva Rodrigues Institute of Chemistry and Biotechnology, Federal University of Alagoas, Campus A.C. Simões, Maceió, Brazil

Alfonso J. Rodriguez-Morales Latin American Network of COVID-19 Research, LANCOVID, Pereira, Risaralda, Colombia;
Grupo de Investigación Biomedicina, Faculty of Medicine, Fundación Universitaria Autónoma de las Américas, Pereira, Risaralda, Colombia

Alessia Ruggieri Department of Infectious Diseases, Molecular Virology, Center for Integrative Infectious Disease Research, University of Heidelberg, Heidelberg, Germany

Paulo Fernando da Silva Santos-Júnior Chemistry and Biotechnology Institute, Federal University of Alagoas, Maceió, AL, Brazil

Radha Saraswathy 120TT Biomedical Genetics Research Lab (BMGRL), Department of Biomedical Sciences, School of Biosciences & Technology, VIT University, Vellore, Tamil Nadu, India

Biswatrish Sarkar Department of Pharmaceutical Sciences & Technology, Birla Institute of Technology, Mesra, Ranchi, Jharkhand, India

Shailendra K. Saxena Centre for Advanced Research (CFAR), Faculty of Medicine, King George's Medical University (KGMU), Lucknow, India

Tanja Schirmeister Institute of Pharmaceutical and Biomedical Science, Johannes Gutenberg-University Mainz, Mainz, Germany

Naveed Shahzad School of Biological Sciences, University of the Punjab, Lahore, Pakistan

Muhammad Sagir Shehu College of Health Technology, Ningi, Bauchi State, Nigeria

Deepak Shilkar Department of Pharmaceutical Sciences & Technology, Birla Institute of Technology, Mesra, Ranchi, Jharkhand, India

Barij Nayan Sinha Department of Pharmaceutical Sciences & Technology, Birla Institute of Technology, Mesra, Ranchi, Jharkhand, India

Ali Syed Department of Internal Medicine, University of Missouri-Columbia, Columbia, MO, USA

Veysel Tahan Associate Professor of Clinical Medicine, Division of Gastroenterology and Hepatology, Department of Internal Medicine, University of Missouri, Columbia, MO, USA

Mohammed Ibrahim Tahir Department of Medical Laboratory Science, Faculty of Allied Health Sciences, College of Medical Sciences, Ahmadu Bello University, Zaria, Nigeria

Fareeda Tasneem Department of Biology, Virtual University of Pakistan, Lahore, Pakistan

Yahaya Usman Department of Medical Laboratory Science, Faculty of Allied Health Sciences, College of Medical Sciences, Ahmadu Bello University, Zaria, Nigeria

María Victoria Vadell CONICET (Consejo Nacional de Investigaciones Científicas y Técnicas, Argentina), Instituto Nacional de Medicina Tropical - ANLIS "Dr. Carlos G. Malbrán", Puerto Iguazú, Misiones, Argentina

Mansi Verma Department of Zoology, Sri Venkateswara College, University of Delhi (South Campus), New Delhi, India

Fenju Wei Department of Medicinal Chemistry, Key Laboratory of Chemical Biology, Ministry of Education, School of Pharmaceutical Sciences, Shandong University, Ji'nan, China

Peng Zhan Department of Medicinal Chemistry, Key Laboratory of Chemical Biology, Ministry of Education, School of Pharmaceutical Sciences, Shandong University, Ji'nan, China

Introduction to Human Viruses

Human Viruses: Infection, Prevention and Potential Target(s) for Therapy – A Comprehensive Review

Shilpi Gupta, Prabhat Kumar, Ankan Mukherjee Das, D. K. Gupta, and Bhudev C. Das

Abstract It is well known that most pathogenic viruses cause acute, chronic and co-infections that lead to pathogenesis and progression and manifestated as various human diseases. Viral diseases mainly AIDS, Zika, Ebola, severe acute respiratory syndrome (SARS), Middle east respiratory syndrome (MERS), influenza and pneumonia of various forms are the biggest cause of mortality and disability in both developed and developing countries. Also certain infectious viruses have the potential to cause cancers in humans. Taken together, the known, unknown and novel viral diseases associated with cancers represent a major global public health challange for social well-being, economic stability, quality human health, productivity and progress. The threats posed by viral diseases mainly depend on the continued emergence, re-emergence of new and novel pathogenic viruses with varied pathogenecity and severity. Chronic infections contribute to as high as 26% of cancer cases in developing countries, only about 11 out of millions of microbes and chemical agents around us have been declared as human carcinogens. In the past few years, several new and highly pathogenic viral infections, that affect humans, have emerged and majority are of zoonotic origin. Thus, monitoring these zoonoses and other novel viruses with unidentified origins required, advanced efforts for increased awarness and the efficient global research co-ordination for

Shilpi Gupta and Prabhat Kumar contributed equally to first authorship.

S. Gupta · P. Kumar · B. C. Das (✉)
Stem Cell and Cancer Research Lab, Amity Institute of Molecular Medicine & Stem Cell Research (AIMMSCR), Amity University Uttar Pradesh, Sector-125, Noida 201313, India
e-mail: bcdas@amity.edu

S. Gupta
Molecular Genetics Laboratory, National Institute of Cancer Prevention and Research (ICMR), I-7, Sector-39, Noida 201301, India

A. Mukherjee Das
Laboratory of Disease Dynamics and Molecular Epidemiology, Amity Institute of Public Health, Amity University Uttar Pradesh, Sector-125, Noida 201313, India

S. Gupta · D. K. Gupta
Felix Hospital, Sector 137, Noida 201305, India

S. I. Ahmad (ed.), *Human Viruses: Diseases, Treatments and Vaccines*, https://doi.org/10.1007/978-3-030-71165-8_1

pursuance of effective approaches for prevention and the control of diseases remains crucial. This chapter presents an update on emerging and re-emerging viruses and viral diseases, highlights their role in causing human diseases including cancer as well as their potential targets for development of new therapeutics and vaccines.

Keywords Infection · DNA virus · RNA virus · HPV · HIV · Influenza · Coronavirus · Covid-19 · Community transmission · Human diseases · SARS · MERS · Pathogenesis · Carcinogenesis · Therapy and vaccine

Abbreviations

AIDS	Acquired immune deficiency syndrome
AP-1	Activator protein 1
ACE2	Angiotensin-converting enzyme 2
ARDS	Acute respiratory distress syndrome
CaCx	Cervical cancer
COVID-19	Coronavirus disease
CRISPR	Clustered regularly interspaced short palindromic repeats
CAS9	CRISPR associated protein 9
CDC	Centres for disease control and prevention
ÇDK	Cyclin-dependent kinases
CD	Cluster of differentiation
CYD-TDV	CYD-tétravalent dengue vaccine
CFR	Case fatality rate
CT	Computed tomography
CMV	Cytomegalovirus
DENV	Dengue virus
dsDNA	Double-strand DNA
EBV	Epstein-Barr virus
ER	Endoplasmicreticulum
ELISA	Enzyme-linked immunosorbent assay
FDA	Food and drug administration
HIV	Human immunodeficiency virus
HPV	Human papillomavirus
HR-HPV	High-risk human papillomavirus
HPyV	Human polyomaviruses
HSV	Herpes simplex virus
HBV	Hepatitis B virus
HCV	Hepatitis C virus
HCQ	Hydroxychloroquine
HCC	Hepatocellular carcinoma
HTLV-1	Human T-cell lymphotropic virus type 1
IFN	Interferon
IL-12	Inter lukin-12

IARC	International agency for research on cancer
ICTV	International committee on taxonomy of viruses
JNK	c-Jun N-terminal kinases
KHSV	Kaposi sarcoma-associated herpesvirus
LR-HPV	Low-risk human papillomavirus
LMIC	Low-middle-income country
MERS-CoV	Middle east respiratory syndrome coronavirus
MCC	Merkel cell carcinomas
MCpyV	Merkel cell polyomavirus
mTOR	Mammalian target of rapamycin
NAAT	Nucleic acid amplification testing
NS1	Non-structural protein 1
NF-kB	Nuclear factor kappa-light-chain-enhancer of activated B cells
ORF	Open reading frame
PML	Progressive multifocal leukoencephalopathy
PIKFYVE	FYVE finger-containing phosphoinositide kinase
PCR	Polymerase chain reaction
R_0	Reproductive number
RT-PCR	Real-time reverse transcription PCR
RSV	Rous sarcoma virus
SARS	Severe acute respiratory syndrome
SARS-CoV-1	Severe acute respiratory syndrome conronavirus-1
SARS-CoV-2	Severe acute respiratory syndrome conronavirus-2
SiRNA	Small interfering RNA
STI	Sexually transmitted infection
STV	Sexually transmitted virus
STD	Sexually transmitted disease
ssDNA	Single-strand DNA
sgRNA	Single guide RNA
STAT3	Signal transducer and activator of transcription 3
TMPRSS2	Transmembrane protease, serine 2
VEGF	Vascular endothelial growth factor
VIA	Visual inspections with acetic acid
WHO	World health organization

1 Introduction

If we think we have greater knowledge of the infectious agents that occupy our planet and affect our health, nature reminds us that our understanding about infectious pathogens is too limited. In the last few decades, several kinds of pathogenic microbes had been discovered to cause various types of human diseases and killed millions of people from almost every continent. The growing challenges to global public health and economic stability to a greater extent can be attributed to various infectious agents including viruses, bacteria, fungi and parasites. These infectious agents can be transmitted indirectly or directly, from human-to-human or animal-to-human or human-to-animal. Though a lot has been done by global health systems to protect and promote human health against infectious diseases, yet its menace is further deepened by the sustained emergence and re-emergence of novel, unrecognized, neglected and long standing infectious agents. At least 28.3% (17 million) of around 60 million deaths that occur globally every year are projected to be due to infectious diseases. Some of these infections (such as pneumonia, AIDS, tuberculosis, diarrhoea, measles and malaria) are more severe and deadly with high incidence and fatality rates (https://www.who.int/healthinfo/global_burden_disease/GBD_report_2004update_part2.pdf). The threats vary extensively in terms of severity, infectivity, morbidity and mortality, as well as their effects on social and global health and economy (https://www.who.int/healthinfo/global_burden_disease/GBD_report_2004update_part2.pdf). Regardless of high morbidity and mortality associated with many of these infections, the attempts in understanding them are restricted to a remarkably small group of biomedical and public health researchers. Whether the global health system as of today can provide effective safety and protection against huge array of infectious diseases is doubtful and this is very clear from by recent outbreaks of SARS-CoV-2 (severe acute respiratory syndrome coronavirus-2) including previous epidemics of Ebola, HIV, Zika, influenza, SARS-CoV-1 and MERS-CoV (Middle East respiratory syndrome coronavirus) along-with looming threat of their increasing resistance (Kim et al. 2017; McNeil and Shetty 2017; https://www.worldometers.info/coronavirus/#countries; Feldmann et al. 2020; Ksiazek et al. 2003; Memish et al. 2020). The concern is magnified by rapid population growth in areas with weak healthcare systems, globalization, urbanization, environmental/cultural variations, ecological factors, civil conflict, global travel along with trade and the altered natural behaviour of pathogens and accelerated spread not only from human-to-human but also from animal-to-human. Human-created epidemics/pandemics originating from laboratory accidents or less possible but planned biological outbreaks are also of increasing concern. These infectious diseases caused by pathogens obviously do not respect human emotions, political and financial boundaries.

Thus, the global research communities have responsibilities to direct adequate attention to studying these diseases, their etiologic agents, and natural history with creating constant awareness and advanced efforts and effective strategies for closly monitoring infectious diseases and their future emergence and re-emergence. In this chapter, we sheds light on infectious agents particularly emerging and re-emerging pathogenic viruses, and briefly introduce its association with human diseases as well as their potential targets for development of new drug therapies and/or vaccines.

2 Infections: Acute and Chronic

Latent or Acute, chronic, recurrent and co-infections may involve in disease pathogenesis and play a serious role in manifestation of disease. It is essential to make a distinction between acute and chronic infections, as they play a different role in the genesis of several diseases. Several acute infections may play a protective role for the development of some diseases while chronic infections appear to be a predisposing factor for the onset of disease.

Acute infections are usually associated with an acute phase reaction with a local inflammatory response, fever and an increased hepatic acute phase protein synthesis. The acute process is characterized by the massive presence of activated monocytes and macrophages in the inflammation area. These inflammatory cells may promote angiogenesis (Corliss et al. 2015) by producing pro-angiogenic factors, such as VEGF (Zhang and Daaka 2011) and prostaglandin E2. Nevertheless, acute infections do not involve new vessel growth. The variety of infection-induced factors (TNF-α, IFN-α/β/γ, IL-12, TGF-β and the acute phase proteins) represent a critical regulator of acute inflammation that could prevent neo-angiogenesis. The observations of spontaneous cancer regression in patients with acute infections, pave the way for studies on the effects of infection and inflammation on cancer (Kucerova and Cervinkova 2016). Acute inflammation can lead to cancer regression as shown by early experiments that artificially infected tumors with erysipelas bacteria cause regression in some cases of incurable cancer (Zhang and Daaka 2011). It is also important to note, that acute infections occurring in childhood, adulthood or old age have different features. Some studies examined the acute infectious disease of childhood and underlined that these infections were associated to a reduced risk of future development of melanoma, ovarian cancer and multiple tumors (Oikonomopoulou et al. 2013). The cancer risk reduction may also be the effect and consequence of acute infectious diseases in childhood. These may provide protection against cancer, but will decrease with age, with the alteration of the immune system. It seems also that acute infections after the infancy may give an important protection against cancer (Jacqueline et al. 2017).

Chronic infections last for many years and are not associated with fever and present variable symptoms. In addition, these kinds of infections can be a consequence of a failed or deficient immune response. In chronic inflammation the active tissue destruction and repair process occur simultaneously with the help of angiogenesis and fibrosis (Kataoka et al. 2003). Chronic infections are a leading public health issue in high-income countries, and they cause a significant toll in low or middle-income countries as well. The onset of the infection and, subsequently, the establishment of the infectious disease and its possible chronic evolution basically depend on host and pathogen factors. These factors depend on biological characteristics of pathogen in terms of pathogenecity, virulence, tenacity, toxinogenesis, infection load, contagiousness and vitality. Instead, the factors inherent the host are represented by sex, age, nutritional state, economic status, social, hygienic cultural and conditions, presence of metabolic disorders and the state of the immune system.

3 Infectious Agents that Can Cause Human Diseases

There are various infectious microbes such as viruses, bacteria, fungi and even parasites have the ability to spread infection from animal-to-human or human-to-human and transmit diseases directly or indirectly from one person to another. Generally, several microorganisms are harmless and live in and on human bodies but under certain conditions, some of them may purport to cause variety of human diseases ranging from acute to chronic including cancers.

Persistent infection of certain pathogenic microorganisms can cause variety of human diseases including cancers (cervical cancer, gastric cancer, liver cancer, oral cancer etc.) in both developing (26%) and developed (8%) countries (IARC 2012; Parkin 2006). Currently, more than 20% of the human malignancies are directly or indirectly caused by pathogenic microbes of which more than 50% of infection associated human malignancies are caused by different strains of human papillomavirus (HPV) (Parkin 2006; zur Hausen 2000). A series of studies showed links between infectious agents and cancer in the last 60 years. Worldwide, it has been estimated that 2.2 million (13%) infection-attributable malignancies (excluding non-melanoma skin cancers) were diagnosed in 2018 alone (de Martel et al. 2019). As per International Agency for Research in Cancer (IARC), about 11 infectious agents that have been classified as group-1 carcinogens, are responsible for more than 90% of infection-associated human cancers globally. These are Helicobacter pylori, high-risk human papillomaviruses (HR-HPVs), hepatitis B and C virus (HBV/HCV), Epstein–Barr virus (EBV), Kaposi sarcoma-associated herpesvirus (KHSV), human T-cell lymphotropic virus type 1 (HTLV-1), Schistosoma haematobium, Opisthorchis viverrini and Clonorchis sinensis. Out of these, *Helicobacter pylori*, HR-HPVs, HBV, and HCV are the most important pathogens which caused $\geq$90% of infection-associated human malignancies world over (Plummer et al. 2016).

In the 21st century, while the world contineus to fight and contain existing diseaases (plague, AIDS/HIV, tuberculosis, hepatitis, Influenza etc.) and newly emerging (Dengue, SARS, MERS, Ebola, Zika, and SARS-CoV-2) the newly emerging agents are infectious wreaking havoc in both developed and developing countries alike when an outbreak as an epidemic or a pandemic occurs. To date, pathogenic diseases and associated fatalities have remained a significant menace throughout the globe. In spite of current advances highlighting the role of infectious agents in human diseases, the global incidence of infection associated diseases is still high, thus an a strong and flexible global health system and focused research on infectious disease surveillance, prevention, control and treatment strategies are required.

4 Viruses and Human Diseases

Viruses are a mystery and can quickly spread explosively around the world. They are "obligate intracellular parasite" that can replicate inside the host cell (Summers 2009). These intracellular parasites can be found in all living organisms such as bacteria, fungi, protozoans, plants, animals and humans. These are mainly those microorganisms that have small genomes, insufficient to encode their own proteins and every step of viral life cycle depends on the host machinery for almost all essential functions. Viruses also can affect human health and cause a wide range of acute to fatal human diseases including variety of cancers.

The World Health Organization (WHO) has reported various human diseases caused by different types of viral infections, and that millions of individuals are at risk globally (https://www.who.int/healthinfo/global_burden_disease/GBD_report_2004update_part2.pdf). The best example is of the recent global pandemic of coronavirus and COVID-19. Though, our understanding of the mechanisms of pathogenic virus -induced diseases has significantly advanced but we are still lacking in-depth knowledge of several existing, emerging and re-emerging viruses that cause severe diseases in humans. Thus, a better in-depth understanding of what types of viruses in which situation pose the highest risk to human health would enable evidence-based targeting of surveillance, treatment and management of these viral diseases.

5 Virus Infection Can Lead to Cancer

Currently, about 15–20% of all human malignancies are associated with persistent infection of several DNA and RNA viruses (Liu and Richardson 2013). There are six major families of animal viruses which have the ability to cause cancer in both humans and animals. Tumor viruses fall into 2 categories; DNA-containing tumor viruses (hepatitis B virus, papillomaviruses, Epstein-Barr virus, Herpesviruses, polyomavirus and Adenoviruses) and RNA-containing tumor viruses (Retroviruses and Hepatitis C virus). Viruses are unable to cause cancer directly, as several other co-factors including host cell factors besides additional exposure to various carcinogenic agents and immune impairment that can influence tumor development and progression.

Originally, tumor causing viruses were identified in animals but in 1964 onwards several human carcinogenic viruses were discovered. In 1911, Francis Peyton Rous showed that it is possible to transmit the tumour by injecting a solution made from extracts of cancer, thus opening the way for studies on the association between virus infections and cancer. Rous Sarcoma Virus (RSV) was the first carcinogenic virus identified; the studies on RSV had a major role in the development of current understanding on cancer. Nowadays, it is well known that infections caused by certain viruses play key roles in the pathogenesis of several cancers. Viruses can subvert the functions of host cell purturting the pathways regulation of growth arrest and apoptosis (Bagga and Bouchard 2014). Viral particles can induce tumor as some of them have oncogenes while others can cause the transformation of a proto-oncogene to oncogene in the host cell. The viral oncogene is a mutated, hyperactive and homologue gene involved in cell proliferation, so it works rapidly in all infected cells. Moreover, the over-expression of the proto-oncogene, responsible for uncontrolled cell proliferation, is caused by the viral promoter or by other factors that regulate the transcription. The carcinogenesis process requires multiple steps, but it seems that virus-associated tumors may have common pathways; it could be, for viruses such as HBV, HPV and EBV, the functional inactivation of p53 by virally encoded oncoproteins (Jiang and Yue 2014; Ko 2015; Kordestani et al. 2014). The viruses associated with tumors are characterized by the capacity of causing a persistent infection, but they are not exclusively carcinogenetic agents; so cancer development may be considered as an accidental event in the course of the infection. Obviously, the host immune status may influence the evolution of the disease. Between the initial infection and the onset of cancer could take several years, and carcinogenesis process in virus associated cancer, can be promoted by direct and indirect mechanisms. The viral carcinogenesis also depends on genome type; therefore, it is necessary to consider the main differences between DNA viruses and RNA viruses and their implications on human cancers (Table 1) and other diseases (Tables 2 and 3).

A. **Pathogenic DNA viruses**

DNA viruses are obligate intracellular parasites and important human pathogens that have the ability to cause several acute and chronic diseases in humans. Their DNA is replicated either by host or virally encoded DNA polymerases (Fig. 1). Viruses lack the internal machinery to synthesize their proteins and their lytic cycle is totally dependent on their capability to control/command host cell genome and thus regulate different signalling pathways. Viruses essentially “hijack” the host cell and its mRNA is translated into proteins and these early proteins are accountable for changing normal cellular actions which in some cases permit the infected lysogenic cells to escape the host restriction systems. The genomes size of DNA viruses varies between 5 and 375 kb. During viral infection, majority of these viruses stimulate host-cell DNA replication, or at least in the early stages of DNA replication, it creates a suitable atmosphere for their own DNA replication.

Currently, 91 human DNA virus species have been identified with 22 genera and 8 families, many of them are associated with human diseases including cancer and are of immense public health importance worldover (Woolhouse and Adair 2013). According to International Committee on Taxonomy of Viruses (ICTV) (Kuhn et al. 2019a) DNA viruses are divided into 3 major categories (i) single-strand DNA (ssDNA) viruses, double-strand DNA (dsDNA) viruses and (iii) para-retroviruses (that replicate their genome through an RNA intermediate; e.g. Hepadnaviruses) (Katakura et al. 2005). There are 22 families of dsDNA viruses and 5 families of ssDNA viruses which have been recognized by ICTV. There are six distinct members of DNA virus families, Hepadnaviridae, Polyomaviridae,

Table 1 Estimated number of different virus-attributable human malignancies in 2018

Virus types	Associated new cases of malignancies in 2018	Total number of new cases due to infectious agents
DNA tumour viruses Human papillomaviruses (HPV)	Cervical (570000), Oropharyngeal (42000), Anal (29000), Penile (18000), Vagina (14000), Vulvar (11000), Oral cavity (5900) and Laryngeal (4100) Carcinoma	694,000 (de Martel et al. 2019)
Hepatitis B virus (HBV)	Hepatocellular carcinoma (HCC) (150000)	360000 (de Martel et al. 2019)
Epstein Barr virus (EBV)	Hodgkin lymphoma (40000), Burkitt's lymphoma (6600) and Nasopharyngeal Carcinoma (110000)	156,600 (de Martel et al. 2019)
Kaposi's sarcoma herpesvirus (KHSV/HHV-8)	Kaposi's sarcoma and primary effusion Lymphoma (12000)	42000 (de Martel et al. 2019)
Merkel cell polyomavirus (MCpyV)	Merkel cell carcinoma (MCC) (United States) (1972)	19,722 (de Martel et al. 2019)
RNA tumourviruses	**Associated malignancies**	
Hepatitis C virus (HCV)	Hepatocellular carcinoma (140000) and Other non-Hodgkin lymphomas (16000)	156,000 (de Martel et al. 2019)
Human immunodeficiency virus (HIV)	AIDS-related Kaposi's sarcoma (KS) (14000)	41,799 (Dalla Pria et al. 2019)
Human T-cell leukemia virus-1 (HTLV-1)	Adult T-cell leukaemia and lymphoma (ATL) (1900)	3600 (de Martel et al. 2019)

Table 2 List of human herpesvirus types and family

Herpes viruses types	Alternate name/Abbreviations	Sub-family
Human herpesvirus 1	Herpes simplex-1 (HSV-1/HHV-1)	α
Human herpesvirus 2	Herpes simplex-2 (HSV-2/HHV-2)	α
Human herpesvirus 3	Varicella-zoster (VZV/HHV-3)	α
Human herpesvirus 4	Epstein-Barr (EBV/HSV-4)	γ
Human herpesvirus 5	Cytomegalovirus (CMV/HHV-5)	β
Human herpesvirus 6	HHV-6	β
Human herpesvirus 7	HHV-7	β
Human herpesvirus 8	KSHV/HHV-8	γ

Table 3 Modes of transmission of most common sexually transmitted viruses (STV) in human and their key features

Common sexually transmitted viruses 	Human Papilloma Virus (HPV) 	Human Immunodeficiency Virus (HIV) 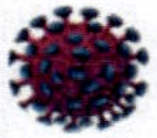	Epstein-Barr virus (EBV) 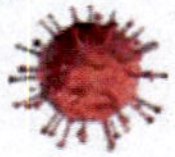	Herpes Simplex Virus (HSV1 and 2) 	Hepatitis B Virus (HBV) 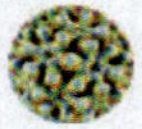
Disease(s)	Genital and skin warts, Cervical, Vaginal, Vulva, Anal, Penile and head and neck cancers	Acquired Immunodeficiency Syndrome (AIDS)	Infectious mononucleosis (glandular fever) Nasopharyngeal carcinoma and Lymphoma	Oral-Herpes and Genital Herpes	Acute and chronic hepatitis, liver cirrhosis and hepatocellular carcinoma
Genomic material	Double stranded DNA	Single-stranded RNA	Double-stranded DNA	Double-stranded DNA	Double-stranded DNA
Incubation period	1–20 months for warts to development. HR-HPV can take 10 years or more to develop cancer	2 weeks to 3 months	~4–6 weeks after initial infection	Symptoms tend to develop 2–20 days after exposure to the virus	~4–24 weeks
Transmission	Transmitted sexually or skin-to-skin contact	Blood, semen, rectal fluid, vaginal fluid, blood transfusion, breast feeding	Body fluids, Saliva, blood and semen during sexual contact, blood transfusions and organ transplantations	Oral-to-oral contact and genital-to-genital contact during sexual intercourse	Transmitted by person-to-person through blood, semen or other body fluids
Common symptoms	Wart; discomfort or pain, though they may itch or feel tender. Cancer;	Systemic disease, rapid weight loss, recurring fever, swollen glands,	Fatigue, fever, sore throat, swollen lymph nodes in the neck and	Mostly asymptomatic or unrecognized but can cause painful blisters or ulcers at	Abdominal pain, Dark urine, Fever, Joint pain, Loss of

(continued)

Table 3 (continued)

Common sexually transmitted viruses 	Human Papilloma Virus (HPV) 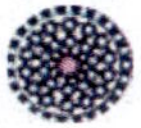	Human Immunodeficiency Virus (HIV) 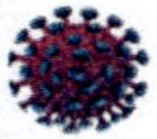	Epstein-Barr virus (EBV) 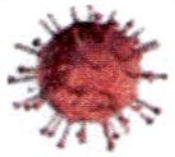	Herpes Simplex Virus (HSV1 and 2) 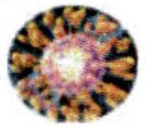	Hepatitis B Virus (HBV) 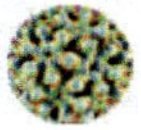
	chronic pain, bleeding, itching, discharge, burning, irritation, changes in color/ thickness of the skin	sores in mouth, anus, genital area, memory loss, depression, pink or purplish blotches on skin	armpits, head and body aches, rash, swollen liver or spleen or both	the site of infection, ranging from mild to moderate to severe	appetite, Nausea and vomiting, Weakness, fatigue and Jaundice
Prevention	Adolescent HPV vaccination, curcumin, neem	PEP, ART	Avoiding radiation exposure, healthy diets	Use of condom, Vitamin C, antioxidant	HBV vaccination of infants & adults
Diagnosis/ testing methods	Pap smear, Colposcopy and acetic acid test, Biopsy, PCR-based HPV DNA test, Southern Blot	ELISA Test, RT-PCR-based DNA test	Physical examination, Blood test; Antibody tests	Clinical evaluation, PCR of cerebrospinal fluid (CSF) and MRI for HSV encephalitis	Blood tests, Liver ultrasound, Liver biopsy
Treatment/ vaccine	FDA approved Vaccines; Cervarix, Gardasil-4 and 9 valent Imiquimod, Podofillox for warts & Surgery, Chemo-radiotherapy for cancerous lesions	No specific treatment and vaccine Supportive treatment-Antiretovirals (AZT, etc.)	No vaccine to protect against infectious mononucleosis, penicillin antibiotics like ampicillin or amoxicillin	No vaccine available, Acyclovir, valacyclovir, famciclovir are commonly used.	Vaccines-Energix B, RecombivaxHB, Drugs-Lamivudine, tenofovir and Interferon alpha

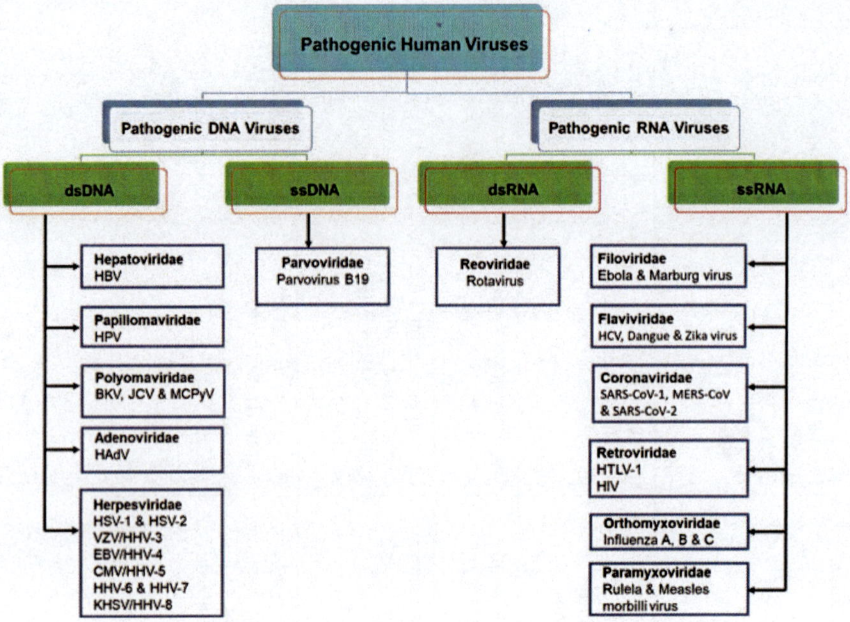

Fig. 1 Classification of most common pathogenic human DNA and RNA viruses

Papillomaviridae, Herpesviridae, Adenoviridae and Poxviridae; they are capable of infecting and causing various acute and chronic diseases in humans. Persistent infections of majority of these DNA viruses are linked with cellular transformation and carcinogenesis.

I. Hepatitis viruses

Viral hepatitis is caused by 5 different types of hepatotropic viruses (HAV, HBV, HCV, HDV and HEV). Both, acute and chronic Hepatitis B (Cho et al. 1973) is caused by the Hepatitis B virus (HBV). HBV is a non-cytopathic, partially (full length negative-sense, partial positive-sense) double stranded with circular DNA belongs to the family *Hepadnaviridae* (Hirschman et al. 1969). Chronic infections caused by HBV represent the main etiological factor for the development of liver diseases and hepatocellular carcinoma (Table 2). HBV infection remains to be a major public health concern globally and in the last two decades the chronically infected people ranged from 240 to 350 million (Lavanchy 2004; Ott et al. 2012), and 8,870,00 deaths were reported in 2015 worldwide (Lozano et al. 2012). It is an endemic in Southeast Asia and Sub-Saharan Africa and more than 8% African and Asian population are chronic carriers of HBV (Maddrey 2001). HBV seropositivity has been estimated to be 3.6%, with the greatest endemicity in Africa (8.8%) and Asia (5.3%). Since India is the second largest populated country in the world, it accounts for a large proportion of the global HBV infection and has about 40–50 million (10–15%) of the entire pool of HBV carriers (Ray 2017).

HBV is mainly found in secretions of semen, vagina, saliva, blood, and menstrual blood of infected persons and transmitted by parenteral, sexual or perinatal mode. HBV infected individuals generally remain asymptomatic or lead to an acute hepatitis after which the infection is cleared by the immune response (Guidotti et al. 1999). However, if HBV infection fails to get cleared, it can lead to chronic infection in a large number of individuals comprising young children and a good percentage of adults.

There are eight HBV genotypes and four serotypes (adv, ayw, adr and ayr) which are characterized by distinct geographic distribution and are clinically important. Several reports about HBV infection showed that genetic characteristics of virus, including genotypes and specific genetic mutations, may lead to the development of hepatocellular carcinoma (HCC) through direct and indirect pathways. The ability of the virus to integrate into the host cell genome leading to alteration of cellular signalling and tumorigenic growth control, assumes a great importance (Chu 2000; Farazi and DePinho 2006). It is interesting to note that integrated HBV DNA in human genome does not replicate independently but can do so along with the human genome and can persist indefinitely. Certain disease status of chronic HBV infection poses a higher risk of liver cancer and the risk is increased in patients with active hepatitis (Farazi and DePinho 2006) and in cirrhotic patients. HBV can cause cancer by integrations into the human genome, even in the absence of cirrhosis. The continuous injury and regeneration of cirrhotic liver lead to increased liver cells turnover, favouring critical genomic mutations, chromosomal rearrangements, activation of oncogenes and inactivation of tumour suppressor genes. The T-cell immune response elicited to combat infection, contributes to hepatocyte necrosis, inflammation and consequently regeneration (Katakura et al. 2005). HBV-induced hepatocellular carcinogenesis, like other malignancies, is considered a multi-step process involving several genetic alterations that ultimately lead to malignant transformation of the hepatocytes. Moreover, an association has been observed between past exposure to HBV and the risk for pancreatic cancer development. This is not surprising, because HBV can replicate within the pancreas and because elevations of serum and urinary level of pancreatic enzymes are often noted in patients with HBV chronic infection (Berrington de Gonzalez et al. 2008).

The HBV vaccine was developed in the late 1970s. This vaccine has provided an effective means to prevent and decrease HBV infection and eventually reduced chronic liver diseases and HBV-associated HCCs, which can therefore be considered the first anti-cancer vaccine against HBV (Hessel and West 2002). Various novel strategies (CRISPR/Cas9, siRNA, entry and secretion inhibitors and immunotherapy) are now being developed that could potentially be used to combat and eliminate HBV infections and prevent virus rebound on therapy discontinuation.

II. **Human Polyomavirus infection**

At present, 14 human polyomavirus (HPyV) species have been identified and some of which are known to cause infection in humans. These viruses are small, non-enveloped circular, dsDNA viruses that belong to the family *Polyomaviridae*. These HPyVs infect a wide range of tissues including skin, kidney, and respiratory tract, usually resulting in a persistent, asymptomatic infection. However, in immunocompromised hosts, HPyVs can cause serious diseases including fatal infection of the brain, kidney and can cause cancer. Recent works identified 8 novel species, MCV, HPyV6, HPyV7, KIPyV, WUPyV, HPyV9, and MW PyV/HPyV10 (Allander et al. 2007; Feltkamp et al. 2013; Feng et al. 2008; Gaynor et al. 2007; Schowalter et al. 2010; Scuda et al. 2011; Siebrasse et al. 2012). Studies have also shown that the JC virus (JCV) and BK virus (BKV), both belonging to HPyVs were discovered in 1971 in patients who were immunosuppressed (Gardner 1971). Evidence suggest that BKV persistently is linked with tissue injury, salivary gland diseases and sclerosis in HIV-positive patients, nephropathy in kidney transplant patients, haemorrhagic cystitis in recipients of bone marrow transplantation and oncogenesis in transplant recipients. JCV causes progressive multifocal leukoencephalopathy (PML) and in immunocompromised/AIDS patients it causes haemorrhagic cystitis and haematological malignancies.

MCPyV is the first polyomavirus directly implicated in more than 80% of aggressive Merkel Cell Carcinomas (MCCs), the second most deadly form of skin cancer after melanoma in elderly white patients. This rare and aggressive neuroectodermal cancer shares clinico-epidemiological features and other aspects with Kaposi sarcoma. MCPyV is an emerging, non-enveloped, fifth tumour polyomavirus virus, which was first described in January 2008 in Pittsburgh, Pennsylvania. About 1,500 new MCC cases are recorded annually in the United States (US), representing a relatively low prevalence compared to other skin cancers; however, its incidence has tripled in the past two decades (Agelli and Clegg 2003; Calder and Smoller 2010; Hodgson 2005).

III. **Human Herpesvirus and associated diseases**

Human herpesviruses (HHVs) are potential human pathogens that belong to the family *Herpesviridae*. There are eight distinct types of herpesviruses (Table 2) and their family is divided in3 subfamilies: Alphaherpesvirinae (α-herpesvirinae), Betaherpesvirinae (β-herpesvirinae) and Gammaherpesvirinae (γ-herpesvirinae). These viruses can cause variety of human diseases including genital or oral herpes, infectious mononucleosis, encephalitis, meningitis, herpes keratitis, neonatal herpes and Kaposi's sarcoma. These viruses are transmitted from the infected area of skin or mucous membrane and as a result of direct contact with the lesions. The distinct types of herpesviruses that infect humans and their key features are summarized in Tables 1 and 3.

HSV-1 and HSV-2: These are the two most common viruses which are generally associated with human oro-facial or genital herpes, encephalitis and conjunctivitis. However, genital herpes may be a consequence of HSV-1 infection and cold sores may also be caused by HSV-2. Once the individual has been infected, reactivation is extremely common in both clinical forms: oral or genital (Whitley 1996; Wilck et al. 2013). HSV infection is untreatable and about 90% of global population is infected with one or both HSVs (Boppana and Fowler 2007). If HSV-encephalitis is left untreated, it can cause >70% mortality (Jereb et al. 2005).

Varicella-zoster virus (HHV-3): It is an exclusively human neurotropic alpha-herpesvirus that causes chickenpox (Varicella) especially in children (1–9 years of age) and reactivation of the dormant virus leading to herpes zoster (shingles) in elderly and immunosuppressed persons (Pergam et al. 2019). In the US, more than 90% of adults suffer from this disease in their childhood, whereas the remaining children and young adults get vaccinated (Gnann 2002).

Epstein-Barr virus (EBV/HSV-4): EBV was identified and fully sequenced for the first time in Burkitt lymphoma biopsy by Denis Burkitt in 1964 in Uganda (Zhang and Daaka 2011). Later it was considered, along-with other genetic and environmental factors (Yu and Zelterman 2002) responsible for the development of benign lesions, infectious mononucleosis, Burkitt lymphoma and other human malignancies, such as nasopharyngeal carcinoma and a series of other human infections and diseases particularly in immunocompromised patients. It also causes Hodgkin and non-Hodgkin lymphoma (NHL), thymic lympho-epithelioma, salivary gland and urogenital carcinoma (Parkin 2006; Javier and Butel 2008). Based on geographical regions, EBV infection alone accounts for 0.5–2% of human cancers (Epstein et al. 1964; Khan and Hashim 2014) and recent epidemiological studies indicate that 1.8% of all cancer-related mortalities were associated with EBV infection in 2010 (Khan and Hashim 2014). Globally, EBV is highly prevalent and more than 90% of younger population are infected with EBV (Young et al. 2008).

Two major types of EBV (EBV-1 and EBV-2) have been identified (Cho et al. 1973) and it has been revealed that the genomes of both the EBV types are very similar except for the regions of the EBV-nuclear antigen (EBNA) genes that have greater ability of type 1 EBV infection to induce B cell proliferation (Tzellos et al. 2014). EBV-2 infection is prevalent in Africa and mostly among homosexual men (Gratama and Ernberg 1995; Higgins et al. 2007; van Baarle et al. 2000). Markedly, percentage of EBV infection differs in different diseases and may have consequences toward the aggressivity of the associated diseases (Jha et al. 2016). Moreover, the presence of autoimmune diseases such as systemic lupus erythematosus, rheumatoid arthritis, and primary Sjögren's syndrome, the condition of transplanted and the consequent iatrogenic immunosuppression and HIV infection are all circumstances in which the EBV infection is strongly associated to leiomyomas, leiomyosarcomas and other smooth muscle neoplasms (Sousa et al. 2008) (Table 3).

Human cytomegalovirus (HCMV/HHV-5): It is a ubiquitous HSV-5 virus with the high morbidity and mortality rates compared to other HSVs worldwide. Infection of this virus inflicts "mononucleosis-like syndrome" (cytomegaly or "cytomegalic inclusion disease) (Davison and Bhella 2007; Liu and Zhou 2007).

Infection risk is higher among young, elderly, immunocompromised, HIV-infected and transplant recipient individuals. It may also cause encephalitis, retinitis, hepatitis, colitis, esophagitis, pneumonia and neonatal infection sequelae. In developed countries, the infection burden of HCMV increases progressively with age, reaching over 70% prevalence by the age 70. The seropositivity rates are higher ($\geq$90%) among lower socioeconomic groups, homosexual men, and in low-income countries (Beam and Razonable 2012; Cannon 2009; Ho 1990; Pass 1985; Razonable 2005; Stagno and Cloud 1990). Clinical management of HCMV infection is challenging due to the absence of 100% protective vaccine or anti-viral drugs against the virus (Griffiths and Boeckh 2007; Heineman 2007; Plotkin and Boppana 2019). The majority of studies have been focusing on the immune response of viral pathogenesis. However, studies on host-genetic interaction of virus may help in identifying molecular pathways that may lead to the discovery of new therapeutics.

HHV-6 and HHV-7: These are ubiquitous viruses and have been linked with lichen planus, pityriasis rosea, hypersensitivity reactions, roseola infantum, graft-vs-host disease, and early febrile infectious syndrome. A large number of adult population ($\geq$95%) is found seropositive to both HHV-6 and HHV-7 infection worldwide. Primary infection commonly occurs in children during infancy. HHV-6 was first identified in 1986 in patients with lymphoproliferative diseases. HHV-7 is one of the causative agents of exanthem subitum (ES) (Tanaka et al. 1994). The peak of HHV-7 seroconversion (Tanaka-Taya et al. 1996) takes place later than seroconversion for HHV-6 (Okuno et al. 1989) and mature CD4 positive T-lymphocytes and epithelial cells of salivary glands seem to be the principal target cells for both the viruses (Dockrell and Paya 2001; Caserta et al. 2001). Further studies on diagnostic techniques and therapy will lead to better detection and targeted anti-viral treatment.

Kaposi's sarcoma herpes virus (KSHV/HHV-8): KHSV is known cause of various Kaposi's sarcoma (KSHV) forms (AIDS-associated, classic, endemic, African, iatrogenic) also known as Human Herpes Virus 8 (HHV8) encodes for several cytokines and their receptors (Cesarman et al. 1995; Soulier et al. 1995). There are four forms of KHSV that have been described; the classic KS, identified by Moritz Kaposi, is found in elderly men of Mediterranean and Eastern European descent (DiGiovanna and Safai 1981). The second type is lymphadenopathic and AIDS associated form of KS present as African endemic KS, which occurs in Eastern and Central African countries (Friedman-Kien and Farthing 1990; Stein et al. 1994). Currently KS is the most common malignancy associated with HIV infection (Table 1). The fourth type of KS is iatrogenic/post-transplant KS, which is associated with the use of immunosuppressive therapy for the prevention of organ transplant rejection (Andreoni et al. 2001; Marcelin et al. 2007). In addition to Kaposi's sarcoma, the viral genome has been found in primary effusion lymphoma (PEL), Castleman disease, angioimmunoblastic lymphadenopathy and in certain cases of reactive lymphadenopathies. Moreover, it is also considered a cause of body-cavity-based lymphomas.

IV. **Parvovirus B19 infection**

Parvovirus B19 infection may play a role in the pathogenesis of rheumatic diseases (Yaghoobi et al. 2015) and acute lymphoblastic and myeloblastic leukaemia (Kerr 2000). Even if Parvovirus B19 is an aetiological agent for certain diseases (Kerr et al. 2003), there is limited literature on Parvovirus B19 involvement in solid tumours. However, a correlation with papillary thyroid carcinoma (PTC) has been hypothesized. In fact, studies have demonstrated the presence of viral DNA and viral proteins in neoplastic epithelium of patients with PTC (Li et al. 2012). The possible role of B19 in pathogenesis of thyroid neoplasm is also suggested by involvement of immune system involving NF-kB, IL-6 and viral proteins. The rapidly activated NF-kB, by tax and tat protein, can play a pivotal role in the HTLV-1-induced acute leukaemia and HIV-induced KS which is also significantly increased and co-localized with the virus DNA in papillary thyroid carcinoma tissues (Chan et al. 2017). Moreover, one of the non-structural protein (NS1) encoded by viral genome is cytotoxic to host cells and linked to NF-kB and IL6. The correlation is probably due to the fact that the non-structural viral protein resembles tax protein of HTLV-1 and tat protein of type 1- HIV. They all play a part in viral propagation and activation of IL6 production through the NF-kB-binding site in the IL6 promoter (Yaghoobi et al. 2015). These findings also reveal a novel link between parvovirus B19 and thyroid carcinoma (Li et al. 2012).

V. **Human adenoviruses (HAdV)**

Human adenoviruses (HAdV) are ubiquitous, non-enveloped, dsDNA virus with $\sim$35 kb of genome size belongs to the Adenoviridae family. HAdV are the first respiratory viruses which were isolated in tissue culture. These viruses are well-recognized as an important human pathogen of upper respiratory tract infections including common cold in children and adults. Presently, $\geq$70 HAdV serotypes are known and these are further divided into 7 species (HAdV-A to HAdV-G). In humans, over fifty adenoviral serotypes have been identified that can cause widespread infections, from mild to life-threatening respiratory infections in young children and in immunocompromised persons (Sun et al. 2018). Severe HAdV infection in young children can be complicated, leading to acute respiratory distress syndrome (Hung and Lin 2015; Stebbing et al. 2020), pleural effusions (Cho et al. 1973), myocarditis (Treacy et al. 2010), respiratory failure (Lai et al. 2013) and central nervous system (CNS) dysfunction (Huang et al. 2013). However, HAdV also can cause variety of lesions of the conjunctiva and gastrointestinal, bladder, central nervous system and genitals. HAdV are tolerant to interferon treatments and only Cidofovir is a choice of drug which gives anti-adenoviral activity in severe disseminated disorders. Unfortunately, no effective vaccine for children and specific anti-adenovirus drugs against HAdV diseases is available (Fu et al. 2019). Majority of viral infections are self-limited but in immunocompromised patients, the burden and of the disease outcome is potentially fatal. This has paved the way for additional work to work identifying novel avenues for immunotherapy and pharmacotherapy.

VI. **The Human papillomaviruses (HPVs)**

HPV being epitheliotropic virus it exclusively infects both, the skin and basalmucosal epithelial tissues of genital, head and neck area and infection can lead to the transformation of basal epithelial cells (Gupta et al. 2012). HPV have been classified into five genera (alpha, beta, gamma, mu and nu (de Villiers 2013; Gupta et al. 2018b). Mucosal HPV infection can cause various human diseases including condyloma acuminata, focal epithelial hyperplasia, cervical neoplasia and cervical and other anogenital and head and neck carcinomas (Table 3). Cutaneous HPVs which infect cutaneous epithelia can cause various types of benign warts (Plantar, and Filiform which may be pigmented), epidermoid cysts and skin cancers (Doorbar et al. 2015). To date >200 HPV genotypes have been discovered and majority of them ($\sim$40 HPV genotypes) infect the anogenital tract (Kocjan et al. 2015). HPVs have been categorized into two main types; (i) high-risk (HR-) and low-risk (LR-); these types are based on their potential to induce malignant lesions. Of these, 15 mucosal HR-HPV types (HPV type 16, 18, 31, 33, 35, 39, 45, 51, 52, 56, 58, 59, 66, 68, 73 and 82) have been identified as oncogenic HPV types and are associated with more than 99% of cervical and other anogenital cancers (de Villiers 2013; Chen and Chou 2019) HR-HPV types 16/18 are considered as the most dominant types and are responsible for >80% of all cervical cancer (CaCx) cases, and 6 other HR-types (HPV 31, 33, 35, 45, 52, and 58) are responsible for an additional 20% (Bosch et al. 1995; de Sanjose et al. 2010).

HPV is considered as one of the most common sexually transmitted virus (STV) and globally $\sim$12% females and 1% males are infected (Chu 2000). It is strongly associated with the pathogenesis of several types of human cancers and accounts for $\sim$10–20% of total human tumors (Table 3). HPV infection transmission is considered through sexual contact, mostly during the vaginal, anal, oral sex, or genital-to-genital contact (Steben and Duarte-Franco 2007). In fact, most of the sexually active partners are exposed at least one strain of HPV at some point in their life-time. Therefore, the difference in the capability of virus to induce malignant transformation is totally depended on oncogenic HPV genotypes and host immune susceptibility. HPV infection is a necessary causative agent for cervical cancers (CaCx), which is one of the leading cause of death in women particularly in low or middle-income countries mainly due to lack of early screening program for CaCx. Other than CaCx, HPV infection also induce cancers of anal (88%), vaginal (78%), penile (50%), vulvar (25%), oropharynx (20–31%), tongue (30%), laryngeal (2.4%), oral cavity (2.2%) and also breast and stomach (Gupta et al. 2018b; Boda et al. 2018; Dunne et al. 2007; Pereira et al. 2015; zur Hausen 2002). Despite available effective prophylactic HPV vaccines, protecting against most prevalent HR- and LR-HPVs, the HPV-associated tumors still remain a major global health problem particularly in low-middle income countries (LMICs). Understanding molecular, mechanisms and conducting large-scale clinicodemiological studies along-with organized HPV screening and vaccination are crucial to completely eradicate HPV-associated disease burden globally.

Genome structure and functions of HPVs. They are non-enveloped, small, circular dsDNA viruses with about 8 kb genomes (Fig. 2). HPV belongs to the family '*Papillomaviridae*'. The viral genome contains three major regions: (i) the early genes (E1, E2, E4, E5, E6 and E7) involved in viral replication, transcription and tumorigenic transformation, (ii) the late genes (L1 and L2) encodes two structural capsid proteins and (iii) the long control region (LCR), a non-coding upstream regulatory region (URR) (Dalla Pria et al. 2019) that controls viral transcription and replication (Fig. 2) (de Villiers 2013; Das et al. 2008). The HPV E6 and E7, which encode oncoproteins consisting of 151 and 98 amino acids, respectively, are largely responsible for the onset and persistence of the malignant phenotype in both anogenital and head and neck cancers (Fig. 2).

Emergence of HPVs as the key human carcinogens for cervical cancer. The existing data highlights that oncogenic HPV infections are the important etiological agents in the genesis of majority of cervical cancers (CaCx). It causes almost all cases of cervical cancer but being an epitheliotropic virus, it also induces a significant variable percentage of certain non-cervical carcinomas. CaCx is a deadly disease among women and a big challenge to the society, as it kills one woman in

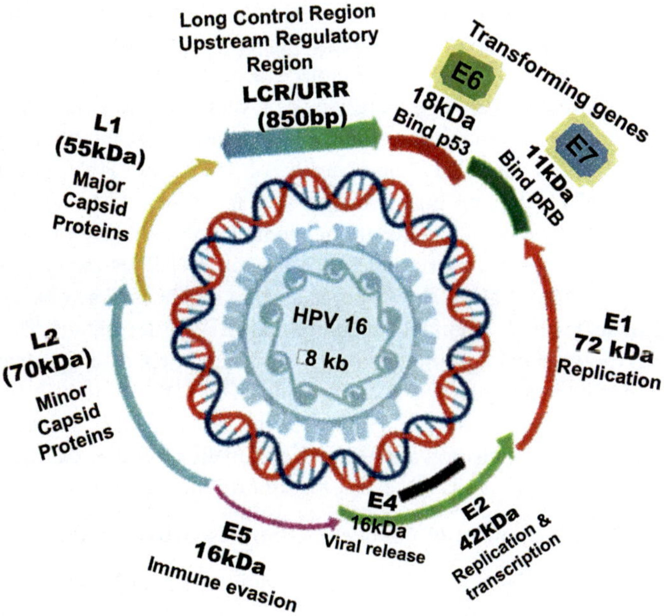

Fig. 2 Genomic organization of HR-HPV type 16. First the early region (E1–E7), second is late capsid (L1–L2) and third is non-coding upstream regulatory region (URR). Most Papillomavirus genomes resemble HPV16 organization in general

every two minutes around the globe (Bray et al. 2015). Globally, CaCx is most frequent gynaecologic cancer with an estimated 570,000 new cases and 311,365 cervical cancer-related deaths recorded in 2018, of which about 90% (9 out of 10) cancer-related deaths were recorded in low-middle income countries (LMIC) (Bray et al. 2015; Ferlay et al. 2018). CaCx is the 3rd most prevalent malignancy among females globally and 2nd leading cause of death in Indian females. Worldwide, it accounts for about 3.3% of all cancer-associated mortalities (Bray et al. 2018).

One-forth of all global cervical cancer cases are diagnosed in LMIC which harbour greater than 21.3% of world's population (Kumar et al. 2019). The substantial differences in CaCx incidence exist within the states, most prominently between urban and rural populations. These figures suggest that there is an urgent need for better prevention and treatment solutions regarding HPV-induced cancers in women. Ironically, regardless of the fact that a cancer control programs being an essential part of the national health strategy, there have been disappointing improvements on HPV-mediated cancer prevention strategies. However, the scenerio in interstingly changing with GAVI and UNDP implementing HPV vaccination programmes in India. More than 90% of all CaCx cases are caused by persistent infections with HR-HPVs which can lead to the genesis of precancer lesions and, eventually, invasive carcinoma (Ferlay et al. 2014). However, the possibility of HPV-DNA persistence and development to lesions is increased by a number of host and environmental factors. The risk of HR-HPV persistence and developing into high-grade cervical lesions increases 5–10 folds due to the host immune impairment (Birkeland et al. 1995). In addition, the relative risk of developing CaCx is increased by several other epidemeological factors such as types of STIs (Smith et al. 2002), continuous use of contraceptives (Moreno et al. 2002), multiple partners, early age marriage (<18 years), poor genital hygiene (Das et al. 2008), high parity, smoking, certain religious practices and ethnicity (Bharti et al. 2009; Wyatt et al. 2001). In majority of the cases, HPV infections are asymptomatic and cleared by host immune system, however, in few cases, persistent infections with specific strains of oncogenic HPVs can develop into invasive cancer (Gupta et al. 2018b). Nevertheless, persistent HR-HPVs are principal cause for more than 90% of CaCx. However, it clearly indicates that CaCx is a preventable disease through regular screening and HPV vaccination. Thus, with these primary and secondary prevention approaches such as HPV vaccination, different cervical screening methods along-with HPV DNA testing in the general population in LMIC will remarkably reduce the risk of HPV-induced cervical cancers as well as will help towards global elimination of cervical cancer in all groups of women (Gupta et al. 2012, 2018b; Bharti et al. 2009).

HPV: molecular mechanism of oncogenesis and target(s) for therapy.

HPV is an epitheliotropic virus that infects stratified squamocolumnar epithelial mucosa at the transformation zone (TZ) of the cervix through micro-abrasions caused during sexual intercourse. Thus, early age of first sexual intercourse, multiple sexual partners and other STIs enhance virus entry, progression of infections and development of lesions in women and progress through a variety of pre-malignant and malignant lesions (Stokley et al. 2014). In majority of patients ($\sim$90%) immune system clears HPV infection thus remaining asymptomatic (Steben and Duarte-Franco 2007), cases of about $\sim$10–12% HR-HPV infections that persist make the differentiating epithelial cells to reach a DNA-synthesis competent state leading to tumorigenic transformation mostly due to up-regulated transcription of E6 and E7 oncoproteins. Therefore, persistent or chronic HR-HPV infection for at least one year or more along-with other endogenous and/or exogenous factors are prerequisite for the development of cellular changes which can lead to aberrant viral oncogene expression and serve as the key factor for the genesis of HPV-induced cancers (Das et al. 2008; Bharti et al. 2009). During productive HPV infection, viral genomes remain in the episomal state in the basal undifferentiated epithelial cells and replicates together with the cellular genome. Apart from genomic predisposition, acute and chronic localized inflammatory co-infections, persistence of HR-HPVs triggers the integration of viral DNA into the host genome. Following this it makes nearly impossible for the premalignant lesions to return back to normal (Bharti et al. 2009; Moody and Laimins 2010), thus considered as a crucial step in the development of high-grade cervical lesions (Cullen et al. 1991; Das et al. 1992a, b; Pirami et al. 1997; Shukla et al. 2010).

HPV-induced tumorigenesis is a complex and multistage process associated with the occurrence of genomic alteration and/or tumorigenic transformation of the cells. Genomic alterations are the hallmark of tumorigenic transformation (Senapati et al. 2016). The molecular mechanism of HPV-induced tumorigenesis can be regulated by two key viral oncoproteins; E6 and E7 that are consistently overexpressed and are critical to the induction and maintenance of the malignant phenotype (IARC 2007; May and May 1999). These oncoproteins are also able to disrupt innate immunity by inhibiting type 1 interferon and alter functioning of several important molecular regulatory signaling pathways such as p53, Rb, EGFR, Notch1, Wnt, JNK, mTOR, STAT, AP-1 and NF-kB-mediated pathways, that are known to play critical role in regulating normal cell growth, differentiation, apoptosis and immune functions (Gupta et al. 2015, 2018a; Kuguyo et al. 2018; Olusola et al. 2019; Tyagi et al. 2017).

For maintenance of tumorigenicity and activation of replication machinery, HR-HPV-E6/E7 oncoprotein interact and bind to two key tumour suppressor proteins, p53 and retinoblastoma (pRb) respectively leading to inactivation of these two essential cell growth regulators and induction of S phase and cell cycle progression (Fig. 3). Furthermore, the E6 oncoprotein binds more efficiently to the p53 leading to the formation of a ternary complex with E6-associated protein (E6AP) causing inactivation of p53 and PDZprotein (post synaptic density protein of Drosophila disk large tumour suppressor-zonula occludens-1 proteins) while stimulating phosphoinositides 3-kinase (PI3K)/protein kinase B (Akt), Wnt/β-catenin and Notch signalling pathways (Gupta et al. 2018b; May and May 1999).

Similarly, E7 binds to unphosphorylated pRb and disrupts phosphorylated state of pRb which ultimately induces the degradation of retinoblastoma (pRb) family members (pRb1, p107 and pRb2). Thereby release of the E2F transcription factor to bind with promoters of cell cycle regulatory genes to translate cell cycle regulatory

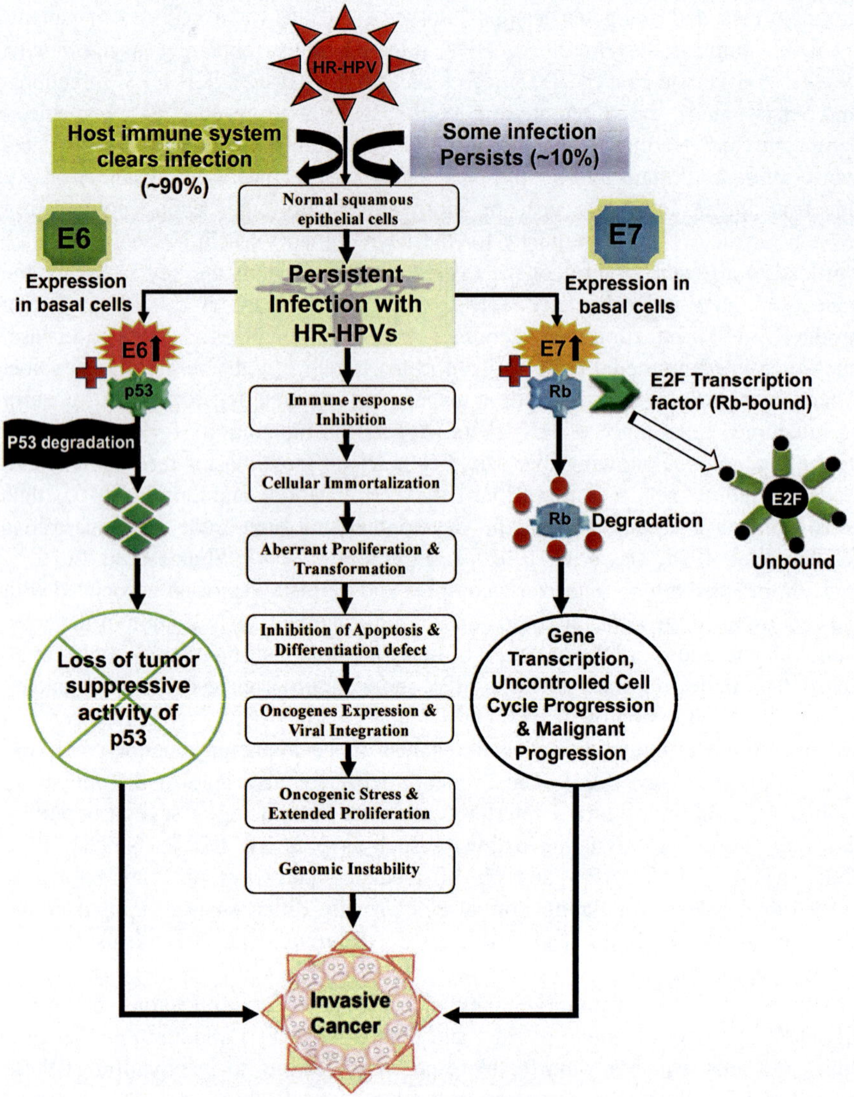

Fig. 3 Schematic diagram of molecular mechanisms of HPV-induced genomic instability and oncogenesis. Viral oncoproteins; E6 & E7 bind to p53 and Rb, respectively and induce their degradation and inactivation leading to uncontrolled cell cycle progression and growth causing genomic instability and chromosome aberrations following HR-HPV infection.

proteins and polymerases required for cell growth and cell cycle progression (Fig. 3). HR-HPV/E7 has ability to induce cell cycle progression by suppressing the activity of p21 and p27 cyclin-dependent kinase (CDK) inhibitors that is induced once p53 is stimulated in response to DNA damage (Cho et al. 2002; Helt et al.

2002; Helt and Galloway 2001). E7 also disrupts p16/pRb pathway and induces cellular immortalization. HR-HPV oncogenes specifically E5, E6 and E7 can abrogate DNA damage responses which may lead to the accumulation of several genomic alterations and activation of diverse signalling pathways and inhibit cancer suppressor genes during HPV-induced tumorigenesis (Demers et al. 1996; Slebos et al. 1994). Further, alterations in regulatory signalling pathways result cell growth, proliferation, inhibition of cell apoptosis and drug resistance. Hence, HR-HPV oncogenes and altered molecular signalling pathways may be excellent molecular biomarkers for developing effective targeted therapeutic approaches for the treatment of HPV-induced cancers (Oikonomopoulou et al. 2013; Parkin 2006). Since transcription of two essential HR-HPV oncogenes E6/E7 is tightly regulated by a host cell transcription factor, activator protein-1 (AP-1) as AP-1 binding to its URR is indispensable for overexpression of E6/E7, AP-1 especially its two-family proteins c-Fos and Fra-1 have been shown to be the potential targets for therapy (Gupta et al. 2018b; Tyagi et al. 2017).

Detection, Diagnosis and treatment of HPV-associated diseases. The comprehensive control of HPV-induced diseases can be achieved by primary prevention (HPV vaccination), secondary prevention (screening and treatment of benign lesions) and tertiary prevention (diagnosis and therapy of invasive HPV-associated cancers) (Gupta et al. 2012; Bharti et al. 2010). It has been established that organised HPV screening programmes or widespread good quality cytology can reduce both incidence and mortality of HPV-infected patients. The introduction of HPV vaccination, development of DNA-based testing for early screening, monitoring and diagnosis of HPV could also effectively reduce the burden of HPV-induced diseases in the next few decades.

HPV-induced tumorigenesis is a multistep, multistage (premalignant lesion to invasive cancer) and complex process which takes 15–20 years to develop invasive cancer thus providing a unique opportunity and an excellent window period for early detection, diagnosis, prevention and treatment of the disease through vaccination, conventional cytology or other methods including cellular and molecular testing and standard treatment approaches. Currently, three FDA-approved prophylactic HPV vaccines including bivalent Cervarix (for HPV types16 and 18), tetravalent Gardasil (for HPV type 6, 11, 16 and 18) and nonavalent Gardasil-9 (for HPV types 6, 11, 16, 18, 31, 33, 45, 52 and 58) are commercially available which provide protection against the most common HR- and LR- types of HPVs. All three vaccines are based on recombinant DNA technology and highly effective and induce high titre of specific antibodies. Further, Pap smear test, visual inspection with acetic acid can also be preferred and employed for early detection of HPV lesions which are suitable for low resource settings (Das et al. 2008). Initially, most of the HPV infected lesions are benign or asymptomatic, and are undetectable by Pap test or VIA, thus molecular HPV-DNA testing can be preferred which provides unequivocal results and are the most reliable methods for early HPV detection (Gupta et al. 2012, 2018b). When premalignant lesions are recognized, cryotherapy, loop electrosurgical excision procedure or cold knife conization can be done depending upon the grade of lesion. However, in-spite of several preventive and treatments methods such as screening, vaccination, surgery, chemo and radiotherapy, the burden of HPV-induced diseases remains high world over. Existing HPV vaccines are providing limited results to eliminate pre-existing HPV infections and having many challenges in its widespread use in developing countries including India. Therefore, new approaches should be made for the development of therapeutic HPV vaccines which can regress pre-existing HPV-associated lesions and will facilitate better control and management of HPV-induced diseases.

B. **Human-infective RNA viruses**

RNA viruses, also referred to as retroviruses, are most common pathogens inducing new human diseases, and 1 to 3 novel RNA viruses are being discovered every year (Rosenberg et al. 2015). Retroviruses exist as more genetically diversified populations than the DNA viruses in that their genetic material is single-stranded or a double stranded RNA and also they have few similar notable features to DNA viruses. Depending on their genome polarities, RNA viruses are divided into 3 groups; (i) positive-strand RNA virus, (Liu and Richardson 2013) (ii) negative-strand RNA virus and (iii) double-strand RNA virus. Presently more than 158 species of human RNA viruses have been discovered which comprise 47 genera and 17 families, only few of them cause human diseases (Parvez et al. 2019). Among all potential pathogenic viruses, RNA viruses are special human etiological agents in particular these have become important zoonotic agents emerging and transmitting from wild animals, occupying about 25–44% of all emerging human infectious diseases (Jones et al. 2009; Morens et al. 2004; Woolhouse and Gowtage-Sequeria 2005). Notably, variant viruses are more often generated in RNA viruses due to their high mutation rate than DNA viruses. Due to this property, RNA viruses are difficult and challenging to treat and control using antiviral therapies. RNA viruses that can cause variety of acute to chronic human disorders involving retroviruses, flaviviruses, filoviruses, orthomyxoviruses, paramyxoviruses and coronaviruses (Fig. 1).

I. **Human retroviruses**

Retroviruses are pathogens that belong to Retroviridae family. Poiesz and colleagues (1980) had isolated the first human retrovirus in a cutaneous T-cell lymphoma patient (Poiesz et al. 1980). These viruses are enveloped and contain about 7–10 kb RNA as their genetic material. Upon infection, their complex replication strategy is characterized by its own enzyme 'reverse transcriptase', which converts their RNA genetic information into a DNA intermediate and produces a double-stranded DNA within a host. Retroviruses can cause several human diseases including AIDS, cancer, auto-immune and neurologic diseases. The Retroviridae family is consisting of 7 subfamilies. Two most pathogenic viruses; human T cell lymphotropic virus (HTLV) type 1 and human immunodeficiency virus (HIV) of which HIV1 are most important pathogens for humans and they have also the ability to induce cancers (zur Hausen 2002; Robertson and Levin 1999).

a. **Human immunodeficiency virus (HIV)**. HIV is a RNA virus of the *lentivirus* genus classified into two strains: HIV-1 and HIV-2, belong to different locations. HIV-1 is a main agent of HIV/AIDS pandemic globally and found in chimpanzees and gorillas living in western Africa, whereas the geographical location of HIV-2 strain is endemic in sooty mangabeys, species of monkey living in western and central Africa (Chen et al. 1996; Gao et al. 1992). The viral infection causes the Acquired Immune Deficiency Syndrome (AIDS), a disease defined as a set of events due to the depletion of T lymphocyte, even if HIV is able to productively infect other cell types such as lymphocytes, macrophages, microglia and dendritic cells. In patients with HIV infections, there is an increased risk to developing cancers such as Kaposi's sarcoma, non-Hodgkin's lymphoma B-cell, primitive brain lymphoma B cells and invasive carcinoma of the uterine cervix (Tables 1 and 3). Other AIDS-related cancers include Hodgkin disease and cancers of the lung, mouth and digestive system; the observation of all these cancers in people who are HIV-positive is so relevant, compared to normal population that it could be considered as an index for AIDS definition. In patients affected by AIDS, the most important causes of death are the opportunistic secondary infections, provoked in majority of cases by viruses, bacteria, fungi and parasites. These microorganisms rarely cause a disease in people with an efficient immune system but in subjects affected by AIDS, they may cause severe diseases which are usually disseminated, difficult to treat and often have relapses. So, HIV infection exposes to a number of opportunistic malignancies and other subsequent infections, some of which in turn are potentially carcinogenic. The chance that an opportunistic infection could develop depends on the deterioration of the immune system and, in the same way, the onset of cancer can be related to the status of patient's immunity. Nowadays, due to the modern and efficient HAART therapy (Highly Active Antiretroviral Therapy), we can observe an increase in survival of patients affected by AIDS. The therapy leads to an enhancement in life expectancy but ironically it can sustain an increased risk of contracting cancers not necessarily linked to HIV infection. Among "no-HIV related" cancers, there are pulmonary, digestive system and liver cancers (Bellan et al. 2003).
b. **Human T-cell leukemia virus type 1 (HTLV-1)**. HTLV-1 is predominantly associated to adult T-cell leukemia (ATL) (Hristov et al. 2010), a disease characterized by uncontrolled growth of CD4+ T-lymphocytes, lymphadenopathy, hypercalcemia, immunodeficiency and weak prognosis (Goncalves et al. 2010). After infection, HTLV-1 has a very long latency period that can last for several decades, but once cancer begins, the progression is rapid (Table 1). Mechanisms leading to ATL are not well understood, but it is possible that a multistep leukemogenic process, in which the main roles are played by viral genes and their products and host immune status (Goncalves et al. 2010). Besides, NF-kB has been reported in ATL cells and is known that it has a regulatory function of immune response to infection and is strongly associated with oncogenesis.

II. **Human flaviviruses**

Flaviviruses belong to the family *Flaviviridae*, which are associated with variety of distinct mosquitoes-transmitted human diseases including dengue, yellow fever, Japanese encephalitis (JE), tick-borne encephalitis (TBE) etc. Flaviviruses are small, enveloped, positive-sense single-stranded vector-borne RNA viruses with approximately 9–12 kb in genome length which share a common genome organization and replication strategy. Viral replication takes place in both the vertebrate host and the insect vector. Flavivirus infections vary from asymptomatic, mild fever, body aches, head-aches, nausea, vomiting joint pain and arthralgia to life threatening and haemorrhagic (Messaoudi et al. 2015). Flaviviruses are also able to persist in patients and may cause long-term morbidities. To date, there is no specific treatment for flaviviruses infections are available (Dovih et al. 2019).

a. **Dengue viruses**. Dengue fever is caused by one of the mosquito-borne virus, called Dengue virus (DENV). The disease is mainly transmitted by biting of Aedes aegypti/or Aedes albopictus mosquitoes. Globally, more than 50 million dengue virus infections have been reported annually in tropical and subtropical regions. More than 0.5 million peoples are hospitalised and more than 3 billion peoples live in dengue endemic regions of the globe (Sanyaolu et al. 2017). DENV is consisting four distinct serotypes; DENV-1, DENV-2, DENV-3 and DENV-4. Exposure to one serotype provides lifelong homologous herd immunity against all other virus's serotype (Gibbons et al. 2007). Symptoms of dengue virus infection range from headache, severe muscle, joint pain, and rashes that usually persists for 7–14 days to potentially life-threatening haemorrhagic fever (bleeding in the skin and gastrointestinal tract) or shock syndrome (Dengue shock syndrome). There is currently no specific treatment available for DENV. Recently, an FDA approved Dengvaxia® (CYD-TDV) dengue vaccine has been developed by Sanofi Pasteur against all four strains of DENV. Dengvaxia® DENV vaccine was first licensed in December 2015 in Mexico. It is recommended for the age group of 9–45 years of individuals in more than 20 countries.
b. **Zika Virus**. The virus is primarily transmitted through the bite of an infected Aedes (Aedes aegypti) mosquito to the human in subtropical and tropical regions. However, viral transmission through sexual contact, blood transfusion, organ transplantation and mother to foetus during pregnancy has also been reported (Wikan and Smith 2016). The virus causing Zika Virus Disease (ZVD), was first discovered in 1947 from the Zika forest in Uganda in rhesus monkeys and was transmitted in humans in 1952 by Aedes aegypti mosquito (Petersen et al. 2016). The first biggest outbreak of Zika virus was reported in in 2007 in Yap Island followed by French Polynesia and Brazil in 2013 and 2015 respectively. By 2018, Zika virus epidemic explosively transmitted in more than 86 countries and territories including America (Musso and Gubler 2016).

In 2016, the WHO declared Zika virusa global public health emergency (McNeil and Shetty 2017). The symptoms of Zika virus disease can last for several days to weeks that are generally of low-grade fever, rash, joint pain, conjunctivitis, malaise, arthralgia and asthenia. Recent studies have suggested that Zika virus has also been associated with serious sequelae like microcephaly and Guillain-Barre syndrome (GBS) in infants of those mothers who were infected with the virus during pregnancy (McNeil and Shetty 2017). To detect Zika virus infection, nucleic acid amplification testing (NAAT) has been used for various specimens such as serum, amniotic fluid, urine, whole blood, semen, cerebrospinal fluid and tissues. Presently, no specific vaccine or antiviral therapy is available for ZVD. Supportive treatment such as analgesics and antipyretics can be used by clinician for viral infection.

III. **Human filoviruses (HFV)**

Human filoviruses are the members of *Filoviridae* family with non-segmented, lipid enveloped, single-stranded negative-sense RNA (-ssRNA). These viruses were first identified as the infectious pathogens of a haemorrhagic fever outbreak in 1967 in Europe. Filovirus forms filamentous thread-like morphology called virions. There are two members of this family that are namely known as the Ebola virus (EV) and Marburg virus (MV) (Martines et al. 2015). Both EV and MV are most lethal human pathogens with epidemic potential that can cause severe haemorrhagic fever illness in humans and non-human primates (Kuhn et al. 2019a, b). Bats are considered the natural reservoir for both EV and MV zoonotic pathogens (Kuhn et al. 2019a, b). HFV transmission in humans occurs from direct human-to-human contact with it infectious body fluids (vomitus, blood, semen, saliva, breast milk stool and tears) through an infected symptomatic patient. Filoviruses are classified as category A infectious agents due to their high CFR, potential aerosol infectivity, direct human-to-human transmission, and absence of specific antiviral therapy and vaccine.

Ebola virus (EV). This viral infection is a most virulent and deadliest that can causes fatal Ebola virus disease (EVD) or Ebola haemorrhagic fever in humans. The first EV infection outbreak was reported in Zaire (now known as Democratic Republic of the Congo) near the Ebola river in 1976, since then the virus is known as Ebola (Messaoudi et al. 2015; Messaoudi and Basler 2015). EVD is a major global public health concern due to its unprecedented outbreak in sub-Saharan African regions in 2013–2016 with high case fatality rate (Feldmann et al. 2020). In 2013–2016, a total number of EVD confirmed cases were more than 28,000 with 11,000 associated deaths due to this endemic. The impact of EV infection has been also observed in US when infected patients who recently travelled from Liberia, West Africa (Chevalier et al. 2014; McCarthy 2014). Recently, three new cases of EVD were reported in the Democratic Republic of the Congo between10 and 14 April 2020 (Feldmann et al. 2020). It is believed that EV is also an animal-borne virus; however bat is the main reservoir and /or nonhuman primates such as monkeys, chimpanzees and apes may be intermediate, or amplifying hosts for this virus (Messaoudi et al. 2015; Siegert et al. 1967). EVD can get through direct close contact with an infected person or animal infected with Ebola virus. It is primarily transmitted by human-to-human through direct contact from the infected patients with infected body fluids and causes severe and acute systemic disease. Ebola virus incubation period is 2–21 days and initially the EVD patients showed with non-specific influenza-like symptoms such as high fever, fatigue, body aches, weakness, stomach disorder, nausea, cough, vomiting and diarrhoea and at later stage the persistent infection may cause breathing difficulties, bloodshot eyes, internal-external bleeding, gastrointestinal dysfunction and multiple organs failure. Currently, no specific treatment or vaccines approved by FDA for EVD and only supportive and symptomatic therapy is the line of treatment. Remdesivir may be considered as the best option for its treatment (Warren et al. 2016).

IV. **Human orthomyxoviruses**

Human Orthomyxoviruses (influenza viruses) contain enveloped, 6–8 segments of negative-sense, single-stranded RNA genome. These viruses belong to the family, *Orthomyxoviridae*. This pathogenic virus causes significant types of diseases both in humans and in animals and comprises five genera on the basis of core proteins: Influenza virus A, Influenza virus B, Thogoto virus and Isa virus. Three surface glycoproteins of influenza virus; nucleoprotein, Hemagglutinin (HA) and neuraminidase (NA) are subdivided to differentiate influenza virus types. These viruses are believed to be transmitted primarily through droplets or respiratory secretions and aerosol from an infected person. Based on environmental conditions and factors such as temperature, humid surfaces, the virus can survive up to many hours (Blut 2009; Scholtissek 1985).

Influenza viruses are the most important members of this family that are major determinant of incidence and mortality and its outbreaks and can cause worldwide epidemics (Blut 2009; Scholtissek 1985). These viruses have ability to cause an acute respiratory disease called influenza. Its global prevalence has affected 5 -15% of adult population and 20–30% children. Pneumonia may develop as a complication and may be fatal, particularly in elderly persons with underlying chronic disease. Influenza type A is highly antigenic and responsible for bird flu, worldwide epidemics and pandemics influenza and types B causes recurring regional epidemics.

In the past 100 years, these viruses have been causing three major pandemics, (i) 1918-'Spanish Flu' (H1N1 strain), (ii) 1957-'Asian Flu' (H2N2 strain) and (iii) 1968-'Hong Kong Flu' (H2N3 strain). 1918-'Spanish Flu' was the worst pandemic associated with highest mortality ($\sim$40 million) worldwide (Table 4). In addition, recent outbreaks of avian subtypes such as H5N1, H9N2, H7N7, H7N3 and H10N7 in birds have caused occasional human diseases in distinct parts of the world (Blut 2009). The common signs and symptoms of influenza disease are sudden onset of fever, sore throat, cough, malaise and headache. The preferred definitive diagnosis is based on rapid influenza virus-specific antibodies detection in the serum via immunostaining tests such as immunofluorescence, enzyme immunoassay (ELISA), hemagglutination inhibition test (HIT) and the neutralisation test (Blut 2009). An inactivated vaccine against influenza virus has been used for ~40 years to prevent the disease. Rimantadine and Amantadine are the choice of drugs being used for treatment of influenza A virus infections.

V. **Human Coronaviruses (HCoV) and their diseases**

The term Coronaviruses (CoVs) named for crown-like spikes on their surface in 1968 and belong to the family *Coronaviridae* was established in 1975 by ICTV. CoVs are enveloped, positive-sense, single-stranded RNA viruses with largest genome size ranged between 26.4 and 31.7 kb that have the ability to infect different animal species including birds and humans. They are emerging and re-emerging viruses which can cause upper respiratory tract, enteric and central nervous system (CNS) illness ranging from mild (common cold like symptoms) to fatal diseases. Coronaviruses include Severe Acute Respiratory Syndrome coronavirus-1 (SARS-CoV-1), Middle East respiratory syndrome coronavirus (MERS) and Severe acute respiratory syndrome Coronavirus-2 (SARS-CoV-2) which causes Coronavirus disease -2019 (COVID-19). There is now few licenced vaccines and antiviral or repurposed drugs available to prevent or treat infections. India was the first country in the world to introduce two successful indigenously developed Covid-19 vaccines, Covaxin (inactivated virus) and covishield (Coronavirus spike protein in Adenovirus vector) in January, 2021. CoV family consists of four genera, α-CoVs, β-CoVs, γ-CoVs, and δ-CoVs. There are 7 known human coronaviruses (HCoVs) species that can infect humans and can cause mild to severe respiratory illness. Four HCoV variants (HCoV-229E, HCoV-NL63, HCoV-OC43 and HCoV-HKU) can cause common mild cold-like symptoms in immunocompromised persons and other 3 (MER-CoV, SARS-CoV-1 and SARS-CoV-2) of the seven HCoVs caused global pandemic and are highly pathogenic with high transmission and fatality rates (Cui et al. 2018). Coronaviruses particularly SARS-CoV-2 is assumed to have transmitted from bat to human first in the animal meat market in Wuhan, China in 2019 (Zhou et al. 2020b).

a. **Severe acute respiratory syndrome coronavirus-1 (SARS-CoV-1)**. Only 12 animal or human coronaviruses identified before the emergence of SARS-CoV-1 in November 2002. The virus is contagious caused by SARS-CoV-1 which was first identified in China and then spread to 4 other countries. The virus potentially infects epithelial cells of the lungs which results in severe and potentially fatal upper respiratory tract illness (Ksiazek et al. 2003). The fatal SARS cases are mainly dominated by diffuse alveolar damage (Peiris et al. 2003). Following its introduction to Hong Kong in 2003, the virus transmitted to more than 30 countries and over 8,422 people were affected with yielding ~10% a global crude fatality rate (CFR) (Table 4). The virus infects humans and has been detected in palm civets and a raccoon-dog in a Southern China market (Lee et al. 2003). The ACE2 (angiotensin converting enzyme 2) transmembrane protein in humans and bats has been identified as an entry receptor into human cells for the SARS-CoV (Donoghue et al. 2000). It is an airborne virus and has ability to transmit from person-to-person and one country to another through infected person's small respiratory droplets. Currently, no

Table 4 Modes of emergence and re-emergence of most common human respiratory transmitted viruses (RTV) and their key characteristics

Respiratory transmitted viruses	SARS-CoV-2	SARS-CoV-1	MERS-CoV	Influenza Viruses			
				H1N1 	H1N1/pdm09 	H2N2 	H5N1 
Disease(s)	COVID-19	SARS	MERS	Spanish flu	Swine Flu/Mexican flu	Asian Flu	Bird/Avian Flu
Known or suspected reservoir(s)	*Bats, Pangolin*	*Bats*	*Dromedary camels*	*Humans and swine*	*Swine, Turkeys, ferrets*	*Bird*	*Birds*
Genomic material	(+) single-stranded RNA	(+) single-stranded RNA	(+) single-stranded RNA	(−) single-stranded RNA	(−) single-stranded RNA	(−) single-stranded RNA	(−) single-stranded RNA
Incubation period	4–14 days	2–7 Days	6 Days	1–7 days	1–4 days	1–4 days	2–5 days up to 17 days
Basic reproduction number (R_o)	2.2	2.8	0.5	1.8	1.75	1.65	1.89
Transmission route	Direct contact via respiratory droplets or aerosols	Direct contact via respiratory droplets or aerosols	Direct contact via respiratory droplets or aerosols	Direct contact through coughing or sneezing from an infected person	Direct contact through coughing or sneezing from an infected person	Direct contact through coughing or sneezing from an infected person	humans contact with infected bird feces, nasal secretions, or secretions from the mouth or eyes
Common symptoms	Fever, cough, and shortness of breath	Fever, malaise, myalgia, headache	Fever, chills, myalgia, and cough	Sore throat, headache and fever	Fever, cough, headache, muscle or joint pain, sore-throat, chills, fatigue	lose weight, fever, sneezing, nasal discharge, pneumonia and seizures	Fever, cough, sore throat, muscle aches, conjunctivitis, breathing problems and pneumonia
Diagnosis	RT-PCR, Serological testing	RT-PCR	RT-PCR, Serological testing	PCR	RT-PCR, Rapid Influenza Diagnostic and Immunofluorescence	PCR	AVantage A/H5N1 flu test kit and RT-PCR

(continued)

Table 4 (continued)

Respiratory transmitted viruses	SARS-CoV-2 	SARS-CoV-1 	MERS-CoV 	Influenza Viruses			
				H1N1 	H1N1/pdm09 	H2N2 	H5N1 
Treatment Vaccine	Supportive No licenced vaccine	Supportive No licenced vaccine	Supportive No licenced vaccine	Traditional medicine No licenced vaccine	Flu Shot, Nasal Spray flu vaccine and monovalent (H1N1)pdm09 vaccine	Supportive No licenced vaccine	Vaccine AUDENZ is Available No highly effective treatment -oseltamivir
Outbreak periods and origin	2019–2020 Wuhan, China	2002–2004 Guangdong, China	2012, 2015, 2018 Saudi Arabia	1918–1920 Spain	2009–2010 Mexico	1957–1958 Guizhou, China	2003–2019 Southeast Asia and Egypt
Outbreak type	Pandemic	Epidemic	Epidemic	Biggest Pandemic	Pandemic	Pandemic	Pandemic
Estimated global cases and deaths	Cases-21,605,614 Deaths-768,226 (as of 15 Aug 2020)	Cases-8,422 Deaths-774	Cases-2499 Deaths-858	Cases-$\sim$500 million Deaths-$\sim$17–50 million	Cases-0.7–1.4 billion Deaths-151700–575400	Cases-$\geq$500 million Deaths-1–4 million	Cases-861 Deaths-455

approved antiviral drugs against SARS-CoV-1 are available. The rapid spread in the human population with high mortality of SARS-CoV-1, its transient re-emergence, and global economic disruptions led to a rush for research on the epidemiological, pathological, molecular, and immunological aspects of the virus and the disease. Due to non-specific clinical illness caused by SARS-CoV-1, diagnosing, management, development of treatment regimens and controlling future infectious diseases will require urgent global coordination to contain the infection and its future outbreak.

b. **The Middle East respiratory syndrome coronavirus (MERS-CoV)**. MERS CoV is a fatal zoonotic virus that cause MERS disease in humans and was first discovered in 2012 in Saudi Arabia and Jordan (Hijawi et al. 2013). From April 2012 and December 2019, a total of 2499 MERS-CoV infection cases and 858 associated deaths (CFR: 34.3%) were recorded in 27 countries of the world. Out of 2499 cases, the majority (2106 cases and 780 related deaths) of them were reported from Saudi Arabia (Arabi et al. 2014). The actual reservoir for MERS-CoV is presently not known but dromedary camels (*Camelus dromedarius*) are supposed to be the animal source of MERS-CoV infection transmitted to humans (see Table 4). Unlike SARS-CoV-1 infection, which was controlled soon after its outbreak, MERS-CoV infection continued to spread and cause human illness worldwide.

For MERS-CoV infection the host cell dipeptidyl-peptidase 4 (DPP4, also known as CD26) was recognized as the receptor for the entry of the virus into the cell. The high case fertility ratio (CFR) in hospital-based and family-based cluster outbreaks was noted especially in co-morbid and immunocompromised patients. The route of human-to-human transmission of MERS-CoV infection could be through direct or indirect close contact, oral routes including consumption of contaminated food items. The incubation period of MERS-CoV infection ranges between 2 and 14 days (median 5.2 days). The clinical presentation of MERS-CoV infection varies from asymptomatic infection to acute to chronic respiratory illness with severe pneumonia, ADR (acute respiratory distress syndrome), respiratory and multiorgan failure leading to patient's fatal outcome. MERS-CoV infection can be detected by PCR and serological tests using different specimens (nasal swabs, saliva, urine, serum, and stool etc.) from MERS patients (Memish et al. 2020). Presently, there is no FDA-approved specific anti-MERS-CoV vaccine or drug are available for the treatment of MERS patients. However, supportive care/treatments are the main clinical management for these patients (Memish et al. 2015).

c. **Severe acute respiratory syndrome coronavirus-2 (SARS-CoV-2)**. Since December 2019, COVID-19 (Coronavirus disease 2019) pandemic has serious ramifications on public health, severely affecting the global healthcare systems, killing millions of people, damaging economy and collapsing of numerous industries. The uncontrollable outbreak of COVID-19 is now wreaking havoc and escalating quickly in nearly 213 countries and territories including the most affected countries: USA, Brazil, India and Russia etc. But now India almost tops the world with its daily toll of coronavirus infection touching 100,000 new cases per day. The novel coronavirus is now known as severe acute respiratory syndrome coronavirus 2 (SARS-CoV-2) which was first originated in Wuhan, Hubei province of China on December 31, 2019, has now pandemic across the globe (https://www.worldometers.info/coronavirus/#countries; Chinazzi et al. 2020). At present (15 September, 2020), the global infection recorded is 29,789,734 and death has reached to 940,362. Cumulative global attempts are ongoing to find effective vaccines to control the viral spread as well treatments to save the life of millions of infected people world over (Table 3) (https://www.worldometers.info/coronavirus/#countries).

COVID-19 has been found to be caused by infection of a very aggressive type of coronavirus; SARS-CoV-2 which has more similarity (88% identity) with two bat-derived SARS-CoVs (bat-SL-CoVZC45 and bat-SL-CoVZXC21) but less sequence homology with SARS-CoV-1 (about 79%) and MERS-CoV (about 50%). (Lu et al. 2020b; Wu 2020). Based on its genomic analysis and phylogenetic relationships, bats have been associated to be most likely primary host for these viruses (Zhou et al. 2020b; Lu et al. 2020a, b; Shang et al. 2020). The virus infection can be asymptomatic, low/mild symptomatic and have strong binding affinity to human respiratory ACE2 receptor (Wan et al. 2020) on the lung epithelial cell surface causing highly lethal pneumonia and acute respiratory distress syndrome (Steben and Duarte-Franco 2007), multi-organs failure and death mainly in elderly people, the majority of whom have co-morbidity factors (Tang et al. 2020).

Genome structure and functions of SARS-CoV-2

SARS-CoV-2 is a spherical, small, enveloped positive-sense single-standard RNA (+ssRNA) virus (see Table 4) with its solar crown-like/club-shaped appearance due to surface spike glycoproteins (Zhou et al. 2020b; Lu et al. 2020a, b; Chan et al. 2020; Wu et al. 2020; Wu and McGoogan 2019). The RNA genome of this virus comprises as many as 29 open reading frames (ORFs) and size of genome is about 29.9 kb. It belongs to the family '*Coronaviridae*' and the genus *Betacoronavirus* (Khailany et al. 2020). SARS-CoV-2 RNA genome contains at least 16 non-structural proteins (nsps) to form the replication-transcription complex (RTC), 4 major structural proteins; nucleocapsid (N) protein, envelope (E) glycoprotein, membrane (M) glycoprotein and spike (S) glycoprotein) encoded by the viral genome on the envelope and 6–7 special structural and accessory proteins (HE protein, 3a/b protein, and 4a/b protein) that are translated from the sgRNAs of virus. Out of four structural proteins, spike (S) glycoprotein is cleaved into two glycosylated subunits (S_1 and S_2) which are responsible for binding to the host human angiotensin-converting enzyme-2 (ACE2) receptor and host-viral cell fusion (Tortorici et al. 2019). The transmembrane S protein has a higher affinity to bind to ACE2 receptor unlike in SARS-CoV-1 allowing the entry of virus into susceptible host cells leading to high chances of human-to-human spread and disease severity (Tortorici et al. 2019). The viral N protein binds to genomic RNA and synthesizes a helical capsid (ribonucleic capsid) which helps in viral genome protection, replication and assembly of virions and also interacts with membrane and non-structural proteins (specifically nsp3). The M protein facilitates virions assembly and budding through enrolment of other SPs to endoplasmic reticulum-golgi-intermediate compartment (ERGIC) (Harapan et al. 2020). It also helps in the interaction with N protein for packaging of RNA into virion and may involve in mitigation of immune response. The last E SARS-CoV-2 protein localizes to ERGIC and involved in assembly, budding and viral pathogenesis (Hoffmann et al. 2020; Mousavizadeh and Ghasemi 2020). For SARS-CoV-1 and SARS-CoV-2, tissue cell culture models are suitable for characterising viral replication cycle, cell tropism, and virus-induced pathogenesis.

Disease pathogenesis and clinical manifestations

The binding of SARS-CoV-2 spike protein to ACE2 receptor allows the virus to enter and infect respiratory mucosa of the host cell (Hoffmann et al. 2020; Mousavizadeh and Ghasemi 2020). Replication cycle of virus mainly found in mucosal epithelium of upper respiratory tract and later further multiplication causes a severe lower respiratory tract infection (Zou et al. 2020). The virus spike proteins help it for the initial attachment and cellular entry through ACE2 receptor and then the virus hijacks the host cell and gain access into the cytoplasm. Studies have reported that this process is commonly accomplished by the cellular surface transmembrane protease, serine 2 (TMPRSS2) followed by membrane fusion with host cells (Hoffmann et al. 2020). Once the receptors binding and fusion completed, viral RNA is released, its genome is translated into polyproteins. Viral nucleocapsid is assembled from genomic RNA and R protein in the cytosol and formation of mature particles by budding into the lumen of the endoplasmic reticulum (ER) (see Fig. 4) (Shereen et al. 2020; Zhou et al. 2020a). The mature virions are then released from the host's infected cell through exocytosis and are transmitted to infect other organs cells of the body such as liver cells, kidney cells, intestine, and lymphocytes as well as lower respiratory system.

Initially, the majority of COVID-19 patients remain asymptomatic and may transmit infection in vulnerable people. Thus, understanding clinical manifestations and following-up of asymptomatic COVID-19 patients is important during coronavirus infection. The sign and symptoms of COVID-19 in patients appear between 5 and 14 days (Wu and McGoogan 2019; Singhal 2020). In rare cases, the incubation period has been found to be as long as 24 days (Harapan et al. 2020; Huang et al. 2020b). The duration of incubation period is based on a number of factors including patient's age, history of co-morbidity, immune system, but majority of infected patients can show symptoms in 11.5 days after virus exposure (World Health Organization (WHO) 2020a; Lauer et al. 2019). SARS-CoV-2 typically causes mild to moderate upper respiratory tract infections but occasionally it can cause severe pneumonia and lower respiratory tract infections with extrapulmonary clinical manifestations in both immunocompromised and older patients. COVID-19 patients usually experience several common symptoms including fever, cough, myalgia, sore throat in conjunction with some other uncommon respiratory problems such as headache, fatigue, diarrhoea, runny nose, dyspnoea, breathing

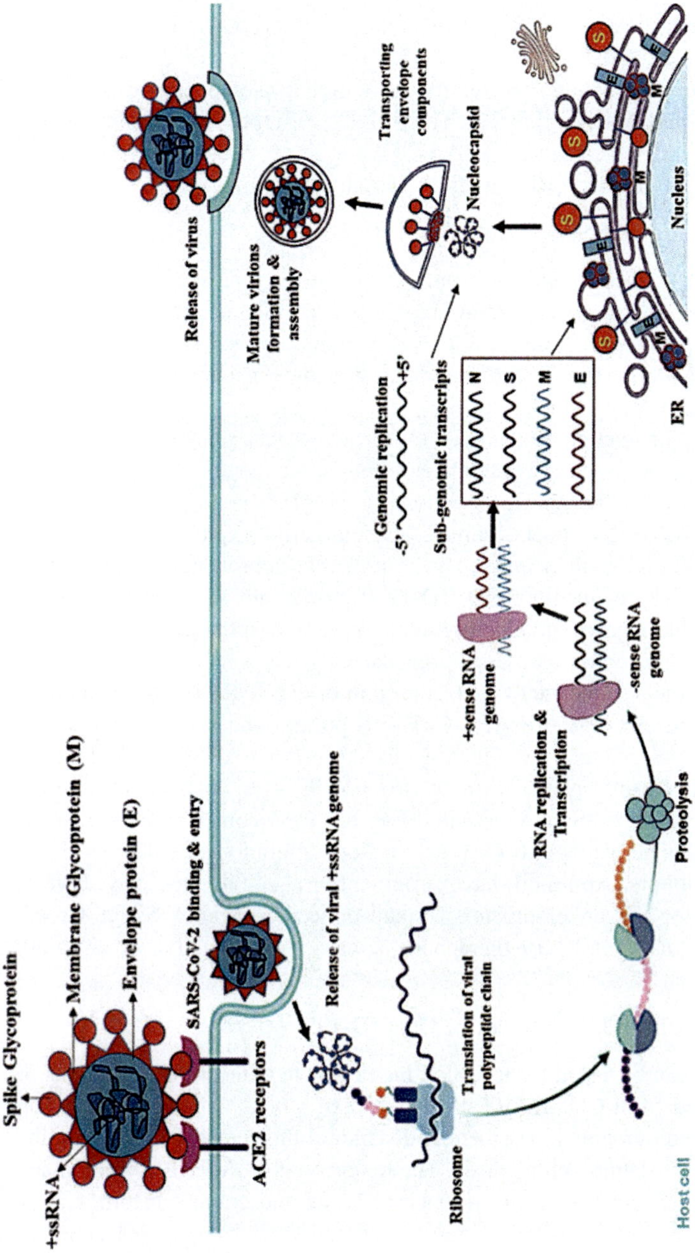

Fig. 4 Novel coronavirus transmission and pathogenesis cycle in the host cell. Spike (S) glycoprotein of SARS-CoV-2 bind to ACE2 receptors on the surface of the target cell membrane allowing virus to enter into the host cell through the endosomal pathway leading to high chances of disease pathogenesis and severity. Next, virus genomic ssRNA unveiled and releases in the cytosol of host susceptible cell. After human ACE2 receptor binding and viral-host cell fusion, virus RNA transcribed and proteins are synthesized in the cytoplasm. Viral helical ribonucleic capsid synthesized, virions assembled at the cell membrane and genomic RNA packaged into virion and mature particle forms by budding into the lumen of the endoplasmic reticulum (ER). The mature virions then released through exocytosis and begin pathogenesis to infect other organ cells

difficulty, pneumonia to acute respiratory distress syndrome (Stebbing et al. 2020), and even death in severe and older patients (Adhikari et al. 2020; Wu et al. 2020; Huang et al. 2020b). The infection of SARS-CoV-2 can be detected in bronchoalveolar-lavage (93%), sputum samples (73%) nasal swab (63%), pharyngeal swab (32%) (Bastola et al. 2020; Wang et al. 2020b, c) and saliva samples (12–50%) (Zhang et al. 2020a, b; To et al. 2020). The overall fatality rate (CFR) depends between country to country but a global rate appears to be around 6.2.% (Abduljalil and Abduljalil 2020). The estimate of reproductive number (R_0) of SARS-CoV-2 varies between 2.2 and 5.7 (Wang et al. 2020a) which is higher than SARS-CoV-1 (Guan and Zhong 2020; Liu et al. 2020).

Potential targets for SARS-CoV-2

Human ACE2 is a surface receptor that protects host from lung injury but it serves as a point of entry for SARS-CoV-2. ACE2 is also identified as binding partner of the SARS-CoV-1 spike (S) protein. Recent functional studies are revealing that SARS-CoV2-ACE2 binding-directed treatment strategies can inhibit the entry of virus into the host cell which can be an ideal antiviral therapeutic target for SARS-CoV2 (Zhang et al. 2020a, b). Cell surface protease enzyme, TMPRSS2 (transmembrane serine protease 2) can cleaves both spike (S) and ACE2 protein of SARS-coronaviruses. Cathepsin L is a lysosomal pH-dependent protease that facilitates SARS-CoV entry via endosomes. Thus, targeting expression and activity of both TMPRSS2 and Cathepsin L may be potential approaches in developing antiviral drug for the treatment of COVID-19 (Lindahl and Li 2020). Furin is a protease enzyme that cleaves envelope proteins of various influenza viruses, HIV, ebolavirus and certain coronaviruses. Furin-like cleavage site has recently been identified in the protein sequence of S glycoprotein of the SARS-CoV-2. Inhibiting Furin can block the cleavage of SARS-CoV-2 S protein and suppress virus production which may serve as a new therapeutic target for SARS-CoV-2. Further, Phosphatidylinositol 3-phosphate 5-kinase (PIKFYVE) is a kinase synthesizes a class of phosphoinositides that are essential for early endosome formation and its activity required for human SARS-CoV-2 infection. Inhibition of PIKfyve kinase using specific inhibitors (Apilimod) have strong antiviral activity and may serve as potential targets for the development of small-molecules against SARS-CoV-2 representing potential COVID-19 therapeutic approach (Bouhaddou et al. 2020; Kang et al. 2020).

Detection, Diagnosis and Treatment of COVID-19

Suspected person can be tested for infection based on the clinical manifestations or who have travelled from COVID-19 affected part of world. Clinical presentation accompanied by radiographic assessment and standard molecular diagnosis are the possible methods for definitive diagnosis. The majority of COVID-19 symptomatic patients show bilateral involvement under chest X-ray and computed tomography (CT) imaging. CT image diagnosis is a non-invasive method that contains a number of X-ray measurements at various angles across a chest to generate cross-sectional images which mainly depends on the stage and onset of symptoms after viral infection. Furthermore, viral RNA can be detected in various respiratory specimens (nasopharyngeal swabs, sputum, bronchoalveolar lavage fluid, blood, deep throat saliva and fibro-bronchoscope brush biopsy) by qRT-PCR method (Xu et al. 2020). Viral protein testing for SARS-CoV-2 infection can be also used for the diagnosing of COVID-19 patients. Serological testing such as ELISA, neutralization assay, chemiluminescent immunoassay and immunochromatographic assays could also be used to check immunologic reaction and which can detect IgM, IgG, or total antibodies against virus. In addition, genome sequencing may be performed to analyse mutational landscape (World Health Organization 2020b, c).

On the basis of clinical presentation, the clinical stages are categorised into 3 phases; (i) the acute phase (pneumonia) (Katakura et al. 2005), (ii) the chronic phase (viremia) and (iii) the recovery phase. If patient's immune system in pneumonia phase functions effectively, and no other diseases exists, the virus can be successfully suppressed and infection can get cleared and enter the recovery phase. If the patient's immunity is weak or the patients are older or associated with comorbidities, the immune system cannot function effectively to control the virus in pneumonia phase and the patient will become critical or go to viremia phase. During the infection period, total leukocyte count in the early stage of the disease is slightly low (lymphopenia) and may gradually decrease when the disease progresses, which may affect antibody production in infected patients (Wang et al. 2020a; Guan and Zhong 2020). Further, a high level of D-Dimer, C-reactive protein, inflammatory cytokines, blood urea nitrogen, creatinine and high prothrombin time are frequently noted among severe patients or non-survivors (Wang et al. 2020a).

Recommendations for admission to critical care units (oxygen-based therapy), strategies for community transmission control, and measures to reduce health care-associated spread are being established (Wax and Christian 2020). More importantly, a thorough understanding of natural history and biological behaviour of the virus and the disease is urgently needed to develop targeted therapies and/or more effective vaccines.

A number of antiviral drugs that have been used for the treatment of other viral diseases have been tried on COVID-19 patients but only with limited success. A recent study showed that effective concentrations of combinational therapy with remdesivir/lopinavir/Ivermectin (protease inhibitors), homorringtonine and emetine against SARS-CoV-2 are proven to have significant improvement on patient's clinical outcome (Choy et al. 2020). Other studies or case reports indicated that combination treatment with anti-HBV (IFN-alpha) and anti-HIV ritonavir-boosted Lopinavir drugs have anti-SARS-CoV activity (National Health Commission of the People's Republic of China 2020). In addition, remdesivir has been shown antiviral activity against SARS-CoV and MERS-CoV both *in vivo* and *in vitro* studies and now used as a potential drug on a compassionate basis for COVID-19 patients treatment (Wu et al. 2020; Holshue et al. 2020; Sheahan et al. 2020). A number of multicentre clinical trials recently conducted in multiple hospitals in various countries to assess safety and efficacy of chloroquine (CQ) or hydroxychloroquine (HCQ) (antimalarial and antirheumatic drugs) for the treatment of COVID-19 patients and these drugs were found to be safe and effective in treating COVID-19 patients (Gao et al. 2020; Savarino et al. 2003). and have potential to inhibit virus entry into the host cells by interfering with the glycosylation of its ACE2 receptor (Huang et al. 2020a). Recently, combination therapy of HCQ and azithromycin have been found to have a significant synergistic effect in decreasing viral load, effective for acute infection and associated with reduced mortality rate in COVID-19 patients (Gautret et al. 2020; Molina et al. 2020). Furthermore, several other potential drugs such as Arbidol, Favipiravir, Umifenovir (Targeting ACE2/S protein), darunavir/cobicistat, Tocilizumab, Camostat mesylate (TMPRSS2 inhibitor), emtricitabine/tenofovir alafenamide and baloxavir are being assessed in clinical trials and now recommended for treatment of COVID-19 patients in various countries including China, Russia and Japan (Harrison 2020). Favipiravir inhibits the SARS-CoV-2 by targeting the catalytic domain of nsp12 and recommended for the treatment of COVID-19. Other immunomodulatory drugs, herbal or bio-active compounds can act as anti-viral agents and used to improve immunity against SARS-CoV-2 in patients (Ingraham et al. 2020; Panyod et al. 2020). Though, these drugs are potentially effective against COVID-19 patients but at this point these are not routinely recommended as standard treatment for SARS-CoV-2. Hence, there is an urgent need for large-scale global coordination for the development of therapeutic vaccine and/or anti-SARS-CoV-2 drug for an effective treatment and control of COVID-19 pandemic. A large number of patients hospitalized with Covid-19 have developed high level of life-threatening blood clots leading to deadly thromboembolic events. Therefore, patients treated with anticoagulants showed improved survival. It is demonstrated that Covid-19 patients given anticoagulants orally or intravenously can prevent possible deadly events such as heart attack, strokes and pulmonary embolism in Covid-19 patients.

Prevention and vaccine approach

A number of features of SARS-Cov-2, including long incubation period, asymptomatic feature and transmission from asymptomatic subjects and recurrent mutations etc. make prevention and management of the disease difficult. Supportive care strategies are critical for dealing with COVID-19 infected patients. Preventive measures such as self-isolation, social/physical distancing, avoiding unnecessary community gathering and use of face mask are being recommended and are mandatory to prevent and control disease spread. Nearly every government in the world are taking precautionary actions such as partial/complete lockdown; public distancing, self-isolation/quarantine (14 days) and fast testing etc. to reduce and control the spread of highly contagious corona virus and in most cases it is working effectively. Quarantine of suspected people helps in reducing the possibility of infection transmission to healthy population. Community lockdown by several nations at the early onset of the outbreak meant limiting possibility of community transmission. Extensive thermal screening of people remains initial step for identification of COVID-19 symptom and mandatory use of face mask and keeping distance (of no less than 2 m) and regular washing of hands with soap or sensitization of hands especially when coming home from outside are ways to protecting and control person-to-person transmission of infection. Healthy diet, fresh air and normal amount of exercise have been found to improve immunity and fight against the infection (Fig. 5). Also, prophylactic use of some medicines such as paracetamol, certain herbal preparation including Homeopathic medicines such as Arsenicum, Rhus toxicodendron and Mercurius solubilis are often recommended.

Until such time the virus exists amongst us and as there is no specific treatment available for COVID-19 patients, except vaccine or administration of analgesic and antipyretic drugs, respiratory support through mechanical ventilation and use of pre-existing antibiotic/antiviral drugs are to control the disease. The most effective long-term approach for prevention and elimination of COVID-19 outbreaks would be the development of an effective SARS-CoV-2 specific vaccine which only can provide immunological protection against this virus. However, currently researchers all over the world are battling for the development of effective SARS-CoV-2 vaccine and its universal access to end the pandemic. Currently, ≥ 100 candidate vaccines are under development globally, and amongst them more than ten were at advanced stage of clinical trials on human subjects (Amanat and Krammer 2020). There are atleast 10 vaccines are already placed in immunization program in different countries (see Table 5). In India, two indigenous vaccines were developed, Covaxin (Bharat Biotech and National Institute of Virology, Pune ICMR), an inactivated Corona vaccine while Covishield (Serum Institute of India in collaboration with AstraZeneca, Oxford UK) is a Corona virus spike protein in adenovirus vector called AZD1222 and ChAdOx1. Both vaccines together have been given to more than 60 millions people in India. The data indicate that the Covishield vaccine has an overall efficacy of 70% with vaccine efficacy at 62.1%. while Covaxin has been r reprted 81% effective in an intrim analysis 2,58,000 participants. Both vaccines were found to be safe and triggerd good immune responses.

Conclusion

Persistent Sars Cov-2 infections may induce a number of cellular and biomolecular modifications in the host, which can evolve over time, leading to an increased risk of variety of human diseases including cancer. One of the first approaches could be large clinico-epidemiological studies to obtain a better understanding of the natural history and biological behavior of virus infection and its relationship with specific human diseases. Some studies focused on the increased use of antipyretics and other drugs available over the pharmacy counter without proper medical prescription, that are able to suppress the symptoms of acute infections regardless of possible effects on the immune system, resulting in an increased risk of disease development. Other priority research could be development of vaccines against various infectious agents to control the disease and associated mortalities. As a matter of fact, the development of vaccines against HBV for young children and HPV vaccines for adolescent girls can be considered as an effective prevention strategy for two most prominent cancers (hepatocellular and genital carcinomas). Further studies are needed to fill our knowledge gaps on the link between emerging, re-emerging and pre-existing viral infections both acute and chronic and the

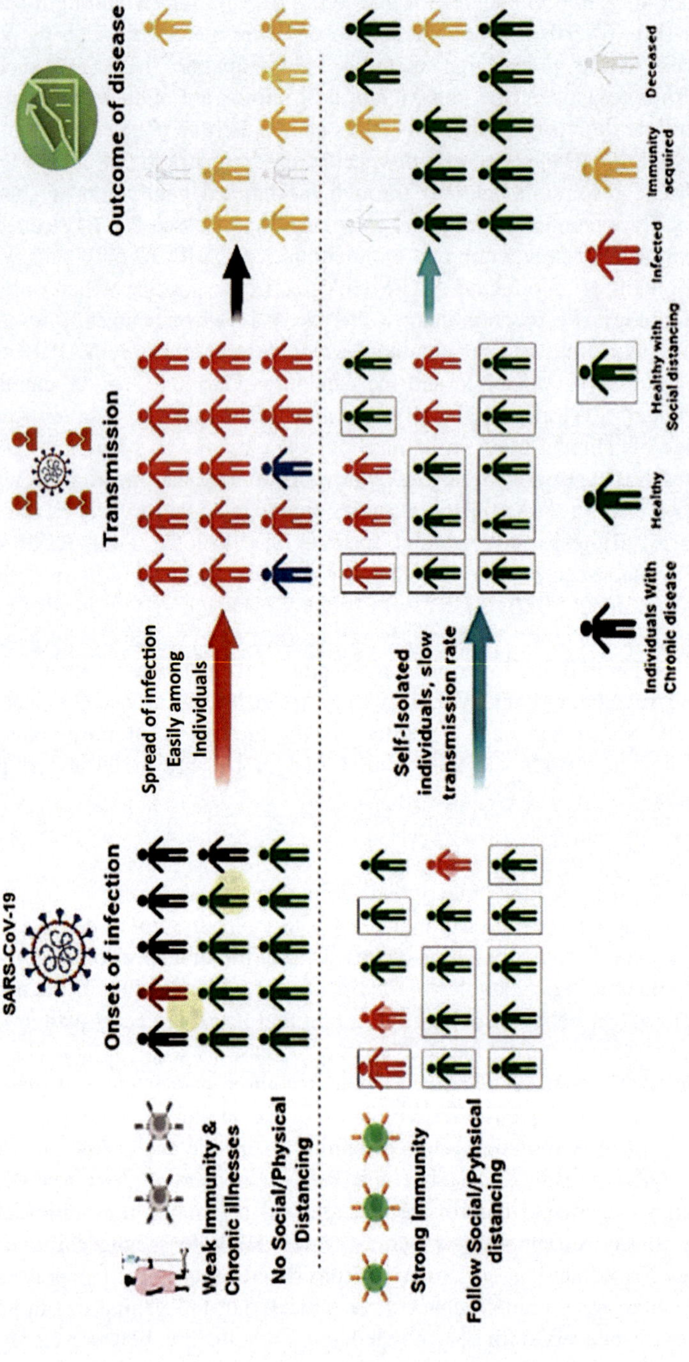

Fig. 5 Social/Physical distancing and strong immunity slow down the rate and extent of community transmission of SARS-CoV-2 infection which eventually leads to reduce disease incidence and mortality

Table 5 List of corona virus vaccines

S. No.	Vaccine Name	Vaccine Type	Primary Developers	Country of Origin
1	Comirnaty (BNT162b2)	mRNA-based vaccine	Pfizer, BioNTech; Fosun Pharma	Multinational
2	Moderna COVID 19 Vaccine (mRNA-1273)	mRNA-based vaccine	Moderna, BARDA, NIAID	USA
3	COVID-19 Vaccine AstraZeneca (AZD1222); also known as Covishield	Adenovirus vaccine	BARDA, OWS	India & UK
4	Sputnik V	Recombinant adenovirus vaccine (rAd26 and rAd5)	Gamaleya Research Institute, Acellena Contract Drug Research and Development	Russia
5	COVID-19 Vaccine Janssen (JNJ-78436735; Ad26.COV2.S)	Non-replicating viral vector	Janssen Vaccines (Johnson & Johnson)	The Netherlands, USA
6	CoronaVac	Inactivated vaccine (formalin with alum adjuvant)	Sinovac	China
7	BBIBP-CorV	Inactivated vaccine	Beijing Institute of Biological Products; China National Pharmaceutical Group (Sinopharm)	China
8	EpiVacCorona	Peptide vaccine	Federal Budgetary Research Institution State Research Center of Virology and Biotechnology	Russia
9	Convidicea (Ad5-nCoV)	Recombinant vaccine (adenovirus type 5 vector)	CanSino Biologics	China
10	Covaxin	Inactivated vaccine	Bharat Biotech, NIV-ICMR	India

development of various viral diseases including cancers in human. The present COVID-19 pandemic has taught us great lesion to conform the continued threat of infectious viral diseases and new infections that may evolve for which world should be prepared to face the future challenges. Whether SARS-CoV-2 or any other highly pathogenic virus will emerge or re-emerge as an epidemic infection is currently not possible to predict. However, constant awareness and viral surveillance studies on animal species around us including bats, rodents, birds, ticks, mosquitoes and livestock, are essential to understand the pathobiology of potential human pathogenic agents that exist around us in the environment before they can spill-over. Therefore, there is a need for research on these potential emerging viruses and human development associated variation in microbiomes and to develop broad-spectrum prophylactic vaccines and therapeutic approaches to control viral diseases and to remain prepared for the present as well as future emerging and re-emerging pathogenic viruses and other microbes.

References

Abduljalil JM, Abduljalil BM. Epidemiology, genome, and clinical features of the pandemic SARS-CoV-2: a recent view. New Microbes New Infect. 2020;35:100672.

Agelli M, Clegg LX. Epidemiology of primary Merkel cell carcinoma in the United States. J Am Acad Dermatol. 2003;49(5):832–41.

Allander T, Andreasson K, Gupta S, Bjerkner A, Bogdanovic G, Persson MA, et al. Identification of a third human polyomavirus. J Virol. 2007;81(8):4130–6.

Amanat F, Krammer F. SARS-CoV-2 vaccines: status report. Immunity. 2020;52(4):583–9.

Andreoni M, Goletti D, Pezzotti P, Pozzetto A, Monini P, Sarmati L, et al. Prevalence, incidence and correlates of HHV-8/KSHV infection and Kaposi's sarcoma in renal and liver transplant recipients. J Infect. 2001;43(3):195–9.

Arabi YM, Arifi AA, Balkhy HH, Najm H, Aldawood AS, Ghabashi A, et al. Clinical course and outcomes of critically ill patients with Middle East respiratory syndrome coronavirus infection. Ann Intern Med. 2014;160(6):389–97.

Bagga S, Bouchard MJ. Cell cycle regulation during viral infection. Methods Mol Biol. 2014;1170:165–227.

Bastola A, Sah R, Rodriguez-Morales AJ, Lal BK, Jha R, Ojha HC, et al. The first 2019 novel coronavirus case in Nepal. Lancet Infect Dis. 2020;20(3):279–80.

Beam E, Razonable RR. Cytomegalovirus in solid organ transplantation: epidemiology, prevention, and treatment. Curr Infect Dis Rep. 2012;14(6):633–41.

Bellan C, De Falco G, Lazzi S, Leoncini L. Pathologic aspects of AIDS malignancies. Oncogene. 2003;22(42):6639–45.

Berrington de Gonzalez A, Yun JE, Lee SY, Klein AP, Jee SH. Pancreatic cancer and factors associated with the insulin resistance syndrome in the Korean cancer prevention study. Cancer Epidemiol Biomarkers Prev. 2008;17(2):359–64.

Bharti AC, Shukla S, Mahata S, Hedau S, Das BC. Anti-human papillomavirus therapeutics: facts & future. Indian J Med Res. 2009;130(3):296–310.

Bharti AC, Shukla S, Mahata S, Hedau S, Das BC. Human papillomavirus and control of cervical cancer in India. 2010.

Birkeland SA, Storm HH, Lamm LU, Barlow L, Blohme I, Forsberg B, et al. Cancer risk after renal transplantation in the Nordic countries, 1964–1986. Int J Cancer. 1995;60(2):183–9.

Blut A. Influenza virus. Transfus Med Hemother. 2009;36(1):32.

Boda D, Docea AO, Calina D, Ilie MA, Caruntu C, Zurac S, et al. Human papilloma virus: apprehending the link with carcinogenesis and unveiling new research avenues (Review). Int J Oncol. 2018;52(3):637–55.

Boppana SB, Fowler KB. Persistence in the population: epidemiology and transmisson. In: Arvin A, Campadelli-Fiume G, Mocarski E, Moore PS, Roizman B, Whitley R, et al., editors. Human herpesviruses: biology, therapy, and immunoprophylaxis. Cambridge; 2007.

Bosch FX, Manos MM, Munoz N, Sherman M, Jansen AM, Peto J, et al. Prevalence of human papillomavirus in cervical cancer: a worldwide perspective. International biological study on cervical cancer (IBSCC) Study Group. J Natl Cancer Inst. 1995;87(11):796–802.

Bouhaddou M, Memon D, Meyer B, White KM, Rezelj VV, Correa Marrero M, et al. The global phosphorylation landscape of SARS-CoV-2 infection. Cell. 2020.

Bray F, Ferlay J, Laversanne M, Brewster DH, Gombe Mbalawa C, Kohler B, et al. Cancer incidence in five continents: inclusion criteria, highlights from Volume X and the global status of cancer registration. Int J Cancer. 2015;137(9):2060–71.

Bray F, Ferlay J, Soerjomataram I, Siegel RL, Torre LA, Jemal A. Global cancer statistics 2018: GLOBOCAN estimates of incidence and mortality worldwide for 36 cancers in 185 countries. CA Cancer J Clin. 2018;68(6):394–424.

Calder KB, Smoller BR. New insights into merkel cell carcinoma. Adv Anat Pathol. 2010;17 (3):155–61.

Cannon MJ. Congenital cytomegalovirus (CMV) epidemiology and awareness. J Clin Virol. 2009;46(Suppl 4):S6–10.

Caserta MT, Mock DJ, Dewhurst S. Human herpesvirus 6. Clin Infect Dis. 2001;33(6):829–33.

Cesarman E, Chang Y, Moore PS, Said JW, Knowles DM. Kaposi's sarcoma-associated herpesvirus-like DNA sequences in AIDS-related body-cavity-based lymphomas. N Engl J Med. 1995;332(18):1186–91.

Chan CP, Kok KH, Jin DY. Human T-cell leukemia virus type 1 infection and adult T-cell leukemia. Adv Exp Med Biol. 2017;1018:147–66.

Chan JF, Kok KH, Zhu Z, Chu H, To KK, Yuan S, et al. Genomic characterization of the 2019 novel human-pathogenic coronavirus isolated from a patient with atypical pneumonia after visiting Wuhan. Emerg Microbes Infect. 2020;9(1):221–36.

Chen CC, Chou YY. Predictive value of the anal cytology for detecting anal intraepithelial neoplasia or worse: a systematic review and meta-analysis. Diagn Cytopathol. 2019;47(4):307–14.

Chen Z, Telfier P, Gettie A, Reed P, Zhang L, Ho DD, et al. Genetic characterization of new West African simian immunodeficiency virus SIVsm: geographic clustering of household-derived SIV strains with human immunodeficiency virus type 2 subtypes and genetically diverse viruses from a single feral sooty mangabey troop. J Virol. 1996;70(6):3617–27.

Chevalier MS, Chung W, Smith J, Weil LM, Hughes SM, Joyner SN, et al. Ebola virus disease cluster in the United States-Dallas County, Texas, 2014. MMWR Morb Mortal Wkly Rep. 2014;63(46):1087–8.

Chinazzi M, Davis JT, Ajelli M, Gioannini C, Litvinova M, Merler S, et al. The effect of travel restrictions on the spread of the 2019 novel coronavirus (COVID-19) outbreak. Science. 2020;368(6489):395–400.

Cho CT, Hiatt WO, Behbehani AM. Pneumonia and massive pleural effusion associated with adenovirus type 7. Am J Dis Child. 1973;126(1):92–4.

Cho NH, Kim YT, Kim JW. Alteration of cell cycle in cervical tumor associated with human papillomavirus: cyclin-dependent kinase inhibitors. Yonsei Med J. 2002;43(6):722–8.

Choy KT, Yin-Lam Wong A, Kaewpreedee P, Sia SF, Chen D, Yan Hui KP, et al. Remdesivir, lopinavir, emetine, and homoharringtonine inhibit SARS-CoV-2 replication in vitro. Antiviral Res. 2020;3:104786.

Chu CM. Natural history of chronic hepatitis B virus infection in adults with emphasis on the occurrence of cirrhosis and hepatocellular carcinoma. J Gastroenterol Hepatol. 2000;15(Suppl): E25–30.

Corliss BA, Azimi MS, Munson JM, Peirce SM, Murfee WL. Macrophages: an inflammatory link between angiogenesis and lymphangiogenesis. Microcirculation. 2015;23(2):95–121.

COVID-19 Coronavirus Pandemic. https://www.worldometers.info/coronavirus/#countries.

Cui J, Li F, Shi ZL. Origin and evolution of pathogenic coronaviruses. Nat Rev Microbiol. 2018;17(3):181–92.

Cullen AP, Reid R, Campion M, Lorincz AT. Analysis of the physical state of different human papillomavirus DNAs in intraepithelial and invasive cervical neoplasm. J Virol. 1991;65(2):606–12.

Dalla Pria A, Pinato DJ, Bracchi M, Bower M. Recent advances in HIV-associated Kaposi sarcoma. F1000Res. 2019;8.

Das BC, Sharma JK, Gopalakrishna V, Luthra UK. Analysis by polymerase chain reaction of the physical state of human papillomavirus type 16 DNA in cervical preneoplastic and neoplastic lesions. J Gen Virol. 1992a;73(Pt 9):2327–36.

Das BC, Sharma JK, Gopalkrishna V, Das DK, Singh V, Gissmann L, et al. A high frequency of human papillomavirus DNA sequences in cervical carcinomas of Indian women as revealed by Southern blot hybridization and polymerase chain reaction. J Med Virol. 1992b;36(4):239–45.

Das BC, Hussain S, Nasare V, Bharadwaj M. Prospects and prejudices of human papillomavirus vaccines in India. Vaccine. 2008;26(22):2669–79.

Davison AJ, Bhella D. Comparative genome and virion structure. In: Arvin A, Campadelli-Fiume G, Mocarski E, Moore PS, Roizman B, Whitley R, et al., editors. Human herpesviruses: biology, therapy, and immunoprophylaxis. Cambridge; 2007.

de Martel C, Georges D, Bray F, Ferlay J, Clifford GM. Global burden of cancer attributable to infections in 2018: a worldwide incidence analysis. Lancet Glob Health. 2019;8(2):e180–90.

de Sanjose S, Quint WG, Alemany L, Geraets DT, Klaustermeier JE, Lloveras B, et al. Human papillomavirus genotype attribution in invasive cervical cancer: a retrospective cross-sectional worldwide study. Lancet Oncol. 2010;11(11):1048–56.

de Villiers EM. Cross-roads in the classification of papillomaviruses. Virology. 2013;445(1–2):2–10.

Demers GW, Espling E, Harry JB, Etscheid BG, Galloway DA. Abrogation of growth arrest signals by human papillomavirus type 16 E7 is mediated by sequences required for transformation. J Virol. 1996;70(10):6862–9.

DiGiovanna JJ, Safai B. Kaposi's sarcoma. Retrospective study of 90 cases with particular emphasis on the familial occurrence, ethnic background and prevalence of other diseases. Am J Med. 1981;71(5):779–83.

Dockrell DH, Paya CV. Human herpesvirus-6 and -7 in transplantation. Rev Med Virol. 2001;11(1):23–36.

Donoghue M, Hsieh F, Baronas E, Godbout K, Gosselin M, Stagliano N, et al. A novel angiotensin-converting enzyme-related carboxypeptidase (ACE2) converts angiotensin I to angiotensin 1-9. Circ Res. 2000;87(5):E1–9.

Doorbar J, Egawa N, Griffin H, Kranjec C, Murakami I. Human papillomavirus molecular biology and disease association. Rev Med Virol. 2015;25(Suppl 1):2–23.

Dovih P, Laing ED, Chen Y, Low DHW, Ansil BR, Yang X, et al. Filovirus-reactive antibodies in humans and bats in Northeast India imply zoonotic spillover. PLoS Negl Trop Dis. 2019;13(10):e0007733.

Dunne EF, Unger ER, Sternberg M, McQuillan G, Swan DC, Patel SS, et al. Prevalence of HPV infection among females in the United States. JAMA. 2007;297(8):813–9.

Epstein MA, Woodall JP, Thomson AD. Lymphoblastic Lymphoma in bone-marrow of African Green Monkeys (Cercopithecus Aethiops) inoculated with biopsy material from a child with Burkitt's Lymphoma. Lancet. 1964;2(7354):288–91.

Farazi PA, DePinho RA. The genetic and environmental basis of hepatocellular carcinoma. Discov Med. 2006;6(35):182–6.

Feldmann H, Sprecher A, Geisbert TW. Ebola. N Engl J Med. 2020;382(19):1832–42.

Feltkamp MCW, Kazem S, van der Meijden E, Lauber C, Gorbalenya AE. From Stockholm to Malawi: recent developments in studying human polyomaviruses. J Gen Virol. 2013;94(Pt 3):482–96.

Feng H, Shuda M, Chang Y, Moore PS. Clonal integration of a polyomavirus in human Merkel cell carcinoma. Science. 2008;319(5866):1096–100.

Ferlay J, Soerjomataram I, Dikshit R, Eser S, Mathers C, Rebelo M, et al. Cancer incidence and mortality worldwide: sources, methods and major patterns in GLOBOCAN 2012. Int J Cancer. 2014;136(5):E359–86.

Ferlay J, Colombet M, Soerjomataram I, Mathers C, Parkin DM, Pineros M, et al. Estimating the global cancer incidence and mortality in 2018: GLOBOCAN sources and methods. Int J Cancer. 2018;144(8):1941–53.

Friedman-Kien AE, Farthing C. Human immunodeficiency virus infection: a survey with special emphasis on mucocutaneous manifestations. Semin Dermatol. 1990;9(2):167–77.

Fu Y, Tang Z, Ye Z, Mo S, Tian X, Ni K, et al. Human adenovirus type 7 infection causes a more severe disease than type 3. BMC Infect Dis. 2019;19(1):36.

Gao F, Yue L, White AT, Pappas PG, Barchue J, Hanson AP, et al. Human infection by genetically diverse SIVSM-related HIV-2 in west Africa. Nature. 1992;358(6386):495–9.

Gao J, Tian Z, Yang X. Breakthrough: Chloroquine phosphate has shown apparent efficacy in treatment of COVID-19 associated pneumonia in clinical studies. Biosci Trends. 2020;14 (1):72–3.

Gardner MB. Current information on feline and canine cancers and relationship or lack of relationship to human cancer. J Natl Cancer Inst. 1971;46(2):281–90.

Gautret P, Lagier JC, Parola P, Hoang VT, Meddeb L, Sevestre J, et al. Clinical and microbiological effect of a combination of hydroxychloroquine and azithromycin in 80 COVID-19 patients with at least a six-day follow up: a pilot observational study. Travel Med Infect Dis. 2020;11:101663.

Gaynor AM, Nissen MD, Whiley DM, Mackay IM, Lambert SB, Wu G, et al. Identification of a novel polyomavirus from patients with acute respiratory tract infections. PLoS Pathog. 2007;3 (5):e64.

Gibbons RV, Kalanarooj S, Jarman RG, Nisalak A, Vaughn DW, Endy TP, et al. Analysis of repeat hospital admissions for dengue to estimate the frequency of third or fourth dengue infections resulting in admissions and dengue hemorrhagic fever, and serotype sequences. Am J Trop Med Hyg. 2007;77(5):910–3.

Gnann JW Jr. Varicella-zoster virus: atypical presentations and unusual complications. J Infect Dis. 2002;15(186 Suppl 1):S91–8.

Goncalves DU, Proietti FA, Ribas JG, Araujo MG, Pinheiro SR, Guedes AC, et al. Epidemiology, treatment, and prevention of human T-cell leukemia virus type 1-associated diseases. Clin Microbiol Rev. 2010;23(3):577–89.

Gratama JW, Ernberg I. Molecular epidemiology of Epstein-Barr virus infection. Adv Cancer Res. 1995;67:197–255.

Griffiths PD, Boeckh M. Antiviral therapy for human cytomegalovirus. In: Arvin A, Campadelli-Fiume G, Mocarski E, Moore PS, Roizman B, Whitley R, et al., editors. Human herpesviruses: biology, therapy, and immunoprophylaxis. Cambridge; 2007.

Guan WJ, Zhong NS. Clinical characteristics of Covid-19 in China. Reply. N Engl J Med. 2020;382(19):1861–2.

Guidotti LG, Borrow P, Brown A, McClary H, Koch R, Chisari FV. Noncytopathic clearance of lymphocytic choriomeningitis virus from the hepatocyte. J Exp Med. 1999;189(10):1555–64.

Gupta S, Kumar P, Kaur H, Sharma N, Gupta S, Saluja D, et al. Constitutive activation and overexpression of NF-κB/c-Rel in conjunction with p50 contribute to aggressive tongue tumorigenesis. 2018a;9(68):33011.

Gupta S, Kumar P, Das BC. HPV: molecular pathways and targets. Curr Probl Cancer. 2018b;42 (2):161–74.

Gupta S, Kumar P, Kaur H, Sharma N, Saluja D, Bharti AC, et al. Selective participation of c-Jun with Fra-2/c-Fos promotes aggressive tumor phenotypes and poor prognosis in tongue cancer. 2015;5:16811.

Gupta S, Kumar P, Tyagi A, Sharma K, Kaur H, Das BC. Modern approaches to prevention and control of cancer of the uterine cervix in women. J Indian Inst Sci. 2012;92(3):353–564.

Harapan H, Itoh N, Yufika A, Winardi W, Keam S, Te H, et al. Coronavirus disease 2019 (COVID-19): a literature review. J Infect Public Health. 2020;13(5):667–73.

Harrison C. Coronavirus puts drug repurposing on the fast track. Nat Biotechnol. 2020;38(4):379–81.

Heineman TC. Human cytomegalovirus vaccines. In: Arvin A, Campadelli-Fiume G, Mocarski E, Moore PS, Roizman B, Whitley R, et al., editors. Human herpesviruses: biology, therapy, and immunoprophylaxis. Cambridge; 2007.

Helt AM, Galloway DA. Destabilization of the retinoblastoma tumor suppressor by human papillomavirus type 16 E7 is not sufficient to overcome cell cycle arrest in human keratinocytes. J Virol. 2001;75(15):6737–47.

Helt AM, Funk JO, Galloway DA. Inactivation of both the retinoblastoma tumor suppressor and p21 by the human papillomavirus type 16 E7 oncoprotein is necessary to inhibit cell cycle arrest in human epithelial cells. J Virol. 2002;76(20):10559–68.

Hessel L, West DJ. Antibody responses to recombinant hepatitis B vaccines. Vaccine. 2002;20 (17–18):2164–5.

Higgins CD, Swerdlow AJ, Macsween KF, Harrison N, Williams H, McAulay K, et al. A study of risk factors for acquisition of Epstein-Barr virus and its subtypes. J Infect Dis. 2007;195 (4):474–82.

Hijawi B, Abdallat M, Sayaydeh A, Alqasrawi S, Haddadin A, Jaarour N, et al. Novel coronavirus infections in Jordan, April 2012: epidemiological findings from a retrospective investigation. East Mediterr Health J. 2013;19(Suppl 1):S12–8.

Hirschman RJ, Shulman NR, Barker LF, Smith KO. Virus-like particles in sera of patients with infectious and serum hepatitis. JAMA. 1969;208(9):1667–70.

Ho M. Epidemiology of cytomegalovirus infections. Rev Infect Dis. 1990;12(Suppl 7):S701–10.

Hodgson NC. Merkel cell carcinoma: changing incidence trends. J Surg Oncol. 2005;89(1):1–4.

Hoffmann M, Kleine-Weber H, Schroeder S, Kruger N, Herrler T, Erichsen S, et al. SARS-CoV-2 cell entry depends on ACE2 and TMPRSS2 and is blocked by a clinically proven protease inhibitor. Cell. 2020;181(2):271–80 e8.

Holshue ML, DeBolt C, Lindquist S, Lofy KH, Wiesman J, Bruce H, et al. First case of 2019 novel coronavirus in the United States. N Engl J Med. 2020;382(10):929–36.

Huang YC, Huang SL, Chen SP, Huang YL, Huang CG, Tsao KC, et al. Adenovirus infection associated with central nervous system dysfunction in children. J Clin Virol. 2013;57(4):300–4.

Huang M, Tang T, Pang P, Li M, Ma R, Lu J, et al. Treating COVID-19 with chloroquine. J Mol Cell Biol. 2020a.

Huang C, Wang Y, Li X, Ren L, Zhao J, Hu Y, et al. Clinical features of patients infected with 2019 novel coronavirus in Wuhan, China. Lancet. 2020b;395(10223):497–506.

Hristov G, Kramer M, Li J, El-Andaloussi N, Mora R, Daeffler L, et al. Through its nonstructural protein NS1, parvovirus H-1 induces apoptosis via accumulation of reactive oxygen species. J Virol. 2010;84(12):5909–22.

Hung KH, Lin LH. Adenovirus pneumonia complicated with acute respiratory distress syndrome: a case report. Medicine (Baltimore). 2015;94(20):e776.

IARC. Smokeless tobacco and some tobacco-specific N-nitrosamines. 2007.

IARC. Biological agents. Volume 100 B. A review of human carcinogens. IARC Monogr Eval Carcinog Risks Hum. 2012;100(Pt B):1–441.

Ingraham NE, Lotfi-Emran S, Thielen BK, Techar K, Morris RS, Holtan SG, et al. Immunomodulation in COVID-19. Lancet Respir Med. 2020;8(6):544–6.

Jacqueline C, Tasiemski A, Sorci G, Ujvari B, Maachi F, Misse D, et al. Infections and cancer: the "fifty shades of immunity" hypothesis. BMC Cancer. 2017;17(1):257.

Javier RT, Butel JS. The history of tumor virology. Cancer Res. 2008;68(19):7693–706.

Jereb M, Lainscak M, Marin J, Popovic M. Herpes simplex virus infection limited to the brainstem. Wien Klin Wochenschr. 2005;117(13–14):495–9.

Jha HC, Pei Y, Robertson ES. Epstein-Barr virus: diseases linked to infection and transformation. Front Microbiol. 2016;7:1602.

Jiang P, Yue Y. Human papillomavirus oncoproteins and apoptosis (Review). Exp Ther Med. 2014;7(1):3–7.

Jones DM, Patel AH, Targett-Adams P, McLauchlan J. The hepatitis C virus NS4B protein can trans-complement viral RNA replication and modulates production of infectious virus. J Virol. 2009;83(5):2163–77.

Kang Y, Chen T, Mui D, Ferrari V, Jagasia D, Scherrer-Crosbie M, et al. Cardiovascular manifestations and treatment considerations in COVID-19. Heart. 2020;106(15):1132–41.

Katakura Y, Yotsuyanagi H, Hashizume K, Okuse C, Okuse N, Nishikawa K, et al. Pancreatic involvement in chronic viral hepatitis. World J Gastroenterol. 2005;11(23):3508–13.

Kataoka H, Tanaka H, Nagaike K, Uchiyama S, Itoh H. Role of cancer cell-stroma interaction in invasive growth of cancer cells. Hum Cell. 2003;16(1):1–14.

Kerr JR. Pathogenesis of human parvovirus B19 in rheumatic disease. Ann Rheum Dis. 2000;59 (9):672–83.

Kerr JR, Barah F, Cunniffe VS, Smith J, Vallely PJ, Will AM, et al. Association of acute parvovirus B19 infection with new onset of acute lymphoblastic and myeloblastic leukaemia. J Clin Pathol. 2003;56(11):873–5.

Khailany RA, Safdar M, Ozaslan M. Genomic characterization of a novel SARS-CoV-2. Gene Rep. 2020;16:100682.

Khan G, Hashim MJ. Global burden of deaths from Epstein-Barr virus attributable malignancies 1990–2010. Infect Agent Cancer. 2014;9(1):38.

Kim KH, Tandi TE, Choi JW, Moon JM, Kim MS. Middle East respiratory syndrome coronavirus (MERS-CoV) outbreak in South Korea, 2015: epidemiology, characteristics and public health implications. J Hosp Infect. 2017;95(2):207–13.

Ko YH. EBV and human cancer. Exp Mol Med. 2015;23(47):e130.

Kocjan BJ, Bzhalava D, Forslund O, Dillner J, Poljak M. Molecular methods for identification and characterization of novel papillomaviruses. Clin Microbiol Infect. 2015;21(9):808–16.

Kordestani R, Mirshafiee H, Hosseini SM, Sharifi Z. Effect of hepatitis B virus X gene on the expression level of p53 gene using Hep G2 cell line. Avicenna J Med Biotechnol. 2014;6(1):3–9.

Ksiazek TG, Erdman D, Goldsmith CS, Zaki SR, Peret T, Emery S, et al. A novel coronavirus associated with severe acute respiratory syndrome. N Engl J Med. 2003;348(20):1953–66.

Kucerova P, Cervinkova M. Spontaneous regression of tumour and the role of microbial infection–possibilities for cancer treatment. Anticancer Drugs. 2016;27(4):269–77.

Kuguyo O, Tsikai N, Thomford NE, Magwali T, Madziyire MG, Nhachi CFB, et al. Genetic Susceptibility for cervical cancer in African populations: what are the host genetic drivers? OMICS. 2018;22(7):468–83.

Kuhn JH, Amarasinghe GK, Basler CF, Bavari S, Bukreyev A, Chandran K, et al. ICTV virus taxonomy profile: Filoviridae. J Gen Virol. 2019a;100(6):911–2.

Kuhn JH, Adachi T, Adhikari NKJ, Arribas JR, Bah IE, Bausch DG, et al. New filovirus disease classification and nomenclature. Nat Rev Microbiol. 2019b;17(5):261–3.

Kumar P, Gupta S, Das AM, Das BC. Towards global elimination of cervical cancer in all groups of women. Lancet Oncol. 2019;20(5):e237.

Lai CY, Lee CJ, Lu CY, Lee PI, Shao PL, Wu ET, et al. Adenovirus serotype 3 and 7 infection with acute respiratory failure in children in Taiwan, 2010–2011. PLoS One. 2013;8(1):e53614.

Lauer SA, Grantz KH, Bi Q, Jones FK, Zheng Q, Meredith HR, et al. The incubation period of coronavirus disease 2019 (COVID-19) from publicly reported confirmed cases: estimation and application. Ann Intern Med. 2020.

Lavanchy D. Hepatitis B virus epidemiology, disease burden, treatment, and current and emerging prevention and control measures. J Viral Hepat. 2004;11(2):97–107.

Lee N, Hui D, Wu A, Chan P, Cameron P, Joynt GM, et al. A major outbreak of severe acute respiratory syndrome in Hong Kong. N Engl J Med. 2003;348(20):1986–94.

Li Y, Dong Y, Jiang J, Yang Y, Liu K, Li Y. High prevelance of human parvovirus infection in patients with malignant tumors. Oncol Lett. 2012;3(3):635–40.

Lindahl U, Li JP. Heparin—an old drug with multiple potential targets in Covid-19 therapy. J Thromb Haemost. 2020.

Liu F-F, Richardson CD. Oncogenic viruses and tumor viruses. Basic Sci Oncol. 2013:117.

Liu F, Zhou ZH. Comparative virion structures of human herpesviruses. In: Arvin A, Campadelli-Fiume G, Mocarski E, Moore PS, Roizman B, Whitley R, et al., editors. Human herpesviruses: biology, therapy, and immunoprophylaxis. Cambridge; 2007.

Liu Y, Gayle AA, Wilder-Smith A, Rocklov J. The reproductive number of COVID-19 is higher compared to SARS coronavirus. J Travel Med. 2020;27(2).

Lozano R, Naghavi M, Foreman K, Lim S, Shibuya K, Aboyans V, et al. Global and regional mortality from 235 causes of death for 20 age groups in 1990 and 2010: a systematic analysis for the Global Burden of Disease Study 2010. Lancet. 2012;380(9859):2095–128.

Lu R, Wu X, Wan Z, Li Y, Zuo L, Qin J, et al. Development of a novel reverse transcription loop-mediated isothermal amplification method for rapid detection of SARS-CoV-2. Virol Sin. 2020a.

Lu R, Zhao X, Li J, Niu P, Yang B, Wu H, et al. Genomic characterisation and epidemiology of 2019 novel coronavirus: implications for virus origins and receptor binding. Lancet. 2020b;395 (10224):565–74.

Maddrey WC. Hepatitis B—an important public health issue. Clin Lab. 2001;47(1–2):51–5.

Marcelin AG, Calvez V, Dussaix E. KSHV after an organ transplant: should we screen? Curr Top Microbiol Immunol. 2007;312:245–62.

Martines RB, Ng DL, Greer PW, Rollin PE, Zaki SR. Tissue and cellular tropism, pathology and pathogenesis of Ebola and Marburg viruses. J Pathol. 2015;235(2):153–74.

May P, May E. Twenty years of p53 research: structural and functional aspects of the p53 protein. Oncogene. 1999;18(53):7621–36.

McCarthy M. US to monitor arrivals from Liberia, Sierra Leone, and Guinea. BMJ. 2014;23(349): g6433.

McNeil CJ, Shetty AK. Zika virus: a serious global health threat. J Trop Pediatr. 2017;63(3):242–8.

Memish ZA, Al-Tawfiq JA, Alhakeem RF, Assiri A, Alharby KD, Almahallawi MS, et al. Middle East respiratory syndrome coronavirus (MERS-CoV): a cluster analysis with implications for global management of suspected cases. Travel Med Infect Dis. 2015;13(4):311–4.

Memish ZA, Perlman S, Van Kerkhove MD, Zumla A. Middle East respiratory syndrome. Lancet. 2020;395(10229):1063–77.

Messaoudi I, Basler CF. Immunological features underlying viral hemorrhagic fevers. Curr Opin Immunol. 2015;36:38–46.

Messaoudi I, Amarasinghe GK, Basler CF. Filovirus pathogenesis and immune evasion: insights from Ebola virus and Marburg virus. Nat Rev Microbiol. 2015;13(11):663–76.

Moderna's Work on a COVID-19 Vaccine. https://www.modernatx.com/modernas-work-potential-vaccine-against-covid-19.

Molina JM, Delaugerre C, Goff JL, Mela-Lima B, Ponscarme D, Goldwirt L, et al. No evidence of rapid antiviral clearance or clinical benefit with the combination of hydroxychloroquine and azithromycin in patients with severe COVID-19 infection. MÃ©decine Mal Infect. 2020.

Moody CA, Laimins LA. Human papillomavirus oncoproteins: pathways to transformation. Nat Rev Cancer. 2010;10(8):550–60.

Moreno V, Bosch FX, Munoz N, Meijer CJ, Shah KV, Walboomers JM, et al. Effect of oral contraceptives on risk of cervical cancer in women with human papillomavirus infection: the IARC multicentric case-control study. Lancet. 2002;359(9312):1085–92.

Morens DM, Folkers GK, Fauci AS. The challenge of emerging and re-emerging infectious diseases. Nature. 2004;430(6996):242–9.

Musso D, Gubler DJ. Zika virus. Clin Microbiol Rev. 2016;29(3):487–524.

Mousavizadeh L, Ghasemi S. Genotype and phenotype of COVID-19: their roles in pathogenesis. J Microbiol Immunol Infect. 2020.

National Health Commission of the People's Republic of China. Notice on printing and distributing the diagnosis and treatment plan of pneumonia with new coronavirus infection (trial version 3). 2020. http://www.nhc.gov.cn/yzygj/s7653p/202001/f492c9153ea9437bb587ce2ffcbee1fa.shtml. Accessed 23 Jan 2020.

Oikonomopoulou K, Brinc D, Kyriacou K, Diamandis EP. Infection and cancer: revaluation of the hygiene hypothesis. Clin Cancer Res. 2013;19(11):2834–41.
Okuno T, Takahashi K, Balachandra K, Shiraki K, Yamanishi K, Takahashi M, et al. Seroepidemiology of human herpesvirus 6 infection in normal children and adults. J Clin Microbiol. 1989;27(4):651–3.
Olusola P, Ousley K, Ndetan H, Singh KP, Banerjee HN, Dasgupta S. Cervical cancer prevention in racially disparate rural populations. Medicines (Basel). 2019;6(3).
Ott JJ, Stevens GA, Groeger J, Wiersma ST. Global epidemiology of hepatitis B virus infection: new estimates of age-specific HBsAg seroprevalence and endemicity. Vaccine. 2012;30 (12):2212–9.
Panyod S, Ho CT, Sheen LY. Dietary therapy and herbal medicine for COVID-19 prevention: a review and perspective. J Tradit Complement Med. 2020;10(4):420–7.
Parkin DM. The global health burden of infection-associated cancers in the year 2002. Int J Cancer. 2006;118(12):3030–44.
Parvez MK, Al-Dosari MS, Alam P, Rehman M, Alajmi MF, Alqahtani AS. The anti-hepatitis B virus therapeutic potential of anthraquinones derived from Aloe vera. Phytother Res. 2019;33 (11):2960–70.
Pass RF. Epidemiology and transmission of cytomegalovirus. J Infect Dis. 1985;152(2):243–8.
Peiris JS, Lai ST, Poon LL, Guan Y, Yam LY, Lim W, et al. Coronavirus as a possible cause of severe acute respiratory syndrome. Lancet. 2003;361(9366):1319–25.
Petersen LR, Jamieson DJ, Powers AM, Honein MA. Zika virus. N Engl J Med. 2016;374 (16):1552–63.
Pereira N, Kucharczyk KM, Estes JL, Gerber RS, Lekovich JP, Elias RT, et al. Human papillomavirus infection, infertility, and assisted reproductive outcomes. J Pathog. 2015;2015:578423.
Pergam SA, Limaye AP, AST Infectious Diseases Community of Practice. Varicella zoster virus in solid organ transplantation: guidelines from the American Society of Transplantation Infectious Diseases Community of Practice. Clin Transplant. 2019;33(9):e13622.
Pirami L, Giache V, Becciolini A. Analysis of HPV16, 18, 31, and 35 DNA in pre-invasive and invasive lesions of the uterine cervix. J Clin Pathol. 1997;50(7):600–4.
Plotkin SA, Boppana SB. Vaccination against the human cytomegalovirus. Vaccine. 2019;37 (50):7437–42.
Plummer M, de Martel C, Vignat J, Ferlay J, Bray F, Franceschi S. Global burden of cancers attributable to infections in 2012: a synthetic analysis. Lancet Glob Health. 2016;4(9):e609–16.
Poiesz BJ, Ruscetti FW, Gazdar AF, Bunn PA, Minna JD, Gallo RC. Detection and isolation of type C retrovirus particles from fresh and cultured lymphocytes of a patient with cutaneous T-cell lymphoma. Proc Natl Acad Sci U S A. 1980;77(12):7415–9.
Ray G. Current scenario of hepatitis B and its treatment in India. J Clin Transl Hepatol. 2017;5 (3):277–96.
Razonable RR. Epidemiology of cytomegalovirus disease in solid organ and hematopoietic stem cell transplant recipients. Am J Health Syst Pharm. 2005;62(8 Suppl 1):S7–13.
Robertson A, Levin ML. AIDS knowledge, condom attitudes, and risk-taking sexual behavior of substance-abusing juvenile offenders on probation or parole. AIDS Educ Prev. 1999;11 (5):450–61.
Rosenberg M, Pettifor A, Van Rie A, Thirumurthy H, Emch M, Miller WC, et al. The relationship between alcohol outlets, HIV risk behavior, and HSV-2 infection among South African young women: a cross-sectional study. PLoS One. 2015;10(5):e0125510.
Sanyaolu A, Okorie C, Badaru O, Adetona K, Ahmed M, Akanbi O, et al. Global epidemiology of dengue hemorrhagic fever: an update. 2017;5(6):00179.
Savarino A, Boelaert JR, Cassone A, Majori G, Cauda R. Effects of chloroquine on viral infections: an old drug against today's diseases? Lancet Infect Dis. 2003;3(11):722–7.
Scholtissek C. Stability of infectious influenza A viruses at low pH and at elevated temperature. Vaccine. 1985;3(3 Suppl):215–8.

Schowalter RM, Pastrana DV, Pumphrey KA, Moyer AL, Buck CB. Merkel cell polyomavirus and two previously unknown polyomaviruses are chronically shed from human skin. Cell Host Microbe. 2010;7(6):509–15.

Scuda N, Hofmann J, Calvignac-Spencer S, Ruprecht K, Liman P, Kuhn J, et al. A novel human polyomavirus closely related to the African green monkey-derived lymphotropic polyomavirus. J Virol. 2011;85(9):4586–90.

Senapati R, Senapati NN, Dwibedi B. Molecular mechanisms of HPV mediated neoplastic progression. Infect Agent Cancer. 2016;11:59.

Shang W, Yang Y, Rao Y, Rao X. The outbreak of SARS-CoV-2 pneumonia calls for viral vaccines. NPJ Vaccines. 2020;5:18.

Sheahan TP, Sims AC, Leist SR, Schafer A, Won J, Brown AJ, et al. Comparative therapeutic efficacy of remdesivir and combination lopinavir, ritonavir, and interferon beta against MERS-CoV. Nat Commun. 2020;11(1):222.

Shereen MA, Khan S, Kazmi A, Bashir N, Siddique R. COVID-19 infection: origin, transmission, and characteristics of human coronaviruses. J Adv Res. 2020;24:91–8.

Shukla S, Bharti AC, Mahata S, Hussain S, Hedau S, Sharma R, et al. Application of a multiplex PCR to cervical cells collected by a paper smear for the simultaneous detection of all mucosal human papillomaviruses (HPVs) and typing of high-risk HPV types 16 and 18. J Med Microbiol. 2010;59(Pt 11):1303–10.

Siebrasse EA, Reyes A, Lim ES, Zhao G, Mkakosya RS, Manary MJ, et al. Identification of MW polyomavirus, a novel polyomavirus in human stool. J Virol. 2012;86(19):10321–6.

Siegert R, Shu HL, Slenczka W, Peters D, Muller G. On the etiology of an unknown human infection originating from monkeys. Dtsch Med Wochenschr. 1967;92(51):2341–3.

Singhal T. A review of coronavirus disease-2019 (COVID-19). Indian J Pediatr. 2020;87(4):281–6.

Slebos RJ, Lee MH, Plunkett BS, Kessis TD, Williams BO, Jacks T, et al. p53-dependent G1 arrest involves pRB-related proteins and is disrupted by the human papillomavirus 16 E7 oncoprotein. Proc Natl Acad Sci U S A. 1994;91(12):5320–4.

Smith JS, Herrero R, Bosetti C, Munoz N, Bosch FX, Eluf-Neto J, et al. Herpes simplex virus-2 as a human papillomavirus cofactor in the etiology of invasive cervical cancer. J Natl Cancer Inst. 2002;94(21):1604–13.

Soulier J, Grollet L, Oksenhendler E, Cacoub P, Cazals-Hatem D, Babinet P, et al. Kaposi's sarcoma-associated herpesvirus-like DNA sequences in multicentric Castleman's disease. Blood. 1995;86(4):1276–80.

Sousa H, Pinto-Correia AL, Medeiros R, Dinis-Ribeiro M. Epstein-Barr virus is associated with gastric carcinoma: the question is what is the significance? World J Gastroenterol. 2008;14(27):4347–51.

Stagno S, Cloud GA. Changes in the epidemiology of cytomegalovirus. Adv Exp Med Biol. 1990;278:93–104.

Stebbing J, Phelan A, Griffin I, Tucker C, Oechsle O, Smith D, et al. COVID-19: combining antiviral and anti-inflammatory treatments. Lancet Infect Dis. 2020;20(4):400–2.

Steben M, Duarte-Franco E. Human papillomavirus infection: epidemiology and pathophysiology. Gynecol Oncol. 2007;107(2 Suppl 1):S2–5.

Stein ME, Kantor A, Spencer D, Bezwoda WR. Classical Kaposi's sarcoma in Caucasians in Africa–experience at the Johannesburg Hospital (1978–1992). Dermatology. 1994;188(3):182–7.

Stokley S, Jeyarajah J, Yankey D, Cano M, Gee J, Roark J, et al. Human papillomavirus vaccination coverage among adolescents, 2007–2013, and postlicensure vaccine safety monitoring, 2006–2014—United States. MMWR Morb Mortal Wkly Rep. 2014;63(29):620–4.

Summers BA. Climate change and animal disease. Vet Pathol. 2009;46(6):1185–6.

Sun J, Xiao Y, Zhang M, Ao T, Lang S, Wang J. Serum inflammatory markers in patients with adenovirus respiratory infection. Med Sci Monit. 2018;7(24):3848–55.

Tanaka S, Komori K, Okadome K, Sugimachi K, Mori R. Detection of active cytomegalovirus infection in inflammatory aortic aneurysms with RNA polymerase chain reaction. J Vasc Surg. 1994;20(2):235–43.

Tanaka-Taya K, Kondo T, Mukai T, Miyoshi H, Yamamoto Y, Okada S, et al. Seroepidemiological study of human herpesvirus-6 and -7 in children of different ages and detection of these two viruses in throat swabs by polymerase chain reaction. J Med Virol. 1996;48(1):88–94.

Tang D, Comish P, Kang R. The hallmarks of COVID-19 disease. PLoS Pathog. 2020;16(5): e1008536.

To KK, Tsang OT, Leung WS, Tam AR, Wu TC, Lung DC, et al. Temporal profiles of viral load in posterior oropharyngeal saliva samples and serum antibody responses during infection by SARS-CoV-2: an observational cohort study. Lancet Infect Dis. 2020;20(5):565–74.

Tortorici MA, Walls AC, Lang Y, Wang C, Li Z, Koerhuis D, et al. Structural basis for human coronavirus attachment to sialic acid receptors. Nat Struct Mol Biol. 2019;26(6):481–9.

Treacy A, Carr MJ, Dunford L, Palacios G, Cannon GA, O'Grady A, et al. First report of sudden death due to myocarditis caused by adenovirus serotype 3. J Clin Microbiol. 2010;48(2):642–5.

Tyagi A, Gupta S, Kuamr P, Bharti AC, Das BC. Emerging role of Fra-1 as a novel target for radiosensitization in HPV-associated cancers. AACR; 2017.

Tzellos S, Correia PB, Karstegl CE, Cancian L, Cano-Flanagan J, McClellan MJ, et al. A single amino acid in EBNA-2 determines superior B lymphoblastoid cell line growth maintenance by Epstein-Barr virus type 1 EBNA-2. J Virol. 2014;88(16):8743–53.

van Baarle D, Hovenkamp E, Dukers NH, Renwick N, Kersten MJ, Goudsmit J, et al. High prevalence of Epstein-Barr virus type 2 among homosexual men is caused by sexual transmission. J Infect Dis. 2000;181(6):2045–9.

Wan Y, Shang J, Graham R, Baric RS, Li F. Receptor recognition by the novel coronavirus from Wuhan: an analysis based on decade-long structural studies of SARS coronavirus. J Virol. 2020;94(7).

Wang D, Hu B, Hu C, Zhu F, Liu X, Zhang J, et al. Clinical characteristics of 138 hospitalized patients with 2019 novel coronavirus-infected pneumonia in Wuhan, China. JAMA. 2020a.

Wang W, Xu Y, Gao R, Lu R, Han K, Wu G, et al. Detection of SARS-CoV-2 in different types of clinical specimens. JAMA. 2020b.

Wang W, Tang J, Wei F. Updated understanding of the outbreak of 2019 novel coronavirus (2019-nCoV) in Wuhan, China. J Med Virol. 2020c;92(4):441–7.

Warren TK, Jordan R, Lo MK, Ray AS, Mackman RL, Soloveva V, et al. Therapeutic efficacy of the small molecule GS-5734 against Ebola virus in rhesus monkeys. Nature. 2016;531 (7594):381–5.

Wax RS, Christian MD. Practical recommendations for critical care and anesthesiology teams caring for novel coronavirus (2019-nCoV) patients. Can J Anaesth. 2020;67(5):568–76.

Whitley RJ. The past as prelude to the future: history, status, and future of antiviral drugs. Ann Pharmacother. 1996;30(9):967–71.

WHO. World Health Organization. Causes of death. https://www.who.int/healthinfo/global_burden_disease/GBD_report_2004update_part2.pdf.

Wikan N, Smith DR. Zika virus: history of a newly emerging arbovirus. Lancet Infect Dis. 2016;16(7):e119–26.

Wilck MB, Zuckerman RA, AST Infectious Diseases Community of Practice. Herpes simplex virus in solid organ transplantation. Am J Transplant. 2013;13 Suppl 4:121–7.

Woolhouse MEJ, Adair K. The diversity of human RNA viruses. Future Virol. 2013;8(2):159–71.

Woolhouse ME, Gowtage-Sequeria S. Host range and emerging and reemerging pathogens. Emerg Infect Dis. 2005;11(12):1842–7.

World Health Organization (WHO). Q&A on coronaviruses (COVID-19): How long is the incubation period for COVID-19?. 2020a. Archived from the original on 20 Jan 2020. Accessed 26 Feb 2020.

World Health Organization. Laboratory testing for coronavirus disease 2019 (COVID-19) in suspected human cases: interim guidance, 2 Mar 2020 (PDF) (Report). 2020b. hdl:10665/331329. WHO/COVID-19/laboratory/2020.4.

World Health Organization. Laboratory testing of 2019 novel coronavirus (2019-nCoV) in suspected human cases: interim guidance, 17 Jan 2020 (PDF) (Report). 2020c.

Wu AHB. Screening the general population for SARS-CoV-2 virus and COVID-19 antibodies: a counterargument. J Appl Lab Med. 2020;5(5):1107–10.

Wu C, Liu Y, Yang Y, Zhang P, Zhong W, Wang Y, et al. Analysis of therapeutic targets for SARS-CoV-2 and discovery of potential drugs by computational methods. Acta Pharm Sin B. 2020.

Wu Z, McGoogan JM. Characteristics of and important lessons from the coronavirus disease 2019 (COVID-19) outbreak in China: summary of a report of 72314 cases from the Chinese Center for Disease Control and Prevention. JAMA. 2020.

Wyatt SW, Lancaster M, Bottorff D, Ross F. History of tobacco use among Kentucky women diagnosed with invasive cervical cancer: 1997–1998. J Ky Med Assoc. 2001;99(12):537–9.

Xu R, Du M, Li L, Zhen Z, Wang H, Hu X. CT imaging of one extended family cluster of corona virus disease 2019 (COVID-19) including adolescent patients and "silent infection". Quant Imaging Med Surg. 2020;10(3):800–4.

Yaghoobi R, Pazyar N, Tavakoli S. Co-existence of multicentric Castleman's disease and Kaposi's sarcoma. Indian J Dermatol. 2015;60(3):323.

Young P, Anderton E, Paschos K, White R, Allday MJ. Epstein-Barr virus nuclear antigen (EBNA) 3A induces the expression of and interacts with a subset of chaperones and co-chaperones. J Gen Virol. 2008;89(Pt 4):866–77.

Yu C, Zelterman D. Statistical inference for familial disease clusters. Biometrics. 2002;58(3):481–91.

Zhang Y, Daaka Y. PGE2 promotes angiogenesis through EP4 and PKA Cgamma pathway. Blood. 2011;118(19):5355–64.

Zhang D, Zhou X, Yan S, Tian R, Su L, Ding X, et al. Correlation between cytokines and coagulation-related parameters in patients with coronavirus disease 2019 admitted to ICU. Clin Chim Acta. 2020a;6(510):47–53.

Zhang W, Du RH, Li B, Zheng XS, Yang XL, Hu B, et al. Molecular and serological investigation of 2019-nCoV infected patients: implication of multiple shedding routes. Emerg Microbes Infect. 2020b;9(1):386–9.

Zhou P, Yang X-L, Wang X-G, Hu B, Zhang L, Zhang W, et al. Discovery of a novel coronavirus associated with the recent pneumonia outbreak in humans and its potential bat origin. BioRxiv. 2020a.

Zhou P, Yang XL, Wang XG, Hu B, Zhang L, Zhang W, et al. A pneumonia outbreak associated with a new coronavirus of probable bat origin. Nature. 2020b;579(7798):270–3.

Zhu FC, Li YH, Guan XH, Hou LH, Wang WJ, Li JX, et al. Safety, tolerability, and immunogenicity of a recombinant adenovirus type-5 vectored COVID-19 vaccine: a dose-escalation, open-label, non-randomised, first-in-human trial. Lancet. 2020.

Zou L, Ruan F, Huang M, Liang L, Huang H, Hong Z, et al. SARS-CoV-2 viral load in upper respiratory specimens of infected patients. N Engl J Med. 2020;382(12):1177–9.

zur Hausen H. Papillomaviruses and cancer: from basic studies to clinical application. Nat Rev Cancer. 2002;2(5):342–50.

zur Hausen H. Papillomaviruses causing cancer: evasion from host-cell control in early events in carcinogenesis. J Natl Cancer Inst. 2000;92(9):690–8.

Dengue Virus

Dengue Preventive Strategies Through Entomological Control, Vaccination and Biotechnology

Idris Nasir Abdullahi, Sharafudeen Dahiru Abubakar, Hafeez Aderinsayo Adekola, Abdurrahman El-fulaty Ahmad, Mohammed Ibrahim Tahir, Abubakar Umar Anka, Muhammad Sagir Shehu, and Yahaya Usman

Abstract Dengue (DEN) is a semi-neglected and dangerously re-emerging, mosquito-borne viral disease of the Flaviviridae family. Dengue virus (DENV) has been rapidly spreading for the past 6 decades, and now affects more than 50% the world's population, especially the tropical and subtropical countries. DENV transmission and geographic expansion is expected to rise due to increased international travels, deforestation, industrialization, urbanization and climatic changes. A major control strategy for dengue is achieved through targeting the mosquito vectors, which put a significant break in the transmission of DENV to humans. For long, most dengue endemic nations relied heavily on chemical control measures, however, there were growing concerns and controversies on cases of multiple insecticide resistance. Therefore, more research interest has shifted toward studies on the use of biological control, biotechnological approaches, as potential alternatives. Indeed, dengue prevention is a key, mainly through continuous search for novel vaccines and/or improves on the potency, effectiveness, efficacy, safety, and affordability of existing ones, in order to assure protection against all serotypes of dengue, regardless of age, sex or race of people. In this chapter, we outlined and elucidated available non-insecticide based dengue control strategies that have been implemented, and/or currently being tested. We also updated information on currently licensed dengue vaccines, those in clinical trials and their public health prospects in the prevention of dengue.

I. N. Abdullahi (✉) · S. D. Abubakar · A. E. Ahmad · M. I. Tahir · A. U. Anka · Y. Usman
Department of Medical Laboratory Science, Faculty of Allied Health Sciences, College of Medical Sciences, Ahmadu Bello University, Zaria, Nigeria
e-mail: inabdullahi@abu.edu.ng

H. A. Adekola
Department of Microbiology, Olabisi Onabanjo University, Ogun State, Nigeria

M. S. Shehu
College of Health Technology, Ningi, Bauchi State, Nigeria

S. I. Ahmad (ed.), *Human Viruses: Diseases, Treatments and Vaccines*, https://doi.org/10.1007/978-3-030-71165-8_2

Keywords Preventive measures · Arbovirus · Dengue vaccines · Epigenetics · Mosquitoes biocontrol · Neglected infection · Re-emerging virus · Zoonosis · Dengue virus

1 Introduction

Dengue is a semi-neglected tropical disease caused by Dengue virus (DENV). DENV has been classified a re-emerging mosquito-borne virus by the World Health Organization (WHO 2019). DENV, the Flaviviridae member, is one of the most widespread mosquito-borne diseases in humans; it is mainly transmitted by the Aedes mosquito, Aedes aegypti and Aedes albopictus (Bhatt et al. 2013). There are four serotypes responsible for dengue fever; Dengue-1, Dengue-2, Dengue-3, and Dengue-4 (Sang et al. 2015). However, a fifth serotype, DENV-5 was isolated in Malaysia in 2013 (Mustafa et al. 2015). Dengue fever is endemic in more than 100 countries in Southeast Asia, the Americas, Western Pacific, Africa, and Eastern Mediterranean regions (Guzman and Harris 2015). Dengue fever has evolved from a sporadic disease to a major public health problem as geographical extension and the number of cases and disease severity are on increase (Guzman and Harris 2015).

The incidence of dengue has changed dramatically over last few decades from a report in only nine countries in the 1970s to an estimated 3.9 billion people at risk in 128 countries in the regions of Africa, the Americas, the Eastern Mediterranean, South-East Asia and the Western Pacific (WHO 2019). Cases across the Americas, South-East Asia and Western Pacific exceeded 1.2 million in 2008 and over 3.34 million in 2016 (based on official data submitted by Member States) with the America, South-East Asia and Western Pacific regions being the most affected (WHO 2019).

India (Murhekar et al. 2019), China (Yue et al. 2019), Indonesia (Harapan et al. 2019), Malaysia (Cohen et al. 2019), Brazil (Cohen et al. 2019; Stolerman et al. 2019), Mexico (Cohen et al. 2019; Macías et al. 2019) and Venezuela (Cabrera and Taylor 2019) have one of the highest rates globally and are considered high-risk countries.

Climate change in combination with several factors, including globalisation and changes of demography, environment and urbanisation, is considered to have an impact on the occurrence and transmission of dengue (Hinz et al. 2019). A Generalized Additive Mixed model (GAMM) framework was used by Cabrera and Taylor (2019) with factors such as growing population density, precarious living conditions, unsuitable water supply, poor well-being policies (Vincenti-Gonzalez et al. 2017), social displacements, demographic dynamic, unplanned urban settlements, unreliable sanitation policies, climatic factors and immunological resistance (Torres et al. 2017) further reveal multidynamics of dengue transmission.

Current trends and statistical mapping techniques predict that 3.83 (3.45–4.09) billion people (roughly 53% of the global population) live in areas that are suitable for dengue transmission, with the vast majority in Asia, followed by Africa and the

Americas (Messina et al. 2019). Unless an effective treatment is found, it is estimated that more people will be at risk of dengue in 2080 compared to 2015, bringing the total population at risk to over 6.1 (4.7–6.9) billion, or 60% of the world populations (Messina et al. 2019).

In the recent years, a lot has been done on the issue of dengue virus control. Although not much could be achieved, there are a few licensed dengue vaccines whereas, ongoing trials prevail towards the production of vaccines and drug measures towards dengue prevention and control. In the absence of any available effective vaccine and preventive drug, controlling the mosquito vectors of dengue is the only way to prevent and control dengue transmission (Alphey et al. 2013).

Mosquitoes control strategies have conventionally focused on killing mosquitoes using a variety of insecticides. Environmental management (through reduction or destruction of mosquito breeding sites) has mainly been used alongside chemical or microbiological ovicides, larvicides, and pupicides in areas endemic with dengue and other arboviral diseases.

1.1 Hosts of Dengue Virus Transmission

The relationship between hosts and parasites is a very complex and interesting one, the parasite develops mechanism to evade host response where as the host initiates an effective immune response against the parasites (Ruiz-Guzman et al. 2016). The ability of arboviruses to infect both, vertebrate and invertebrate hosts, has led to concerns about how the viruses could survive using different cellular machineries and overcome different antiviral responses (Villordo et al. 2015). Dengue transmission occurs as a result of interactions between the natural hosts (e.g. humans and non-human primates), intermediate hosts (mosquitoes) and environmental factors (Guzman et al. 2016) (Fig. 1).

The mosquito Aedes aegypti and to some extent Aedes albopictus are vectors serving as intermediate hosts for DEN (Fig. 2). Spread of virus directly relates to the geographic distribution of these vectors which thrives in close proximity to humans (Back and Lundkvist 2013). Monkeys and dogs have been reported to be reservoir hosts for dengue virus, even though humans are natural hosts for dengue, pathogenesis of the virus is dependent on various factors host genetics, virus genetics, prior immune exposure in humans and gut microbiota of mosquitoes (Sim and Hibbered 2016).

Originally, dengue was transmitted in sylvatic cycles in Asia and Africa between the Aedes mosquito and nonhuman primates, the global spread however has supported the evolution of the virus from the sylvatic (enzootic) cycles to urban (epidemic) cycles, the primary lifecycle of the virus is now exclusively between humans serving as natural host and Aedes mosquitoes serving as intermediate hosts (Virginia Tech 2010). Vertical transmission between vector species has been observed but more research is still needed to determine the reservoir status of animals such as dogs (Thongyuan and Kittayapong 2017).

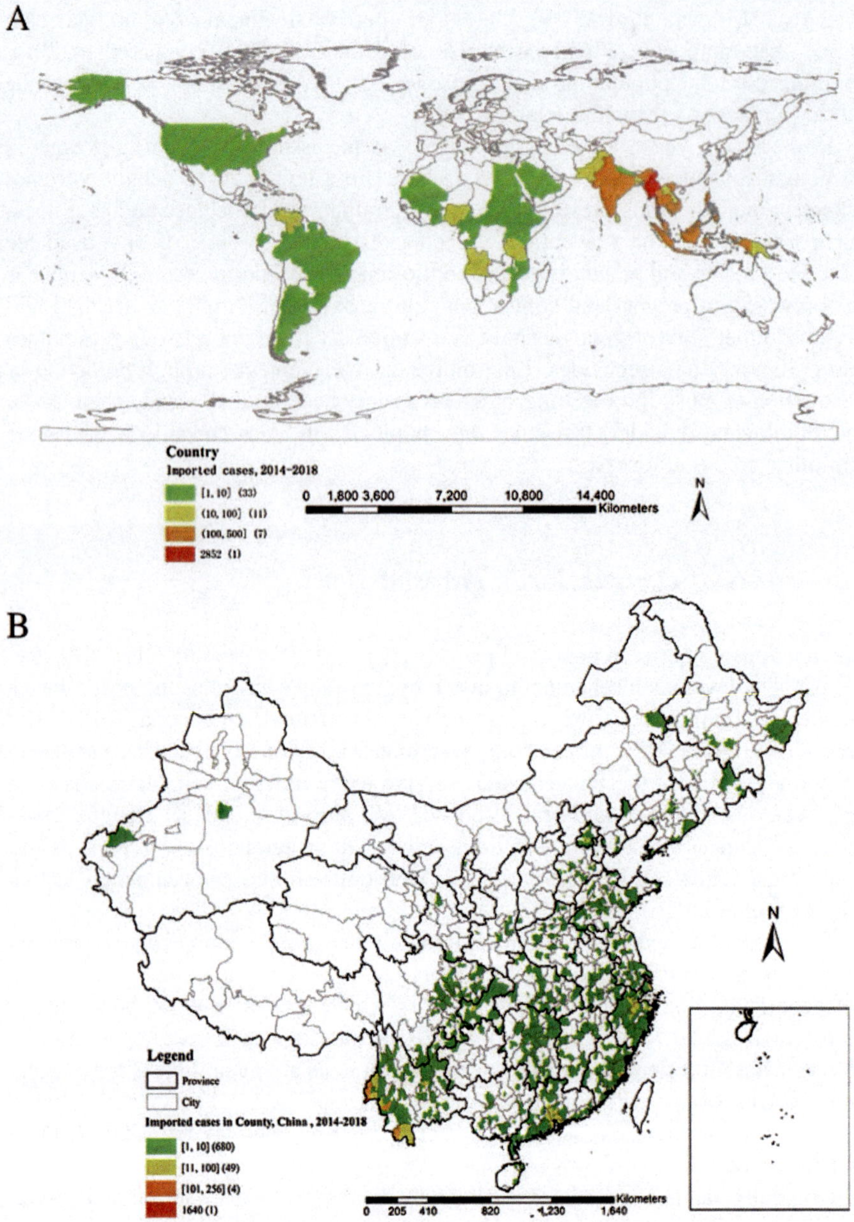

Fig. 1 Global distribution of dengue imported cases

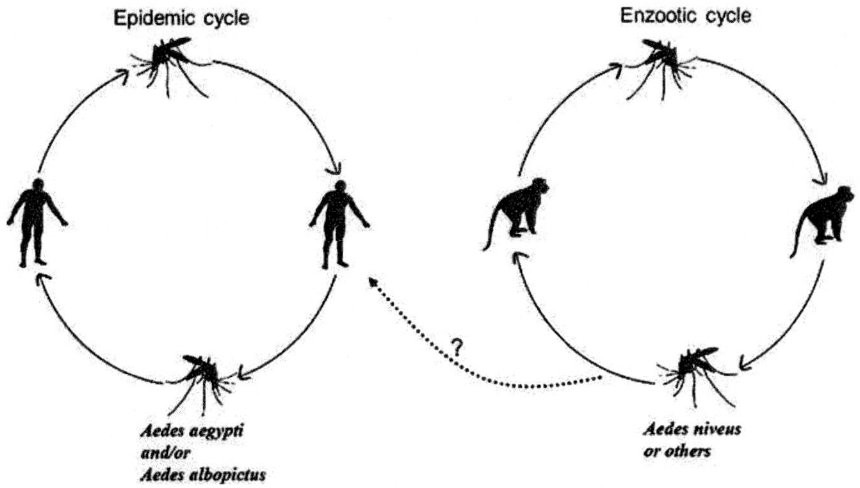

Fig. 2 On the left side (epidemic cycle), dengue virus is transmitted and maintained between humans and Aedes aegypti and Aedes albopictus in urban settlements, whereas on the right hand (enzootic cycle), dengue virus circulates between nonhuman primate (such as monkeys, baboons and chimpanzees) and Aedes niveus and other related species in the jungle

2 Control of Dengue Virus Transmission

A key control strategy in major mosquito-borne diseases such as malaria, yellow fever, dengue, Chikungunya fever, and Zika virus infection is by targeting mosquito vectors, which put a break in the transmission of the diseases to human. The effective control strategy of dengue will depend on the suppression of Aedes aegypti and Ae albopictus as the two most important vectors for transmission of the dengue. Vector control is an important approach to dengue prevention as there is lack of a universally effective vaccine against the disease.

2.1 Chemical Approach

About three decades after the discovery of the insecticidal properties of Dichlorodiphenyltrichloroethane (DDT), it was banned in the United States in 1972 for its mammalian toxicity, poisoning risks to nontarget organisms, persistence in the biosphere surface, and an accumulation in food-chains. Synthetic pyrethroids compounds were afterward added to the battery of public health insecticides. There are growing concerns and controversies on the multiple insecticide resistance mechanisms and therefore, more research interest is shifted toward studies on biological control, transgenic and paratransgenic approaches as potential alternatives, or complements to current chemical strategies.

2.2 Biological Approach

Biological control is the deliberate use of natural enemies to reduce the number of vector organisms. Natural enemies are the predators, parasites, parasitoids, and pathogens of the vectors. Biological control includes no chemical contamination of the environment and has specificity against target vectors.

2.3 Predators

Natural enemies feeding on mosquito larvae and pupae in aquatic environments can play an important role in reducing Culicidae populations. These larvae and pupae are being preyed by many aquatic organisms such as fish, amphibians, copepods, odonate young instars, water bugs, and even larvae of other mosquito species (Sarwar 2015). Fishes that prey on mosquito larvae include goldfish, guppies, bass, bluegill and catfish. But the most important fish predator is the Gambusia affinis, commonly known as the mosquito fish (Sarwar 2015).

Larvivorous fish have been established to be very effective at suppressing mosquito larval populations in different habitats. Birds have also been demonstrated to play vital role in remarkable reduction of adult and aquatic stages of mosquitoes. These birds include purple martins, swallows, waterfowl (geese, terns and ducks) and migratory songbirds (Sarwar 2015).

In addition, dragonflies are also arthropod predators that prey on mosquitoes and often referred to as mosquito hawks. Larvae of species of mosquitoes in the genus Toxorhynchites have been reported to be predators for larvae of medically important mosquitoes. Toxorhynchites are autogenous, and lack the need for blood feeding, which makes this species of mosquito ideal for release in the environment for biological control without increasing the risk of disease transmission (Huang et al. 2017).

2.4 Parasitoid

Insect parasitoids are natural enemies and have an immature life stage that develops on or within a single insect host which eventually kill the host. Adult parasitoids are free-living and may be predatory (Sarwar 2015). Parasitoids have specificity for their host and usually smaller than the host. This precise recognition of the host by the parasitoids is important in the biological control strategy. Ascogregarina culicis is a gregarine protozoon parasite and has been detected from the larvae of A. aegypti which can harmfully affect the natural population of mosquito (Sarwar 2015).

2.5 *Plants*

Plants produce compounds to protect themselves from insects, and these compounds can affect insect development in many ways (Ong 2016). These plant-borne molecules are effective at a few parts per million (ppm) against Aedes (Ae.), Anopheles (An.) and Culex (Cx.) young larval instars (Rodríguez-Pérez et al. 2012). Hundreds of plant species have been tested for their effects against mosquitoes. The neem tree (Azadirachta indica) is a well-known medicinal plant that has been widely tested against mosquitoes (Rodríguez-Pérez et al. 2012).

3 Application of Biotechnology in Dengue Control

Dengue Virus (DENV) control can be stratified into various categories depending on the type of approach. This section will focus on the use of gene modifications on Aedes Mosquito as a means of control dengue virus infections.

From Epigenetic perspective, dengue control can be achieved through vertical (mating-based) transmission of heritable elements which are specie specific (Aedes spp) (Alphey et al. 2013). This approach is most attractive due to its environment friendly nature. It exquisitely targets the vector species (Aedes Mosquito). Some genetic strategies have been developed overtime using classical genetics such as Sterile Insect Technique (SIT) which is termed as "Population suppression strategy" aimed at achieving sterile male population of the Aedes Mosquito. Through recombinant DNA methods, several Aedes Mosquitoes have been transformed using either transposon vectors or by artificial infection with various Wolbachia (Alphey et al. 2013).

The intracellular bacteria, Wolbachia can induce a form of sterility known as Cytoplasmic Incompatibility (CI), where embryos of uninfected females are fertilized by sperm from infected males, and the fertilized eggs fail to develop and eventually die. However, when infected males mate with infected females, the eggs can develop by a method called "Incompatible Insect Technique" (IIT). Death of offsprings of incompatible crosses is also achieved by Wolbachia through IIT process, though the biochemical and the genetic mechanism is unknown (Alphey et al. 2013).

Sterility or death of most or all offsprings can also be achieved through the introduction of dominant lethal allele into the vector insect genome by recombinant DNA methods, rather than irradiation in sterilizing vector male population. Although, the SIT-like methods described here are self-limiting, self-sustaining population suppression strategies can be applied using sequence-specific nucleases called homing endonucleases (HEGs) where reduced-fitness trait are introduced into the target population using the super-Mendelian inheritance property of HEGs, which can drive a population or even a specie into extinction. With the development of SIT-like system called RIDL (Release of Insect carrying a Dominant Lethal), a dominant lethal transgene is inserted, but its expression is artificially repressed to allow the insect vector to be reared. The advantage of this approach over the use of DNA damage or CI technique is the ability to select the time death of the offspring.

However, the DENV can also be controlled through suppression of replication by the RNAi mechanism in the mid-gut or saliva of the Aedes Mosquito. Transgene-based expression of a hairpin RNA corresponding to part of the DENV2 virus in the mid-gut or salivary gland of the host vector has shown to provide a strong block to virus transmission. Though for the midgut expressing-line, anti-DENV2 hairpin expression and the associated refractory phenotype were reported to be lost after 13 generations (Aguiar et al. 2019). This could be as result of the unusual inverted repeats involved which may be subject to some form of epigenetic silencing.

The "gene drive" technique has also incorporated Wolbachia as result of its ability to reduce susceptibility of infected mosquito to various pathogens. Wolbachia are capable of spreading through insect populations as a heritable modification by manipulating the host's reproductive biology- in other words, "gene drive system". This brought about the concept of Gene-drive-plus-refractory-gene package. The attention was focused on wMel, a strain of Wolbachia from Drosophila melanogaster, because its infection has a similar dengue-blocking effect in Ae. Albopictus. The introduction of a single infected female can lead to Wolbachia invading that population, especially if the effective population size is low. The public perception has generally been positive about the use of Wolbachia, reason being that it is natural, and has avoided public concerns relating to the use of recombinant DNA methods.

Human miRNA pathway, GRP75 protein and hsa-mir-126 have been reported to have regulatory role in dengue life-cycle. Though this experiment was performed at cell line level, it has shown positive result on the ability of such genes to regulate DENV replication, which is a positive step toward the search for dengue virus control (Kakumani et al. 2016). Targeting these genes will generate positive outcomes in the efforts to control dengue infection by arresting its replication.

Bacillus thuringiensis H-14

Bacillus thuringiensis H-14 is a gram-positive, sporulating bacterium that releases insecticidal toxins and virulence factors that selectively target the larval stages of insects (Ong 2016). It effectively kills the larval stage of mosquitoes such as Aedes aegypti and Aedes albopictus without undesirable effects on the environment and non-target organisms. Ingestion of the sporulated bacterial cells by mosquito larvae lead to ingestion of crystals being activated in the naturally alkaline environment of the larval mid-gut. This cause the proteolytic enzymes that existed in the mid-gut of larval to break down the endotoxin of the crystallized bacteria, as the result, a polypeptide toxin fraction was released by the bacteria and triggers the killing mechanism (Ong 2016).

Three of the four toxins (Cry4A, Cry4B, and Cry11A) produced by Bacillus thuringiensis are homologous in their structures and utilize similar mechanisms of killing with other members in the Cry toxin family (Huang et al. 2017). Despite the success recorded in use this bacterium to biologically control the dengue vector, reports demonstrated that Ae. albopictus is capable of developing resistance to B. thuringiensis H-14 and this indicate the application of B. thuringiensis H-14 might not be sustainable for a long-term control program because of its potential in developing resistance (Ong 2016).

3.1 Genetically-Modified Aedes Aegypti

The use of genetically engineered mosquitoes for biological control of dengue is borne from the idea of using the genetically engineered mosquitoes to replace the population of competent mosquitoes in nature (Ong 2016). Aedes aegypti is genetically modified using a technology known as RIDL (Release of Insects carrying Dominant Lethal).

The science behind RIDL is known as 'repressible lethality' which involves introduction of a specific DNA construct into Aedes aegypti eggs through micro-injection so that the transformed mosquito is destined to die at larval or pupal stage unless it is provided with a nutritional supplement otherwise antidote (tetracycline) in the rearing medium. The tetracycline represses the lethal gene and hence allows the mosquito larvae to grow normally into adults when they are reared in a laboratory or rearing facility (Ong 2016).

The engineered males are released into the natural environment which then mates with wild females, producing offspring that die at the larval stage in the absence of tetracycline which is unstable in the environment (Vythilingam and Wan-Yusoff 2017). This approach has the advantage of being species-specific and has no long-lasting effects on the target species as the aim is to eliminate the population in the released area.

4 Vaccination and Vaccine Development Against Dengue

Currently, there are no licensed dengue-specific therapeutics, nor are any candidate drugs in late-stage development (Hadinegoro et al. 2015). Management options are often helpful but could be difficult to implement in remote or resource-poor areas (Stanaway et al. 2016). Public Health methods such as vector control can be effective at preventing infections but these measures remain challenging because the mosquito vectors rapidly evolve to avoid chemo-prevention and expand their ability to live in harsh environments (World Health Organization 2019). All these factors point to the utmost need to develop a vaccine candidate that will prevent against all the Dengue serotypes (Whitehead and Subbarao 2018; Fauci et al. 2019).

Laboratory evidence has led to the widely accepted hypothesis that heterotypic antibodies bind to DENVs and increase their uptake into cells of the monocytic lineage via their FcG receptors (FcGR) and drive up virus load, leading to severe dengue. In this context, it has been discovered that antibodies to the pre-membrane (prM) protein and the fusion loop (FL) epitope of the envelope (E) protein, primarily play a role in enhancement of DENV infection (Swaminathan and Khanna 2019). The occurrence of antibody-dependent enhancement (ADE) of dengue disease strongly supports the rationale underlying current dengue vaccine development efforts, which is to provide balanced and durable immunity to all four DENV serotypes (Raut et al. 2019).

The quest for a safe and effective DENV vaccine has been ongoing for nearly 70 years (Ngono and Shresta 2019). A live attenuated vaccine (LAV), chimeric yellow fever/dengue tetravalent dengue vaccine (CYD-TDV), developed by Sanofi (Hadinegoro et al. 2015) and marketed as Dengvaxia, was first licensed in 2015 and is now licensed in several dengue-endemic countries. Two major drawbacks of this vaccine are suboptimal antibody response against different serotypes and its relative ineffectiveness in children younger than 9 years and naïve individuals (Ngono and Shresta 2019). Several vaccine candidates are being designed and are undergoing both preclinical and clinical trials.

4.1 Concept of Chimerism in Dengue Vaccinology

In DENV endemic areas, multiple serotypes of dengue viruses (DENV-1 to DENV-4) are in co-circulation. Infection by one serotype of DENV confers lasting protection against the disease, and possible infection, following re-exposure to the same serotype. Secondary exposure to any of the other three serotypes only experiences transient protection (Li et al. 2013). Therefore, dengue vaccine development only focuses on a vaccine that simultaneously provides long lasting protection against all the DENV serotypes. This takes the approach of developing a tetravalent DENV vaccine.

DENV have a single stranded positive-sense RNA genome that contains a single Open Reading Frame (ORF) flanked by two untranslated regions (5' and 3' UTRs) (Huang et al. 2006). The ORF encodes a single polyprotein (proteolytically processed into three structural proteins, C, prM and E) and seven non-structural proteins (NS1, NS2A, NS2B, NS3, NS4A, NS4B, and NS5). Induction of protective immunity against viral infection relies on the E glycoprotein antigen, with co-expression of prM as a requirement to acquire its native conformation. The non-structural proteins are actively involved in viral genome replication, translation, and regulation (Yang et al. 2014). The high degree of similarity in the *flaviviruses* genome organization, replication and translation strategy makes the viruses viable chimeric viruses. Viable chimeric viruses have been rationally designed and generated by interchanging various genes among *flaviviruses* by using reverse genetics. In recent years, this kind of chimerism approach is been used in dengue vaccine development. Example of vaccine that is product of this kind of technology is the ChimeriVax-Dengue (CYD), which was created with yellow fever vaccine strain, 17D (YF-17D) this vaccine is currently in phase 2b trial in Thailand (Li et al.

2013). Attenuated DENV-2 strain PDK53 is another chimeric dengue vaccine on clinical trial. Additionally, recombinant DENV-4 with 30 nucleotides deletion in 3' UTR (DENV4Δ30) has been utilized successfully as a backbone for the development of chimeric dengue vaccine (Yang et al. 2018).

Chimeric DENV has been developed in recent years by using reverse genetics using the full-length infectious cDNA clone of Japanese encephalitis virus (JEV) strain, SA14-14-2 as a backbone, named ChinDENV. The ChinDENV (novel chimeric DENV) was designed and constructed using JEV strain SA14-14-2, replacing the pre-membrane (prM) and envelop genes (E) of DENV-2. The chimeric virus showed growth and plaque properties similar to those of parental DENV in mammalian and mosquito cells.

5 Conclusion

Global dengue virus re-emergence and expansion appear to be very difficult to contain anytime soon. Being an arthropod-borne infection, the best that could be done is to re-enforce vector control measures, preferably using novel biotechnological and molecular approaches viz-a- viz, continue with the search for novel vaccines and/or improve on the potency, effectiveness, efficacy, safety, and affordability of existing ones, in order to assure protection against all serotypes of dengue, regardless of age, sex or race of people.

References

Aguiar GPCG, Leite CMGDS, Dias B, Vasconcelos SMM, Moraes RA, Moraes MEA, Miyajima F et al. Evidence for host epigenetic signatures arising from arbovirus infections: a systematic review. Front Immunol. 2019;10:1207.

Alphey L, McKemey A, Nimmo D, Neira Oviedo M, Lacroix R, Matzen K, Beech C. Genetic control of Aedes mosquitoes. Pathog Glob Health. 2013;107(4):170–9.

Back A, Lundkvist A. Dengue viruses—an overview. Infect Ecol Epidemiol. 2013;3:1–21.

Bhatt S, Gething PW, Brady OJ, Messina JP, Farlow AW, Moyes CL, Drake JM, Brownstein JS, Hoen AG, Sankoh O, Myers MF. The global distribution and burden of dengue. Nature. 2013;496(7446):504.

Cabrera M, Taylor G. Modelling spatio-temporal data of dengue fever using generalized additive mixed models. Spat Spatio-Tempor Epidemiol. 2019;28:1–13.

Cohen C, Moreira ED Jr, Nañez H, Nachiappan JP, Arvinder-Singh HS, Huoi C, Nealon J, Sarti E, Puentes-Rosas E, Moureau A, Khromava A. Incidence rates of neurotropic-like and viscerotropic-like disease in three dengue-endemic countries: Mexico, Brazil, and Malaysia. Vaccine. 2019;37(13):1868–75.

Fauci A, Erbelding E, Whitehead S, Cassetti MC, Handley FG, Gupta R. Dengue vaccine clinical trials in India—an opportunity to inform the global response to a re-emerging disease challenge. Int J Infect Dis. 2019;84S:S4–6.

Guzman MG, Harris E. Dengue. The Lancet. 2015;385(9966):453–65.

Guzman MG, Gubler DJ, Izquierdo A, Martinez E, Halstead SB. Dengue infection. Nat Rev. 2016;2:1–25

Hadinegoro SR, Arredondo-García JL, Capeding MR, Deseda C, Chotpitayasunondh T, Dietze R, Hj Muhammad Ismail HI, Reynales H, Limkittikul K, Rivera-Medina et al. Efficacy and long-term safety of a dengue vaccine in regions of endemic disease. N Engl J Med. 2015;373(13):1195–206.

Harapan H, Michie A, Yohan B, Shu PY, Mudatsir M, Sasmono RT, Imrie A. Dengue viruses circulating in Indonesia: A systematic review and phylogeneticanalysis of data from five decades. Rev Med Virol. 2019;29(4):e2037. https://doi.org/10.1002/rmv.2037

Hinz R, Frickmann H, Krüger A. Climate change and infectious diseases. In:International Climate Protection. Cham: Springer; 2019. p. 269–276.

Huang KJ, Yang KC, Lin YS, Huang JH, Liu HS, Yeh TM, Chen SH, Liu CC, Lei HY. The dual-specific binding of dengue virus and target cells for the antibodydependentenhancement of dengue virus infection. J Immunol. 2006;176:2825–2832

Huang YJ, Higgs S, Vanlandingham D. Biological control strategies for mosquito vectors of arboviruses. Insects. 2017;8(1):21.

Kakumani PK, Medigeshi GR, Kaur I, Malhotra P, Mukherjee SK, Bhatnagar RK. Role of human GRP75 in miRNA mediated regulation of dengue virus replication. Gene. 2016;586(1):7–11.

Li XF, Deng YQ, Yang HQ, Zhao H, Jiang T, Yu XD, et al. A chimeric dengue virus vaccine using Japanese encephalitis virus vaccine strain SA14-14-2 as backbone is immunogenic and protective against either parental virus in mice and nonhuman primates. J Virol. 2013;87(24):13694–705.

Macías AE, Puentes-Rosas E, Velandia D, Morley D, Coudeville L, Baurin N. Real-world evidence of dengue burden on hospitals in Mexico: insights from the Automated Subsystem of Hospital Discharges (Saeh) database. Revista de Investigación Clínica. 2019;71(3):168–77.

Messina JP, Brady OJ, Golding N, Kraemer MU, Wint GW, Ray SE, Pigott DM, Shearer FM, Johnson K, Earl L, Marczak LB. The current and future global distribution and population at risk of dengue. Nat Microbiol. 2019;4:1508–15.

Murhekar M, Joshua V, Kanagasabai K, Shete V, Ravi M, Ramachandran R, Sabarinathan R, Kirubakaran B, Gupta N, Mehendale S. Epidemiology of dengue fever in India, based on laboratory surveillance data, 2014–2017. Int J Infect Dis. 2019;84(Supplement):S10–4.

Mustafa MS, Rasotgi V, Jain S, Gupta V. Discovery of fifth serotype of dengue virus (DENV-5): a new public health dilemma in dengue control. Med J Armed Forces India. 2015;71(1):67–70.

Ngono AE, Shresta S. Cross-reactive T cell immunity to dengue and Zika viruses: new insights into vaccine development. Front Immunol. 2019;10:1316. https://doi.org/10.3389/fimmu.2019.01316.

Ong SQ. Dengue vector control in Malaysia: a review for current and alternative strategies. Sains Malaysiana. 2016;45(5):777–85.

Raut R, Corbett KS, Tennekoon RN, Premawansa S, Wijewickrama A, Premawansa G, Mieczkowski P, Rückert C, Ebel GD, De Silva AD, de Silva AM. Dengue type 1 viruses circulating in humans are highly infectious and poorly neutralized by human antibodies. Proc Natl Acad Sci. 2019;116(1):227–32.

Rodríguez-Pérez MA, Howard AF, Reyes-Villanueva F. Biological control of dengue vectors. Integrated Pest Management and Pest Control—Current and future Tactics. Feb 24; 2012.

Ruiz-Guzman G, Ramos-Castaneda J, Hernandez-Quintero A, Contreras-Garduno J. Cost and benefits of vertical and horizontal transmission of dengue virus. J Exp Biol. 2016;219:3865–9.

Sang S, Gu S, Bi P, Yang WZ, Xu L, Yang J, Chu C. Predicting unprecedented dengue outbreak using imported cases and climatic factors in Guangzhou,2014. PLoS Negl Trop Dis. 2015;9:e0003808

Sarwar M. Reducing dengue fever through biological control of disease carrier Aedes Mosquitoes (Diptera: Culicidae). Int J Prevent Med Res. 2015;1(3):161–6.

Sim S, Hibbered M. Genomic approaches for understanding dengue: insights from the virus, vector and host. Genome Biol. 2016;17(38):1–15.

Stanaway JD, Shepard DS, Undurraga EA, Halasa YA, Coffeng LE, Brady OJ, Hay SI, Bedi N, Bensenor IM, Castañeda-Orjuela CA, Chuang TW. The global burden of dengue: an analysis from the Global Burden of Disease Study 2013. Lancet Infect Dis. 2016;16(6):712–23.

Stolerman LM, Maia PD, Kutz JN. Forecasting dengue fever in Brazil: an assessment of climate conditions. PLoS ONE. 2019;14(8):e0220106.

Swaminathan S, Khanna N. Dengue vaccine development: Global and Indian scenarios. Int J Infect Dis. 2019;84(Supplement):S80–6.

Thongyuan S, Kittayapong P. First Evidence of dengue infection in domestic dogs living in different ecological settings in Thailand. PLoS ONE. 2017;12(8):e0180013.

Torres JR, Orduna TA, Piña-Pozas M, Vázquez-Vega D, Sarti E. Epidemiological characteristics of dengue disease in latin America and in the Caribbean: a systematic review of the literature. J Trop Med. 2017;3:1–18. https://doi.org/10.1155/2017/8045435.

Villordo S, et al. Dengue Virus RNA structure Speialization Facilitates Host Adaptation. Plos Pathog. 2015;11(1):e1004604.

Vincenti-Gonzalez MF, Grillet ME, Velasco-Salas ZI, Lizarazo EF, Amarista MA, Sierra GM, Comach G, Tami A. Spatial analysis of dengue seroprevalence and modeling of transmission risk factors in a dengue hyperendemic city of venezuela. PLoS Negl Trop Dis. 2017;11(1): e0005317.

Virginia Tech. Dengue Virus. 2010. [Online] http://ci.vbi.vt.edu/pathinfo/pathogens/Dengue1.html [Accessed 20 September 2019].

Vythilingam I, Wan-Yusoff WS. Dengue vector control in Malaysia: Are we moving in the right direction? Trop Biomed. 2017;34(4):746–58.

World Health Organization. Dengue and severe dengue fact sheet. World Health Organization, Geneva, Switzerland. 2019. https://www.who.in.

Whitehead SS, Subbarao K. Which dengue vaccine approach is the most promising, and should we be concerned about enhanced disease after vaccination? The risks of incomplete immunity to dengue virus revealed by vaccination. Cold Spring Harb Perspect Biol. 2018;10(6):a028811.

Yang C, Hou J, Chen T, Chen W. Discriminable roles of Aedes aegypti and Aedes albopictus in establishment of dengue outbreaks in Taiwan. Acta Trop. 2014;130:17–23.

Yang H, Yang H, Li Z, Liu L, Wang W, He T, et al. Japanese encephalitis virus/yellow fever virus chimera is safe and confers full protection against yellow fever virus in intracerebrally challenged mice. Vaccine. 2018;36(18):2450–5.

Yue Y, Liu X, Xu M, Ren D, Liu Q. Epidemiological dynamics of dengue fever in mainland China, 2014–2018. Int J Infect Dis. 2019;86:82–93.

Dengue Fever Epidemic in Pakistan and Its Control Measures: Where Are We Moving?

Ali Ahmed and Gul Majid Khan

Abstract Dengue fever is one of the significant health issues throughout the world, especially in tropical and subtropical territories. The cause of dengue fever is the dengue virus, which is a single-stranded RNA virus that is part of Flaviviridaeviruses family and has four distinct serotypes: DENV-1, DENV-2, DENV-3, and DENV-4. Dengue virus is spread to humans by the bite of the mosquitoes *Aedes aegypti* and *Aedes albopictus*. Clinically dengue fever symptoms range from mild (headache, nausea, vomited, pain in muscles and bones) to severe form such as dengue haemorrhagic fever and dengue shock syndrome. Since 1994 Pakistan has suffered from dengue endemic. Nevertheless, since 2006, the world has faced the worst dengue attack situation in which the disease has infected thousands of people and hundreds of people have been killed. DENV-2, DENV-3 and DENV-1 are most predominant serotypes in Pakistan.Popular diagnostic techniques being used in Pakistan are Enzyme-Linked Immunosorbent Assay (ELISA), polymerase chain reaction and rapid diagnostic tests, during differential diagnosis. Critical issues with dengue diagnosis include shortcomings in screening tests and a weak health-care system. The major factors responsible for dengue epidemics in Pakistan are favourable climatic conditions, unplanned urbanisation, population growth, commuting and many socioeconomic factors etc. This chapter offers updates on Pakistan's dengue epidemic and explains how to strengthen the region against dengue virus.

Keywords DENV-1 · DENV-2 · DENV-3 · DENV-4 · Emergence · Immunity · Serotypes · Endemic · Aedes aegypti · Ae. Albopictus · N, N-diethyl-m-toluamide · *Bacillus thuringiensis*

A. Ahmed (✉)
School of Pharmacy, Monash University, Jalan Lagoon Selatan, 47500 Bandar Sunway, Malaysia
e-mail: ali.ahmed@monash.edu

G. M. Khan
Department of Pharmacy, Quaid I Azam University Islamabad, Islamabad 15320, Pakistan
e-mail: gmkhan@qau.edu.pk

S. I. Ahmad (ed.), *Human Viruses: Diseases, Treatments and Vaccines*, https://doi.org/10.1007/978-3-030-71165-8_3

1 Introduction

Mosquito-born disease caused by the dengue virus is dengue fever. Symptoms of dengue fever are associated with flu-like symptoms which develop after three to fourteen days of viral infection (Kularatne 2015). Other symptoms may include high fever, nausea, vomiting, skin rashes, muscles and joints pain. Recovery generally takes place within 7 to 10 days.

In some cases dengue fever becomes severe, such as the occurrence of low platelets, bleeding and blood plasma leakage; while in dengue shock syndrome, blood pressure of patient becomes dangerously low, hepatomegaly and shock (Heydari et al. 2018). Dengue, which is an enveloped single-stranded RNA virus, belongs to the genus Flavivirus and family Flaviviridae (Ahmad and Poh 2019). The virus is transmitted to humans via the bite of an infected female mosquito of the genus Aedes, usually *Aedes aegypti* (Ae Aegypti) and rarely *Ae. Albopictus*. The Ae Aegypti grow in stagnant water collected at different areas in tropical and subtropical regions, and it is a day-biting mosquito. These mosquitos carrying the virus, their biting to the person can transmit the virus to a non-infected person. Dengue can also be transmitted by infected blood products and organ donations (Dean et al. 2018). Vertical transmission from mother to foetus during pregnancy has also been reported (Wiwanitkit 2010). According to World Health Organization (WHO), dengue fever is the fastest spreading dengue-endemic disease affecting the health of peoples of tropical and sub-tropical countries of the world. Dengue is the primary cause of hospitalisation, and it is estimated that annually 500,000 people infected with severe dengue required hospitalization in which children are dominant. In contrast, about 2.5% of the affected people die annually (Yousaf et al. 2018).

2 Dengue Serotypes

There are four serotypes of dengue viruses: DEN-1, DEN-2, DEN-3, and DEN-4. They are antigenically different but closely related serotypes of dengue show 65–70% sequence homology (Ali et al. 2016). Each serotype has different genotype with extensive genetic variability, which causes vaccine development difficulties against all four dengue serotypes. If a person has been diagnosed with dengue from one serotype and is later recovered from dengue it will keep him immune from this serotype throughout life. While cross-immunity to the different dengue, serotype is short-lived (Ali et al. 2016). Re-infection of dengue with new serotype can present a high risk of creating dengue hemorrhagic fever due to antibody-dependent change, a miracle where specific antibodies to infection strengthen infectious disease, and sometimes infection replication within monocytes/macrophages and granulocytic cells can aid (Wahala and De Silva 2011). Infection caused by all four dengue serotypes is analogous, but several studies have shown that DEN-2 and DEN-3 are related to severe dengue infection, whereas DEN-1 is causing mild disease (Passos et al. 2001).

3 Prevalence of Dengue Serotypes in Pakistan

Dengue was first discovered in Karachi, a congested city of Pakistan with a seaport causing the importation of eggs of infected mosquitoes usually with tyres from foreign countries (Yousaf et al. 2018). This virus is endemic in Pakistan mostly in the post-monsoon period and can stay dormant for several years, and several outbreaks have been reported in different regions of this country at a different time (Table 1).

Why dengue endemic in Pakistan?

Pakistan is suffering from dengue endemic since 1994, and now this virus has become a significant public health concern. Although started in 1994 but from 2006 it has taken status of the epidemic in which the disease has infected thousands of people and hundreds of people lost their lives. From 1995 to 2019, there have been around 1,47,200 instances of dengue infection and over 800 deaths. Dengue serotype 2 is the most usual circulating serotype in Pakistan, with few reported cases of serotype 3 (Fatima 2019).

Table 1 Distribution of dengue serotypes in various geographical areas of Pakistan

Year	Serotype	Region	References
1985	DEN-2	Karachi	(Cobelens et al. 2002)
1994	DEN-2	Karachi	(Chan et al. 1994)
1995	DEN-2	Baluchistan	(Paul et al. 1998)
1997	DEN-2	Karachi	(Qureshi et al. 1997)
1998	DEN-1, DEN-2	Karachi	(Akram et al. 1998)
2003	DEN-2	Nowshehra, Haripur, Khushab	(Khanani et al. 2011)
2005	DEN-3	Karachi	(Jamil et al. 2007)
2006	DEN-2, DEN-3	Karachi	(Khan et al. 2008)
2007	DEN-2, DEN-3	Lahore	(Fatima et al. 2011)
2008	DEN-2, DEN-3	Lahore	(Fatima et al. 2011)
2009	DEN-2, DEN-3	Lahore	(Fatima et al. 2011)
2010	DEN-1, DEN-2	Gujranwala, Lahore, Sheikhupura	(Mahmood et al. 2012)
2011	DEN-2	Punjab	(Fatima et al. 2012)
2013	DEN-2, DEN-3	Swat	(Khan 2013)

There are several factors involved in spreading of this virus; the most important is the favorable climate. As the climate in this country is most beneficial to the mosquito to grow and reproduce, especially in the post-monsoon period when there are hot and humid conditions. Another factor supporting the growth and reproduction of Ae Aegypti is the moderate temperature in which vector replication and maturation are enhanced (Ahmed et al. 2017; Rasheed et al. 2013). Another factor favoring mosquito growth is the unregulated and unplanned urbanization which provide favorable breeding ground to Ae Aegypti. Other reason includes the inability of environmental sanitation to clean up the plethora of waste containers (Pulford et al. 2012). Overpopulation, lack of fresh consuming water, inadequate mosquito control, air travel, poor socio-economic conditions, lack of public health and lack of awareness are other vital reasons for the dengue epidemic (Ahmed et al. 2017; Tahir et al. 2015).

4 Seasonal Effects of Dengue Outbreak

Cases of dengue infection increased during and after rainy seasons as compared to winter and summer seasons. high humidity, optimum temperature and rainfalls remain significant predictors of dengue incidence in Pakistan. The surge of cases occurs between July and October as shown in Fig. 1 (Shaikh et al. 2014).

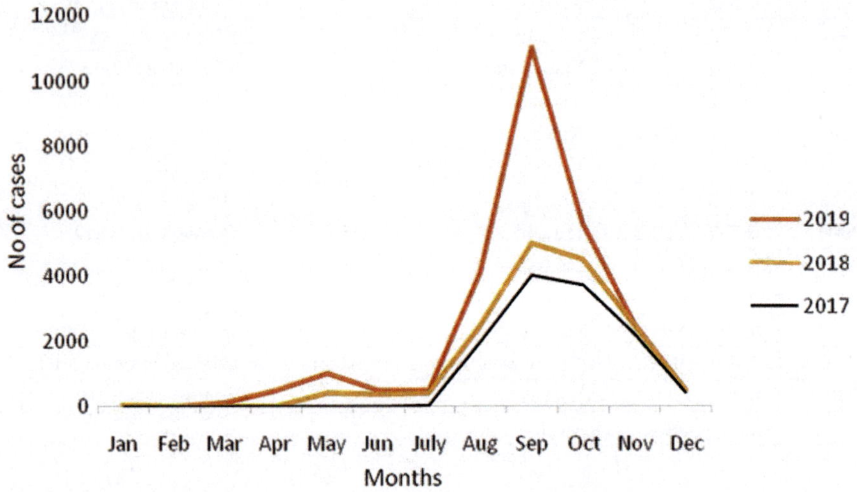

Fig. 1 Dengue fever cases in Pakistan: monthly data collected from 2017 to September 2019 (n = 28,328) (Internal data source 2014)

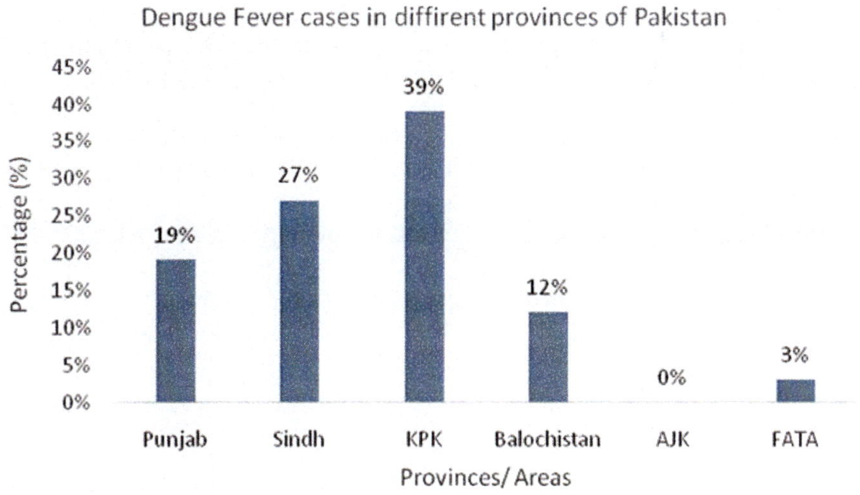

Fig. 2 Represents the dengue fever cases in all provinces of Pakistan from 2017 to September 2019 (n = 28,328)

5 Climate Change Adds to Risk Factors

The key risk factors for dengue-endemic are urban population, indoor storage of water with neglected storage facilities, and poor sanitary conditions. But, an unmarked condition in this equation is climate change. The unplanned vector surveillance cannot be predicted for the post-monsoon, unexpected rain, where water drainage facilities are of poor quality, time after time. The open pockets of rainwater in the urban areas results in suitable nurseries for the dengue mosquito hence result in failures of most measures in controlling the vector population, by the end of the monsoon period.

6 Dengue Infection Statistics in 2019

According to the data of WHO, around the world, per 100 million people there are around 20,000 people die due to dengue virus infection (Guzman et al. 2016). According to weekly epidemiology report of National Institute of Health (NIH) Pakistan in November 2019: "there was a total of 24,336 cases including 2,686 cases in Baluchistan, 2847 cases in Islamabad and 5214 cases in Khyber Pakhtunkhwa (KPK). According to the report for Punjab and Sindh province, there were 8,770 and 2,937 cases respectively (Internal data 2014).

Dawn (*18 November 2019)* a Pakistan national media news channel claimed in 2019 reported a total 49,587 cases from the whole country. Thirteen thousand two hundred fifty-one cases from Sindh, 13,173 from Islamabad, 9,855 from Punjab,

7,776 cases from KPK, 3,217 cases from Baluchistan, and least number of cases 1,690 reported from Azad Jammu Kashmir (AJK). Other 625 cases were placed in 'other' categories such as the cases whose origin could not be determined. No deaths from KPK and Gilgit-Baltistan were recorded, while in the federal capital Islamabad, as many as 22 dengue patients died. Of the remaining 57 cases, 33 people died in the province of Sindh, 20 in Punjab, 3 in Baluchistan and 1 died in AJK (Countrywide 2019).

In Pakistan, several hospitals running the facilities to treat patients infected with dengue virus include Pakistan Institute of Medical Sciences (PIMS), Islamabad, Mayo, Jinnah and Ganga Ram Hospitals, Lahore, Civil, Jinnah and Agha Khan, Hospital, Karachi, and Allied Hospital, Faisalabad, Benazir Bhutto Hospital, Rawalpindi (Fig. 2).

7 Way Forward

In Pakistan control of dengue and other vector-borne diseases is extremely challenging due to insufficient expert entomologist, lack of quality assurance standards, adequate monitoring, assessment system and the lack of designated authority for dengue vector control interventions. Prevention of dengue in this country can be enhanced by adopting the following measures.

1. **Health care system up-gradation**

In Pakistan, weak health care systems may be responsible for the high mortality rate from dengue infection. Mortality rate can lessen by improving staff training and clinical management, early clinical and laboratory diagnosis, intravenous rehydration, and reorganizing hospital procedures. Pakistan's health ministry should organize health promotion programs to disseminate community information to eradicate mosquito breeding sites. Better hygiene procedures, personal safety measures, encouragement of the use of larvicidal agents, and treatment of unwanted stagnant water should be the objectives of the health care staff alongside the campaigns.

2. **Strengthening of surveillance system in Pakistan**

Surveillance is a critical step for any dengue prevention and control as it offers the required information for risk management and policy guidance, including disease response and policy evaluation. The weak dengue surveillance system in Pakistan is one of the main reasons of unsatisfactory dengue prevention. Application of functional and continuous dengue surveillance at every level is the primary requirement of dengue control in Pakistan, and it should be a part of the national health care program. Passive monitoring, active surveillance, and event-based data collection should be the valuable component of our monitoring program to assess dengue transmission, spread of serotypes, and investigate unexplained health incidents, including unknown etiology of fevers and case clustering.

3. **Personal protection**

Protective garb can reduce the risk of mosquito biting and hence it is recommended that socks, shirts and trousers be worn in full sleeves. Mosquito mats, coils, pressurized canned products and repellents are generally used for domestic safety. Plant extricates inclusive of Neem oil and chemical such as DEET (N, N-diethyl-m-toluamide) are repellents which offer protection against mosquitoes. Certain insecticides are available for those who use skin-tight clothing; their clothes can be treated with the chemical to avoid mosquito biting through skin-tight clothing. Their clothes ought to be treated with the recommended dose to prevent irritation on the skin. Precautions must be taken while the usage of repellents both natural and chemical and use of mats and coils (the locally produced insect repellent which when burnt slowly produce vapours acting as mosquito repellents) should be avoided to be used in tightly closed rooms.

4. **Environmental management**

Environmental management can also be effectively used to prevent mosquito growth and infection. Environmental management refers to the modifications in the environment to reduce man-vector contact and consequent transmission hazard. Quality waste management enhanced and effective water supply network, enough drainage system, maintenance of containers for domestic water storage, cleaning of flower vases and recycling of worn out tyres, are all part of environmental management. In Pakistan, government will enhance environmental protection to be helped by the society in controlling mosquito spread. Improper sanitation is the key problem in Pakistan, and the country should concentrate on street sweeping, improved drainage network and stagnant water removal.

5. **Biological control**

Biological control agents can be used for the prevention of dengue. Larvivorous fish were commonly used for management of Ae. Aegypti in a large container of water and its output depends on the type of container or body of water being used similarly copepods have played a role in regulating Ae. Agypti (Guzmán and Kouri 2004). According to national dengue control guidelines in Pakistan, mosquito fish should only be used in limited, clean water bodies and should be released after comprehensive surveys of mosquito breeding sites. Fungi such as Beauveria bassiana and Metarhizium anisopliae have also been proposed as potential biological control agent of Ae. Aegypti. Also, bacterium *Bacillus thuringiensis* sub-sp. *israelensis* is useful to control mosquitoes. These bacteria should be applied only during first three instars of the larval stage. The use of endosymbionts such as *Wolbachia pipientis*, a gram-negative bacterium, has also been reported to control of mosquito-borne transmission of pathogens (Tahir et al. 2015).

6. **Chemical control**

Certain larvicides have also been found to be effective in controlling the dengue vectors by using larvicides. The agents are mostly used as a surface spray, limited to domestic use and to be applied ona short term basis. Several different larvicides include pyriproxyfen, H-14 and temephos. Space spraying can also kill the mosquito by small droplets of insecticides in air, and it has been the primary method used in several countries to control dengue fever/DHF (Dengue Hemorrhagic Fever). Larvicides are occasionally applied at mosquito's breeding sites to control vector population, but precautions must be taken after proper and careful breeding sites assessment assays carried out.

7. **Research and development**

Research is Pakistan is also essential for dengue control. It should address the cost-effective enhancement of new control methods. Pakistan's research and development organizations such as Pakistan Science Foundation (PSF) and Pakistan Council of Scientific and Industrial Research (PCSIR) should focus on developing specific, sensitive and rapid test devices for dengue diagnosis, and development of effective vaccines against all serotypes and non-insecticidal methods to control Dengue virus.

8. **Dengue vaccines**

Sanofi Pasteur's first dengue vaccine, Dengvaxia (CYD-TDV), was reported late in 2015 and early in 2016. Generic name CYD-TDV and brand name Dengvaxia is a tetravalent chimeric, live attenuated vaccine. Other tetravalent live-attenuated vaccines such as TAK003, and Dengvaxia were developed by Takeda in Japan, have completed the phase II clinical trial in children of age 2-17 years in Asia and Latin America (Sáez-Llorens et al. 2017). Other vaccine candidates like V180, TDENV, and PIV developed by Merck from dengue virus envelope protein ((DEN-80E) effective against all serotypes of dengue have completed the first phase of clinical trials and shown promising results in some countries including Pakistan (Yousaf et al. 2018; Manoff et al. 2019). Vaccines remain the most effective method for combating infectious viral diseases. A secure, effective, and affordable dengue vaccine against all serotypes of dengue would result in considerable improvement in dengue management.

8 Treatment for Dengue Virus Infection

In terms of the medicines required for the treatment of dengue infection, intravenous fluids such as crystalloids or colloids may be given. Also, intravenous infusion employing antipyretics and oral rehydration salts are usually available at most hospitals. Other drugs to treat severe cases are injectable vitamin K1, calcium gluconate, sodium bicarbonate, glucose, furosemide, potassium chloride solution, vasopressors, and inotropes. But vasopressors, vasopressin and inotropes, Diltiazepam and Verapamil are not available in injectable form in Pakistan. These injectable drugs are highly essential to savethe lives of critically ill patients (Lo et al. 2017).

9 Conclusion

This chapter showed that dengue is widespread throughout Pakistan. Dengue growth in Pakistan is due to multiple factors that may consist of climate change alternatives, virus evolution, and social elements such as rapid urbanization, population boom, development, socioeconomic influences, and global tour and trade. For areas where humans interact with vector organisms, more efficient control steps for dengue mosquitoes are critical. Dengue is spreading to non-endemic regions globally. To minimize the spread and effect of this disease, the Worldwide Plan for the prevention and control of dengue as outlined by WHO must be enforced.

Recommended dengue management initiatives may include vector population abolition by using eco-friendly control methods, use of air conditioning, window/door screening in homes and workplaces, improved water garage practices, and waste cloth disposal, infrastructure that may minimize dengue vector breeding. There is dire need to revisit the policies regarding dengue control in Pakistan.

References

Ahmad Z, Poh CL. The conserved molecular determinants of virulence in dengue virus. Int J Med Sci. 2019;16(3):355.

Ahmed A, Tanveer M, Khan G, Imran M. Dengue fever again in Pakistan: Are we going in the right direction. Public Health. 2017;152:153–6.

Akram DS, Igarashi A, Takasu T. Dengue virus infection among children with undifferentiated fever in Karachi. Ind J Pediatr. 1998;65(5):735–40.

Ali A, Ahmad H, Idrees M, Zahir F, Ali I. Circulating serotypes of dengue virus and their incursion into non-endemic areas of Pakistan; a serious threat. Virol J. 2016;13(1):144.

Chan Y, Tan H, Seah C, Li J, Chow V, Salahuddin N, et al. Dengue haemorrhagic fever outbreak in Karachi, Pakistan, 1994. 1995.

Cobelens FG, Groen J, Osterhaus AD, Leentvaar-Kuipers A, Wertheim-van Dillen PM, Kager PA. Incidence and risk factors of probable dengue virus infection among Dutch travellers to Asia. Tropical Med Int Health. 2002;7(4):331–8.

Junaidi I. Countrywide dengue cases nearing 50,000 Dawn: Dawn; 2019 https://www.dawn.com/news/1517320.

Dean CL, Wadc J, Roback JD. Transfusion-transmitted infections: an update on product screening, diagnostic techniques, and the path ahead. J Clin Microbiol. 2018;56(7):e00352–18.

Fatima Z. Dengue infection in Pakistan: not an isolated problem. Lancet Infect Dis. 2019;19(12):1287–8.

Fatima Z, Idrees M, Bajwa MA, Tahir Z, Ullah O, Zia MQ, et al. Serotype and genotype analysis of dengue virus by sequencing followed by phylogenetic analysis using samples from three mini outbreaks-2007-2009 in Pakistan. BMC Microbiol. 2011;11(1):200.

Fatima Z, Akram M, Raza SM, Ali L, Hussain A, Amin I, et al. Dengue virus serotype 2 (DEN-2): the causative agent of 2011-dengue epidemic in Pakistan. Am J Biomed Sci. 2012;4(4):307–15.

Guzmán MG, Kourı G. Dengue diagnosis, advances and challenges. Int J Infect Dis. 2004;8(2):69–80.

Guzman MG, Gubler DJ, Izquierdo A, Martinez E, Halstead SB. Dengue infection. Nat Rev Dis Primers. 2016;2(1):1–25.

Heydari M, Metanat M, Rouzbeh-Far MA, Tabatabaei SM, Rakhshani M, Sepehri-Rad N, et al. Dengue fever as an emerging infection in Southeast Iran. Am J Trop Med Hyg. 2018;98 (5):1469–71.

Internal data source: Surveillance Data: Field Epidemiology & Disease Surveillance Division [Internet]. https://www.nih.org.pk/. 2014–2019.

Jamil B, Hasan R, Zafar A, Bewley K, Chamberlain J, Mioulet V, et al. Dengue virus serotype 3, Karachi, Pakistan. Emerg Infect Dis. 2007;13(1):182.

Khan W. Dengue virus serotype 2 and 3 causing high morbidity and mortality in Swat, Pakistan. 2013.

Khan E, Hasan R, Mehraj V, Nasir A, Siddiqui J, Hewson R. Co-circulations of two genotypes of dengue virus in 2006 out-break of dengue hemorrhagic fever in Karachi, Pakistan. J Clin Virol. 2008;43(2):176–9.

Khanani MR, Arif A, Shaikh R. Dengue in Pakistan: Journey from a disease free to a hyper endemic nation. J Dow Univ Health Sci. 2011;5(3):81–4.

Kularatne SA. Dengue fever. Bmj. 2015;351:h4661.

Lo Y-C, Huang I-H, Ho T-C, Chien Y-W, Perng GC. Antiviral drugs and other therapeutic options for dengue virus infection. Curr Treat Options Infect Dis. 2017;9(2):185–93.

Mahmood N, Rana MY, Qureshi Z, Mujtaba G, Shaukat U. Prevalence and molecular characterization of dengue viruses serotypes in 2010 epidemic. Am J Med Sci. 2012;343 (1):61–4.

Manoff SB, Sausser M, Falk Russell A, Martin J, Radley D, Hyatt D, et al. Immunogenicity and safety of an investigational tetravalent recombinant subunit vaccine for dengue: results of a Phase I randomized clinical trial in flavivirus-Naïve adults. Human Vaccines Immunother. 2019;15(9):2195–204.

Passos MNP, Santos LMJG, Pereira MRR, Casali CG, Fortes BdPM, Ortiz VLI, et al. Clinical differences observed in patients with dengue caused by different serotypes in the epidemic of 2001/2002, occurred in Rio de Janeiro. Revista da Sociedade Brasileira de Medicina Tropical. 2004;37(4).

Paul RE, Patel AY, Mirza S, Fisher-Hoch SP, Luby SP. Expansion of epidemic dengue viral infections to Pakistan. Int J Infect Dis. 1998;2(4):197–201.

Pulford J, Oakiva T, Angwin A, Bryant M, Mueller I, Hetzel MW. Indifferent to disease: a qualitative investigation of the reasons why some Papua New Guineans who own mosquito nets choose not to use them. Soc Sci Med. 2012;75(12):2283–90.

Qureshi JA, Notta NJ, Salahuddin N, Zaman V, Khan JA. An epidemic of dengue fever in Karachi-associated clinical manifestations. J Pak Med Assoc. 1997;47:178–80.

Rasheed S, Butlin R, Boots M. A review of dengue as an emerging disease in Pakistan. Public Health. 2013;127(1):11–7.

Sáez-Llorens X, Tricou V, Yu D, Rivera L, Tuboi S, Garbes P, et al. Safety and immunogenicity of one versus two doses of Takeda's tetravalent dengue vaccine in children in Asia and Latin America: interim results from a phase 2, randomised, placebo-controlled study. Lancet Infect Dis. 2017;17(6):615–25.

Shaikh S, Kazmi SJH, Qureshi S. Monitoring the diversity of malaria and dengue vector in Karachi: studying variation of genera and subgenera of mosquitoes under different ecological conditions. Ecol Process. 2014;3(1):12.

Tahir U, Khan UH, Zubair MS. Wolbachia pipientis: A potential candidate for combating and eradicating dengue epidemics in Pakistan. Asia Pac J Trop Med. 2015;8(12):989–98.

Wahala WM, De Silva AM. The human antibody response to dengue virus infection. Viruses. 2011;3(12):2374–95.

Wiwanitkit V. Unusual mode of transmission of dengue. J Infect Dev Ctries. 2010;4(01):051–4.

Yousaf MZ, Siddique A, Ashfaq UA, Ali M. Scenario of dengue infection & its control in Pakistan: An up—date and way forward. Asia Pac J Tropical Med. 2018;11(1):15.

Employing Geographic Information System and Spatiotemporal Analysis of Dengue Outbreaks in a Metropolitan Area in Pakistan

Shakeel Mahmood and Ahtisham Irshad

Abstract Dengue is endemic in Pakistan with its usual peak incidence in the post monsoon period. In the recent past dengue outbreaks have occurred in major urban areas particularly Karachi, Lahore and Peshawar affecting large number of population. This study is an attempt to analyze the spatial and temporal variation of dengue fever (DF) in Samanabad Town, district Lahore. The study is based on primary and secondary data. Primary data is acquired through semi-structured questionnaire. Secondary data have been acquired from concerned Government departments. Geographic information system (GIS) is used to perform spatial analysis. It has been found that temporally DF prevalence varies from month to month and year to year. Similarly spatial variation has been observed. Analysis reveals that DF is still a major threat to the area as socio-economic and geographic conditions favors mosquitoes breeding and transfers of disease from one person/place to another. This study presents useful information related to the dengue outbreak spatio-temporal patterns and may bring the attention of public health departments to plan strategies to control the spread of disease. Risk zones of DF have been delineated using Inverse Distance Weighted (IDW) technique of spatial interpolation. The methodology is general for spatio-temporal analysis and can be applied for other infectious diseases as well.

Keywords Dengue fever · Dengue hemorrhagic fever Aedes Albopictus · Aedes Aegypti · Spatial distribution · Temporal pattern · Risk zones · Geographic information system · Karachi · Lahore · Disease transmission

S. Mahmood (✉) · A. Irshad
Department of Geography, GC University Lahore, Lahore, Pakistan
e-mail: shakeelmahmoodkhan@gmail.com

S. I. Ahmad (ed.), *Human Viruses: Diseases, Treatments and Vaccines*, https://doi.org/10.1007/978-3-030-71165-8_4

1 Introduction

Globally, dengue fever (DF) has become one of the most prominent mosquito-borne infectious diseases (Shaikh et al. 2014; Senthil et al. 2014). The global incidence of DF has been mounting and making large human population vulnerable particularly in tropical and sub tropical regions (Lin and Wen 2011; Jeefoo et al. 2011; Li et al. 2011). In the past 50 years, DF has amplified 33-fold with an increase of 50 million cases per annum (Fareed et al. 2016). Thousands of DF and dengue hemorrhagic fever (DHF) cases have been reported each year and became a leading factor of morbidity and mortality (Amarsinghe et al. 2011).This has made 2.5 billion people at risk in about 120 counteries (Raheel et al. 2010; Arshad et al. 2011; Mulligan et al. 2012; Naqvi et al. 2015). Dengue virus is transmitted to human by mosquito that is *Aedes Aegypti* and A*edes Albopictus* (Li et al. 2011).

The spatial distribution of DF is directly influenced by prevailing weather and seasonal conditions because its vector is sensitive to it (Hales et al. 2002). The temperature conditions have significant influence over mosquito feeding habits, virus development and population dynamics. Similarly, the developmental stages of mosquito life are controlled by temperature. The chance of female mosquito fertilization reduces when temperature falls below 20 °C and feeding activity ceases when temperature falls below 15 °C. Exposure to high temperatures results in increased mortality of adult mosquitoes. The temperature of water plays vital role in breeding and reproduction of mosquitoes (Balmaseda et al. 2006; Mahmood et al. 2019). *Aedes Aegypti* laid eggs in cooler water having shaded container. The *Aedes Aegypti* continues to live in a wide range of temperatures and humidity while *Aedes Albopictus* does not. This is the reason that *Aedes Aegypti* is dominant in urban environment and *Aedes Albopictus* likes better Peri-urban areas (Juliano et al. 2002). Rainfall makes available the favorable habitat for mosquito initial stage of life. The combine impact of higher temperature and rainfall increases humidity. The higher humidity causes higher feeding rates and better development of *Aedes Aegypti* (Scott et al. 2000).

In Pakistan, DF is endemic with its usual peak in the post-monsoon season affecting urban and sub-urban areas (Mahmood et al. 2019). The very first DF outbreak was reported in 1994–1995 in Karachi with 4500 registered cases (Khan and Hassan 2011; Khnanani et al. 2011). During 2005–2006, there was an unprecedented increase in DF with more than 3640 reported cases and 40 deaths (Shakoor et al. 2012). In Lahore, few DF cases were reported in 2006–2007 and epidemic occurred in 2008 (Jahan 2011). The total number of reported cases was 17,493 with 290 deaths in year (Rasheed et al. 2013). The dengue epedmic has been occuring every year since 2006 and extended to most of the cities in Pakistan (Rasheed et al. 2013). It has badly affected the major cities of Karachi, Lahore, Multan, Fasilabad and Rawalpendi (Laghari et al. 2015). Naqvi (2015) found that transportation is the main factor of dengue spread from Karachi to Lahore.

Currently, in Peshawar (The capital city of Khyber Pakhtunkhwa) the registered cases of DF is maximum in the entire country and is spreading to the surrounding districts very rapidly.

Geographic information system (GIS) is a strong geographical tool for spatial and temporal analysis (Mahmood et al. 2019) and have been applied in epidemiological and public health studies for many years (Mondini and Chiaravalloti-Neto 2008) to assess and identify potential risk factors involved in disease transmission (Twumasi and Merem 2005; Jeefoo et al. 2011). This study involved micro-level detailed investigation of DF spatial pattern at union council (UC; the smallest electoral unit) which made it different from past studies. Similarly, there are few studies regarding the dengue outbreak and it spatial pattern in the study area. The main objective of this research is to employ GIS and Spatiotemporal Analysis of Dengue Outbreaks in a Metropolitan Area in Pakistan.

2 Study Area

Lahore is the second largest city of Pakistan and capital of the province Punjab (Shirazi and Kazmi 2014). Geographically, it is extended from is 31° 15′ to 31° 45′ N latitude and 74° 01′ to 74° 39′ E longitude with total area of 2260 km^2. The population density is 2778 person/km^2 (Shirazi and Kazmi 2013). Administratively, Lahore is divided into nine towns with one cantonment area. Iqbal Town is the largest covering 476.79 km^2 and Shalimar town is the smallest with 26.88 km^2 area. Climate of the study area is sub-tropical. In January and February the temperature remains low and in June and July temperature remains high (Shirazi 2012). It is located in the lucky monsoon region of south asia with plenty of rainfall which is significant factor for dengue growth. In July and August the study recienves maximum rainfall (Naqvi et al. 2015).

Samanabad Town has been selected for detail and micro-level investigation regarding the spatio-temporal trend of DF. It was severely affected by dengue outbreak in 2008. Out of the total reported cases in Lahore, 62% were registered from the study area (Fig. 1). The severely affected Union Council (UC; smallest electoral unit) were Samanabad, Ichra and Gulshan-e-Ravi (Mushtaq et al. 2010). In the Northwest of the study area River Ravi is flowing. Waste water drain (locally Ghanda Nalla) is flowing from north to south of the town and make it favorable for the breeding and growth of mosquito.

2.1 Data Acquisition and Analysis

To achieve objective of the study, seven UCs were selected as sample sites, out of nineteen in the study area by random means of sampling namely Babu Sabu, Abu Bakar Siddque, Gulshan-e-Ravi, Samanabad, New Samanabad, Ichra, and Sikander Block (Fig. 1).

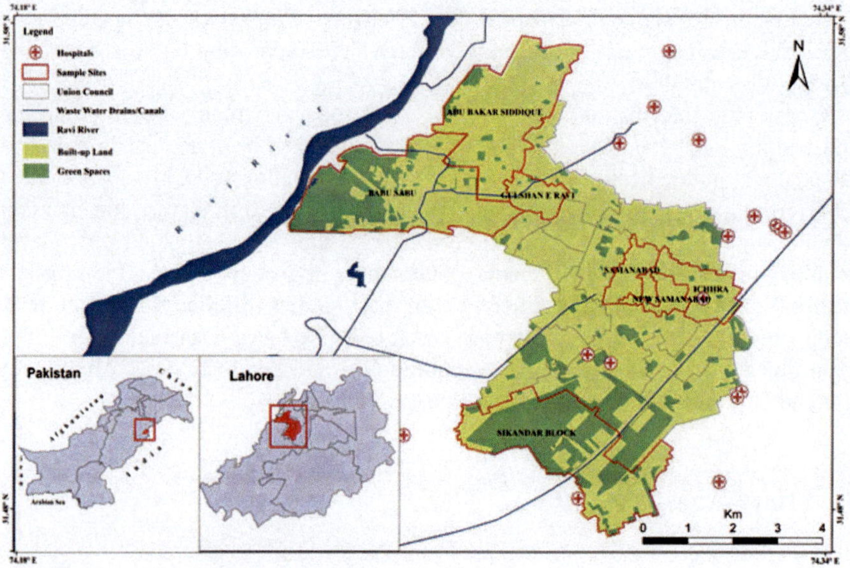

Fig. 1 Location of the study area and sample UCs

The required data were collected from concerned Government Departments. Population record was collected from Pakistan Statistical Bureau, Lahore. Rainfall, temperature and humidity data were collected from Pakistan Meteorological Department (PMD), Lahore. Data regarding the confirmed DF cases for the years 2012–15 were obtained from Governments Hospitals and Health Departments, district Lahore. The data comprised of patient's age, gender and postal address. Google Earth image was downloaded as base layer to develop land use and land cover map of the selected area. Spatial data regarding the administrative boundaries of the study area was collected from Punjab Provincial Disaster Management Authority (PPDMA), Lahore.

Relative location of the dengue affected person was converted to absolute location by applying Point Level Geo-coding technique to visualize the record on map. Then location data was converted KML using ArcGIS10.2 and geo-visualize the record on Google Earth for validation purpose. The data were validated using GPS-based questionnaire surveys using stratified purposive sampling techniques with sample size of 22%. The validity of the data was 83%. Simultaneously, GPS survey was also conducted to acquire the geographic location of health units including hospitals and Basic Health Units (BHU). MS excel was used for data processing and descriptive statistical analysis.

3 Results and Discussion

Analysis reveals that out of the 7 sample sites (Fig. 1), *Samanabad, Ichhra and Gulshan-e-Ravi* were the most affected with high prevalence. Sikandar Block and Abu Bakkar Siddique were less affected with low prevalence (Table 1). DF, DHF, and DSS are three serotypes with 394 registered cases in 2012, 388 had DF, 5 has DHF and 1 had DSS, 905 cases were registered in 2013, 887 had DF, 17 had DHF and 1 had DSS while in the 2014 registered cases were 130 of only DF. In the selected time DF cases were maximum and 2013 was severely affected by DF outbreak (Table 2). The total registered cases of dengue from 2007–15 were 2793.

Samanabad, Ichhra and Gulshan-e-Ravi are the densely populated sample sites. Most of the sample households have medium (7–9 members) family size. Similarly 44% of the surveyed households have monthly income ranges from 16 to 50 thousand (Pakistani Rupees). The age of confirmed DF patient ranged from 3 to 84 years with a median of 28 years. There were 725 males and 311 females out of the total affected persons and among them 131 (12.6%) were children (15 years or below), 579 (55.8%) were of 16–30 years; this was the most affected age group in the selected spatial and temporal dimension. Approximately 67%, 83% and 71% of these cases were confirmed during post Monsoon season in 2012, 2013 and 2014 respectively. In the mentioned season the rainfall decreases and relative humidity increases which favor the breeding and growth of mosquitoes. The number of DF cases increases from July onward, reached to its highest recorded peak in September, and then decreases. Out of the total patients, male patients were more than female (Fig. 2). In selected dengue outbreaks registered male patients were about 60%. Based on age, 16–30 years age group is most affected where as the elder persons are least affected (Fig. 3).

Temporal analysis reveals that dengue distribution varies from month to month and year to year. Year wise 2013 was severely affected with maximum reported cases (905), in 2012 the reported cases were 394 and in 2014 of 130. Monthly distribution of dengue is also variable. January to May the incidence is minimum then from Jun incidence rises, in September maximum cases were recorded in 2012 and 2013. Incidence of dengue decreases from October to November. In the year 2013 two outbreaks occurred in September and November (Fig. 4).

Spatial analysis reveals that dengue distribution is not uniform. Samanabad, Ichhra and Gulshan-e-Ravi were the most affected sites with incidence rate; number of patients/1000 persons of 5.8, 4.2 and 2 respectively. The dengue incidence were high in Ichara followed by Samanabad in 2012 and 2013 whereas in Kashmir

Table 1 Serotype prevalence of Dengue in study area

Dengue type	Year 2012	Year 2013	Year 2014
DF	388	887	130
DHF	5	17	0
DSS	1	1	0

Source Health department and Arfa Kareem tower, Lahore

Table 2 Spatio-temporal distribution of dengue incidences in studied area

Sample UC	2012	2013	2014
Abu Bakkar Siddique	6	8	2
Gulshan-e-Ravi	49	109	16
Babu Sabu	17	11	2
Ichhra	89	301	34
New Samanabad	16	61	2
Samanabad	98	213	24
Sikandar block	0	1	2

Source Health department, Lahore

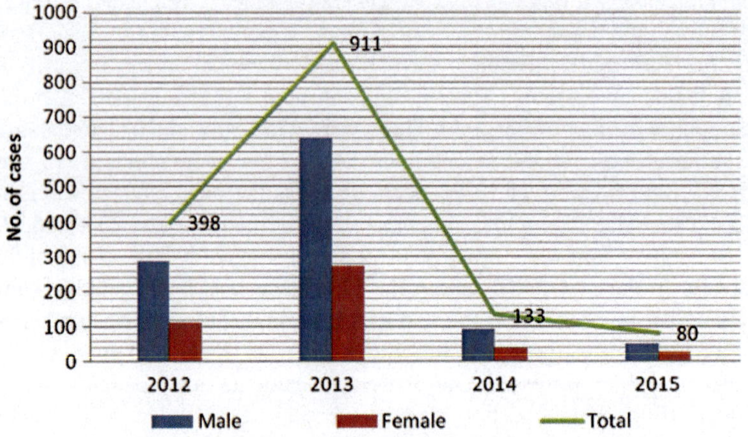

Fig. 2 Annual distribution of dengue occurrence in study area. *Source* Health department, Lahore

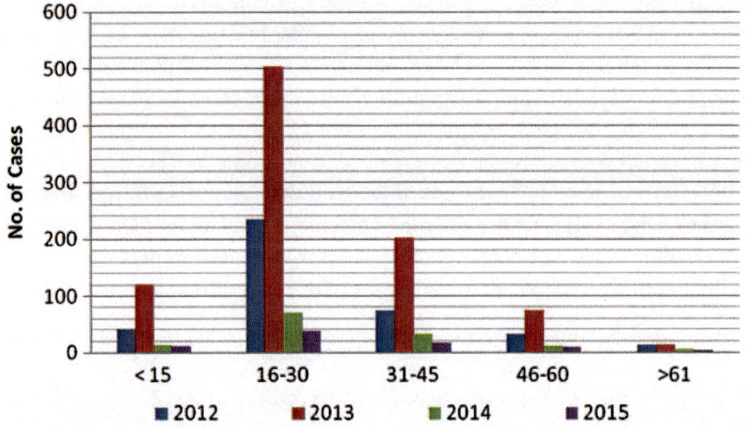

Fig. 3 Age wise distribution of dengue occurrence in study area. *Source* Health department, Lahore

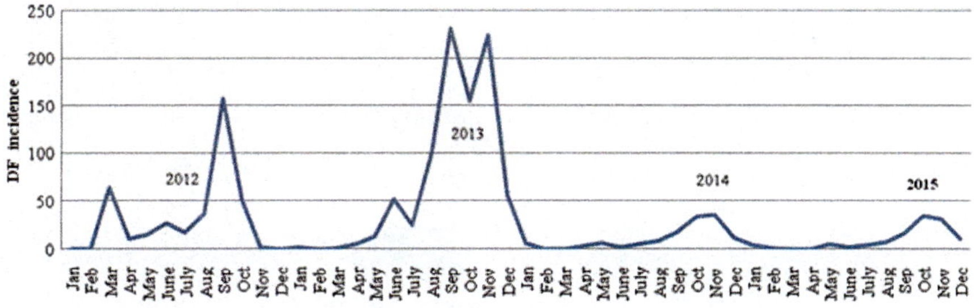

Fig. 4 Dengue fever outbreaks in district Lahore after Punjab Health Department, 2015

Block, Paki Thatti and Abu Bakar Siddique Colony the incidence was lowest (Fig. 5). Dengue incidence maps were developed for the year 2012, 2013 and 2014 using the layers of DF, population density and landuse. Different themes were combined in one layout to compare and identify most affected sample site (spatial) and year/moth (temporal) as well as to identify incidence relationship with study area attributes. Spatio-temporal analysis of the dengue reveals that prevalence is high in *Samanabad, Ichhra and Gulshan-e-Ravi,* where more than 90% land is built up with high population density while prevalence was low in *Babu Sabu, Abu Bakar Siddque, New Samanabad, and Sikander Block* where green spaces are more with comparatively low population density (Fig. 6). High risk zone are spatially located in north-western parts of the study area because over there no vegetation cover and high population density (Fig. 7).

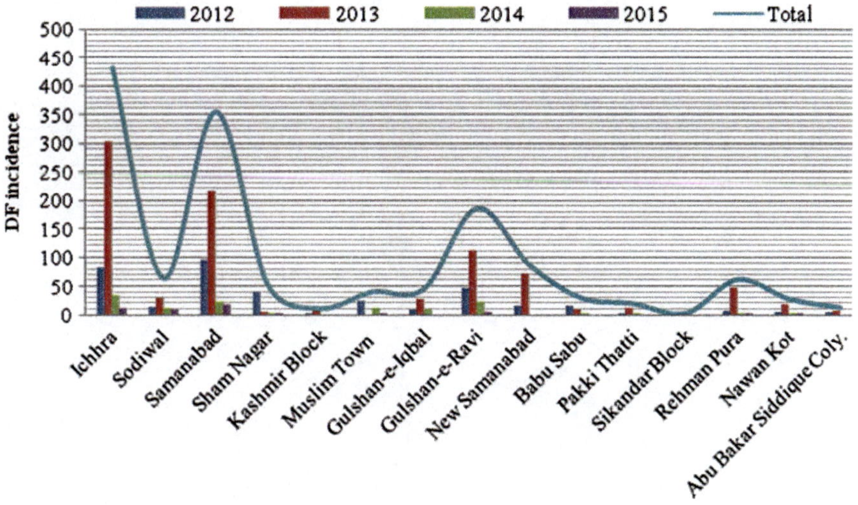

Fig. 5 Spatio-Temporal distribution of dengue incidences in studied area

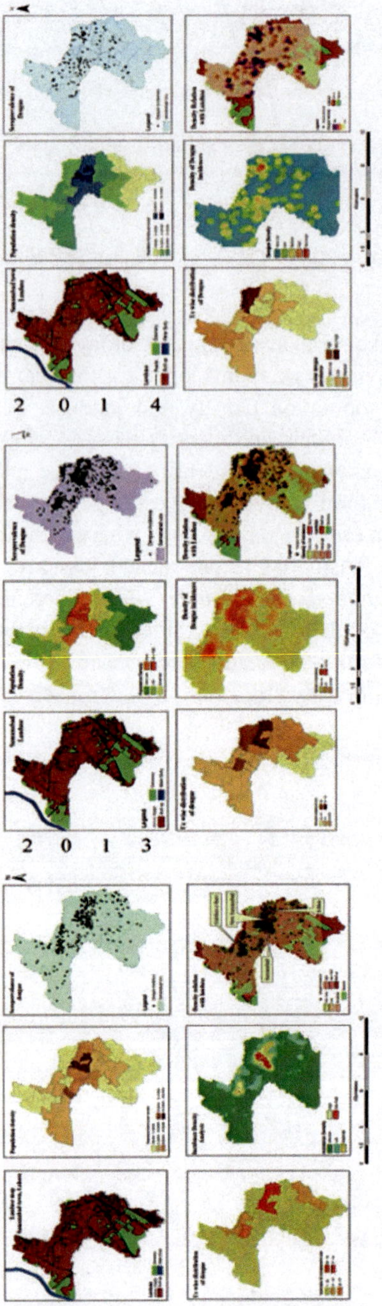

Fig. 6 Land use, population density, dengue prevalence from 2012–14

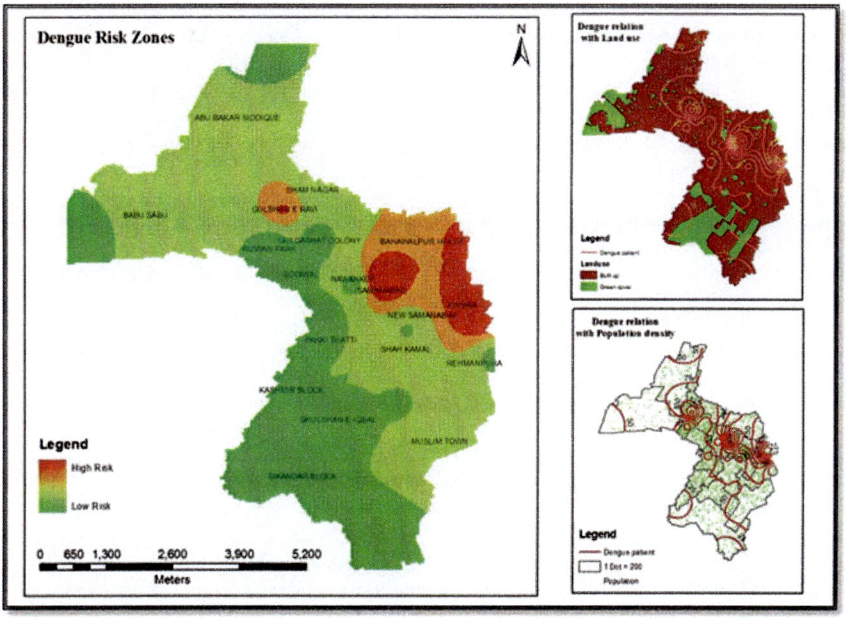

Fig. 7 Risk zones of Dengue

4 Conclusion

The study concludes that temporally dengue distribution varies from month to month and year to year. Dengue prevalence was high in 2013 with 905 reported cases followed by 2012 with 394 cases. Monthly prevalence also varies; from December to May minimum incidence recorded and from Jun it gets momentum and then in September attains recorded peak. In 2013 two outbreaks occurred in September and November. Spatio analysis reveals that prevalence is high in those UCs where green spaces are rare and more than 90% land is built up with high population density. On other hand where green spaces are more prevalence the incidence is low. Spatially, high risk zone is located in the North-Western areas. Finally, it is concluded that dengue fever infected person are more in those areas where population density is high with no green spaces or greenery.

References

Amarasinghe A, Kuritsk JN, Letson GW, Margolis HS. Dengue virus infection in Africa. Emerg Infect Dis. 2011;17(8):1349–54.

Arshad I, Malik F, Hussain A, Shah SA. Dengue fever; clinco-pathologic coreation and their associations with poor outcomes. Prof Med J. 2011;18(1):57–63.

Balmaseda A, Hammond SN, Pérez L, Tellez Y, Indirasaborío S, Mercado JC, et al. Serotype-specific differences in clinical manifestations of dengue. Am J Trop Med Hyg. 2006;74(3):449–56.

Fareed N, Ghaffar A, Malik TS. Spatio-temporal extension and spatial analyses of dengue from Rawalpindi, Islamabad and Swat during 2010–2014. Climate. 2016;4(2):23.

Hales S, De Wet N, Maindonald J, Woodward A. Potential effect of population and climate changes on global distribution of dengue fever: an empirical model. The Lancet. 2002;360 (9336):830–4.

Halstead SB. Pathogenesis of dengue: challenges to molecular biology. Science. 1988; 239 (4839):476–81.

Jahan F. Dengue fever (DF) in Pakistan. Asia Pac Family Med. 2011;10(1):1. https://doi.org/10.1186/1447-056X-10-1.

Jeefoo P, Tripathi NK, Souris M. Spatio-temporal diffusion pattern and hotspot detection of dengue in Chachoengsao province, Thailand. Int J Environ Res Public Health. 2011; 8(1):51–74. https://doi.org/10.3390/ijerph8010051.

Juliano SA, O'Meara GF, Morrill JR, Cutwa MM. Desiccation and thermal tolerance of eggs and the coexistence of competing mosquitoes. Oecologia. 2002;130:458–69.

Khanani MR, Arif A, Shaikh R. DENGUE IN PAKISTAN: Journey from a Disease free to a Hyper Endemic Nation. J Dow Univ Health Sci Ii Karachi. 2011;5(3):81–4.

Khan E, Hasan R. Dengue infection in Asia; a regional concern. J Postgrad Med Inst (Peshawar-Pakistan). 2011; 26:1–6.

Laghari TM, Memon AA, Mustufa MA, Memon AW, Aishwani MK, Otho MS. Assessment of mass level public awareness campaigns regarding dengue among parents visiting tertiary care children hospital Karachi, Pakistan. JLUMHS. 2015;14(3):124–8.

Lin CH, Wen TH. Using geographically weighted regression (GWR) to explore spatial varying relationships of immature mosquitoes and human densities with the incidence of dengue. Int J Environ Res Public Health. 2011;8(7):2798–815.

Li S, Fang M, Zhou B, Ni H, Shen Q, Zhang H, Cao G. Simultaneous detection and differentiation of dengue virus serotypes 1–4, Japanese encephalitis virus, and West Nile virus by a combined reverse-transcription loop-mediated isothermal amplification assay. Virol J 2011; 8.

Mahmood S, Irshad A, Nasir JM, Sharif F, Farooqi SH. Spatiotemporal analysis of dengue outbreaks in Samanabad town, Lahore metropolitan area, using geospatial techniques. Environ Monit Assess. 2019;191(2):55–69.

Mondini A, Chiaravalloti-Neto F. Spatial correlation of incidence of dengue with socioeconomic, demographic and environmental variables in a Brazilian city. Sci Total Environ. 2008;393 (2):241–8.

Mulligan K, Elliott SJ, Schuster-Wallace C. The place of health and the health of place: dengue fever and urban governance in Putrajaya, Malaysia. Health Place. 2012;18(3):613–20.

Naqvi SAA, Kazmi SJ, Shaikh S, amd M. Akram. Evaluation of prevalence patterns of dengue fever in Lahore district through geo-spatial techniques. J Basic Appl Sci. 2015;11:20–30.

Raheel U, Faheem M, Riaz MN, Kanwal N, Javed F, Qadri I. Dengue fever in the Indian subcontinent: an overview. The J Infect Developing Countries. 2010;5(04):239–47.

Rasheed SB, Butlin RK, Boots M. A review of dengue as an emerging disease in Pakistan. Public Health. 2013;127:11–7.

Scott TW, Amerasinghe PH, Morrison AC, Lorenz LH, Clark GG, Strickman D. Longitudinal studies of Aedes Aegypti (diptera: culicidae) in Thailand and Puerto Rico: blood feeding frequency. J Med Entomol. 2000;37:89–101.

Senthil J, Rajeswari S, Vadivel S, Malarvizhi F. Scenario of Dengue and Healthcare Utilization in Nagapattinam Town, Tamilnadu, India. Int J Dev Res. 2014;4(11):2547–52.

Shaikh K, Memon KN, Sarah B, Akhtar R, Memon M, Memon S. Dengue fever; an audit of risk factors among patients reporting at a tertiary care hospital in Hyderabad. Professional Med J. 2014;21(3):455–9.

Shakoor MT, Ayub S, Ayub Z. Dengue fever: Pakistan's worst nightmare. WHO South-East Asia J Public Health. 2012;1(3):229–31.

Shirazi SA. Spatial analysis of NDVI and density of population: a case study of Lahore Pakistan. Sci Intl. 2012;24(3):323–8.

Shirazi SA, Kazmi SJ. Appraisal of the Change in spatio-Temporal Patterns of Lulc and its Impacts on the Vegetation of Lahore, Pakistan. Pakistan Vision. 2013; 14(1).

Shirazi SA, Kazmi SJ. Analysis of population growth and urban development in Lahore-Pakistan using geospatial techniques: suggesting some future options. South Asian Stud. 2014;29 (1):269–80.

Twumasi YA, Merem EC. GIS applications in land management: The loss of high quality land to development in central Mississippi from 1987–2002. Int J Environ Res Public Health. 2005;2 (2):234–44.

Identification of Dengue NS2B-NS3 Protease Inhibitors Through High-Throughput Virtual Screening—Impacts on Drug Development Against the Dengue Virus

Sheikh Murtuja, Deepak Shilkar, Biswatrish Sarkar, Barij Nayan Sinha, and Venkatesan Jayaprakash

Abstract Amongst the neglected tropical diseases, dengue remains a disease of concern, and the rise of global temperatures every year puts additional liabilities on the infrastructure of the health sector of various underdeveloped and developing countries. Increasing incidence puts additional stress on an already burdened healthcare infrastructures. This calls for an urgent need of discovering an inhibitor for Dengue virus (DENV). The success in targeting HIV and HCV proteases has driven considerable attention of researchers to explore the same towards the DENV NS2B-NS3 protease. Although most attempts yet remains futile, advancing computational power and biotechnology can certainly play a pivotal role in speeding up the drug discovery process. High-throughput Virtual Screening is one such frontier that could help screening of large libraries of chemical entities, making the in vitro pipeline efficient and cost effective. This chapter presents the DENV NS2B-NS3 protease inhibitors identified through High-throughput Virtual Screening and its impact on drug development against the Dengue virus.

Keywords Flavivirus · NS2B-NS3 · DENV · HTVS · Protease inhibitors · Dengue · In silico · Crystal structure · Catalytic triad · Docking · Hydrophobic interactions · Virtual screening

1 Introduction

Dengue has emerged as one of the most threatening arthropod-borne viral diseases in the past 50 years (WHO 2012) affecting nearly 390 million people (Bhatt et al. 2013) in tropical and the subtropical regions of the globe (Guzman et al. 2010).

S. Murtuja · D. Shilkar · B. Sarkar · B. N. Sinha · V. Jayaprakash (✉)
Department of Pharmaceutical Sciences & Technology, Birla Institute of Technology, Mesra, Ranchi, Jharkhand 835215, India
e-mail: venkatesanj@bitmesra.ac.in

S. I. Ahmad (ed.), *Human Viruses: Diseases, Treatments and Vaccines*, https://doi.org/10.1007/978-3-030-71165-8_5

Dengue fever is caused by the Dengue virus (DENV), which belongs to the genus *flavivirus* of the *Flaviviridae* family. DENV is mainly transmitted by the mosquito species *Aedes aegypti* and *Aedes albopictus.* DENV has four serotypes, namely DENV-1, DENV-2, DENV-3, and DENV-4, whose clinical complexities in humans range from a milder dengue fever (DF) to fatal dengue hemorrhagic fever (DHF) and dengue shock syndrome (DSS) (Ebrahim 1993; Guzman et al. 2010). Interestingly, it was noted that being infected from one serotype did not provide immunity on cross-infection with other serotypes. A subsequent infection from a different serotype posed a potential risk of developing fatal dengue. The phenomenon is called the "antibody-dependent enhancement" effect (ADE) (Goncalvez et al. 2007). Vaccine development programs have suffered serious setbacks. The first vaccine developed by Sanofi Pasteur in 2015, despite being approved in twenty countries, carries a risk of causing ADE and has been limited for people in the age group of 9–45 years and who have had a previous viral infection at least once (Guy et al. 2017).

The dengue virus genome is a single strand of RNA, often referred to as *positive-sense RNA,* since it can be directly translated into proteins. The viral genome encodes ten genes. The genome is translated as a single, long polypeptide and then cut into ten proteins, three structural proteins, capsid (C), envelope (E), and membrane (M) proteins, and seven nonstructural proteins (NS1, NS2A, NS2B, NS3, NS4A, NS4B, and NS5). These nonstructural proteins play roles in viral replication and assembly (Mukhopadhyay et al. 2005).

With the establishment of HCV (Wyles 2013) and HIV (De Clercq 2009) protease as promising drug targets, researchers worldwide began investigating DENVNS3 protease to develop a possible inhibitor. NS3 protease is a trypsin-like serine protease harboring a classical serine protease catalytic triad (His51, Asp75, and Ser135) (Falgout et al. 1991). It resides at the N-terminal end of NS3 protein (Nestorowicz et al. 1994). The proteolytic activity of NS3 protease is witnessed in the presence of a cofactor NS2B, improves the catalytic function of NS3 (Falgout et al. 1991; Yusof et al. 2000). The NS2B-NS3 protease complex cleaves DENV polyprotein into its components, thereby promoting viral replication. Due to its crucial role in cleavage and assembling of viral proteins, it appears as a promising target for the design and development of anti-dengue drugs (Phong et al. 2011).

High Throughput Virtual Screening has now become a fundamental component of in silico studies. This approach uses computational algorithms to filter in potential bioactive molecules from large chemical compound libraries. A typical HTVS methodology involves docking and scoring millions of compounds from chemical databases against a protein binding site of interest. The binding site and the protein are sourced and prepared from crystal structures available in protein databases or by developing homology models. The chemical hits determined from this campaign are then screened in vitro using biochemical and biophysical assays. The in silico virtual screening approach has its limitations, e.g., when screening an extensive library, it is often required to ignore protein flexibility and other factors such as binding entropy and desolvation. However, a large number of studies have shown that HTVS had a significant impact on lead discovery.

In this chapter, we have compiled every attempts made by various groups searching for an anti-dengue therapeutics through the HTVS approach while stressing on factors such as uniqueness of the approach, merits, demerits and success metrics.

2 Structure of Dengue NS2B-NS3 Protease

Several protein structures of DENV NS2B-NS3 protease have so far been solved. Among these, only a few have so far been used for DENV protease in silico studies. Among these, the most prominent ones include 2FOM, 2VBC, and 3U1I.

Erbel et al. (2006) published the first reliable and enzymatically active, high resolution (1.5 Å) NS2B-NS3 protease structure that contained the catalytically important part of NS2B (47-residue core region) linked to the N-terminal of NS3 protease via a glycine linker. The NS3 protease domain adopted chymotrypsin-like folds with two β-barrels forming 6 β-strands. The two β-barrels formed a cleft where catalytic triad His51-Asp75-Ser135 were located. The catalytic triad is a classical feature of serine proteases responsible for their mechanism of action. This research held phenomenal importance in the understanding of the role of NS2B as a cofactor. The authors uncovered that, while a construct with NS2B residues 49–66 could produce a soluble enzyme, an active enzyme was only produced if the entire NS2B fragment was involved. When the structure was studied in the presence and absence of the ligand, the C terminus folds of NS2B adopted markedly different configurations. It was also noted that the integration of residues Arg78-Leu87 from the β-loop of NS2B into the protease-cofactor complex greatly influenced the active site. In summary, the association of 40 amino acid NS2B fragment with NS3protease is key to protease activity. The cofactor actively aids in S2 and S3 pocket formation in the protease active site.

A ligand-bound crystal structure of DENV3NS2B-NS3 protease published, where NS3 adopted a closed conformation and was wrapped by NS2B, which included the β-hairpin. They further studied the crystal structure in comparison with WNV protease. This study revealed the amino acid level changes that influence binding pocket and binding affinities (Noble et al. 2012).

3 High-Throughput Virtual Screening Against Dengue NS2B-NS3 Protease

Ganesh et al. (2005) were amongst the first to perform virtual screening of compounds from the available chemical directory (ACD, MDL system, Foster city CA) against DENV2NS3 protease crystal structure (PDB ID:1BEF). The same structure was also used to model the West Nile Virus (WNV) protease using Modeller 4.0.

Three biguanides and seventeen single guanidine containing groups were evaluated for their inhibitory activity against DENV and WNV protease, and five non-substrate-based inhibitors were identified. In their study, they observed how the modifications of the bridge group between two biguanidine arms affected the K_i value. Also, the effect of a single guanidino group on K_i was observed. As shown in Fig. 1, three compounds had K_i value below 50μM against both the proteases and thus served as effective lead for further modifications. Molecular modelling studies of compound 1showed that it existed in two confirmations (U shape and Z shape). Further RMS data (0.68 Å) on deviation and centroid to centroid distance (7.7 Å) between the two guanidine groups indicated that the inhibitor acted best in the U confirmation (Fig. 3). Residues Asp129, Tyr150, Ser163 showed electrostatic interactions with compound 1 (Figs. 2 and 3). Modeling data of Compound 2 & 3, which had a single guanidine group, showed that amide oxygen of the indolin ring in compound 2 and Oxygen atom of the phosphonic group in compound 3 could form a hydrogen bond with active site Ser135 of DENV NS3 protease, resulting in good inhibitor protease interaction.

Tomlinson et al. (2009) identified two small molecule inhibitors of DENV2 NS3 protease through Structure-based drug design. The chemical library was an in-house database of small molecules from Mayo Clinic that contained 2.5 million three-dimensional structure of compounds. Two filters were employed to screen the molecules, one being the selection of only non-zwitterionic compounds at physiological pH and the other being easy availability from reputed vendors. Further EUDOC program was used to computationally screen the filtered library against DENV2 NS3 pro (PDB ID: 1BEF, 1DF9, and 2FOM). Top 20 hits, identified from Molecular modeling studies were purchased and evaluated for in vitro inhibitory activities against NS2B-NS3 protease. Two compounds also inhibited viral replication in cell culture experiments. ARDP0006 (Fig. 4) was most efficacious in inhibiting DEN2V replication with EC_{50} of 4.2 ± 1.9 μM. Modeling studies were done, interactions were identified, and ARDP0006 was predicted to interact with the residues, Gln35 (a residue near active site), His51, and Ser135 of the active site and residues Gly151 and Gly153 of the P1 pocket. Other observed interactions were between the inhibitor and residues Ser131, Pro132, Gly133, Thr134, Asn152, and Val155 (Fig. 5). Images were generated with SWISS PDB.

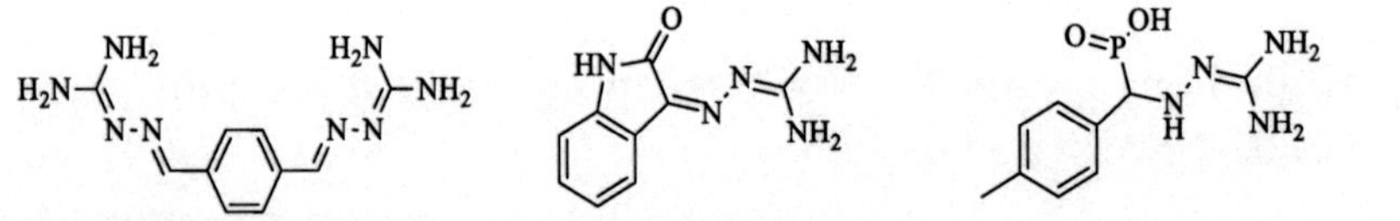

K_i= 44± 5μM (DENV NS2B(H):NS3)
K_i=35± 5μM (WNV NS2B(H):NS3)
Compound 1

K_i= 23±2μM (DENV NS2B(H):NS3)
K_i=16±2μM (WNV NS2B(H):NS3)
Compound 2

K_i= 14±2μM (DENV NS2B(H):NS3)
K_i=13±2μM (WNV NS2B(H):NS3)
Compound 3

Fig. 1 The dissociation constant (K_i) of the compounds against DENV and WNV NS2B-NS3 protease

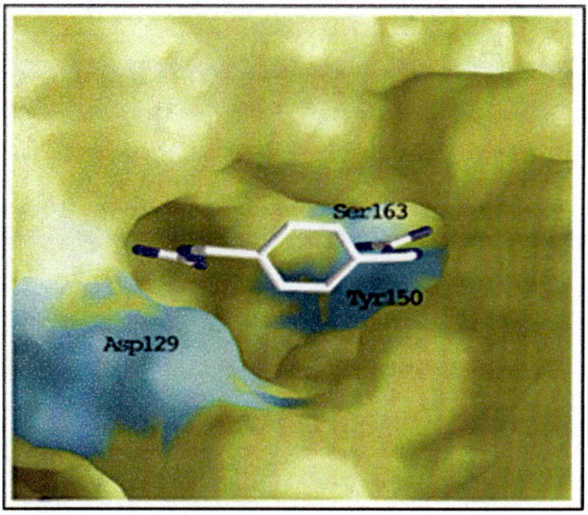

Fig. 2 DEN2NS3 protease-Compound 1 interaction (Ganesh et al. 2005). Reprinted with permission, Copyright © 2004 Elsevier Ltd.

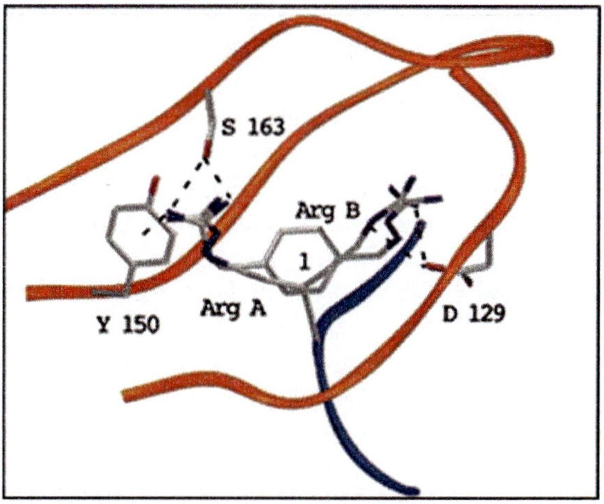

Fig. 3 U-conformation of compound 1 with ribbon drawing model of DEN2 NS3 protease (Ganesh et al. 2005). Reprinted with permission, Copyright © 2004 Elsevier Ltd.

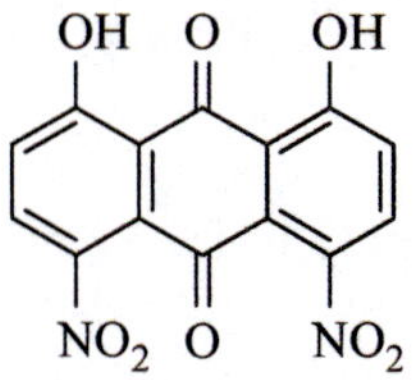

Compound code:ARDP0006
EC_{50}=4.2±1.9μM

Fig. 4 The most potent compound of Tomlinson et al.

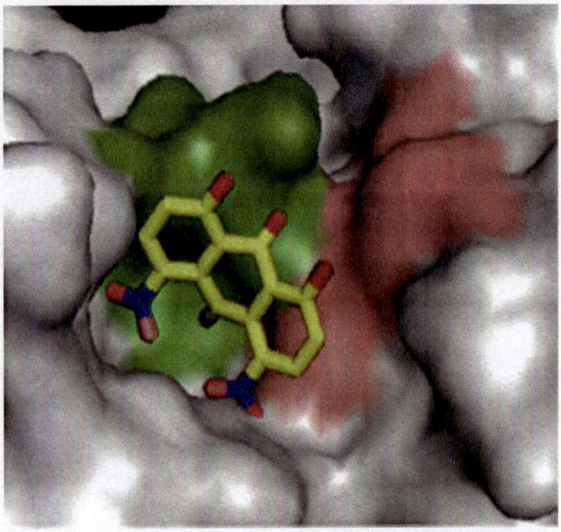

Fig. 5 Model of inhibitor-protease interaction (Tomlinson et al. 2009). Reprinted with permission, Copyright © 2009 Elsevier B.V.

Knehans et al. (2011) demonstrated the successful application of a structure-guided fragment-based in silico drug design approach for DENV protease inhibitor (Fig. 6). Applying retrosynthetic combinatorial synthetic analysis procedure, a library of the molecular fragment was derived from ZINC database. This library of about 1,50,000 fragments was docked to DENV NS2B-NS3 protease, which was developed by homology modeling. High scoring fragments were assembled and through an implicit linking approach, with a focus on interactions of the fragments with S1 and S2 pocket of the protease and similarity search methods, twenty-three compounds were selected, and inhibition assay was performed. The best compound obtained had an IC_{50} value of 7.7 μM. Figure 7, shows the predicted binding pose of the most potent compound. The inhibitor is shown in magenta color, and using GOLD (Version 4.1), it was docked into the homology modeled DENV2 substrate binding site. Docking study of the best inhibitor showed that the inhibitor was engaged in hydrogen bonding and electrostatic interactions with Asp129 of the S1 pocket, a π-π interaction was predicted between the phenyl guanidine group and Tyr161 and this arene-arene interaction was supported by hydrophobic interaction with Pro132 and thus Tyr161 and Pro132 served as a hydrophobic clamp in the S1 pocket. The opposite phenyl moiety interacted with His51 of the catalytic triad through π-π stacking interaction, and its guanidine group formed a hydrogen bond with Asn152 of the S2 pocket. Since S2 pocket has a negatively charged environment, the positive charge of the guanidine group compensates the same. Thus, this approach explains how a combination of homology modeling, fragment docking, chemical similarity and structural filters could pave the way for hit generation.

In another study, Deng et al. (2012) utilized the crystal structure of the DENV NS2B-NS3 complex (PDB ID: 2FOM) and screened ~600,000 compounds in the

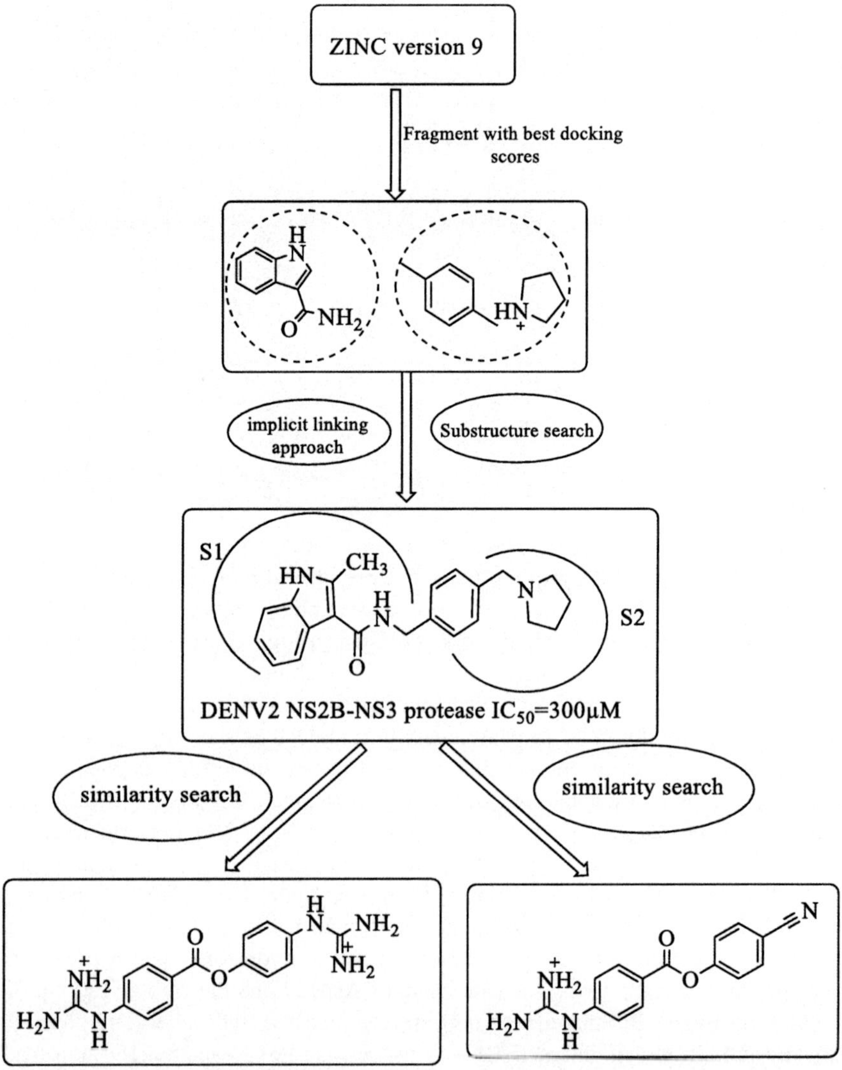

DENV2 NS2B-NS3 protease IC_{50}=7.7μM
competitive mode of inhibition

DENV2 NS2B-NS3 protease IC_{50}=37.9μM
competitive mode of inhibition

Fig. 6 Schematic representation of the evolution of potent compounds of Knehans et al.

ACD database by molecular docking. Twenty-seven hits were purchased and evaluated for DENV2 NS2B-NS3 protease inhibition. As a result, three different scaffolds small molecules were found to be promising. One scaffold was chosen as a starting molecule for modification based on various parameters, including synthetic ease and its analogues were synthesized and evaluated (Table 1). Further, the scaffold hopping technique mediated structural modifications were done;

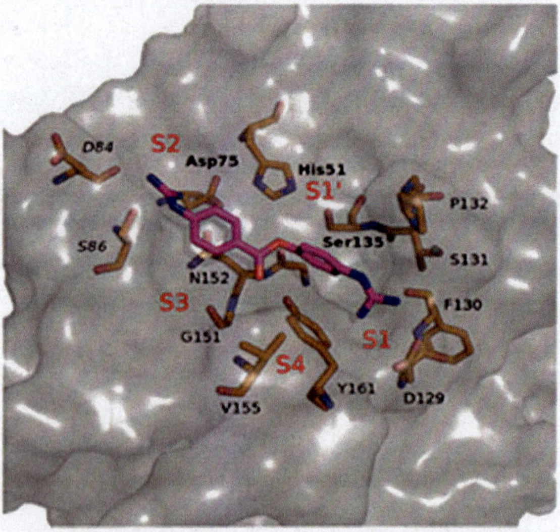

Fig. 7 Predicted binding pose of the most potent compound. Reprinted with permission, Copyright © 2011 Springer Science +Business Media B.V

eventually, the quinoline moiety replaced the core phenyl moiety and the resulting molecules were evaluated for DENV protease inhibitory activity (Table 2). A schematic representation of their work is shown in Fig. 8. Altogether thirty-five compounds were synthesized, and their inhibitory abilities against DENV2 NS2B-NS3 protease were evaluated. Of the thirty-five molecules, seventeen were potent inhibitors of protease, and eight molecules showed good antiviral activity against DENV2 in the cell line assay. The best molecule designed (compound A) had an IC_{50} value of 7.46 ± 1.15 μM against DENV2 NS2B-NS3 protease. For the Molecular modeling studies, screening was performed using DOCK 4.0 and docking was done using GOLD3.0. Visualization was done using PyMol. The quinoline ring of compound B showed hydrophobic interactions with Leu76 and Ile 165. The nitrogen of quinoline formed hydrogen bond with the side chain of Asn152. The first and the second nitrogen of hydrazone were involved in hydrogen bond formation with the side chain of Asn152 and the carbonyl group of Lys73, respectively. Bromobenzene portion occupied the hydrophobic region, P4 and showed hydrophobic interactions with the residue Ile123 and Val154. Further, hydrogen-bonding interaction was also predicted between Lys 74 and the carbonyl group of the amide linking the hydrazine on the right side of the structure and between amide-hydazine nitrogen and the main chain of Ile165. The remaining one phenyl portion was occupied between the alkyl portion of Glu88 side chain and the side chain of Ala166 (Fig. 10). The better inhibitory activity of compound A was attributed to the sulfonamide group present in its structure, which enabled more hydrogen-bonding interactions in the P4 region of it and the residues involved were Thr120 and Lys73 (Fig. 9).

Yet in another study Pambudi et al. (2013), using structure based screening and cell-based viral replication assay identified a small molecule inhibitor, SK12, that interfered with the interaction between NS2B and NS3. They performed HTVS of a

Table 1 Inhibition constant of compounds obtained by structural modification

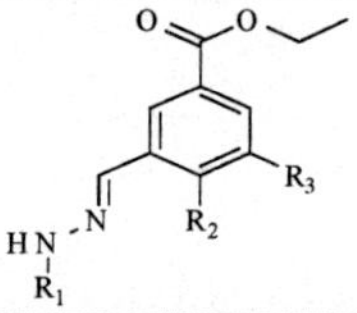

R_1	R_2	R_3	IC_{50} (μM) DENV2 NS2B-NS3 protease
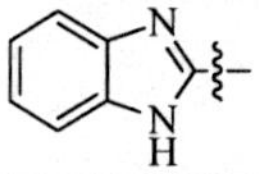	—OH	$-CH_3$	48.59 ± 3.46
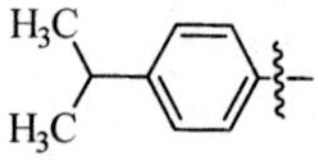	—OH	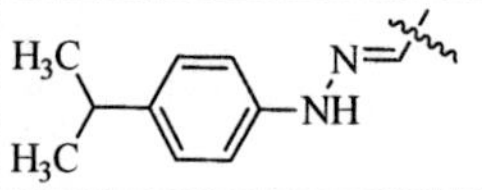	39.46 ± 1.43
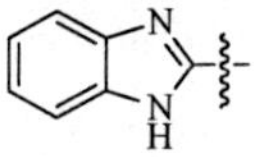	—H	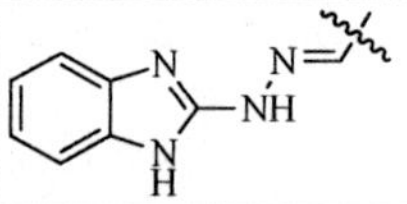	29.53 ± 2.15
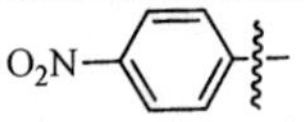	—OH	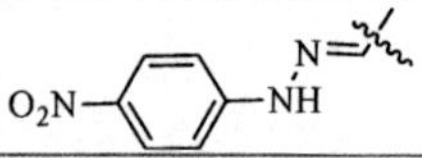	14.58 ± 2.06
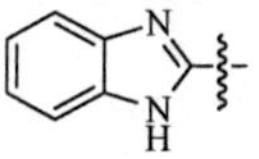	—OH	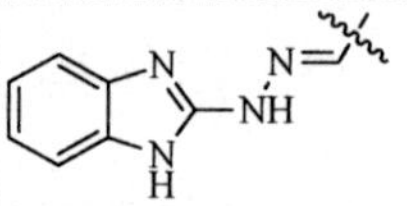	13.12 ± 1.03

library of 661,417 compounds (derived from molecular operating environment lead-like database) against the X-ray crystal structure of DENV2 NS2B–NS3 protease (PDB ID: 2FOM). MOE site finder identified the docking region of the NS2B-NS3 interaction site. Thirty-nine compounds having the top score were identified and subjected to antiviral cell-based assay. Inhibitor SK12 (Fig. 11) was found to inhibit all four DENV serotypes with EC_{50} ranging between (0.74–2.43 μM). Insilico studies predicted that SK12 preoccupied the NS2B binding site of NS3, and it interacted with the NS2B binding site of NS3 through a hydrogen bond, and the residues involved were Lys26, Gln27, Met59, and His60. The ligand interaction studies were done using the MOE program. Figure 12, shows stereo view of SK-12 bound to NS3 (represented in pink color). Figure 13, shows the interaction of the inhibitor with the residues. Interestingly NS2B was also predicted to interact with NS3 through Arg24, Lys26, Gln27, Met59, and His 60 residues. Hence SK12 here interferes with the interaction between NS2B and NS3. Further, it was established that SK12 interfered with DENV protease non-competitively. Modification in the structure of SK12 to reduce the cytotoxicity and increase the inhibitory potential could give us a good lead for future drug development and a pan DENV agent.

Table 2 Inhibition constant of compounds obtained by scaffold hopping

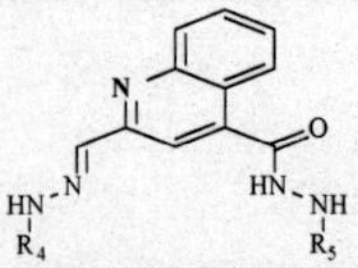

R_4	R_5	IC_{50}(μM) DENV2 NS2B-NS3 protease
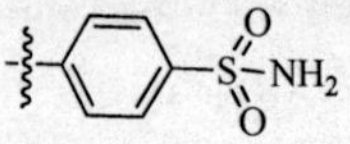	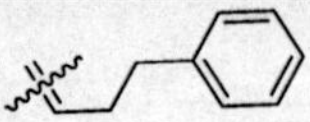	36.02 ± 3.05
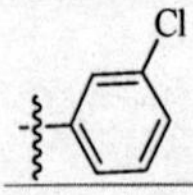	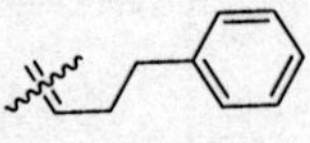	29.04 ± 1.78
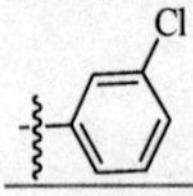	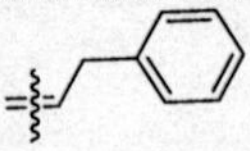	28.12 ± 1.96
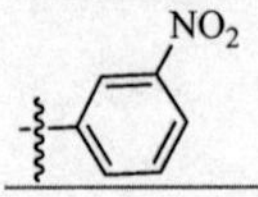	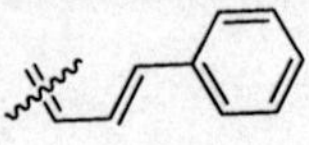	21.96 ± 2.05
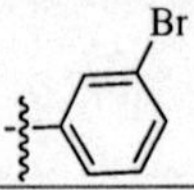	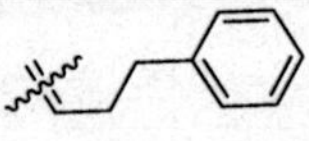	19.93 ± 0.98
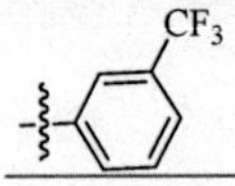	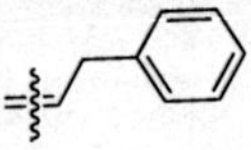	19.8 ± 1.15
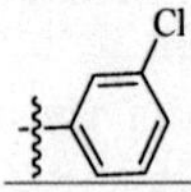	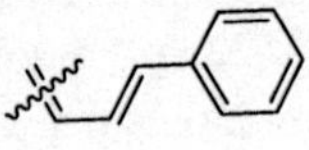	14.32 ± 2.49
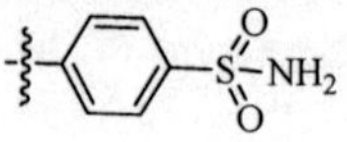	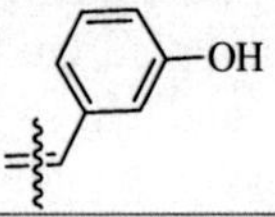	7.83 ± 0.94

Additional experiments performed, employing High throughput virtual screening (HTVS) of about 300,000 compounds (obtained from ChemDiv Inc., San Diego, CA, USA) against DENV4 NS2B-NS3 proteases (PDB ID: 2VBC) identified few novel inhibitors of DENV4 NS2B-NS3 proteases. AutoDock 3.0.5 was used for computational molecular docking simulation. Grid Application Platform Virtual Screening Services (GVSS) for the batch process was employed for computation. Hydrophobic and hydrogen bond interactions with the active site residues were predicted for the active molecules. Of the thirty-six compounds selected for in vitro DENV NS2B-NS3 protease screening, seven molecules were identified as novel DENV4 protease inhibitors with IC_{50} in the micromolar concentration range. The best molecule identified, shown in Fig. 14, (Chemdiv ID: K286-0036), had

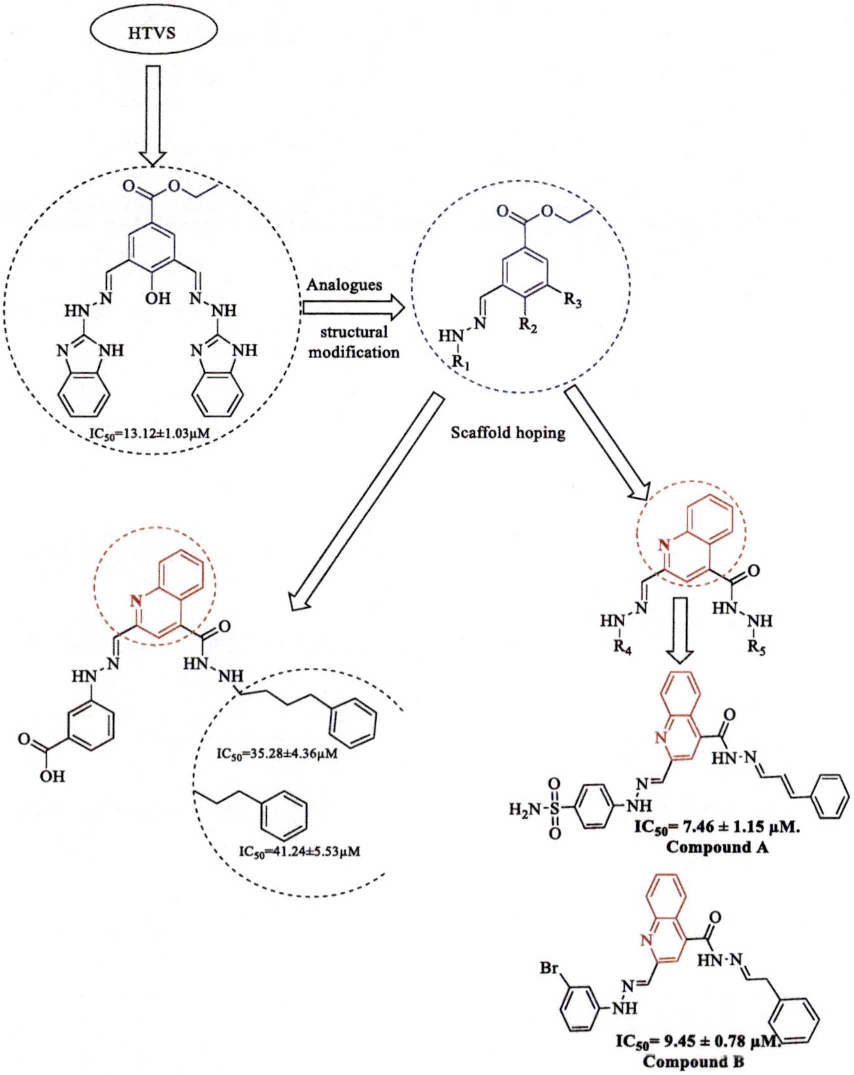

Fig. 8 Schematic representation of Deng et al. work

binding affinity −11.9 kcal/mol and IC_{50} of 3.9 ± 0.6 μM against DENV4 NS2B-NS3 protease, and the mode of inhibition identified was competitive (K_i value of 3.4 ± 0.4 μM). Molecular docking studies predicted that the inhibitor (magenta color) was bound to the NS3 protease active site pocket (Fig. 15), hydrogen bonding interaction and hydrophobic interaction were responsible for the stabilization of the inhibitor (Fig. 16). The inhibitor formed hydrophobic interactions with Trp50, His51, Arg54, Arg73, Asn74, Asp75, Asn152, and Gly153. The

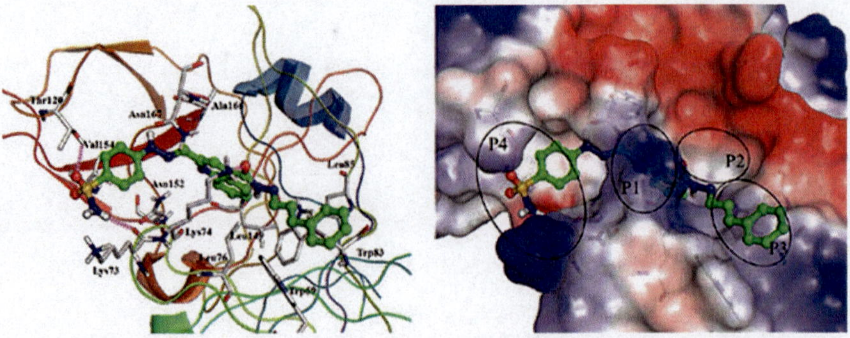

Fig. 9 3D Interaction of Docked pose of compound A with DENV2 NS2B-NS3 protease. Reprinted with permission from (Deng et al. 2012), Copyright © 2012 American Chemical Society

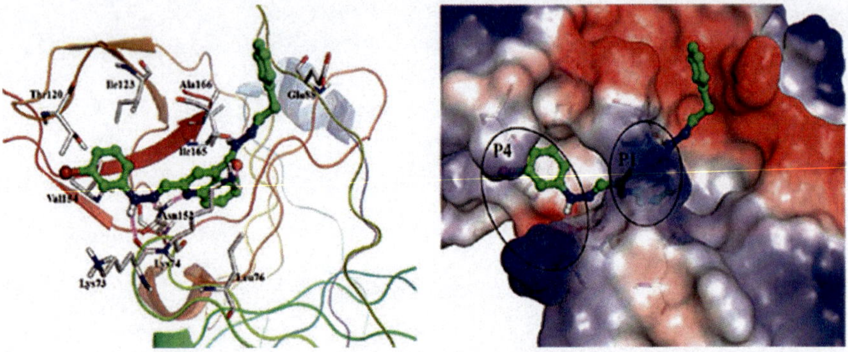

Fig. 10 3D Interaction of Docked pose of compound B with DENV2 NS2B-NS3 protease. Reprinted with permission from (Deng et al. 2012). Copyright © 2012 American Chemical Society

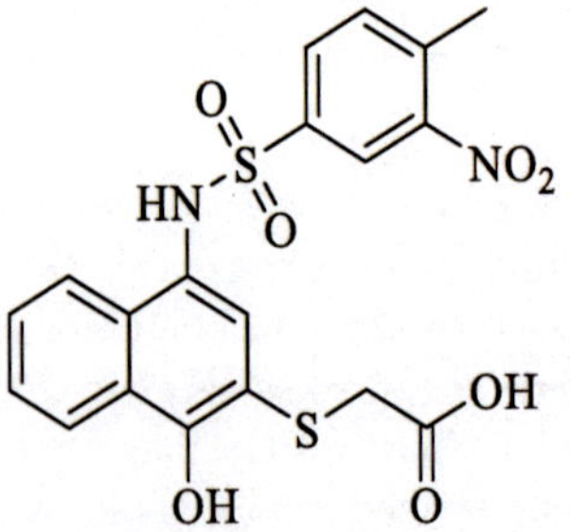

EC_{50}=0.97 ± 0.420 μM (DENV1)
EC_{50}=0.98 ± 0.39 μM (DENV2)
EC_{50}=2.43 ± 0.630 μM (DENV3)
EC_{50}=0.74 ± 0.48 μM (DENV4)

Fig. 11 Structure of SK12 and its reported EC_{50} values against DENV serotypes

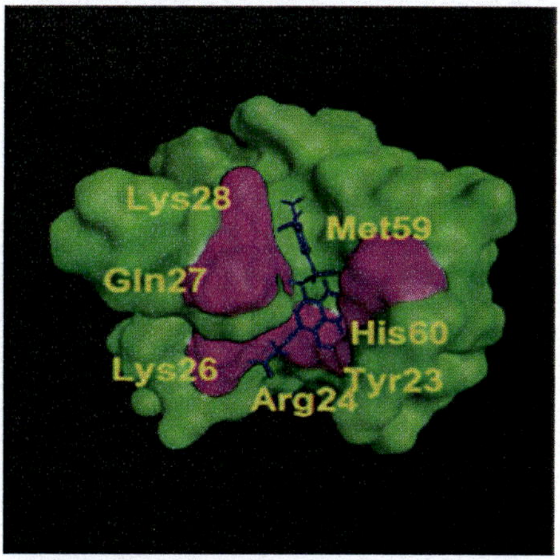

Fig. 12 Stereo view of SK12-NS3 interaction (Pambudi et al. 2013). Reprinted with permission, Copyright © 2013 Elsevier

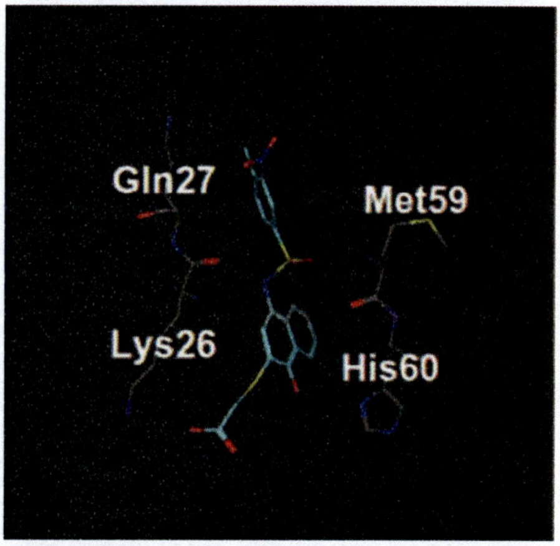

Fig. 13 SK12-NS3 residue interaction (Pambudi et al. 2013). Reprinted with permission, Copyright © 2013 Elsevier.

pyrimidine group oxygen atom formed two hydrogen bonds, one with the hydroxyl group of Ser135 and the other with an amine group of Asn152. Thus here an attempt was made to correlate the interaction results obtained through molecular modeling studies to the in vitro assay result (Nguyen et al. 2013).

In another study DENV NS2B-NS3 protease inhibitors were identified using the rational drug discovery approach. Using Modeller 9.11 software and WNV protease as a template, the homology model design of DENV NS2B-NS3 protease was realized. Pinostorbin was used as the standard reference ligand. AutoDock4.2 was used for the virtual screening of 13,341 molecules from the ZINC database, and the top hits were screened for in-vitro protease assay resulting in the identification of four small molecules as non-competitive inhibitors of DENV2 NS2B-NS3 protease. The best molecule (Fig. 17) had a binding affinity −6.17 kcal/mol and K_i of 69 μM against DENV2 NS2B-NS3 protease. Molecular modeling data (Figs. 18 and 19)

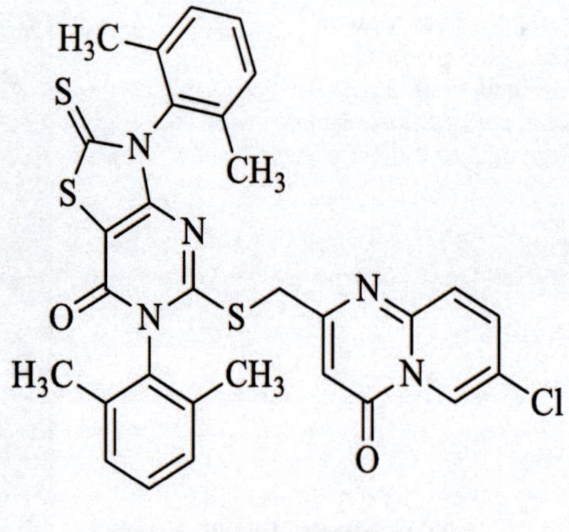

Fig. 14 Most potent molecule of Nguyen et al.

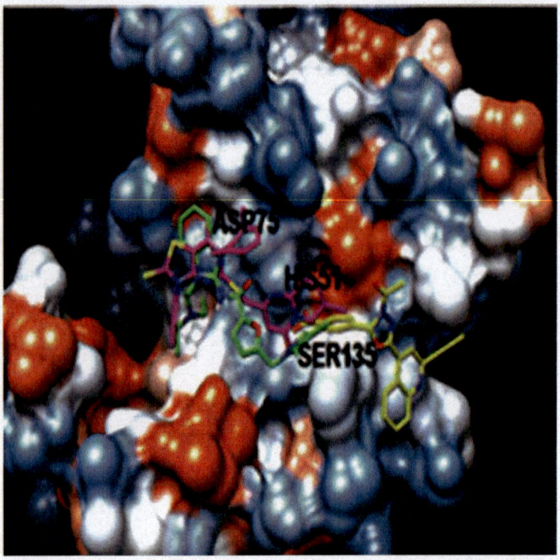

Fig. 15 Inhibitor-protease Interaction (3D) (Nguyen et al. 2013), (*Open Access*)

predicted hydrogen bonding interaction of inhibitor with Asn167 (shown in pink color), hydrophobic interaction of the inhibitor was predicted with Ile78, Lys73 and Ile123 residues (shown in green color) these results were prepared using Ligplot and further Discovery Studio Visualizer 3.1 predicted pi-cation interaction between the inhibitor and Lys74 residue (shown in orange color) (Heh et al. 2013).

Viswanathan et al. (2014) introduced their newly constructed web-based drug discovery portal (DrugDiscovery@TACC) for structure-based drug discovery. In their HTVS study, screening was done using both the inhibitor bound protease structure (PDB ID: 3U1I & 3U1J) and inhibitor-free protease structure (PDB ID:

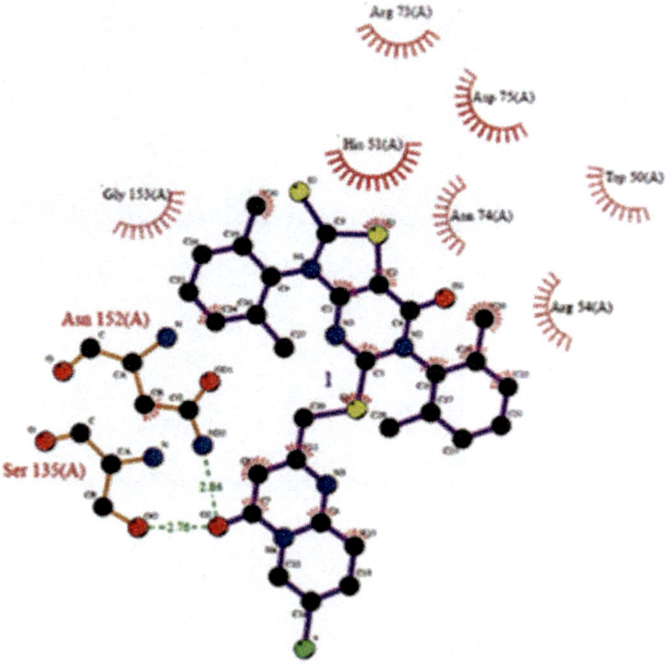

Fig. 16 Inhibitor-protease interactions (2D) (Nguyen et al. 2013), (*Open Access*).

Fig. 17 Most potent molecule of Heh et al.

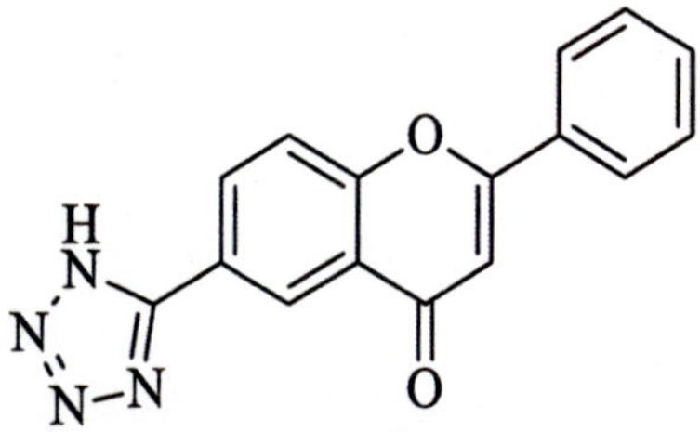

K_i=69 μM

2FOM). Further, two virtual libraries were compiled, a library of 642,769 molecules, which was a subset of ZINC database and was named "collective" library; it was obtained after applying the Lipinski rule of 5 filters along with other filters. The second library, known as the focused library, obtained by applying the ClogP filter, included 45,458 small molecules. An independent Virtual screening was performed against DENV2 NS2B-NS3 protease (PDB ID: 2FOM) and DENV NS2B-NS3 protease (PDB ID: 3U1I, 3U1J). NS3 proteases of 2FOM and 3U1I, 3U1J share approximately 70% sequence identity and active site residues of the same shared approximately 90% sequence identity. Virtual screening yielded 179 hits from the

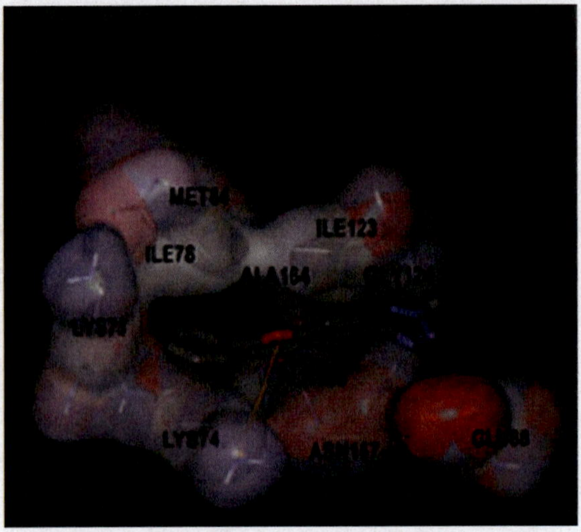

Fig. 18 3D model of Interactions (Heh et al. 2013). Reproduced with permission, Copyright © 2013 John Wiley & Sons A/S

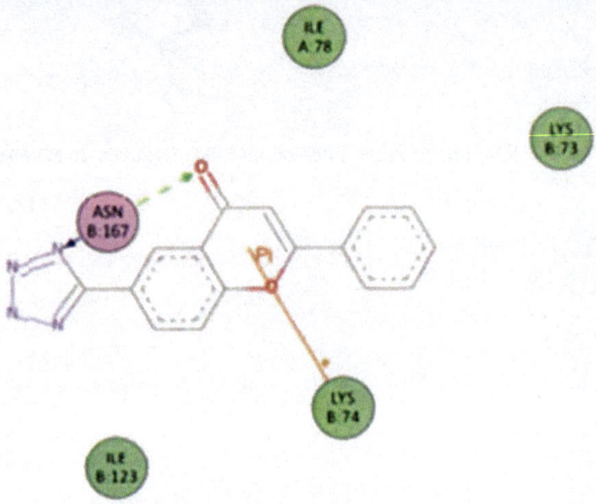

Fig. 19 2D model of Interactions (Heh et al. 2013). Reproduced with permission, Copyright © 2013 John Wiley & Sons A/S

collective library and 117 hits from the focused library. Thus, with its computational resources, this platform provided an option to screen multiple ligands for similar valid targets. Further, these hits were narrowed down by applying several filters, which included.

- Discard of hits which did not form any hydrogen bond with catalytic side residue.
- Discarding the one which had poor predicted solubility.
- Discard of hits with fewer hetero atoms.
- Discarding the polar hits or which contained formal charges.

Finally, they presented a proof of concept result for a validated compound ZINC04321905 (Fig. 20), which was a mixed non-competitive inhibitor and had K_i of 7 μM against DENV2 NS2B-NS3 protease. Drug-like properties of ZINC04321905 were calculated using ORIS property calculator, which indicated no adverse mutagenic, irritant effect, or tumorigenic effect associated with this pharmacophore. However, the cyclohexene ring was associated with an adverse reproductive effect. Hence future analog development approach would be to replace this unwanted fragment and substitute it with more drug-like molecules. For molecular modeling studies, the Vina docking program was used to calculate the orientation of the inhibitor to the DENV3 NS2B-NS3 protease. Figure 22 was generated using PyMol, and it showed the inhibitor (in magenta color) bound to the protease while the catalytic site was shown in green color. Three hydrogen bonding interactions (shown by green dashed lines) of the inhibitor with the amides of Gly133, Thr134, and Ser135 can be seen in Fig. 21. The hydrophobic interactions were shown by red hashed lines and residues involved were Val152, Val136, His51, Pro132, Asp129, Phe130, Tyr161, and Tyr150. The image was prepared using LigPlot+. An interesting study done by this group saw a comparison of NS3 protease sequence from 2211 strains spanning to the four DENV serotypes, and about 40% of the residues were conserved, more interestingly the active site residue conservation was 66% for all the 4 serotypes. Also, of the 11 residues, involved in the inhibitor ZINC04321905-protease interaction, 10 residues were conserved in all the 2211 strains of DENV examined, hence this molecule could be further exploited for the development of pan-dengue inhibitor (Viswanathan et al. 2014).

Subsequently in 2015, Li et al. attempted a structure-guided discovery of a small non-peptide as an inhibitor of DENV2 NS2B-NS3 protease. They carried out a multistep virtual screening campaign on a library of 5 million compounds obtained from 4 different commercial sources (Chembridge, Enamine, Life chemicals, and Maybridge). The pharmacophore-based preliminary filter narrowed the compounds to a few hundred, to this rigid docking, and induced fit docking was applied using AutoDock 4.2.5. Fourteen compounds based on docking scores were identified and subjected to biological screening. The protease inhibition assay was performed, and the EC_{50} of the top hit was around 5.0 μM in BHK21 cells (Fig. 23). Docking

Fig. 20 Mixed non-competitive inhibitor

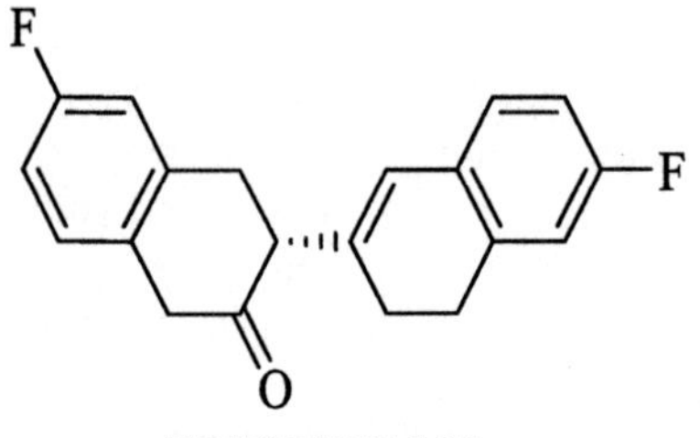

ZINC04321905

K_i=7.0 μM

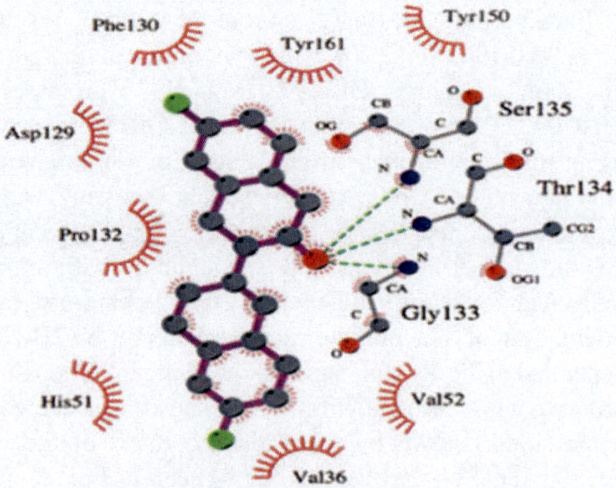

Fig. 21 2D representation of inhibitor-protease interaction. Reprinted with permission from (Viswanathan et al. 2014). Copyright © 2014 American Chemical Society

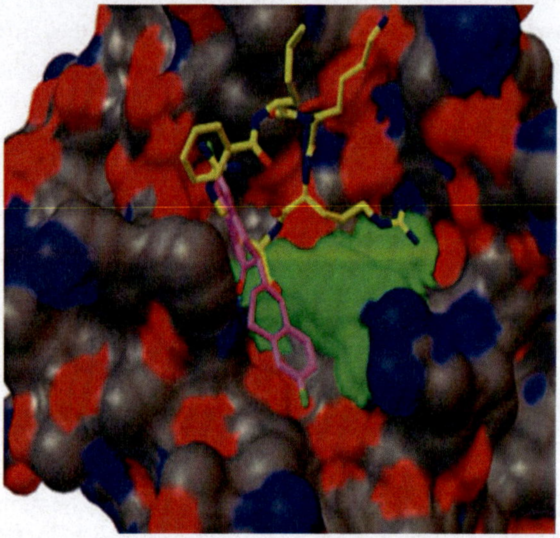

Fig. 22 3D representation of inhibitor-protease interactions. Reprinted with permission from (Viswanathan et al. 2014). Copyright © 2014 American Chemical Society

study of this compound further supported their concept of pharmacophore-based design as its results were in good agreement with the interactions seen in the known crystal structure (PDB ID: 3U1I) and the binding free energy being −10.65 kcal/mole. This hit occupied the pockets of the binding site (Fig. 24). The benzyl ring of the compound occupied the hydrophobic region between Pro 132 and Val 155. Also, Tyr 161 showed interaction with the lone pair of N atom, and an additional π-π interaction could also be seen with it. Further stabilization of the inhibitor-protease was seen when an additional water group linked the amino group of the inhibitor with Met184 and Gly153.

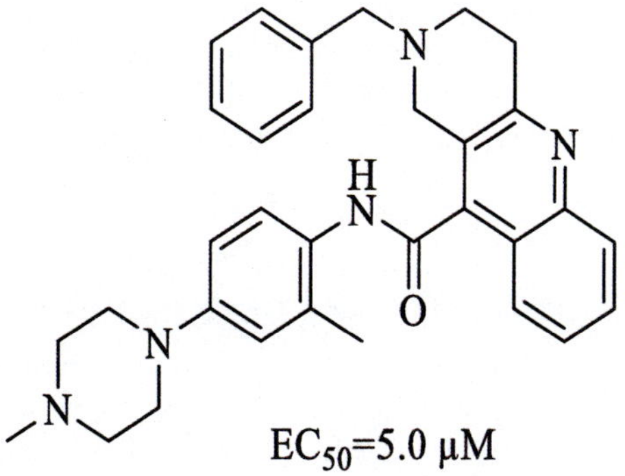

Fig. 23 Top Hit of Li et al.

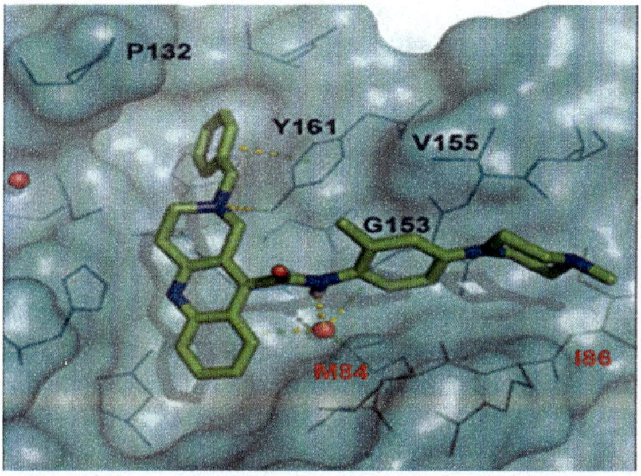

Fig. 24 Binding mode of inhibitor (Li et al. 2015). Reprinted with permission, Copyright © 2014 John Wiley & Sons A/S

Subsequently, Timiri et al. (2015) performed HTVS of ZINC 8 database to identify an inhibitor of DENV2 NS2B-NS3 protease. Top hundred molecules were manually analyzed and they decided to synthesize phthalimide-sulphonamide analogues as they were encouraged by protease (HIV/HCV) (Lampa et al. 2014; Davis et al. 2017) inhibition activity of sulphonamides and antiviral (Selvam et al. 2012), antimycobacterial (Akgün et al. 2012) and anticancer (Matsushita et al. 2015) activities reported by phthalimide derivatives. Twenty derivatives of 4-(1,3-dioxo-2,3-dihydro-1H-isoindol-2-yl) benzene- 1-sulphonamide were synthesized and screened for DENV2 protease inhibitory activity. The compound shown in Fig. 25 had an IC_{50} value of 48.2 μM against DENV2 NS2B-NS3 protease. Molecular docking and molecular dynamic simulation studies were carried out to understand the mechanism of action of the inhibitor. Using Modeller 9.15 and 3U1I as a template, the structure of DENV2 NS2B-NS3 protease was modeled and Induced–fit docking was performed using GLIDE of Maestro 9.2. The docking score of the most potent compound reported was −5.12. The interactions which were attributed for the DENV protease inhibition activities were π-π stacking interactions between ring A and Tyr200. Hydrogen bonding interaction between O

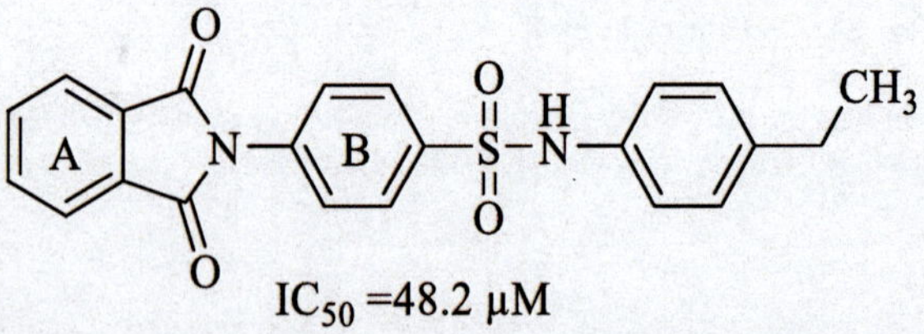

Fig. 25 Most potent compound of Timiri et al.

atom of ring A and N atom of imidazole ring in Hip 90, hydrophobic interactions of =CH group of ring A with Asp168, Phe169, Pro171, Tyr189, Gly190 residues. Another hydrophobic interaction between 4-ethyl phenyl group attached to sulphonamide and Trp89 and Val111, A hydrogen bond interaction between N atom attached to sulphonamide and O atom of carbonyl group of Asn191, and π-cation interactions between electron cloud of ring B and cationic N of imidazole ring in Hip90 (Fig. 26). MD simulations revealed that NS2B/NS3 protease was more stable when it was bound with the active compound.

Subsequently Pelliccia et al. (2017). adopted a virtual screening approach to identify allosteric inhibitors against DENV NS2B-NS3 protease. The virtual screening mainly focused on the allosteric site of the enzyme. Based on their findings, they designed and synthesized compounds and evaluated their inhibitory potential by performing the cell-based and enzyme-based assay against DENV NS2B-NS3 protease. In the cell-based assay, the best compound showed an EC_{50} value of 6.7 ± 0.78 μM and in the enzyme-based assay, the EC_{50} value obtained was 4.7 ± 0.3 μM (Fig. 27). For the molecular modeling studies, crystal structure of DENV2 NS2B-NS3 protease (PDBID: 2FOM) was used. Docking was performed using PLANTS and scoring was done using ChemPLP scoring function (Korb 2009). Figure 28, shows the binding mode of the inhibitor (Green color) with the protease.

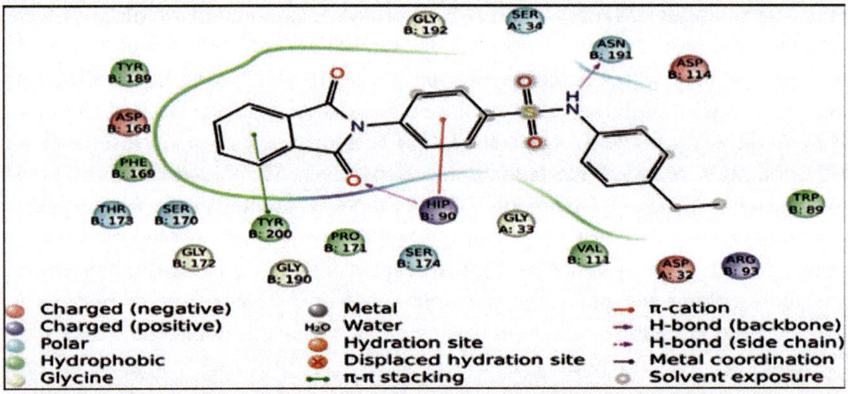

Fig. 26 2D representation of Inhibitor-modelled DENV2 NS2B-NS3 protease interaction (Timiri et al. 2015). Reprinted with permission, Copyright © 2015 Elsevier Inc.

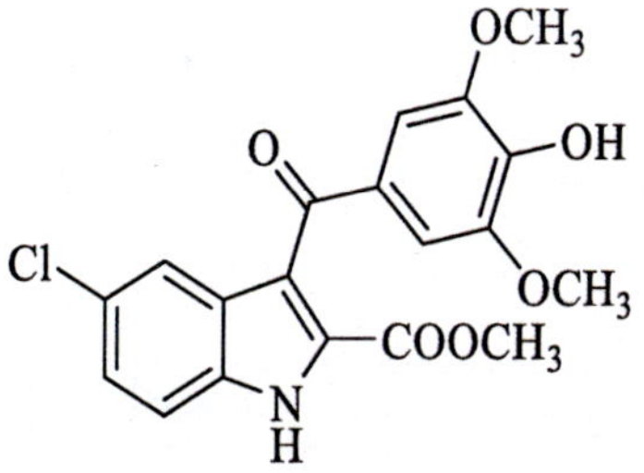

EC50= 6.7 ± 0.78μM (Cell-based assay)
EC50= 4.7 ± 0.3μM (Enzyme-based assay)

Fig. 27 Potent molecule of Pelliccia et al.

PyMol was used to generate this structure. The carbonyl portion of the ester formed H-bond with Asp152 side chain and the ester was well-positioned inside the binding pocket, further hydrogen bond was predicted between indole, NH and Asn167. Residues Val146, Leu149 and Leu76 were predicted to form hydrophobic interaction with the indole ring, the methoxyphenyl portion also showed hydrophobic interactions with Trp83 and a pi-cation interaction with Lys74.

These studies were followed by, Hariono et al. (2019). efforts wherein they identified a hit (D0713) against DENV2 NS2B-NS3 protease by virtual screening of a library from the National Cancer Institute database. The binding free energy and the experimental IC_{50} against DENV2 NS2B-NS3 protease of this hit was −7.10 kcal/mol and 62 μM, respectively (Fig. 29). AutoDockVina (www.autodock.scripps.edu) was used for this virtual screening exercise. This hit had a thioguanine scaffold in it, and upon its modification, the best molecule obtained had an IC_{50} value of 0.38 μM against DENV2 NS2B-NS3 protease. Further, this molecule was docked to the DENV2 NS2B-NS3 protease model using AutoDock4.2 and the binding free energy obtained was −7.48 kcal/mol. Further, interactions associated with the inhibitory activity were identified, and residues Ser135, Tyr161, and Gly153 were predicted to form hydrogen bonding interactions

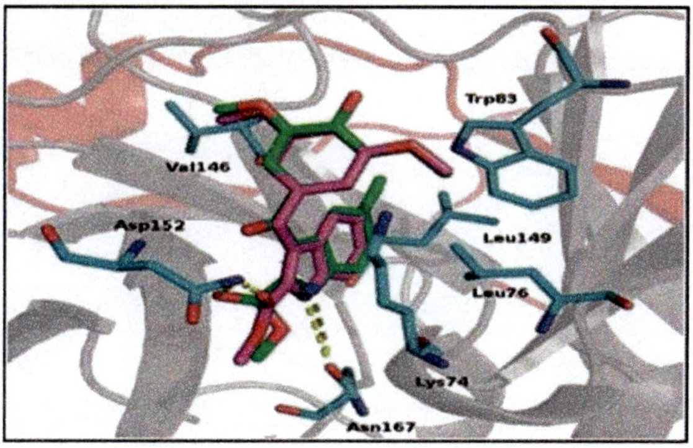

Fig. 28 Inhibitor-protease interactions. (Pelliccia et al. 2017), (*Open Access*)

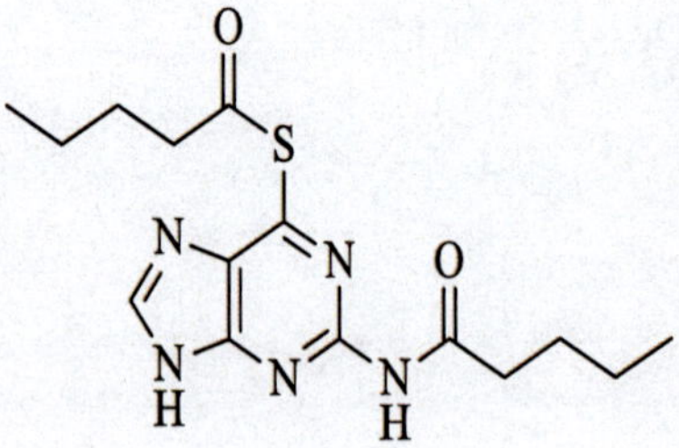

IC50=0.38μM

Fig. 29 The most potent compound of Hariono et al. (2019)

with the inhibitor; other hydrophobic interactions identified across the length of the inhibitor were Pro132, Tyr154, Thr134, Ser131, Gly151, His51, Gly52, Asp51, Met34, Val72, Asp75, Tyr50, Asn152, Ser53 (Figs. 30 and 31). To get a better insight this inhibitor–protease interaction, Amber 14 (Le Grand et al. 2013) was used to perform molecular dynamics simulations, and key residues identified for hydrogen binding interactions were His51, Asp129, Ser135, Asn152, and Tyr 161.

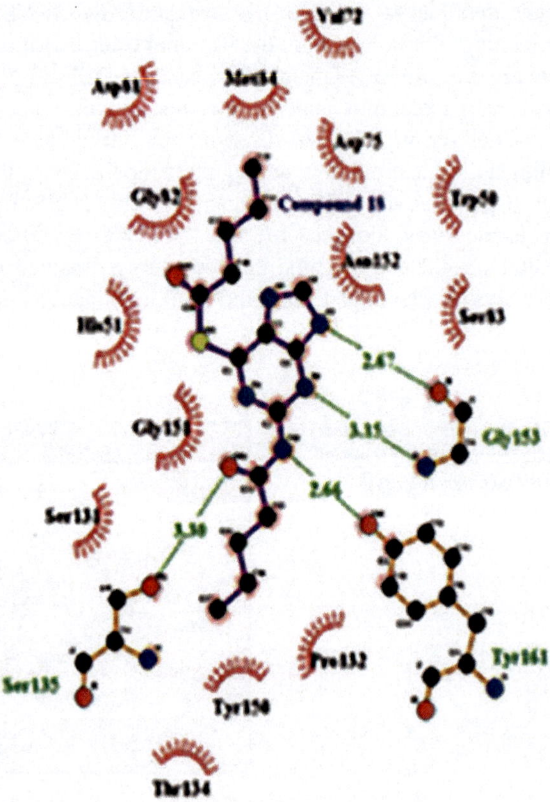

Fig. 30 2D view of inhibitor-protease interaction. (Hariono et al. 2019), (*Open Access*)

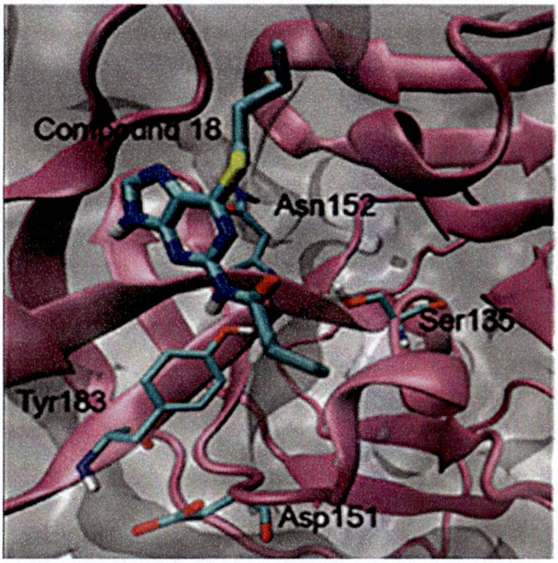

Fig. 31 3D view of inhibitor-protease interaction (Hariono et al. 2019), (*Open Access*)

MMPBSA.py module of AMBER 14 was used to perform MM/PBSA calculations of this molecule, and the free binding energy obtained was −16.10 kcal/mol. MM/PBSA calculations indicated the role of polar and non-polar interactions contributing to this molecule's low IC_{50} value.

In a very recent study, a multistep virtual screening of the Asinex database containing 590,859 compounds was performed against DENV2 NS2B-NS3 protease using crystal structure 2FOM. The first step of the virtual screening was done with the AutoDock Vina program. The set criteria narrowed the compound numbers to 3645 compounds. Further, using the virtual screening workflow of Schrodinger Suite and MM/GBSA based binding free energy estimation, the compounds were narrowed down to 102. In silico pharmacokinetic assessment further narrowed down this number and five compounds were proposed as potential NS2B-NS3 protease inhibitors. The best molecule identified had a docking score of −7.619 kcal/mol (Fig. 32) and the binding free energy of this molecule, calculated from MD simulation trajectories using MM/GBSA approach was 60.666 kcal/mol, which was better than the standard inhibitor (MB21) chosen for this study. MB 21 had binding free energy of −40.207 kcal/mol. The binding interaction of this inhibitor is shown in Fig. 33. Residues Lys73, Gly153, and Asn167 were predicted to form four Hydrogen bonds with the inhibitor. Lys74 and Leu76 showed hydrophobic interactions with the inhibitor and the only π-cation interaction was between Lys74 and the inhibitor. Figure 34, shows the binding pose of the inhibitor in 3D space (Bhowmick et al. 2020).

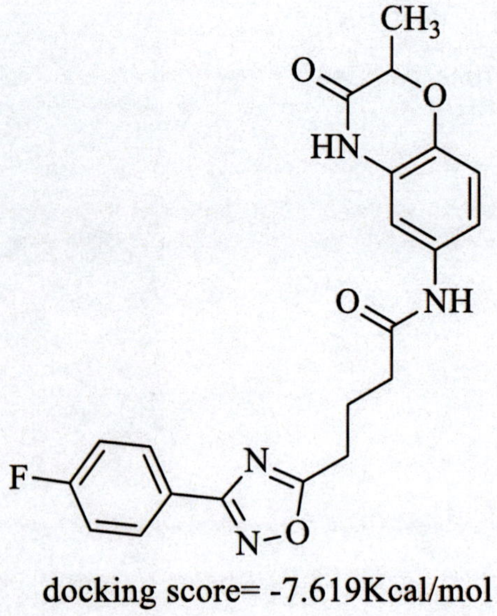

Fig. 32 Lowest binding energy molecule

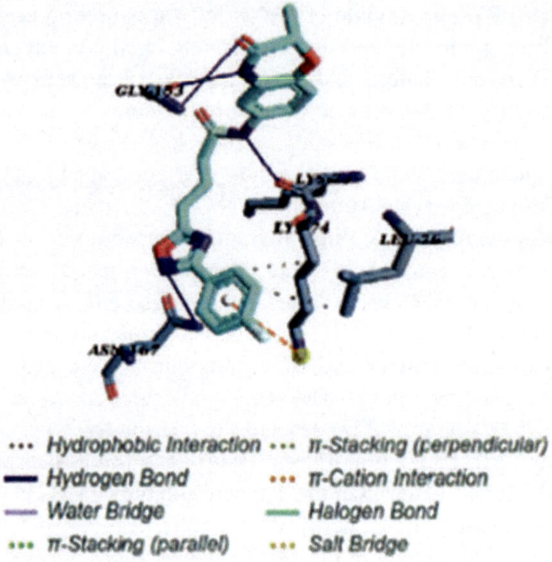

Fig. 33 Inhibitor-protease binding interaction (Bhowmick et al. 2020). Reprinted with permission, Copyright © 2020 John Wiley & Sons Ltd.

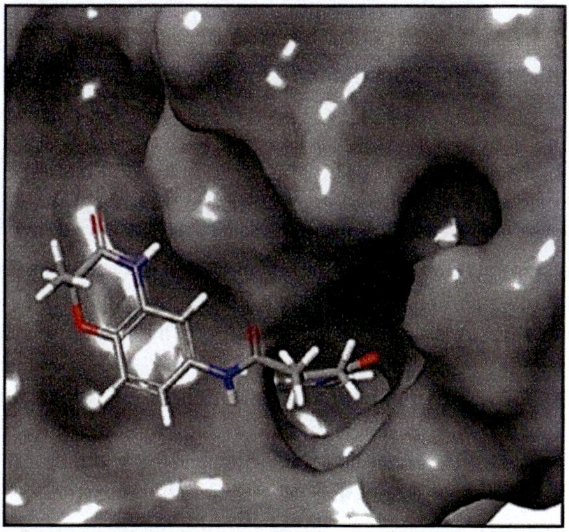

Fig. 34 3D view of inhibitor-protease interaction (Bhowmick et al. 2020). Reprinted with permission, Copyright © 2020 John Wiley & Sons Ltd.

4 Conclusion

As conclusion here we summarise the the important interactions chronologically reported by various groups from their HTVS studies. Ganesh et al. (2005) reported important electrostatic interaction involving residues Asp129, Tyr150, Ser163. Also they noted that, Ser135 was involved in hydrogen bonding. Tomlinson et al. (2009) reported P1 pocket interactions, which included residues Gly151, Gly153, some residues near to the catalytic triad (Gln35) and also the residues of the catalytic triad (His51, Ser135). Besides this, some other interactions involving residue Asn152 were also identified. Knehans et al. (2011) showed the engagement of residues Asp129 and Asn152 in hydrogen bonding in the S1 and S2 pocket of protease, respectively. His51 of the catalytic triad was involved in π-π interaction, Tyr161 was predicted to show pi-pi stacking interactions and Pro132 was involved in hydrophobic interaction. Deng et al. (2012) reported hydrogen bonding by residues Thr120, Lys73, Lys74, Asn152, and Ile165 with their potent compounds. Pambudi et al. (2013) reported hydrogen bonding interactions involving residues Lys26, Gln27, Met59, and His60. Nguyen et al. (2013) identified residues Ser135 and Asn152 to be involved in hydrogen bonding, and residues His51 and Asn152, along with few other residues, were found to be engaged in hydrophobic interactions. Heh et al. (2013) predicted hydrogen bond formation by residue Asn167 and a pi-cation interaction by Lys74. Besides this, Lys73, along with other residues, contributed to hydrophobic interactions. Li et al. (2015) reported important residue Tyr161 along with others for hydrophobic interactions. Timri et al. (2015) found residues Hip90 and Asn191 to be engaged in hydrogen bonding interactions and again residue, Hip90 was found to be involved in pi-cation interaction. Pelliccia et al. (2017) reported the involvement of residues Asp152 and

Asn167 in hydrogen bonding and Lys74 in pi-cation interaction. This was followed by a report by Hariono et al. (2019) who found residue Tyr161 to be involved in hydrogen bonding interaction. Besides this, Asn152 and His51, along with other residues were involved in hydrophobic interactions. Finally, Bhowmick et al. (2020) reported residues, Lys73, Gly153, and Asn167 to be involved in hydrogen bonding, and Lys74 showing a pi-cation interaction.

Looking into these interactions, a commonality was exhibited by few residues pertaining to certain interactions, viz involvement of Lys74 in pi-cation interaction, the involvement of Asn152, and Asn167 in hydrogen bonding interaction, and involvement of Tyr161 in pi-pi stacking interactions. Understanding the role of these residues can certainly help in the designing of a suitable pharmacophore model.

Hence, it could be inferred from the above observations that HTVS has certainly helped the researchers in making a rational approach towards dengue drug discovery. It has given an option of speedily evaluating the big libraries available from different databases and has tactically provided a smarter approach for the drugdiscovery process. It may not be a surprise if in the near future we discover a pan-dengue inhibitor from the HTVS studies.

Acknowledgements The Authors thank DST-SERB, Govt. of India for providing financial support through their project (EMR/2016/005711/) dated 7th August 2017 and Birla Institute of Technology, Mesra, Ranchi, India for providing the necessary infrastructure support.

References

Akgün H, Karamelekoglu I, Berk B, et al. Synthesis and antimycobacterial activity of some phthalimide derivatives. Bioorganic Med Chem. 2012;20:4149–54. https://doi.org/10.1016/j.bmc.2012.04.060.

Bhatt S, Gething PW, Brady OJ, et al. The global distribution and burden of dengue. Nature. 2013;496:504–7. https://doi.org/10.1038/nature12060.

Bhowmick S, Alissa SA, Wabaidur SM, et al. Structure-guided screening of chemical database to identify NS3-NS2B inhibitors for effective therapeutic application in dengue infection. J Mol Recognit. 2020;33:1–18. https://doi.org/10.1002/jmr.2838.

Davis BW, Diep JT, Jose S. (12) United States Patent 2. 2017.

De Clercq E. Anti-HIV drugs: 25 compounds approved within 25 years after the discovery of HIV. Int J Antimicrob Agents. 2009;33:307–20. https://doi.org/10.1016/j.ijantimicag.2008.10.010.

Deng J, Li N, Liu H, et al. Discovery of novel small molecule inhibitors of dengue viral NS2B-NS3 protease using virtual screening and scaffold hopping. J Med Chem. 2012;55:6278–93. https://doi.org/10.1021/jm300146f.

Ebrahim GJ. Dengue and dengue haemorrhagic fever. J Trop Pediatr. 1993;39:262–3. https://doi.org/10.1093/tropej/39.5.262.

Erbel P, Schiering N, D'Arcy A, et al. Structural basis for the activation of flaviviral NS3 proteases from dengue and West Nile virus. Nat Struct Mol Biol. 2006;13:372–3. https://doi.org/10.1038/nsmb1073.

Falgout B, Pethel M, Zhang YM, Lai CJ. Both nonstructural proteins NS2B and NS3 are required for the proteolytic processing of dengue virus nonstructural proteins. J Virol. 1991;65:2467–75. https://doi.org/10.1128/jvi.65.5.2467-2475.1991.

Ganesh VK, Muller N, Judge K, et al. Identification and characterization of nonsubstrate based inhibitors of the essential dengue and West Nile virus proteases. Bioorganic Med Chem. 2005;13:257–64. https://doi.org/10.1016/j.bmc.2004.09.036.

Goncalvez AP, Engle RE, St. Claire M, et al. Monoclonal antibody-mediated enhancement of dengue virus infection in vitro and in vivo and strategies for prevention. Proc Natl Acad Sci USA. 2007;104:9422–7. https://doi.org/10.1073/pnas.0703498104.

Guy B, Noriega F, Ochiai RL, et al. A recombinant live attenuated tetravalent vaccine for the prevention of dengue. Expert Rev Vaccines. 2017;16:671–83. https://doi.org/10.1080/14760584.2017.1335201.

Guzman MG, Halstead SB, Artsob H, et al. Dengue: a continuing global threat. Nat Rev Microbiol. 2010;8:S7–16. https://doi.org/10.1038/nrmicro2460.

Hariono M, Choi SB, Roslim RF, et al. Thioguanine-based DENV-2 NS2B/NS3 protease inhibitors: virtual screening, synthesis, biological evaluation and molecular modelling. PLoS ONE. 2019;14:1–21. https://doi.org/10.1371/journal.pone.0210869.

Heh CH, Othman R, Buckle MJC, et al. Rational discovery of dengue type 2 non-competitive inhibitors. Chem Biol Drug Des. 2013;82:1–11. https://doi.org/10.1111/cbdd.12122.

Knehans T, Schüller A, Doan DN, et al. Structure-guided fragment-based in silico drug design of dengue protease inhibitors. J Comput Aided Mol Des. 2011;25:263–74. https://doi.org/10.1007/s10822-011-9418-0.

Korb O. Efficient ant colony optimization algorithms for structure- and ligand-based drug design. Chem Cent J. 2009;3:247–58. https://doi.org/10.1186/1752-153X-3-S1-O10.

Lampa AK, Bergman SM, Gustafsson SS, et al. Novel peptidomimetic hepatitis C virus NS3/4A protease inhibitors spanning the P2–P1′ region. ACS Med Chem Lett. 2014;5:249–54. https://doi.org/10.1021/ml400217r.

Le Grand S, Götz AW, Walker RC. SPFP: Speed without compromise—a mixed precision model for GPU accelerated molecular dynamics simulations. Comput Phys Commun. 2013;184:374–80. https://doi.org/10.1016/j.cpc.2012.09.022.

Li L, Basavannacharya C, Chan KWK, et al. Structure-guided discovery of a novel non-peptide inhibitor of dengue virus NS2B-NS3 protease. Chem Biol Drug Des. 2015;86:255–64. https://doi.org/10.1111/cbdd.12500.

Matsushita M, Ozaki Y, Hasegawa Y, et al. A novel phthalimide derivative, TC11, has preclinical effects on high-risk myeloma cells and osteoclasts. PLoS ONE. 2015;10:1–15. https://doi.org/10.1371/journal.pone.0116135.

Mukhopadhyay S, Kuhn RJ, Rossmann MG. A structural perspective of the Flavivirus life cycle. Nat Rev Microbiol. 2005;3:13–22. https://doi.org/10.1038/nrmicro1067.

Nestorowicz A, Chambers TJ, Rice CM. Mutagenesis of the yellow fever virus NS2A/2B cleavage site: effects on proteolytic processing, viral replication, and evidence for alternative processing of the NS2A protein. Virology. 1994;199:114–23. https://doi.org/10.1006/viro.1994.1103.

Nguyen TTH, Lee S, Wang HK, et al. In vitro evaluation of novel inhibitors against the NS2B-NS3 protease of dengue fever virus type 4. Molecules. 2013;18:15600–12. https://doi.org/10.3390/molecules181215600.

Noble CG, Seh CC, Chao AT, Shi PY. Ligand-bound structures of the dengue virus protease reveal the active conformation. J Virol. 2012;86:438–46. https://doi.org/10.1128/JVI.06225-11.

Pambudi S, Kawashita N, Phanthanawiboon S, et al. A small compound targeting the interaction between nonstructural proteins 2B and 3 inhibits dengue virus replication. Biochem Biophys Res Commun. 2013;440:393–8. https://doi.org/10.1016/j.bbrc.2013.09.078.

Pelliccia S, Wu YH, Coluccia A, et al. Inhibition of dengue virus replication by novel inhibitors of RNA-dependent RNA polymerase and protease activities. J Enzyme Inhib Med Chem. 2017;32:1091–101. https://doi.org/10.1080/14756366.2017.1355791.

Phong WY, Moreland NJ, Siew P, et al. Dengue protease activity: the structural integrity and interaction of NS2B with NS3 protease and its potential as a drug target. Biosci Rep. 2011;31:399–409. https://doi.org/10.1042/BSR20100142.

Selvam P, Lakra DR, Pannecouque C, De Clercq E. Synthesis, antiviral and cytotoxicity studies of novel N-substituted indophenazine derivatives. Indian J Pharm Sci. 2012;74:274–8. https://doi.org/10.4103/0250-474X.106077.

Timiri AK, Subasri S, Kesherwani M, et al. Synthesis and molecular modelling studies of novel sulphonamide derivatives as dengue virus 2 protease inhibitors. Bioorg Chem. 2015;62:74–82. https://doi.org/10.1016/j.bioorg.2015.07.005.

Tomlinson SM, Malmstrom RD, Russo A, et al. Structure-based discovery of dengue virus protease inhibitors. Antiviral Res. 2009;82:110–4. https://doi.org/10.1016/j.antiviral.2009.02.190.

Viswanathan U, Tomlinson SM, Fonner JM, et al. Identification of a novel inhibitor of dengue virus protease through use of a virtual screening drug discovery web portal. J Chem Inf Model. 2014;54:2816–25. https://doi.org/10.1021/ci500531r.

WHO. Treatment, prevention and control global strategy for dengue prevention and control 2. 2012.

Wyles DL. Antiviral resistance and the future landscape of hepatitis C virus infection therapy. J Infect Dis. 2013;207:33–9. https://doi.org/10.1093/infdis/jis761.

Yusof R, Clum S, Wetzel M, et al. Purified NS2B/NS3 serine protease of dengue virus type 2 exhibits cofactor NS2B dependence for cleavage of substrates with dibasic amino acids in vitro. J Biol Chem. 2000;275:9963–9. https://doi.org/10.1074/jbc.275.14.9963.

Ebola Virus

Ebola Virus: Overview, Genome Analysis and Its Antagonists

Sahar Qazi, Ayesha Khanam, and Khalid Raza

Abstract Ebola is an inexorable, negative-sense, single-strand, non-segmented RNA virus which rapidly develops alterations through error-prone replication and is known to be one of the most fatal diseases in humans. A member of Filoviridae, under the genus Ebolavirus, has found to be a cataclysmic and an apex threat to human health worldwide. In 1976 we have had epidemic outbursts, including Nzara, Sudan, and in Yambuku and the Democratic Republic of Congo to be the first two consecutive ones. Fruit bats are mainly considered to be the primary agents of spreading and carrying the zoonotic infection of Ebola, especially the Hypsignathus Monstrosus, Epomopsfranqueti, and Myonycteris Torquata. Possible sources of spread can be plants, arthropods, rodents, and aves. This spread of zoonotic infection can be understood by factors such as biological, environmental, and social factors, and therefore, the infectious interaction between bats and the ecosystem is considered the janitor for the outbreak. Ebola causes a chronic and lethal hemorrhagic fever called the Ebola Virus Disease (EVD). In previous bursts of EVD, many people tasted lethality in less than a week's infection incubation period. Thus, EBOV is the main cynosure for researchers as it has increased interests in developing new models for virus evolution and therapies. The virus affiliates with the immune system of the host which aids to determine the dynamics as to how the virus functions and affects the immune system. After sequencing the Ebola virus genome, it has attracted the researchers, especially from Bioinformatics, Genomics, Transcriptomics, Proteomics and Metabolomics to analyze it and infer fruitful knowledge which can be used for efficient and robust drugs for EVD treatment approaches. This chapter presents a brief review about the Ebola virus and discusses its genome analysis, the antagonist and control in the future.

Keywords Ebola · Ebola Virus Disease (EVD) · World Health Organization (WHO) · Zoonosis · System Biology (SB) · EBOV glycoprotein (GP) · Pathogenesis · Transmission · Genome analysis · Niemann-Pick C1 · Next Generation Sequencing (NGS)

S. Qazi · A. Khanam · K. Raza (✉)
Department of Computer Science, Jamia Millia Islamia, New Delhi 110025, India
e-mail: kraza@jmi.ac.in

S. I. Ahmad (ed.), *Human Viruses: Diseases, Treatments and Vaccines*, https://doi.org/10.1007/978-3-030-71165-8_6

1 Introduction

1.1 Ebola virus (EBOV), is an inexorable, negative-sense single-strand, nonsegmented RNA ((−)ssRNA) virus with a 19 kb genome, rapidly develop alterations through error-prone replication and is known to be one of the most fatal diseases in humans (Goeijenbier et al. 2014). Humans have experienced Ebola virus disease (EVD) outbreaks in the past, mostly in socio-economically deprived countries, which had a mortality rate of approximately 90%. It was first observed in 1976 in two consecutive epidemics, in Nzara, Sudan, and in Yambuku, Democratic Republic of Congo (Fig. 1) (Murphy, WHO, Ghazanfar et al. 2015). There are mainly four strains of the EBOV, such as Bundibugyo virus (BDBV), Sudan virus (SUDV), Taï Forest virus (TAFV) and Zaire Ebola virus and are collectively of the 'Marburg Viruses' (Sanchez et al. 1993). Genetic studies on EBOV have discerned that the infectious viral particles of such filoviridae are composed of seven structural proteins that are observed from seven different genes which have been discerned from the complete nucleotide sequence of the Marburg genome (Feldmann et al. 1992). In case of EBOV these proteins are: (i) RNA Dependent RNA polymerase (L, 180 kDa), (ii) glycoprotein (GP; 125 kDa), (iii) nucleoproteins (NP), 104 kDal and a possible second nucleoprotein (VP30), (iv) matrix protein (VP40), (v) encased protein (VP241), (vi) and protein WP35. They have been suggested to have a pivotal role in other P-proteins of such viruses. Geyer et al. (1992) analyzed the carbohydrate residue of the Marburg glycoprotein stated that the glycoprotein of EBOV (Feldmann et al. 1991) and other filoviridae contain a large amount of N-linked and Olinked glycans (Geyer et al. 1992) while, Feldmann and colleagues in 1991–93, deduced the same in their lectin binding studies (Feldmann et al. 1993).

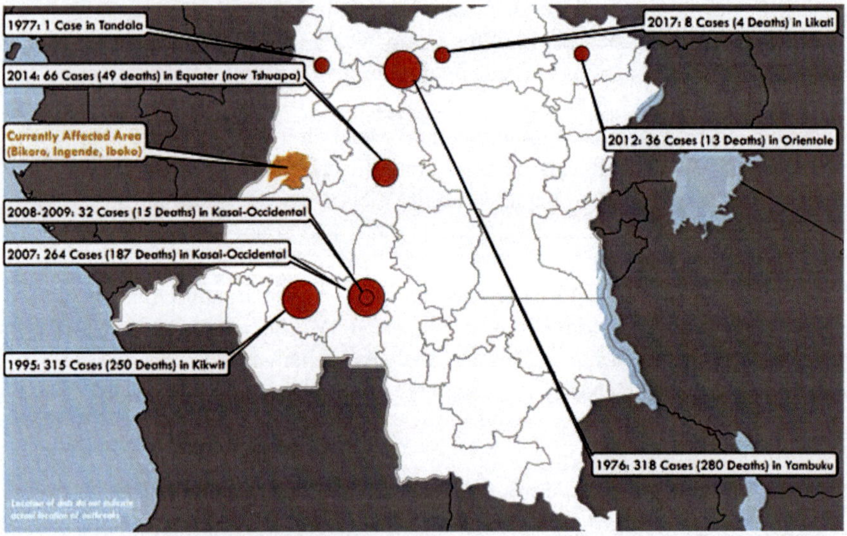

Fig. 1 Pictorial representation of the past outbreaks of EVD in DRC

Drug discovery approaches have developed commendable accomplishments in antiviral medications. Studies based on System biology have aided virtuously in the recognition of new antiviral nominees. According to Cheng et al. (Cheng et al. 2015), an interspersed approach is required to discern new antiviral competitors for the existing medicines by analyzing drug-gene patterns into the virus-host interactome. Using this analogy, they have discerned three active drugs, Ajmaline, Piroxicam, and Azlocillin, as the new drug candidates for anti-Ebola virus treatment. Moreover, they have also suggested four the doohickey representing how these drugs can inhibit the virus, thus blocking the infection (Khan et al. 2017).

1.2 How does it spread? It is suggested that Ebola infection spreads only through direct contact between people, either with blood or any other body fluids of an infected person to a healthy one (Drazen et al. 2014; CDC 2014). Saliva, mucus, puke, feculent, sweat, breast milk, tears, urine, and semen are the bodily fluids that are the transmission cause of the infection from an infected individual to a healthy one (CDC 2019a, b). According to the World Health Organization (WHO) (WHO 2018a, b, c, d, e), individuals who are severely infected are the main vectors of disease spread via saliva, blood, defecation and vomit (McNeil 2014). The access points for EBOV are the nose, mouth, eyes, open wounds and lesions (CDC 2019a, b).

Most viral infection is usually spread through sneezing and coughing in which large amount of infectious droplets that comes out remains in the environment. Ebola virus too spreads through these large infectious droplets that are discharged when an infected person sneezes or coughs (CDC 2019a, b). Contaminated syringes and needles are other culprits behind the transmission of the virus (Chowell and Nishiura 2014). In a dry state, the EBOV can survive for hours, even days within the bodily fluids of an individual. The virus can also survive on objects for a few hours or for a few days in a dried state.

1.3 Signs and Symptoms: Once a person is infected, the symptoms start to develop between two to three weeks such as fever, sore throat, muscle pain, headaches, vomiting, diarrhea, rashes, reduced functioning of the liver and kidneys, internal and external bleeding. The coalescence of relentless puking and diarrhea generally leads to obdurate dehydration in patients, which usually turns out to be catastrophic (Sharma and Cappell 2015). Further symptoms such as shortness of breath and pain in the chest may also develop along with swelling, inconvenience, and disorientation (Alan 2013). In almost half of the cases, the skin of the infected patient may foster a maculopapular rash viz. a flat red area covered with little confluent bumps occurring five to seven days after the major symptoms begin (Goeijenbier et al. 2014; Hoenen et al. 2006; Osterholm et al. 2015).

A decreased blood clotting is the common observation in people infected with EBOV (Hoenen et al. 2006). In almost 40–50% of cases bleeding occurs from the mucous membranes or from sites of needle punctures (King 2012) which may be the reason behind the presence of blood in puke, cough or in feculence (Ministry of Health and Long-Term Care 2019). Skin bleeding may result in development of petechiae, purpura, hematomas (Osterholm et al. 2015) moreover bleeding of eyes and gastrointestinal tract also occurs (Shantha et al. 2016; West and von Saint André-von Arnim 2014). Ebola has been tagged as the mortal zoonosis disease killing about 50% of infected people. Mortality is often observed due to low blood pressure because of reduced fluid content.

2 Harking Down the Past Lanes, Recapitulation About Ebola Virus Disease

To understand the origin of outbreaks EVD, its studies become mandatory; its cataclysm started in the Sub-Saharan region of Africa. It was first identified in 1976 and through that time till 2013, the WHO confirmed almost 1,716 cases of people affected by EVD (WHO 2018a, b, c, d, e). The highest number of epidemics of EVD till date was in West Africa, which caused a large number of mortalities in Guinea, Sierra Leone, and Liberia (CDC 2018a, b, c).

2.1 History of EVD, epidemics during 1976–2018: The itinerary of the course of outbreaks starts from the year of 1976. Two consecutive epidemics were observed, namely—Sudan and Zaire outbreaks (Baize et al., 2014; Yozwiak et al. 2016). The first-ever known spread of EVD determined was between June and November 1976 in Nzara, South Sudan (Hoenen et al. 2012) caused due to Sudan virus (SUDV) and the first EVD case was of a storekeeper in a cotton factory from whom slow and steady spread killed 151 people and infected about 284 (Feldmann and Geisbert 2011). It was 26 August 1976, the second outbreak of EVD initiated in Yambuku, a small rural village in the Mongala District in northern Zaire (the Democratic Republic of the Congo) (Barry and Bonnie 2007) caused by Zaire Ebolavirus, a different member of the genus Ebolavirus. The first person infected with the disease was the village school's pedagogue named Mabalo Lokela who died 14 days after he began displaying symptoms (CDC 1995). The second biggest epidemic in Zaire occurred in 1995, killing 254 and infecting almost 315 people. In 2000, SUDV was again the culprit in the epidemic observed in Uganda affecting 425 and taking the lives of almost 224 (WHO 2018a, b, c, d, e). Back then in the year 2003, a widespread epidemic in the Republic of the Congo latched 143 and had a mortality rate of almost 90% (Formenty et al. 2003). A Russian boffin tasted death in 2004 after injecting her with a contaminated Ebola needle (Miller 2004).

In 2007, a febrile outbreak due to EVD was confirmed in four regions of the Democratic Republic of the Congo, killing 187 and infecting 264 individuals (CBC 2007). Furthermore, the Uganda Ministry of Health also confirmed the Ebola invasion in Bundibugyo district; this strain was named as—Bundibugyo (UN News Service 2007). Two small epidemics caused by SUDV strain were again observed in Uganda in the year 2012 which were confirmed by the WHO, which initially infected 7 and the death toll was 4 people, while, the second outbreak had a death toll of 17 people (WHO 2018a, b, c, d, e). The same year, the Ministry of Health of the Democratic Republic of the Congo (DRC) reported an outbreak of the Bundibugyo Ebola variant (VOA 2012) in the eastern region (WHO 2012).

The genome sequencing proved that the 7th, 2014 Ebola outbreak of (DRC) (Holmes et al. 2016) was not related to the 2014–15 which occurred in West Africa Zaire Ebola strain (WHO 2014a, b). In 2015, Guinea was declared free of Ebola transmission after the last person was tested negative against Ebola (Dudas and Rambaut, 2014; CBC 2015). Then in 2016, a new case was identified in Sierra Leone (Subissi et al. 2018; WHO 2016). On 11 May 2017, the DRC Ministry of Public Health notified the WHO about an outbreak of Ebola. Four people died, and other four survived; five of these eight cases were laboratory-confirmed. A total of 583 contacts was monitored. On 2 July 2017, the WHO declared the end of the outbreak. In 2018, the WHO stated that 39 people confirmed of Ebola in DRC, including death toll of 19 (WHO 2017) and this spread revolved around regions of Bikoro, Iboko, and Wangata areas in the Equateur region including Mbandaka. The 2018 outbreak in the DRC was discerned as over on July 24, 2018 (WHO 2018a, b, c, d, e). Apart from this attack, the 10th Ebola virus epidemic was noticed in North Kivu, which was centered in a military conflict area constituting of many refugees (Beaubien 2018; Specia 2018).

3 Factors: Reservoir of Ebola Zoonotic Viruses

Natural repository yet to be confirmed, but fruit bats are hypothesised to be responsible because of the first cases of EVD in 1976–79 and also in Marburg virus spread in 1975 and 1980 involved this bat (Chowell and Nishiura 2014; Pourrut et al. 2005). The three types of fruit bats viz.,—Hypsignathusmonstrosus, Epomopsfranqueti and Myonycteris Torquata have been observed to play a vital role in the EVD transmission. These bat species actually are like a 'catalyst' where they carry the virus and spread it, without themselves being affected or sick (Laupland and Valliquette 2014). Other sources like plants, arthropods, rodents, and birds are the possible repositories of this virus and its dispersal (Sharma and Cappell 2015).

3.1 Pathogenesis: The zoonotic EBOV multiplies rapidly in cells leading to the production of humongous concentrations of virus in monocytes, macrophages, liver cells, fibroblasts, dendrites, adrenal glands (Ansari 2014) where its multiplication activates inflammatory signals leading to a septic state in the patients affected (Tosh and Sampath kumar 2014). Once an individual gets infected, the endothelial cells (usually within three days of infection) (Chippaux 2014), liver cells and other cells are the first sites of attack, which is then followed by immune cells, which carry this EBOV to lymph nodes where the virus grows and multiplies further. Once the virus is established, it shifts from the lymph nodes to the main bloodstream of the patient and eventually is omnipresent in the entire body (Funk and Kumar 2015). The first group of immune cells, affected in the body, is the macrophages (Chippaux 2014) and cells like the white blood cells (WBC) which ultimately results in apoptosis (programmed cell death) leading to weakening of the immunity of the individual (Funk and Kumar 2015).

The EBOV glycoprotein (GP) can be tagged as the janitor of the disruption of the endothelial cells along with the breakdown of the blood vessels, reducing the accessibility of specific integrins, responsible for cell attachment to the intracellular structure. Their generation leads to liver damage and improper blood clotting, causing swelling (Smith 2005) and this excessive bleeding and clotting increase the activation of the external pathway of the coagulation cascade occurring due to extra tissue factor generation by macrophages and monocytes (Goeijenbier et al. 2014).

Once the virus has infected an individual, a small-sized, soluble GP is developed which forms a dimeric complex and disturbs the signalling pathway of neutrophils, a type of WBC which in turn helps the virus to attack the immune system. EBOV multiplication overpowers protein generation of infected cells and the host immunity defenses. The GP forms a trimeric complex, which unleashes the virus to the endothelial cells (Goeijenbier et al. 2014).

3.2 Transmission of EBOV: As mentioned above as to how does Ebola spreads, viz. by direct contact with the blood or other body fluids of an infected person. The virus is able to exist for more than 3 months in the semen (Soka et al. 2016) leading to infections via sexual intimacy. It also occurs in the colostrum (breast milk) of women after recovery, which is actually not sure when it is safe to breastfeed again (CDC 2018a, b, c). The capacity for the epidemic of such infections is low, where and in such situations, tele-based health care is usually preferred over the traditional practice (Qazi et al. 2019). Carcasses of infected people tend to remain contagious and it becomes very tedious for people to bury such bodies in accordance with traditional rituals and practices (CDC 2014) as the chance of healthy people becoming infected is very high. Furthermore, clinicians, staff and nurses, who are in direct contact with the infected person, are at a higher risk of getting themselves contaminated with the EBOV (CDC 2019a, b). This occurs mainly when they are not wearing appropriate ocular gears, masks, gown, clothing or do not handle these according to instructions. Airborne transmission of EVD is only observed in contaminated milieu transmission or animal-to-human but not in direct human-to-human contact (Funk and Kumar 2015). Transmission of the virus by water/food apart from bush meat has not been observed; also spread by mosquitoes or other insects have not been observed (Ansari 2014). Other possible methods for EVD transmission are being studied (Osterholm et al. 2015).

4 Pathophysiology of EBOV: Immune System Under Attack!

EBOV zoonotic virus also disturbs the appropriate functioning of the innate immune system (Misasi and Sullivan 2015). Its proteins disturb the immune system's reaction in favor of viral infections by meddling with the cells' potential to develop and react to interferon proteins, interferon-alpha, interferon-beta, and interferon gamma. The structural proteins of EBOV, VP24 and VP35, have a vital role playing in this disturbance and interference. The receptors presence in the cytoplasm such as—RIG-I and MDA5, or receptors on the outside of the cytosol like Toll-like receptor3 (TLR), TLR7, TLR8, and TLR9 identify the infectious molecules which are affiliated with the EBOV whenever a healthy cell gets infected with the virus (Kühl and Pöhlmann 2012). Once the TLR get triggered positive, IRF3 (Interferon regulatory factor) and IRF7 activate a signalling avalanche which exhibits the expression of type 1 interferons (Kühl and Pöhlmann 2012). The IFNAR1 and IFNAR2 get attached to the type 1 interferons, usually expressed on the surface of their neighboring cells and the interferon binds to its receptors on the neighboring cells, thus the signalling proteins STAT1 and STAT2 are triggered positively and shift to the cell's nucleus (Kühl and Pöhlmann 2012) which activates the expression of interferon-activated genes, coding for proteins with antiviral properties. The structural protein of EBOV V24 inhibits the generation of these antiviral proteins by stopping the STAT1 to enter in the neighboring cell via the nucleus (Kühl and Pöhlmann 2012). The VP35 protein directly blocks the generation of interferon-beta (Ramanan et al. 2011). The EBOV can rapidly spread the entire body if these immune responses are blocked (WHO 2014a, b).

5 Diagnosis and Prognosis

The diagnosis is usually carried out at two-level categories: (a) Laboratory Examinations and (b) Differential Diagnosis respectively. Both these are done consecutively [Sichtig, FDA; Carneiro and Pereira, 2016; Swetha et al. 2015; Whitmer et al. 2016].

(a) **Laboratory Examinations**: Laboratory examinations are composed of tests for low platelet count, decreased and increased WBC count, high levels of liver enzymes- Alanine aminotransferase (ALT) and Aspartate aminotransferase (AST), and difficulty in blood clotting often regular with disseminated intravascular coagulation (DIC) such as a prolonged prothrombin time, partial thromboplastin time, and bleeding time (Kortepeter et al. 2011). Electron microscopy can also be very helpful in identification of the unique EBOV structures in cell culture (Goldsmith and Miller 2009). The main procedure for diagnosing EBOV is done by isolating the virus, identifying its RNA or proteins, or rectifying antibodies against the virus in a patient's blood. Isolate the virus by cell culture, observing the viral RNA by real-time polymerase chain reaction (RT-PCR) (Goeijenbier et al. 2014) and detecting proteins by Enzyme-Linked Immunosorbent Assay (ELISA) are the two best methods employed in the initial stages of the zoonotic disease. Rectifying antibodies against the virus is better in the later stages of the disease and in those persons who recovered (CDC 2018a, b, c). IgM antibodies are perceptible only two days after symptom initiation and IgG antibodies can be observed 6–18 days after symptom occurrence (Goeijenbier et al. 2014). Isolation of the virus is not possible during the epidemic of EVD but, in emergency cases, mobile hospitals can come handy which facilitate in procedures such as RT-PCR and ELISA (Grolla et al. 2005). Results soon retrieved with the help of newly deployed diagnostic machinery in Libya after the 2014 outbreak (WHO 2014a, b). The WHO approved a fast antigen examination diagnosis back in 2015 which produces results in just 15 min and is able to attest EVD in 92% infected patients (WHO 2018a, b, c, d, e).

(b) **Differential diagnosis**: The initial signs and symptoms of Ebola are quite similar to malaria and dengue (Gatherer 2014), also to those of Marburg virus disease, Lassa fever, and Crimean Congo hemorrhagic fever accordingly (Beeching et al. 2014). Differential diagnosis is tedious and involves the consideration of myriad infectious diseases such as—typhoid fever, cholera, sepsis, plague, Q fever, trypanosomiasis, measles, shigellosis, leishmaniasis, etc. (Dulaurier et al. 2016). There are non-infectious/contagious diseases as well, which may end up signs similar to those of the zoonotic disease (Grolla et al. 2005; Gear 1989; Bogomolov 1998; Gear et al. 1978). For instance, acute promyelocytic leukemia, hemolytic uremic syndrome, blood clotting deficiencies, thrombotic thrombocytopenic purpura, Kawasaki disease, and warfarin poisoning.

Prognosis: Ebola has had a high mortality rate (~25–90%) in its past outbreaks (Murphy, WHO). The average risk of mortality among those infected was 50% back in 2014. The highest risk of death was about 90% during 2002–2003 in DRC (BBC 2014). Due to low blood pressure and dehydration, a person can die soon after the symptoms start to appear (Ruzek 2014). In order to prevent mortality in patients, the only best and quick way is to re-hydrate them. Once the patient survives, his/her recovery can be rapid. Patients who are severely affected with the virus have a longer course of treatment and are very tedious to treat them (Goeijenbier et al. 2014; Shantha et al. 2017). Swelling of the uveal layer of the eye is the most common ocular complexity in survivors of Ebola (Shantha et al. 2017). Ocular signs like photosensitivity, over tearing, and eyesight loss have been observed (Wiwanitkit and Wiwanitkit 2015). The Ebola virus can stay in certain body parts such as the eyes, breasts, and testicles after infection (Wiwanitkit and Wiwanitkit 2015; Varkey et al. 2015; Mackay and Arden 2015). Furthermore, transmission of the virus has also been discerned after a sexual intimacy between the infected and a normal person (Rogstad and Tunbridge 2015). The Ebola outbreak study in Sierra Leone showed around 44 survivors which had musculoskeletal ache (70%), headache (48%) and ocular related issues (14%) (Scott et al. 2016).

6 Ebola Virus Genome: Explanation and Comparison

6.1 Organization and genetic elements: When a genome-wide screening was conducted to identify differentially expressed RNA during replication and transcription of long non-coding RNAs (lncRNAs), circular RNAs (circRNAs), microRNAs (miRNAs) and mRNAs using a tetracistronic transcription and replication component virus-like particle (trVLP) system which models the life cycle of EBOV in 293T cells (Duy et al. 2018). The result shows that CLDN18 gene and cirRNA19 are potentially important in EBOV pathogenesis (Wang et al. 2017).

For the identification of siRNA activity in EBOV genome replication and transcription, a genome-wide siRNA screening carried out consisting of 64,755 individual siRNA against 21,566 human genes (Martin et al. 2018). Results show that de novo pyramiding synthesis pathway has antiviral activity against EBOV infection as well as other nonsegmented negative-sense RNA viruses (Martin et al. 2018).

A gene, Niemann-Pick C1 (NPC1), works as a promoter for the entry of Marburg and EBOV into the cytoplasm and when these viruses have been internalized, it works as an intraluminal receptor (Muhlberger et al. 1999). Since blocking the virus entry into the target cell leads to suppression of viral infection and it would be an attractive antiviral strategy. Thus, studies conducted on small molecule inhibitors with NPC1 to inhibit the interaction of EBOV and glycoprotein which mediates viral entry into the cell showed that NPC1 protein shares sequence homology with tumor suppressor i.e. protein patched homolog 1 and 2 (Wang et al. 2017). The NPC1 showed differentially expressed in different cells, and oocytes express it far more than the other cells. Therefore, it is recommended that there is a need to check for embryonic and oocyte toxicity while exploring the inhibitors to inhibit NPC1 as well to explore the effect of NPC1 inhibitors on cancer (Cherian and Kuppanan 2018).

Certain structural proteins play important role in attachment, entry, stability, gene expression and pathogenesis of EBOV within the host (Shawan et al. 2015), and various studies show that glycoprotein and VP40 are stimulus that leads to the activation of EBOV infection (Ning et al. 2018; Wahl et al. 2011). Monocytes and Macrophages are the first targets of EBOV which leads to the release of proinflammatory cytokines and chemokines. Chemokines are the most regulated genetically regulated protein at EBOV infection (Tong et al. 2015) . Host proteins such as Tsg101 and Nedd4 physically and functionally interacts with EBOV VP40 and lead to the release of VLPs (virus-like particles) and infectious virus results to TLR4 which is signalling pathway in host causes production of proinflammatory cytokines (Okumura et al. 2015) as shown in Fig. 2.

On studying the host gene expression, infected by EBOV, it was found that genes are differentially expressed within the initial hour of infection (Menicucci et al. 2017; Wahl et al. 2011). In the EBOV virus pathogenesis, there is a major role of deregulation of normal host immune response (consisting of destruction in lymphocytes, increase in circulating cytokine level and development of coagulation abnormalities) (Yen et al. 2010). Due to small RNA genome, EBOV requires host factors for replication for pathogenesis (Cilloniz et al. 2011; Belyi et al. 2010;

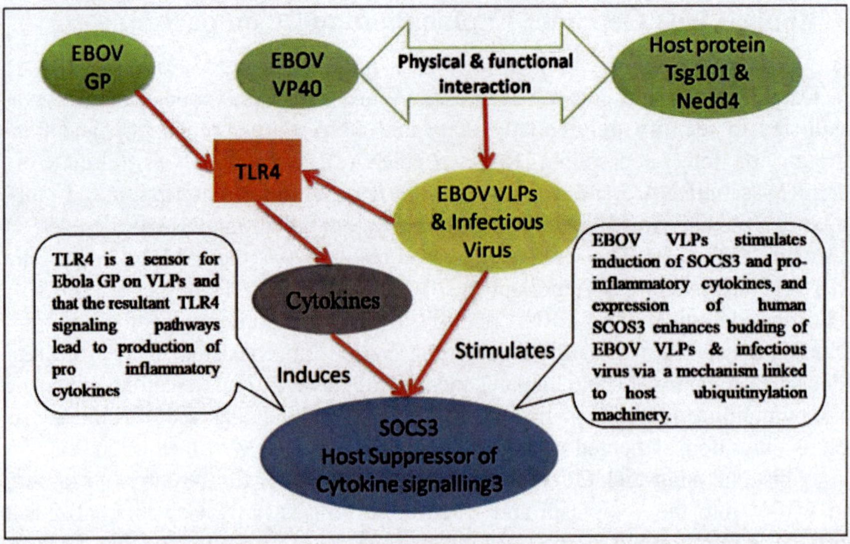

Fig. 2 Diagrammatic representation of the function of VP40, GP in EBOV pathogenesis

Bosworth et al. 2017). Polyamine synthesis is one of the factors which is essential for viral replication/infection as shown in Fig. 3.

Spermidine is a polyamine which is important of hypusination of eukaryotic initiation factor (EIF5A). EBOV gene expression and replication can be inhibited by blocking the hypusination of EIF5A. Interaction between EIF5A and VP30 can help to fight with EBOV (Olsen et al. 2016).

Importance of non-coding region in EBOV was discovered through a highly complex interaction between different cis-acting elements that modulate transcription and shows a specific combination of IR (intergenic regions) and UTR

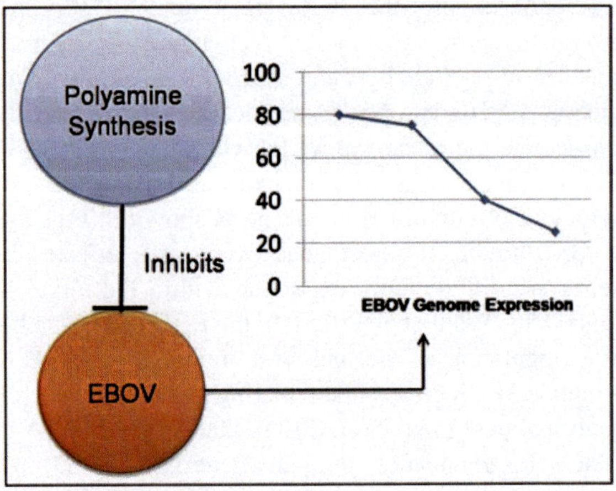

Fig. 3 At the inhibition of polyamine synthesis, the EBOV gene expression is decreased

(Brauburger et al. 2016). According to transcriptomic analysis of the host response to pathogen infection, there is a strong upregulation interferon signalling and acute phase response in patients who died compared with the survivor (Liu et al. 2017).

There is a profound effect of EBOV IRADs (innate response antagonizing domains) on the host immune adaptive response through huge transcriptional down-regulation of DC (dendritic cell). The combined effect of VP35 VP24, IRAD, and change in dendritic cell transcriptome associated with an individual prevents the maturation of dendritic cell which leads to T cell response less effective as shown in Fig. 4. VP35 protein blocks the expression of DC maturation markers by targeting the identified suppressive pathway which could result in functional adaptive immune response; that leads to the possibility of treatment of EBOV infection (Ilinykh et al. 2015).

Transcription pattern of EBOV and Marburg Virus was analyzed using NGS (next generation sequencing) method in two types of cells i.e. Huh7 (human liver cell) and Mpg (human macrophages) infected with four wild type EBOV: Ebola (EBOV), Sudan (SUDV), Bundibugyo (BDBV) and Reston (RSTV). The NGS study showed a similar type of transcription pattern with a minor difference in the two types of cell among these four viruses. The expression level of genes is like—NP, VP35, VP40 and GP mRNAs, highest level of expression was in VP30, in VP24 an intermediate level and gene L (RNA dependent RNA polymerase) showed lower levels of expression. It was found that the pattern of all the gene expression was almost similar in both types of cell lines Huh7 and Mpg which are infected by EBOV but there was a small increase in expression level of gene VP30 and VP24 in case of Mpg cell line.

NGS was applied in Marburg virus to analyze the transcription pattern and its recombination. In the human Mpg cell infected with Marburg virus showed that rMARV/GFP had reduced growth in comparison with wild type, rMARV. The reduced expression of downstream genes and expression of GFP into the NP-VP35

Fig. 4 Different effects of VP24 and VP35 on IRAD

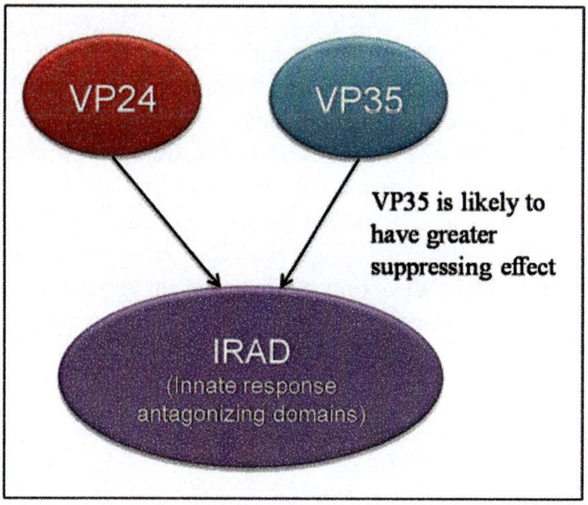

intergenic region because of the insertion of an additional transcription unit (ATU) would be the reason behind this phenotype. It was thought that a decrease in the level of viral protein with an innate immune inflammatory function like VP30 and VP40 and perturbation in the natural order of transcription was due to the insertion of an additional transcription unit. Through the use of genomic design, similar to other MARV viruses a new recombinant MARV expressing ZsGreen (rMARV/ZsG) was made to correct the abnormal phenotype of the recombinant MARV expressing a reporter gene. Encoded proteins are released through the action of P2A by fusing a reporter gene to the self-cleaving peptide P2A and to NP. The growth level of rMARV/ZsG was higher than the rMARV/GFP but lower than the wild type Rmarv and observed through the analysis of growth kinetics of three recombinant viruses (Albarino et al. 2018).

7 Computational Genome Sequencing Solves EBOLA Mystery

Various genetic changes in EBOV result in each EVD outbreak. Thus, genome sequencing can play a major role in pathogen determination and its transmission dynamics in order to develop an accurate diagnosis (Gire et al. 2014). For the detection of pathogen and biosurveillance studies pathogen, specific ELISA and Real-time PCR, are the techniques that are commonly used. Since Real-time PCR is specific and highly sensitive, screening of multiple samples using Real-time PCR for an unknown pathogen would be time-consuming and expensive too (Seifert et al. 2018). It also has small sample volume, thus it is limited to broad pathogen screening studies. For this, a multiplexed real-time PCR assay is needed that can allow multiple targets to be assessed in a single sample. But this multiplexed technique was undeveloped as there are various issues like inherent specificity for the targeted agents, the number of discrete dyes for separating multiplexed result and limited number of assays which was capable of being multiplexed (Quick et al. 2016).

NGS is an appropriate technique for screening a large number of pathogens and it can identify both types of pathogens that would be known or unknown from both clinical and environmental samples. It can be applied to RNA or DNA and generates a sequence that is opposite to pathogen sequence and these sequencing data are non-target metagenomic sequence or just like an exclusively host sequence (Koehler et al. 2014). The major advantage of this technique is that there is no limitation of target sequences. Viral transmission and evolution of genes can be understood by detection of mutation or variation in the genome which is easily determined through full genome sequencing using NGS (Brigid et al. 2018; Jabeen et al. 2019).

8 Discussion, Conclusion, and Research Directions

Ebola virus disease (EVD) is one of the most catastrophic forms of zoonotic infection, which can take a pernicious form if not diagnosed at the earliest and not treated appropriately. Researchers around the globe have been on their tenterhooks and working out hard in order to decode its molecular mechanism and trying to develop specific personalized therapeutics which can be helpful in treating the disease. EVD is a cataclysmic infectious disease to human health worldwide caused by Ebola Virus (EBOV) which belongs to the family Filoviridae, and is encased zoonosis virus comprising of 18.8–19 kB negative-sense single-stranded RNA which codes for 7 proteins. It also has a surface glycoprotein for executing the specific roles of—attachment, fusion and cell entry, for human immune attack and pathogenesis of disease accordingly. Fruit bats are thought to be the primary suspects of spreading and carrying the zoonotic infection of Ebola, especially—Hypsignathus Monstrosus, Epomopsfranqueti, and Myonycteris Torquata. Other possible sources of zoonotic infection spread are—plants, arthropods, rodents, and aves respectively. The spread of zoonotic infection occurs from bats to humans, but the dynamical mechanism of the lethal EBOV can be understood by factors such as —Biological Factors, Environmental Factors, and Social Factors, and therefore, the infectious interaction between bats and the ecosystem is considered the janitor for the outbreak. There are many ways through which Ebola infection can spread in a healthy individual, such as, saliva, mucus, puke, feculent, sweat, breast milk, tears, urine, and semen are the bodily fluids.

As per the WHO, people who are majorly affected are the main transmission initiators of the disease outbreak mainly due to saliva, blood, defecation, and puke. The entry points for EBOV are the nose, mouth, eyes, abrasions, and lesions. The EBOV has the most intriguing path of multiplication as it multiplies in the cells progressing to the generation of huge amounts of virus in monocytes, macrophages, liver cells, fibroblasts, dendrites, adrenal glands. The macrophages, WBC are the cells which are initially attacked and harmed, which in turn, results in apoptosis, thus reducing the immunity of the individual. This division of EBOV triggers the inflammatory signals which in turn move ahead to the septic state in the individual accordingly.

When a person gets affected by EBOV, in the first three days of infection, the endothelial cells, liver cells, and other cells are damaged followed by the immune cells carrying the EBOV infection to the lymph nodes, the site where the virus breeds further. From the lymph nodes the virus shifts to the main bloodstream of an individual and thus is ubiquitous in the entire body of the individual. Various studies have shown that CLDN18 gene and cirRNA19 are potentially mandatory in EBOV zoonotic transmission and pathogenesis. Determining the activity of siRNA in EBOV genome multiplication and transcription analysis, it was discerned that de novo pyrimidine synthesis pathway has antiviral activity against EBOV infection as well as some other non-segmented negative-sense RNA viruses. Niemann—Pick C1 (NPC1), acts as a promoter for the accession of Marburg and EBOLA into the cytosol, which then works as an intraluminal receptor in an individual. Direct inhibition of the virus accession in the target cell progresses to curbing of the viral infection, which can be employed as a fantastic antiviral strategy for the same. Studies have proved that NPC1 protein shares sequence homology with tumor suppressor protein patched homolog 1 and 2. Thus, it is suggested by researchers and medical fraternity to check for embryonic and oocyte toxicity when analyzing the inhibitors to halt the activity of NPC1. Furthermore, there is a need to explore the effect of NPC1 inhibitors on cancer as well.

There is an apex clinical challenge in the treatment of EVD as there is no approved therapeutics, but some of the diagnostic methods such as Enzyme-Linked Immunosorbent Assay (ELISA) and Reverse Real Time-Polymerase Chain Reaction (rRT-PCR) are commonly employed. Additionally, a number of vaccines and antiviral drugs are under the development against EVD which are still under the tedious and time taking processes of the clinical trials. The only lack in the development of specific diagnosis and prognosis is the employment of rational, robust super-powered technological strategies. Our century has seen a drastic revolution in terms of scientific experimentation, analysis, and interpretation. Next-Generation Sequencing (NGS) (Jabeen et al. 2016; Raza and Ahmad 2019) and Nanopore sequencing (Raza and Qazi 2019) is among the revolutionized techniques, which have proved to successfully provide novel solutions for the identification of viruses for both—simultaneous and unbiased examinations for different infectious agents. One of the biggest advantages of it is that many samples can be multiplexed in a single sequencing procedure, giving an edge to the computational analysis from the traditional ones. Moreover, the introduction and generation of real-time sequencing-based platforms, have aided in rapid processing and interpretation of individualization within a short period of time. The super-powered technique is applicable to both RNA and DNA, thus, making it an unbiased approach (Jabeen et al. 2018). The basic modus operandi of NGS is simple and it takes into account sequences which are opposite to the zoonotic pathogen sequences so that a comparative study can be executed without any bias. It is also helpful in detecting mutations, alterations in the entire genome. Further, evolution of genes and pathogenesis of the virus can also be understood by employing NGS studies. There are still some loopholes in the computational analysis approaches for EVD, but researchers are working hard to increase the sensitivity of NGS-employed diagnostic studies by focusing on the enhancement of the virus and betterment of the sequencing libraries, which is inclusive of amplicon sequencing, PCR- produced baits, and some capture techniques accordingly. Recently, Nanopore Sequencing has been very successful in studying virus outbreaks due to its several important features such as small-size, portability, high throughput, decentralized sequencing, low-cost and can be used on-site to generate data in real-time and suitable for genomic surveillance of epidemics such as Zika and Ebola (Quick et al. 2016; Hoenen et al. 2016; Raza and Qazi 2019). It has been discerned that during infectious epidemics where traditional PCR studies are seemingly impossible to execute, NGS and Nanopore sequencing approaches can be easily employed, which can rapidly process information about the pathogenic agent and control management and measures.

The future holds a bright Anschluss for both the wet-lab traditional experimentation and computational intelligence. To deal with infectious lethal epidemics, such as EVD, it is high time to collaborate with the new technologies in order to learn and understand the lethal and the catastrophic mechanism of such pathogens. The amalgamations of traditional and novel technologies are the best ingredients for curbing and protecting our world.

Acknowledgements Sahar Qazi is supported by DST-INSPIRE fellowship by the Department of Science and Technology, Government of India.

Conflict of Interest There is no conflict of interest.

References

Alan M. Hunter's tropical medicine and emerging infectious diseases. 9th ed. New York: Saunders; 2013. p. 332.

Albarino CG, Wiggleton Guerrero L, Chakrabarti AK, et al. Transcriptional analysis of viral mRNAs reveals common transcription patterns in cells infected by five different filoviruses. PLos One. 2018;13(8):e0201827.

Ansari AA. Clinical features and pathobiology of Ebolavirus infection. J Autoimmun. 2014;55:1–9.

Baize S, Pannetier D, Oestereich L, et al. Emergence of Zaire Ebola virus disease in Guinea. N Engl J Med. 2014;371:14181–425. https://doi.org/10.1056/NEJM1404505.

Barry H, Bonnie H. Ebola, culture and politics: the anthropology of an emerging disease. Cengage Learning; 2007. p. 103.

BBC. How deadly is Ebola. 2014. https://www.bbc.co.uk/programmes/p0244jdt.

Beaubien J. Ebola in a conflict zone. National Public Radio. 2018. https://www.npr.org/sections/goatsandsoda/2018/08/02/635016034/ebola-in-a-conflictzone.

Beeching NJ, Fenech M, Houlihan CF. Ebola virus disease. BMJ. 2014;10(349):g7438.

Belyi VA, Levine AJ, Skalka AM, et al. Unexpected inheritance: multiple integration of ancient Bornavirus and Ebola virus/Marburg virus sequences in vertebrate. PLos Pathog. 2010;6(7): e1001030.

Bogomolov BP. Differential diagnosis of infectious diseases with hemorrhagic syndrome. Terapevticheskii Arkhiv. 1998;70(4):63–8.

Bosworth A, Dowall SD, Garcia-Dorival I, et al. A comparison of host gene expression signatures associated with infection in vitro by the Makona and Ecran (Mayinga) variants of Ebola virus. Sci Rep. 2017;7:43144.

Brauburger K, Boehmann Y, Krahling V, et al. Transcriptional regulation in Ebola virus: effect of gene border structure and regulatory elements on gene expression and polymerase scanning behavior. J Virol. 2016;90(4):1898–909.

Brigid MO, Li Y, Yang T, et al. Comprehensive viral enrichment enables sensitive respiratory virus genomic identification and analysis by next generation sequencing. Genome Res. 2018;28:869–77.

Carneiro J, Pereira F. EbolaID: an online database of informative genomic regions for Ebola identification and treatment. PLos Negl Trop Dis. 2016;10(7):e004757.

CBC News. Ebola gone from Guinea. CBC News. 2015. https://www.cbc.ca/news/health/ebola-guinea-1.3382945.

CBC News. Ebola Outbreak in Congo. CBC. 2007. https://www.cbc.ca/news/world/ebolaoutbreak-in-congo-1.648593.

CDC. 2014. Ebola (Ebola Virus Disease) Transmission. Archived from the original on 6 Nov 2014.

CDC. 2014–2016 Ebola Outbreak in West Africa. 2018a. https://www.cdc.gov/vhf/ebola/history/2014-2016-outbreak/index.html CDC.

CDC. Ebola: Breastfeeding infant feeding Ebola. 2018b. https://www.cdc.gov/vhf/ebola/hcp/recommendations-breastfeeding-infant-feedingebola.html.

CDC. EVD: Diagnosis. 2018c. https://www.cdc.gov/vhf/ebola/diagnosis/index.html.

CDC Telebriefing on Ebola outbreak in West Africa. 2014. https://www.cdc.gov/media/releases/2014/t0728-ebola.html.

CDC. Ebola: Ebola Virus Disease: Transmission. 2019a. https://www.cdc.gov/vhf/ebola/transmission/index.html.

CDC. EVD: Transmission. 2019b. https://www.cdc.gov/vhf/ebola/transmission/.

Centers for Disease Control Prevention (CDC). Outbreak of Ebola Viral Hemorrhagic Fever—Zaire, 1995. Morb Mortal Wkly Rep. 1995;44(19):381–2.

Cheng H, Lear-Rooney CM, et al. Inhibition of Ebola and Marburg virus entry by G protein coupled receptor antagonists. J Virol. 2015;89(19):9932–8.

Cherian PV, Kuppanan S. Predicted side effects of NPC1 protein inhibitors. Int J Life Sci. 2018; A11:73–6 (special issue).

Chippaux JP. Outbreaks of Ebola virus disease in Africa: the beginnings of a tragic saga. J Venom Anim Toxins Incl Trop Dis. 2014;20:44.

Chowell G, Nishiura H. Transmission dynamics and control of Ebola virus disease (EVD): a review. BMC Med. 2014;12:1–196.

Cilloniz C, Ebihara H, Ni C, et al. Functional genomics reveals the induction of inflammatory response and metalloproteinase gene expression during lethal Ebola virus infection. J Virol. 2011;85(17):9060–8.

Drazen JM, Kanapathipillai R, et al. Ebola and quarantine. N Engl J Med. 2014;371(21):2029–30.

Dudas G, Rambaut A. Phylogenetic analysis of Guinea 2014 EBOV Ebola virus outbreak. PLos Curr Outbreaks. 2014, 1st ed. https://doi.org/10.1371/currents.Outbreaks.84eefe5ce43ec9dc0bf0670f7b8b417d.

Dulaurier M, Moyer K, Wallihan R. Ebola virus disease: epidemiology, clinical presentation, and diagnostic and therapeutic modalities. 2016; EB Medicine.

Duy J, Honko AN, Altamura LA, et al. Virus encoded miRNAs in Ebola virus disease. Sci Rep. 2018;8.

Feldmann H, Geisbert TW. Ebola hemorraghic fever. Lancet. 2011;377(9768):849–62.

Feldmann H, Klenk, HD, Sanchez A. Molecular biology and evolution of filoviruses. Arch Virol. 1993 (in Press).

Feldmann H, Miihlberger E, et al. Marburg virus, a filovirus: messenger RNAs, gene order, and regulatory elements of the replication cycle. Virus Res. 1992;24:1–19.

Feldmann H, Will C, et al. Glycosylation and oligomerization of the spike protein of Marburg virus. Virology. 1991;182:353–6.

Formenty P, Libama F, et al. Outbreak of Ebola hemorrhagic fever in the Republic of the Congo, 2003: a new strategy?. Med Trop (Mars). 2003;63(3):291–5 (in French).

Funk DJ, Kumar A. Ebola virus disease: an update for anesthesiologists and intensivists. Can J Anaesth. 2015;62(1):80–91.

Gatherer D. The 2014 Ebola virus disease outbreak in West Africa. J Gen Virol. 2014;95(8):1619–24.

Gear JH. Clinical aspects of African viral hemorrhagic fevers. Rev Infect Dis. 1989;11(4):S777–82.

Gear JH, Ryan J, Rossouw E. A consideration of the diagnosis of dangerous infectious fevers in South Africa. S Afr Med J. 1978;53(7):235–7.

Geyer H, Will C, et al. Carbohydrate structure of Marburg virus glycoprotein. Glycobiology. 1992;2:299–312.

Ghazanfar H, Orooj F, et al. Ebola, the killer virus. Infect Dis Poverty. 2015;4(1):15.

Gire SK, Goba A, Andersen KG, et al. Genomic surveillance elucidates Ebola virus origin and transmission during the 2014 outbreak. Science. 2014;345(6202):1369–72.

Goeijenbier M, van Kampen JJ, et al. Ebola virus disease: a review on epidemiology, symptoms, treatment and pathogenesis. Neth J Med. 2014;72(9):442–8.

Goldsmith CS, Miller SE. Modern uses of electron microscopy for detection of viruses. Clin Microbiol Rev. 2009;22(4):552–63.

Grolla A, Lucht A, et al. Laboratory diagnosis of Ebola and Marburg hemorrhagic fever. Bull Soc Pathol Exot. 2005;98(3):205–9.

Hoenen T, Groseth A, Feldmann H. Current Ebola vaccines. Expert Opin Biol Ther. 2012;12 (7):859–72.

Hoenen T, Groseth A, et al. Ebola virus: unravelling pathogenesis to combat a deadly disease. Trends Mol Med. 2006;12(5):206–15.

Hoenen T, Groseth A, et al. Nanopore sequencing as a rapidly deployable Ebola outbreak tool. Emerg Infect Dis. 2016:22(2):331.

Holmes EC, Dudas G, et al. The evolution of Ebola virus: insights from the 2013–2016 epidemic. Nature. 2016;538(7624):193–200.

Ilinykh PA, Lubaki NM, Widen SG, et al. Different temporal effects of Ebola virus VP35 and VP24 proteins on global gene expression in human dendritic cells. J Virol. 2015;89(15):7567–83.

Jabeen A, Imam N, Ahmad N, Raza K. Next generation sequencing: a new platform for disease study. JMI Int J Math Sci. 2016;7:50–61.

Jabeen A, Ahmad N, Raza K. Differential expression analysis of ZIKV infected human RNA sequence reveals potential genetic biomarkers. Lect Notes Bioinform. 2019;11465:1–12.

Jabeen A, Ahmad N, Raza K. Machine learning-based state-of-the-art methods for the classification of RNA-Seq data. In: Dey N, Ashour A, Borra S, editors. Classification in BioApps. Lecture notes in computational vision and biomechanics, vol. 26. Springer; 2018. p. 133–72.

Khan FN, Qazi S, Tanveer K, et al. A review on the antagonist Ebola: a prophylactic approach. Biomed Pharmacother. 2017;96:1513–26 (Elsevier).

King JW. Ebola Virus, Clinical Presentation. Medscape. Archived from the original on 1 Jan 2012. https://emedicine.medscape.com/article/216288-clinical.

Koehler JW, Hall AT, Rolfe PA, et al. Development and evaluation of a panel of Filovirus sequence capture probes for pathogen detection by next generation sequencing. PLoS One. 2014;9(9):e107007.

Kortepeter MG, Bausch DG, Bray M. Basic clinical and laboratory features of filoviral hemorrhagic fever. J Infect Dis. 2011;204(3):S810–6.

Kühl A, Pöhlmann S. How Ebola virus counters the interferon system. Zoonoses Public Health. 2012;59(Suppl 2):116–31.

Laupland KB, Valliquette L. Ebola virus disease. Can J Infect Dis Med Microbiol. 2014;25 (3):128–9.

Liu X, Speranza E, Munoz-Fontela C, et al. Transcriptomic signature differentiate survival from fatal outcomes in human infected with Ebola virus. Genome Biol. 2017;18:4.

Mackay IM, Arden KE. Ebola virus in the semen of convalescent men. Lancet Infect Dis. 2015;15 (2):149–50.

Martin S, Chiramel AI, Schmidt ML, et al. A genome wide siRNA screen identifies a druggable host pathway essential for the EBOLA virus life cycle. Genome Med. 2018;10(1):58.

McNeil DG Jr. Ask Well: How Does Ebola Spread? How Long Can the Virus Survive?. The New York Times. Archived from the original on 24 Oct 2014.

McNeil, D. Nigeria's actions seem to contain ebola outbreak. The New York times (2014). https://www.nytimes.com/2014/10/01/health/ebola-outbreak-in-nigeria-appears-to-be-over.html

Menicucci AR, Versteeg K, Woolsey C, et al. Transcriptome analysis of circulating immune cell subset highlight the role of monocytes in Zaire ebola virus Makona pathogenesis. Front Immunol. 2017;8:1372.

Ministry of Health and Long-Term Care. Appendix A: Disease-Specific Chapters—Chapter: Hemorrhagic fevers caused by: i) Ebola virus and ii) Marburg virus and iii) Other viral causes including bunyaviruses, arenaviruses, and flaviviruses (PDF). 2019. http://www.health.gov.on.ca/en/pro/programs/publichealth/oph_standards/docs/hemorrhagic_fevers_chapter.pdf.

Miller J. Russian Scientist Dies in Ebola Accident at Former Weapons Lab. The New York Times. 2004. https://www.nytimes.com/2004/05/25/world/russian-scientist-dies-inebola-accident-at-former-weapons-lab.html.

Misasi J, Sullivan NJ. Camouflage and misdirection: the full on assault of Ebola virus disease. Cell. 2015;159(3):477–86. 24.

Muhlberger E, Weik M, Volchkov VE, et al. Comparison of the transcription and replication strategies of Marburg virus and Ebola virus by using artificial replication systems. J Virol. 1999;73(3):2333–42.

Murphy FA. Ebola Virus Disease: An Introduction. World Health Organization. http://www.searo.who.int/entity/emerging_diseases/ebola/ebola_virus_disease_intro.pdf?ua=1.

Ning Y-J, Kang Z, Xing J, et al. Ebola virus mucin like glycoprotein (Emuc) induces remarkable acute inflammation and tissue injury: evidence for Emuc pathogenicity in vivo. Protein Cell. 2018;9:389–93.

Okumura A, Rasmussen AL, Halfmann P, et al. Suppressor of cytokine signaling 3 is an inducible host factor that regulates virus egress during Ebola virus infection. J Virol. 2015;89 (20):10399–406.

Olsen ME, Filone, CM, Rozelle, D, et al. Polyamines and hypusination are required for Ebola virus gene expression and replication. mBio. 2016;7(4):e00882–16.

Osterholm MT, Moore KA, et al. Transmission of Ebola viruses: what we know and what we do not know. MBio. 2015;6(2):e00137.

Pourrut X, Kumulungui B, et al. The natural history of Ebola virus in Africa. Microbes Infect. 2005;7(7–8):1005–14.

Qazi S, Tanveer K, El-bahnasy K, Raza K. From telediagnosis to teletreatment: the role of computational biology and bioinformatics in tele-based healthcare. In: Telemedicine technologies. Academic Press; 2019. p. 153–69.

Quick J, Loman NJ, Duraffour S, et al. Real-time, portable genome sequencing for Ebola surveillance. Nature. 2016;530(7589):228–32.

Ramanan P, Shabman RS, et al. Filoviral immune evasion mechanics. Viruses. 2011;3(9):1634–49.

Raza K, Ahmad S. Recent advancement in next-generation sequencing techniques and its computational analysis. Int J Bioinform Res Appl. 2019;15(3):191–220.

Raza K, Qazi S. Nanopore sequencing technology and internet of living things: a big hope for U-healthcare. Sens Health Monit. 2019;5:95–116.

Rogstad KE, Tunbridge A. Ebola virus as a sexually transmitted infection. Curr Opin Infect Dis. 2015;28(1):83–5.

Ruzek D. Viral hemorrhagic fevers. Boca Raton: CRC Press, Taylor & Francis Group; 2014. p. 444.

Sanchez A, Kiley MP, et al. Sequence analysis of the Ebola virus genome: organization, genetic elements, and comparison with the genome of Marburg virus. Virus Res. 1993;29(3):215–40.

Scott JT, Sesay FR, Massaquoi TA, et al. Post-Ebola syndrome, Sierra Leone. Emerg Infect Dis. 2016;22(4):641–6.

Seifert SN, Schulz JE, Matson MJ, et al. Long range polymerase chain reaction method for sequencing the Ebola virus genome from ecological and clinical samples. J Infect Dis. 2018;128:301–4.

Shantha JG, Yeh S, Nguyen QD. Ebola virus disease and the eye. Curr Opin Ophthalmol. 2016;27 (6):538–44.

Shantha JG, Crozier I, Yeh S. An update on ocular complications of Ebola virus disease. Curr Opin Ophthalmol. 2017;28(6):600–6.

Sharma N, Cappell MS. Gastrointestinal and hepatic manifestations of Ebola virus infection. Dig Dis Sci. 2015;60(9):2590–603.

Shawan MMAK, Hossain MM, Hasan MM, et al. In silico characterisation and investigation of putative promoter motifs in Ebola virus genome. J Glob Biosci. 2015;4:1747–57.
Sichtig. Decoding Ebola: Next-Generation Sequencing of the Ebola Genome for the FDAARGOS Database. U.S. Food & Drug Administration. https://www.fda.gov/emergencypreparedness/counterterrorism/medicalcountermeasures/mcmregulatoryscience/ucm452650.htm.
Smith T. Ebola (deadly diseases and epidemics). Chelsea House Publications; 2005.
Soka MJ, Choi MJ, et al. Prevention of sexual transmission of Ebola in Liberia through a national semen testing and counselling programme for survivors: an analysis of Ebola virus RNA results and behavioural data. Lancet Glob Health. 2016;4:e736–43.
Specia M. The Latest Ebola Outbreak is centered in a war zone. New York Times. 2018. https://www.nytimes.com/2018/08/02/world/africa/ebola-outbreak-congo-warzone.html.
Subissi L, Keita M, Mesfin S, et al. Ebola virus transmission caused by persistently infected survivors of the 2014–2016 outbreak in west Africa. J Infect Dis. 2018;218:287–91.
Swetha RG, Ramaiah S, Anbarasu A, et al. Ebola virus database: gene and protein information resource for Ebola viruses. Adv Bioinform. 2015;2016:1–4.
Tong Y-G, Shi W-F, Liu D, et al. Genetic diversity and evolutionary dynamics of Ebola virus in Sierra Leone. Nature. 2015;524(7563):93–6.
Tosh PK, Sampath kumar P. What clinicians should know about the 2014 Ebola outbreak. Mayo Clin Proc. 2014;89(12):1710–17.
UN. Uganda: Deadly Ebola Outbreak Confirmed—UN. UN News Service. 2007. https://allafrica.com/stories/200711301070.html.
Varkey JB, Shantha JG, et al. Persistence of Ebola virus in ocular fluid during convalescence. N Engl J Med. 2015;372(25):2423–7.
VOA. DRC confirms Ebola Outbreak. 2012. https://www.voanews.com/a/Ebola-drcoutbreak/1492233.html.
Wahl JV, Kurz S, Feldmann F, et al. Ebola virion attachment and entry into human macrophages profoundly effects early cellular gene expression. PLoS Negl Trop Dis. 2011;5(10):e1359.
Wang ZY, Guo ZD, Li JM, et al. Genome wide search for competing endogenous RNAs responsible for the effects induced by EBOLA virus replication and transcription using a trVLP system. Front Cell Infect Microbiol. 2017;7:479.
West TE, von Saint André-von Arnim A. Clinical presentation and management of severe Ebola virus disease. Ann Am Thorac Soc (Rev). 2014;11(9):1341–50.
Whitmer SLM, Albarino C, Shepard SS, et al. Preliminary evaluation of the effect of investigational Ebola virus disease treatment on viral genome sequences. J Infect Dis. 2016;214(S3):333–41.
Wiwanitkit S, Wiwanitkit V. Ocular problem in Ebola virus infection: a short review. Saudi J Ophthalmol. 2015;29(3):225–6.
World Health Organization, Liberia: New Ebola mobile lab speeds up diagnosis and improves care. 2014a. http://www.who.int/features/2014/liberia-mobile-ebola-lab/en/.
World Health Organization. Media Centre. 2014b. http://www.who.int/mediacentre/news/ebola/2-september-2014/en/.
World Health Organization. 19 dead, 39 infected so far in Congo: Ebola Outbreak. 2017. https://www.msn.com/en-us/health/medical/who-says-19-dead-39-infected-so-far-incongo-ebola-outbreak/ar-AAxfhX9?ocid=ientp.
World Health Organization. Ebola outbreak in DRC ends: WHO calls for international efforts to stop other deadly outbreaks in the country. 2018a. http://www.who.int/newsroom/detail/24-07-2018-ebola-outbreak-in-drc-ends–who-calls-for-internationalefforts-to-stop-other-deadly-outbreaks-in-the-country.
World Health Organization. Ebola Virus Disease. 2018b. http://www.who.int/en/newsroom/fact-sheets/detail/ebola-virus-disease.
World Health Organization. Emergencies preparedness, responses. 2018c. http://www.who.int/csr/don/31-august-2018-ebola-drc/en/.
World Health Organization. Emergencies, preparedness, responses. EVD: Democratic Republic of Congo. 2018d. http://www.who.int/csr/don/23-may-2018-ebola-drc/en/.

World Health Organization. First Antigen Rapid Test for Ebola through Emergency Assessment and Eligible for Procurement. 2018e. http://www.who.int/medicines/ebolatreatment/1st_antigen_RT_Ebola/en/.

World Health Organization. Emergencies preparedness, responses. 2012. http://www.who.int/csr/don/2012_08_18/en/.

World Health Organization. Situation report: ebola virus disease - 10th June 2016. (2016). https://apps.who.int/iris/bitstream/handle/10665/208883/ebolasitrep_10Jun2016_eng.pdf;jsessionid=11FD565E392F555E0D5C29BB602A79F2?sequence=1

Yozwiak NL, Happi CT, et al. Roots, not parachutes: research collaborations combat outbreaks. Cell. 2016; 166(1):5–8

Global Outbreaks of Ebola Virus Disease and Its Preventive Strategies

Ameer Khusro and Chirom Aarti

Abstract Ebola virus disease (EVD) has always been a challenge for humankind since its discovery. Ebola virus is transmitted from human-to-human via direct contact with the body fluids of infected people, and with surfaces and materials contaminated with these fluids. Fever, profound weakness, diarrhoea, cramping, abdominal pain, nausea, and vomiting are typical symptoms of EVD. The African fruit bat (*Rousettus aegyptiacus)* is considered the natural reservoir for EVD. Ebola virus epidemic emphasized the necessity for developing auspicious antiviral drugs and vaccines to combat the disease. Over the past two decades, few effective vaccines have been developed using DNA, virus-like particles, recombinant viral vectors, and recombinant proteins. These vaccines were tested in rodents and non-human primates which showed high efficacy towards the prevention of EVD. Currently, rVSV-ZEBOV vaccine is the only approved vaccine for EVD based on its high efficiency rate. In addition, few vaccine candidates are in preclinical stage. Considering the dearth of promising antiviral drugs and approval of limited vaccines for humans, prevention is a leading strategy to eradicate EVD at present. This chapter highlights not only the biology and pathogenesis of EVD but also discusses the epidemiology, disparate vaccines development, and other common preventive strategies to combat EVD outbreak in future.

Keywords Ebola virus · Ebola virus biology · Ebola virus disease · Epidemic · Non-human primates · Outbreak · Pathogenesis · Prevention · Transmission · Vaccination

---

A. Khusro (✉) · C. Aarti
Research Department of Plant Biology and Biotechnology, Loyola College, Chennai 600034, TN, India

S. I. Ahmad (ed.), *Human Viruses: Diseases, Treatments and Vaccines*, https://doi.org/10.1007/978-3-030-71165-8_7

1 Introduction

Ebola virus disease (EVD) is considered as one of the deadliest diseases to humankind because of its high mortality (Wiwanitkit 2014). This disease was first discovered in 1976 by Dr. Peter Piotin in Zaire, Africa (now Democratic Republic of Congo) from the blood of a catholic nun who was a suspected case of yellow fever (Centres for Disease Control and Prevention 2015). It was formerly called as Ebola haemorrhagic fever and has been a colossal challenge since then for communities worldwide. It is a zoonotic disease which is transmitted accidentally due to direct contact with infected live or dead animals. Ebola virus contains a lipid-enveloped negatively stranded ribonucleic acid (RNA) virus that belongs to the family Filoviridae (World Health Organization [WHO] 2015). Ebola haemorrhagic fever is caused by any of five genetically different members of the Filoviridae family: *Zaire ebolavirus* (ZEBOV), *Sudan ebolavirus* (SEBOV), *Côte d'Ivoire ebolavirus*, *Bundibugyo ebolavirus* (BDBV), and *Reston ebolavirus* (REBOV). Le Guenno et al. 1995 in the Côte-d'Ivoire had isolated a new strain of Ebola *Côte d'Ivoire ebolavirus* from a non-fatal human case infected during the autopsy of a wild chimpanzee. It was suggested that this animal has been responsible for the outbreaks of haemorrhagic syndromes. This is the first time that a human infection has been connected to naturally-infected monkeys in Africa. REBOV infects non-human primates (NHPs) only (Barrette et al. 2009). ZEBOV, SEBOV, and BDBV account for most of the Ebola haemorrhagic fever epidemic but ZEBOV is threat to both human and NHPs in sub-Saharan Africa (Groseth et al. 2007). Recent EVD outbreak began in the republic of Guinea in February 2014 and spread to the republic of Liberia and the Sierra Leone (Enserink 2014).

Now we understand that the virus is transmitted to people from wild animals including fruit bats, porcupines, and NHPs, and then spreads in the human population through direct contact with the secretions such as, saliva, mucus, vomit, faeces, sweat, tears, blood, breast milk, urine, and semen or other bodily fluids of infected people, and with surfaces and materials (e.g. bedding, clothing) contaminated with these fluids (Moghadam et al. 2015).

The African fruit bat (*Rousettus aegyptiacus)* is considered the natural reservoir for Ebola virus. These bats transmit the virus to apes, monkeys, and other animals' species of forest. Humans living in these areas come in contact with such infected animals and transmit the infection (Rajak et al. 2015) (Fig. 1).

The incubation period of the virus is 2 to 21 days. Symptoms of EVD include fever, fatigue, muscle pain, headache, and sore throat followed by vomiting, diarrhoea, rash, impaired kidney and liver function, low white blood cell and platelet counts, and elevated liver enzymes (Fig. 2).

Unfortunately, the symptoms of EVD are similar to malaria, typhoid fever, and meningitis which make it difficult to diagnose in the primary stage of infection. The present chapter summarizes not only various essential aspects of EVD such as virus biology, pathogenesis, and epidemiology but also role of vaccination towards its prime preventive strategies in future outbreak.

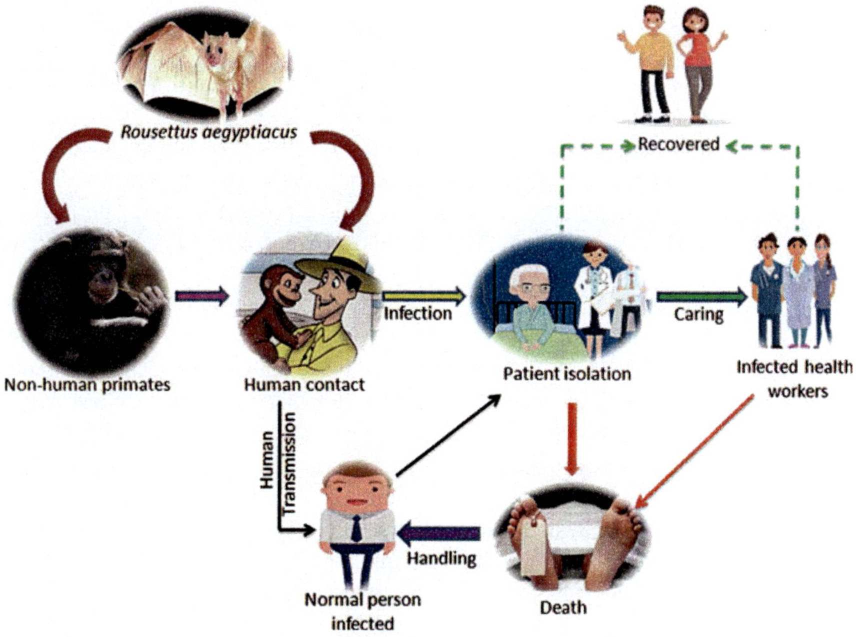

Fig. 1 Transmission cycle of EVD

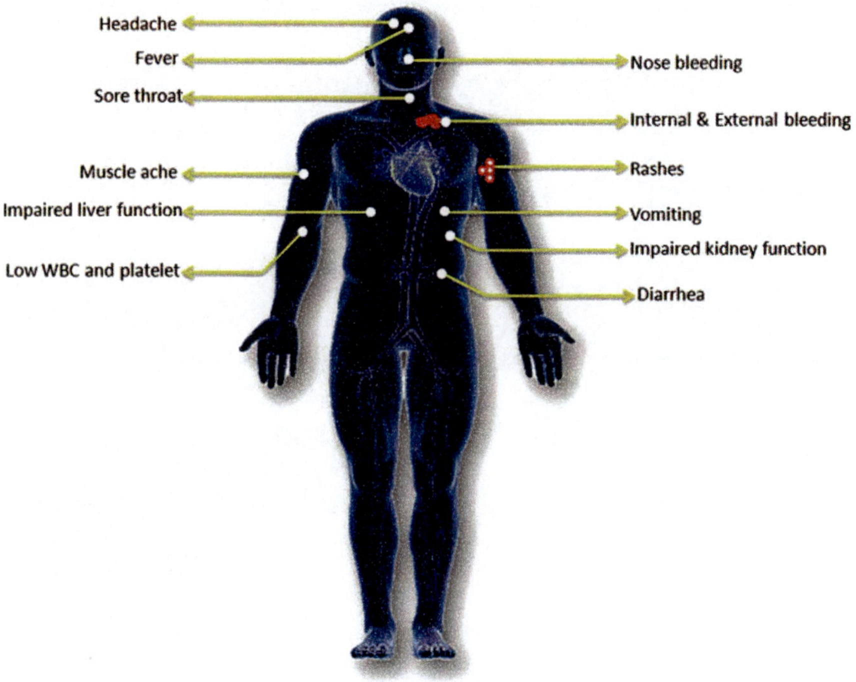

Fig. 2 Symptoms of EVD

2 Ebola Virus Biology

Ebola virus is an enveloped virus containing non-segmented negative-strand ribonucleic acid (RNA) genome of approximately 19 kb.

The viral genome encodes seven different genes which are known to encode nine proteins: nucleoprotein (NP), polymerase cofactor (VP35), matrix protein (VP40), glycoprotein (GP), soluble GP (sGP), small soluble GP (ssGP), transcription activator (VP30), minor matrix protein (VP24), and RNA-dependent RNA polymerase (L) (Fig. 3).

The genome shows following structure: 3'-leader→*nucleoprotein (NP)* gene→*viral protein (VP) 35* gene→*VP40* gene→*glycoprotein (GP)* gene→*VP30* gene→*VP24* gene→*polymerase (L)* gene→5'-trailer. Genome replication promoters and packaging signals are present in conserved leader and trailer regions. Each gene is flanked by 3' and 5' untranslated regions (UTRs) including conserved transcriptional start and stop signals. In spite of overlapping of few genes in UTRs, most genes are separated by intergenic regions of varied lengths. All genes are monocistronic except *GP*, which encodes three GP. The sGP is the preliminary product of *GP*. The virus down-regulates transmembrane GP expression produces a second small ssGP through RNA editing by polymerase complex (Furuyama and Marzi 2019) (Fig. 4).

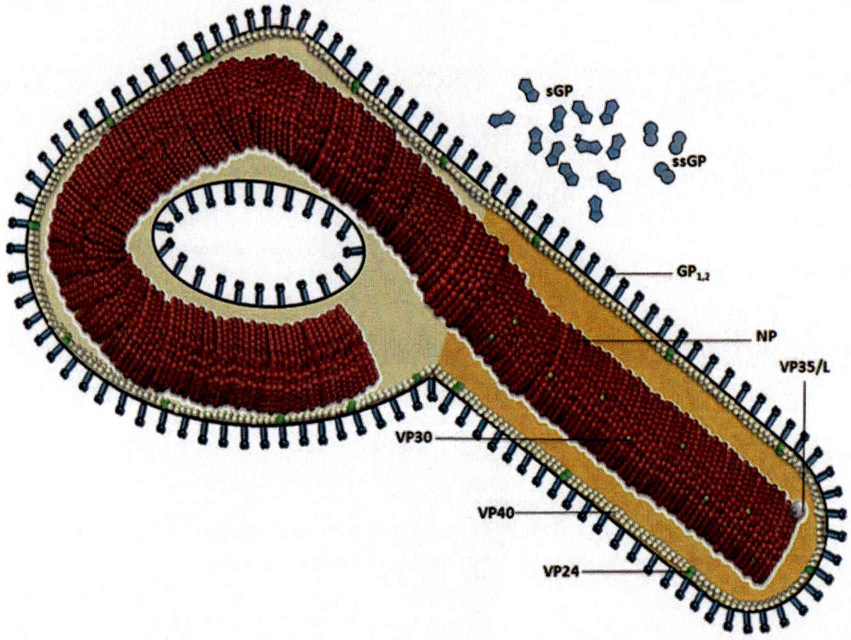

Fig. 3 Structure of Ebola virus showing various proteins

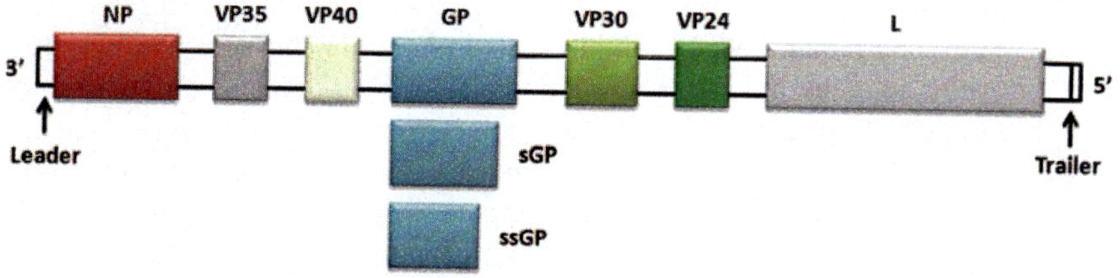

Fig. 4 Ebola virus genome displaying 3'-leader→*nucleoprotein (NP)* gene→*viral protein (VP) 35* gene→*VP40* gene→*glycoprotein (GP)* gene→*VP30* gene→*VP24* gene→*polymerase (L)* gene→5'trailer

The size of Ebola virus particle i.e. virion varies from 50 to 80 nm. The length of virus ranges from 10000 to 14000 nm. The most common shape of the virus is filamentous. Both the sGP are non-structural proteins and are secreted from infected cells. The membrane-associated matrix protein (VP40) forms the filamentous structure of viral particles and connects the nucleocapsid with the host-derived lipid membrane (Hoenen and Feldmann 2014). The virus proteins perform distinct function in the replication cycle (Banadyga et al. 2016).

Nucleoprotein is a prime constituent of ribonucleoprotein complex (RNP) which is known to encapsidate the genome and protects RNA degradation. It is composed of three major domains: N-terminal, core (located in the centre), and C-terminal domains. N and C-terminal as well as lobes of the core domain recognize and clamp to RNA binding groove regions. NP oligomerization and NP tube structure formation are mainly carried out by N-terminal portion (Rojas et al. 2020).

VP35 (a polymerase cofactor) is a phosphoprotein which interferes with the early innate response of host. VP35 interacts with NP and forms VP35-NP complex. This complex is known to prevent premature NP oligomerization and causes variations in NP in viral RNA synthesis machinery. VP30 is a transcriptional activator which supports primary transcription and RNA editing. It plays pivotal role in the stimulation of RNA synthesis. Leucine-zipper motif is present in N-terminal region which helps in homo-oligomerization (Hartlieb et al. 2007). The basic amino acid cluster of C-terminal region interacts with NP and control VP30 activity (Kirchdoerfer et al. 2016). This protein is not essential for the replication of virus.

The RNA-dependent RNA polymerase L carries the enzymatic functional domains for genome transcription and replication. It also causes activation of RNA synthesis and edits messenger RNA (mRNA). It binds to genomic RNA at 3′ leader promoter and alters virus negative-sense RNA into positive-sense messenger RNA for producing new virions. The L protein is responsible for mRNA polyadenylation and capping too because of its catalytic nature (Martin et al. 2017). The editing trait of L protein leads to generate GP transcript and regulates proteins expression (Volchkova et al. 2011).

VP24 plays an important role in RNP assembly. The protein-protein interaction is carried out by N-terminal region of this protein and leads to the formation of capsid. Protein aggregation occurs due to the mutation in this particular region (Han et al. 2003). C-terminal deletion inhibits the formation of nucleocapsid-like structure formation, thereby indicating the importance of both domains in the assembly of nucleocapsid (Han et al. 2003). VP24 interacts with NP via conserved loop by depending on VP24 amino acids (V170 and N171) that plays important role in nucleocapsid condensation, thereby producing new virions (Banadyga et al. 2017). VP24 determines the virulence property of virus and inhibits IFN-α/β/γ trigger by sequestering proteins belonging to the karyopherin α (KPNA) family viz. KPNA α1, α5, and α6 proteins, thus, inhibiting activated signal transducer and activator of transcription 1 (STAT1) (Reid et al. 2006). VP24 also binds to STAT1 and inhibits its phosphorylation, nuclear transport, and interferon (IFN)-associated gene activation (Zhang et al. 2012). VP24 is also known to inhibit the production of IFN via both Interferon regulatory factor 3 (IRF3) and early tumour necrosis factor (TNF)-induced nuclear factor kappa-light-chain-enhancer of activated B cells (NF-κB) signalling (Guito et al. 2017).

The coordination of virion assembly, maintenance of structural integrity, and budding of viral particles from infected cell is performed by VP40 (Pleet et al. 2017). The N-terminal domain of VP40 helps in dimerization and oligomerization in the cytoplasm and translocates to the plasma membrane. The C-terminal domain causes the interaction with the plasma membrane. Additionally, this protein induces exosomes formation (Madara et al. 2015).

Glycoprotein is essential in viral pathogenesis. It is the only observable protein on the surface of virus. It initiates the entry of virus into the host by fusion with receptor (Lee and Saphire 2009). The GP gene encodes three different proteins: GP with two subunits—GP1 (for receptor binding) and GP2 (for viral fusion); sGP lacks transmembrane domain and is formed from the unedited transcript. The third protein is called as ssGP and its activity is undetermined (Mehedi et al. 2011). The fusion between virus and target cell is generally mediated by GP1. GP1 contains three domains: the receptor binding domain (RBD) (interacts with one or more receptors), the glycan cap (protects the RBD from antibody recognition and interacts with GP2 to prevent premature fusion), and the O-linked glycosylated mucin-like domain (protects RBD from immune recognition and has a cytotoxic influence through extracellular signal-regulated kinase and mitogen-activated protein kinase pathways) (Zampieri et al. 2007). GP2 causes the fusion of membrane. It contains five domains: the N-terminal heptad repeat region (HRR), C-terminal HRR, fusion loop, the transmembrane region, and the cytoplasmic tail (Mohan et al. 2012). The membrane fusion is carried out by fusion loop due to the presence of a hydrophobic sequence. Subsequently, GP2 HRR forms a transmembrane six-helix bundle that serves as an opening in the membrane (Malashkevich et al. 1999). sGP is known for evading the immune system (Mohan et al. 2012).

3 Ebola Virus Pathogenesis

Ebola virus enters into the host cells by interacting with fluids of infected body through mucous membranes or skin lesions (Rojas et al. 2020). In vivo studies in NHPs models revealed that endothelial cells, adrenal cells, immature dendritic cells (DCs), monocytes, macrophages, and Kupffer cells are mainly infected by virus in the liver because of the potentiality of virus GP1 to interact with varied host-cell proteins (Hensley et al. 2011). Several attachment factors have been reported and may cause binding of the virus.

Lectins viz. C-type lectins dendritic cell-Specific intercellular adhesion molecule-3-grabbing non-integrin (DC-SIGN), and L-SIGN are prime receptors for virion attachment (Matsuno et al. 2010). The increased infectivity of Ebola virus may be because of an increase in the glycan extension on the viral GP from the virus, which shows high affinity for the "macrophage galactose-type calcium-type lectin" (Fujihira et al. 2018). Other co-receptors such as β1-integrin (Schornberg et al. 2009), folate receptor-α (Chan et al. 2001), glycosaminoglycans (O'Hearn et al. 2015), Tyro3, Axl, and Mer (TAM) receptor tyrosine kinases (Dahlmann et al. 2015), and T-cell immunoglobulin and mucin 1 (TIM-1) (Dahlmann et al. 2015) are also responsible for virus entry. After binding, the virion is taken up into the target cells by macro-pinocytosis (a non-specific endocytosis mechanism) (Aleksandrowicz et al. 2011). The host cell membrane interacts with the phosphatidylserine present on the virion and stimulates the reorganization of cytoskeleton, triggers the ruffling of the plasma membrane, and invagination of the virion with further internalization into the early endosome (Moller-Tank and Maury 2015). After host-cell membrane fusion process, the nucleocapsid is released into the cytoplasm. At this stage, mRNA is formed from virus RNA via translation mechanism. Virus proteins viz. GP, NP, VP24, VP35, VP40, and L are then translated from viral mRNA by the host machinery. After that, NP, VP35, and L proteins act as a viral polymerase complex and allow the replication of viral genome (Muhlberger 2007). The new virion is released due to the interaction of VP40 with the host machinery (Jasenosky and Kawaoka 2004) (Fig. 5).

The secretion of pro-inflammatory cytokines such as interleukin 6 (IL-6), TNF, and IFN-β as well as the release of tissue factor and vasoactive peptides are enhanced due to the infection of Ebola virus (Rogers and Maury 2018). Additionally, the virus infection shows impairment in DC activation by the action of VP35 and VP24. Ebola virus-infected cells lack type-I IFNs production and does not show maturation process (Bosio et al. 2003). Ebola virus hijacks IFN-I response by inhibiting its production and by blocking its signalling mechanism, and thus, causes impairment in the maturation of DCs, thereby inactivation of adaptive immune response. Ebola virus primarily targets lymphoid organs. As a matter of fact, the virus initially infects DCs and macrophages, and these immune cells guide the virus towards the proximal lymph nodes and infect those (Steele et al. 2009). As per in vivo NHPs models study, the virus destructs lymphocytes, follicular

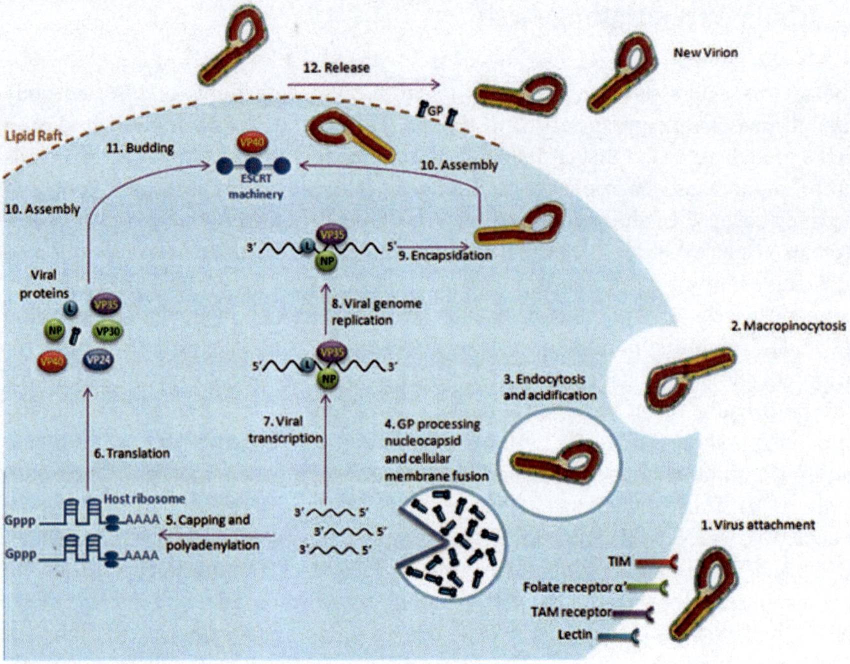

Fig. 5 Ebola virus replication cycle showing the binding of the host cell through different co-receptors, followed by its invasion into the host cytoplasm and transcription of viral RNA into protein. Further, the viral genome is replicated, followed by encapsidation, assembly, and release of a new virion. [*Note* ESCRT: Endosomal sorting complexes required for transport; GP: Glycoprotein; TIM-1: T-cell immunoglobulin and mucin domain 1; TAM receptors: Tyro3, Axl, and Mer receptors; VP: Virion protein; and NP: Nucleoprotein]

structure, and parenchyma (Twenhafel et al. 2013). The virus is known to disrupt vascular endothelium too, causing bleeding, partial blood circulation, and intravascular coagulation blood-tissue barrier permeability (Rasmussen 2017).

4 Epidemiology of Ebola Virus Disease

From 1976 to 2013, the WHO reported 2387 confirmed cases with 1590 deaths. The Ebola virus epidemic in West Africa is considered to be the largest outbreak, which caused thousands of mortalities in Guinea, Sierra Leone, and Liberia. In March 2014, the WHO reported a major EVD outbreak in Guinea, a West African nation. The disease rapidly spread to the neighbouring countries of Liberia and Sierra Leone. It was the largest EVD outbreak ever documented, and the first recorded in the region. On 8 August 2014, the WHO declared this epidemic an international public health emergency (Moghadam et al. 2015). On 29 December

2015, Guinea was declared free of Ebola virus transmission. On 14 January 2016, Ebola virus was detected again in Sierra Leone but emergency was not declared. On 11 May 2017, the DRC Ministry of Public Health notified the WHO about an outbreak of Ebola which killed 4 people. On 2 July 2017, the WHO again declared the end of the outbreak. On 1 August 2018, the world's 10th EVD outbreak was declared in North Kivu province of the Democratic Republic of the Congo. By November 2018, nearly 200 Congolese died of Ebola. By March 2019, this became the second largest EVD outbreak ever recorded, with more than 1000 cases. As of 4 June 2019, the WHO reported 2025 confirmed and probable cases with 1357 deaths. In June 2019, 2 people died of EVD in neighbouring Uganda. On 17 July 2019, the WHO declared the EVD outbreak a global health emergency (World Health Organization 2019).

From 13 to 19 January 2020, 15 new confirmed cases of EVD were reported from North Kivu Province in the Democratic Republic of the Congo. From 30 December 2019 to 19 January 2020, 37 confirmed cases were reported from 12 of the 87 health areas in 6 active health zones in North Kivu and Ituri Provinces. As of 19 January 2020, a total of 3414 EVD cases, including 3295 confirmed and 119 probable cases have been reported, of which 2237 patients died. Of the total confirmed and probable cases, 56% were female, 28% were children aged less than 18 years, and 5% were health workers. As of 10 March 2020, a total of 3444 EVD cases were reported, including 3310 confirmed and 134 probable cases, of which 2264 patients died. Of the total confirmed and probable cases, 56% were female, 28% were children aged less than 18 years, and 5% were health care workers. As of 30 March, 2020, a total of 3453 cases (3310 confirmed and 143 suspected) were reported which includes 2264 deaths, 1169 survivors, and rest of the patients are still under care (World Health Organization 2020).

5 Ebola Virus Vaccination

Over the past few centuries, vaccination has been considered as a prime strategy to prevent and control outbreaks of certain deadly human diseases. The development of efficacious vaccine for EVD was started in the late 1970s. The inactivated virus based vaccine was the first vaccine developed against EVD, which prevented guinea pigs from this deadly infection (Lupton et al. 1980). Since then, a variety of vaccines has been developed against EVD including deoxyribonucleic acid (DNA), virus-like particles (VLPs), recombinant viral vectors, and recombinant proteins (Reynolds and Marzi 2017). The effectiveness of those vaccines was assessed in rodents or NHPs (Reynolds and Marzi 2017). At present, 8 vaccine candidates are in clinical trials stages which target viral GP (Table 1).

Plasmid encoding sGP and GP were used for the development of first DNA vaccine for EVD. The vaccine was known to elicit humoral and T cell responses (Xu et al. 1998). This vaccine is safer to use, produce, and shows advantages over live attenuated vaccines. Previous study revealed that the administration of DNA

Table 1 Vaccine candidates in clinical trial phases

Vaccines	Class	Immunogen	Trial phases
DNA	DNA	GP and NP	Phase 1
Protein	Subunit	GP	Phase 1
Ad26	Replication-deficient virus expressing EBOV protein	GP	Ad26/cAd3—phase I Ad26/MVA—phases I, II, III
Ad5	Replication-deficient virus expressing EBOV protein	GP	Ad5—phases I, II Ad5/VSV—phases I, II, IV
chAd3	Replication-deficient virus expressing EBOV protein	GP	chAd3—phases I, II chAd3/MVA—phase I
MVA	Replication-deficient virus expressing filovirus protein	GP and NP	Ad26/MVA—phases I, II, III chAd3/MVA—phase I
VSV	Replication-competent virus expressing EBOV protein	GP	Phases I, II, III
HPIV3	Replication-competent virus expressing EBOV protein	GP	Phase I

Abbreviation: Ad—adenovirus; chAd—chimpanzee adenovirus; GP—glycoprotein; HPIV3—Human parainfluenza virus 3; MVA—modified vaccinia Ankara; NP—nucleoprotein; VSV—vesicular stomatitis virus

vaccine encoding viral GP and NP protected all the mice from EVD (Vanderzanden et al. 1998). In another study, the vaccination of DNA vaccine with optimized antigen expression showed 83% efficiency in NHPs (Grant-Klein et al. 2015). A combination of DNA prime together with an adenovirus boost (both encoding viral GP) exhibited 100% protection in NHPs (Sullivan et al. 2000). The Ebola virus GP-based DNA vaccination depicted protection to NHPs (Patel et al. 2018).

Ebola VLPs have also been used for the development of vaccine by co-expressing Ebola virus GP and VP40 in transfected cells. The VLP vaccination stimulated NK cells for protecting against the virus (Warfield et al. 2004). Additionally, VLP vaccination induces host responses through Toll-like receptor and type I IFN signalling, thereby leading to the initiation of early innate protective immune responses (Ayithan et al. 2014). VLPs containing VP40 and GP showed 100% protection against viral infection in rodents (Swenson et al. 2005). The vaccination of VLPs along with RIBI adjuvant stimulated viral GP-specific antibodies and strong T cell responses (Warfield et al. 2007).

The replication-deficient Ebola virus lacking the VP30 gene (EBOV_VP30) is a subunit vaccine. Study showed that the administration of mice with EBOV_VP30 showed robust Ebola virus GP-specific antibody and Ebola virus NP-specific T cell responses with high survival rate of animals. Guinea pigs immunized twice with EBOV_VP30 showed protection from Ebola virus (Halfmann et al. 2009). In another report, the immunization of NHPs with EBOV_VP30 showed protection from the infection (Marzi et al. 2015b).

The viral vectors encoding Ebola virus GP is another EVD vaccine in preclinical stage. A replicon system based on Venezuelan equine encephalitis virus (VEEV) is effective against Ebola virus. VEEV replicons expressing Ebola virus GP protected guinea pigs and mice from Ebola virus infection (Pushko et al. 2000).

Human adenovirus serotype 1, serotype 2, and serotype 5 (Ad5) cause mild upper respiratory tract infections. Replication-deficient human Ad5 has been used as a vaccine candidate for EVD. Ad5 stimulates cellular and humoral immunity against the encoded antigen. Replication-deficient recombinant Ad5 revealed protection in NHPs against viral infection (Sullivan et al. 2003).

The vaccination of Ad26-Ebolavirus showed effectiveness against Ebola virus in NHPs. Further, its effectiveness was enhanced when used with an Ad35-Ebola virus boost (Geisbert et al. 2011). In another study, a single dose of recombinant chAd3 expressing Ebola virus GP depicted 100% protection in NHPs against EVD. Unfortunately, the protection reduced to 50% when animals were challenged 10 months post vaccination. Hence, a chAd3- Ebola virus prime was boosted with a modified vaccinia Ankara (MVA) vector encoding the Ebola virus GP (MVA-BN Filo) for improving the protection (Stanley et al. 2014).

Vesicular stomatitis virus (VSV) belongs to *Rhabdoviridae* family and infects livestock as well as other animals. It causes mild infection in humans (Letchworth et al. 1999). Thus, a VSV-based vaccine is considered crucial for protection against EVD. The VSV-G coding sequence was replaced in the wild-type VSV genome with a sequence encoding Ebola virus GP for engineering recombinant VSV-Ebola virus vaccine (also called as rVSV-ZEBOV) (Mire et al. 2012). rVSV-ZEBOV showed high protection in NHPs (Marzi et al. 2015a). Headache, fatigue, and muscle pain were commonly reported as adverse events in this period across all age groups. Overall, rVSV-ZEBOV offered substantial protection against EVD.

Recombinant rabies virus (RABV) has also been used for developing EVD vaccine. The vaccine showed complete prevention against Ebola virus in NHPs (Blaney et al. 2013). However, a replication-deficient or an inactivated version of this vaccine candidate showed only 50% protection (Blaney et al. 2013). The inactivated RABV vaccine vector was improved using a codon-optimized antigen which showed complete prevention against Ebola virus (Willet et al. 2015).

The EVD vaccine was also developed using recombinant paramyxovirus-based vectors. Multiple GPs were inserted into the genome of Human parainfluenza virus 3 (HPIV-3) (Skiadopoulos et al. 2002). Recombinant HPIV-3 expressing viral GP alone or along with NP prevented guinea pigs against EVD (Bukreyev et al. 2006). Two-dose immunization process using this very vaccine showed protection in NHPs (Bukreyev et al. 2007). Mucosal (IN/intratracheal) administration of the HPIV-3-Ebolavirus GP vaccine prevented NHPs from EVD (Meyer et al. 2015).

Prime-boost regimen with a combination of the Ad26-Ebolavirus and MVA-BN Filo vaccines are in clinical trial phases. In one phase I trial, immunization with Ad26-Ebolavirus or MVA-BN Filo did not show any side effect (Milligan et al. 2016). GamEvac-Combi, the combination vaccine of VSV-Ebola virus and

Ad5-Ebola virus was developed in Russia. The vaccine induced viral GP-specific antibodies in 95% of the volunteers, and neutralizing antibodies were detected in 93.1%. Furthermore, CD4+ and CD8+ T cells were detected in 82.8% and 58.6% of participants, respectively (Dolzhikova et al. 2017). This vaccine is being tested in an ongoing phase IV clinical trial in Guinea and Russia.

After EVD epidemic, several human clinical trials with VSV-Ebola virus are ongoing or have been completed in North America, Europe, and Africa (Feldmann et al. 2018), and still others are planned. Ten phase I clinical trials of VSV-Ebola virus were evaluated to determine the safety and immunogenicity of the vaccine as well as to identify doses and regimes for further clinical trial phases (Agnandji et al. 2017). Five phase II trials using the VSV-Ebola virus vaccine in healthy adults, HIV-positive adults, and adults at risk for Ebola virus infection were completed in the United States, Canada, Liberia, Sierra Leone, and Guinea (Kennedy et al. 2017). The efficacy of the VSV-Ebola virus vaccine was tested to prevent EVD which showed serious side effects (Henao-Restrepo et al. 2017). Further trials suggested the administration of VSV-Ebola virus as an emergency vaccine for the prevention of EVD (Halperin et al. 2017; Carter et al. 2018). During the recent outbreak in North Kivu of the Democratic Republic of the Congo (DRC), VSV-Ebola virus was administered to over 93000 individuals for preventing EVD (World Health Organization 2019).

In December 2019, the U.S. Food and Drug Administration (FDA) approved 'Ervebo' or rVSV-ZEBOV, the first FDA-approved vaccine for the prevention of EVD, caused by Zaire Ebola virus in individuals 18 years of age and older (Fig. 6). As discussed earlier, it is a live attenuated vaccine that has been genetically engineered to contain a protein from the Zaire Ebola virus. The approval of this vaccine was granted to Merck & Co., Inc. (Centers for Disease Control and Prevention 2019).

According to Anna Abram, FDA Deputy Commissioner for Policy, Legislation, and International Affairs, "While the risk of EVD in the U.S. remains low, the U.S. government remains deeply committed to fight devastating Ebola outbreaks in Africa, including its current outbreak in the Democratic Republic of the Congo. Today's approval is an important step in our continuing efforts to fight EVD in close coordination with our partners across the U.S. Department of Health and Human Services, as well as our international partners, such as the WHO. These efforts, including today's landmark approval, reflect the FDA's unwavering dedication to leveraging our expertise to facilitate the development and availability of safe and effective medical products to address urgent public health needs and fight infectious diseases, as part of our vital public health mission" (Centers for Disease Control and Prevention 2019).

The approval of 'Ervebo' is a colossal step towards the prevention of EVD outbreaks. As a matter of fact, the vaccine is an outcome of the study conducted in Guinea during the 2014–2016 outbreak in individuals, 18 years of age and older. The safety of 'Ervebo' was assessed in approximately 15,000 individuals in Africa, Europe, and North America. This vaccine is administered as a single-dose injection. However, side effects such as pain, swelling and redness at the injection site,

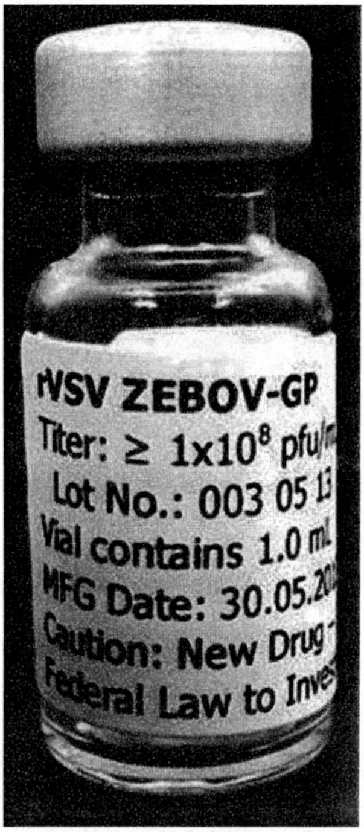

Fig. 6 FDA approved only vaccine for prevention of Ebola virus infection

headache, fever, joint and muscle aches, and fatigue were reported (Centers for Disease Control and Prevention 2019).

6 rVSV Technology for EVD Vaccination

VSV is an attenuated vector backbone which is used to develop vaccines. It causes asymptomatic infections in humans. VSV as a vector has several advantages such as its reduced prevalence of immunity to the vector in most populations targeted for immunization, lack of viral RNA integration into the host, expression of large foreign transgenes, and pseudo-typing of virus with heterologous viral glycoproteins (Monath et al. 2019). VSV-Indiana and VSV New Jersey are two serotypes of VSV. VSV-Indiana (VSV-I) is the basis for current vaccine candidates. The VSV genome contains 11000 nucleotides which encode five major proteins. The VSV glycoprotein (G) is present in the envelope of virus which causes cellular attachment, fusion with endosomal membranes, and release of viral RNA into the cytoplasm (Monath et al. 2019). This protein provides protective immunity against VSV. VSV vectors are constructed expressing genes from diversified species such as Ebola virus, Marburg virus, Lassa fever virus, and influenza virus, bacteria (Zhang et al. 2017), and tumour antigens (Blanchard et al. 2015) using reverse genetic process. Sometimes, a portion of the VSV G protein is retained for expression as well as enabling fusion and internalization of the recombinant virus (Clarke et al. 2016). VSV vectors lacking the VSV G gene (VSVDG) must reconstitute the attachment, fusion, and budding functions with one or more proteins encoded by the heterologous envelope gene. The rVSVDG-ZEBOV-GP produced by reverse genetic system constitutes plasmids containing the VSV genome with G deleted and replaced with ZEBOV GP, together with helper plasmids expressing the VSV N, P, and L genes (Huttner et al. 2015). The rVSVDG-ZEBOV-GP is constructed with full-length GP. On the other hand, native ZEBOV expresses sGP (de La Vega et al. 2015). rVSVDG-ZEBOV-GP is more efficiently neutralized by antibody than wild-type ZEBOV due to lack of sGP (Ilinykh et al. 2016). The full length heterologous GP is incorporated into the rVSV particle, showing bullet shaped morphology. ZEBOV and rVSV pseudo-typed with ZEBOV GP enter the cell by macropinocytosis in a GP protein-dependent manner (Nanbo et al. 2010). The deletion of VSV G protein and replacement by the heterologous transgene in rVSVDG-ZEBOV-GP shows highly attenuated phenotype (Monath et al. 2019).

7 Preventive Approaches Other Than Vaccination

Plethora of strategies such as case management, surveillance, and laboratory testing at wide scale are essential to prevent any outbreak. Raising awareness of risk factors for EVD can certainly be an effective approach to reduce human-to-human transmission. Reducing transmission of EVD relies on diversified factors as mentioned below:

- Contact with infected fruit bats, monkeys, apes, forest antelope or porcupines should be avoided maximally in order to reduce wildlife-to-human transmission. Animals should be handled with gloves and other appropriate protective clothing. Animal's raw products (blood and meat) should not be consumed, and if unavoidable, cooked thoroughly before consumption.
- Close contact with EVD infected people or people with EVD symptoms should be avoided in order to reduce human-to-human transmission. Contact with the body fluids of infected people should be avoided. Gloves and appropriate personal protective equipment should be worn when taking care of ill patients. Regular hand washing is required after visiting patients in hospital, as well as after taking care of patients at home.
- Safe and dignified burial of the dead person should be carried out.
- People who may have been in contact with someone infected with EVD should be identified and monitored for 21 days during quarantine.
- Risk of possible sexual transmission should be reduced. Male survivors of EVD should practice safer sex and hygiene for at least a year from onset of symptoms or until their semen tests negative twice for Ebola virus.

- Pregnant women who have survived EVD need to be monitored.
- Health-care workers should always take standard precautions while caring the patients. The health-care workers should wear face protective mask, goggles, and gloves.
- Samples received for Ebola virus confirmation should be handled by trained staff and processed in equipped laboratories.
- After sex, used condoms should be handled safely, and safely disposed of in order to prevent contact with seminal fluids.

8 Conclusion

Since the discovery of Ebola virus, the EVD has been a serious threat to humankind due to its human-to-human transmission feature. The epidemics, unavailability of effective antiviral drug, and reduced research activities makes this disease fatal and colossal health concern of recent times. Few vaccine candidates have shown tremendous outcomes in rodents and NHPs at experimental stages. The use of those vaccines in human based on NHPs results will not only be a time consuming approach but also expensive process. Currently, rVSV-vector based vaccine is the only FDA approved vaccine for EVD based on its high efficacy rate. Hence, at present, the best approach to reduce the epidemic of EVD is the prevention only. The quarantine and care of the infected patients, avoiding close contact with NHPs and infected persons, early diagnosis based on sign and symptoms, and safe burial of the dead can be prime preventive options to stop the human-to-human transmission. The disease can be eradicated by initiating awareness programmes at large scale. The rapid, cheap, and simple diagnostic kits should be made available for health care workers in nearby medical centres and hospitals. A significant effort should be undertaken by the government to test the suspected cases free of cost in order to examine maximum number of people. The research funding should be provided to prestigious institutes or research centres for the development of ideal vaccine and antiviral drugs against Ebola virus. Most importantly, the government and community should clarify any misconception among people regarding role of any religious factor or groups towards the outbreak of EVD.

Conflict of Interest None declared.

References

Agnandji ST, Fernandes JF, Bache EB, ObiangMba RM, Brosnahan JS, Kabwende L, et al. Safety and immunogenicity of rVSV_G-ZEBOV-GP Ebola vaccine in adults and children in Lambarene, Gabon: a phase I randomised trial. PLOS Med. 2017;14.

Aleksandrowicz P, Marzi A, Biedenkopf N, Beimforde N, Becker S, Hoenen T, et al. Ebola virus enters host cells by macropinocytosis and clathrin-mediated endocytosis. J Infect Dis. 2011;204:S957–67.

Ayithan N, Bradfute SB, Anthony SM, Stuthman KS, Dye JM, Bavari S, et al. Ebola virus-like particles stimulate type I interferons and proinflammatory cytokine expression through the toll-like receptor and interferon signaling pathways. J Interferon Cytokine Res. 2014;34:79–89.

Banadyga L, Dolan MA, Ebihara H. Rodent-adapted filoviruses and the molecular basis of pathogenesis. J Mol Biol. 2016;428:3449–66.

Banadyga L, Hoenen T, Ambroggio X, Dunham E, Groseth A, Ebihara H. Ebola virus VP24 interacts with NP to facilitate nucleocapsid assembly and genome packaging. Sci Rep. 2017;7:7698.

Barrette RW, Metwally SA, Rowland JM, Xu L, Zaki SR, Nichol ST, et al. Discovery of swine as a host for the *Reston ebolavirus*. Science. 2009;325:204–6.

Warfield KL, Swenson DL, Olinger GG, Kalina WV, AmanMJ, Bavari S. Ebola virus-like particle based vaccine protects nonhuman primates against lethal Ebola virus challenge. J Infect Dis. 2007;196:S430–437.

Blanchard AM, Shim KG, Grams MP, Rajani K, Diaz RM, Furutani KM, et al. Definitive management of oligometastatic melanoma in a murine model using combined ablative radiation therapy and viral immunotherapy. Int J Radiat Oncol Biol Phys. 2015;93:577–87.

Blaney JE, Marzi A, Willet M, Papaneri AB, Wirblich C, Feldmann F, et al. Antibody quality and protection from lethal Ebola virus challenge in nonhuman primates immunized with rabies virus based bivalent vaccine. PLoS Pathog. 2013;9.

Bosio CM, Aman MJ, Grogan C, Hogan R, Ruthel G, Negley D. Ebola and Marburg viruses replicate in monocyte-derived dendritic cells without inducing the production of cytokines and full maturation. J Infect Dis. 2003;188:1630–8.

Bukreyev A, Yang L, Zaki SR, Shieh WJ, Rollin PE, Murphy BR, et al. A single intranasal inoculation with a paramyxovirus-vectored vaccine protects guinea pigs against a lethal-dose Ebola virus challenge. J Virol. 2006;80:2267–79.

Bukreyev A, Rollin PE, Tate MK, Yang L, Zaki SR, Shieh W-J, et al. Successful topical respiratory tract immunization of primates against Ebola virus. J Virol. 2007;81:6379–88.

Carter RJ, Idriss A, Widdowson MA, Samai M, Schrag SJ, Legardy-Williams JK, et al. Implementing a multisite clinical trial in the midst of an Ebola outbreak: lessons learned from the Sierra Leone trial to introduce a vaccine against Ebola. J Infect Dis. 2018;217:S16–23.

Centers for Disease Control and Prevention. Ebola in Liberia. Atlanta: Centers for Disease Control and Prevention; 2015. http://wwwnc.cdc.gov/travel/notices/warning/ebola-liberia. Accessed 4 Feb 2015.

Centers for Disease Control and Prevention. Ebola virus disease Prevention and Vaccine. Atlanta: Centers for Disease Control and Prevention; 2019. https://www.cdc.gov/vhf/ebola/prevention/index.html. Accessed 15 Nov 2019.

Chan SY, Empig CJ, Welte FJ, Speck RF, Schmaljohn A, Kreisberg JF, et al. Folate receptor-alpha is a cofactor for cellular entry by Marburg and Ebola viruses. Cell. 2001;106:117–26.

Clarke DK, Hendry RM, Singh V, Rose JK, Seligman SJ, Klug B, et al. Live virus vaccines based on a vesicular stomatitis virus (VSV) backbone: standardized template with key considerations for a risk/benefit assessment. Vaccine. 2016;34:6597–609.

Dahlmann F, Biedenkopf N, Babler A, Jahnen-Dechent W, Karsten CB, Gnirs K, et al. Analysis of ebola virus entry into macrophages. J Infect Dis. 2015;212:S247–57.

de La Vega MA, Wong G, Kobinger GP, Qiu X. The multiple roles of sGP in Ebola pathogenesis. Viral Immunol. 2015;28:3–9.

Dolzhikova IV, Zubkova OV, Tukhvatulin AI, Dzharullaeva AS, Tukhvatulina NM, Shcheblyakov DV, et al. Safety and immunogenicity of GamEvac-Combi, a heterologous VSV- and Ad5-vectored Ebola vaccine: an open phase I/II trial in healthy adults in Russia. Hum Vaccines Immunother. 2017;13:613–20.

Enserink M. Infectious diseases: ebola drugs still stuck in lab. Science 2014;345:364–365.

Feldmann H, Feldmann F, Marzi A. Ebola: lessons on vaccine development. Annu Rev Microbiol. 2018;72:423–46.

Fujihira H, Usami K, Matsuno K, Takeuchi H, Denda-Nagai K, Furukawa J, et al. A critical domain of ebolavirus envelope glycoprotein determines glycoform and infectivity. Sci Rep. 2018;8:5495.

Furuyama W, Marzi A. Ebola virus: pathogenesis and countermeasure development. Annu Rev Virol. 2019;6:435–58.

Geisbert TW, Bailey M, Hensley L, Asiedu C, Geisbert J, Stanley D, et al. Recombinant adenovirus serotype 26 (Ad26) and Ad35 vaccine vectors bypass immunity to Ad5 and protect nonhuman primates against ebolavirus challenge. J Virol. 2011;85:4222–33.

Grant-Klein RJ, Altamura LA, Badger CV, Bounds CE, Van Deusen NM, Kwilas SA, et al. Codon-optimized filovirus DNA vaccines delivered by intramuscular electroporation protect cynomolgus macaques from lethal Ebola and Marburg virus challenges. Hum Vaccines Immunother. 2015;2015(11):1991–2004.

Groseth A, Feldmann H, Strong JE. The ecology of Ebola virus. Trends Microbiol. 2007;15:408–16.

Guito JC, Albarino CG, Chakrabarti AK, Towner JS. Novel activities by Ebola virus and Marburg virus interferon antagonists revealed using a standardized in vitro reporter system. Virology. 2017;501:147–65.

Halfmann P, Ebihara H, Marzi A, Hatta Y, Watanabe S, Suresh M, et al. Replication-deficient Ebola virus as a vaccine candidate. J Virol. 2009;83:3810–5.

Halperin SA, Arribas JR, Rupp R, Andrews CP, Chu L, Das R, et al. Six-month safety data of recombinant vesicular stomatitis virus–Zaire Ebola virus envelope glycoprotein vaccine in a phase 3 double-blind, placebo-controlled randomized study in healthy adults. J Infect Dis. 2017;215:1789–98.

Han Z, Boshra H, Sunyer JO, Zwiers SH, Paragas J, Harty RN. Biochemical and functional characterization of the Ebola virus VP24 protein: implications for a role in virus assembly and budding. J Virol. 2003;77:1793–800.

Hartlieb B, Muziol T, Weissenhorn W, Becker S. Crystal structure of the C-terminal domain of Ebola virus VP30 reveals a role in transcription and nucleocapsid association. Proc Natl Acad Sci USA. 2007;104:624–9.

Henao-Restrepo AM, Camacho A, Longini IM, Watson CH, Edmunds WJ, Egger M, et al. Efficacy and effectiveness of an rVSV-vectored vaccine in preventing Ebola virus disease: final results from the Guinea ring vaccination, open-label, cluster-randomised trial. Lancet. 2017;389:505–18.

Hensley LE, Alves DA, Geisbert JB, Fritz EA, Reed C, Larsen T, et al. Pathogenesis of Marburg hemorrhagic fever in cynomolgus macaques. J Infect Dis. 2011;204:S1021–31.

Hoenen T, Feldmann H. Reverse genetics systems as tools for the development of novel therapies against filoviruses. Expert Rev Anti Infect Ther. 2014;12:1253–63.

Huttner A, Dayer JA, Yerly S, Combescure C, Auderset F, Desmeules J, et al. The effect of dose on the safety and immunogenicity of the VSV Ebola candidate vaccine: a randomised doubleblind, placebo-controlled phase 1/2 trial. Lancet Infect Dis. 2015;15:1156–66.

Ilinykh PA, Shen X, Flyak AI, Kuzmina N, Ksiazek TG, Crowe JE Jr, et al. Chimeric filoviruses for identification and characterization of monoclonal antibodies. J Virol. 2016;90:3890–901.

Jasenosky LD, Kawaoka Y. Filovirus budding. Virus Res. 2004;106:181–8.

Kennedy SB, Bolay F, KiehM, Grandits G, Badio M, Ballou R, et al. Phase 2 placebo-controlled trial of two vaccines to prevent Ebola in Liberia. N Engl J Med. 2017;377:1438–1447.

Kirchdoerfer RN, Moyer CL, Abelson DM, Saphire EO. The ebola virus VP30-NP interaction is a regulator of viral RNA synthesis. PLoS Pathog. 2016;12.

Le Guenno B, Formenty P, Wyers M, Gounon P, Walker F, Boesch C. Isolation and partial characterisation of a new strain of Ebola virus. Lancet. 1995;345:1271–4.

Lee JE, Saphire EO. Ebola virus glycoprotein structure and mechanism of entry. Future Virol. 2009;4:621–35.

Letchworth GJ, Rodriguez LL, Del C, Barrera J. Vesicular stomatitis. Vet J. 1999;157:239–60.

Lupton HW, Lambert RD, Bumgardner DL, Moe JB, Eddy GA. Inactivated vaccine for Ebola virus efficacious in guineapig model. Lancet. 1980;2:1294–5.

Madara JJ, Han Z, Ruthel G, Freedman BD, Harty RN. The multifunctional Ebola virus VP40 matrix protein is a promising therapeutic target. Future Virol. 2015;10:537–46.

Malashkevich VN, Schneider BJ, McNally ML, Milhollen MA, Pang JX, Kim PS. Core structure of the envelope glycoprotein GP2 from Ebola virus at 1.9-A resolution. Proc Natl Acad Sci USA. 1999;96:2662–2667.

Martin B, Canard B, Decroly E. Filovirus proteins for antiviral drug discovery: structure/function bases of the replication cycle. Antivir Res. 2017;141:48–61.

Marzi A, Halfmann P, Hill-Batorski L, Feldmann F, Shupert WL, Neumann G, et al. An Ebola whole-virus vaccine is protective in nonhuman primates. Science. 2015a;348:439–42.

Marzi A, Robertson SJ, Haddock E, Feldmann F, Hanley PW, Scott DP, et al. VSV-EBOV rapidly protects macaques against infection with the 2014/15 Ebola virus outbreak strain. Science. 2015b;349:739–42.

Matsuno K, Kishida N, Usami K, Igarashi M, Yoshida R, Nakayama E, et al. Different potential of C-type lectin-mediated entry between Marburg virus strains. J Virol. 2010;84:5140–7.

Mehedi M, Falzarano D, Seebach J, Hu X, Carpenter MS, Schnittler H-J, et al. A new Ebola virus nonstructural glycoprotein expressed through RNA editing. J Virol. 2011;85:5406–14.

Meyer M, Garron T, Lubaki NM, Mire CE, Fenton KA, Klages C, et al. Aerosolized Ebola vaccine protects primates and elicits lung-resident T cell responses. J Clin Invest. 2015;125:3241–55.

Milligan ID, Gibani MM, Sewell R, Clutterbuck EA, Campbell D, Plested E, et al. Safety and immunogenicity of novel adenovirus type 26- and modified vaccinia Ankara-vectored Ebola vaccines: a randomized clinical trial. JAMA. 2016;315:1610–23.

Mire CE, Miller AD, Carville A, Westmoreland SV, Geisbert JB, Mansfield KG, et al. Recombinant vesicular stomatitis virus vaccine vectors expressing filovirus glycoproteins lack neurovirulence in nonhuman primates. PLOS Negl Trop Dis. 2012;6.

Moghadam SRJ, Omidi N, Bayrami S, Moghadam SJ, Alinaghi SAS. Ebola viral disease: a review literature. Asian Pac J Trop Biomed. 2015;5:260–7.

Mohan GS, Li W, Ye L, Compans RW, Yang C. Antigenic subversion: a novel mechanism of host immune evasion by Ebola virus. PLoS Pathog. 2012;8.

Moller-Tank S, Maury W. Ebola virus entry: a curious and complex series of events. PLoS Pathog. 2015;11.

Monath TP, Fast PE, Modjarrad K, Clarke DK, Martin BK, Fusco J, et al. rVSVΔG-ZEBOV-GP (also designated V920) recombinant vesicular stomatitis virus pseudotyped with Ebola Zaire Glycoprotein: standardized template with key considerations for a risk/benefit assessment. Vaccine X. 2019;1.

Muhlberger E. Filovirus replication and transcription. Future Virol. 2007;2:205–15.

Nanbo A, Imai M, Watanabe S, Noda T, Takahashi K, Neumann G, et al. Ebolavirus is internalized into host cells via macropinocytosis in a viral glycoprotein-dependent manner. PLoS Pathog. 2010;6.

O'Hearn A, Wang M, Cheng H, Lear-Rooney CM, Koning K, Rumschlag-Booms E, et al. Role of EXT1 and glycosaminoglycans in the early stage of filovirus entry. J Virol. 2015;89:5441–9.

Patel A, Reuschel EL, Kraynyak KA, Racine T, Park DH, Scott VL, et al. Protective efficacy and long-term immunogenicity in cynomolgus macaques by Ebola virus glycoprotein synthetic DNA vaccines. J Infect Dis. 2018;219:544–55.

Pleet ML, DeMarino C, Lepene B, Aman MJ, Kashanchi F. The role of exosomal VP40 in ebola virus disease. DNA Cell Biol. 2017;36:243–8.

Pushko P, Bray M, Ludwig GV, Parker M, Schmaljohn A, Sanchez A, et al. Recombinant RNA replicons derived from attenuated Venezuelan equine encephalitis virus protect guinea pigs and mice from Ebola hemorrhagic fever virus. Vaccine. 2000;19:142–53.

Rajak H, Jain DK, Singh A, Sharma AK, Dixit A. Ebola virus disease: past, present and future. Asian Pac J Trop Biomed. 2015;5:337–43.

Rasmussen AL. Host factors involved in ebola virus replication. Curr Top Microbiol Immunol. 2017;419:113–50.

Reid SP, Leung LW, Hartman AL, Martinez O, Shaw ML, Carbonnelle C, et al. Ebola virus VP24 binds karyopherin alpha1 and blocks STAT1 nuclear accumulation. J Virol. 2006;80:5156–67.

Reynolds P, Marzi A. Ebola and Marburg virus vaccines. Virus Genes. 2017;53:501–15.

Rogers KJ, Maury W. The role of mononuclear phagocytes in Ebola virus infection. J Leukoc Biol. 2018;104:717–27.

Rojas M, Monsalve DM, Pacheco Y, Acosta-Ampudia Y, Ramirez-Santana C, Ansari AA, et al. Ebola virus disease: an emerging and re-emerging viral threat. J Autoimmun. 2020;106.

Schornberg KL, Shoemaker CJ, Dube D, Abshire MY, Delos SE, Bouton AH. Alpha5beta1-integrin controls ebolavirus entry by regulating endosomal cathepsins. Proc Natl Acad Sci USA. 2009;106:8003–8.

Skiadopoulos MH, Surman SR, Riggs JM, Orvell C, Collins PL, Murphy BR. Evaluation of the replication and immunogenicity of recombinant human parainfluenza virus type 3 vectors expressing up to three foreign glycoproteins. Virology. 2002;297:136–52.

Stanley DA, Honko AN, Asiedu C, Trefry JC, Lau-Kilby AW, Johnson JC, et al. Chimpanzee adenovirus vaccine generates acute and durable protective immunity against ebolavirus challenge. Nat Med. 2014;20:1126–9.

Steele KE, Anderson AO, Mohamadzadeh M. Fibroblastic reticular cell infection by hemorrhagic fever viruses. Immunotherapy. 2009;1:187–97.

Sullivan NJ, Sanchez A, Rollin PE, Yang ZY, Nabel GJ. Development of a preventive vaccine for Ebola virus infection in primates. Nature. 2000;408:605–9.

Sullivan NJ, Geisbert TW, Geisbert JB, Xu L, Yang ZY, Roederer M, et al. Accelerated vaccination for Ebola virus haemorrhagic fever in non-human primates. Nature. 2003;424: 681–4.

Swenson DL, Warfield KL, Negley DL, Schmaljohn A, Aman MJ, Bavari S. Virus-like particles exhibit potential as a pan-filovirus vaccine for both Ebola and Marburg viral infections. Vaccine. 2005;23:3033–42.

Twenhafel NA, Mattix ME, Johnson JC, Robinson CG, Pratt WD, Cashman KA. Pathology of experimental aerosol Zaire ebolavirus infection in rhesus macaques. Vet Pathol. 2013;50: 514–29.

Vanderzanden L, Bray M, Fuller D, Roberts T, Custer D, K Spik, et al. DNA vaccines expressing either the GP or NP genes of Ebola virus protect mice from lethal challenge. Virology. 1998;246:134–144.

Volchkova VA, Dolnik O, Martinez MJ, Reynard O, Volchkov VE. Genomic RNA editing and its impact on Ebola virus adaptation during serial passages in cell culture and infection of Guinea pigs. J Infect Dis. 2011;204:S941–6.

Warfield KL, Perkins JG, Swenson DL, Deal EM, Bosio CM, Aman MJ, et al. Role of natural killer cells in innate protection against lethal ebola virus infection. J Exp Med. 2004;200: 169–79.

Willet M, Kurup D, Papaneri A, Wirblich C, Hooper JW, Kwilas SA, et al. Preclinical development of inactivated rabies virus-based polyvalent vaccine against rabies and filoviruses. J Infect Dis. 2015;2015(212):S414–24.

Wiwanitkit V. Unprecedented scale ebola epidemic in Guinea: what we should know. Asian Pac J Trop Biomed. 2014;4:675.

World Health Organization. Ebola situation report-2 April 2019. Geneva, Switz: World Health Organ; 2019.

World Health Organization. Ebola situation reports: Democratic Republic of the Congo. World Health Organization; 2020. https://www.who.int/csr/don/06-february-2020-ebola-drc/en/. Accessed 6 Feb 2020.

World Health Organization. WHO guidelines for epidemic preparedness and response to measles outbreaks. Geneva: World Health Organization; 2015. http://www.who.int/csr/resources/publications/measles/whocdscsrisr991.pdf. Accessed 4 Feb 2015.

Xu L, Sanchez A, Yang Z, Zaki SR, Nabel EG, Nichol ST, et al. Immunization for Ebola virus infection. Nat Med. 1998;4:37–42.

Zampieri CA, Fortin J-F, Nolan GP, Nabel GJ. The ERK mitogen-activated protein kinase pathway contributes to Ebola virus glycoprotein-induced cytotoxicity. J Virol. 2007;81:1230–40.

Zhang M, Dong C, Xiong S. Vesicular stomatitis virus-vectored multi-antigen tuberculosis vaccine limits bacterial proliferation in mice following a single intranasal dose. Front Cell Infect Microbiol. 2017;7.

Zhang APP, Bornholdt ZA, Liu T, Abelson DM, Lee DE, Li S, et al. The ebola virus interferon antagonist VP24 directly binds STAT1 and has a novel, pyramidal fold. PLoS Pathog. 2012;8.

Chikungunya Virus

Chikungunya Disease: A Concise Review and Its Transmission Model for India

Rajasekhar Mopuri, Satya Ganesh Kakarla, and Srinivasa Rao Mutheneni

Abstract Chikungunya is a major public health problem globally and the disease spreads mainly in tropical and subtropical countries. Since its re-emergence in the islands of the Indian Ocean in 2005, the numbers of Chikungunya cases are gradually increasing across the world. In 2016, the epidemic of Chikungunya occurred in National Capital Territory (NCT) of Delhi, India, reported more than 12,000 infection cases. Subsequent years this disease spread all over the India causing a large number of infections. Besides presenting a concise review in this chapter we have devised and make use of R_0 mathematical model to understand the potential transmission of this virus in India. The R_0 model is a mechanistic transmission model and it is driven by the climatic factors such as minimum, maximum and mean temperatures. The temperature data show that the disease gradually increases over the Southern region of India. The study in this chapter shows that the viral transmission takes place between 20 and 34 °C whereas the peak transmission occurs at 29 °C. Likewise, it also identifies the possible regions in India which are more vulnerable for the virus transmission and disease outbreak. The outcome of this study should help the public health authorities to take care at the beginning and spreading of the infection well in advance and try to implement appropriate measures including vector control operations and management in India more so in the high risk zones.

Keywords Chikungunya virus · Mosquitoes · Transmission · Infections · India · Climate impacts · Reproductive rate · Transmission potential

R. Mopuri · S. G. Kakarla · S. R. Mutheneni (✉)
ENVIS Resource Partner on Climate Change & Public Health, Applied Biology Division, CSIR-Indian Institute of Chemical Technology (CSIR-IICT), Tarnaka, Hyderabad 500007, Telangana, India
e-mail: msrinivas@iict.res.in

R. Mopuri · S. G. Kakarla · S. R. Mutheneni
Academy of Scientific & Innovative Research (AcSIR), Ghaziabad 201002, Uttar Pradesh, India

S. I. Ahmad (ed.), *Human Viruses: Diseases, Treatments and Vaccines*, https://doi.org/10.1007/978-3-030-71165-8_8

1 Introduction

Chikungunya is a viral disease transmitted to humans by the infected mosquitoes (World Health Organization (WHO) 2020). Its outbreaks occur mostly during the monsoon season when the mosquito vector population reaches to its peak. According to the World Health Organization (WHO) report, the vector is the primary reason accountable for millions of cases globally every year (World Health Organization. World health report 1996). Although malaria and dengue have attracted the public attention as widely transmitted mosquito borne diseases, the past decades have experienced epidemics caused by a group of several arthritogenic *Alphaviruses* which belong to the *Togaviridae* family (Suhrbier et al. 2012). Chikungunya virus (CHIKV) is one of the *Alphaviruses* which poses a serious public health threat by causing a disease condition called Chikungunya fever or Chikungunya infection.

Chikungunya was first discovered in 1952 in Makonde, United Republic of Tanzania, and described by Robinson and Lumsden in 1953 (Robinson 1955; Lumsden 1952). The word Chikungunya is derived from the African dialect Swahili or Makonde, which refers both the virus and disease, which reflect the meaning of "to walk bent over" like a chicken, in Swahili (Pialoux et al. 2007). The disease is transmitted by the mosquito *Aedes aegypti*, the principal vector for transmission in urban areas of Asia and *Aedes albopictus*, another predominant vector, responsible for latest large scale epidemics in Asia (Jupp and McIntosh 1988; Parola et al. 2006). Intermittently *Culex* and *Anopheles* have also been reported as vectors for Chikungunya virus transmission globally (Kakarla et al. 2019). Apart from these, the other species of mosquitoes include *A. africanus* in East Africa, *A. furcifer, A. taylori, A. dalzieli, A. luteocephalus,* and *A. vittatus* in West Africa and *A. taylori* and *A. cordellieri* in South Africa serve as vectors for Chikungunya virus (Zeller et al. 2016). Besides human, monkeys, rodents, and other certain unidentified vertebrates can also serve as reservoir for the virus.

2 History of Outbreaks and Geographical Distribution

Chikungunya disease may be followed back as far as in 1779, in a *knokkel-koorts* febrile epidemic, as mentioned in the logbook of Dutch physician, Dr. David Bylon, in Batavia (present day Jakarta, Indonesia) (Metz 2013; Carey 1971). Later the cases started reporting from India during 1824–1825, West Indies and Southern states of the USA during 1826–1828, London in 1869, Tanzania, Zanzibar during 1871–1872, China in 1872 and Africa in 1917–1948. During initial periods it was erroneously identified as "dengue"; this confusion was mainly due to the two diseases were transmitted by the *Aedes* vectors, whereas prior to 1940 there was no proper diagnostic techniques to distinguish between dengue and Chikungunya viruses (Kuno 2015).

The outbreak of Chikungunya virus reported to occurred throughout the World was: South Africa (1956 and 1975–1977); Zimbabwe (1957, 1961–1962, and 1971); Democratic Republic of Congo (1958 and 1960); Zambia (1959); Senegal (1960, and 1996–1997); Uganda (1961–1962 and 1968); Nigeria (1964, 1969, and 1974); Angola (1970–1971); and Central African Republic (1978–1979) (Jupp and McIntosh 1988; Zeller et al. 2016; Powers and Logue 2007; Desdouits et al. 2015). In Asia, the first Chikungunya virus outbreak was reported in Bangkok, Thailand in 1958 by *A. aegypti* vector (Weaver and Forrester 2015). In India, Chikungunya outbreak was first reported in Calcutta in 1963, followed by, epidemics in Chennai (previous known as Madras), Puducherry and Vellore of Tamilnadu (reported 400,000 cases) in 1964. Similarly, Vishakapatnam, Rajahmudry, Kakinada (Andhra Pradesh), Nagpur (Maharashtra) in 1965; and then remerged in Barsi village (Solapur District, Maharashtra) in 1973 (Parashar et al. 2015). Several other epidemics occurred on all of the South-western Indian Ocean islands except Madagascar and Europe during 2005–2007. The emergence of Chikungunya virus occurred in the Americas, in December 2013 and it spread to 17 countries of the South America until December 2014. Since it was first reported in the Saint Martin, the autochthonous transmission of Chikungunya virus was identified in 45 countries in the Caribbean, North America, South America, and Central America (World Health Organization (WHO) 2020; Leparc-Goffart et al. 2014).

The resurgence and global distribution of Chikungunya in different geographical regions is associated with various factors such as globalization of economy, social and cultural aspects, geographical expansion of vectors, loss of herd immunity, urbanization, and international travel and climate change (Pialoux et al. 2007). Globally, about 2.5 million Chikungunya cases were recorded in the last decade, of which majority of the cases reported from Asia and African regions (Powers 2014). As of September 2019, the autochthonous cases of Chikungunya were found in 114 countries and territories, including 32 countries in Africa, 21 countries in Asia, 47 countries in the Americas, two countries in Europe and twelve countries in the Oceania/Pacific islands (Centers for Disease control and prevention 2019).

3 Chikungunya in India

Since the occurrence of one of the largest outbreak of Chikungunya virus in Indian Ocean countries, in December 2005 cases of Chikungunya were reported from several Indian states including Andhra Pradesh, Karnataka, Maharashtra, Tamil Nadu, Madhya Pradesh and Gujarat states in which more than a million cases of Chikungunya fever suffered (Yergolkar et al. 2006; Lahariya and Pradhan 2006). The African genotype of Chikungunya virus carried by the *Aedes* vector was responsible for this outbreak (Yergolkar et al. 2006). By 2010, more than 3.7 million people were infected and the virus spread to more than 18 states and Union

Territories of India (Ministry of Health and Family Welfare 2010). The National Capital Territory (NCT), Delhi was reported to be worst affected (Telle et al. 2016). During the dengue outbreak in 2006, certain sera samples were found positive for both dengue and Chikungunya viruses. This combined outbreak followed in NCT of Delhi during October to December 2010 and again in 2016 (Chahar et al. 2009; Singh et al. 2012).

Chikungunya cases from 2006 to 2019 were analysed and a total of 590,756 cases were reported in India. During this period the highest incidence rate (84.6 per million populations) was reported in 2008 (Fig. 1). Subsequently the incidence rate gradually decreased from 2008 to 2012 but started again, showing increasing trend, from 2013 to 2019. The average incidence rate of the cases was unequally distributed among the different states, and the highest average incidence rates were observed from West Bengal in east, Goa in west, NCT of Delhi in North, Karnataka, Kerala, Pondicherry and Telangana in south, and Arunachal Pradesh, Meghalaya, Tripura, Mizoram and Sikkim in North East states of India (Fig. 2). During the outbreak of Chikungunya virus in 2016, the most affected regions were NCT of Delhi in which were 12,279 cases (incidence rate: 698.10) and in Karnataka were 15,666 cases (incidence rate: 247.32).

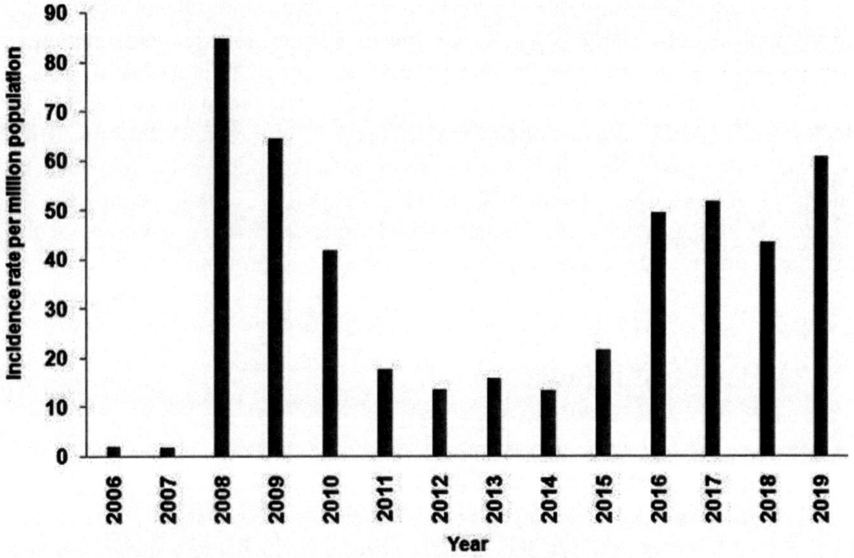

Fig. 1 Time series of Chikungunya case Incidence rate (per million population) in India from 2006 to 2019 (*Data source* NVBDCP, Govt. of India)

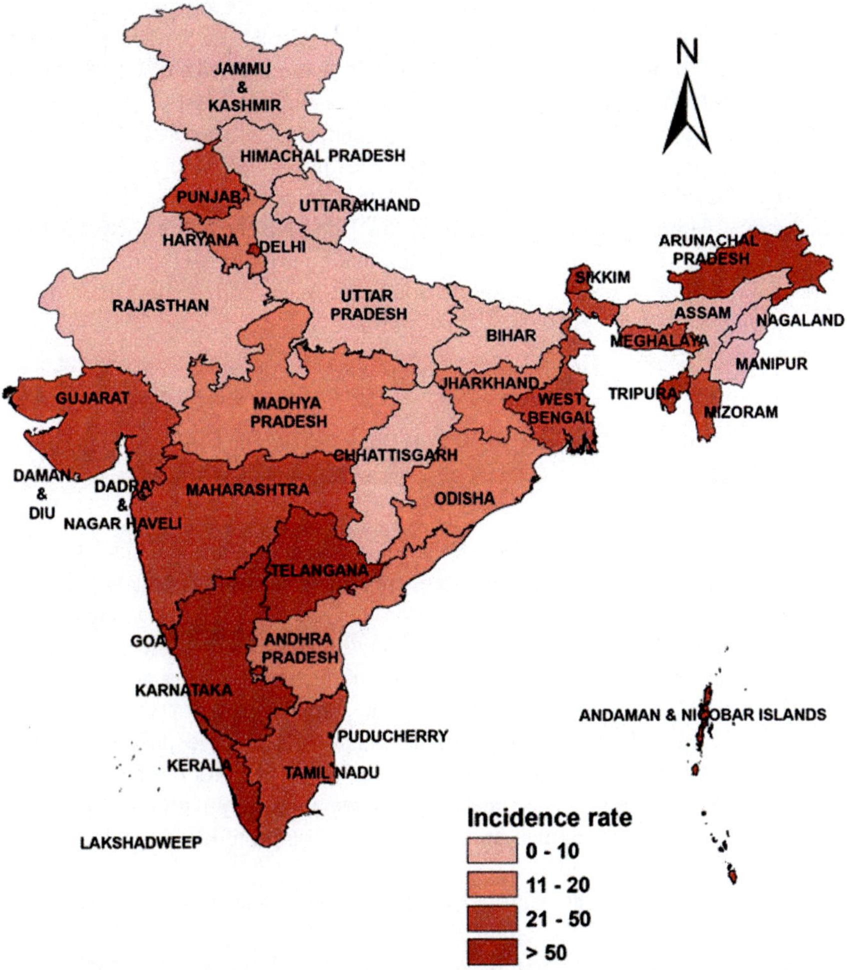

Fig. 2 Spatial distribution Chikungunya incidence rate (per million population) in India from 2006 to 2019 (*Data source* NVBDCP, Govt. of India)

4 Phylogenetic Analysis of Chikungunya Virus

The phylogenetic studies have identified three genotypes of Chikungunya virus as categorised by their geographic locations. The spread of Chikungunya virus occurred from Africa with two different strains: East/Central/South African (ECSA) and the West African strains. Subsequently the ECSA strain spread to Asiatic region (Li et al. 2010). The ECSA strain of Chikungunya had the largest spread and beside its spread in East/Central/South Africa, during 1963–1973 it entered in Europe, Australia, Japan and India. During 2005–2007 outbreaks this Asian strain found in South-east Asia and India and the West African strain prevalent in Western Africa. The ECSA was further grouped into ECSA enzootic, ECSA Brazil and IOL (Indian Ocean lineage) (Volk et al. 2010; Caglioti et al. 2013). Then the Indian Ocean lineage (IOL) strain of Chikungunya virus arose from the ECSA lineages in Indian Ocean outbreak. From the sequencing studies, it is revealed that the IOL strain was actively replaced the existing Asian strains. The phylogenetic and epidemiological studies of IOL strain indicated that infested *Ae. albopictus* always carried an alanine-to-valine substitution in the E1 envelope glycoprotein (E1-A226V) as a key evolutionary conversion that contributed to the re-emergence and spatial distribution of CHIKV in the Southeast Asia (Volk et al. 2010; Powers et al. 2000).

5 Life Cycle of Chikungunya Virus Vectors—The Mosquitoes

Essentially low mosquito population has been associated with dry and cool conditions and increase when the temperature increase and stay (bellow 32 °C) and wet season commences. During onset of monsoon the vector *Aedes aegypti* and *Aedes albopictus* number increases by creating the new breeding habitat. It is a domesticated tropical species that prefers to breed in fresh water, primarily found in artificial water storage containers. Likewise, the stagnant water in and around the locality will serve as artificial breeding sites (such as artificial containers, bottles, buckets, tanks, flower pot, air coolers, coconut shell, tires etc.) for these mosquitoes. *Aedes* mosquitoes are endophagic and lives in close proximity with humans due to its feeding requirement of human and during each gonotrophic cycle which facilitate effective growth of virus.

The vector mosquitoes are diurnal feeder, have four stages of their life cycle: egg, larva, pupa and adult. To complete from egg to adult stage, the insect requires approximately 8–10 days (https://www.cdc.gov/dengue/resources/factSheets/MosquitoLifecycleFINAL.pdf). The adult *Aedes* mosquito shows distinct patterns of white lines on body, a lyre-shaped line in the thorax and palps are half white. The proboscis is generally black and its peak activity is known at mid-morning and late afternoon. It has a short flight range (less than 200 m).

Adult female mosquito lays eggs on moist wall of the container with water. A gravid female can lay about 100 eggs at a time. *Aedes* eggs are hard and can stick to walls of container with glue like substance and can survive up to 8 months. At suitable temperature that is >13° C and availability of stagnated water, within four days the eggs can hatch out into larvae. The larvae then moults three times and passes through four larval stages (Fig. 3), during which the larvae feed on organic particulate matters, microorganisms such as small algae and other phytoplanktons and become pupa. Pupa is a non-feeding motile stage and within two days it develops into adult mosquito.

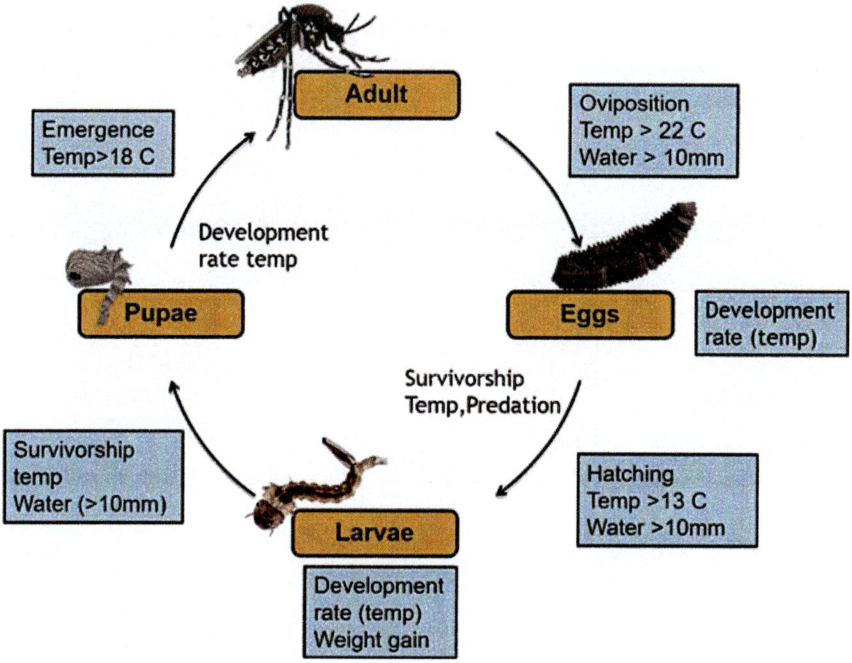

Fig. 3 Mosquito life cycle (eggs, larvae, pupae and adult) and its climate dependency

6 Molecular Structure and Genome Organisation of Chikungunya Virus

Chikungunya virus is a small enveloped microbe with icosahedral symmetry shape which came from the Old World Semliki Forest virus, a group of arthritogenic *alphaviruses* within the *Togaviridae* family (Kuhn 2013). It has a single-stranded, positive-sense RNA genome which is approximately 11,800 nucleotides with a 5′ 7-methylguanosine cap and a 3′ poly-A tail (Khan et al. 2002). The 12 KB Chikungunya virus RNA genome encodes two open reading frames (ORFs) present alienated by a non-coding junction and 5′- and 3′-untranslated regions (Yap et al. 2017). Four non-structural proteins namely nsP1, nsP2, nsP3, and nsP4 constitute the RNA replicase are encoded by the 5′ end of the genome. The 3′ ORF translated from a subgenomic mRNA encodes three major structural proteins namely capsid (C) envelope glycoprotein 1 (E1), envelope glycoprotein 2 (E2) and three peptides {(6 K, transframe (TF), and E3)} (Fig. 4a). The structural proteins are synthesized as a long polyprotein, which is then posttranslationally cleaved into C, E1, 6 K, and p62 proteins (Silva and Dermody 2017).

The 240 copies of capsid protein C form a nucleocapsid shell and encapsidation of genomic RNA Occurs. During release from the host cell, the new virus particles become coated with the envelope glycoproteins E1 and E2. The heterodimers of

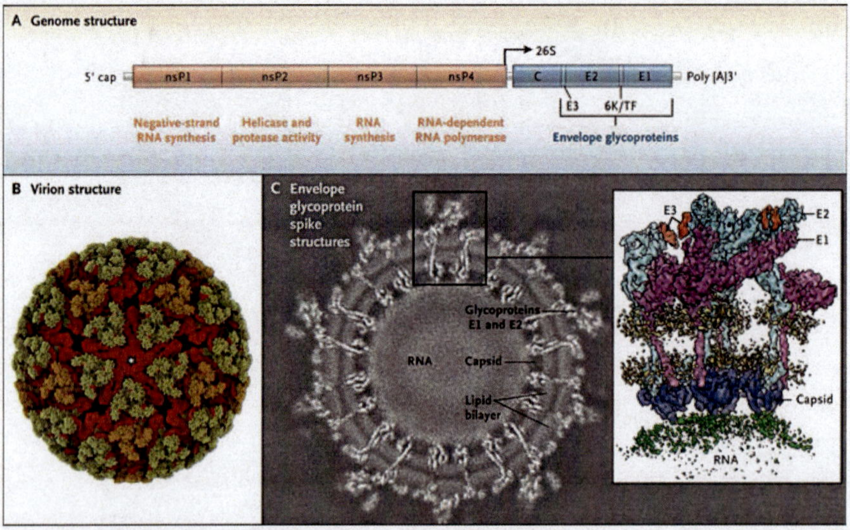

Fig. 4 Chikungunya Virus Genetic and Physical Structure. (*Source* Weaver, S.C. et al., N Engl J Med 2015; 372:1231–1239. https://doi.org/10.1056/nejmra1406035)

E1-E2 glycoproteins arranged in trimers form trimeric spikes on the viral surface form an icosahedral lattice (Fig. 4b, c). The glycoprotein E2 facilitates the entry into the new host cell through binding to yet unidentified receptors. Subsequently at low pH the glycoprotein E1 exposes and precedes the cell fusion and release of the nucleocapsid (Voss et al. 2010). Whereas glycoprotein E3 facilitates E1-p62 heterodimerization and prevents the exposure of the E1 fusion loops from premature fusogenic activation. The envelope glycoproteins of Chikungunya virus are the primary antigenic targets for the host immune response (Skehel and Wiley 2000; Eckert and Kim 2001).

The 20 trimeric spikes of the virus located on the icosahedral three-fold axes and another 60 trimeric spikes in general positions that obey $T = 4$ quasi-symmetry (Voss et al. 2010). The icosahedral capsid of viral particle is surrounded by a lipid envelope, and measures approximately 70 nm in diameter. This lipid bilayer is acquired from the host cell membrane during the process of maturation, assembly, and budding of the virus. Subsequently, the infection requires a fusion of the lipid membrane of virus with the plasma membrane of host cell during the process of a new cycle. During the cell–cell fusion delivers the genetic materials of the virus into the cytoplasm of the infected host. The lipid bilayer of virus contains transmembrane proteins those are involved in the membrane fusion and also for binding to the cellular receptors (Harrison 2005; Kielian and Rey 2006; Weissenhorn et al. 2007).

7 Pathogenesis of Chikungunya Virus

The acute phase of infection is characterized by painful polyarthralgia, high fever, asthenia, headache, body ache, backache, vomiting, rash, and myalgia. In the chronic phase, incapacitating arthralgia persists for months. Neurological syndromes includes encephalitis, encephalopathy, and myelopathy or myeloneuropathy and non-neurological systemic syndromes are renal, hepatic, respiratory, cardiac and haematological manifestations including lymphadenopathy, oral ulcers and encephalitis petechiae. Other specific symptom of Chikungunya infection is prolonged joint pain which occurs over a period of months to years (Suryawanshi et al. 2009; Kannan et al. 2009). The virus can also affect the eyes in several ways such as conjunctivitis, retinitis, optic neuritis photophobia and retro-orbital pains are often seen in the acute phase of fever (Mahendradas et al. 2013). During Chikungunya infection any joint of the body can be affected, commonly reported are the distal extremities such as wrists, metacarpal and interphalangeal joints, as well as the ankles and metatarsophalangeal joints, and knees are commonly affected. Generally these manifestations of arthralgia will resolve within 1–4 weeks after the initial onset but, mortality due to infection is rare (Chang et al. 2014; Goupil and Mores 2016).

Few more observations also reported that Chikungunya virus infection may leads to abortion in pregnant woman. However, if the infection persists the virus can be transmitted from mother-to-child (Ramful et al. 2007). In infants Chikungunya fever manifests with certain differences; fever is commonly present which is associated with some constitutional symptoms such as lethargy or irritability and crying excessively. Acrocyanosis may be prominent followed by symmetrical superficial vesicobullous lesions, erythematous asymmetrical macules, and patches, which later is progressed to morbiliform rashes. The face and oral cavity are usually spared (Mohan et al. 2010; Sebastian et al. 2009).

7.1 Immune Response to Chikungunya Virus

Chikungunya virus infection elicits strong systemic innate immune responses, by producing IFN-α as well as several pro-inflammatory cytokines, chemokines, and growth factors (Wauquier et al. 2011). This is followed by the activation of the adaptive immunity through activation sand proliferation of $CD8^+$ T cells in the early stages of the disease. Followed by $CD4^+$ T-cell response and anti-inflammatory proteins like IL-1RA and IL-2RA produces during acute phase of infection. Chikungunya infection induces a strong inflammatory response that is possibly coordinated by the production of IL-16, IL-17, monocyte chemo-attractant protein 1 (MCP-1), IP-10, and macrophage inflammatory protein (MIP-1α) (Wauquier et al. 2011). BST-2 (bone marrow stromal antigen-2) may protect lymph tissues and regulate CHIKV-induced host inflammatory responses (Mahauad-Fernandez et al. 2014). Similarly, several other cytokines such as TNF-α,

IL-6, and IL-1 are associated with infection also promote osteoclast activity and it has been associated with osteoclastogenesis (Chen et al. 2015).

During the end of acute phase of infection is characterized by the production of proinflammatory Macrophage Migration Inhibitory Factor (MIF), MIP-1β, stromal cell-derived factor (SDF-1α), and IL-6 and IL-8. Similarly, Chemokine (C-C motif) ligand 5 (CCL5: RANTES) levels also high in all patients during the first week after symptom onset (Wauquier et al. 2011). CCL5, MCP-1, IP-10, MIP-1β, and IL-8 are produced by activated macrophages that are susceptible to CHIKV infection (Sourisseau et al. 2007). These chemokines play a major role in leukocyte engage at sites of infection, coordinating the deployment of efficient antiviral defences.

Chikungunya virus infection induces strong cellular immune response. Higher plasma levels of IFN-γ, IL-4, IL-7, and IL-12p40 and cytokines promote the adaptive immunity (Wauquier et al. 2011). The B cell-promoting cytokines IL-4 and in some cases IL-10, were also up regulated in the first few days after symptom onset probably initiating the production of CHIKV-specific IgG. In addition to this, $CD4^+$ T lymphocytes are also involved to maintain the humoral response strongly to finish the acute phase (Wauquier et al. 2011). IgG antibodies can be noticed after the first week of Chikungunya virus infection, indicating rapid seroconversion and high levels of antibody responses among CHIKV-infected individuals (Sourisseau et al. 2007).

8 Diagnosis of Chikungunya Virus

There are several methods to diagnose Chikungunya virus infection. Reverse Transcription-Polymerase Chain Reaction (RT-PCR) method is useful during the initial viremia phase that is first week after the onset of symptoms. The real-time PCRs have been used to amplify nsP1, nsP2, or even envelope protein genes E3, E2, or E1 regions of the CHIKV genome. ELISA (Enzyme Linked Immunosorbent Assay) is considered to be most appropriate test to be used for detection of high levels of IgM and IgG anti-Chikungunya antibodies. IgM antibody levels peaks at 3–5 weeks after the onset of illness and persist for about 2 months. Similarly, high levels of IgG can be detected in convalescent samples and it persists for several years (World Health Organization (WHO) 2020; Kam et al. 2012; Álvarez-Argüelles et al. 2019).

MAC-ELISA (IgM Antibody Capture-ELISA) is another commonly used test for detection of Chikungunya virus. The sensitivity and specificity of serological tests are poor because MAC-ELISA may show cross-reaction with other flavivirus antibodies such as Ross River virus, Barmah Forest virus, Sindbis virus etc. (Marta et al. 2019). Hence, the commercially available rapid diagnostic kits often do not provide as reliable results as RT-PCR because they all tend to detect host-derived anti-CHIKV IgM antibodies. Furthermore, since IgM antibodies are produced later in the course of infection than the antigen, this is a less sensitive test and as such can delay diagnosis and reduce the effectiveness of disease management (Rianthavorn et al. 2010).

9 Treatment of Disease

Currently there is no known vaccine available for Chikungunya virus; however a few candidate vaccines are under development and some at clinical trial stages. These vaccines are making use of various techniques such as attenuation of live virus, inactivated viral vaccine, recombinant viral vaccines, chimeric-alphavirus vaccines, DNA vaccines and virus-like particles (VLPs) (https://www.who.int/blueprint/what/norms-standards/meeting-report.pdf?ua=1).

Chikungunya treatment is generally focused on relieving the disease symptoms. These include taking full rest, intake of fluids; taking an appropriate analgesic and, antipyretics which may relieve symptoms of fever and aching. Movement and mild exercise may improve stiffness and joint pains. In unresolved arthritis that does not respond to an analgesic a non-steroidal anti-inflammatory drug is given (Ramful et al. 2007). Infected persons should be protected from further mosquito exposure (staying indoors and/or under a mosquito net) so that they cannot contribute to further transmission cycle.

10 Transmission of Chikungunya Virus

The Chikungunya virus is mainly transmitted through the bite of infected mosquitoes of *Aedes* species (subgenus *Stegomyia*). The infected female *Aedes* mosquitoes inject the Chikungunya virus into the blood stream of a susceptible human host during their blood meal. The human host becomes infected with the Chikungunya virus and the virus particles replicate within the body causing the Chikungunya fever. The subsequent blood meal by another susceptible mosquito pickups the virus from infected human host and transfer the virus from one human to the next. The behaviour and ecology of *Aedes aegypti* favours the epidemic transmission of the virus since these species have ready access to feed on human hosts and obtain several partial blood meals during a single gonotrophic cycle (Monteiro et al. 2019).

11 Control of Mosquito Breeding

A. aegypti is a container habitat species and breeds primarily in clay and damp containers and vessels. The following steps should be taken to control mosquito breeding.

- Take all possible care that no containers including air cooler fridge etc. where water is standing for long time making it stagnant and infected should be left. Fill the ditches and clean the blocked drains.

- If any kind of accumulation of mosquito found, ask health department to spray the area with insecticidal agents.
- Introduce Larvivorous fish like *Gambusia affinis*, *lebister* or some *cyprinids* which can effectively reduce mosquito larval population in mosquito breeding ponds.

12 Climate Driven Chikungunya Transmission Potential Model

12.1 Basic Reproduction Rate (R_0) Model

Like other vector-borne diseases, the life cycle of Chikungunya consists of an epidemiological triad this includes host, vector and virus. The association between host vector and virus is strongly influenced by various climatic factors such as temperature and rainfall. Rainfall creates breeding habitats for vectors and temperature plays a critical role in development of mosquitoes and viral transmission. In the present study different entomological and epidemiological factors related to disease transmission mechanism such as mosquito biting rate, vector to host transmission rate of virus, host to vector transmission rate, extrinsic incubation period and mortality rate were simulated as functions of temperature and which were used to assess the transmission intensity of Chikungunya through basic reproductive rate (R_0) model. The basic reproductive rate (R_0) is helpful to understand the risk of spread of an infectious disease as it provides an index for transmission intensity and its threshold levels. The R_0 can be defined as the average number of secondary infections which arises from a primary infection in a susceptible population. If R_0 is greater than one it indicates that the number of people will be infected is more and it could spread further transmission, which leads to an epidemic, however, if R_0 is less than one, virus transmission goes to reduce and disease outbreak may not appear in the population (Smith et al. 2007; Van den Driessche and Watmough 2002; Caminade et al. 2017).

The transmission rate is modelled as the basic reproduction rate R_0. The R_0 model denotes that, the relationships of parameters are required for an infection to spread within a population ($R_0 > 1$) as opposed to dying out ($R_0 < 1$). R_0 helps the public health authorities to quantify the intensity of control interventions required to prevent an outbreak. R_0 can be expressed by following equation (Degallier et al. 2010):

$$R_0(t) = m\,a^2\,bc\;exp\,(-\mu\tau_e)/\mu r \tag{1}$$

Here R_0(t) represents the reproductive rate as the function temperature t, m is relative density of mosquitoes per human, a is number of bites per day, b indicates vector to host infection probability, c indicates host to vector infection probability,

μ is daily mortality rate of vector, $_e$ is the extrinsic incubation period (EIP) in days, γ viremia in days. The parameters which were included in the model were derived from temperature dependent mathematical equations. For visualization of data, standard anomalies of R_0 were calculated using following equation:

$$R_0\,stdanomaly = \frac{R_0 - mean(R_0)}{std(R_0)} \tag{2}$$

Biting rate: The average blood meal frequency (a) of female *Ae.aegypti* mosquito linearly increased with increasing temperature estimated per day using the following equation (Liu-Helmersson et al. 2014; Scott et al. 2000):

$$a(T) = 0.0943 + 0.0043T\ 21°\mathrm{C} \leq \mathrm{T} \leq 32°\mathrm{C} \tag{3}$$

Human to mosquito transmission: The average number of hosts infected by the introduction of an infective vector in a susceptible population is given by the transmission probability of mosquito to human per bite (b) represented as (Focks et al. 1995):

$$b(T) = -0.9037 + 0.0729T \tag{4}$$

Mosquito to human transmission probability is temperature dependent and increases linearly with temperature. The temperature range is (26.1 °C < T < 32.5 °C) equal to 1 and the transmission probability was '0' if the temperature was < 12.4°C. This relationship is piecewise linear, with increasing probability starting at 12.4 °C and a constant probability of one above 26.1 °C (Focks et al. 1995).

Mosquito to Human transmission: The number of vectors directly infected from the single infective host is given by the transmission probability of human to mosquito per bite (c). It is a linear relation with temperature (T) and it increases when 12.4 °C < T < 28 °C decreases sharply when T > 28 °C and equal to zero at T > 32.5 °C (Focks et al. 1995):

$$c(T) = 0.001044T(\mathrm{T} - 12.286)\sqrt{32.461 - T} \tag{5}$$

Extrinsic incubation period: To investigate the effect of temperature on the extrinsic incubation period (EIP) of Chikungunya virus in India, we used the model described by Focks et al., for dengue virus (Focks et al. 1995). This model relates the EIP with temperature as a covariate range between 12 and 36 °C (McLean et al. 1974; Watts et al. 1987). The EIP (τ_e, in days) estimated for each state on the basis of temperature (T, in °C), by using the following equation:

$$\tau_e(\mathrm{T}) = 4 + \exp(5.15 - 0.123T) \quad (6)$$

Mortality rate: The mortality rate (μ) per day of adult mosquitoes was computed using temperature as a function. The mortality rates were assessed based on an experimental study on female *Ae. aegypti* mosquitoes over the temperature range of 10.54 °C $\leq T \leq$ 33.41 °C (Yang et al. 2009). The mortality rates were increasing with increasing temperature.

$$\mu(T) = 0.8692 - 0.1590T + 0.01116T^2 - 3.408 \times 10^{-4}T^3 + 3.809 \times 10^{-6}T^4 \quad (7)$$

Constant parameter: Based on existing literature, the viremia (γ) period was reported as six days and the viral load peak during commencement of illness followed by it gradually decline (Ng et al. 2017; Riswari et al. 2016). Many modelling studies estimated the mosquito density (m) to be between 1 and 3 per person and also found that the high vector density can be found during the outbreak period. Hence the study considered three mosquitoes per person for model development (Christofferson et al. 2014; Johansson et al. 2014). The above parameters are sensitive to temperature and hence, R_0 is also highly sensitive to climate. Among these derived parameters, the biting rate, and vector to host and host to vector transmission of Chikungunya is directly proportional to the R_0 whereas EIP and mosquito mortality rate are inversely proportional to R_0.

13 Climatic Conditions in India

India displays a wide range of climatic conditions due to its vast geographical area and other factors like latitude, longitude which influence local climatic conditions. The temperatures between 1948–1981 and 1982–2016 across the India show that there was an increase (1–2 °C) in average temperatures over South India and foothills of the Himalaya region. The average maximum temperatures increased over South India and average minimum temperatures doesn't show much differences in other parts of India. Similarly, a significant decrease ($\sim$20 mm) of rainfall was observed over Western Ghats region and some parts of Uttar Pradesh.

14 Temperature Dependent Chikungunya Transmission Potential Model

Figure 5 describes the climate suitability for Chikungunya transmission potential at different temperatures such as the Chikungunya transmission was calculated between 15 and 35 °C. The simulated R_0 shows that the optimal temperature for

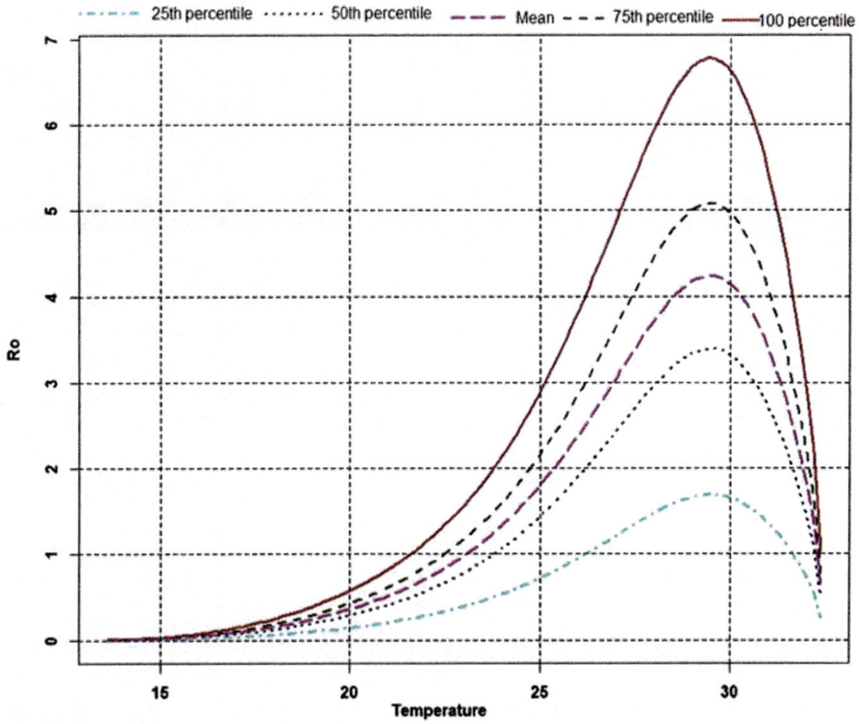

Fig. 5 The R_0 scenario at different constant temperature at 25th, 50th, 75th, 100th and mean percentile

peak Chikungunya transmission is 29 °C and the R_0 was reported as zero when the temperature was below <17 °C and above >34 °C. Based on mean R_0 values it is assessed that the Chikungunya transmission is unsuitable when temperatures below <21.5 °C and above >34 °C (mean $R_0 < 0.5$) and highly suitable between >24 °C and <34 °C (mean $R_0 > 1$). The R_0 thermal response curve shows that the predicted temperatures are better suited for *Aedes* vectors.

15 R_0 Model for India

The average mean annual R_0 values were simulated for the period of 1948–1981 and 1982–2016 by using monthly mean temperature for all the states of India. This reveals that the central, north and northeast states have low R_0 values, but East and West coastal states have high R_0 values. The high altitude regions of north and north-eastern states are situated in the foothills of the Himalayas, hence, generally report low temperatures, which leads to low observation of R_0. The states Gujarat, Maharashtra, Karnataka, Kerala, Tamil Nadu, Andhra Pradesh (united), Odisha and

West Bengal have shown high R_0 values. Similarly these states are relatively endemic for Chikungunya and high incidence rate is observed. However the Chikungunya transmission pattern is expanding during 1982–2016 period mainly in Southern, Northern and Western region in comparison with 1948–1981periods. This increase may be due to increasing of temperature which results low EIP days and mortality rates and high biting rate. Among all months the high Chikungunya transmission and high R_0 values was reported during the month of September across the India.

Similarly, the present study also assessed the seasonal pattern of Chikungunya transmission in India, which shows that high R_0 values was reported in summer due to increased temperatures, followed by monsoon period. The monsoon period, rainfall provides ample breeding habitats and temperature helps the lifecycle to complete faster in both vector and virus. Similarly, the temperature dependent parameters shows that monsoon period has high biting rate, low EIPs days and mortality rates which leads the higher R_0.

16 Conclusions

It has been long since the Chikungunya virus infection in human started and causing a major public health issue globally. The mathematical model is developed and presented her is based on simulations of pathogen traits and entomological factors which help to understand the effects of the temperature range for optimal transmission of Chikungunya in India. The study also predicts the regions that are more vulnerable for the viral transmission or outbreak in India. The model showed that the R_0 peaked at 29 °C ($R_0 > 6.5$) which indicates that it is most suitable temperature for transmission of Chikungunya virus. The model indicates that the coastal areas and Southern region of India has the environment more suitable for Chikungunya transmission. Whereas the Northeast, central region and few geographic regions of North India show low suitability for Chikungunya transmission, this may be due to low-temperature.

In India most of the historical outbreaks of Chikungunya were associated with the monsoon and post-monsoon periods. The first outbreak of Chikungunya (1963 July to December) in India occurred during the monsoon and post monsoon period (Ray et al. 2012). Followed by the outbreaks in Orissa in 2006–2007, and in NCT of Delhi in monsoon and post monsoon periods (Dwibedi et al. 2011). Therefore the Chikungunya transmission and outbreaks in India follows a strong seasonality predominantly in monsoon period where rainfall and temperature provides ambient conditions for both vector and virus. Our model predicts the risk of Chikungunya transmission and outbreaks during the monsoon period. Similarly, the population of vectors such as *Aedes albopictus* and *Aedes aegypti* is very high in many parts of the India and they are not only vectors for Chikungunya but they can also transmit dengue, yellow fever and Zika virus (Mutheneni et al. 2017). A comprehensive vector/larval control operations should be implemented on priority basis and the study also emphasize that, most of the states are climatologically suitable for disease transmission hence a continuous surveillance studies can be taken to understand the disease burden.

Acknowledgements The authors Rajasekhar Mopuri, Satya Ganesh Kakarla and Srinivasa Rao Mutheneni are grateful to the Director of the Council of Scientific and Industrial Research-Indian Institute of Chemical Technology, Hyderabad for his encouragement and support. Srinivasa Rao Mutheneni acknowledges Ministry of Environment, Forest & Climate Change (MoEF & CC), Government of India for funding the project environmental information system (ENVIS: Resource Partner on Climate Change and Public Health). CSIR-IICT communication number of the book chapter is IICT/Pubs./2020/165.

References

Caglioti C, Lalle E, Castilletti C, Carletti F, Capobianchi MR, Bordi L. Chikungunya virus infection: an overview. New Microbiol. 2013;36(3):211–27. https://doi.org/10.1016/j.rmu.2015.06.001.

Caminade C, Turner J, Metelmann S, et al. Global risk model for vector-borne transmission of Zika virus reveals the role of El Niño 2015. Proc Natl Acad Sci U S A. 2017;114(1):119–24. https://doi.org/10.1073/pnas.1614303114.

Carey DE. Chikungunya and dengue: a case of mistaken identity? J Hist Med Allied Sci. 1971;26(3):243–62. https://doi.org/10.1093/jhmas/xxvi.3.243.

Centers for Disease control and prevention. Geographic distribution of Chikungunya virus: Countries with reported current or previous local transmission of Chikungunya virus (as of September 17, 2019) Atlanta, GA: National Center for Emerging and Zoonotic Infectious Diseases Division of Vector-Borne Diseases; 2019.

Chahar HS, Bharaj P, Dar L, Guleria R, Kabra SK, Broor S. Co-infections with chikungunya virus and dengue virus in Delhi, India. Emerg Infect Dis. 2009;15(7):1077–80. https://doi.org/10.3201/eid1507.080638.

Chang LJ, Dowd KA, Mendoza FH, et al. Safety and tolerability of chikungunya virus-like particle vaccine in healthy adults: a phase 1 dose-escalation trial. Lancet. 2014;384(9959):2046–52. https://doi.org/10.1016/S0140-6736(14)61185-5.

Chen W, Foo SS, Sims NA, Herrero LJ, Walsh NC, Mahalingam S. Arthritogenic alphaviruses: new insights into arthritis and bone pathology. Trends Microbiol. 2015;23(1):35–43. https://doi.org/10.1016/j.tim.2014.09.005.

Christofferson RC, Chisenhall DM, Wearing HJ, Mores CN. Chikungunya viral fitness measures within the vector and subsequent transmission potential. PLoS ONE. 2014;9(10): https://doi.org/10.1371/journal.pone.0110538.

Degallier N, Favier C, Menkes C, Lengaigne M. Toward an early warning system for dengue prevention: modeling climate impact on dengue transmission. Clim Change. 2010;98:581–92.

Desdouits M, Kamgang B, Berthet N, et al. Genetic characterization of Chikungunya virus in the Central African Republic. Infect Genet Evol. 2015;33:25–31. https://doi.org/10.1016/j.meegid.2015.04.006.

Dwibedi B, Sabat J, Mahapatra N, et al. Rapid spread of chikungunya virus infection in Orissa: India. Indian J Med Res. 2011;133(3):316–21.

Eckert DM, Kim PS. Mechanisms of viral membrane fusion and its inhibition. Annu Rev Biochem. 2001;70:777–810. https://doi.org/10.1146/annurev.biochem.70.1.777.

Focks DA, Daniels E, Haile DG, Keesling JE. A simulation model of the epidemiology of urban dengue fever: literature analysis, model development, preliminary validation, and samples of simulation results. Am J Trop Med Hyg. 1995;53(5):489–506. https://doi.org/10.4269/ajtmh.1995.53.489.

Goupil BA, Mores CN. A review of chikungunya virus-induced arthralgia: clinical manifestations, therapeutics, and pathogenesis. Open Rheumatol J. 2016;10:129–40. https://doi.org/10.2174/1874312901610010129.

Harrison SC. Mechanism of membrane fusion by viral envelop proteins. Adv Virus Res. 2005;64:231–61. https://doi.org/10.1016/S0065-3527(05)64007-9.

Kuhn RJ. Togaviridae. In: Knipe DM, Howley PM, eds. Fields Virology. Philadelphia, Pennsylvania, USA: Lippincott, Williams, and Wilkins; 2013:629–650.

Johansson MA, Powers AM, Pesik N, Cohen NJ, Staples JE. Nowcasting the spread of chikungunya virus in the Americas. PLoS ONE. 2014;9(8): https://doi.org/10.1371/journal.pone.0104915.

Jupp P, McIntosh B. Chikungunya virus disease. In: Monath T, editor. The arboviruses: epidemiology and ecology. Boca Raton Florida: CRC Press; 1988. p. 137–57.

Kakarla SG, Mopuri R, Mutheneni SR, et al. Temperature dependent transmission potential model for chikungunya in India. Sci Total Environ. 2019;647:66–74. https://doi.org/10.1016/j.scitotenv.2018.07.461.

Kam YW, Lum FM, Teo TH, et al. Early neutralizing IgG response to Chikungunya virus in infected patients targets a dominant linear epitope on the E2 glycoprotein. EMBO Mol Med. 2012;4(4):330–43. https://doi.org/10.1002/emmm.201200213.

Kannan M, Rajendran R, Sunish IP, et al. A study on chikungunya outbreak during 2007 in Kerala, south India. Indian J Med Res. 2009;129(3):311–5.

Khan AH, Morita K, Parquet MDC, Hasebe F, Mathenge EGM, Igarashi A. Complete nucleotide sequence of chikungunya virus and evidence for an internal polyadenylation site. J Gen Virol. 2002;83(Pt12):3075–84. https://doi.org/10.1099/0022-1317-83-12-3075.

Kielian M, Rey FA. Virus membrane-fusion proteins: more than one way to make a hairpin. Nat Rev Microbiol. 2006;4(1):67–76. https://doi.org/10.1038/nrmicro1326.

Kuno G. A re-examination of the history of etiologic confusion between dengue and chikungunya. PLoS Negl Trop Dis. 2015;9(11): https://doi.org/10.1371/journal.pntd.0004101.

Lahariya C, Pradhan SK. Emergence of chikungunya virus in Indian subcontinent after 32 years: a review. J Vector Borne Dis. 2006;43(4):151–60.

Leparc-Goffart I, Nougairede A, Cassadou S, Prat C, De Lamballerie X. Chikungunya in the Americas. Lancet. 2014;383:514. https://doi.org/10.1016/S0140-6736(14)60185-9.

Li L, Jose J, Xiang Y, Kuhn RJ, Rossmann MG. Structural changes of envelope proteins during alphavirus fusion. Nature. 2010;468(7324):705–8. https://doi.org/10.1038/nature09546.

Liu-Helmersson J, Stenlund H, Wilder-Smith A, Rocklöv J. Vectorial capacity of *Aedes aegypti*: effects of temperature and implications for global dengue epidemic potential. PLoS ONE. 2014;9(3): https://doi.org/10.1371/journal.pone.0089783.

Lumsden WH. An epidemic of virus disease in Southern Province, Tanganyika Territory, in 1952–53. II. General description and epidemiology. Trans R Soc Trop Med Hyg. 1955;49(1):33–57. https://doi.org/10.1016/0035-9203(55)90081-x.

Mahauad-Fernandez WD, Jones PH, Okeoma CM. Critical role for bone marrow stromal antigen 2 in acute Chikungunya virus infection. J Gen Virol. 2014;95:2450–61. https://doi.org/10.1099/vir.0.068643-0.

Mahendradas P, Avadhani K, Shetty R. Chikungunya and the eye: a review. J Ophthalmic Inflamm Infect. 2013;3(1):35. https://doi.org/10.1186/1869-5760-3-35.

Marta E. Álvarez-Argüelles, Susana Rojo Alba, Mercedes Rodríguez Pérez, Jose Antonio Boga Riveiro and Santiago Melón García (June 13th 2019). Diagnosis and Molecular Characterization of Chikungunya Virus Infections, Current Topics in Neglected Tropical Diseases, Alfonso J. Rodriguez-Morales, IntechOpen. 10.5772/intechopen.86957.

McLean DM, Clarke AM, Coleman JC, et al. Vector capability of *Aedes aegypti* mosquitoes for California encephalitis and dengue viruses at various temperatures. Can J Microbiol. 1974;20 (2):255–62. https://doi.org/10.1139/m74-040.

Metz SWH. Chikungunya virus-like particle vaccine. Wageningen, NL: Wageningen University; 2013.

Ministry of Health and Family Welfare. Annual Report 2009–10. Government of India: Ministry of Health and Family Welfare; 2010.

Mohan A, Kiran DH, Manohar IC, Kumar DP. Epidemiology, clinical manifestations, and diagnosis of Chikungunya fever: lessons learned from the re-emerging epidemic. Indian J Dermatol. 2010;55(1):54–63. https://doi.org/10.4103/0019-5154.60355.

Monteiro VVS, Navegantes-Lima KC, de Lemos AB, et al. Aedes-Chikungunya virus interaction: key role of vector midguts microbiota and its saliva in the host infection. Front Microbiol. 2019;10:492. https://doi.org/10.3389/fmicb.2019.00492.

Mutheneni SR, Morse AP, Caminade C, Upadhyayula SM. Dengue burden in India: recent trends and importance of climatic parameters. Emerg Microbes Infect. 2017;6(8): https://doi.org/10.1038/emi.2017.57.

Ng V, Fazil A, Gachon P, et al. Assessment of the probability of autochthonous transmission of chikungunya virus in Canada under recent and projected climate change. Environ Health Perspect. 2017;125(6): https://doi.org/10.1289/EHP669.

Parashar D, Amdekar S, More A, Patil P, More R, Babu VR. Chikungunya fever outbreak in Guntur, Andhra Pradesh, India. Indian J Med Res. 2015;142 Suppl (Suppl 1): S111–S115. https://doi.org/10.4103/0971-5916.176639.

Parola P, de Lamballerie X, Jourdan J, et al. Novel chikungunya virus variant in travellers returning from Indian Ocean islands. Emerg Infect Dis. 2006;12(10):1493–9. https://doi.org/10.3201/eid1210.060610.

Pialoux G, Gaüzère BA, Jauréguiberry S, Strobel M. Chikungunya, an epidemic arbovirosis. Lancet Infect Dis. 2007;7(5):319–27. https://doi.org/10.1016/S1473-3099(07)70107-X.

Powers AM. Chikungunya virus control: is a vaccine on the horizon? Lancet. 2014;384 (9959):2008–9. https://doi.org/10.1016/S0140-6736(14)61290-3.

Powers AM, Logue CH. Changing patterns of chikungunya virus: re-emergence of a zoonotic arbovirus. J Gen Virol. 2007;88(Pt 9):2363–77. https://doi.org/10.1099/vir.0.82858-0.

Powers AM, Brault AC, Tesh RB, Weaver SC. Re-emergence of Chikungunya and O'nyong-nyong viruses: evidence for distinct geographical lineages and distant evolutionary relationships. J Gen Virol. 2000;81(Pt 2):471–9. https://doi.org/10.1099/0022-1317-81-2-471.

Ramful D, Carbonnier M, Pasquet M, et al. Mother-to-child transmission of Chikungunya virus infection. Pediatr Infect Dis J. 2007;26(9):811–5. https://doi.org/10.1097/INF.0b013e3180616d4f.

Ray P, Ratagiri VH, Kabra SK, et al. Chikungunya infection in India: results of a prospective hospital based multi-centric study. PLoS ONE. 2012;7(2): https://doi.org/10.1371/journal.pone.0030025.

Rianthavorn P, Wuttirattanakowit N, Prianantathavorn K, Limpaphayom N, Theamboonlers A, Poovorawan Y. Evaluation of a rapid assay for detection of IgM antibodies to chikungunya. Southeast Asian J Trop Med Public Health. 2010;41(1):92–6.

Riswari SF, Ma'roef CN, Djauhari H, et al. Study of viremic profile in febrile specimens of chikungunya in Bandung, Indonesia. J Clin Virol. 2016;74:61–5. https://doi.org/10.1016/j.jcv.2015.11.017.

Robinson MC. An epidemic of virus disease in Southern Province, Tanganyika Territory, in 1952–53. I. Clinical features. Trans R Soc Trop Med Hyg. 1955;49(1):28–32. https://doi.org/10.1016/0035-9203(55)90080-8.

Scott TW, Amerasinghe PH, Morrison AC, et al. Longitudinal studies of *Aedes aegypti* (Diptera: Culicidae) in Thailand and Puerto Rico: blood feeding frequency. J Med Entomol. 2000;37 (1):89–101. https://doi.org/10.1603/0022-2585-37.1.89.

Sebastian MR, Lodha R, Kabra SK. Chikungunya infection in children. Indian J Pediatr. 2009;76 (2):185–9. https://doi.org/10.1007/s12098-009-0049-6.

Silva LA, Dermody TS. Chikungunya virus: epidemiology, replication, disease mechanisms, and prospective intervention strategies. J Clin Invest. 2017;127(3):737–49. https://doi.org/10.1172/JCI84417.

Singh P, Mittal V, Rizvi MM, et al. The first dominant co-circulation of both dengue and chikungunya viruses during the post-monsoon period of 2010 in Delhi, India. Epidemiol Infect. 2012;140(7):1337–42. https://doi.org/10.1017/S0950268811001671.

Skehel JJ, Wiley DC. Receptor binding and membrane fusion in virus entry: the influenza hemagglutinin. Annu Rev Biochem. 2000;69:531–69. https://doi.org/10.1146/annurev.biochem.69.1.531.

Smith DL, McKenzie FE, Snow RW, Hay SI. Revisiting the basic reproductive number for malaria and its implications for malaria control. PLoS Biol. 2007;5(3): https://doi.org/10.1371/journal.pbio.0050042.

Sourisseau M, Schilte C, Casartelli N, et al. Characterization of reemerging chikungunya virus. PLoS Pathog. 2007;3(6): https://doi.org/10.1371/journal.ppat.0030089.

Suhrbier A, Jaffar-Bandjee MC, Gasque P. Arthritogenic alphaviruses–an overview. Nat Rev Rheumatol. 2012;8(7):420–9. https://doi.org/10.1038/nrrheum.2012.64.

Suryawanshi SD, Dube AH, Khadse RK, et al. Clinical profile of chikungunya fever in patients in a tertiary care centre in Maharashtra, India. Indian J Med Res. 2009;129(4):438–41.

Telle O, Vaguet A, Yadav NK, et al. The spread of dengue in an endemic Urban Milieu—the case of Delhi, India. PLoS ONE. 2016;11(1): https://doi.org/10.1371/journal.pone.0146539.

Van den Driessche P, Watmough J. Reproduction numbers and sub-threshold endemic equilibria for compartmental models of disease transmission. Math Biosci. 2002;180:29–48. https://doi.org/10.1016/s0025-5564(02)00108-6.

Volk SM, Chen R, Tsetsarkin KA, et al. Genome-scale phylogenetic analyses of chikungunya virus reveal independent emergences of recent epidemics and various evolutionary rates. J Virol. 2010;84:6497–504. https://doi.org/10.1128/JVI.01603-09.

Voss JE, Vaney MC, Duquerroy S, et al. Glycoprotein organization of Chikungunya virus particles revealed by X-ray crystallography. Nature. 2010;468(7324):709–12. https://doi.org/10.1038/nature09555.

Watts DM, Burke DS, Harrison BA, Whitmire RE, Nisalak A. Effect of temperature on the vector efficiency of *Aedes aegypti* for dengue 2 virus. Am J Trop Med Hyg. 1987;36(1):143–52. https://doi.org/10.4269/ajtmh.1987.36.143.

Wauquier N, Becquart P, Nkoghe D, Padilla C, Ndjoyi-Mbiguino A, Leroy EM. The acute phase of Chikungunya virus infection in humans is associated with strong innate immunity and T CD8 cell activation. J Infect Dis. 2011;204(1):115–23. https://doi.org/10.1093/infdis/jiq006.

Weaver SC, Forrester NL. Chikungunya: Evolutionary history and recent epidemic spread. Antiviral Res. 2015;120:32–9. https://doi.org/10.1016/j.antiviral.2015.04.016.

Weissenhorn W, Hinz A, Gaudin Y. Virus membrane fusion. FEBS Lett. 2007;581(11):2150–5. https://doi.org/10.1016/j.febslet.2007.01.093.

World Health Organization (WHO). 2020. https://www.who.int/news-room/fact-sheets/detail/chikungunya. Accessed 08 Sept 2020.

World Health Organization (WHO). https://www.who.int/blueprint/what/norms-standards/meeting-report.pdf?ua=1. Accessed 13 March 2020.

World Health Organization. World health report 1996: Executive summary of insect borne diseases. 1996.

Centre for Disease Control and Prevention. https://www.cdc.gov/dengue/resources/factSheets/MosquitoLifecycleFINAL.pdf.

Yang HM, Macoris ML, Galvani KC, Andrighetti MT, Wanderley DM. Assessing the effects of temperature on the population of *Aedes aegypti*, the vector of dengue. Epidemiol Infect. 2009;137(8):1188–202. https://doi.org/10.1017/S0950268809002040.

Yap ML, Klose T, Urakami A, Hasan SS, Akahata W, Rossmann MG. Structural studies of Chikungunya virus maturation. Proc Natl Acad Sci U S A. 2017;114(52):13703–7. https://doi.org/10.1073/pnas.1713166114.

Yergolkar PN, Tandale BV, Arankalle VA, et al. Chikungunya outbreaks caused by African genotype, India. Emerg Infect Dis. 2006;12(10):1580–3. https://doi.org/10.3201/eid1210.060529.

Zeller H, Van Bortel W, Sudre B. Chikungunya: Its history in Africa and Asia and its spread to new regions in 2013–2014. J Infect Dis. 2016;214(suppl 5):S436–40. https://doi.org/10.1093/infdis/jiw391.

Effect of Chikungunya Viral Infection on the Auditory System

Prashanth Prabhu and Shezeen Abdul Gafoor

Abstract It is plausible that auditory symptoms can become comorbidity with Chikungunya (CHIK) virus infections. The exact etiology is not known due to the lack of full literature as well as the lower number of reports regarding the incidence of hearing related issues in the affected persons. Moreover, in the developing and under-developed countries (which is where the infection is more prevalent), hearing difficulties are usually not reported until it is significantly noticeable by the patient or affecting quality of life (which is dependent on the listening needs of the individual). It has been reported that the neurotropic nature of the CHIK virus can have effect on the auditory and associated neurons. As demyelination neuropathy is mostly reported in patients with CHIKinfection (Chandak et al., Neurol India 57:177–181, 2009), it is possible that a similar demyelination can be the cause of the auditory symptom, such as in case of Auditory Neuropathy Spectrum Disorder (ANSD) presentations. It is also likely that, if the virus crosses and reaches the cochlea it can have similar effect of other viral infections in the cochlea such as rubella and herpes simplex, which can cause varying degrees of atrophy of organ of corti, striavascularis and/or tectorial membrane.

Keywords Chikungunya · Viral infections · Neuropathy · Neuro-virulence · Spectrum disorder · Auditory system · Auditory nerve · Demyelination · Atrophy

1 Introduction

Chikungunya (CHIK) virus is gradually rising to be a "global public health menace", that affects millions of people from across the world, majorly in the tropical and sub-tropical countries (Petersen and Powers 2016). The CHIK, a single stranded RNA virus, is mostly transmitted by vector mosquitoes such as

P. Prabhu (✉) · S. A. Gafoor
Department of Audiology, All India Institute of Speech and Hearing, Manasagangothri, Naimisham Campus, Mysore 570006, Karnataka, India

S. I. Ahmad (ed.), *Human Viruses: Diseases, Treatments and Vaccines*, https://doi.org/10.1007/978-3-030-71165-8_9

Aedesaegypti and Aedesalbopictus, belonging to the Alpha virus genus. In some cases, vertical maternal fetal transmission have also been reported (Robillard et al. 2006). Once infected, the virus utilizes an incubation period of up to a week before the symptoms start appearing. The CHIK infection spreads to wherever the vectors move and are more common in tropical countries. Increased travel from one place to the other by humans and the capacity of CHIK to produce high levels of infection, and mutation also contribute to its fast spread. The most common symptoms include elevated body temperatures, joint pain and swelling, rash, headache, tiredness etc.

According to the World Health Organization, this viral infection, first identified in Tanzania in 1952, has spread to at least 40 countries including Asia, Africa, Europe and America (World Health Organisation, 2017). Phylogenetic analyses carried out on CHIK virus identifies its four different genotypes—Asian/Caribbean, West African, Eastern/Central/South African and Indian Ocean Line (Moizeis et al. 2018). The major difference between these genotypes lies in their transmission cycles.

The CHIK virus is detected through laboratory testing (microscopic presence of CHIK virus and its Immunoglobulin M analyses). A reverse transcription polymerase chain reaction can also be used; however its sensitivity remains weak (Johnson et al. 2016).

2 Clinical Presentations

CHIK virus fever widely varies according to clinical manifestations. Samples were collected from the residents of a CHIK virus hit area in Southern Thailand and tested for the presence of the virus (Nakkhara et al. 2013). Although 61.9% of the samples tested were positive for the virus, 47.1% of the positive tested were asymptomatic. The authors concluded that even though the transmission is based on the environmental conditions, symptoms are dependent on "personal biologic factors and social setting". Individuals aged above 30 years with Rh positive blood groups were found to be more susceptible to the virus. In addition, males were 1.3 times more likely to be affected by the disease than females (Kumar et al. 2010). Once infected, the individuals may be debilitated to work as there can be a reduction in dexterity, mobility and reaction.

The symptomatic cases are mostly characterized by a triad of symptoms which include ocular manifestations like conjunctivitis, retinitis etc. In some cases, the CHIK virus infection also exhibits neurological complications such as encephalitis, bulbar palsy, acute disseminated encephalomyelitis, cerebellitis, myelopathy, radiculoneuropathy, carpel tunnel syndrome and tremors were reported (Anand et al. 2019). Other signs like peripheral neuropathy and myeloneuropathy have also been presented with CHIK virus infection. The neuropathy was mostly caused by the demyelination of the neurons. Studies carried out in South India found that individuals with CHIK virus infection also had some ocular signs such as anterior uveitis, optic neuritis, retrobulbar neuritis and dendritic lesions (Chandak et al. 2009; Lalitha et al. 2007). The prognosis for the above mentioned visual difficulties were reported to be good and most patients recovered completely (Lalitha et al. 2007). Some other atypical manifestations like cardiovascular, respiratory, vascular, renal, hepatic, gastrointestinal and adrenal system have also been noticed by some authors (Mehta et al. 2018; Anand et al. 2019; Chandak et al. 2009).

3 Chikungunya Virus Effect on the Auditory System

Amongst CHIK virus infected persons, 60% reported facing some kind of comorbidities, which was more in individuals who had the onset of disease at an older age (mean age: 54.2 years). Such patients also had lesser recovery rates. Of this, six percent of the individuals reported sustained severe to profound hearing loss (Couturier et al. 2012). There are other instances where bilateral mild to severe sensorineural hearing loss has been reported with CHIK virus infection (Dutta et al. 2011).

Therefore, development of hearing difficulty or hearing related problems, though less encountered, are prevailing amongst neurological complication post CHIK virus infection. In a case study carried out in Colombia, it was observed that a child acquired mild hearing impairment, with altered auditory evoked potential, after 13 months of CHIK virus infection (Alvarado-Socarras et al. 2016).

There are also instances where sudden sensorineural hearing loss was reported one week post onset of CHIK virus fever, which was preceded by tinnitus. The hearing difficulty was associated with other aural symptoms such as fullness in the ear, severe giddiness, nausea and vomiting. Upon audiological testing, the patient was diagnosed as having profound hearing loss. The probable etiology was routed to "atrophy of organ of corti, striavascularis and tectorial membrane with variable loss of neuronal population" during the virus attack (Bhavana et al. 2008).

4 Peripheral Neuropathy

Peripheral neuropathy has been associated with CHIK virus infected patients. Its symptoms usually are linked with the conditions where it is gradually progressive. When this happens in the auditory system, the condition is termed as auditory neuropathy spectrum disorder (ANSD). In ANSD (which is a "cochlear afferent disorder"), it is sound transduction that is majorly affected. This is mostly accounted to primary demyelination of neurons and axonal loss (Starr et al. 1996). It is likely that the CHIK virus follows similar pathology within the inner ear and causes the various auditory symptoms.

ANSD can happen at three different levels—pre-synaptic, synaptic (at the ribbon synapse between inner hair cells and the auditory nerve) and post synaptic where it causes a disruption in the VIII cranial nerve activity. This disruption mostly arises either due deafferentiation (reduced number of activated auditory nerve fibers) which in most cases of axonal type of neuropathy or due to reduced neural synchrony occurs mostly in demyelinating neuropathies (Rance and Starr 2015). Since CHIK virus has been associated with demyelination of nerve cells and its neurotoxic nature established, ANSD is very likely neurological comorbidity in individuals with CHIK fever, if the virus enters the auditory system.

The possibility exists that neural hearing loss may be present in individuals having CHIK virus infection. Prabhu (2015), observed such findings in a 14 year old boy with complaint of a decline in his hearing and increase in humming tinnitus (catastrophic) in both his ears along with difficulty in understanding speech onset after the attack of CHIK fever. These were also associated with vertigo and Tullio phenomenon. The boy also had mild form of depression. Upon detailed investigation, he was found to have bilateral mild sensorineural hearing loss (rising configuration) with poor speech identification scores. Abnormal or absent auditory brainstem response and cervical vestibular evoked myogenic potential recordings were obtained for this patient. This along with the findings in otoacoustic emissions indicated the presence of retrocochlear pathology like ANSD.

Hence, a test battery approach is mandatory to diagnose and habilitate individuals with hearing impairment post CHIK virus infection. The audiological test battery for post CHIK virus infestation should include pure tone audiometry and speech audiometry to understand the degree of loss of sensitivity. Speech audiometry results also can help in predicting the neuronal ability to conduct speech sounds. An immittance audiometry need to be carried out which includes both tympanometry as well as reflexometry to understand the middle ear status of the individual. Otoacoustic emissions can be measured to obtain information about the cochlear status and the outer hair cell functioning. Auditory evoked responses such as brainstem evoked response audiometry, vestibular evoked myogenic potentials etc. need to be administered to rule out retrocochlear pathology such as ANSD. Further testing can be decided based on the findings from these mentioned tests. The rehabilitative options suggested to the patient will also vary depending on these test results.

5 Conclusion

It is intriguing that such a small single stranded RNA virus can cause such a varied genre of symptoms in humans. It is quite evident that CHIK virus can lead to auditory deficit, probably as a result of demyelination of neurons leading to neuropathy or toxicity caused in the inner ear. However, there is dearth of literature to point out exact signs and symptoms or prognostic factors associated with this viral infection in the auditory system. It is likely that the symptoms vary depending on the extent of the spread and severity of the infection.

If a person reports with hearing loss onset after CHIK virus infection, it is important that a complete audio-vestibular test battery approach be administered.

A single test cannot give a clear information about the patient's audio-vestibular profile. Multiple tests have to be carried out and correlated before reaching a diagnostic conclusion.

Hence, a test battery approach is mandatory to diagnose and habilitate individuals with hearing impairment post CHIK virus infection. The audiological test battery for individuals post CHIK virus infestation should include pure tone audiometry and speech audiometry to understand the degree of loss of sensitivity. Speech audiometry results also can help in predicting the neuronal ability to conduct speech sounds. An immittance audiometry need to be carried out which includes both tympanometry as well as reflexometry to understand the middle ear status of the person. Otoacoustic emissions can be measured to obtain information about the cochlear status and the outer hair cell functioning. Auditory evoked responses such as brainstem evoked response audiometry, vestibular evoked myogenic potentials etc. need to be administered to rule out retrocochlear pathology such as ANSD. Further testing can be decided based on the findings from the mentioned tests. The rehabilitative options suggested to the patient will also vary depending on these test results.

References

Alvarado-Socarras JL, Ocampo-Gonzalez M, Vargas-Soler JA, Rodriguez-Morales A, Franco-Paredes C. Congenital and Neonatal Chikungunya in Colombia. J Pediatr Infect Dis Soc. 2016;5(3):17–20. https://doi.org/10.1093/jpids/piw021.

Anand KS, Agrawal AK, Garg J, Dhamija RK, Mahajan RK. Spectrum of neurological complications in chikungunya fever: experience at a tertiary care centre and review of literature. Troplical Doctor. 2019;28–9. https://doi.org/10.1177/0049475518825219.

Bhavana K, Tyagi I, Kapila RK. Chikungunya virus induced sudden sensorineural hearing loss. Int J Pediatr Otorhinolaryngol. 2008;72:257–9. https://doi.org/10.1016/j.ijporl.2007.09.022.

Chandak NH, Kashyap RS, Kabra D, Karandikar P, Saha SS, Morey SH, Daginawala HF. Neurological complications of Chikungunya virus infection. Neurol India. 2009;57(2):177–81. https://doi.org/10.4103/0028-3886.51289.

Couturier E, Guillemin F, Mura M, Le L, Valk HD, Virion J, Simon F. Impaired quality of life after chikungunya virus infection: a 2-year follow-up study. Rheumatology. 2012;51 (March):1315–22. https://doi.org/10.1093/rheumatology/kes015.

Dutta P, Khan SA, Khan AM, Borah J, Chowdhury P, Mahanta J. First evidence of chikungunya virus infection in Assam, Northeast India. Trans R Soc Trop Med Hyg. 2011;105(6):355–7. https://doi.org/10.1016/j.trstmh.2011.02.004.

Johnson BW, Russell BJ, Goodman CH. Laboratory diagnosis of chikungunya virus infections and commercial sources for diagnostic assays. J Infect Dis. 2016;214:471–4. https://doi.org/10.1093/infdis/jiw274.Laboratory.

Kumar NCVM, Nadimpalli M, Vardhan VR, Gopal SDVR. Association of ABO blood groups with Chikungunya virus. Virol J. 2010;7(1).

Lalitha P, Rathinam S, Banushree K, Maheshkumar S, Vijayakumar R, Sathe P. Ocular involvement associated with an epidemic outbreak of chikungunya virus infection. Am J Ophthalmol. 2007;144(4):552–6. https://doi.org/10.1016/j.ajo.2007.06.002.

Mehta R, Gerardin P, Alexandre C, Brito A De, Solomon T, Brito ML. The neurological complications of chikungunya virus: a systematic review. Rev Med Virol. 2018;28(3). https://doi.org/10.1002/rmv.1978

Moizeis RNC, Fernandes TAA de M, Guedes PM da M, Pereira HWB, Lanza DCF, Azevedo JWV de, Fernandes JV, et al. Chikungunya fever: a threat to global public health. Pathog Glob Health. 2018;112(4), 182–94. https://doi.org/10.1080/20477724.2018.1478777

Nakkhara P, Chongsuvivatwong V, Thammapalo S. Risk factors for symptomatic and asymptomatic chikungunya infection. Trans R Soc Trop Med Hyg. 2013;107 (September):789–96. https://doi.org/10.1093/trstmh/trt083.

Petersen LR, Powers AM. Chikungunya: epidemiology. F1000Research. 2016;5, 1–8. https://doi.org/10.12688/f1000research.7171.1

Prabhu P. Acquired auditory neuropathy spectrum disorder after an attack of chikungunya: case study. Eur Arch. 2015;273(1):257–61. https://doi.org/10.1007/s00405-015-3578-9.

Rance G, Starr A. Pathophysiological mechanisms and functional hearing consequences of auditory neuropathy. Brain. 2015;138(11):3141–58. https://doi.org/10.1093/brain/awv270.

Robillard P, Boumahni B, Gerardin P, Michault A, Fourmaintraux A, Schuffenecker I, Barau G. Vertical maternal fetal transmission of the Chikungunya virus. Presse Med. 2006;35 (May):785–8.

Starr A, Picton TW, Kim R. Pathophysiology of auditory neuropathy. In Auditory neuropathy; 1996. p. 741–53.

World Health Organisation. WHO | Chikungunya. WHO. 2017. https://doi.org/10.1016/S1726-4901(10)70043-5.

Epstein-Barr Viruses

Brief Introduction of Epstein-Barr Virus and Lymphoma

Daniel Esau

Abstract Epstein-Barr virus (EBV) is a double-stranded DNA virus of the γ-herpesvirus subfamily that is implicated in a heterogeneous array of pathologies ranging from benign to malignant. Several lymphomas have been associated with latent infection of EBV, including Burkitt lymphoma (BL), Hodgkin lymphoma, diffuse large B-cell lymphoma, and T-cell lymphoma. The role of EBV in the pathogenesis of lymphoma is thought to be related to the induction of T-cell independent immunoglobulin class switching, antigenic stimulation of the B-cell receptor, and promotion of genetic instability of B-cells. In this chapter, the viral particle, route of infection, and proposed pathogenesis of EBV-related lymphomas are briefly discussed. The history of the discovery of EBV and BL, the first human malignancy proven to be caused by a virus, is presented.

Keywords Lymphoma · Epstein-Barr virus · Burkitt lymphoma · Hodgkin's lymphoma · Diffuse large B-Cell lymphoma · Oncology · Oncogenesis

1 Epstein-Barr Virus

Epstein-Barr virus (EBV) is a double-stranded DNA virus of the γ-herpesvirus subfamily that encodes for nearly 100 proteins (Jenson 2011; Minarovits and Niller 2017). There are two types of EBV, type 1 and type 2 (also known as type A and type B, respectively), which are homologous in 70–85% of their genome. There are no identified differences in disease or presentation between the two types of EBV, but EBV-1 is more prevalent worldwide while EBV-2 is more prevalent in Africa (Jenson 2011). Although it can be transmitted through

D. Esau (✉)
Department of Medicine, The University of British Columbia, Vancouver, BC V6T 1Z4, Canada
e-mail: desau@alumni.ubc.ca

Royal Jubilee Hospital, Victoria, BC V8R 1J8, Canada

S. I. Ahmad (ed.), *Human Viruses: Diseases, Treatments and Vaccines*, https://doi.org/10.1007/978-3-030-71165-8_10

genital secretions and sexual contact, it is transmitted primarily by saliva, which often occurs as a child through handling toys or as an adolescent through kissing. Following primary infection, EBV is shed in saliva at high concentrations for 6 months, as well as intermittently throughout the remainder of life at lower concentrations (Jenson 2011; Vockerodt et al. 2015).

In low- and middle-income countries transmission of EBV occurs largely during infancy, while in high-income countries approximately one third of transmissions occur in later childhood and adolescence – presumably because of differences in hygiene (Jenson 2011). Infections in infancy and early childhood are more likely to be asymptomatic or mild while around 50% of adolescence and adults will experience symptoms and signs of infectious mononucleosis with primary infection (Jenson 2011).

EBV primarily infects the B-lymphocyte and results in lifelong infection by establishing latency in the host genome (Jenson 2011; Minarovits and Niller 2017; Vockerodt et al. 2015). Binding of the viral envelope glycoprotein, gp350 and gp42, to the CD21 receptor and to human leukocyte antigen class II molecules, respectively, on the surface of B cells begins the process of EBV entrance into B cells (Borza and Hutt-Fletcher 2002; Nemerow et al. 1987). Infection of other cell types is possible but much less efficient (Borza and Hutt-Fletcher 2002). In vitro, B cells infected with EBV almost always transform into proliferating lymphoblastoid cells and unregulated polyclonal expansion occurs (Diehl et al. 1968). In vivo, the virus rarely achieves full lytic replication (Young and Rickinson 2004) instead leads to latent infection of lymphoblastoid cells. Latently infected cells express a limited set of viral gene products made up of six EBV nuclear antigens (EBNAs 1, 2, 3A, 3B, 3C and LP) and three latent membrane proteins (LMPs 1, 2A, and 2B). Non-coding viral RNA transcripts, including *EBER1* and *EBER2*, are also produced from latently infected cells, although the function of these transcripts is not clear (Kieff and Rickinson 2001). The constitutive expression of these proteins without progression to lytic replication is a hallmark of latent infection and allows EBV (and other herpesviruses) to achieve life-long infection in a host.

Because it is transferred primarily by saliva and leads to lifelong infection; EBV is an extremely common pathogen, with greater than 90% of adult humans being seropositive and latently infected (Jenson 2011; Henle and Henle 1966). The persistent, latent infection that follows primary EBV infection is in itself not known to cause any symptoms or clinical manifestation, however primary infection with or re-activation of EBV can cause a wide spectrum of clinical manifestations. Infectious mononucleosis, usually a benign disease in otherwise healthy individuals, is a widely recognized manifestation of EBV infection, and presents classically with fevers, malaise, lymphadenopathy, sore throat, and hepatosplenomegaly (Jenson 2011; Henle et al. 1968). In immunocompromised individuals, EBV infection can cause non-malignant manifestations including lymphoid interstitial pneumonitis, oral hairy leukoplakia, and virus-associated hemophagocytic syndrome (Jenson 2011; Minarovits and Niller 2017); several malignant disorders including primary central nervous system lymphoma, primary effusion lymphoma, plasmablastic lymphoma of the oral cavity (Carbone et al. 2009), and other lymphoproliferative disorders (Thomas et al. 1990) may also occur in the immunocompromised host. In both cases immunocompetent and immunocompromised individuals, EBV infection is associated with several malignancies, most classically with Burkitt lymphoma (Zur Hausen et al. 1970; Epstein et al. 1964), but also with nasopharyngeal carcinoma (Zur Hausen et al. 1970), Hodgkin lymphoma (Vockerodt et al. 2015), diffuse large B-cell lymphoma (Vockerodt et al. 2015), and T-cell lymphoma (Vockerodt et al. 2015). Through molecular mimicry, EBV has been postulated to play a role in several auto-immune conditions, including systemic lupus erythematosus, rheumatoid arthritis, and multiple sclerosis (Evans et al. 1971; Poole et al. 2008; Lunemann et al. 2008; 2008; Rojas et al. 2018). The ability of EBV to induce certain cancers and induce or exacerbate inflammatory and autoimmune conditions is thought to be related to the induction of T-cell independent immunoglobulin class switching, which leads to unregulated and aberrant IgG, IgA, and/or IgE production (He et al. 2003).

In addition to its importance as a human pathogen causing a variety of diseases, EBV holds an important place in medical research and virology. EBV was the first virus found to cause cancer in humans (Greaves 2016; Harford 2012; Magrath 2012), and its discovery stimulated further study in the fields of human oncoviruses, molecular virology, and viral diagnostics (Niller and Bauer 2017).

2 Historic Context

The history of EBV is tied to a unique malignancy we now term Burkitt Lymphoma (BL). Prevalent along equatorial Africa, BL often presented with rapidly growing and aggressive jaw tumours in children. Traditional wooden masks from equatorial Africa often show proptosis and facial deformities, which may be a sign that the typical facial involvement of BL was noted well before the modern era (Pulvertaft 1965). Albert Cook, who was likely the earliest European physician to devote his practice to Uganda (IARC Working Group 1997), noted in 1901 that cancer was prevalent in Uganda, with jaw tumours being especially common (Davies et al. 1897). Several other reports of jaw tumours surfaced from equatorial African countries (Edington 1956; Smith and Elmes 1934), but it was not until the late 1950s that a clinically unique syndrome could be described.

In 1957 Denis Burkitt, a medical officer with the Colonial office stationed in Uganda, was able to examine two children with tumours of the jaw. When both of his patients died, he began to search in local records for similar cases of tumours in young children (Wright 2012) . Over several years Burkitt and Gregory O'Conor, the pathologists at Mulago Hospital, gathered information on tumour cases from Uganda and neighbouring countries. In 1961, they published their findings of a unique lymphoma syndrome characterized by rapidly growing and often fatal tumours occurring in the jaw, abdomen, salivary gland, bone, or spinal column (Burkitt and O'Conor 1961). Already, there were hints of a viral aetiology; O'Conor noted that the malignant lymphoma syndrome was clinically similar in presentation to lymphocytic bovine leukemia, a disease in cows known at that time to have a viral cause (O'Conor 1961). The prevalence of this newly described lymphoma appeared to be clustered along central Africa in a "lymphoma belt", the margins of which Burkitt and two colleagues demarcated in 1961 by visiting around 57 hospitals in 8 countries during a 10-week span, travelling around 10 000 miles of rough terrain in a 1954 Ford station wagon (Magrath 2012). Through this epidemiological reconnaissance, Burkitt was able to describe the lymphoma as being most common in the tropic regions of Africa and rare in colder, dry, elevated regions, raising the possibility of an insect-vectored virus as the cause of the lymphoma (Burkitt 1962, 1969)—a hypothesis that was eventually proven false when it was determined that EBV was transmitted by saliva (Klein 2009). Several previous studies had examined the possibility that malaria, transmitted through the anopheles mosquito, played a role in the pathogenesis of what had by then been dubbed Burkitt Lymphoma (Dalldorf et al. 1964; Edington et al. 1963), and in 1969 Burkitt updated his hypothesis by stating that malaria may act as a cofactor with EBV to promote oncogenesis, and the unique distribution of BL was thus explained by the habitat of the anopheles mosquito (Burkitt 1969). The role of malaria and EBV in causing endemic BL is complex and continues to be an active area of research to this day (Magrath 2012; Current Cancer Research 2013).

The discovery of the viral cause of BL began in 1961 when Burkitt, back in the United Kingdom for home leave, presented on the newly described lymphoma at a lecture at Middlesex Hospital Medical School. Anthony Epstein, a medical virologist whose research focused on chicken tumour viruses, was in the audience to hear Burkitt's speech (Epstein and Eastwood 1955). Epstein had used electron microscopy, still a relatively novel research tool, to demonstrate that the Rous fowl sarcoma virus was a RNA virus and not a DNA virus (Epstein 1958; Epstein and Holt 1958 June). Burkitt's description of the strange malignant tumour syndrome with a geographical distribution immediately had Epstein thinking about a human cancer-causing virus as a potential cause, and Epstein shifted his research focus to isolating a viral cause for BL (Smith 2012; Epstein 2015). After the lecture, he met with Burkitt and the two entered in a collaborative research (Epstein 2015). Epstein then had tumour cells from BL patients in Uganda flown to his laboratory in London for study (Epstein 2015). Epstein was able to secure a grant from the US National Cancer Institute, which allowed him to recruit the pathologist, Bert Achong, to use electron microscopy, and PhD student Yvonne Barr, to assist with tissue cultures (Epstein 2015). After several years of attempting to grow human lymphoma cells with disappointing results, a breakthrough was finally made when the airplane bringing biopsy samples from Kampala to London was diverted due to fog. When the samples finally arrived, Epstein noted cells growing in the transit fluid, and thus a method of culturing free-floating lymphocytes in fluid was discovered (Epstein 2015). These cell lines, when examined under electron microscopy, showed evidence of a herpes-like virus, and Epstein, Achong, and Barr were able to publish this result in 1964 (Epstein et al. 1964). However, years of further studies would be needed before the virus particles, later named Epstein-Barr virus, could be branded as oncogenic.

To confirm these findings at an independent laboratory, Epstein contacted Werner and Gertrude Henle, husband and wife virologists working at the Children's Hospital in Philadelphia. In addition to confirming the presence of a viral particle in lymphoblast, taken from BL patients (Epstein et al. 1965), the Henle contributed much of the early research into the characterization and oncogenic potential of EBV. They worked collaboratively with many researchers and clinicians, including Harald zur Hausen, who eventually won the Nobel Prize for the discovery of human Papilloma virus (Nobelprize 2008), another human oncovirus. At that time, it was unclear if EBV was an innocent bystander or an oncogenic virus: although the newly described virus could be detected in nearly all cases of African BL, it was also prevalent in asymptomatic and healthy individuals (Henle and Henle 1966). However, Henle and others showed that BL patients expressed antibody titres against EBV that were, on average, 8-fold higher than controls (Henle et al. 1969), supporting an association between EBV and BL. Henle then showed that co-cultivation of irradiated lysed BL cells with healthy control leukocytes often lead to proliferation of hematopoietic cells, indicating that the oncogenic potential that may be present within the contents of BL cells (Henle et al. 1967; Henle 1968). In 1970 they published the discovery of EBV DNA within cells taken from BL biopsies, supporting the association of EBV with BL (Zur Hausen et al. 1970). When a technician working in the Henle laboratory serendipitously became ill with infectious mononucleosis, it was noticed that the technician developed antibodies to EBV over the course of the illness (Henle and Henle 1979). Further study was then devoted to the role of EBV in the development of mononucleosis, and the ability of EBV to cause several distinct clinical pathologies were described (Henle et al. 1968; Henle and Henle 1979). Through the work of Epstein, Barr, Achong, the Henle group, and many other researchers, by the late 1970s enough evidence was gathered to prove that EBV played a carcinogenic role in humans (Epstein and Achong 1979), and EBV became widely recognized as the first human oncovirus.

3 Lymphomas Associated With EBV Infection

The malignancies that have been associated with EBV infection can be classified into two broad categories: those that arise from B-cell origin (and thus are a by-product of the persistence of EBV in the B-cell system) and those that arise from non-B-cell origin (and are thought to be the result of EBV accessing unusual target cell types) (Chapman and Rickinson 1998). The former category includes BL, Hodgkin disease, diffuse large B-cell lymphoma (DLBCL), and primary effusion lymphoma, while the latter includes T-cell lymphoma, nasopharyngeal carcinoma, gastric carcinoma, and leiomyosarcoma.

Because EBV infection is ubiquitously found, it is sometimes challenging to determine if EBV infection truly plays a pathogenic role in any specific disease in which it has been associated with, or whether it is an innocent bystander (Thorley-Lawson and Gross 2004). The discovery that isolated the virus from different tumours, expressing different latent genes has been used to explain various roles of EBV in lymphomagenesis (Vockerodt et al. 2015), and provides evidence in several cases against an “innocent bystander” role. Furthermore, EBV has been shown to interact with memory B cells causing B cell hyperplasia and triggering T cell-independent class switch recombination and somatic hypermutation of B cells (He et al. 2003; Magrath 2012; Epeldegui et al. 2007). Errors associated with unregulated class switch recombination and somatic hypermutation can lead to oncogene mutation or translocation (Epeldegui et al. 2007), which may provide a mechanistic explanation for the role of EBV in the development of B cell lymphomas.

4 Role of EBV In Oncogenesis Of Burkitt Lymphoma

Burkitt Lymphoma, as the first human cancer proven to be caused by a virus, has been extensively studied and can be used as an example of some of the mechanisms by which EBV causes lymphomas. Despite the extensive research into EBV and BL, there are still gaps in our knowledge of the role of EBV in the oncogenesis of BL (Bornkamm 2009). An overview of the current knowledge is presented below.

The World Health Organization (WHO) classifies BL into three broad groups (Swedlow et al. 2016). ‘Endemic’ BL refers to cases of BL in high incidence areas including equatorial Africa and Papua New Guinea (Magrath 2012); ‘Sporadic’ BL occurs (at much lower incidence rates) throughout the rest of the world while immunodeficiency-associated BL, first described in patients with HIV (Ziegler et al. 1982), occurs in those who are immunocompromised (Vockerodt et al. 2015). Nearly all of endemic BL is EBV-positive (Henle and Henle 1966), while only about 10–30% of sporadic (IARC Working Group 1997; Frick et al. 2012) and 40% of immunodeficiency-associated cases are EBV-positive (Frick et al. 2012). EBV causes antigenic stimulation of the B-cell receptor (BCR) and promotes genetic instability of B-cells (He et al. 2003; Epeldegui et al. 2007; Bornkamm 2009).

Malaria is an important cofactor in the pathogenesis of endemic BL. In addition to causing hyper-gammaglobulinemia and polyclonal B cell activation, *Plasmodium falciparum* has been shown to directly induce EBV reactivation and increase viral loads (Chene et al. 2009). This effect intensifies the B cell hyperplasia, CD40-idenpendent class switch recombination, and somatic hypermutation of B cells that occurs with EBV infection (He et al. 2003; Magrath 2012; Epeldegui et al. 2007). It is the occasional errors in the cellular machinery in class switch recombination that leads to oncogene mutations or translocations, genetic instability, and deregulated expression of the MYC gene thought to underlie the development of endemic BL (Magrath 2012; Torgbor et al. 2014).

Early studies showed that EBV serum antibodies were more prevalent and of higher titre in cases of endemic BL than in controls (Henle et al. 1969; Klein et al.

1970). Soon after it was shown that nearly 100% of tumour samples taken from endemic BL cases contain multiple EBV genomes (Zur Hausen et al. 1970), and that generally the full set of latent EBV genes (including nuclear antigens and latent membrane proteins) were expressed when BL cells are cultured artificially (Magrath 2012; Reedman and Klein 1973). In contrast, studies examining sporadic cases of BL found levels of EBV serum antibody titres that, although higher than age-matched controls, were much lower than endemic BL cases (Levine et al. 1972). In addition, compared to endemic BL, sporadic BL cases have higher rates of mutations of TCF3/ID3 genes, which cause intrinsic activation of the BCR-pathway independently of the activation of this pathway caused by EBV and malaria (Amato et al. 2016). These findings support the hypothesis that significantly different pathways exist for the pathogenesis of endemic and sporadic BL, with EBV (and other cofactors like malaria) playing a larger role in endemic BL while sporadic BL relies more on intrinsic mutations.

5 Conclusion

The current understanding of the role of EBV in the pathogenesis of several heterogeneous malignancies, though incomplete, is being used to guide future treatment options. One exciting area of research is the use of cellular immunotherapies to target EBV antigens on infected cells. Since 1995 this technique has been used on a small number of patients with a variety of EBV-related diseases including post-transplant lymphoproliferative disease, Hodgkin's lymphoma, and nasopharyngeal carcinoma (Merlo et al. 2011; Pfeffermann et al. 2018), although success has been somewhat variable.

From humble beginnings with a missionary surgeon in Uganda and a chicken-virus researcher in England to new cellular immunotherapies created through the combined knowledge of 40 years of research, researchers are still revealing new information about the oncogenesis of EBV-related malignancies. Burkitt Lymphoma and other EBV-related lymphomas continue to cause significant mortality and worldwide, and research has begun to shift towards creating new treatment options to prolong survival and perhaps, in some cases, provide cure.

References

Amato T, Abate F, Piccalugo P, Iacono M, Fallerini C, Renieri A, et al. Clonality analysis of immunoglobulin gene rearrangement by next-generation sequencing in endemic Burkitt lymphoma suggests antigen drive activation of BCR as opposed to sporadic Burkitt lymphoma. Am J Clin Pathol. 2016;145(1):116–27.

Bornkamm GW. Epstein-Barr virus and its role in the pathogenesis of Burkitt's lymphoma: an unresolved issue. Semin Cancer Biol. 2009;19(6):351–65.

Borza CM, Hutt-Fletcher LM. Alternate replication in B cells and epithelial cells switches tropism of Epstein-Barr virus. Nat Med. 2002;8(6):594–9.

Burkitt D. A children's cancer dependent on climatic factors. Nature. 1962; 194.

Burkitt DP. Etiology of Burkitt's lymphoma—an alternative hypothesis to a vectored virus. J Nat Cancer Inst. 1969; 42(1).

Burkitt D, O'Conor GT. Malignant lymphoma in African children I. A clinical syndrome. Cancer. 1961;14:258–69.

Carbone A, Cesarman E, Spina M, Gloghini A, Schulz TF. HIV-associated lymphomas and gamma-herpesviruses. Blood. 2009;113(6):1213–24.

Chapman ALN, Rickinson AB. Epstein-Barr virus in Hodgkin's disease. Ann Oncol. 1998;9(5): 5–16.

Chene A, Donati D, Orem J, Bjorkman A, Mbidde ER, Kironde F, et al. Endemic Burkitt's lymphoma as a polymicrobial disease: new insights on the interaction between Plasmodium falciparum and Epstein-Barr virus. Semin Cancer Biol. 2009;19(6):411–20.

Current Cancer Research. Burkitt's Lymphoma Robertson ES, editor. New York: Springer; 2013.

Dalldorf G, Linsell CA, Barnhart FE, Martyn R. An epidemiological approach to the lymphomas of African children and Burkitt's sarcoma of the jaws. Perspect Biol Med. 1964;7:435–49.

Davies JNP, Elmes S, Hutt MSR, Mtimavalye LAR, Owor R, Shaper L. Cancer in an African community, 1897–1956: an analysis of the records of Mengo Hospital, Kampala, Uganda: part 1. Br Med J. 1964; 1(5379): 336–341.

Diehl V, Henle G, Henle W, Kohn G. Demonstration of a herpes group virus in cultures of peripheral leukocytes from patients with infectious mononucleosis. J Virol. 1968;2(7):663–9.

Epstein A, Eastwood MA. Denis Parsons Burkitt. 28 February 1911–23 March 1993. Biographical Memoirs of Fellows of the Royal Society. 1995 November; 41: 88-102.

Edington GM, MacLean CMU, Okubadejo OA. One-Hundred one necropsies on tumours of the reticulo-endothelial system in Ibadan, Nigeria, with special reference to childhood lymphosarcoma Roulet FC, ed. Basel: Karger; 1963.

Edington GM. Malignant disease in the Gold Coast. Br J Cancer. 1956; 10.

Epeldegui M, Hung YP, McQuay A, Ambinder RF, Martinez-Maza O. Infection of human B cells with Epstein-Barr virus results in the ezpression of somatic hypermutation-inducing molecules and in the accrual of oncogene mutations. Mol Immunol. 2007;44(5):934–42.

Epstein MA. Composition of the Rous Virus Nucleoid. Nature. 1958;181:1808.

Epstein A. Why and How Epstein-Barr Virus was discovered 50 years ago. In: Münz C, editor. Epstein barr virus voluma 1, one herpes virus: many diseases. Zürich: Springer; 2015. p. 3–15.

Epstein MA, Achong BG. Chapter 14: The relationship of the Virus to Burkitt's lymphoma. In: Epstein MA, Achong BG, editors. The Epstein-Barr virus. Berlin: Springer; 1979. p. 321–37.

Epstein MA, Holt SJ. Observations on the Rous virus; integrated electron microscopical and cytochemical studies of fluorocarbon purified preparations. Br J Cancer. 1958;12:363–9.

Epstein MA, Achong BG, Barr YM. Virus particles in cultured lymphoblasts from Burkitt's lymphoma. Lancet. 1964;1(7335):702–3.

Epstein MA, Henle G, Achong BG, Barr YM. Morphological and biological studies on a virus in cultured lymphoblasts from Burkitt's lymphoma. J Exp Med. 1965;121(5):761–70.

Evans AS, Rothfield NF, Niederman JC. Raised antibody titres to E.B. virus in systemic lupus erythematosus. The Lancet. 1971; 297(7691):167–168.

Frick M, Dorken B, Lenz G. New insights into the biology of molecular subtypes of diffuse large B-cell lymphoma and Burkitt lymphoma. Best Prac Res Clin Haematol. 2012; 25(1).

Greaves M. Leukaemia 'firsts' in cancer research and treatment. Nat Rev Cancer. 2016;16:163–72.

Harford JB. Viral infections and human cancers: the legacy of denis Burkitt. Br J Haematol. 2012;156:709–18.

He B, Raab-Traub N, Casali P, Cerutti A. EBV-encoded latent membrane protein 1 cooperates with BAFF/BLyS and April to induce T cell-independent Ig heavy chain class switching. J Immunol. 2003;171(10):5215–24.

Henle G, Henle W. Immunofluorescence in cells derived from Burkitt's lymphoma. J Bacteriol. 1966;91(3):1248–56.

Henle G, Henle W. Chapter 13: the virus as the etiologic agent of infectious mononucleosis. In: Epstein MA, Achong BG, editors. The Epstein-Barr virus. Berlin: Springer; 1979. p. 297–320.
Henle W, Diehl V, Kohn G, zur Hausen H, Henle G. Herpes-type virus and chromosome marker in normal leukocytes after growth with irradiated Burkitt cells. Science. 1967; 157(3792): 1064–1065.
Henle W. Evidence for viruses in acute leukemia and Burkitt's tumor. Cancer. 1968; 21(4).
Henle G, Henle W, Diehl V. Relation of Burkitt tumor associated herpes-type virus to infectious mononucleosis. Proc Natl Acad Sci USA. 1968;59:94–101.
Henle G, Henle W, Clifford P, Diehl V, Kafuko GW, Kirya BG, et al. Antibodies to Epstein-Barr virus in Burkitt's lymphoma and control groups. J Natl Cancer Inst. 1969;43(5):1147–57.
IARC Working Group on the Evaluation of Carcinogenic Risk to Humans. Studies of Cancer in Humans. In Epstein-Barr Virus and Kaposi's Sarcoma Herpesvirus/Human Herpesvirus 8. Lyon (FR): International Agency for Research on Cancer; 1997.
Jenson HB. Epstein-Barr Virus. Pediatr Rev. 2011;32(9):375–83.
Kieff E, Rickinson AB. Epstein-Barr Virus and its replication. In: Knipe DM, Howley PM, editors. Fields virology. Philadelphia: Williams and Wilkins; 2001. p. 2511–74.
Klein G. Burkitt lymphoma—a stalking horse for cancer research? Semin Cancer Biol. 2009;19:347–50.
Klein G, Geering G, Old LJ, Henle G, Henle W, Clifford P. Comparison of the anti-EBV titer and the EBV-associated membrane reactive and precipitating antibody levels in the sera of Burkitt lymphoma and nasopharyngeal carcinoma patients and controls. Int J Cancer. 1970;5(2): 185–94.
Levine PH, O'Conor GT, Berard CW. Antibodies to Epstein-Barr Virus (EBV) in American patients with Burkitt's lymphoma. Cancer. 1972;30(3):610–5.
Lunemann JD, Jelcic I, Roberts S, Lutterotti A, Tackenberg B, Martin R, et al. EBNA1-specific T cells from patients with multiple sclerosis cross react with myelin antigens and co-produce IFN-γ and IL-2. J Exp Med. 2008a;205(8):1763–73.
Lunemann JD, Frey O, Eidner T, Baier M, Roberts S, Sashihara J, et al. Increased frwequency of EBV-specific effector memory CD8 + T cells correlates with higher viral load in rheumatoid arthritis. J Immunol. 2008b;181(2):991–1000.
Magrath I. Epidemiology: clues to the pathogenesis of Burkitt lymphoma. Br J Haematol. 2012;156:744–56.
Merlo A, Turrini R, Dolcetti R, Zanovello P, Rosato A. Immunotherapy for EBV-associated malignancies. International Journal of Hematology. 2011; 93(3): 281–93.
Minarovits J, Niller HH. Current trends and alternative scenarios in EBV research. In Minarovits J, Niller HH, eds. Epstein barr virus. Methods in Molecular Biology, vol 1532. New York, NY: Humana Press; 2017. p. 1–32.
Wright D. Nailing BurkittLymphoma. Br J Haematol. 2012;156:780–2.
Nemerow GR, Mold C, Schwend VK, Tollefson V, Cooper NR. Identification of gp350 as the viral glycoprotein mediating attachment of the Epstein-Barr virus (EBV) to the EBV/C3d receptor of B cells: sequence homology of gp350 and C3 complement fragment C3d. J Virol. 1987;61(5):1416–20.
Niller HH, Bauer G. Epstein-Barr virus: clinical diagnostics. In Minarovits J, Niller H, eds. Epstein barr virus. Methods in Molecular Biology, vol 1532. New York, NY: Humana Press; 2017. p. 33–55.
Nobelprize.org. The Nobel Prize in Physiology or Medicine 2008/2018. Retrieved 5 Dec 2018 from https://www.nobelprize.org/prizes/medicine/2008/summary/.
O'Conor GT. Malignant lymphoma in African children II: a pathological entity. Cancer. 1961; 14 (2): p. 270–283.
Pfeffermann LM, Pfirrmann V, Huenecke S, Bremm M, Bonig H, Kvasnicka HM, et al. Epstein-Barr virus-specific cytokine-induced killer cells for treatment of Epstein-Barr virus-related malignant lymphoma. Cytotherapy. 2018;20(6):839–50.

Poole BD, Gross T, Maier S, Harley JB, James JA. Lupus-like autoantibody development in rabbits and mice after immunization with EBNA-1 fragments. J Autoimmun. 2008;31(4): 362–71.

Pulvertaft RJV. A study of malignant tumours in Nigeria by short-term tissue culture. J Clin Path. 1965;18:261–73.

Reedman BM, Klein G. Cellular localization of an Epstein-Barr virus (EBV)-associated complement-fixing antigen in producer and non-producer lymphoblastoid cell lines. Int J Cancer. 1973;28(11):499–520.

Rojas M, Restrepo-Jimenez P, Monsalve DM, Pacheco Y, Acosta-Ampudia Y, Remirez-Santana C, et al. Molecular mimicry and autoimmunity. J Autoimmun. 2018;95:100–23.

Smith O. Denis Parsons Burkitt CMG, MD, DSc, FRS, FRCS, FTCD (1911-93) Irish by birth, Trinity by the grace of God. Br J Haematol. 2012;156:770–6.

Smith EC, Elmes BGT. Malignant disease in natives of Nigeria. Ann Trop Med Parasitol. 1934;28:461–512.

Swedlow SH, Campo E, Pileri SA, Harris NL, Stein H, Siebert R, et al. The 2016 revision of the World Health Organization classification of lymphoid neoplasms. Blood. 2016;127(20): 2375–90.

Thomas JA, Hotchin NA, Allday MJ, Amlot P, Rose M, Yacoub M, et al. Immunohistology of Epstein-Barr virus-associated antigens in B call disorders from immunocompromised individuals. Transplantation. 1990;49(5):944–53.

Thorley-Lawson DA, Gross A. Persistence of the epstein-barr virus and the origins of associated lymphomas. N Engl J Med. 2004; 350(13).

Torgbor C, Awuah P, Deitsch K, Kalantari P, Duca KA, Thorley-Lawson DA. A multifactorial role for P. falcipatum malaria in endemic Burkitt's lymphoma pathogenesis. PLoS pathogens. 2014; 10(5).

Vockerodt M, Yap LF, Shannon-Lowe C, Curley H, Wei W, Vrzalikova K, et al. The Epstein-Barr virus and the pathogenesis of lymphoma. J Pathol. 2015;235:312–22.

Young LS, Rickinson AB. Epstein-Barr virus: 40 years on. Nat Rev. 2004;4:757–68.

Ziegler JL, Miner RC, Rosenbaum E, Lennette ET, Shillitoe E, Casavant C, et al. Outbreak of Burkitt's-Like Lymphoma in Homosexual Men. The Lancet. 1982;320(8299):631–3.

Zur Hausen H, Schulte-Holthausen H, Klein G, Henle G, Henle W, Clifford P, et al. EBV DNA in biopsies of Burkitt tumours and anaplastic carcinomas of the nasopharynx. Nature. 1970; 228 (5276).

Crimean-Congo Hemorrhagic Fever

Diagnosis, Prognosis and Clinical Trial in Crimean-Congo Hemorrhagic Fever

Seyit Ali Büyüktuna and Halef Okan Doğan

Abstract Crimean-Congo hemorrhagic fever (CCHF) is one of the most common zoonotic diseases of viral hemorrhagic fever.The causative agent is a single-stranded RNA virus belonging to the genus Orthonairovirus of the Nairoviridae family. The virus is transmitted to humans by tick bites, or through direct contact with blood or tissues of a viremic patient or infected animal. Information on the pathogenesis is very limited. The disease has a wide geographical distribution, with significant numbers of cases reported from the Republic of South Africa, the Democratic Republic of Congo, Mauritania, Tanzania, Iraq, Pakistan, Oman, China, and Senegal. CCHF typically has four different clinical stages, including incubation, pre-hemorrhagic, hemorrhagic, and recovery periods. It is difficult to diagnose because the clinical manifestations of the disease can be confused with many other diseases. This fact increases the need for new biomarkers to help diagnose CCHF. Routine clinical laboratory parameters including platelet, AST, ALT, first step coagulation tests, LDH creatinine, and fibrinogen have been considered as laboratory indicators of CCHF. As disease-specific treatment options are limited, supportive approaches are often used in the treatment of the disease.

Keywords Crimean-Congo hemorrhagic fever virus · Tick-borne · Biomarkers · Transmission · Orthonairovirus · Viremia · Alanine aminotransferase · Lactate dehydrogenase · Favipiravir · Humoral immunity

S. A. Büyüktuna
Department of Infectious Disease, School of Medicine, Cumhuriyet University, Sivas, Turkey

H. O. Doğan (✉)
Department of Biochemistry, School of Medicine, Cumhuriyet University, Sivas, Turkey
e-mail: halefokan@gmail.com

S. I. Ahmad (ed.), *Human Viruses: Diseases, Treatments and Vaccines*, https://doi.org/10.1007/978-3-030-71165-8_11

1 History and Geographical Distribution

CCHF was first described in the twelfth century in Tajikistan. Different nomenclatures have been used for the disease in different parts of the world, such as Central Asian hemorrhagic fever, Crimean hemorrhagic fever, Congo fever, Asian Ebola, Khungribta (blood taking), Khunymuny (nose bleeding), and Karakhalak (black death). In modern literature, the disease was first recognized in the 1940s, with an outbreak of a hemorrhagic febrile disease in 200 soldiers working to increase agricultural production in Crimea. This disease, which resulted in mortality in 10% of cases, was called Crimean fever. Subsequently, the virus isolated from the blood of a patient in Belgian Congo in 1956 was found to be antigenically equivalent to the causative agent of the Crimean disease. The current name of Crimean-Congo hemorrhagic disease was applied as a result of this connection between the two place names. Before the 1970s, most cases were reported from the Soviet Union and the Democratic Republic of Congo, and from then until 2000, detailed studies and significant numbers of cases were reported from the Republic of South Africa, the Democratic Republic of Congo, Mauritania, Tanzania, Iraq, Pakistan, Oman and China, Senegal. Since 2000, new cases have been reported from Bulgaria, Iran, Turkey, Greece, Albania and Kosovo (Chinikar et al. 2010; Hoogstraal 1979; Whitehouse 2004; Ince et al. 2014; Morikawa et al. 2007; Papa 2019). Important case reports were published in France (cases from Senegal), Germany (from Afghanistan) and the UK (in Afghanistan and Bulgaria). Following a case from Spain in 2016 it was suggested that the main vector represents a change in the geographical distribution (Spengler et al. 2019). The most commonly reported countries for CCHF cases, between 1998 and 2013, were Turkey, Russia, Iran, Pakistan and Afghanistan (Ince et al. 2014; Morikawa et al. 2007).

2 Causative Agent

The causative agent is a single-stranded RNA virus of the Orthonairovirus genus of the Nairoviridae family. The single-stranded RNA consists of 3 different segments called L, M and S, which encode 4 different proteins. The S segment encodes the viral nucleocapsid, the M segment encodes structural proteins (Gn and Gc) and the L segment encodes the viral polymerase. Eight different genetic lineages have been reported for viral RNA. The first lineage comprises the viruses in the region of Eastern Europe. The second lineage of viruses is the so-called AP92, is mostly found in Greece. The third lineage comprises the viruses in the region of Central Asia, the fourth lineage comprises viruses in the region of Pakistan, Madagascar and some parts of Iran, and the fifth lineage comprises the viruses in the region of Iran and Senegal and Mauritania. The other lineages comprise viruses in the regions of Africa (Whitehouse 2004; Spengler et al. 2019; Adams et al. 2017; Shayan et al. 2015; Ergönül 2006).

The virus is sensitive to environmental conditions and can be inactivated with disinfectants and detergents. The CCHF virus (CCHFV) can be inactivated by disinfectants, including 1% sodium hypochlorite, 70% alcohol, 2% glutaraldehyde, hydrogen peroxide and peracetic acid. With this treatment the virus can be inactivated in 30 min at 56 °C (Appannanavar and Mishra 2011).

3 Seasonal Distribution

It occurs in the Northern Hemisphere between April and September and reaches its highest level in June and July. However, this may vary depending on the region; and cases have even been reported also in January. Climatic characteristics such as warm winters, reduced precipitation and increased temperatures in spring and summer can cause an increase in the number of ticks of the genus Hyalomma, and this is associated with increased cases of human CCHF (Duygu et al. 2018b). The geographical distribution of CCHF patients and Hyalomma ticks is very similar. In addition to Hyalomma, CCHFV has been detected in tick species such as Rhipicephalus, Ornithodoros, Boophilus, Dermacentor and Ixodes (Telmadarraiy et al. 2015).

4 Risk Groups

The risk of exposure to the virus is high among agricultural workers, livestock farmers, shepherds, slaughterhouse workers, meat and meat products workers, veterinarians, leather factory workers, rural campers, hunters, soldiers, health workers and those travelling to endemic areas. Risk factors for seropositivity include a history of tick bites, livestock or farming, manual removal of ticks from animals, and living in rural areas (Whitehouse 2004; Mourya et al. 2019; Gunes et al. 2009).

5 Transmission

The virus is transmitted to humans through tick bites or by direct contact with infected animal blood. Due to their role in the life cycle of ticks, some animals play an important role in the transmission and multiplication of the virus. Viremia or antibodies against the virus have been detected in a wide variety of wild and domestic mammals. Viremia in mammals can last up to two weeks with no clinical manifestations. Although there data are scarce and further research is required, it should be kept in mind that contact with body fluids or tissues of viremic animals may cause an infection risk for slaughterhouse workers, veterinarians, shepherds and farmers. High risk activities such as slaughtering, transporting animals and inspecting livestock without protective measures and the unauthorized slaughter of animals can facilitate the transmission of the disease. In addition, human-to-human transmission can be seen by direct or indirect contact of body fluids, skin or mucous membranes of patients (Gozel et al. 2014; Gunes et al. 2011). Until 2016, 158 CCHF hospital infection cases had been reported from 20 countries in Africa, Asia and Europe. Most of these cases were reported in healthcare workers (86.1%). This was followed by visitors (12.7%) and hospitalized patients (1.3%). Nurses (44.9%) and doctors (32.3%) and the healthcare workers most affected by CCHF. It has been reported that percutaneous contact (34.3%) and cutaneous contact (22.2%) are the most common routes of transmission in healthcare workers (Tsergouli et al. 2019). Uncertainty remains about the risk of the transmission of the CCHF agent by breast milk and sexually transmission. If appropriate precautions are taken in the laboratory, the risk of laboratory-related contamination is low (Erbay et al. 2008; Leblebicioglu et al. 2016b). Possible causes of CCHF outbreaks also include the transport of viruses by birds or other animals from endemic to non-endemic areas. In addition, changing agricultural practices, climate change, the presence of vulnerable animal populations and wildlife increase the risk of contact with vectors (Bente et al. 2013).

6 Clinical Findings

The disease may present with different clinical manifestations ranging from asymptomatic cases to severe forms (Akinci et al. 2016a). The clinical manifestations of CCHF are non-specific and may be confused with many other diseases. Fever, headache, retro-orbital pain, weakness, muscle pain, sore throat, dizziness, conjunctivitis, photophobia, abdominal pain, nausea and vomiting are common findings of CCHF (Fillâtre et al. 2019). CCHF typically has four different clinical stages, including incubation, pre-hemorrhagic, hemorrhagic and recovery periods (Ergönül 2006). The incubation time may vary depending on the viral load and the route of transmission of the virus. After a tick bite, the incubation period is usually between 1 and 3 days (maximum 12 days). The incubation period is usually between 5 and 6 days following contact with blood and body fluids, although it can be up to 13 days (Vorou et al. 2007; Kaya et al. 2011).

The pre-hemorrhagic period lasts between 1 and 7 days. This period is characterized by sudden onset of fever, headache, myalgia, dizziness and hyperemia in the face and neck region and conjunctivitis. During this period, the fever continues for an average of 4–5 days, after which the hemorrhagic period begins, which is a very troublesome period for the patient. This is a short-term period (usually 2–3 days) and develops rapidly. It usually begins on the third to fifth day of the disease. In the hemorrhagic period, different types of bleeding can be seen, such as petechiae, ecchymoses, mucosal hemorrhages, large hematomas of the skin and intraparenchymal hemorrhages. Cerebral and intramuscular hemorrhages can be seen in CCHF patients, in addition to typical bleeding symptoms such as nosebleed, gingival hemorrhage, hemoptysis, hematemesis, melana, menometrorrhagia and hematuria (Hoogstraal 1979; Ergönül 2006).

Death usually occurs during the hemorrhagic period. Somnolence, gross hematuria and melanaare associated with mortality in this period (Fillâtre et al. 2019). The above-mentioned findings are associated with cytokine storm, endothelial cell activation and increased vascular permeability (Papa et al. 2016). The recovery period is between 10 and 20 days after the onset of clinical symptoms. The length of hospital stay is 7 $\pm$ 2.6 days, which varies depending on the severity of the disease. It has been reported that surviving patients may experience health problems such as weakness, hair loss, appetite disorders, polyneuritis, hearing loss, and memory impairment (Ergönül 2006; Leblebicioglu et al. 2016a).

Various scores are used to predict the clinical course of the disease. Two scoring systems have been developed in Turkey, where the disease is endemic. These scores can be used to predict the clinical outcomes in countries where the disease is endemic (Bakır et al. 2015; Dokuzoguz et al. 2013). Although not included in clinical scoring systems, particularly high viral load exceeding 10^8 copies/mL has been independently associated with poor prognosis (Duh et al. 2007).

Mortality rates in endemic countries range from 5 to 30%. The presence of bleeding such as gastrointestinal bleeding and hematuria, impaired consciousness, splenomegaly, thrombocytopenia, leukocytosis, increased alanine aminotransferase, aspartate aminotransferase, lactate dehydrogenase activity and prolonged activated partial thromboplastin time and decreased fibrinogen levels are among the independent predictors of mortality. No relapse has been reported to date (Bente et al. 2013; Fillâtre et al. 2019).

7 Laboratory Findings

Common laboratory findings in CCHF patients include leukopenia, thrombocytopenia, increased activity of alanine aminotransferase (ALT), aspartate aminotransferase (AST) and lactate dehydrogenase (LDH), high creatine kinase (CK) level, prolonged prothrombin time (PT) and activated partial thromboplastin time (PTT). Kupffer cells, hepatic endothelium and hepatocytes are the main targets in CCHFV. Necrosis of hepatocytes causes an increase in liver enzymes. As a result of virus-induced endothelial damage, the coagulation cascade is activated and platelet counts and functions are reduced. As a result, DIC and multiple organ failure are observed (Shayan et al. 2015).

8 Diagnosis

The clinical diagnosis of CCHF is difficult because it can be confused with many other diseases. Therefore, finding a rapid and reliable laboratory diagnosis is necessary. The virus can be isolated from blood, plasma or tissues. Diagnosis can be made by detecting the genomic RNA of the virus using reverse transcription-PCR, demonstrating the viral antigen by immunohistochemical tests, or isolating the virus in cell culture in the acute phase of the disease. Whether the patient is in the recovery phase is confirmed by showing the virus-specific antibody response. Enzyme-Linked ImmunoSorbent Assay (ELISA) or IFA can detect IgGand IgM in the serum. IgM levels are detected in serum by IFA on the 7th day of the diagnosis of the disease. IgM antibodies may persist for up to 4 months. IgG levels are determined on days 7–9 of the disease. Although they gradually reduce in the following periods still they can be detected for up to 5 years. The presence of viral RNA persists until the 16th day of the disease. CCHP is classified as biosafety level 4 (BSL-4) pathogen due to the lack of a safe vaccine or specific treatment (Shayan et al. 2015; Tezer and Polat 2015).

9 Differential Diagnosis

Sepsis, influenza, malaria, leptospirosis, Q fever, typhoid fever, septicemic plague, rickettsial infections, meningococcemia and viral hepatitis should be considered in the differential diagnosis. In addition, Alkhurma and Rift Valley fever in the Middle East, Omsk hemorrhagic fever in Russia, Kyasanur forest disease in India, Hantavirus in Europe and Asia, Lassa, Ebola, Margburg and dengue hemorrhagic fever in Africa should be considered in the differential diagnosis. Non-infectious causes including hematological or solid organ malignancies, ITP, liver cirrhosis and collagen tissue disorders should be considered in the differential diagnosis (Tanyel et al.).

10 Treatment

Treatment options for CCHF have been investigated for the last 30 years. However, disease-specific treatment options are limited. Therefore supportive approaches are often used in the treatment of the disease. Patients should be monitored closely for fluid and electrolytes. As this disease occurs mostly in rural areas, it should be investigated for potential coinfections, which have been reported in the literature as brucellosis, malaria and leptospirosis. An analgesic such as Paracetamol can be used for the management of febrile and painful periods. Platelet suspension, fresh frozen plasma and erythrocyte suspension form the basis of supportive treatment for the control of hemostasis. The criteria for platelet transfusion have been defined as (Leblebicioglu et al. 2012);

- Platelet counts lower than 10,000/mm^3 in patients with no fever or abnormalities in coagulation parameters,
- Platelet counts lower than 50,000/mm^3 in patients with bleeding or who have undergone an invasive intervention,
- Platelet counts lower than 20,000/mm^3 with fever or systemic hemostatic defect.

The criteria for fresh frozen plasma transfusion has been defined as (Leblebicioglu et al. 2012);

- PT/INR 1.5 times the upper limit of normal, or aPTTat the upper limit of normal

The need for whole blood transfusion should be evaluated according to hemoglobin level and general clinical condition. Unnecessary invasive procedures should be avoided to minimize the risk of bleeding (Leblebicioglu et al. 2012).

Nucleoside analogs are a class of drugs that have been used for many years to treat viral hemorrhagic infections, and some are currently being developed for therapeutic use in emerging infections. Ribavirin, a purine nucleoside analogue, is an agent with effective antiviral activity on several RNA viruses. The main mechanism of action of this drug is to disrupt RNA synthesis by competing with cellular nucleotides (Espy et al. 2018). Ribavirin is recommended by the World Health Organization for the treatment of CCHF. In retrospective studies conducted in Turkey and Iran it was stated that the use of ribavirin (within 4 days after the onset of symptoms) may decrease the death rate and accelerate healing (Ergönül et al. 2004; Mardani et al. 2003).

However, clinical data supporting the use of ribavirin to treat CCHF is inconsistent. In-vivo and in-vitro studies have shown that ribavirin has an inhibitory effect on virus replication in cell culture but has no effect on reducing mortality. Furthermore, since most of the studies evaluating the efficacy of ribavirin are non-randomized, the information on its efficacy is limited and further studies are needed (Johnson et al. 2018). Furthermore, due to the ethical problems of placebo-controlled studies, such studies are extremely difficult to be performed. Therefore, obtaining definitive results regarding the efficacy of ribavirin in CCHF patients is problematic. Another broad-spectrum RNA virus inhibitor is favipiravir, which is a drug approved for the treatment of influenza virus infections in Japan. The use of this drug is promising against other highly pathogenic RNA viruses, including Ebola and Lassa. It has also been shown to be effective against CCHF virus in in-vivo and in-vitro studies, with the suggestion that in-vivo efficacy is higher than that of ribavirin (Oestereich et al. 2014).

11 Vaccine

Vaccine studies for CCHF began in the 1960s, and the only vaccine tested in humans to date is the inactivated vaccine formulated with aluminum hydroxide adjuvant, which was developed in Bulgaria. The vaccine activates both cellular and humoral immunity. Neutralization analyses of vaccinated individuals have shown that immunity to the virus has developed. However, neutralizing antibody responses have been reported to be low and require additional vaccinations to improve the immune response. To date, there have been no controlled efficacy studies of the vaccine, and it has not been licensed by the European Drug Agency or the US Food and Drug Administration. Although this has accelerated the studies on vaccines in order to control the disease at a global level, there is still no licensed vaccine available (Dowall et al. 2016).

12 Novel Biomarkers in CCHF

Early diagnosis and prediction of clinical course are very important factors in CCHF for the patient's survival (Akinci et al. 2016b). Although the gold standard for the diagnosis of CCHF is virus isolation, it cannot be used in routine laboratory tests because of the need for biosafety level 4. Therefore, many laboratories use IgM and IgG ELISA and/or RT-PCR to determine the presence or absence of CCHFV infection (Vanhomwegen et al. 2012). The sensitivities of these tests have been reported to be 87.8–93.9%, 80.4–86.1% and 79.6–83.3%, respectively. Therefore several biomarkers which have high sensitivity and specificity have been evaluated to determine whether they can be used in early diagnosis and prediction of prognosis in CCHF.

12.1 Diagnostic and Prognostic Biomarkers

Routine clinical laboratory parameters including platelet, AST, ALT, first step coagulation tests, LDH creatinine, fibrinogen have been considered as laboratory indicators of CCHF (Cevik et al. 2008). According to our laboratory and clinical experience, these parameters do not always change in every case. This fact increases the need for new biomarkers to help diagnose CCHF. In a previous study by the current authors, serum NADPH oxidases (NOX) levels were determined in CCHF patients and healthy controls.

Higher median NOX-1 and NOX-5 levels were found in patients compared to the control group. Increased median serum NOX-5 levels were found in the low-grade disease group compared to the intermediate-high grade disease group according to two different severity scores. Despite low r values, negative correlations were also found between the serum NOX-5 levels and the afore mentioned severity scores. The area under the curve (AUC) values for the NOX-1 and NOX-5 were 0.67 and 0.99, respectively. Lower NOX-5 levels were found in patients receiving thrombocyte suspension. It was therefore concluded that NOX-5 might have a protective effect on CCHF patients and the measurement of serum NOX-5 levels might be used as a novel biochemical test in the diagnosis of CCHF (Büyüktuna et al. 2018). Altay et al. (2016) investigated members of triggering receptor expressed on myeloid cells-1 (TREM-1). In that study, sera of 39 CCHF patients on admission and in the recovery period were compared with healthy control subjects. TREM-1 levels were found to be higher in patients than in the control group. In addition, TREM-1 levels were found to be lower in the recovery period than on admission to hospital. The specificity and sensitivity of TREM-1 in discriminating healthy control subjects and CCHF patients were found to be 94.9% and 87.5%, respectively (Altay et al. 2016). In a recent study by Kayadibi et al. (2019), a novel index entitled the Hitit index was reported as a novel diagnostic tool to identify patients with CCHF in endemic regions (Kayadibi et al. 2019). In addition, organ-specific biomarkers have also been investigated in CCHF. It has been reported that total protein and urine neutrophil gelatinase-associated lipocalin could be useful biomarkers to monitor renal involvement in CCHF (Deveci et al. 2013).

CCHF is characterized by vascular endothelial damage, disseminated intravascular coagulation, thrombocytopenia, liver damage, coagulation abnormalities and hemorrhagia including ecchymosis, gingival bleeding, epistaxis and gastrointestinal bleeding, which are prominent findings for the estimation of the case fatality rate (Ergonul et al. 2006). Several studies have been performed to find reliable biomarkers for estimating the prognosis of patients with CCHF.

In our recent study, different routine laboratory parameters including ferritin, activated partial thromboplastin time, alanine aminotransferase, aspartate aminotransferase and lactate dehydrogenase were evaluated in 40 CCHF patients to discriminate bleeding, and a cut-off value of 149 U/L for AST was determined to discriminate the bleeding condition in CCHF patients (Büyüktuna et al. 2019). Duygu et al. (2018a, b) reported that platelet distribution width (PDW) could be beneficial in predicting haemorrhage and mortality in CCHF (Duygu et al. 2018a). In a study of patients with CCHF and Hemorrhagic fever with renal syndrome, fatal outcomes have been associated with interleukin-6 (IL-6), interferon-alpha2 and monocyte chemo-attractive protein-1(MCP-1) (Korva et al. 2019). Sagmak Tartar et al. (2019) retrospectively investigated the laboratory findings of CCHF patients to find indicators of the clinical course. In their study, ALP and GGT were associated with prognosis of CCHF (Sagmak Tartar et al. 2019). Kerget et al. classified 60 CCHF patients according to severity criteria and compared the groups in terms of endothelin-1 (ET-1), endothelial cell-specific receptor tyrosine kinase (Tie-2) and angiopoietins 2 (Ang-2). Lower ET-1 and Tie-2 and higher Ang-2 were reported in surviving patients (Kerget et al. 2019). It has been reported that PDW and troponin might be used in discriminating severe cases in CCHF (Yilmaz et al. 2017, 2016). It has been shown that monocyte chemo-attractive protein-1 (MCP-1), IL-8 and IL-6 are indicators of fatality in the early period of the disease (Ergönül et al. 2017). Levels of IL-6, tumor necrosis factor alpha and IL-10 have been associated with the disseminated intravascular coagulation scores in CCHF patients (Ergonul et al. 2006; Kaya et al. 2014). Bakir et al. reported that soluble fms-like tyrosine kinase-1 receptor, which is a vascular growth factor receptor and cell free DNA might be used to predict CCHF prognosis (Bakir et al. 2016, 2013). In a study performed on 69 CCHF patients, prolonged aPTTwas considered as an independent risk factor for mortality (Cevik et al. 2008). Biochemical biomarker genetic studies have also been performed on CCHF to determine biomarkers which can be used to estimate prognosis (Arslan et al. 2019; Kızıldağ et al. 2018). MicroRNAs (miRNAs) have been investigated in CCHF patients and miR-625-5p,miR-342-5p and miR-320e have been found to be associated with mortality (Arslan et al. 2019). Kızıldağ et al. (2018) reported that toll-like receptor 10 polymorphisms might be an important biomarker to determine susceptibility and mortality rate in CCHF (Kızıldağ et al. 2018).

13 Conclusion

As CCHF shows a course in different clinical tables varying from an asymptomatic form to a severe clinical form, it remains a mystery to clinicians. The emergence of cytokine storm and endothelial dysfunction at different levels in patients has led to the need to investigate various biomarkers to explain the physiopathology of the disease in clinical, biochemical and genetic aspects. In this paper the results have been evaluated of biomarker studies in several areas.

References

Adams MJ, Lefkowitz EJ, King AMQ, Harrach B, Harrison RL, Knowles NJ, Kropinski AM, Krupovic M, Kuhn JH, Mushegian AR, Nibert M, Sabanadzovic S, Sanfaçon H, Siddell SG, Simmonds P, Varsani A, Zerbini FM, Gorbalenya AE, Davison AJ. 2017. Changes to taxonomy and the international code of virus classification and nomenclature ratified by the international committee on taxonomy of viruses. Arch Virol. 2017;162:2505–38.

Akinci E, Bodur H, Sunbul M, Leblebicioglu H. Prognostic factors, pathophysiology and novel biomarkers in Crimean-Congo hemorrhagic fever. Antiviral Res. 2016;132:233–43.

Akinci E, Bodur H, Sunbul M, Leblebicioglu H. Prognostic factors, pathophysiology and novel biomarkers in Crimean-Congo hemorrhagic fever. Antiviral Res. 2016b;132:233–43.

Altay FA, Elaldi N, Şentürk GÇ, Altin N, Gözel MG, Albayrak Y, Şencan İ. Serum sTREM-1 level is quite higher in Crimean Congo hemorrhagic fever, a viral infection. J Med Virol. 2016;88:1473–8.

Appannanavar SB, Mishra B. An update on Crimean Congo hemorrhagic fever. J Glob Infect Dis. 2011;3:285–92.

Arslan S, Engin A, Aydemir EI, Sahin NO, Bayyurt B, Sari I, Cosgun Y, Bakir M. Identification of potential microRNA markers related to Crimean-Congo hemorrhagic fever disease. J Cell Biochem. 2019;120:15506–17.

Bakir M, Bakir S, Sari I, Celik VK, Gozel MG, Engin A. Evaluation of the relationship between serum levels of VEGF and sVEGFR1 with mortality and prognosis in patients with Crimean-Congo hemorrhagic fever. J Med Virol. 2013;85:1794–801.

Bakir M, Engin A, Kuskucu MA, Bakir S, Gündag O, Midilli K. Relationship of plasma cell-free DNA level with mortality and prognosis in patients with Crimean-Congo hemorrhagic fever. J Med Virol. 2016;88:1152–8.

Bakır M, Gözel MG, Köksal I, Aşık Z, Günal Ö, Yılmaz H, But A, Yılmaz G, Engin A. Validation of a severity grading score (SGS) system for predicting the course of disease and mortality in patients with Crimean-Congo hemorrhagic fever (CCHF). Eur J Clin Microbiol Infect Dis. 2015;34:325–30.

Bente DA, Forrester NL, Watts DM, McAuley AJ, Whitehouse CA, Bray M. Crimean-Congo hemorrhagic fever: history, epidemiology, pathogenesis, clinical syndrome and genetic diversity. Antiviral Res. 2013;100:159–89.

Büyüktuna SA, Doğan HO, Bakir M, Elaldi N, Gözel MG, Engin A. The protective effect and diagnostic performance of NOX-5 in Crimean-Congo haemorrhagic fever patients. J Med Microbiol. 2018;67:543–8.

Büyüktuna SA, Doğan HO, Unlusavuran M, Bakir M. An evaluation of the different biomarkers to discriminate bleeding in Crimean-Congo Hemorrhagic Fever. Ticks Tick Borne Dis. 2019;10:997–1002.

Cevik MA, Erbay A, Bodur H, Gülderen E, Baştuğ A, Kubar A, Akinci E. Clinical and laboratory features of Crimean-Congo hemorrhagic fever: predictors of fatality. Int J Infect Dis. 2008;12:374–9.

Chinikar S, Ghiasi SM, Hewson R, Moradi M, Haeri A. Crimean-Congo hemorrhagic fever in Iran and neighboring countries. J Clin Virol. 2010;47:110–4.

Deveci K, Uysal EB, Kaya A, Sancakdar E, Alkan F. Evaluation of Renal involvement in children with Crimean-Congo hemorrhagic fever. Jpn J Infect Dis. 2013;66:493–6.

Dokuzoguz B, Celikbas AK, Gök ŞE, Baykam N, Eroglu MN, Ergönül Ö. Severity scoring index for Crimean-Congo hemorrhagic fever and the impact of ribavirin and corticosteroids on fatality. Clin Infect Dis. 2013;57:1270–4.

Dowall SD, Buttigieg KR, Findlay-Wilson SJD, Rayner E, Pearson G, Miloszewska A, Graham VA, Carroll MW, Hewson R. A Crimean-Congo hemorrhagic fever (CCHF) viral vaccine expressing nucleoprotein is immunogenic but fails to confer protection against lethal disease. Hum Vaccin Immunother. 2016;12:519–27.

Duh D, Saksida A, Petrovec M, Ahmeti S, Dedushaj I, Panning M, Drosten C, Avsic-Zupanc T. Viral load as predictor of Crimean-Congo hemorrhagic fever outcome. Emerg Infect Dis. 2007;13:1769–72.

Duygu F, Sari T, Celik H. Effects of platelet function on the haemorrhagic manifestations and mortality in Crimean-Congo haemorrhagic fever. Le Infez Med. 2018;26:341–6.

Duygu F, Sari T, Kaya T, Tavsan O, Naci M. The relationship between Crimean-Congo hemorrhagic fever and climate: does climate affect the number of patients? Acta Clin Croat. 2018;57:443–8.

Erbay A, Cevik MA, Onguru P, Gözel G, Akinci E, Kubar A, Bodur H. Breastfeeding in Crimean-Congo haemorrhagic fever. Scand J Infect Dis. 2008;40:186–8.

Ergönül O. Crimean-Congo haemorrhagic fever. Lancet Infect Dis. 2006;6:203–14.

Ergönül O, Celikbaş A, Dokuzoguz B, Eren S, Baykam N, Esener H. Characteristics of patients with Crimean-Congo hemorrhagic fever in a recent outbreak in Turkey and impact of oral ribavirin therapy. Clin Infect Dis. 2004;39:284–7.

Ergönül Ö, Şeref C, Eren Ş, Çelikbaş A, Baykam N, Dokuzoğuz B, Gönen M, Can F. Cytokine response in crimean-congo hemorrhagic fever virus infection. J Med Virol. 2017;89:1707–13.

Ergonul O, Tuncbilek S, Baykam N, Celikbas A, Dokuzoguz B. Evaluation of serum levels of interleukin (IL)-6, IL-10, and tumor necrosis factor-alpha in patients with Crimean-Congo hemorrhagic fever. J Infect Dis. 2006;193:941–4.

Espy N, Pérez-Sautu U, Ramírez de Arellano E, Negredo A, Wiley MR, Bavari S, Díaz Menendez M, Sánchez-Seco MP, Palacios G. Ribavirin Had Demonstrable Effects on the Crimean-Congo Hemorrhagic Fever Virus (CCHFV) Population and Load in a Patient With CCHF Infection. J Infect Dis. 2018;217:1952–6.

Fillâtre P, Revest M, Tattevin P. Crimean-Congo hemorrhagic fever: an update. Médecine Mal Infect. 2019;49:574–85.

Gozel MG, Bakir M, Oztop AY, Engin A, Dokmetas I, Elaldi N. Investigation of Crimean-Congo hemorrhagic fever virus transmission from patients to relatives: a prospective contact tracing study. Am J Trop Med Hyg. 2014;90:160–2.

Gunes T, Engin A, Poyraz O, Elaldi N, Kaya S, Dokmetas I, Bakir M, Cinar Z. Crimean-Congo hemorrhagic fever virus in high-risk population Turkey. Emerg Infect Dis. 2009;15:461–4.

Gunes T, Poyraz O, Vatansever Z. Crimean-Congo hemorrhagic fever virus in ticks collected from humans, livestock, and picnic sites in the hyperendemic region of Turkey. Vector Borne Zoonotic Dis. 2011;11:1411–6.

Hoogstraal H. The epidemiology of tick-borne Crimean-Congo hemorrhagic fever in Asia, Europe, and Africa. J Med Entomol. 1979;15:307–417.

Ince Y, Yasa C, Metin M, Sonmez M, Meram E, Benkli B, Ergonul O. Crimean-Congo hemorrhagic fever infections reported by ProMED. Int J Infect Dis. 2014;26:44–6.

Johnson S, Henschke N, Maayan N, Mills I, Buckley BS, Kakourou A, Marshall R. Ribavirin for treating Crimean Congo haemorrhagic fever. Cochrane Database Syst Rev. 2018;6:CD012713.

Kaya A, Engin A, Güven AS, Içağasıoğlu FD, Cevit O, Elaldı N, Gültürk A. Crimean-Congo hemorrhagic fever disease due to tick bite with very long incubation periods. Int J Infect Dis. 2011;15:e449–52.

Kaya S, Elaldi N, Kubar A, Gursoy N, Yilmaz M, Karakus G, Gunes T, Polat Z, Gozel MG, Engin A, Dokmetas I, Bakir M, Yilmaz N, Sencan M. Sequential determination of serum viral titers, virus-specific IgG antibodies, and TNF-α, IL-6, IL-10, and IFN-γ levels in patients with Crimean-Congo hemorrhagic fever. BMC Infect Dis. 2014;14:416.

Kayadibi H, Yapar D, Akdogan O, Ulusu NN, Baykam N. Hitit Index to distinguish patients with and without Crimean-Congo hemorrhagic fever. Ticks Tick Borne Dis. 2019;10:1035–40.

Kerget F, Öztürk Z, Öztürk N, Yılmaz S. The relationship with clinical course and prognosis of serum endothelin-1, angiopoietin-2, and tie-2 levels in Crimean-Congo hemorrhagic fever. Turkish J Med Sci. 2019;49:1192–7.

Kızıldağ S, Arslan S, Özbilüm N, Engin A, Bakır M. Effect of TLR10 (2322A/G, 720A/C, and 992T/A) polymorphisms on the pathogenesis of Crimean Congo hemorrhagic fever disease. J Med Virol. 2018;90:19–25.

Korva M, Rus KR, Pavletič M, Saksida A, Knap N, Jelovšek M, Smrdel KS, Jakupi X, Humolli I, Dedushaj J, Petrovec M, Avšič-Županc T. Characterization of Biomarker Levels in Crimean-Congo Hemorrhagic Fever and Hantavirus Fever with Renal Syndrome. Viruses. 2019;11.

Leblebicioglu H, Bodur H, Dokuzoguz B, Elaldi N, Guner R, Koksal I, Kurt H, Senturk GC. Case management and supportive treatment for patients with Crimean-Congo hemorrhagic fever. Vector Borne Zoonotic Dis. 2012;12:805–11.

Leblebicioglu H, Sunbul M, Barut S, Buyuktuna SA, Ozkurt Z, Yapar D, Yilmaz G, Guner R, But A, Cicek Senturk G, Murat N, Ozaras R, Crimean Congo Hemorrhagic Fever Research Network of Turkey. Multi-center prospective evaluation of discharge criteria for hospitalized patients with Crimean-Congo Hemorrhagic Fever. Antiviral Res. 2016a;133:9–13.

Leblebicioglu H, Sunbul M, Guner R, Bodur H, Bulut C, Duygu F, Elaldi N, Cicek Senturk G, Ozkurt Z, Yilmaz G, Fletcher TE, Beeching NJ. Healthcare-associated Crimean-Congo haemorrhagic fever in Turkey, 2002–2014: a multicentre retrospective cross-sectional study. Clin Microbiol Infect. 2016b;22:387.e1–e4.

Mardani M, Jahromi MK, Naieni KH, Zeinali M. The efficacy of oral ribavirin in the treatment of crimean-congo hemorrhagic fever in Iran. Clin Infect Dis. 2003;36:1613–8.

Morikawa S, Saijo M, Kurane I. Recent progress in molecular biology of Crimean-Congo hemorrhagic fever. Comp Immunol Microbiol Infect Dis. 2007;30:375–89.

Mourya DT, Yadav PD, Gurav YK, Pardeshi PG, Shete AM, Jain R, Raval DD, Upadhyay KJ, Patil DY. Crimean Congo hemorrhagic fever serosurvey in humans for identifying high-risk populations and high-risk areas in the endemic state of Gujarat India. BMC Infect Dis. 2019;19:104.

Oestereich L, Rieger T, Neumann M, Bernreuther C, Lehmann M, Krasemann S, Wurr S, Emmerich P, de Lamballerie X, Ölschläger S, Günther S. Evaluation of antiviral efficacy of ribavirin, arbidol, and T-705 (favipiravir) in a mouse model for Crimean-Congo hemorrhagic fever. PLoS Negl Trop Dis. 2014;8:e2804.

Papa A. Diagnostic approaches for Crimean-Congo hemorrhagic fever virus. Expert Rev Mol Diagn. 2019;19:531–6.

Papa A, Tsergouli K, Çağlayık DY, Bino S, Como N, Uyar Y, Korukluoglu G. Cytokines as biomarkers of Crimean-Congo hemorrhagic fever. J Med Virol. 2016;88:21–7.

Sagmak Tartar A, Özer Balin S, Çatak Z, Akbulut A, Demirdag K. Importance of biochemical parameters in order to predict clinical severity in patients diagnosed with Crimean-Congo haemorrhagic fever. Turkish J. Biochem. 2019;44:177–81.

Shayan S, Bokaean M, Shahrivar MR, Chinikar S. Crimean-Congo hemorrhagic fever. Lab Med. 2015;46:180–9.

Spengler JR, Bergeron É, Spiropoulou CF. Crimean-Congo hemorrhagic fever and expansion from endemic regions. Curr Opin Virol. 2019;34:70–8.

Tanyel E, Sunbul M, Fletcher TE, Leblebicioglu H. Aetiology of PCR negative suspected Crimean-Congo hemorrhagic fever cases in an endemic area. Pathog. Glob. Health 110:173–7.

Telmadarraiy Z, Chinikar S, Vatandoost H, Faghihi F, Hosseini-Chegeni A. Vectors of Crimean Congo Hemorrhagic Fever Virus in Iran. J Arthropod Borne Dis. 2015;9:137–47.

Tezer H, Polat M. Diagnosis of Crimean-Congo hemorrhagic fever. Expert Rev Anti Infect Ther. 2015;13:555–66.

Tsergouli K, Karampatakis T, Haidich A-B, Metallidis S, Papa A. Nosocomial infections caused by Crimean-Congo haemorrhagic fever virus. J Hosp Infect. 2019.

Vanhomwegen J, Alves MJ, Zupanc TA, Bino S, Chinikar S, Karlberg H, Korukluoğlu G, Korva M, Mardani M, Mirazimi A, Mousavi M, Papa A, Saksida A, Sharifi-Mood B, Sidira P, Tsergouli K, Wölfel R, Zeller H, Dubois P. Diagnostic assays for Crimean-Congo hemorrhagic fever. Emerg Infect Dis. 2012;18:1958–65.

Vorou R, Pierroutsakos IN, Maltezou HC. Crimean-Congo hemorrhagic fever. Curr Opin Infect Dis. 2007;20:495–500.

Whitehouse CA. Crimean-Congo hemorrhagic fever. Antiviral Res. 2004;64:145–60.

Yilmaz H, Yilmaz G, Kostakoğlu U, Yaman H, Örem A, Köksal İ. The prognostic significance of serum troponin T levels in Crimean-Congo hemorrhagic fever patients. J Med Virol. 2017;89:408–12.

Yilmaz H, Yilmaz G, Menteşe A, Kostakoğlu U, Karahan SC, Köksal İ. Prognostic impact of platelet distribution width in patients with Crimean-Congo hemorrhagic fever. J Med Virol. 2016;88:1862–6.

Zika Virus

Zika Virus Disease: Progress and Prospects

Swatantra Kumar, Rajni Nyodu, Vimal K. Maurya, and Shailendra K. Saxena

Abstract Zika virus (ZIKV) is a mosquito borne flavivirus that exhibits wide range of transmission routes including mosquito, transplacental, sexual contact, blood transfusion and organ transplantation. ZIKV is highly neurotropic which infects central nervous system (CNS) in adult and developing brain. As of 2021, at least 86 countries and territories have been reported with autochthonous transmission of ZIKV. Preexisting flavivirus specific sero-cross reactivity antibodies may results in antibody dependent enhancement (ADE) of ZIKV infection with significant higher viral load and severe disease outcomes. However, the mechanism of ADE has not been completely understood. Upon ZIKV infection, innate immune system, humoral immune response and cellular mediated response get activated. Even though, the virus has developed several immune escape strategies for successful infection. Currently there is no specific treatment available for ZIKV infection and the treatment relies on the symptomatic relief of the patients. However, several of the drug candidates have been discovered and are currently in the clinical trials. Molecules have been targeted against potential viral proteins including entry inhibitors, protease inhibitors and replication inhibitors. In spite of progress made in this area, the absence of clinically approved drug or vaccine increases the chances of recurrent outbreak in endemic regions and therefore needs utmost attention in this issue.

Keywords Zika virus · Antibody dependent enhancement · Zika syndrome · Antivirals · Guillain-Barré syndrome · DENV infection · STAT2 protein · Galidesivir · Non-steroidal anti-inflammatory drugs

---

S. Kumar · R. Nyodu · V. K. Maurya · S. K. Saxena (✉)
Centre for Advanced Research (CFAR), Faculty of Medicine, King George's Medical University (KGMU), Lucknow 226003, India
e-mail: shailen@kgmcindia.edu

S. I. Ahmad (ed.), *Human Viruses: Diseases, Treatments and Vaccines*, https://doi.org/10.1007/978-3-030-71165-8_12

1 Introduction

Zika virus (ZIKV) is a flavivirus which gets transmitted primarily via bite of infected *Aedes aegypti* mosquito. ZIKV is also transmitted from mother to fetus via crossing transplacental barrier, through sexual contact, (Musso et al. 2019) blood transfusion (Joob and Wiwanitkit 2019) and organ transplantation (Fig. 1) (Nogueira et al. 2017). The incubation period of ZIKV infection is 3–14 days from the exposure to the virus. Most of the individuals infected with ZIKV remain asymptomatic. However, mild symptoms for 2–7 days may arise in few individuals which ranges from fever, joint pain, rash, headache and malaise (Baud et al. 2017). During pregnancy, ZIKV infection in mother may cause intrauterine growth restriction, preterm birth and miscarriage as well as ZIKV may be transmitted to fetus which causes infants to be born with Zika syndrome (Teixeira et al. 2020) including cerebral atrophy, cortical and/or periventricular calcifications, corpus callosum abnormalities, ventriculomegaly, including congenital malformations like microcephaly (Ribeiro et al. 2017).

Adults and children infected with ZIKV are at the risk of neurological complications including neuropathy, myelitis, encephalomyelitis, encephalitis,

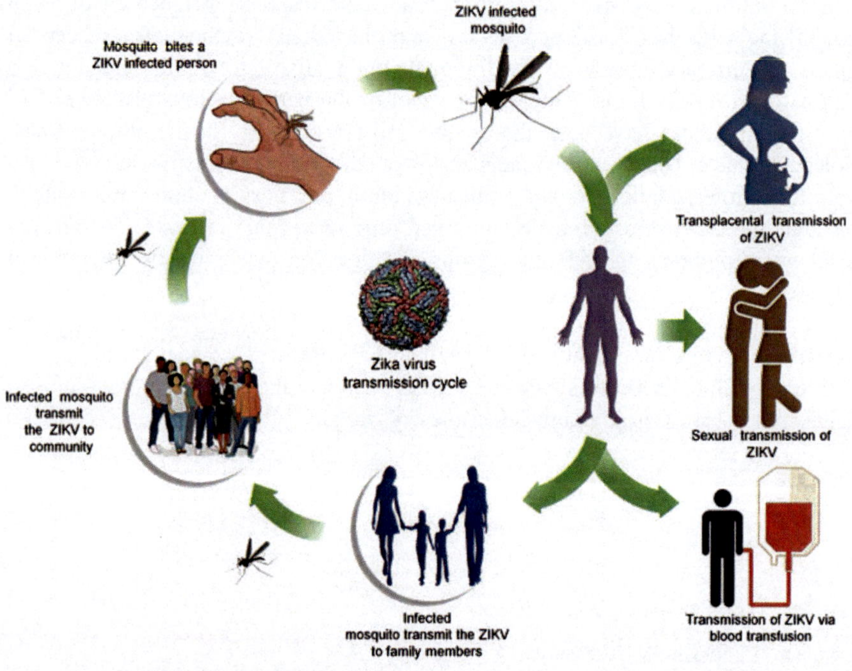

Fig. 1 ZIKV transmission cycle. ZIKV transmission initiate when an infected mosquito bites to an individual, family members or community for its transmission. The transmission cycle also involve the transplacental transmission of ZIKV to fetus, sexual contact and via blood transfusion

meningoencephalitis, sensory polyneuropathy and Guillain-Barré syndrome (Muñoz et al. 2018). Neuropathological investigations of ZIKV infection have shown the presence of ZIKV in brain and cerebrospinal fluid samples (Fig. 2). These findings suggest that ZIKV is highly neurotropic which infects central nervous system (CNS) in adult and developing brain (Chimelli et al. 2017). ZIKV replication follows the similar mode of replication cycle as other flaviviruses. Dendritic cell-specific intercellular adhesion molecule 3-grabbing nonintegrin (DC-SIGN) is the primary cellular receptor involve in the receptor mediated internalization of ZIKV (Hamel et al. 2015). The acidic environment inside the endosomal compartment induces the fusion of viral envelope protein with endosomal membrane that results in release of ZIKV RNA into the cytoplasm (Agrelli et al. 2019). The positive sense RNA genome of ZIKV is a 10.8 Kb which comprised of $\sim$100 nucleotides of 5′ untranslated region (UTR), a single open reading frame (ORF) of $\sim$10 Kb and a 3′ UTR region of $\sim$420 nucleotides (Wang et al. 2017). The single ORF encodes for a single polyprotein that gets processed into structural proteins as capsid (C), precursor membrane protein (prM), envelop (E) and non-structural proteins as NS1, NS2a, NS2b, NS3, NS4a, NS4b, and NS5 (Rossignol et al. 2017). These proteins assist in the replication of viral RNA at the endoplasmic reticulum surface where viral RNA and proteins assembled to generate immature virions. Upon furin-mediated cleavage of prM to M protein, these virions get mature in the trans-Golgi network and results in the process of egression (Rossignol et al. 2017).

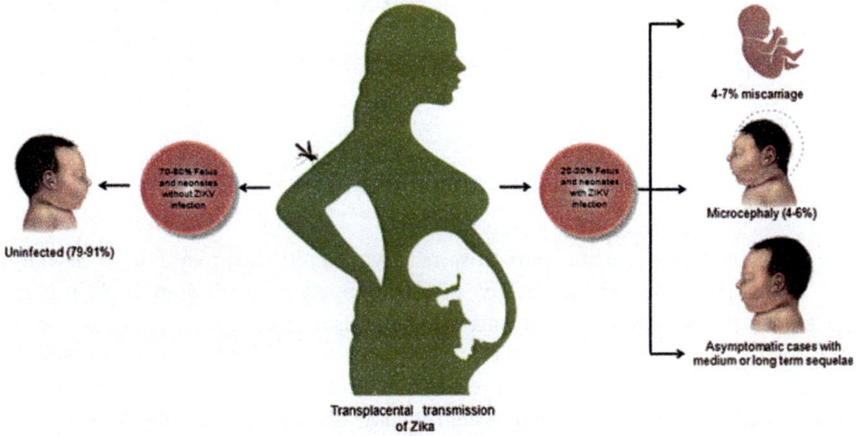

Fig. 2 Transplacental transmission of ZIKV. 70–80% of ZIKV infection during pregnancy may not transmit the virus to fetus. Whereas, 20–30% of ZIKV infection during pregnancy may transmit the ZIKV to fetus and may cause preterm birth and miscarriage (4–7%) and may be transmitted to fetus which results in microcephaly (4–6%) or asymptomatic cases with medium or long term sequelae

2 Epidemiology of Zika Virus Infection

In 1947, ZIKV is first identified in sentinel rhesus monkey in Uganda and later in 1952, identified in humans in Uganda and United Republic of Tanzania. However, little attention was paid until 2007, when a major outbreak of ZIKV reported in the Yap State of Micronesia. Consequently, in year 2013 outbreaks were reported in French Polynesia and subsequently in other Pacific Islands. Till date, the largest outbreak of ZIKV has been reported in Brazil in year 2015–2016 with over 200,000 cases where over 8000 babies were born with malformations caused by ZIKV (Hills et al. 2017). Recent epidemiological data helps in understanding of global Zika transmission and its associated challenges. As of 2021, at least 86 countries and territories have been reported with autochthonous transmission of ZIKV that includes Africa, America, South-East Asia and Western Pacific regions (Fig. 3) (https://www.who.int/health-topics/zika-virus-disease#tab=tab_1). Several factors including prevalence of mosquito vectors, global trade and travel, tropical/subtropical climate, poor waste management has causes recent emergence of Zika virus in countries like China, India, Indonesia, Maldives, Thailand, and Pakistan in recent years. Due to co-circulation of dengue in endemic tropical and sub-tropical region, it has been estimated that people living in 2 million square kilometer are at the highest risk of getting ZIKV infection (Messina et al. 2016). In these individuals, people may be diagnosed with co-infection with Zika and dengue both. A retrospective population based serological survey in Indonesia revealed that by the age of 5 years approximately 9% of the children had previous ZIKV infection. Seasonal patterns of ZIKV transmission in Thailand demonstrated that transmission coincided with dengue virus by sharing common mosquito vectors. In year 2018, ZIKV outbreak has been reported in Rajasthan, India (Saxena et al. 2019).

3 Antibody-Dependent Enhancement

Preexisting flavivirus specific sero-cross reactivity antibodies may results in antibody dependent enhancement (ADE) of ZIKV infection with significant higher viral load and severe disease outcomes (Dejnirattisai et al. 2016). ADE has been studied extensively specifically with respect to secondary dengue infection, which is attributed by the observation that pre-existing antibodies that results in secondary infection such as DSS and/or DHF which is more severe form of DENV infection (Guzman et al. 2013). The cross-reactive antibodies facilitates the viral uptake that increases the severity of infection by forming a complex between the cross-reactive Abs and the viral particles and, hence, helping these particles to get internalized via Fc gamma receptor (Moi et al. 2010). Considerable degree of genetic similarities and structural homology among ZIKV and other flaviviruses, results in a generation of sero cross-reactivity antibodies (Rathore and St John 2020). Additionally, in vitro studies shows human monoclonal Abs (mAbs) to the DENV fusion loop

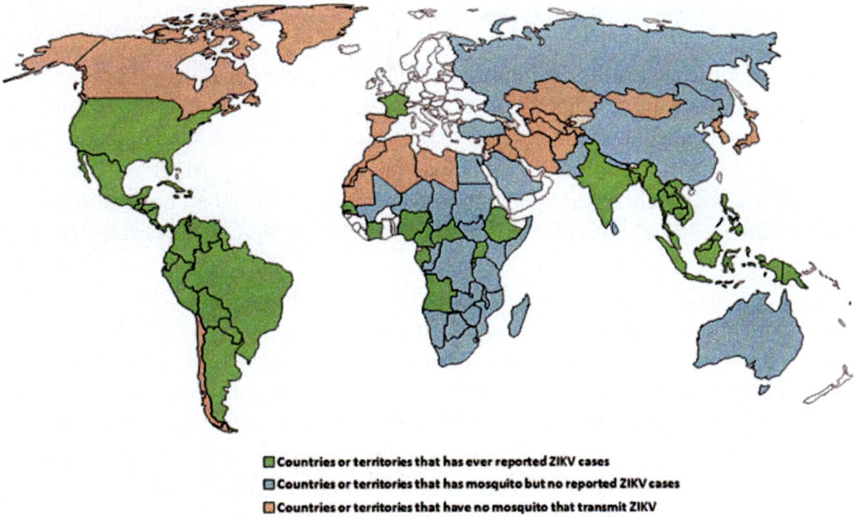

Fig. 3 Epidemiology of ZIKV showing the countries or territories that have reported with ZIKV infection as well as areas where the mosquito are present or absent with no evidence of ZIKV transmission as per Centers for Disease Control and Prevention (https://wwwnc.cdc.gov/travel/files/zika-areas-of-risk.pdf)

epitope causing ADE of ZIKV in Fc receptor bearing cell lines which is shown to be facilitating ZIKV infection (Priyamvada et al. 2016). On the other hand, DENV patient derived mAbs to EDE1 (E dimer-epitope) which neutralizes all the four serotypes of DENV, in vitro, neutralizes ZIKV too, with a high potency, whereas EDE2 binds, but does not significantly neutralizes ZIKV (Fernandez et al. 2020). Similarly, when the convalescent plasma from DENV/WNV infected samples was transferred in vitro, enhancement of ZIKV infection has been found which is mediated through Fc gamma receptors (Swanstrom et al. 2016). Although, WNV mediated Abs shows less cross-reactivity to ZIKV in comparison to DENV-mediated Abs, but when studied, human polyclonal Abs against WNV is reported to be cross reactive towards ZIKV resulting in enhancement of ZIKV infection both in vivo and in vitro (Bardina et al. 2017). So far, ADE response during ZIKV infection has not been clearly understood.

4 Understanding the Host-Immune Response

Nucleic acid receptors such as RIG-I-like receptors (RLRs) activate upon binding to the viral RNA and initiate the release of interferon (IFN) via stimulating IFN receptor subunits that activates the JAK-STAT pathways and thereby establishes an antiviral state (Rehwinkel and Gack 2020). ZIKV NS1 protein has been shown to involve in the immune escape mechanism by inhibiting the type 1 IFN signaling pathway (Xia et al. 2018). In addition, NS5 protein has been shown to degrade the STAT2 protein that results in inhibition of IFN signaling pathway (Grant et al. 2016). Humoral immune response confers first-line of protection against flaviviral infection. B cell mediated antibody response is crucial in any viral infection and may results in the development of effective vaccines. The E, prM and NS1 protein has been found to elicit stronger antibody response which has been utilized in several studies that shows protection against ZIKV infection in various models (Amrun et al. 2020). Strong B cell response suggests a crucial role in ZIKV infection that helps us to identify B cell epitopes identified specifically from NS1 protein which have been found to be preventive upon challenged with ZIKV (Kam et al. 2019). Similarly, envelope has been found to contain B cell epitopes especially from the domain III region and has been found to protect mice upon challenged with ZIKV infection (Yang et al. 2017). Cell mediated immune response also plays a crucial role in prevention of viral infections. Both $CD8^+$ and $CD4^+$ T cells are involved in cell mediated immune responses during ZIKV infection (Winkler et al. 2017). Protective role $CD8^+$ T cells in during ZIKV infection has been validated by observing susceptibility of infected mice towards the ZIKV infection in its absence (Elong Ngono et al. 2017). In addition, adoptive transfer of ZIKV specific $CD8^+$ T cells has been found to protect mice challenged with ZIKV infection by reducing the viral load (Huang et al. 2017).

5 Investigational Medicines for the Treatment of Zika

Currently there is no specific treatment available for ZIKV infection. However, the treatment relies on the symptomatic relief of the patients. According to CDC, the recommended treatment for ZIKV infection is to take plenty of rest, enough fluid intakes to prevent dehydration, acetaminophen for reducing fever and pain and not to use aspirin or other non-steroidal anti-inflammatory drugs (NSAIDs) until dengue has been ruled out (https://www.cdc.gov/zika/symptoms/treatment.html). Considering the replication cycle of ZIKV, several of the potential drugs targets have been identified for the development of effective antiviral drugs. Inhibitors that can interfere with the interaction of viral envelope protein with host cell receptors may be used for the development of potential entry inhibitors. One of the natural product nanchangmycin derived from *Streptomyces nanchangensis*, has been shown to act via inhibition of AXL receptors and abolishment of viral internalization process via receptor mediated endocytosis (Rausch et al. 2017). Compound screening of entry inhibitor has revealed ZINC33683341 molecule as a potential entry inhibitor during ZIKV infection (Fernando et al. 2016). RNA dependent RNA polymerases have been well characterized and explored drug targets due to its essential role in virus replication and production of progeny virus. Nucleoside analogues including its phosphoramidate product have been shown be a promising RdRp inhibitors. Several of the potential drug candidates have been investigated against ZIKV. 3′-O-methylribonucleosides and 2′-C-methylated ribonucleoside has been found to exhibits inhibitory effect on ZIKV replication with less toxicity (Boehr et al. 2019). Adenosine analogues named as NITD008 has been shown as a potent anti-ZIKV activity in cell culture model where in it reduces the viremia and prevent virus induced cell death (Deng et al. 2016). This compound has been also found to inhibit DENV replication. However, due to associated toxicity, the NITD008 has been discontinued from the pipeline of drug discovery for ZIKV. One of the nucleoside analogue named as 7DMA which has been designed for the treatment of HCV has been found as a potent inhibitor of ZIKV infection in cell culture and animal model (Zmurko et al. 2016). However, it has not been successful in clinical trial. Similarly, the clinically approved HCV drug, sofosbuvir has been extensively studied for its activity against ZIKV infection. In both cell culture and model, sofosbuvir has been found to exhibit potent antiviral activity (Bullard-Feibelman et al. 2017). FDA has considered the sofosbuvir as a pregnancy class B drug for the treatment of HCV infection and therefore it might be an ideal drug candidate for ZIKV treatment in pregnant women. In addition, ribavirin which has been also developed as an anti-HCV drug, found to exhibit antiviral properties against ZIKV infection (Kamiyama et al. 2017). Galidesivir which is an adenosine analogue drug developed for filoviruses and YFV has been shown to be effective against ZIKV infection (Lim et al. 2017). In addition to nucleoside based analogues drugs, several of the non-nucleoside drug candidates have been tested. NS2B-NS3 protease of ZIKV is the most targeted ZIKV proteins for the development of antivirals due to its crucial role in replication and maturation. A boric acid containing dipeptides inhibitors have been designed which showed strong antiviral activity. This compound has been found to exhibits broad spectrum anti-flaviviral activity (Lei et al. 2016). Similarly, a pyrazole ester derivative has been developed as a covalent inhibitor of protease activity (Li et al. 2018). Apart from these drug candidates, complementary and alternative medicine in alliance with conventional medicine may be used for Zika therapeutics and prevention (Saxena et al. 2016).

6 Conclusions

Zika virus is an emerging mosquito-borne flavivirus infection that causes Zika syndrome and various neurological complications in adults. The absence of clinically approved drug or vaccine increases the chances of recurrent outbreak in endemic regions. Several of the vaccine candidates have been developed while understanding the immunogenicity of ZIKV proteins and host cellular response during infection. Presence of flavivirus sero-cross reactivity antibodies may results in antibody dependent enhancement (ADE) of ZIKV infection that significantly increases the viral load with severe disease outcomes. Currently there is no specific treatment available for ZIKV infection and the treatment relies on the symptomatic relief of the patients. However, several of the drug candidates have been discovered and are currently in the clinical trials.

7 Future Perspectives

Intense research is required for the better understanding of ZIKV pathogenesis in various models focusing on developmental biology, transplacental transmission and neurological complications. In addition, basic research long-term cohort studies needs to be conducted for the better understanding of the disease impact as well as assessment of the risk for coming future. In this regard, leading heath organizations should establish a surveillance monitoring system in endemic areas of flaviviruses for co-infection cases that will help us to understand more about ADE of ZIKV infection. Due to increased risk associated with pregnant women and related congenital abnormalities, non-essential travelling to the affected areas needs to be avoided by travelers. Due to absence of clinically approved vaccine or drug, personal protective measures need to be compelled.

Acknowledgements The authors are grateful to the Vice Chancellor, King George's Medical University (KGMU), Lucknow, India, for the encouragement of this work. The authors have no other relevant affiliations or financial involvement with any organization or entity with a financial interest in or financial conflict with the subject matter or materials discussed in the manuscript apart from those disclosed.

References

Agrelli A, de Moura RR, Crovella S, Brandão LAC. ZIKA virus entry mechanisms in human cells. Infect Genet Evol. 2019;69:22–9. https://doi.org/10.1016/j.meegid.2019.01.018.

Amrun SN, Yee WX, Abu Bakar F, et al. Novel differential linear B-cell epitopes to identify Zika and dengue virus infections in patients [published correction appears in Clin Transl Immunology. 2020 Feb 19;9(2):e01118]. Clin Transl Immunol. 2019;8(7):e1066. Published 26 July 2019. https://doi.org/10.1002/cti2.1066.

Bardina SV, Bunduc P, Tripathi S, et al. Enhancement of Zika virus pathogenesis by preexisting antiflavivirus immunity. Science. 2017;356(6334):175–80. https://doi.org/10.1126/science.aal4365.

Baud D, Gubler DJ, Schaub B, Lanteri MC, Musso D. An update on Zika virus infection. Lancet. 2017;390(10107):2099–109. https://doi.org/10.1016/S0140-6736(17)31450-2.

Boehr AK, Arnold JJ, Oh HS, Cameron CE, Boehr DD. 2'-C-methylated nucleotides terminate virus RNA synthesis by preventing active site closure of the viral RNA-dependent RNA polymerase. J Biol Chem. 2019;294(45):16897–907. https://doi.org/10.1074/jbc.RA119.010214.

Bullard-Feibelman KM, Govero J, Zhu Z, et al. The FDA-approved drug sofosbuvir inhibits Zika virus infection. Antiviral Res. 2017;137:134–40. https://doi.org/10.1016/j.antiviral.2016.11.023.

Chimelli L, Melo ASO, Avvad-Portari E, et al. The spectrum of neuropathological changes associated with congenital Zika virus infection. Acta Neuropathol. 2017;133(6):983–99. https://doi.org/10.1007/s00401-017-1699-5.

Dejnirattisai W, Supasa P, Wongwiwat W, et al. Dengue virus sero-cross-reactivity drives antibody-dependent enhancement of infection with Zika virus. Nat Immunol. 2016;17(9):1102–8. https://doi.org/10.1038/ni.3515.

Deng YQ, Zhang NN, Li CF, et al. Adenosine analog NITD008 is a potent inhibitor of Zika virus. Open Forum Infect Dis. 2016;3(4):ofw175. Published 30 Aug 2016. https://doi.org/10.1093/ofid/ofw175.

Elong Ngono A, Vizcarra EA, Tang WW, et al. Mapping and role of the CD8+ T cell response during primary Zika virus infection in mice. Cell Host Microbe. 2017;21(1):35–46. https://doi.org/10.1016/j.chom.2016.12.010.

Fernandez E, Dejnirattisai W, Cao B, et al. Human antibodies to the dengue virus E-dimer epitope have therapeutic activity against Zika virus infection [published correction appears in Nat Immunol. 2020 Mar;21(3):354]. Nat Immunol. 2017;18(11):1261–9. https://doi.org/10.1038/ni.3849.

Fernando S, Fernando T, Stefanik M, Eyer L, Ruzek D. An approach for Zika virus inhibition using homology structure of the envelope protein. Mol Biotechnol. 2016;58(12):801–6. https://doi.org/10.1007/s12033-016-9979-1.

Grant A, Ponia SS, Tripathi S, et al. Zika virus targets human STAT2 to inhibit type I interferon signaling. Cell Host Microbe. 2016;19(6):882–90. https://doi.org/10.1016/j.chom.2016.05.009.

Guzman MG, Alvarez M, Halstead SB. Secondary infection as a risk factor for dengue hemorrhagic fever/dengue shock syndrome: an historical perspective and role of antibody-dependent enhancement of infection. Arch Virol. 2013;158(7):1445–59. https://doi.org/10.1007/s00705-013-1645-3.

Hamel R, Dejarnac O, Wichit S, et al. Biology of Zika virus infection in human skin cells. J Virol. 2015;89(17):8880–96. https://doi.org/10.1128/JVI.00354-15.

Hills SL, Fischer M, Petersen LR. Epidemiology of Zika virus infection. J Infect Dis. 2017;216 (suppl_10):S868–74. https://doi.org/10.1093/infdis/jix434.

Huang H, Li S, Zhang Y, et al. CD8+ T cell immune response in immunocompetent mice during Zika virus infection. J Virol. 2017;91(22):e00900-17. Published 27 Oct 2017. https://doi.org/10.1128/jvi.00900-17.

Joob B, Wiwanitkit V. Zika virus, blood donation and transfusion transmission risk. Hematol Transfus Cell Ther. 2019;41(3):278. https://doi.org/10.1016/j.htct.2018.11.001.

Kam YW, Leite JA, Amrun SN, et al. ZIKV-specific NS1 epitopes as serological markers of acute Zika virus infection. J Infect Dis. 2019;220(2):203–12. https://doi.org/10.1093/infdis/jiz092.

Kamiyama N, Soma R, Hidano S, et al. Ribavirin inhibits Zika virus (ZIKV) replication in vitro and suppresses viremia in ZIKV-infected STAT1-deficient mice. Antiviral Res. 2017;146:1–11. https://doi.org/10.1016/j.antiviral.2017.08.007.

Lei J, Hansen G, Nitsche C, Klein CD, Zhang L, Hilgenfeld R. Crystal structure of Zika virus NS2B-NS3 protease in complex with a boronate inhibitor. Science. 2016;353(6298):503–5. https://doi.org/10.1126/science.aag2419.

Li Y, Zhang Z, Phoo WW, et al. Structural insights into the inhibition of Zika virus NS2B-NS3 protease by a small-molecule inhibitor. Structure. 2018;26(4):555–64.e3. https://doi.org/10.1016/j.str.2018.02.005.

Lim SY, Osuna C, Lakritz J, Chen E, Yoon G, Taylor R, MacLennan S, Leonard M, Giuliano E, Mathis A, Berger E. Galidesivir, a direct-acting antiviral drug, abrogates viremia in rhesus macaques challenged with Zika virus. In: Open forum infectious diseases. Oxford University Press; 2017.

Messina JP, Kraemer MU, Brady OJ, et al. Mapping global environmental suitability for Zika virus. Elife. 2016;5:e15272. https://doi.org/10.7554/eLife.15272.

Moi ML, Lim CK, Takasaki T, Kurane I. Involvement of the Fc gamma receptor IIA cytoplasmic domain in antibody-dependent enhancement of dengue virus infection. J Gen Virol. 2010;91(Pt 1):103–11. https://doi.org/10.1099/vir.0.014829-0.

Muñoz LS, Parra B, Pardo CA. Neuroviruses emerging in the Americas study. Neurological implications of Zika virus infection in adults. J Infect Dis. 2018; 217(8):1334.

Musso D, Ko AI, Baud D. Zika virus infection—after the pandemic. N Engl J Med. 2019;381 (15):1444–57. https://doi.org/10.1056/NEJMra1808246.

Nogueira ML, Estofolete CF, Terzian AC, et al. Zika virus infection and solid organ transplantation: a new challenge. Am J Transplant. 2017;17(3):791–5. https://doi.org/10.1111/ajt.14047.

Priyamvada L, Quicke KM, Hudson WH, et al. Human antibody responses after dengue virus infection are highly cross-reactive to Zika virus. Proc Natl Acad Sci U S A. 2016;113 (28):7852–7. https://doi.org/10.1073/pnas.1607931113.

Rathore APS, St John AL. Cross-reactive immunity among flaviviruses. Front Immunol. 2020;11:334. Published 26 Feb 2020. https://doi.org/10.3389/fimmu.2020.00334.

Rausch K, Hackett BA, Weinbren NL, et al. Screening bioactives reveals nanchangmycin as a broad spectrum antiviral active against Zika virus. Cell Rep. 2017;18(3):804–15. https://doi.org/10.1016/j.celrep.2016.12.068.

Rehwinkel J, Gack MU. RIG-I-like receptors: their regulation and roles in RNA sensing [published online ahead of print, 13 Mar 2020]. Nat Rev Immunol. 2020;1–15. https://doi.org/10.1038/s41577-020-0288-3.

Ribeiro BG, Werner H, Lopes FPPL, et al. Central nervous system effects of intrauterine Zika virus infection: a pictorial review. Radiographics. 2017;37(6):1840–50. https://doi.org/10.1148/rg.2017170023.

Rossignol ED, Peters KN, Connor JH, Bullitt E. Zika virus induced cellular remodelling. Cell Microbiol. 2017;19(8). https://doi.org/10.1111/cmi.12740.

Saxena SK, Elahi A, Gadugu S, Prasad AK. Zika virus outbreak: an overview of the experimental therapeutics and treatment. Virusdisease. 2016;27(2):111–5. https://doi.org/10.1007/s13337-016-0307-y.

Saxena SK, Kumar S, Sharma R, Maurya VK, Dandu HR, Bhatt MLB. Zika virus disease in India—update October 2018. Travel Med Infect Dis. 2019;27:121–2. https://doi.org/10.1016/j.tmaid.2018.10.022.

Swanstrom JA, Plante JA, Plante KS, et al. Dengue virus envelope dimer epitope monoclonal antibodies isolated from dengue patients are protective against Zika virus. mBio. 2016;7(4): e01123-16. Published 19 July 2016. https://doi.org/10.1128/mbio.01123-16.

Teixeira FME, Pietrobon AJ, Oliveira LM, Oliveira LMDS, Sato MN. Maternal-fetal interplay in Zika virus infection and adverse perinatal outcomes. Front Immunol. 2020; 11:175. Published 14 Feb 2020. https://doi.org/10.3389/fimmu.2020.00175.

Wang A, Thurmond S, Islas L, Hui K, Hai R. Zika virus genome biology and molecular pathogenesis. Emerg Microbes Infect. 2017;6(3):e13. Published 22 Mar 2017. https://doi.org/10.1038/emi.2016.141.

Winkler CW, Myers LM, Woods TA, et al. Adaptive immune responses to Zika virus are important for controlling virus infection and preventing infection in brain and testes. J Immunol. 2017;198(9):3526–35. https://doi.org/10.4049/jimmunol.1601949.

Xia H, Luo H, Shan C, et al. An evolutionary NS1 mutation enhances Zika virus evasion of host interferon induction. Nat Commun. 2018;9(1):414. Published 29 Jan 2018. https://doi.org/10.1038/s41467-017-02816-2.

Yang M, Dent M, Lai H, Sun H, Chen Q. Immunization of Zika virus envelope protein domain III induces specific and neutralizing immune responses against Zika virus. Vaccine. 2017;35 (33):4287–94. https://doi.org/10.1016/j.vaccine.2017.04.052.

Zika epidemiology update, World Health Organization. https://www.who.int/emergencies/diseases/zika/zika-epidemiology-update-july-2019.pdf?ua=1. Accessed May 2020.

Zika virus Treatment. Centre for Disease control and Prevention. https://www.cdc.gov/zika/symptoms/treatment.html. Accessed May 2020.

Zmurko J, Marques RE, Schols D, Verbeken E, Kaptein SJ, Neyts J. The viral polymerase inhibitor 7-Deaza-2′-C-Methyladenosine is a potent inhibitor of in vitro Zika virus replication and delays disease progression in a robust mouse infection model. PLoS Negl Trop Dis. 2016;10(5):e0004695. Published 10 May 2016. https://doi.org/10.1371/journal.pntd.0004695.

The Medicinal Chemistry of Zika Virus

Érica Erlanny da Silva Rodrigues, Hannah Maus, Stefan Josef Hammerschmidt, Alessia Ruggieri, Elane Conceição dos Santos, Ênio José Bassi, Leticia Anderson, Pedro Gregório Vieira Aquino, João Xavier de Araújo-Júnior, Fenju Wei, Xinyong Liu, Peng Zhan, Tanja Schirmeister, and Edeildo Ferreira da Silva-Júnior

Abstract Arthropod-borne viruses, also known as arboviruses, are transmitted by bites of infected mosquito or tick vectors. In this context, the *Flavivirus* genus is mainly transmitted by mosquitoes from the *Aedes* genus, being the *Ae. africanus, Ae. aegypti*, and *Ae. Albopictus* species are responsible for transmitting the Zika virus (ZIKV). It is a lipid-enveloped virus constitute of an RNA genome, which is translated into a polyprotein encoding three structural proteins {(capsid (C), membrane (M), and envelope (E)} and seven non-structural proteins (NS1, NS2A, NS2B, NS3, NS4A, NS4B, and NS5). Several biological targets have been identified for developing antiviral agents against ZIKV, which could prevent virus

É. Erlanny da S. Rodrigues · J. X. de Araújo-Júnior · E. F. da Silva-Júnior (✉)
Institute of Chemistry and Biotechnology, Federal University of Alagoas, Campus A.C. Simões, Lourival Melo Mota Avenue, Maceió 57072-970, Brazil
e-mail: edeildo.junior@iqb.ufal.br

H. Maus · S. J. Hammerschmidt · T. Schirmeister
Institute of Pharmaceutical and Biomedical Science, Johannes Gutenberg-University Mainz, Staudingerweg 5, 55128 Mainz, Germany

A. Ruggieri
Department of Infectious Diseases, Molecular Virology, Center for Integrative Infectious Disease Research, University of Heidelberg, Im Neuenheimer Feld 344, 69120 Heidelberg, Germany

E. C. dos Santos · Ê.J. Bassi
IMUNOREG – Immunoregulation Research Group, Laboratory of Research in Virology and Immunology, Institute of Biological Sciences and Health, Federal University of Alagoas, Campus A.C. Simões, Lourival Melo Mota Avenue, Maceió 57072-970, Brazil

L. Anderson
CESMAC University Center, Cônego Machado Street, Maceió 57051-160, Brazil

P. G. V. Aquino
Environment Laboratory, Federal University of Agreste From Pernambuco, Bom Pastor Avenue, Boa Vista, Garanhuns 55292-270, Brazil

F. Wei · X. Liu · P. Zhan
Department of Medicinal Chemistry, Key Laboratory of Chemical Biology, Ministry of Education, School of Pharmaceutical Sciences, Shandong University, Ji'nan 250012, China

S. I. Ahmad (ed.), *Human Viruses: Diseases, Treatments and Vaccines*, https://doi.org/10.1007/978-3-030-71165-8_13

entry, assembly, or release of new virion particles, by interactions with structural proteins. Drugs targeting non-structural proteins could inhibit the ZIKV-replication cycle. Nature has provided excellent compounds for designing new potent analogs. The medicinal chemistry emerges as an impressive, rationally design and to develop new antiviral agents. Additionally, virtual and high-throughput screenings have contributed greatly to the identification of previously approved-drugs that could be repurposed for ZIKV, as antiviral agents. In this chapter, we offer a literature review about the most relevant and recent advances made on the medicinal chemistry of ZIKV research. We focus on natural, nature-based, semi-synthetic, and synthetic antiviral compounds, as well as repurposed drugs and other inhibitors targeting ZIKV.

Keywords Zika virus · Natural compounds · Nature-based · Synthetic compounds · Drug repurposing · Virtual screening · Antivirals · Medicinal chemistry

1 Introduction

The *Flavivirus* genus consists of about 70 known viruses (King et al. 2012), some of them are associated with human diseases which cause important health problems worldwide (Schmaljohn and McClain 1996). Most of them are arboviruses (or arthropod-borne viruses) transmitted by mosquito or tick vectors, causing diseases in humans ranging from mild to sub-clinical infections to lethal hemorrhagic fever or encephalitis (Holbrook 2017). The group of mosquito-borne Flaviviruses includes Japanese encephalitis virus (JEV), Dengue virus (DENV), West Nile virus, Yellow fever virus (YFV), and Zika virus (ZIKV). Typically, the most common clinical manifestations of Flaviviruses infection include fever, rashes, encephalitis, and hemorrhagic fever.

In 1947, ZIKV was isolated from the blood of *Rhesus* monkeys (*Macaca mulatta*) during a study on YFV, in Uganda at the Zika forest. Later it was isolated from *Aedes africanus* mosquitoes collected in the same forest (Dick 1952). The first case of human ZIKV infection was reported in Uganda and Nigeria, where the patients presented analogous symptoms of YFV infections (MacNamara 1954).

1.1 Zika Virus Infection

The ZIKV infection is characterized by an acute exanthematic febrile illness lasting for a period of up to 7 days. It is estimated that approximately 20% of infections in humans result in clinical manifestations and symptoms thus presenting a low hospitalization rate (Hayes 2009; Duffy et al. 2009). Symptoms may include skin rash, fever, arthralgia, non-purulent conjunctivitis, headache, non-intense myalgia mainly in the hands and feet joints, edema, and vomiting (Zanluca et al. 2015).

The viral transmission is mainly mediated by mosquitoes from the *Aedes* genus and is initiated by bite of a ZIKV-infected female mosquito at the course of its hematophagy. ZIKV is injected into human skin through the mosquito proboscis (Guerbois et al. 2016; Briant et al. 2014) and can infect fibroblasts, keratinocytes, and dendritic cells (DCs), activating an immune response (Hamel et al. 2015). Moreover, it was shown that amniotic epithelial, fetal mesenchymal (Van Der Eijk et al. 2016), neural (Cumberworth et al. 2017), trophoblasts (Aagaard et al. 2017), fibroblasts (placental, uterine, and pulmonary) (Chen et al. 2016), endothelial, and peripheral blood mononuclear (Foo et al. 2017) cells are susceptible to the virus, suggesting that ZIKV is a pantropic virus, which interacts with distinct cells from different tissues.

During the ZIKV outbreak in Brazil, the reported cases have suggested a possible association between the virus infection in pregnant patients and birth defects, with a 20-fold increase in microcephaly cases in the Northeast of the country (ECDC and Assessment 2015). Further, a ZIKV-microcephaly association was noticed in French Polynesia and Brazil outbreaks (Cauchemez et al. 2016; de Araújo et al. 2016). The virus was detected in the 28th week of pregnancy in the amniotic fluid of two pregnant women who presented previous clinical manifestations of ZIKV infection (fever, myalgia, and rash), suggesting that the virus crosses the placental barrier (Calvet et al. 2016). In another study, it was possible to detect the ZIKV in fetal brain tissue with microcephaly after fetal autopsy (Mlakar et al. 2016). Besides the microcephaly, the virus was associated with a set of neurological disorders in newborns, including intellectual disability, cerebral palsy, epilepsy, visual, auditory, and behavioral disorders, being described as the congenital Zika syndrome (Ashwal et al. 2009).

Guillain-Barré syndrome (GBS) is a serious neurological disorder that affects the nervous system and causes demyelination of the neuron's axon, resulting in respiratory failure characterized by muscle weakness and paralysis (Burns 2008). Recently, ZIKV infection has been pointed out as one of the causes associated with GBS, and 1,708 cases were reported amid the ZIKV outbreak in Brazil (Pan American Health Organisation 2015).

1.2 Epidemiology of Zika Virus

After the first discovery of ZIKV, few sporadic cases of its disease were registered in the African and Asian continents (Olson et al. 1981). ZIKV cases were restricted to African and Asian continents, having been reported in Nigeria (MacNamara 1954), Malaysia (Marchette et al. 1969), Sierra Leone (Robin and Mouchet 1975), Senegal (Renaudet et al. 1978; Bres et al. 1963), and Indonesia (Olson et al. 1981). Nonetheless, in 2007, the first outbreak occurred on Yap Island, Micronesia, and the ZIKV spread to French Polynesia, New Caledonia, Cook Island, Easter Island, and Brazil (Tognarelli et al. 2016; Grard et al. 2014; Duffy et al. 2009).

The ZIKV has been isolated from different mosquitoes species in Africa and Asia, including the *Aedes africanus* (Dick 1952), *Ae. aegypti* (Marchette et al. 1969), and *Ae. albopictus* (Grard et al. 2014). Furthermore, ZIKV was also isolated from the serum of *Rhesus* monkeys in Uganda (Dick 1952) and anti-ZIKV antibodies were detected in bats in Ethiopia, birds in Morocco, and rodents in the Central African Republic (Andral et al. 1968). Vector competence was analyzed with different *Aedes* populations infected with ZIKV and both *Ae. aegypti* and *Ae. albopictus* were shown to succeed in the transmission of the virus (Chouin-Carneiro et al. 2016).

In addition to vector-borne transmission (Guerbois et al. 2016), several studies showed that ZIKV is potentially sexually transmitted and persists in genital secretions for a prolonged period after symptom onset as previously reviewed (Moreira et al. 2017). High viral RNA load was detected in infected patients' semen and urine (Musso et al. 2015). Interestingly, although the sexual transmission of ZIKV has been described and a high viral load has been found in the semen, its presence has been shown to inhibit ZIKV and other Flaviviruses infection in several cell lines, in primary lines and tissues endometrial and vaginal, suggesting that semen acts as a protector factor of infection (Müller et al. 2018).

Vertical transmission was observed amid the Brazilian outbreak (Oliveira Melo et al. 2016; Calvet et al. 2016; Mlakar et al. 2016). In another study was shown the presence of infective ZIKV particles in breast milk with meaningful viral loads (Dupont-Rouzeyrol et al. 2016). ZIKV can also be potentially transmitted by blood transfusions (Motta et al. 2016).

2 Zika Virus Structure

ZIKV is a lipid-enveloped Flavivirus having an RNA genome of about 10.8 kilobases (kb) with a 5′ cap (Fig. 1). The viral RNA is translated into a polyprotein (3,423 amino acids) encoding three structural proteins, capsid (C), membrane (M) which is generated from its precursor-membrane (prM), and envelope (E), as well as seven non-structural proteins (NS1, NS2A, NS2B, NS3, NS4A, NS4B, and NS5) (Sirohi and Kuhn 2017). The structural proteins are responsible for virus entry, assembly, and release of new virions from the host cell. The non-structural proteins are involved in replication, assembly, and regulation of the host immune pathways to viral infection.

The structure of mature ZIKV has been solved at 3.8 Å resolution at room temperature (Sirohi et al. 2016), 3.7 Å at 40 °C (Kostyuchenko et al. 2016), and 3.1 Å resolution by cryo-electron microscopy (cryo-EM) single-particle reconstruction (Sevvana et al. 2018). The mature ZIKV particle has 500 Å diameter containing a single positive-strand RNA complexed with multiple copies of the C protein to form the nucleocapsid core; and there are 180 copies of both E ($\sim$500 amino acids) and M ($\sim$75 amino acids) proteins on the virus surface embedded in the host-derived bilayer lipid membrane via their transmembrane regions (Kostyuchenko et al. 2016), (Fig. 1).

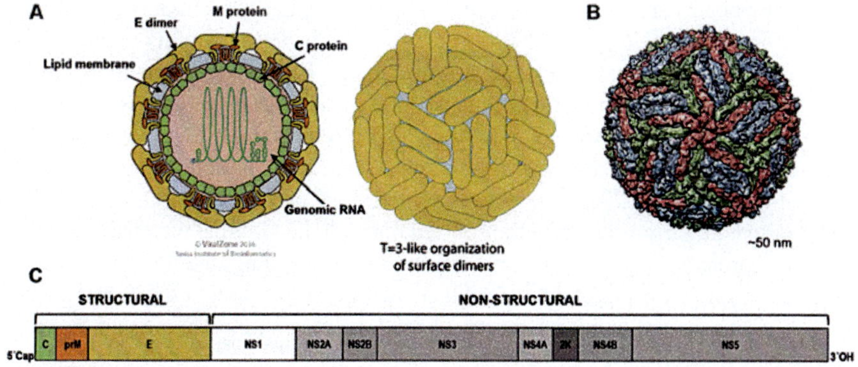

Fig. 1 ZIKV structure and organization. **a** Scheme of ZIKV organization showing cross-section and surface perspectives; **b** Cryo-EM structure (PDB ID: 5IRE) of ZIKV with the envelope proteins in trimeric arrangement on the surface; **c** Viral genome is a single-stranded positive-sense RNA (+) ssRNA of ~10.8 kb organized in a single open-reading frame (ORF) encoding non- and structural proteins Figure **a** *Source* : SIB—Swiss Institute of Bioinformatics (https://viralzone.expasy.org/6756)

2.1 Zika Virus Replication Cycle

ZIKV replication cycle is similar to other known Flaviviruses. Firstly, the E proteins interact with host-cell receptors, initiating cell infection. Diverse host cell receptors were described as entry and/or adhesion factors for ZIKV, including AXL, DC-SIGN, Tyro3, and TIM-1 (Hamel et al. 2015). ZIKV entry is mediated by phosphatidylserine receptors as TIM-1 and TAM family members (Axl and Tyro3), and dendritic cell-specific adhesion receptor DC-SIGN, acting as adhesion factors and participating in viral internalization (Hastings et al. 2019; Hamel et al. 2015). The receptors blockage by neutralizing antibodies were already pointed out as a mechanism to inhibit the viral endocytosis, reducing ZIKV infection (Meertens et al. 2017; Hamel et al. 2015; Hastings et al. 2019). AXL, a member of the TAM receptor family of cell surface receptor tyrosine kinases, is the primary ZIKV entry cofactor on human umbilical vein endothelial cells (HUVECs) (Richard et al. 2017) and glial cells (Meertens et al. 2017), like others. ZIKV has been shown to infect different cells, including skin keratinocytes, human dermal fibroblasts, neuronal, placental, ocular tissues, immune, and reproductive tract cells, as previously reviewed (Miner and Diamond 2017).

After virus E protein attachment on the host cell membrane, ZIKV is internalized via endocytosis. Into the endosome, the low pH environment triggers the trimerization of the E protein thus resulting in the fusion of viral and cell membranes. The viral positive-strand RNA is released into the host cytoplasm followed by its translation into the polyprotein with the endoplasmic reticulum (ER)-localization signal, promoting the association of ribosome to the ER membrane. The polyprotein presents various transmembrane sequences traversing the ER membrane bilayer and it is proteolytically cleaved by host and viral proteases into structural (C, prM, and E) and non-structural (NS1 to NS5) proteins. Thereafter, the virus assembly is initiated, and "immature viruses" are formed in the ER lumen, containing the E, prM, lipid membrane, and nucleocapsid (Hasan et al. 2018). The cryo-EM of immature ZIKV has been solved and it was shown to be similar to the other known immature Flavivirus, although a less prominent capsid protein shell has been observed (Prasad et al. 2017).

The immature virions are then transported through the Golgi apparatus where the host furin protease cleaves the prM protein thus yielding the pr peptide and membrane-anchored M protein to produce mature virus particles (Stadler et al. 1997). The virions are then transported to the host plasma membrane and exit the cell via exocytosis (Fig. 2).

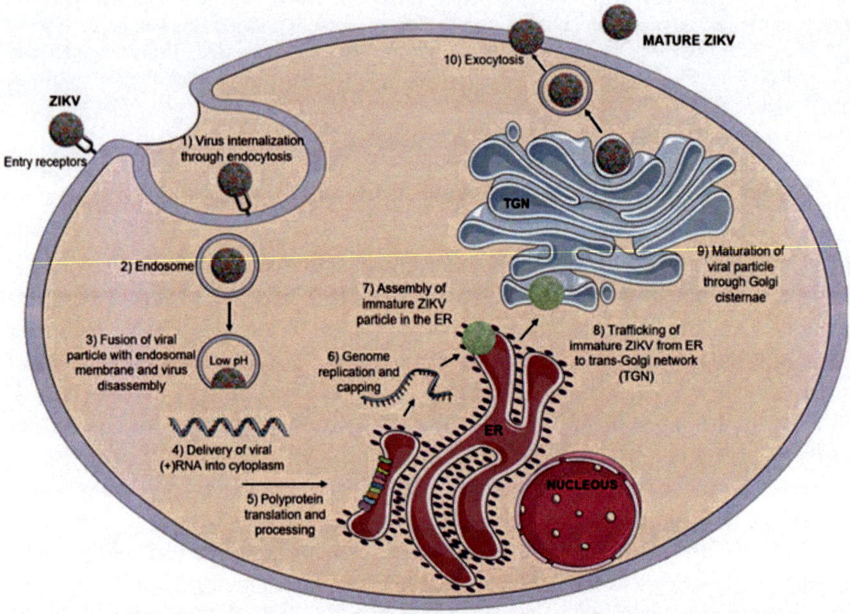

Fig. 2 Schematic overview for ZIKV replication cycle. E protein interacts with host-cell receptors and the virus enters cells by receptor-mediated endocytosis. Several ZIKV receptors were identified including AXL, Tyro3, DC-SIGN, and TIM-1. Into the endosome occurs the fusion of viral particles with the endosomal membrane by and acidic-pH-triggered mechanism followed by virus disassembly and release of viral RNA into the cytoplasm. The (+)ssRNA is translated into a polyprotein further processed and cleaved into all structural (C, prM, and E) and non-structural proteins (NS1 to NS5). The replication process occurs at the surface of the endoplasmic reticulum (ER). A dsRNA genome is synthesized from the genomic (+)ssRNA. Transcription and replication of the dsRNA genome provide viral mRNAs and new (+)ssRNA. Virus assembly occurs at the ER and leads to the formation of immature virus particles that are transported to the Golgi apparatus. The prM protein is cleaved in the trans-Golgi network (TGN) by host furin protease, completing the viral particle maturation. Finally, mature new virions are released from the host-cell by exocytosis

2.2 Targets at Zika virus for Developing Antiviral Agents

Viral protein structure analyses with refined resolution enhance the identification of targets for antiviral therapeutic agents. The replicative cycle of ZIKV depends on structural proteins to form the virus particle and non-structural proteins to form the replication complex inside the host cell, and both could be targeted by antiviral agents. Antiviral compounds targeting the structural proteins could prevent the virus entry, assembly, and release of new virion particles. Otherwise, antiviral agents targeting non-structural proteins (some of them with enzymatic activities) could interrupt the replication cycle (Sinigaglia et al. 2018).

The ZIKA glycoprotein E is involved in the host cell receptor binding, membrane fusion, and host immune recognition. This protein comprises four domains: the stem-transmembrane domain (E-Stem and E-TM), responsible for the membrane anchor, and three domains of the surface portion of the protein: a β-barrel-shaped domain I, an elongated finger-like domain II responsible for dimerization of E protein, and a *C*-terminal immunoglobulin-like domain III (Dai et al. 2016; Shi and Gao 2017). The structure of E protein presents a pocket between domains I and II, determined as βOG pocket with conserved residues, making it a remarkable antiviral target at ZIKV and other Flaviviruses (Wispelaere et al. 2018).

The prM protein plays a role in the assembly of mature virions and might be implicated in the ZIKV neurovirulence. The prM might act as a chaperone-like role for the folding and assembly of the E proteins. During the virus maturation, pr peptide is cleaved off from the prM protein into the mature M. The M protein anchors the lipid membrane and it is hidden under the E protein layer lying on the virus surface. This rearrangement of E and prM proteins is critical to form the infective virus particle (Shi and Gao 2017; Hasan et al. 2018; Dai et al. 2016).

The flavivirus C protein interacts with the viral RNA to produce the nucleocapsid and modulates diverse cellular processes, such as metabolism, apoptosis, and immune response. ZIKV C protein after polyprotein translation remains on the cytoplasmic side of the ER where they form homodimers and attach to the TM regions of the prM/E tripods. Capsid gene encodes a protein with five α-helices with a pre-α1 loop, critical for membrane association, and a cleavage site at helix α4-α5 to NS2B/NS3 protease (Shang et al. 2018). Recently, the cryo-EM structure revealed in the mature virions a mixture of capsid proteins with and without helix α5, suggesting incomplete NS2B/NS3 cleavage (Tan et al. 2020). The C protein interacts with lipid droplets in ZIKV infected cells and possesses the broad binding capability to RNA or DNA (Shang et al. 2018) playing a central role in the virus assembly process (Tan et al. 2020).

The NS3 protein together with the transmembrane NS2B cofactor is a chymotrypsin-like serine protease NS2B-NS3, which plays an essential role in viral replication; also, an attractive target for drug design and development of antivirals. NS3 is a multifunctional protein with an *N*-terminal serine protease domain ($NS3^{Pro}$) that cleaves the viral polyprotein, and a *C*-terminal region that acts as RNA helicase (NS3hel) unwinding RNA duplex (Zhang et al. 2016; Phoo et al.

2018). NS2B cofactor stabilizes the correct protein fold of the core structure and participates structurally as part of the substrate-binding pocket. NS3 catalytic site is composed of His^{51}, Asp^{75}, and Ser^{135} as a catalytic triad, in which serine is the nucleophilic amino acid {(Ser(O^{-})}. Differently of DENV, the catalytic site of ZIKV NS3 protease has an S3 pocket with a Gly instead of a Lys, allowing the design of ZIKV specific inhibitors (Kang et al. 2017; Zhu et al. 2019). NS3hel is activated by GTPγS (triphosphate) at the conserved triphosphate pocket with NTPase activity to unwind and translocate the RNA during replication accommodated in the positive-charged tunnel (Tian et al. 2016).

The largest non-structural protein from ZIKV is NS5, responsible for the replication of the viral genome and host interferon suppression. Similar to other Flaviviruses, NS5 has an *N*-terminal RNA methyltransferase (MTase) domain (responsible for 5′ viral RNA capping) fused with a *C*-terminal RNA-dependent RNA polymerase (RdRp) domain that replicates the RNA viral genome (Upadhyay et al. 2017). Additionally, flaviviral NS5 is the most conserved amongst the viral proteins and is not present in mammalian hosts, also a promising target for drug discovery and development (Lim et al. 2015). These fused flaviviral domains act synergistically to RNA polymerase and 5′-capping functions, meanwhile, the RNA template fits in the RdRp channel and GTP (guanosine-5′-triphosphate) along with SAM (S-adenosyl-L-methionine) cofactor binds MTase domain (Ferrero et al. 2019). NS5 RdRp structure has a canonical right-hand conformation comprising fingers, palm, and thumb subdomains, which the channel has a single-strand RNA (ssRNA) and NTP entry, and an exit region of the double-strand RNA (dsRNA). The priming loop in the RdRp fit the 3′-RNA template binding at the active site is suitable for the discovery of nucleoside and non-nucleoside inhibitors. The priming loop of the ZIKV provides a smaller and closed binding pocket with different amino acids compared to DENV2, thus affecting the drug design (Godoy et al. 2017).

The major host-interaction Flavivirus protein is the NS1, playing a role in replication, pathogenesis, and modulation of the host immune response. Flavivirus NS1 structure has three domains, an *N*-terminal β-roll, an epitope-rich ring, and a *C*-terminal β-ladder and it is included in the replication complex at the ER luminal side (Brown et al. 2016). The crystallographic structure of full-length ZIKV NS1 protein showed an elongated hydrophobic surface for membrane association and a polar surface that varies among Flaviviruses, and two *N*-linked glycosylation sites (Shi and Gao 2017; Brown et al. 2016). Likewise, the NS1 ZIKV protein structure reveals the two faces of dimer: inner face with a hydrophobic protrusion for membrane interaction and a polar outer face with glycosylation residues (Shi and Gao 2017). It is known that flaviviral secreted NS1 (sNS1) as a hexameric lipoprotein particle interacts with components and factors from host innate and adaptive immune systems, as an antigenic marker for ZIKV infection (Hilgenfeld 2016).

The structures of non-structural NS2A, NS4A, and NS4B proteins from ZIKV remain unsolved, having an absence of information about the replication complex

Table 1 Summary of the main ZIKV targets to antiviral development

ZIKV Target	Activity	Function	Structural characteristics
Envelope protein dimer	Virus entry	Viral entry into host cells and membrane fusion	Stem-transmembrane domains (E-Stem and E-TM), three ectodomains and βOG pocket
Capsid	Pack the virus particle	Viral replication	Hydrophobic helices $\alpha 1$ to interact with viral membrane/lipid droplets and positively charged helices $\alpha 4$ to interact with viral RNA
NS3 helicase domain	Unwind RNA duplex	Viral replication	RNA binding site positively charged and triphosphate pocket with NTPase activity
NS2B-NS3 protease complex	Cleavage of the viral polyprotein	Viral replication	Serine protease domain
NS5 MTase domain	5′-capping of nascent RNA	RNA viral replication	Binding pocket for cofactor SAM/SAH and a biding site for GTP and RNA cap
NS5 RdRp domain	RNA-template binding and polymerization	RNA viral replication	Active site in RNA channel to ssRNA and NTP entry, and dsRNA exit region
NS1	Viral RNA replication and Particle assembly	RNA viral replication and host immune modulation	Hydrophobic surface for membrane interaction and polar region with glycosylation sites

comprising all the NS proteins. This information will provide important data to elucidate protein domains, active sites, and perform studies focused on the rational design and development of new antiviral agents.

In summary, main ZIKV proteins as targets for antiviral design and development are shown in Table 1.

3 Medicinal Chemistry of Zika Virus

Below we discuss recent advances in the medicinal chemistry of active substances against ZIKV, focusing on natural, nature-based, semi- and synthetic compounds, and repurposed drugs, as well as, other inhibitors used to combat this infectious disease. Additionally, for the study of those inhibitors, directed toward the NS2B-NS3 protease, we suggest three literature reviews recently presented by Silva-Júnior et al. (2018), Silva-Júnior and Araújo-Júnior (2019), Nitsche (2019), and Voss and Nitsche (2020).

3.1 Natural Compounds

Most studies for screening natural compounds as inhibitors for ZIKV infections focus on the detection of substances that interfere with virus entry and release steps. They contain several newly identified agents that exhibit antiviral activities by targeting the virus envelope (E) protein.

In a study, Carneiro et al. (2016) identified the molecule (−)-epigallocatechin gallate (EGCG, Fig. 3), a polyphenol from the green tea (*Camellia sinensis*), as a ZIKV entry inhibitor with an EC_{50} value of 21.4 μM and negligible toxicity at 200 μM concentration. Intriguingly, Uganda strain (MR766) was more sensitive to EGCG compared to a Brazilian strain (not specified) shown by a two-fold decrease in EC_{50} value (Carneiro et al. 2016). The proposed mechanism of action is analogous to the anti-human immunodeficiency virus (HIV) activity by destroying the envelope phospholipids, thus destabilizing the virus particles (Yamaguchi et al. 2002). Given the low metabolic stability of EGCG (Steinmann et al. 2013), chemical modification or encapsulation methods need to be pursued (Lambert et al. 2006; Fangueiro et al. 2016).

In the following works, Sharma et al. (2017), as well as Hengphasatporn et al. (2019), investigated the binding mode of EGCG on the dimeric E protein using molecular docking studies and found conflicting results. While Sharma et al. described the binding site to be in the flexible linker between domains I and III, Hengphasatporn et al. (2019). Found the most likely binding site to be in domain II, a highly conserved region in flaviviral E proteins (Hengphasatporn et al. 2019; Sharma et al. 2017). However, these new structural insights may be utilized to further improve E protein targeting substances as ZIKV entry inhibitors.

After identifying protoporphyrins IX (PPIX) as antiviral active agents against Yellow fever virus (YFV) and Dengue virus (DENV), Neris et al. (2018) showed, that the heme and its cobalt and tin analogs CoPPIX and SnPPIX (Fig. 4) also inhibit various other enveloped viruses, including ZIKV by interfering with virus adsorption and entry with IC_{50} values in the single-digit micromolar range (Assunção-Miranda et al. 2016; Neris et al. 2018). Interestingly, while both heme and CoPPIX did not show significant changes in activity, SnPPIX displays an improvement of its activity under light stimulation by a nearly 40-fold decrease of

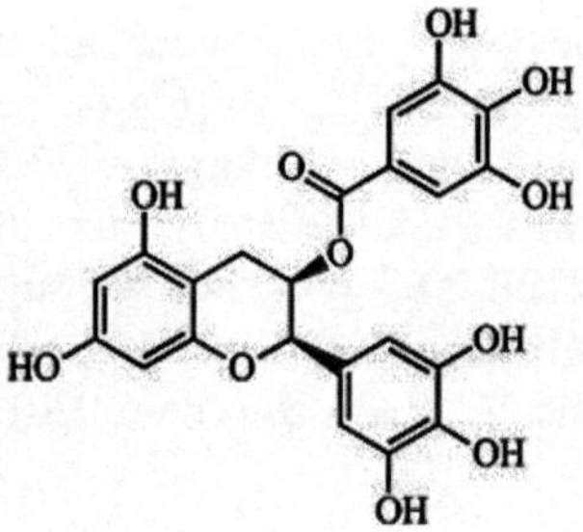

Fig. 3 The chemical structure of EGCG identified as a Zika entry inhibitor by Carneiro et al. (2016)

Fig. 4 Chemical structures of heme and its analogs with activity against Zika virus

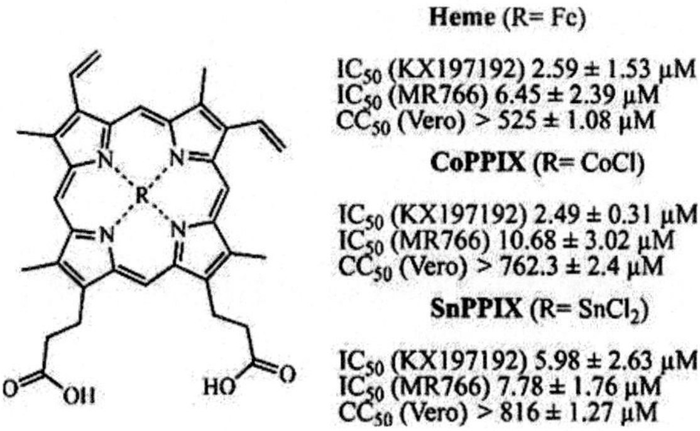

its IC_{50} and CC_{50} values against ZIKV (EC_{50} = 5.98 μM vs. 0.16 μM, CC_{50} = 816 μM vs. 20.7 μM).

Mounce et al. (2017) investigated the antiviral activities of *Curcuma longa* derived curcumin, its natural derivatives bisdemethoxycurcumin and demethoxycurcumin, as well as two synthetic analogs, namely EF-24 and tetramethylcurcumin (FLLL31, Fig. 5), utilized to study either curcumin effects as IκB kinase inhibitor or as JAK/STAT3 inhibitor (Mounce et al. 2017; Vassiliou et al. 2013; Yuan et al. 2014). Despite a large number of other postulated mechanisms of action (Dutta et al. 2009; Cells et al. 2012; Kim et al. 2010; Anggakusuma et al. 2014; Mazumder et al. 1995; Si et al. 2007; Chen et al. 2013; Mounce et al. 2017) provided evidence that curcumin and its derivatives are virus entry inhibitors, possibly by influencing membrane fluidity of either the enveloped virus itself or of their host cell membranes, preventing binding and fusion. These suggestions are supported by finding that enveloped viruses as ZIKV and West Nile virus (WNV) were sensitive to curcumin treatment, without effects against the non-enveloped Coxsackie B virus. Another possibility is the interference with receptor function by altering envelope dynamics on the virus itself. A finding, supporting this mechanism of action is that viral particle formation of enveloped viruses was not impeded if virus entry was bypassed. The high antiviral activities of the tested curcumin derivatives in the low

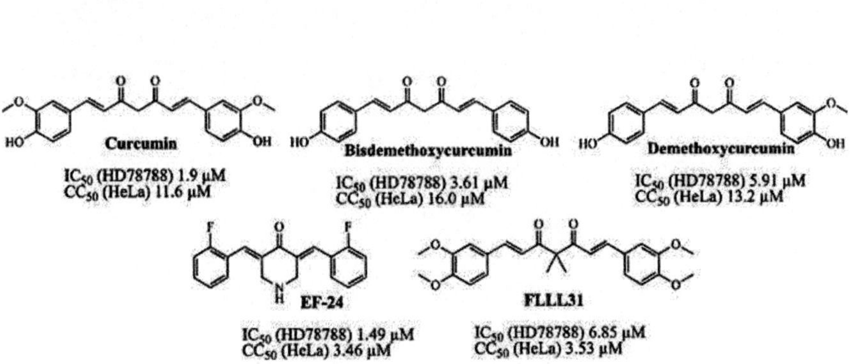

Fig. 5 Curcumin and its derivatives as ZIKV antiviral agents

micromolar range are contrasted with high toxicities resulting in poor selectivity indices ranging from 2 to 6.

In another approach to identify natural E protein binding compounds as lead structures for the development as ZIKV entry inhibitors, Priya et al. (2018) screened 25 natural phyto-compounds including flavones, alkaloids, and polyphenols of 25 herbs with known antiviral activities, identified by Ganjhu et al. (2015). To identify the corresponding binding site of the E protein, Priya et al. (2018) referred to a publication, reviewing the supporting role of unfolded protein responses (UPR) of host cells in viral replication (Blázquez et al. 2014). Assuming a hepatitis C virus (HCV) analog UPR-inducing effect of the ZIKV E-protein, Priya et al. (2018) then performed protein interaction studies of five UPR involved proteins with ZIKV E protein and concluded that the main interactions are formed with the hydrophobic residues in domain II. To find compounds that interfere with this interaction, they performed docking studies of herbal compounds to this fusion site of the E protein and identified tannic acid (Fig. 6) from *Terminalia arjuna* to achieve best scores (Priya et al. 2018; Ganjhu et al. 2015).

A potent ZIKV inhibitor that acts at a similar mechanism as chloroquine but with higher potency, 6-deoxyglucose-diphyllin (DGP or patentiflorin A, Fig. 7), was recently discovered and isolated from *Justicia gendarussa* to interfere with virus particle pre- or fusion steps (Martinez-Lopez et al. 2019; Persaud et al. 2018; Delvecchio et al. 2016; Shiryaev et al. 2017). They found that it prevents acidification of endosomal compartments as previously described for diphyllin via affecting the expression of vacuolar ATPase (Sørensen et al. 2007). The patentiflorin A was tested against different ZIKV strains as well as other members of the *Flaviviridae* family as DENV1, Japanese Encephalitis virus (JEV), Tick-borne Encephalitis virus (TBEV), WNV, and Ebola Virus (EBOV) in African green

Fig. 6 The tannic acid from *Terminalia arjuna* as a promising compound as anti-Zika virus

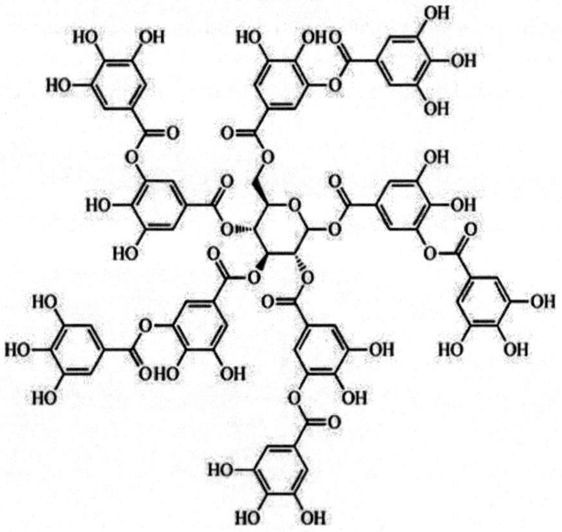

Fig. 7 Chemical structure of patentiflorin A, active against different flavivirus

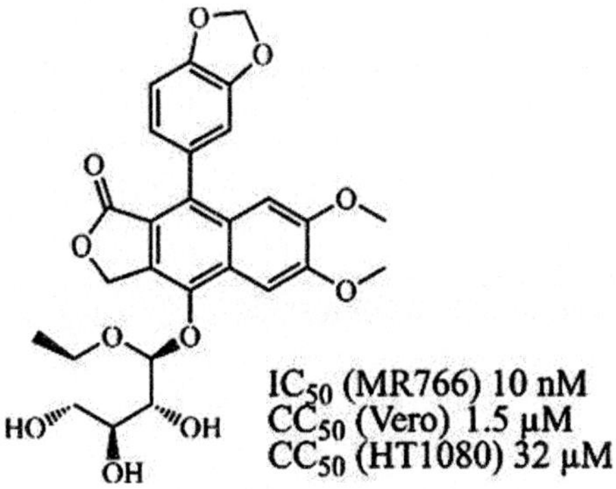

monkey kidney epithelial cells (*Vero*), human fibroblast (*HT1080*), and human microglial cells (*CHME3*), exhibiting IC_{50} values varying from 10 to 70 nM with good selectivity indices especially for human cells (CC_{50} = 1.5–32 μM).

Because of their abundance in food, most flavonoids are considered to possess low toxicities, thus representing promising substances for medical applications. Lee et al. (2019) carried out a high-throughput screening with 483 flavonoids using a viral plaque reduction assays and identified pinocembrin (Fig. 8), mostly isolated from *Pinus* heartwood and *Eucalyptus*, with an EC_{50} value of 17.4 μM to be the best flavonoid to protect human placental *JEG-3* cells from ZIKV infection (Lee et al. 2019). Pinocembrin showed low toxicity (CC_{50} = 251 μM) as expected since it is found in honey, tea, and red wine. Time-course studies allowed deducing that pinocembrin acts on post-entry steps of the ZIKV replication cycle, however, the exact mechanism of action remains unclear.

Based on a study, where it was reported that the ZIKV E protein binds to human placental chondroitin sulfate as a potential host cell surface receptor for ZIKV binding (Kim et al. 2017), Kim et al. (2019) characterized the composition of glycosaminoglycan (GAG) disaccharide of ZIKV infection-related tissues of human and *Aedes* mosquito hosts (Kim et al. 2019). Besides that, they investigated the binding effects of heparin (HP) and its derivatives to the E protein on ZIKV. As a control, they used DENV2 with known reported inhibition by HP (Chen et al. 1997). Intriguingly, the tested HP, low-molecular-weight heparin (LMWH), heparin dodecasaccharide (HPdp12), and non-anticoagulant heparin (s-NACH) inhibited DENV2 as expected, but promoted ZIKV replication in a significant and dose-dependent manner in *Vero* cells. Ghezzi et al. (2017) reported HP to moderately inhibit ZIKV replication in human neural progenitor cells; hence it seems

Fig. 8 Pinocembrin isolated from *Pinus* heartwood and *Eucalyptus* genus

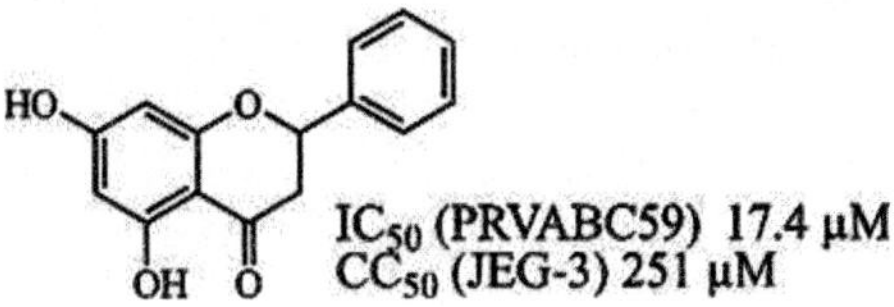

possible, that GAGs may exhibit tissue-dependent and alternative roles in ZIKV pathogenesis mediated via other mechanisms (Ghezzi et al. 2017).

Liu et al. (2012) reported that cholesterol-25-hydroxylase (CH25H) is an interferon-stimulated protein, induced in response to ZIKV infection (Liu et al. 2012). In the following experiments, they tested the viral titer reduction of its enzymatic product, 25-hydroxycholesterol (25-HC, Fig. 9) against ZIKV, DENV, YFV, and WNV in *Vero* cells. The 25-HC was most potent against ZIKV and reduced viral titers with an EC_{50} value of 188 nM. In baby hamster kidney cells (*BHK-21*), they also showed that 25-HC acts as a ZIKV entry inhibitor. In the following experiments, Li and Deng (2017) successfully provided additional data about the protective effects of 25-HC against microcephaly in a mouse model, in *Rhesus* monkeys and human cortical organoids, demonstrating its potential as an antiviral agent to treat flaviviral infections, especially ZIKV (Li and Deng 2017).

In *silico* studies concerning molecular docking and molecular dynamics simulations of natural compounds from ZINC library and small drug-like compounds from a helicase focused library (HFL) into the NTPase site from NS3 ZIKV (PDB ID: 5GJC) showed that the compounds ZINC03860720 (Shikimic acid) and CHEMBL1540619 (Fig. 10) have highest docking scores (−10.3 and −11.73 kcal/mol, respectively) and these promote stable interactions with amino acids at the NTPase site from the NS3(hel). Wherefore, these compounds modulate the side-chain fluctuations and interact with the P-loop residues (Lys^{200}, Thr^{201}, and Glu^{286}), suggesting further interference in the helicase function (Kumar et al. 2019).

In an in vitro study performed by using a dataset of 2,000 ligands, containing 1,000 FDA-approved drugs it was possible to identify the natural product from the microbial origin nanchangmycin (Fig. 11), capable of blocking the ZIKV entry in various cell types, including primary cells, with an IC_{50} value of 0.1 μM and CC_{50} of 7.0 μM for human osteosarcoma cells (*U2OS*), IC_{50} of 0.4 μM and $CC_{50} > 10$ μM for human brain microvascular endothelial cells (*HBMVEC*), and an IC_{50} value of 0.97 and CC_{50} of 6.10 μM for human *Jec-3* cells. Furthermore, it was verified that nanchangmycin blocks the virus entry via inhibition of the clathrin-mediated

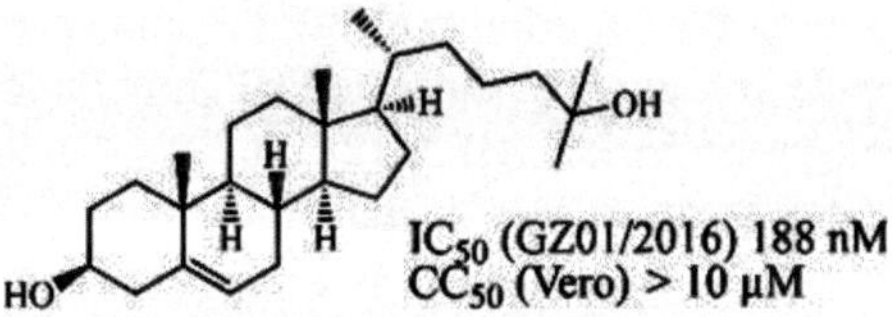

Fig. 9 A cholesterol-derived (25-HC) with significant activity against different flaviviruses

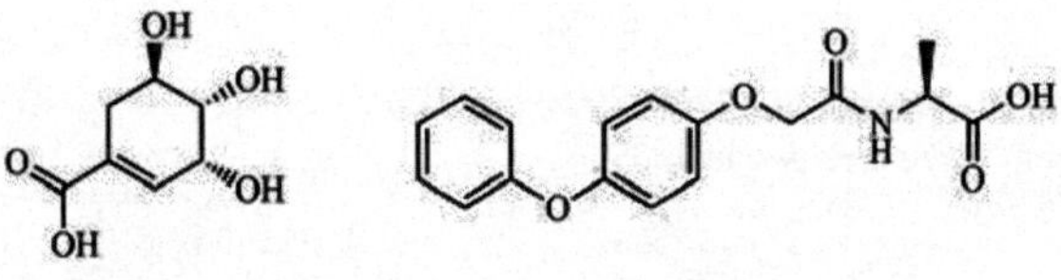

ZINC03860720 CHEMBL1540619

Fig. 10 Natural compounds identified in silico against NS3 (hel) Zika virus

Fig. 11 Chemical structure of nanchangmycin

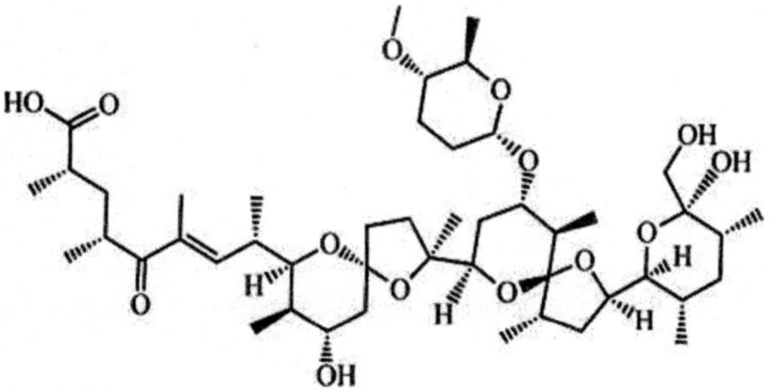

endocytosis, by blocking some specific aspect of the virus since the blocking of endocytosis of other substances by this route did not occur. These results were obtained by investigations upon other flaviviruses, alphaviruses, and influenza virus (Rausch et al. 2017).

Applying Quantitative Structure–Activity Relationship (QSAR) method, the potential activity of 62 molecules from natural origin against *Aedes aegypti* larvae was evaluated, aiming to establish a set of molecular characteristics needed for the activity, being the model useful for the development of new substances for vector control (Saavedra et al. 2018).

QSAR strategy was also used to describe the mode of binding of flavonoids to the ZIKV NS2B-NS3 protease. As requirements were identified the presence of sp^2 or sp^3 oxygen atoms connected to carbon, oxygen connected to the ring via a double bond, oxygen with branching and presence of oxygen in the absence of nitrogen, sulfur or phosphorus atoms, being the presence of sp^2 carbon atom at a ring with branching, and the presence of a glucose system deleterious for activity. Based on this, it was possible to identify amentoflavone, fisetin, isorhamnetin, and theaflavin-3-gallate (Fig. 12) as the most promising inhibitors, which were validated by docking studies (Bhargava et al. 2019).

In total, 38 different flavonoids were studied and analyzed. It was revealed that ECGC-7-*O*-α-glucopyranoside, isoquercetin, rutin, and ECGC-4′-*O*-α-glucopyranoside (Fig. 13) are able to efficiently interact with the catalytic triad (His^{51}, Asp^{75}, and Ser^{135}) from this protease, being good candidates for future studies (Yadav et al. 2020).

Knowing that flavanone naringenin (Fig. 14) has activity against several viruses, its potential for inhibiting ZIKV in vitro was evaluated. Cataneo et al. (2019) described that this flavonoid is capable of preventing infection in human *A549* cells and also in primary human monocyte-derived dendritic cells in a concentration-dependent manner, and regardless of the viral lineage, probably acting in the assembly phase of new viral particles. Moreover, it was found that its maximum non-toxic concentration against *A549* cells is 125 μM and that its IC_{50} value is 58.79 μM, with a CC_{50} value of 693.6 μM and selectivity index (SI) of 11.79. Finally, a molecular docking study suggested that naringenin probably acts interacting with NS2B-NS3 protease (Cataneo et al. 2019).

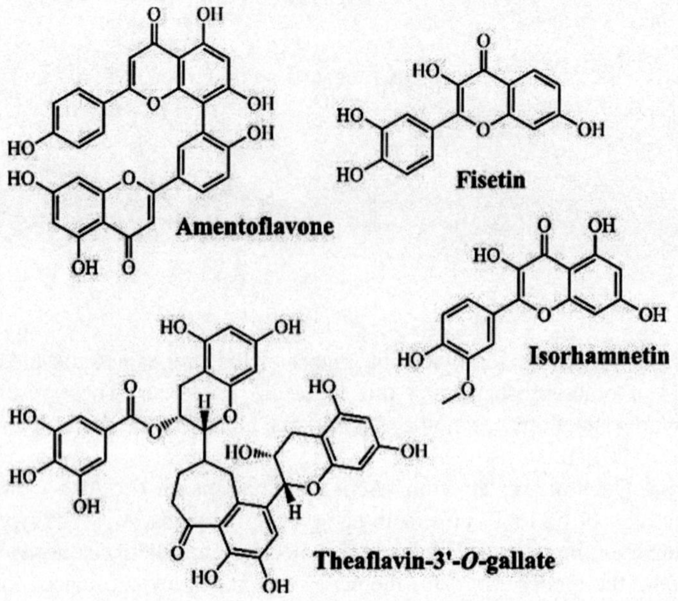

Fig. 12 Natural compounds identified by using a Quantitative Structure–Activity Relationship (QSAR)

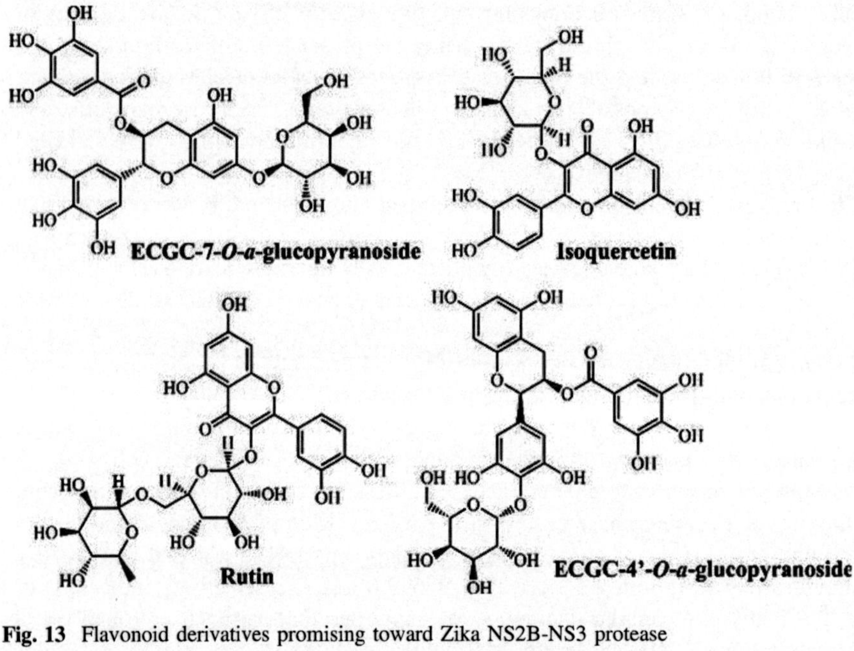

Fig. 13 Flavonoid derivatives promising toward Zika NS2B-NS3 protease

Fig. 14 Chemical structure of the flavanone naringenin

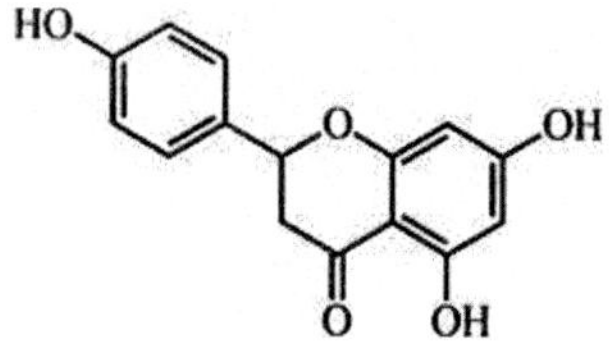

3.2 *Nature-Based Compounds*

The synthetic tetra-*O*-acetylated houttuynoid B (TK1023, Fig. 15) is structurally related to flavonoid glycosides that can be isolated from *Houttuynia cordata.* Basic et al. (2019) showed that this molecule interferes with the virus entry into the cell, it reduces the infection. Besides, they observed that TK1023 impairs the release of the internalized viral particles in the cytoplasm. In contrast, the synthesized parental houttuynoid B (TK1024, Fig. 15), known as a naturally occurring antiviral agent, has no activity against ZIKV. Therefore, it was concluded that the "prodrug" TK1023 could be translocated better across the cellular membrane than the more hydrophilic TK1024, being metabolized into an active substance by enzymatic deacetylation. The half-maximal effective concentration (EC_{50}) for TK1023 and TK1024 was determined by plaque assay. The detected EC_{50} values for French Polynesia (PF13/251,013-18) and Uganda 976 strains were at 1.675, and 1.552 μM for TK1023. The TK1024 did not inhibit PF13/251,013-18 and displayed an EC_{50} value of 25.8 μM for Uganda 976 (Basic et al. 2019).

Pitts et al. (2017) showed that the synthetic retinoid *N*-(4-hydroxyphenyl)-retinamide (4-HPR, Fig. 16) potently inhibits ZIKV and DENV2 *New Guinea* strain in mammalian cells and significantly reduced both serum viremia and brain burden in a murine model of ZIKV infection. The reduction was associated with a major decrease in the rate of virus RNA synthesis that did not result from the direct inhibition of the viral replicase activity but likely through an effect on the host. The

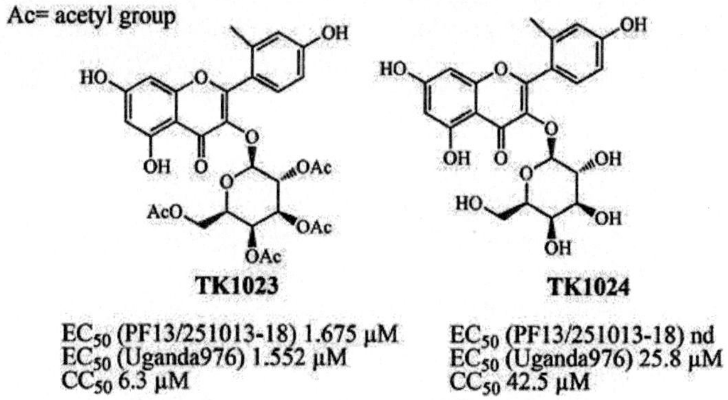

Fig. 15 Houttuynoid B analogs with meaningful activity against Zika virus

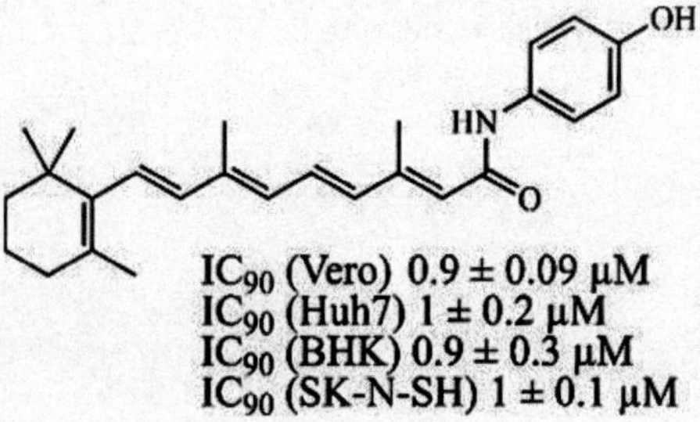

Fig. 16 Chemical structure of *N*-(4-hydroxyphenyl)-retinamide

4-HPR inhibits the ZIKV MR766 as well as PF13-251013-18. IC_{90} values were closed to 1 μM concentration in different mammalian cell lines including *Vero* cells, human hepatocarcinoma *Huh7*, *BHK-21*, and human neuroblastoma *SK-N-SH* cells (Pitts et al. 2017).

Chen et al. (2017a) investigated the antiviral effect of peptides derived from the JEV E protein. The protein has four domains: a stem-transmembrane domain which anchors the protein to the membrane, a central domain that connects to the extended domain, and the globular domain via short flexible loops (Sirohi et al. 2016).

Peptides derived from helix 2 of the stem e.g. P4 (AWDFGSIGGVFNSIGKA VHQVFGGAFRTL) and P5 (AWDFGSIGGVFNSIGKAVHQV) showed IC_{50} values in low nanomolar range against JEV. Based on the high degree of stem sequence conservation among flaviviruses, they tested their cross-inhibition potential against ZIKV and measured an IC_{50} of 3.27 μM for P5 against ZIKV infected *Vero* cells. In contrast, the homolog peptide from ZIKV ZP2 (AWDFGSVGGALNSLGKGIHQIF) did not prevent ZIKV infection. Additionally, ZP1 (AWDFGSVGGALNSLGKGIHQIFGAAFKSL), which is similar to P4, had an IC_{50} of 1.32 μM. The group concluded that the inhibition of ZIKV infection might be described in a two-step mechanism i.e. a relatively non-specific and hydrophobic membrane-binding step followed by specific interaction with E protein during low-pH-triggered conformational change and fusion. Moreover, P5 reduced ZIKV infection-associated histopathological damages in brain and testes as well as provided protection against JEV infection-induced lethality in type I and II interferon receptor-deficient (AG129) mice (Chen et al. 2017a).

Synthetic carbohydrate receptors (SCR) are small molecules alternative to lectin-based therapeutic agents whose development failed due to high molecular weight and their peptidic nature. The aim of the development of those compounds is the disruption of the viral docking process by carbohydrate-binding. Palanichamy et al. (2019) synthesized a series of 16 SCRs based on tetrayltetrakis(methylene) tetrakis (Fig. 17) which is known as a mannose selective SCR (Rieth et al. 2013). The most potent SCR was tetrakistriazole-derived (Fig. 17) with IC_{50} values of 0.16 and 0.24 μM in *Vero* and *HeLa* cells, respectively. Applying time-of-addition studies, it was shown that the SCRs inhibit the early stages of the virus infection.

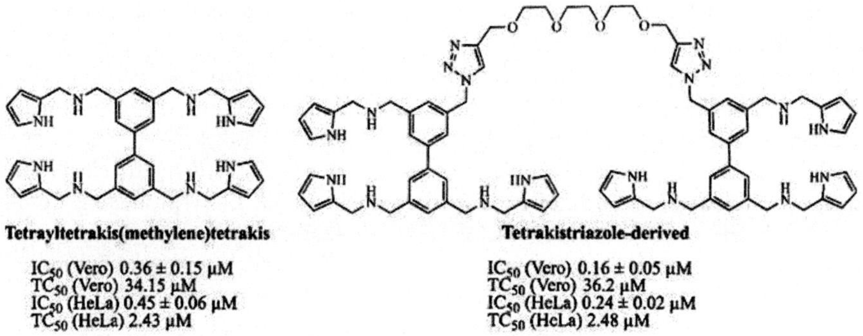

Fig. 17 Tetrakis derivatives with activity against Zika virus

Therefore, they assumed that SCRs inhibit the binding between the virus and cell surface glycans and thus prevent viral entry into the cell (Palanichamy et al. 2019).

Sinefungin (SIN), a competitive inhibitor of several methyltransferases (Yebra et al. 1991; Barbés et al. 1990), and its analog EPZ004777 are potent anti-fungal antibiotics isolated from *Streptomyces griseolus*. Tao et al. (2018) designed and synthesized a series of new SIN derivatives and also their respective deprotected intermediates, with IC_{50} values ranging from 4.56 to 20.16 μM. The guanidine adenosine analog (Fig. 18) showed better activity than EPZ004777, with an IC_{50} value of 35.19 μM, while SIN had a value above 50 μM concentration. The most promising SIN derivative, regarding activity and cytotoxicity, was adenosine derivative (Fig. 18, Cyclohexyl adenosine), substituted by X = chloro and W = cyclohexyl. Among the synthesized intermediates two—i.e., *N*-propyladenosine and *N*-hydropropyladenosine (Fig. 18)—displayed anti-ZIKV activity with the promising capability to be lead compounds for further development. The antiviral activity of these compounds was determined in an infection-based cell

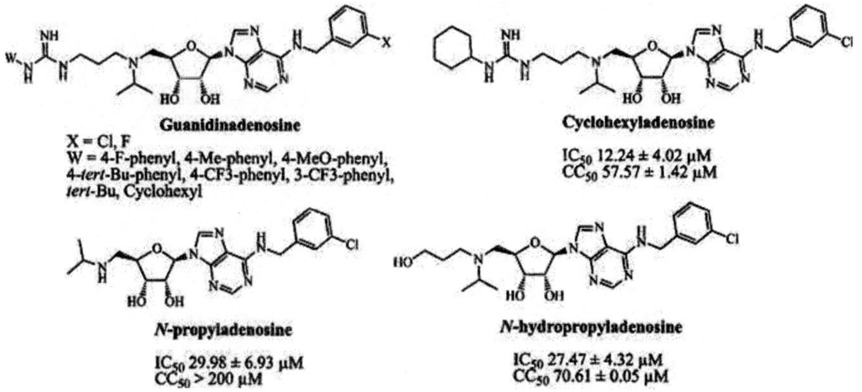

Fig. 18 Adenosine derivatives with activity against Zika virus

culture model and the authors did not provide a potential mechanism of action (Tao et al. 2018).

(−)-Aristeromycin is a naturally occurring carbocyclic nucleoside, which can be derived from *Streptomyces citricolor* (Kusaka et al. 1968). In order to obtain dual-target compounds that address both ZIKV RNA-dependent RNA polymerase (RdRp, NS5) and the host cell *S*-adenosyl-L-homocysteine (SAH) hydrolase, Yoon et al. (2019) designed 6′-fluorinated aristeromycins. The most promising compound, an adenosine-4,4-difluoro analog (Fig. 19), inhibits the SAH hydrolase with an IC_{50} value of 1.06 μM. In addition to ZIKV RdRp, which is inhibited with an EC_{50} value of 0.26 μM, replication of several other RdRp of positive-stranded RNA viruses, such as Severe Acute Respiratory Syndrome Coronavirus (SARS-CoV), Middle-East Respiratory Syndrome Coronavirus (MERS-CoV), and Chikungunya virus (CHIKV), is inhibited in low micromolar range by adenosine-4,4-difluoro derivative. The corresponding phosphoramidate-prodrug (Fig. 19) did not affect the SAH hydrolase but inhibited the replication of ZIKV with an IC_{50} value of 1.75 μM (Yoon et al. 2019).

Suroengrit et al. (2017) developed halogenated chrysins, 6,8-dibromochrysin, and 6,8-diiodochrysin (Fig. 20), two flavonoid derivatives that are potent against ZIKV and all DENV serotypes (DENV1–4). Both inhibitors showed EC_{50} values in the low micromolar range (1.65 ± 0.86 and 1.39 ± 0.11 μM). Both affected post-attachment and -replication steps probably by acting of host factors. Nevertheless, specific drug targets are still unknown (Suroengrit et al. 2017).

Galidesivir (BCX4430, Fig. 21) is a broad-spectrum antiviral that is potently active against YFV, EBOV, and Marburg Virus (MARV) in in vivo experiments (Warren et al. 2014). Julander et al. (2017) tested BCX4430 in an AG129 mouse model after exposure with an Asian ZIKV strain (P6-740) and characterized the course of ZIKV-induced disease. BCX4430 significantly improved the outcome of the infected mice and was found to be the first treatment that protects from ZIKV-associated mortality in a mouse model (Julander et al. 2017).

In a molecular dynamics study, Chuang et al. (2018) identified different substructures of the Zika NS5 methyltransferase domain that might be responsible for

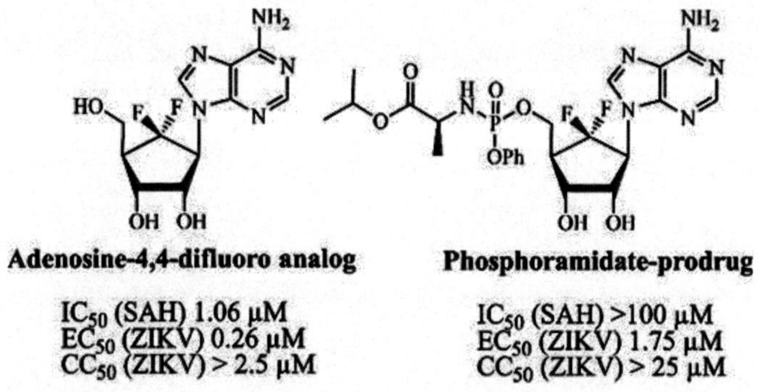

Fig. 19 Adenosine derivatives studied by Yoon et al. (2019)

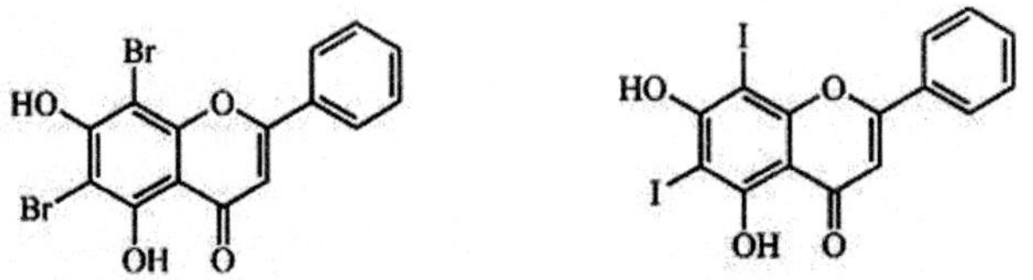

6,8-Dibromochrysin

EC_{50} (ZIKV) 1.65 ± 0.86 μM
CC_{50} (LLC/MK2) 44.28 ± 2.9 μM

6,8-Diiodochrysin

EC_{50} (ZIKV) 1.39 ± 0.11 μM
CC_{50} (LLC/MK2) 42.51 ± 2.53 μM

Fig. 20 Halogenated chrysin analogs with activity against Dengue and Zika viruses

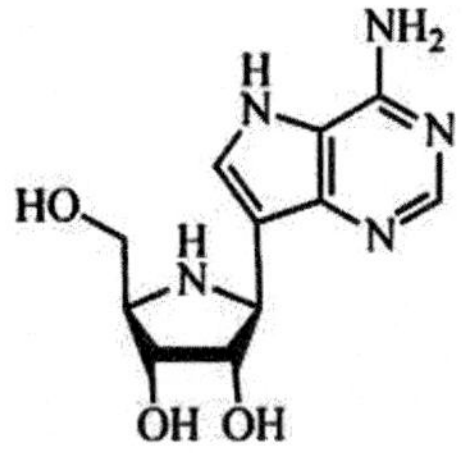

Fig. 21 Chemical structure of galidesivir (BCX4430), a nucleoside analog with anti-Zika virus activity

the selectivity of the drugs *S*-adenosyl-L-homocysteine (SAH) and 7-methylguanosine-5′-triphosphate (m7GTP). IC_{50} values were previously determined to be 0.43 μM (SAH) and 184 μM (m7GTP) by Coutard et al. (2017). For the ZIKV NS5-SAH complex, the research group determined four NS5 substructures (residue orders: 101–112, 54–86, 127–136, and 146–161) and Ser^{56}, Gly^{81}, Arg^{84}, Trp^{87}, Thr^{104}, Gly^{106}, Gly^{107}, His^{110}, Asp^{146}, Ile^{147}, and Gly^{148} residues affecting the ability of SAH to bind with full-length ZIKV NS5. For ZIKV NS5-m7GTP complex, three ZIKV NS5 substructures (residue orders: 11–31, 146–161, and 207–218) and Asn^{17}, Phe^{24}, Lys^{28}, Lys^{29}, Ser^{150}, Arg^{213}, and Ser^{215} residues influenced ZIKV NS5, and m7GTP binding through the entrapment of this RNA analog (Chuang et al. 2018).

3.3 Semi-Synthetic Compounds

Li et al. (2020) investigated derivatives of andrographolide, a compound found in the herbaceous plant *Andrographis panicula* Nees, for their antiviral activity against the ZIKV PRVABC59 strain, first isolated in Puerto Rico, in *Vero* and human glioblastoma cells (*SNB-19*). The group tested more than 20 derivatives of andrographolide and concluded that the combination of andrographolide scaffold and quinoline moiety is a potential anti-Zika strategy. With 19-acetylated 14*a*-(5′,7′-dichloro-8′-quinolyloxy) and 14*b*-(8′-quinolyloxy)-3,19-diol derivatives (Fig. 22)

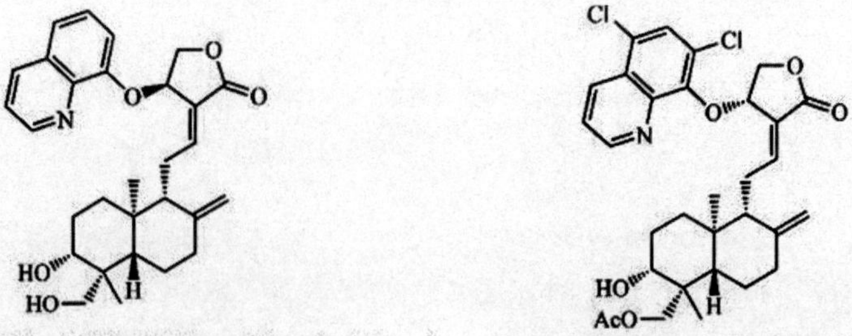

19-Acetylated 14*a*-(5',7'-dichloro-8'-quinolyloxy)

EC_{50} (PRVABC 59) 1.3 ± 0.1 μM
CC_{50} (SNB-19) 22.7 ± 1.1 μM
CC_{50} (Vero) 20.8 ± 0.5 μM

14*b*-(8'-Quinolyloxy)-3,19-diol

EC_{50} (PRVABC 59) 4.5 ± 0.2 μM
CC_{50} (SNB-19) 88.7 ± 1.1 μM
CC_{50} (Vero) 85.0 ± 1.6 μM

Fig. 22 Quinolyloxyl analogs with antiviral activity in the low micromolar range

they found leads with EC_{50} values in the low micromolar range and selectivity indices (SI) around 20 (Li et al. 2020).

3.4 Synthetic Compounds

Hrd1 complex were identified in several screens as host factors required for flavivirus replication (Krishnan et al. 2008; Dang et al. 2015; Scaturro et al. 2018; Mairiang et al. 2013). Ruan et al. (2019) tested 1,696 compounds with a null Hong Kong mutant form of α-1-antitrypsin (NHK)-dislocation assay in which the dislocation of NHK is monitored by the reconstitution of GFP in the cytosol (Zhong and Fang 2012), which possibly target Hrd1. Among these, CP26 (Fig. 23) inhibited the dislocation of NHK with an IC_{50} value of 4.45 μM. In *Huh7* cells, CP26 exhibited a broad-spectrum activity against all DENV serotypes and three ZIKV strains (MR766, PRVABC59, and P2-740). ZIKV replication and virus production were reduced by about 80% at 20 nM of CP26 via an unclear mechanism (Ruan et al. 2019).

To develop drugs that potentially exhibit prophylactic activity against DENV and ZIKV infections, Chen et al. (2020) targeted the host calcium/

Fig. 23 Chemical structure of compound CP26

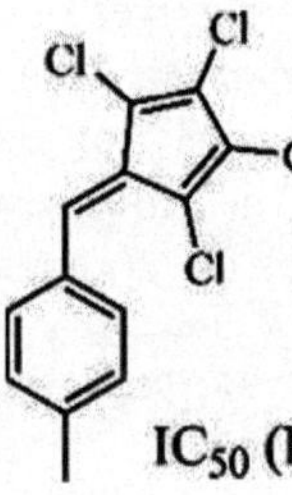

IC_{50} (NHK dislocation) 4.45 μM

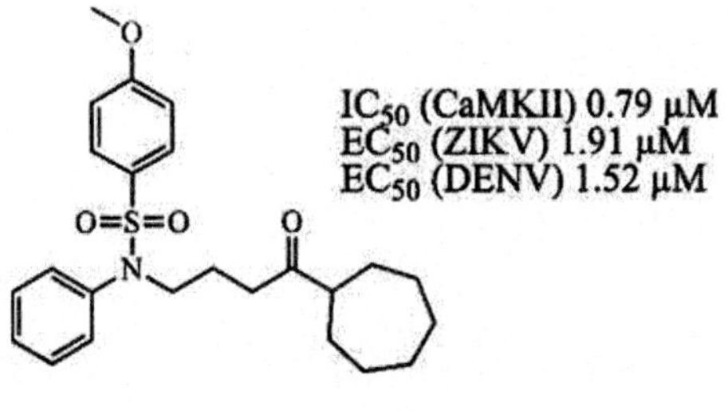

Fig. 24 Chemical structure of *N*-(4-cycloheptyl-4-oxobutyl)-4-methoxy-*N*-phenylbenzenesulphonamide

calmodulin-dependent kinase II (CaMKII). In a previous study, the group observed that JEV activated CaMKII via dopamine D2 receptor-phospholipase C signaling to increase the surface expression of JEV binding/entry molecule in neuronal cells (Simanjuntak et al. 2017). Since members of *Flaviviridae* share life cycles and host–pathogen interactions (Leyssen et al. 2000), they proposed that CaMKII can serve as a target for prophylactic drugs against flaviviral infections. Besides, they designed several benzenesulfonamides of which *N*-(4-cycloheptyl-4-oxobutyl)-4-methoxy-*N*-phenylbenzenesulphonamide (Fig. 24) was the most promising compound. CaMKII was inhibited by an IC_{50} value of 0.79 μM. ZIKV E protein expression and viral progeny production were inhibited with an EC_{50} value of 1.91 μM. Furthermore, it significantly reduced the viremia level and increased animal survival time in mouse-challenge models (Chen et al. 2020).

Quek et al. (2020) used fragment-based drug design to identify fragment scaffolds, which can serve as a starting point for lead compound development. Using a primary thermal shift assay, they screened a library of fragment compounds and found twenty-two fragments that bind to the NS2B-NS3 protease from ZIKV. The X-ray structure of two hits, 4-(hydroxymethyl)benzoic acid and 2-thioxothiazolidin-4-one (Fig. 25) were determined and showed that both fragments bind to the S1 pocket (Quek et al. 2020).

To find new antiviral inhibitors Yang et al. (2017) screened a focused library of about 200 biaryl-substituted quinolones (Yang et al. 2017). After EC_{50} value determination and summary of Structure–Activity Relationship (SAR), new molecules for hit optimization were synthesized. The most potent molecule was RYL-634 (Fig. 26) with an EC_{50} value of 6 nM (determined by RT-PCR and luciferase assays). The antiviral profile of RYL-634 was revealed, as well as its potency against several RNA viruses, such as DENV, enterovirus 71 (EV71), HIV, CHIKV, respiratory syncytial virus (RSV), severe thrombocytopenia syndrome

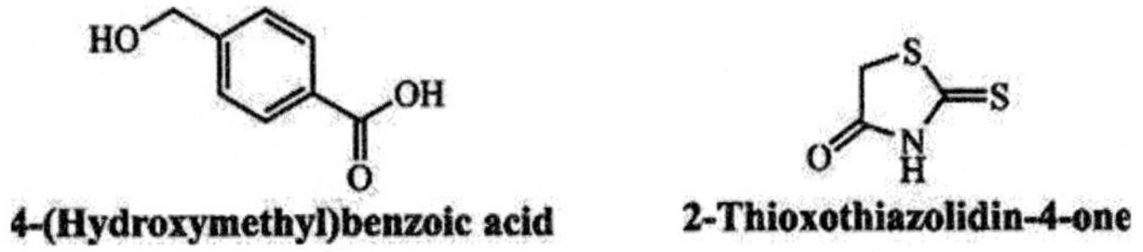

Fig. 25 Chemical structures of inhibitors of NS2B-NS3 from Zika virus

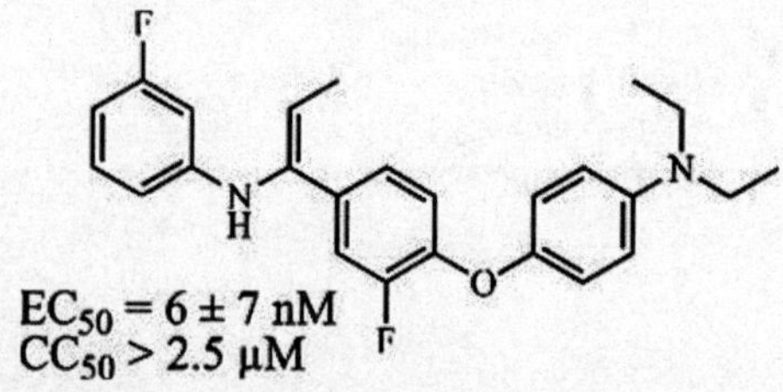

Fig. 26 Chemical structure of RYL-634

virus (SFTSV), MERS-CoV, influenza virus, and HCV. Using activity-based protein profiling, the research group identified different human proteins as possible targets. By using bioinformatics analysis and SAR-guided selection in reverse docking, the authors validated dihydroorotate dehydrogenase (DHODH) as the major target for RYL-634 (Yang et al. 2019).

Lian et al. (2018) investigated the E protein from flaviviruses as a drug target for preventing infections. They developed a proximity-based luminescent assay screen and HTS to identify small-molecule inhibitors of DENV2 E protein. In total, 21,271 compounds were screened and IC_{50} values of 50 hits were determined. Among these, 35 efficiently competed with GNF2-biotin, a 4,6-disubstituted pyrimidine previously identified (Wispelaere et al. 2018; Clark et al. 2016), for binding to DENV soluble dimeric form of the E protein (sE_2) with IC_{50} values lower than 10 μM. Thereafter, 12 compounds inhibited DENV2 infectivity by 90% or higher, at 10 μM concentration. Lastly, eight small molecules were identified to inhibit DENV infection by binding to the E protein and preventing membrane fusion. Among them, five compounds A–E (Fig. 27) showed antiviral activity against ZIKV in a plaque reduction assay lower than 10 μM concentration (Lian et al. 2018).

Other approaches employed in *silico* methods to identify new anti-ZIKV active substances. In order to target the ZIKV RdRp, Pattnaik et al. (2018) screened

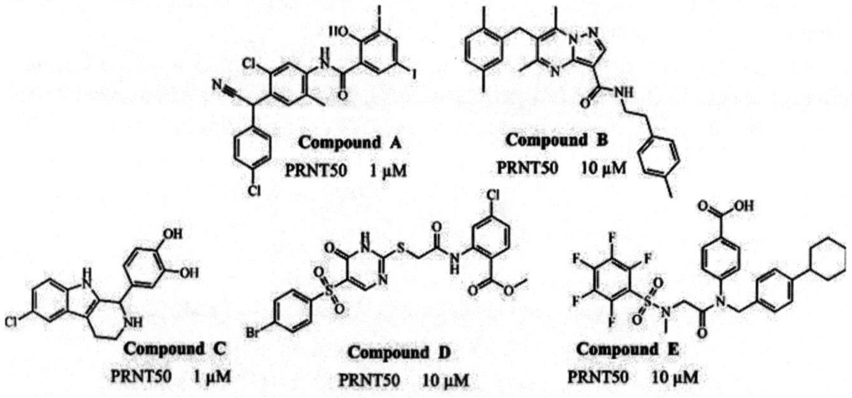

Fig. 27 Zika virus inhibitors identified by Lian et al. (2018)

100,000 small molecules in *silico* and tested the top ten for their ability to inhibit the viral replication. At 1 μM concentration of the non-nucleoside inhibitor 3-chloro-*N*-[({4-[4-(2-thienylcarbonyl)-1-piperazinyl]phenyl}amino)carbonothioyl]-1-benzothiophene-2-carboxamide (TPB) (Fig. 28) inhibited the replication of ZIKV PRVABC59 in *Vero* cells by 99% compared to the control. The EC_{50} value for TPB is about 94 nM and was determined in a plaque reduction assay. In in vivo studies, TPB significantly reduced ZIKV viremia (Pattnaik et al. 2018).

Singh and Jana (2017) modeled the catalytic domain of the ZIKV RdRp with special attention to conserved aspartate residues in the palm domain. Commercially available inhibitors available in the ZINC database (Irwin and Shoichet 2005) were virtually screened against ZIKV modeled RdRp. After refinement based on absorption, distribution, metabolism, excretion, and toxicity (ADMET) parameters, one promising compound, ZINC50166190 (Fig. 29), was selected. This compound showed a remarkable docking profile against the conserved aspartate residues. However, validation of this predicted inhibitor using a cell-based assay is still to be performed (Singh and Jana 2017).

Using structure-based virtual screening and dynamics simulation Santos et al. (2020a) identified a promising inhibitor targeting the NS5 methyltransferase. The group virtually screened 42,390 structures from the Development Therapeutics Program AIDS Antiviral Screen Database, while ZINC1652386 (Fig. 30) stood out due to its high affinity in comparison to the co-crystallized ligand (Santos et al. 2020a).

Onawole et al. (2017) used consensus scoring to identify new inhibitors for the mature particle of ZIKV. They screened 36 million compounds from the MCULE database (Kiss et al. 2012) and the results of the 25 top-ranked molecules from MCULE and Drug Score Xtended (DSX) were combined. Two hit compounds,

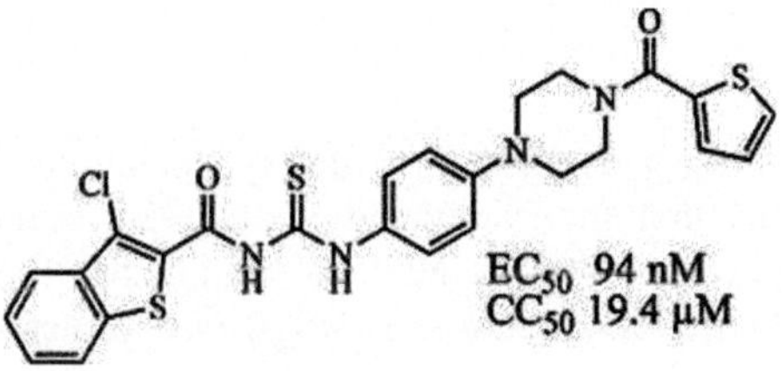

Fig. 28. 3-Chloro-*N*-[({4-[4-(2-thienylcarbonyl)-1-piperazinyl]phenyl}amino) carbonothioyl]-1-benzothiophene-2-carboxamide (TPB)

Fig. 29 Chemical structure of ZINC50166190

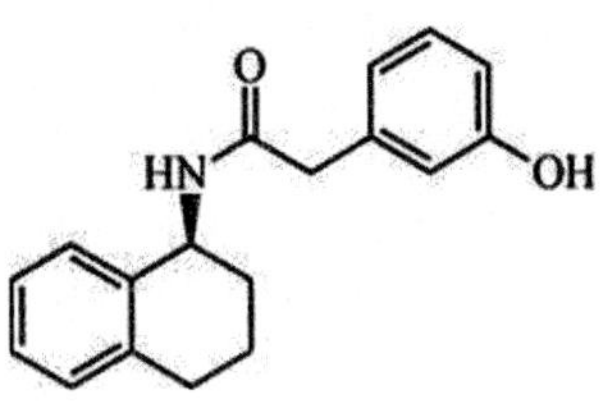

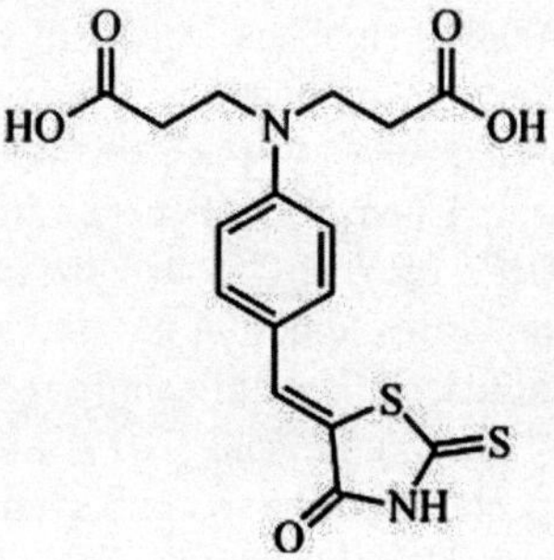

Fig. 30 Chemical structure of ZINC1652386

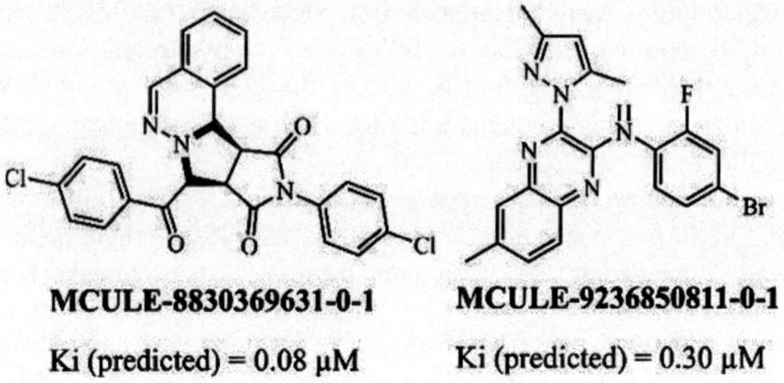

Fig. 31 Chemical structures of inhibitors identified by Onawole et al. (2017)

MCULE-8830369631-0-1 and MCULE-9236850811-0-1 (Fig. 31), were selected and their K_i values were estimated by the docking score values. Both hit compounds have similar predicted binding modes indicated by interactions with the same amino acids. Finally, physicochemical properties, including ADMET, were predicted using MCULE's property calculator and Molinspiration tool (Onawole et al. 2017).

An in *silico* screening performed on a library containing 6,265 protease inhibitors (PI's) revealed that the compound 9B, a C2-symmetric HIV protease inhibitor, emerges as a potential ZIKV NS2B-NS3 inhibitor. Hence, compound 9B showed a meaningful affinity value towards NS2B-NS3 protease (PDB ID: 5LC0) (Fig. 32). By performing molecular dynamics, it was verified that this compound forms a stable complex with this target, in the course of 40 ns simulation time. Analyzes revealed that compound 9B mainly interacts with the ZIKV NS3 domain, more specifically with His^{51} and Ser^{135} residues at the catalytic triad. Lastly, in vitro experiments unveiled that compound 9B inhibits NS2B-NS3 protease with an IC_{50} value of 143.25 ± 5.45 µM, corroborating the in *silico* screening (Akaberi et al. 2019).

The original Fig. 32 has later been replaced by this figure by the author. Kindly help me to bring to normal size for publication.

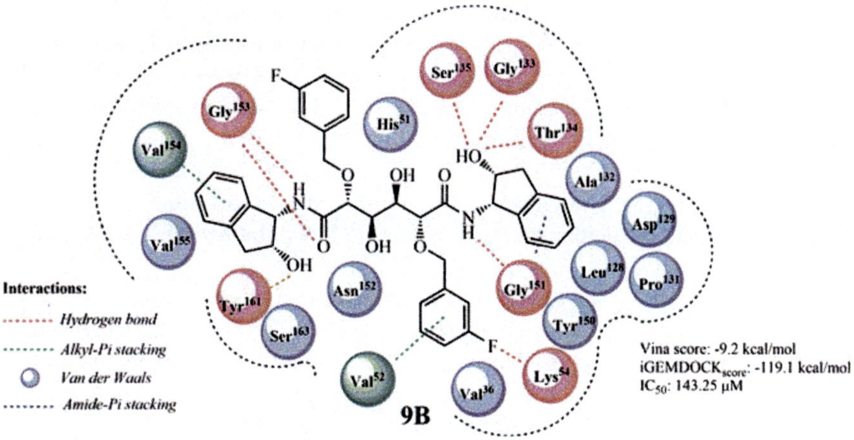

Fig. 32 Structure of compound 9B and its interactions with NS3 protease from the Zika virus

Through molecular docking, molecular dynamics simulations, using implicit solvation models, some molecules were identified to have significant affinities forward dimerization site from the Axl receptor with the Gas6 protein, wherefore interfering with the virus entry into neural cells. Thereupon, R428 and its analogs, compounds 1′ and 2′ (Fig. 33), exhibited a greater affinity for the Axl receptor, with binding energy values of −10.5, −11.9, and −13.06 kcal/mol, respectively. Thusly,

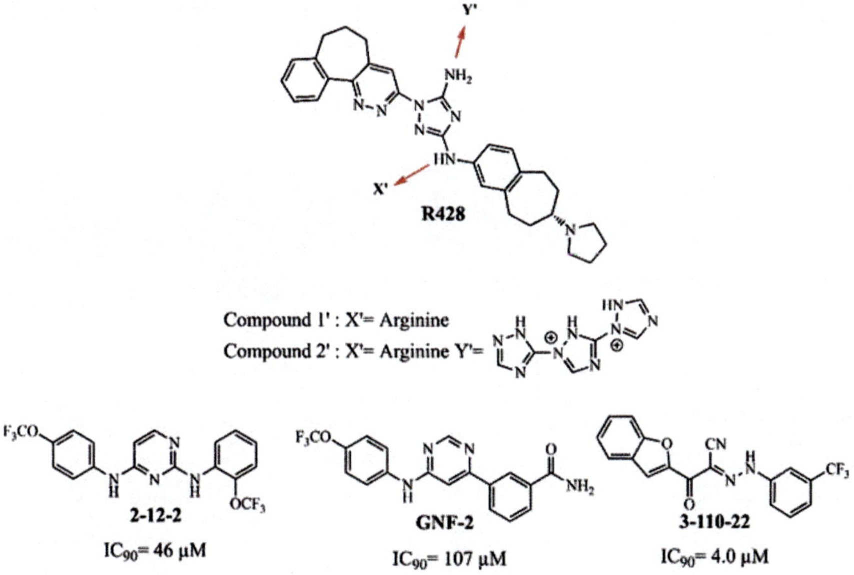

Fig. 33 Dengue virus entry inhibitors identified by in silico and in vitro experiments

compound 2′ showed a greater affinity for the binding pocket, and it was recommended as a lead compound for synthesis and pre-clinical trials (Sarukhanyan et al. 2018). Alternative studies reported that compounds 2-12-2, GNF-2, and 3-110-22, those are 2,4-, 4,6-disubstituted pyrimidines, and a cyanohydrazone analog, respectively (Fig. 33), inhibit DENV fusion by interactions with βOG pocket from prefusion form of the E protein from DENV (Schmidt et al. 2012; Clark et al. 2016; Wispelaere et al. 2018). Notwithstanding with the conserved residues at the βOG pocket, these analogs demonstrate in vitro ZIKV inhibition with IC_{90} values ranging from 4 to 107 μM (Wispelaere et al. 2018).

An interesting molecule for ZIKV control is the molecular tweezer CLR01 (Fig. 34), which had already been found to inactivate human immunodeficiency virus 1 (HIV-1) and herpes virus, acting via viral envelope membrane disruption. This molecule was shown to be able to reduce the ZIKV cytopathic effect (CPE) in *Vero E6* cells, with an IC_{50} value of 8.2 μM, and also prevented ZIKV infection on cells. Finally, it has been shown that CLR01 loses its effect in the presence of serum but not urine, saliva, semen, or cerebrospinal fluid (Röcker et al. 2018).

Further, acting on the viral envelope, but at the stage of binding to host cells, it could be highlighted the cyanohydrazone, as 3-110-22 (see Fig. 33), capable of reducing viral infection with an IC_{50} value of 4.2 μM. However, it has demonstrated high in vivo toxicity. Somewhat to reduce this toxicity, structural optimizations were performing, yielding the compound JBJ-01-162-04 (Fig. 35), which was tested only in a DENV infection model, where exhibited an IC_{90} value of 1.5 μM. Furthermore, it demonstrated in vivo antiviral activity when intraperitoneally administered in mice at a dose of 40 mg/kg. Such results are encouraging for the therapy of ZIKV infection since the region in the viral envelope protein, in which this class of molecules binds, is well conserved among the different flaviviruses (Li et al. 2019).

Fig. 34 Chemical structure of the molecular tweezer CLR01

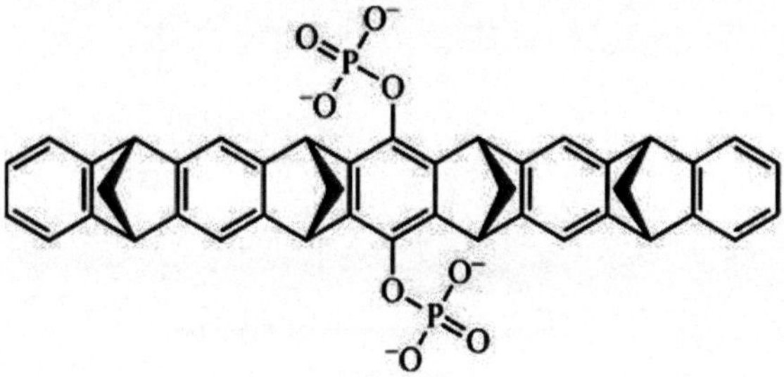

Fig. 35 Chemical structure of cyanohydrazone JBJ-01-162-4 with promising activity against Zika virus

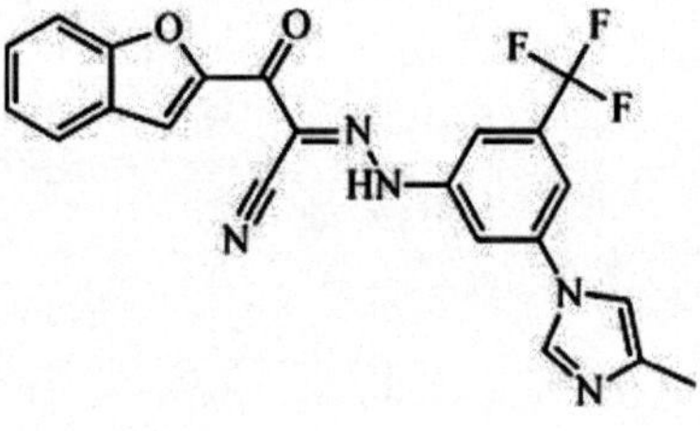

An artificial nucleotide, NITD008 (Fig. 36), was used as an application model for a ZIKV RNA-dependent RNA polymerase (ZIKV RdRp) model, corresponding to the *C*-terminal portion of NS5, essential for viral replication. This is a very detailed model with the potential to guide future work on rational drug design against ZIKV (Šebera et al. 2018). Through structure-based virtual screening (SBVS), it was possible to identify inhibitors of viral replication targeting NS5 and RdRp proteins, such as compounds ZINC64717952 and ZINC39563464 (Fig. 36), whose these could be used as promising scaffolds for the future development of new drugs (Ramharack and Soliman 2018).

Based on the idea that ZIKV uses extracellular vesicles (EVs) as a mode of escaping the immune system and to assist its propagation, and which such vesicles are produced via ceramide formation from sphingomyelin by sphingomyelinase, it has been proposed that the inhibition of this route could be a mechanism for controlling the viral spread. Notwithstanding this information, a neutral sphingomyelinase-2 (nSMase2) inhibitor (GW4869, Fig. 37) was found to be able to inhibit viral propagation in astrocytes in a dose-dependent manner, already at 2 μM concentration (Huang et al. 2018).

Deeming the host's metabolic pathways used by the virus, a celecoxib derivative kinase inhibitor AR-12 (Fig. 38), designed to act downregulating the PI3K/Akt pathway (phosphatidylinositol 3 kinase/protein kinase B) was able to inhibit viral replication in *Huh-7* cells with IC_{50} and CC_{50} values ranging from 0.82 to 0.88 and 7.01 μM, respectively (depending on the viral lineage), and SI of 7.97–8.55. AR-12

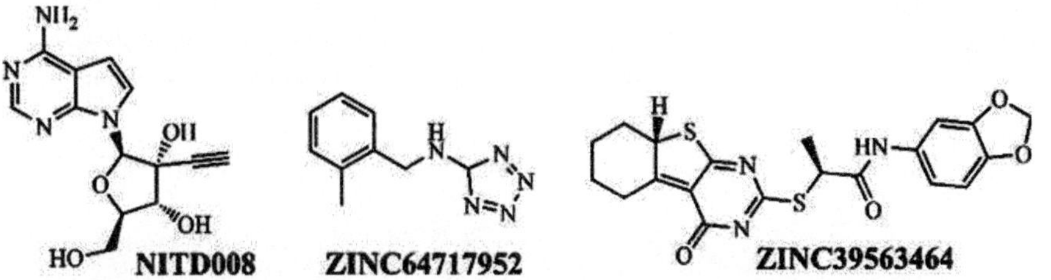

Fig. 36 Compounds identified by structure-based virtual screening

Fig. 37 Chemical structure of neutral sphingomyelinase-2 inhibitor, GW4869

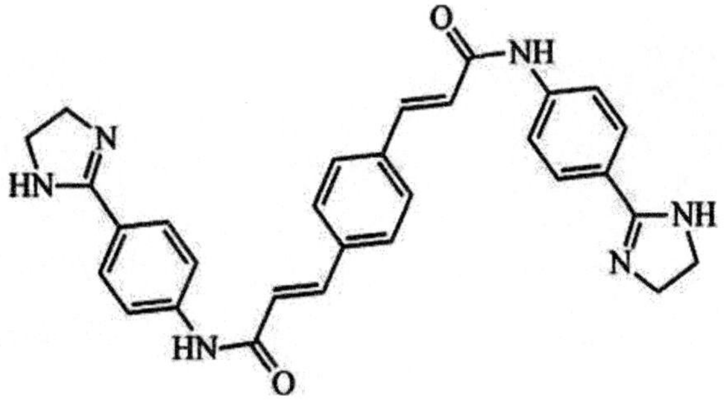

Fig. 38 Chemical structure of AR-12

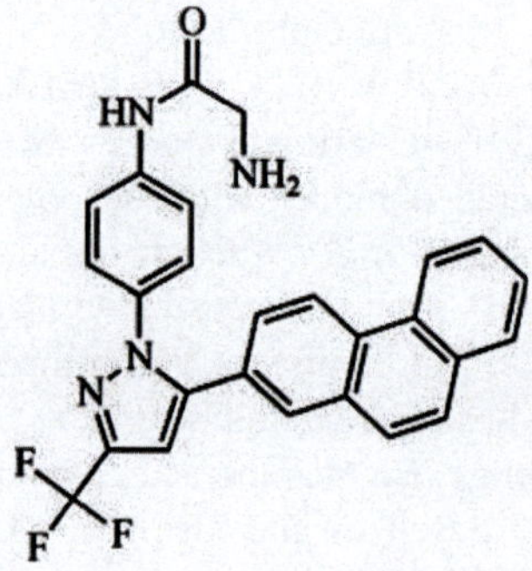

was also tested in neuronal cell lines (*U251* and *SF268*), showing IC_{50} and CC_{50} values of 0.84 and 1.18, and 6.99 and 11 μM, and SI values of 8.32 and 9.32 for *U251* and *SF268* cells, respectively. Furthermore, it has seen that AR-12 inhibited the protein expression of the envelope and NS1 proteins. Finally, AR-12 was found to be active in vivo, leading to a survival rate from 0 to 83.3% when comparing infection in untreated or AR-12-treated mice at a dose of 25 mg/kg intraperitoneally. Studying the route of administration, the researchers demonstrated that the effect was only achieved with a 200 mg/kg AR-12, when the substance was administered orally (Chan et al. 2018).

Yang et al. (2018a) aimed to enhance the antiviral potency of AR-12 (see Fig. 38), a celecoxib analog which has been described as a favorable antiviral agent against some relevant viruses, such as DENV, Lassa virus (LASV), EBOV, among others (Chen et al. 2017b; Mohr et al. 2015). Deeming compound AR-12, glycine-containing pyrazole and pyrrole derivatives were evaluated towards ZIKV. From this, compounds P12-23 and P12-34 (Fig. 39a) displayed the best antiviral

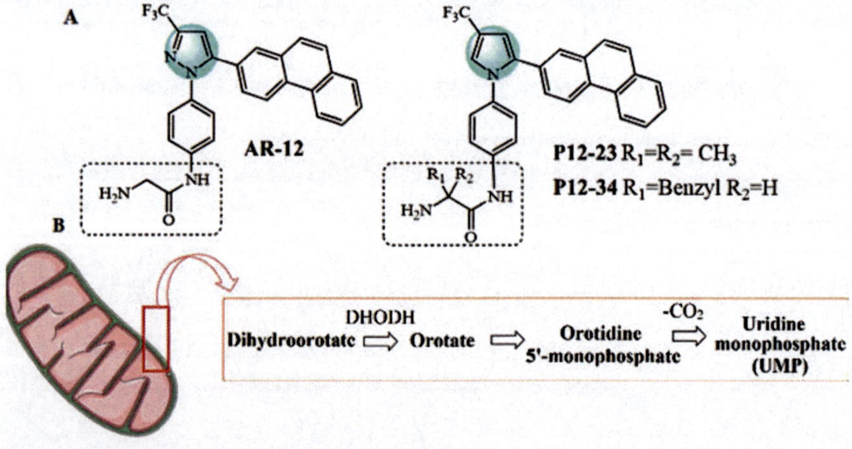

Fig. 39 Structure of AR-12 and its derivatives (**a**) and the mechanism of viral inhibition by dihydroorotate dehydrogenase blocking (**b**)

effects and suppressed the ZIKV replication in *A549* cells at nanomolar concentrations (IC_{50} = 130.3 and 118.6 nM, respectively), being about tenfold more potent than AR-12 (IC_{50} = 1.373 nM). Further, antiviral effects against DENV-2 and JEV were experimentally observed. Surprisingly, P12-34 was effective against the four DENV serotypes (DENV-1-4), with IC_{50} values ranging from 62.2 to 98 nM concentrations. Compounds P12-23 and P12-34 act by interrupting DENV RNA replication, without interfering on the viral binding, entry, or protein translation (Yang et al. 2018a). Therefore, their mechanism of inhibition of the viral replication is associated with the interference into the cytochrome bc1 complex and, consequently the inhibition of DHODH. Finally, it is a mitochondrial enzyme responsible for catalyzing the oxidation of the dihydroorotate to orotate and inhibitors targeting it contribute to inhibiting De Novo pyrimidine biosynthesis (Fig. 39b) (Reis et al. 2017; Yang et al. 2018a).

Another metabolic pathway used by ZIKV in its replication involves proteins of the Hrd1 complex, which is involved in the degradation of misfolded proteins, mediating the transport of proteins from the endoplasmic reticulum to the cytosol, a process that can be inhibited by the synthetic triterpenoid CDDO-Me (Fig. 40). Thereafter, it was described that CDDO-Me could also bind to the grp94 chaperone, inhibiting viral replication with an EC_{50} value of 15 nM and protecting cells from viral cytopathic effect, at the same concentration range (Rothan et al. 2019).

It has been proposed that heat-shock protein 70 (HSP70) is involved in the replication process of the various flaviviruses, and can function as a target for antiviral therapy. In this perspective, a series of different HSP70 inhibitors was evaluated, and it was described that apoptozole (Fig. 41) had antiviral potential against ZIKV in vitro and in vivo. This molecule was able to inhibit the viral cytopathic effect at 3 μM concentration. Additionally, it inhibits viral replication in

Fig. 40 Chemical structure of triterpenoid CDDO-Me

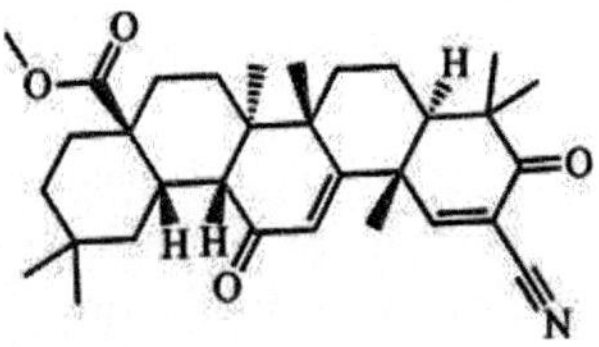

Fig. 41 Chemical structure of apoptozole

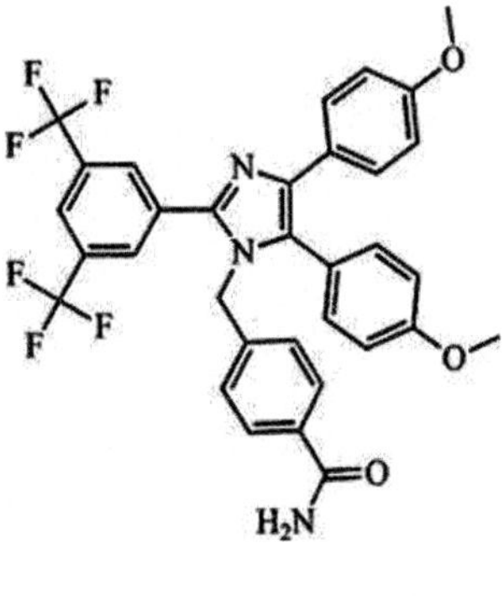

a dose-dependent manner in *Vero*, *BHK*, *A549*, *Huh7*, and *Huh7.5* cells, ranging from 1 to 10 μM, without toxicity. Also, it inhibits the production of viral proteins, acting mainly after the virus entry. In vivo, the substance was intraperitoneally administered at a dose of 10 mg/kg in mice, being able to reduce viremia via an increase in the animal's innate immune response (Yang et al. 2020).

Another strategy to attack the virus is through the technology of short interfering RNA (siRNA), which blocks the expression of genes essential for viral replication. An *in silico* analysis was proposed to design candidates for this type of therapy. Through the alignment of 347 available ZIKV genomes, the researchers selected 13 conserved regions from among the various genomes for designing 20 siRNA molecules. Virtual screening was then carried out to identify and to discard which of these molecules could bind to immune epitopes (susceptible to mutation), which could bind to targets different from that planned and those which could interfere with normal human cell processes, selecting S09 sequences (CAAGUAUAUGACUUUUUGG) and S12/13 (UAUUCCGGCAGAUGCAACCUG) as a promising molecule (Giulietti et al. 2018).

3.5 *Drug Repurposing for Zika Virus*

Since flaviviruses require a large number of human host factors that could already be effectively targeted with approved drugs, it is a promising strategy to test them in high-throughput screenings (HTS) against several flaviviruses. Antivirals identified by using this strategy have previously shown advantageous risk–benefit ratios and safety profiles, allowing them to quickly enter clinical trials.

3.5.1 Antiviral Drugs

Ribavirin (Fig. 42), a synthetic guanosine nucleoside pharmacologically utilized in the treatment of hepatitis C virus (HCV), has shown effects on the replication of other well-known arboviruses (Rattanaburee et al. 2015; Abdel-Hady et al. 2014; Silva-Júnior et al. 2017). Notwithstanding the previous advances, Kamiyama et al.

Fig. 42 Chemical structure of ribavirin

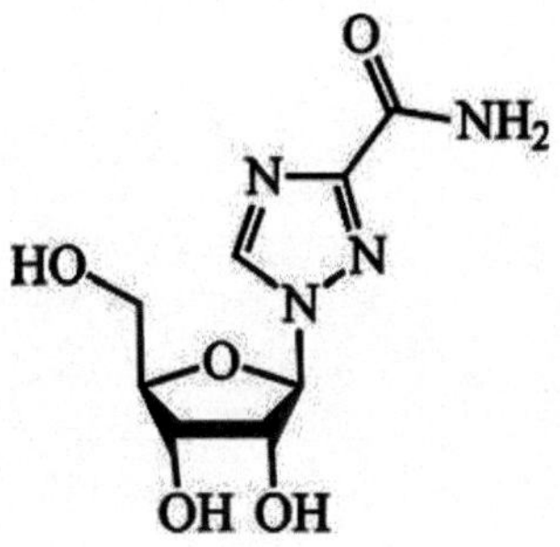

(2017) demonstrated that ribavirin 40 μg/mL significantly inhibited ZIKV replication. Eventually, its association with IFN-β 2.5 U/mL was able to completely inhibit the ZIKV replication in *Vero* cells. Furthermore, the inhibitory effect on the ZIKV replication was observed in SH-SY5Y and C6/36 cells lineages, when treated with ribavirin at 10 μg/mL concentration. Thereafter, it was noticed that ribavirin prevents apoptosis and ZIKV-induced death on *Vero* cells. Lastly, high doses of ribavirin inhibit the early stages of ZIKV viremia in STAT1-deficient mice (Kamiyama et al. 2017).

Another HTS study, focused on previously known broad-spectrum antiviral compounds and substances from the NIH Clinical Compound Collection Library was published by Adcock et al. (2017) shortly after. Amongst the broad-spectrum antivirals, they found the vacuolar ATPase inhibitor saliphenylhalamide (SaliPhe) and innate immune response agonist CID 91632869 (Fig. 43) to be effective against ZIKV MR766 and PRVABC59 infection in *Vero* cells. Interestingly, the previously described entry inhibitor chloroquine showed no anti-ZIKV activity in their assays. As an explanation for that, Adcock et al. (2017) hypothesized that the lack of antiviral activity was probably due to the nature of their assay, where readout was performed after multiple replication cycles of the virus. Uridine analog 6-azauridine (Fig. 43) exhibited antiviral activity in the low micromolar range. In contrast to what was observed for other pyrimidine-synthesis inhibitors like brequinar, 6-azauridine did not decrease viral titers in infected *Vero* cells. Adcock et al. (2017) also found the 5-α-reductase inhibitor finasteride and (HMG-CoA) reductase inhibitor mevastatin (Fig. 43) to act against ZIKV infections, whereas the latter showed inversed effects on ZIKV replication at concentrations above 5 μM

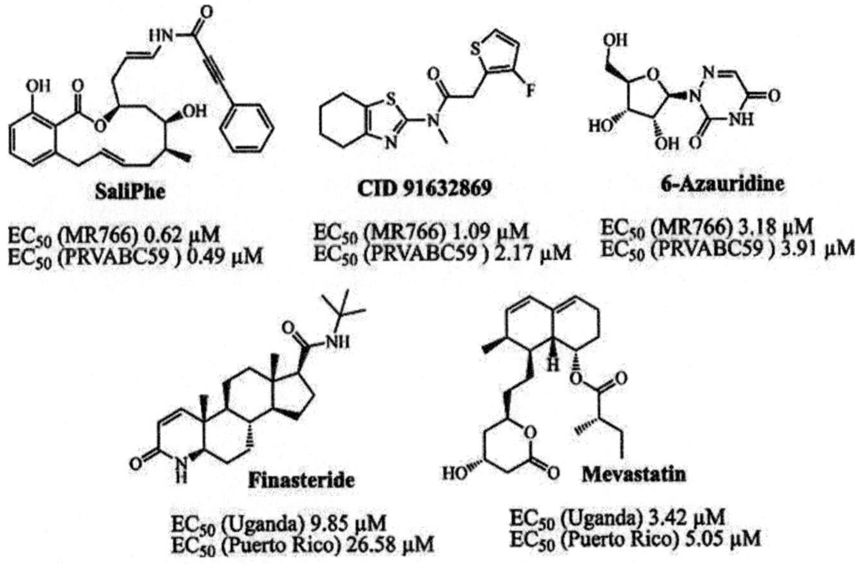

Fig. 43 Repurposed drugs by Adcock et al. (2017)

(Adcock et al. 2017). The nucleoside analog NITD008, well-known for its antiviral activity against DENV by Yin et al. (2009), showed activity against ZIKV MR766 and PRVABC59, with EC_{50} values of 0.51 and 0.56 μM, respectively (Adcock et al. 2017).

Sofosbuvir (Fig. 44) is an RNA polymerase inhibitor (RdRp) approved for the treatment of HCV flavivirus infections, one of the most targeted virus among *Flaviviridae* family members. In this context, sofosbuvir was tested in models of ZIKV cell infection, such as *Huh-7*, neuroblastoma (SH-Sy5y), neural stem (NSC) cells, and brain organoids. In the direct inhibition assay of ZIKV RdRp, sofosbuvir showed an IC_{50} value of 0.38 μM, slightly higher than the positive control (Ribavirin-triphosphate, IC_{50} = 0.21 μM). In cellular tests, the substance showed EC_{50} values ranging from 0.41 to 1.9 μM when working with a multiplicity of infection (MOI) of 1.0, and ranging from 0.12 to 1.7 μM when working with MOI of 0.5, all lower than the positive control (Ribavirin). Finally, the substance was shown to be able to drastically reduce the ZIKV production in brain organoids, in addition to causing mutations of the A-to-G type in the virus in doses lower than those required for antiviral action, suggesting the possibility of a second pharmacological effect (Sacramento et al. 2017). In another study, sofosbuvir showed EC_{50} value ranging from 1 to 5 μM in *Huh-7* (human hepatocellular carcinoma cell) and *Jar* (human placental choriocarcinoma cell) infection models, with MOI 0.1 from several ZIKV strains, in addition to showing a protective effect when orally administered to mice at a dose of 33 mg/kg/day (Bullard-Feibelman et al. 2017).

Considering nucleoside analogs, such as 6-methylmercaptopurine riboside (6MMPr, Fig. 45), which is capable of inhibiting the replication of a series of flaviviruses, it was evaluated in vitro against ZIKV and using *Vero* and human neuroblastoma cells (*SH-Sy5y*) as infection models. 6MMPr showed IC_{50} values of 24.5 and 20.3 μM against infection in *Vero* and *SH-Sy5y* cells, respectively, but with a quite different SI (11.9 and 22.7, respectively). It was suggested that 6MMPr is much less toxic for *SH-Sy5y* cells than for *Vero* cells (Carvalho et al. 2017).

Arbidol (Fig. 46) is an antiviral used for decades in the treatment of influenza that has been shown to inhibit six different ZIKV isolates from various cell lines, including primary vaginal and human cervical epithelial lines. A study showed that the substance is more effective when administered before virus infection, exhibiting an IC_{50} value of 11 μM if administered 12 h before infection and 15 μM if

Fig. 44 Chemical structure of sofosbuvir

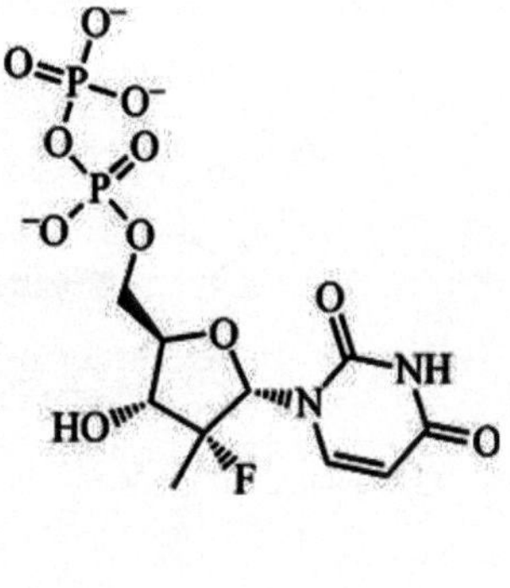

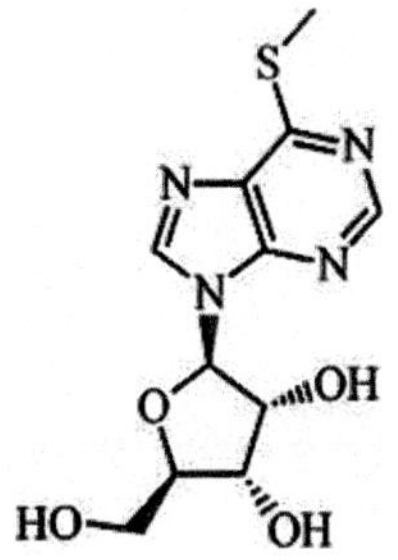

Fig. 45 Chemical structure of 6-methylmercaptopurin ribose

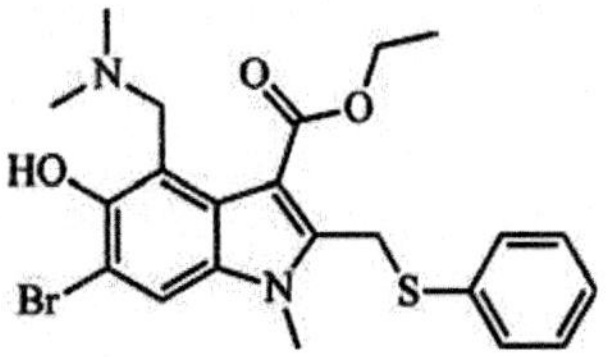

Fig. 46 Chemical structure of arbidol

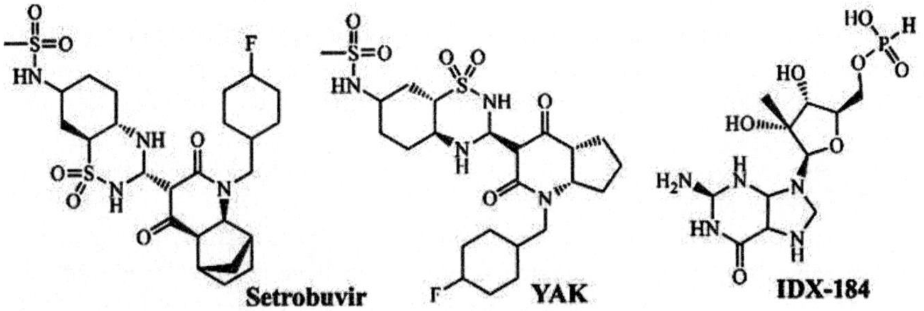

Fig. 47 Repurposed drugs by Elfiky and Ismail (2018) for Zika virus

administered 1 h before, with $CC_{50} > 40$ μM in both cases. This information corroborates its proposed mechanism of action, which is to prevent from virus entering the cell (Fink et al. 2018).

The in *silico* approach was also used to identify candidates for drug repositioning, where a virtual screening including 16 ligands studied in the last 20 years as antivirals allowed to identify potential inhibitors of ZIKV RdRp, being identified the compounds setrobuvir, YAK, and IDX-184 (Fig. 47) as promising anti-ZIKV agents (Elfiky and Ismail 2018).

3.5.2 Lipid-Lowering Drugs and Nutraceuticals

Ianevski et al. (2018) developed an in vitro investigational study involving approved, under-analysis and experimental antiviral agents against Rift Valley fever virus (RVFV), Echovirus 1 (EV1), Herpes Simplex virus type 1 (HSV-1), ZIKV,

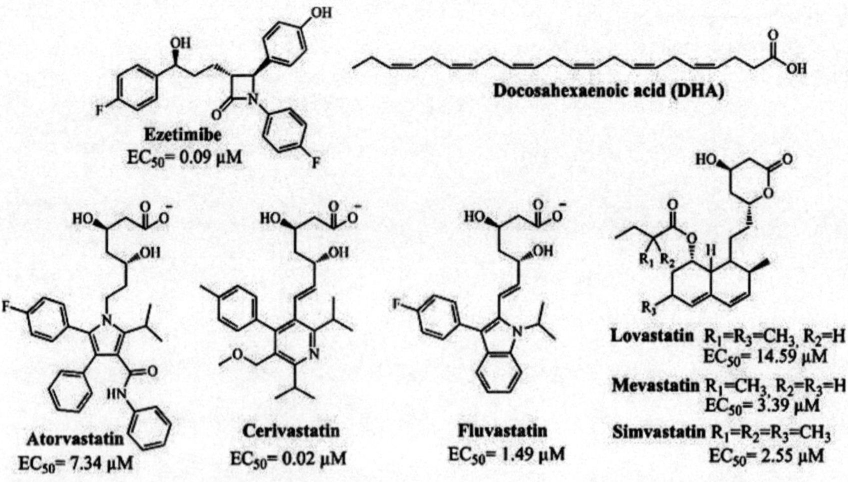

Fig. 48 Hypolipidemic drugs and nutraceuticals with effects against Zika virus replication

HIV-1, CHIKV, Influenza A virus (FLUAV), and Ross River virus (RRV). Interestingly, the authors identified novel antiviral activities. Surprisingly, ezetimibe (Fig. 48), a drug for reducing cholesterol absorption in the small intestine, has been identified as a promising anti-ZIKV agent, exhibiting an EC_{50} value of 0.09 μM and selectivity index (SI) high than 333 in retinal pigment epithelium (*RPE*) cells (Ianevski et al. 2018).

Neuroprotective effect of docosahexaenoic acid (DHA) (Fig. 48) in ZIKV infection, an omega-3 polyunsaturated fatty acid (PUFA), has been analyzed forwards *SH-Sy5y* cells. Likewise, it has been reported that pre-treatment with DHA increases the viability, proliferation, and decreases ZIKV-induced cell apoptosis. Furthermore, it modulates the inflammatory response by decreasing pro-inflammatory cytokines levels (IL-6 and MCP-1) and increased secretion of anti-inflammatory cytokine (TGF-β); inhibits the production of reactive species and prevents ZIKV-induced mitochondrial dysfunction; it reduces ZIKV viral load and infectivity. Nevertheless, pre-treatment with DHA exhibited neuroprotective and anti-inflammatory effects on neuronal cells infected with ZIKV (Braz-De-Melo et al. 2019).

The in vitro activity of statins against ZIKV, a class of drugs that reversibly inhibit HMG-CoA reductase enzyme, which is a key enzyme involved in cholesterol synthesis, was recently investigated for their potential antiviral activity (Moghadasian 1999; Españo et al. 2019). Thereupon, it was detected that lipophilic statins such as atorvastatin, cerivastatin, fluvastatin, lovastatin, mevastatin, and simvastatin (Fig. 48) could inhibit ZIKV infection in *Vero* cells, exhibiting EC_{50} values ranging from 0.02 to 14.59 μM, and also SI values varying from 2.66 to 18.19. In general, it is proposed that these drugs reduced the production of ZIKV infectious particles, reaching up to 98% of inhibition by using lovastatin, for

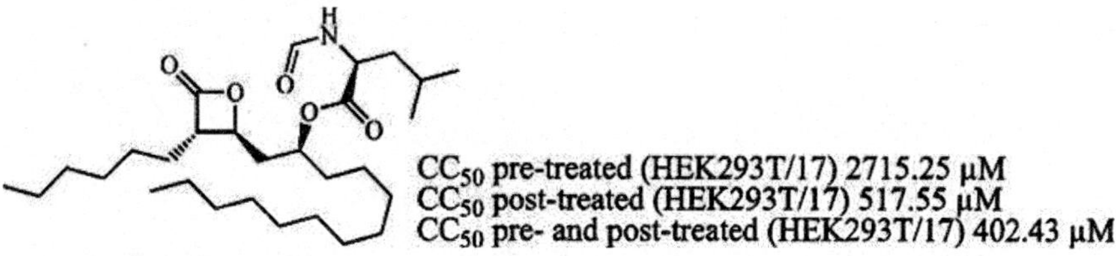

Fig. 49 Chemical structure of orlistat

example. Finally, these results highlighted the role of cholesterol and other mevalonic acid products in ZIKV replication (Españo et al. 2019).

Lipids are required at multiple stages of the viral replication cycle (Martín-Acebes et al. 2016). Hitakarun et al. (2020) investigated the potential of orlistat (Fig. 49, tetrahydrolipstatin) an FDA-approved drug that inhibits the fatty acid synthase (FASN), as broad-spectrum anti-flaviviral and -alphaviral agent. Similar to orlistat, another FASN inhibitor, namely C75, was identified previously to block DENV replication by Samsa et al. (2009). Therefore, they screened orlistat at 100 μM concentration against different strains of DENV, ZIKV, JEV, YFV, and CHIKV in three treatment regiments of infected *HEK 293T/17* cells (pre-, post-infection, and combined treatment). For ZIKV (Uganda and Asian SV0010/15), very moderate change in the level of infection was observed, although virus titer and genome copy numbers were reduced (Hitakarun et al. 2020; Martín-Acebes et al. 2016).

3.5.3 Antiasthmatic Drugs

An in *silico* screening was accomplished by analyzing of 1,597 drugs toward viral proteins from DENV (NS5), DENV2 (E, C, NS2B/NS3), DENV3 (NS5), ZIKV (E, C, MTase, NS1, NS3, NS5, and NS2B/NS3), and CHIKV (E1, E2, and nsP2). Consequently, five compounds were selected for in vitro testing. In this context, pranlukast (Fig. 50), an antiasthmatic drug, exhibited a better antiviral activity upon infected human monocytic cells with DENV and ZIKV, with IC_{50} values of 11.85 and 22.28 μM, respectively. Finally, pranlukast blocked DENV attachment on human monocytic cells surface and interacts with E, NS1, and C proteins from

Fig. 50 Pranlukast, an antiasthmatic drug repurposed for the Zika virus

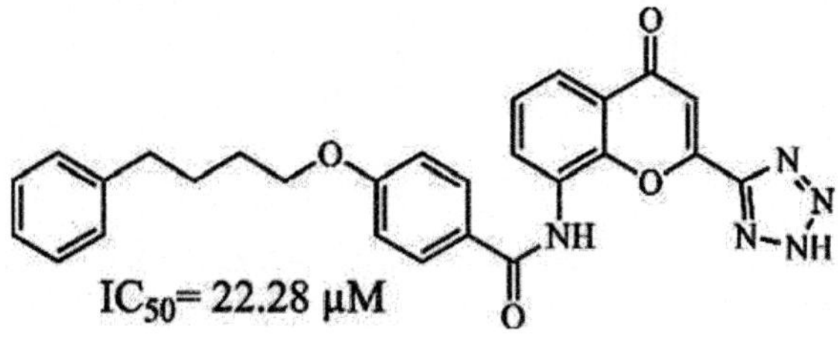

DENV, suggesting that its antiviral effect may be related to these essential proteins (Montes-Grajales et al. 2020).

3.5.4 Antimalarial Drugs

Chloroquine is a 4-aminoquinoline used to treat malaria infections (Fig. 51). Interestingly, it has exhibited antiviral effects against different flaviviruses, such as DENV and JEV (Browning 2014; Zhu et al. 2012; Farias et al. 2013). Additionally, it has demonstrated anti-ZIKV effects in *Vero*, *hBMEC*, and *hNSC* cells, with EC_{50} values ranging from 9.82 to 14.2 μM, and SI from 7.68 to 13.70 (Delvecchio et al. 2016). Based on its remarkable antiviral effects, 7-chloroquinoline-containing analogs have been synthesized and screened for antiviral properties (Fig. 51). In this context, some *N*-(2-(arylmethylimino)ethyl)-7-chloroquinolin-4-amine derivatives (Fig. 51, *a–i*) should be highlighted due to their inhibitory effects on ZIKV replication (> 75%), which exhibited activity about tenfold more potent than chloroquine, except for compound (*f*). Ultimately, the best derivative (compound ***h***) showed EC_{50} and SI values of 0.8 μM and 515, respectively (Barbosa-Lima et al. 2017).

Li et al. (2017) found an antiviral activity in cell assays and showed that it protected mice from ZIKV-induced mortality and reduced microcephaly in

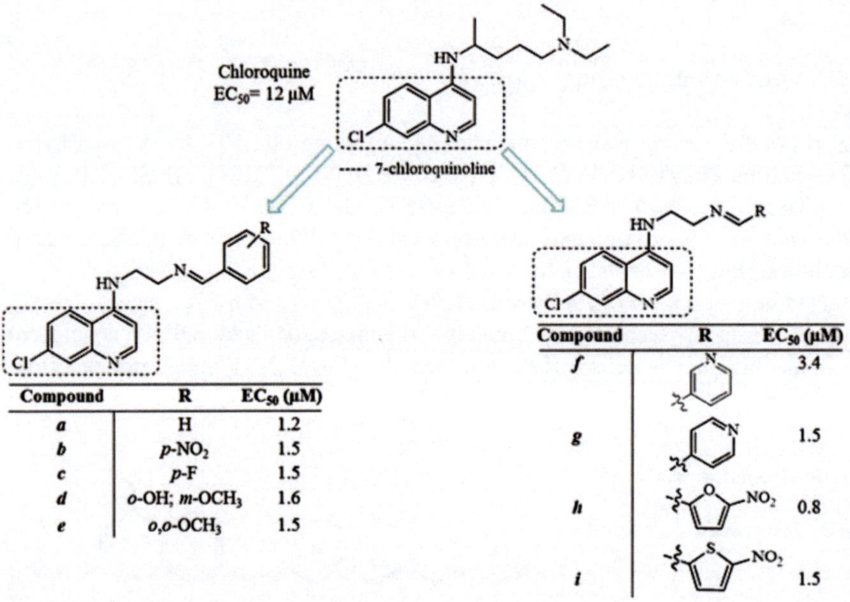

Compound	R	EC_{50} (μM)
a	H	1.2
b	*p*-NO_2	1.5
c	*p*-F	1.5
d	*o*-OH; *m*-OCH_3	1.6
e	*o,o*-OCH_3	1.5

Compound	R	EC_{50} (μM)
f		3.4
g		1.5
h	NO_2	0.8
i	NO_2	1.5

Fig. 51 Structure and EC_{50} values for chloroquine and its *N*-(2-(arylmethylimino)ethyl)-7-chloroquinolin-4-amine derivatives against Zika virus replication

newborns. The IC_{50} value was determined to be in the low single-digit micromolar range with no detectable toxicity at 10 μM concentration. They showed that chloroquine inhibited virus internalization but neither interfered with particle binding nor viral replication (Li et al. 2017). Zhang et al. (2019) further investigated the antiviral effects of chloroquine in different cell types as well as in a newly established maternal to fetal transmission model in mice (Zhang et al. 2019) and found that chloroquine inhibited ZIKV infection by at least two mechanisms, with EC_{50} values of 1.72 and 2.72 μM in *GZ01* and *FSS13025* cells, respectively. Besides, it presented a CC_{50} value higher than 10 μM in both *Vero* and *Huh-7* cells. It affects RNA release from virus internalized endosomes and autophagy-dependent viral replication as proposed by Gorshkov et al. (2019) and Delvecchio et al. (2016). Deeming these findings, they support that more research should be performed to evaluate chloroquine to be considered for the treatment of ZIKV infection, especially for reducing the risk of microcephaly in pregnant women.

Balasubramanian et al. (2017) continued their previous work (Boonyasuppayakorn et al. 2015), in which the antimalarial amodiaquine (Fig. 52) and analogs were found to have anti-DENV-2 and -WNV activities, and calculated stereo-electronic profiles. Based on these profiles, they developed a computational model that identified three additional compounds, quinacrine, mefloquine, and GSK369796 (Fig. 52) that act against DENV2, as well as ZIKV in a plaque reduction assay with similar EC_{50} values of 2.27, 3.95, and 2.57 μM (Balasubramanian et al. 2017). Using a fluorescence-based enzyme activity assay with purified and activated cathepsin B, a lysosomal protein involved in the

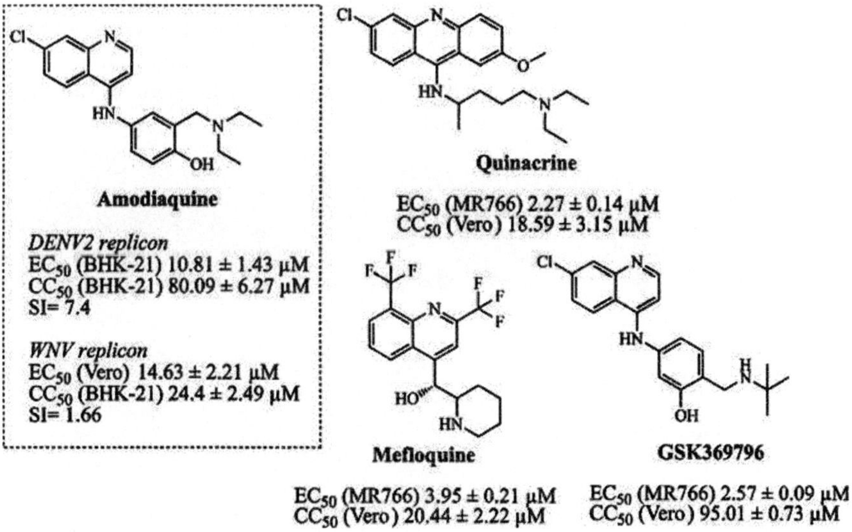

Fig. 52 Repurposed drugs by Balasubramanian et al. (2017) and Zilbermintz et al. (2015)

autophagic flux, they demonstrated that amodiaquine, mefloquine, and GSK369796 act as previously described for amodiaquine (Zilbermintz et al. 2015) by directly binding and inhibiting cathepsin B. As showed by Ha et al. (2010) cathepsin B mediates the fusion of lysosome with endosomes which, therefore, may be important for viral release into the cytoplasm (Ha et al. 2010).

3.5.5 Anthelmintic Drugs

Niclosamide (Fig. 53) is an FDA-approved drug for the treatment of parasitic infections by helminths. Further, it has been reported as an inhibitor of the replication of three different ZIKV strains in a screening upon *SNB-19* cells. In this context, niclosamide presented antiviral activity against ZIKV, with an IC_{50} value of 0.37 μM (Xu et al. 2016). Ivermectin (Fig. 53), another antiparasitic drug, was able to inhibit ZIKV infection in an experimental screening on different cell lines (Barrows et al. 2017). Nevertheless, it has not shown effectiveness in a limited study in Ifnar1 knockout mice (Ketkar et al. 2019). Recently, a study has shown the effectiveness of synthetic poly(lactide-*co*-glycolide)-*b*-polyethylene glycol (PLGA-*b*-PEG) ivermectin-containing nanoparticles, with IgG Fc conjugated on top of the nanoparticles. These nanoparticles were able to increase plasma levels of ivermectin and also showed reduced toxicity when compared to free ivermectin. Furthermore, these nanoparticles showed the capability of reducing the expression of the ZIKV NS1 protein and could be a therapeutic alternative against this virus (Surnar et al. 2019).

In an approach for repurposing drugs for utilization against ZIKV infection, Barrows et al. (2016) screened 774 US Food and Drug Administration (FDA)-approved drugs in a *Huh7* cell-based assay, where the infection rate was assessed. Therefore, only drugs that interfere with viral entry, translation, or replication could be identified. Amongst drugs with already known antiviral activity, they successfully identified the antiparasitic drug ivermectin (see Fig. 53) (EC_{50} ranging from 1

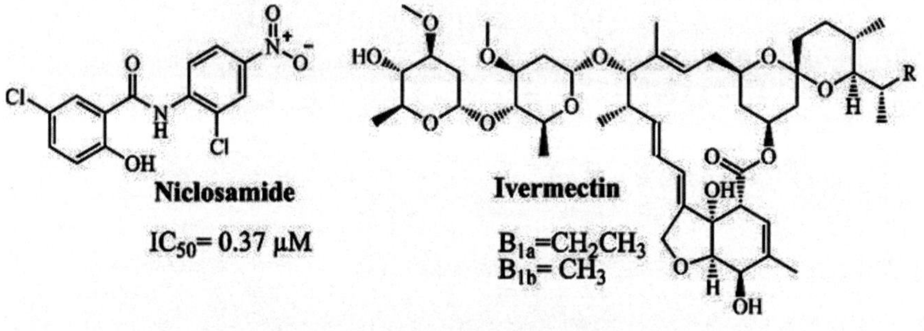

Fig. 53 Anthelmintic drugs—niclosamide and ivermectin with antiviral effects against the Zika virus

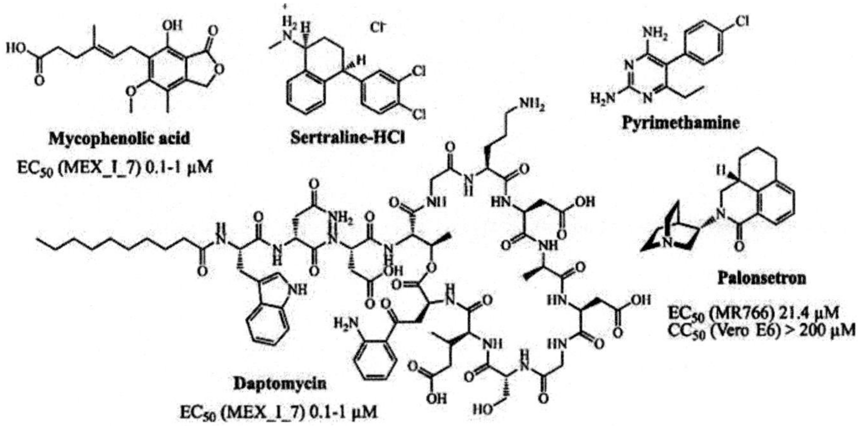

Fig. 54 Repurposed drugs by Barrow et al. (2016)

to 10 μM), the immunosuppressant mycophenolic acid (EC_{50} varying from 0.1 to 1 μM), and the antibiotic daptomycin (Fig. 54) (EC_{50} ranging from 0.1 to 1 μM) as new anti-ZIKV active substances. Drugs that also showed antiviral activity included the antidepressant sertraline-HCl, the anti-protozoal drug pyrimethamine, and the antiemetic palonosetron-HCl (Fig. 54). In successive experiments with *HeLa* and *JEG-3* cell lines, deviations regarding toxicity and antiviral activity could be observed depending on the used cell type (Barrows et al. 2016).

3.5.6 Antihistamine Drugs

In a virtual screening (VS) campaign, using a co-crystal structure of ZIKV NS2B-NS3 (PDB ID: 5H4), Santos et al. (2020b) screened 1,861 FDA-approved drugs available in DrugBank and applied the Glide software as well as DOCK6 in a consensus docking approach to minimize false positive hits. After selecting the four drugs chlorcyclizine, clozapine, methazolamide, and levonordefrin for hit validation in antiviral assays in *Vero* cells, only chlorcyclizine (Fig. 55) showed activity with an EC_{50} value of 69 μM and CC_{50} of 130 μM (Santos et al. 2020b).

Fig. 55 Chemical structure of chlorcyclizine

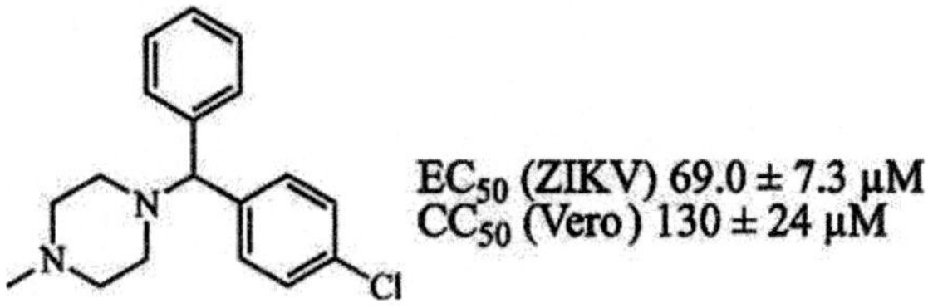

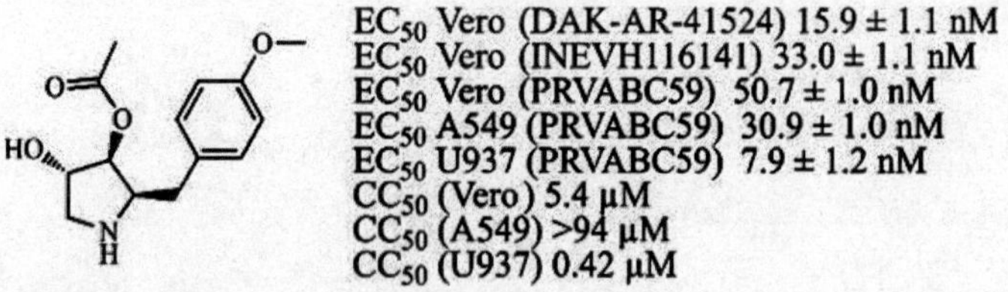

Fig. 56 Chemical structure of anisomycin

3.5.7 Antimicrobial Drugs

The antiviral mechanism of action of the antibiotic alkaloid anisomycin (Fig. 56) against ZIKV and DENV is still uncovered. However, Quintana et al. (2020) revealed that anisomycin to impair post-entry stages of viral replication. Its EC_{50} values ranging from 7.9 to 50.7 nM were determined for several ZIKV strains on *Vero*, human epithelial *A549*, and human myeloid leukemia *U937* cells. In a mouse AG129 model, low doses (4 mg/kg/d) of anisomycin were moderately beneficial in terms of mortality, whereas higher dosages (100 mg/kg/d) reversed this effect (Quintana et al. 2020).

3.5.8 Anticancer Drugs

In order to identify substances that interact with factors from the host to inhibit viral replication, a series of ligands with antiviral activity were tested against several viruses in the human retinal pigment epithelial (RPE) cells model, which are ZIKV targeted. As a result, SaliPhe (see Fig. 43), obatoclax, and gemcitabine (Fig. 57) were identified as capable of rescuing cells from ZIKV-mediated death, with EC_{50} values of 0.04, 0.05, and 0.01 μM and SI of 65, >200 and >1000, respectively. Finally, it was proposed that SaliPhe and obatoclax inhibit virus endocytosis, and gemcitabine interferes with viral RNA transcription (Kuivanen et al. 2017).

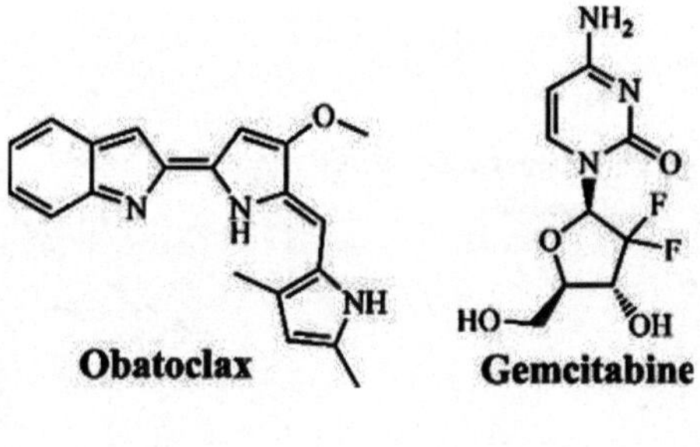

Fig. 57 Drugs repurposed by Kuivane et al. (2017) for Zika virus

3.5.9 Other Repurposed Drugs

A phenotypic HTS, based on immunofluorescence detection of ZIKV E-protein in infected *JEG-3* cells was performed by Loe et al. (2019) for 1,172 FDA-approved drugs. Out of 29 positive hits, four drugs were selected for further studies and validation (candesartan cilexetil, emetine, primaquine diphosphate, and calcifediol (Fig. 58)). All of them showed antiviral activity in the sub-micromolar range. The candesartan cilexetil, an angiotensin II receptor subtype 1 (AT1) antagonist with so far, no reported antiviral activity, inhibited ZIKV with an IC_{50} value of 3.363 nM and a CC_{50} value of 22.33 μM, resulting in a high SI of 6640. Further studies revealed a post-entry inhibitor mechanism and reduction of viral replication and consequently protein synthesis (Loe et al. 2019). Anti-protozoal (primaquine diphosphate) and emetic (emetine) drugs (Su, Yin, Low et al. 2009) were previously described as an antiviral drug against ZIKV infections in a work that also elucidated its mechanism of action against ZIKV and EBOV (Yang et al. 2018b).

Candesartan cilexetil inhibited the ZIKV NS5 RdRp in a cell-free assay with a 20-fold higher IC_{50} value compared to the cell-based viral infection assay probably caused by an enrichment of candesartan cilexetil in cells (Khandelwal et al. 2017).

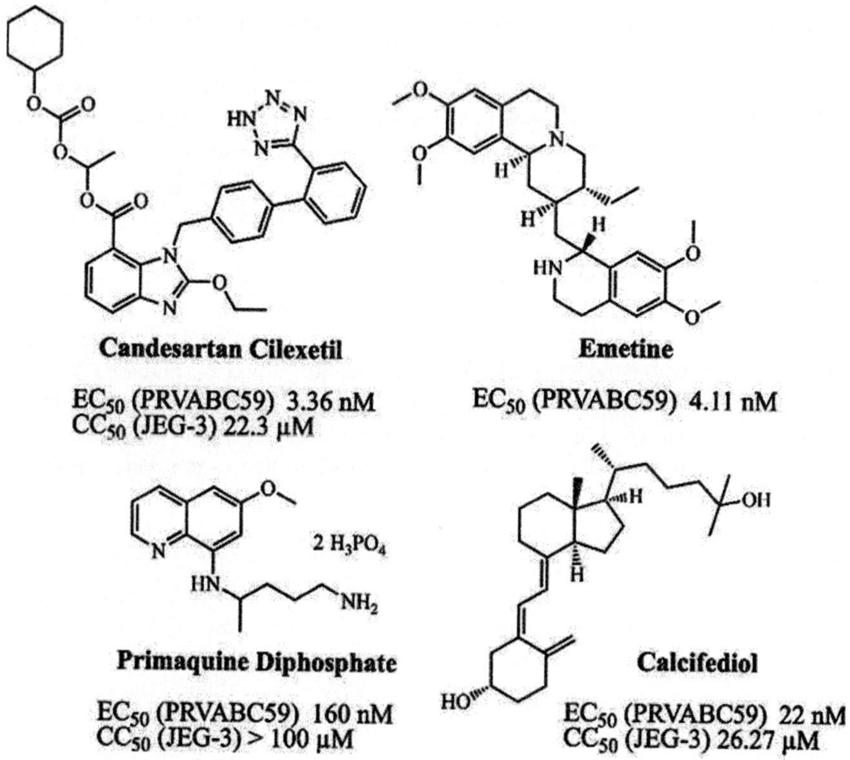

Fig. 58 Repurposed drugs against Zika and Ebola viruses

Another proposed mechanism is that candesartan cilexetil affects the Niemann-Pick type C1 cholesterol transporter, possibly leading to reduced autophagy, reducing the virus binding (as reported for EBOV) and dose-dependent lipid accumulation in lysosomes (Carette et al. 2012; Martín-Acebes et al. 2016; Yang et al. 2018b).

3.6 *Other Inhibitors Against Zika Virus*

3.6.1 Free Form Amino Acid (FFAAP)

Vasireddi et al. (2019) showed that ZIKV replication in human placental cells is dependent on intracellular glutathione levels. Cells that were treated with a specific free form amino acid (FFAAP) formulation comprising cysteine, glycine, and a glutamate source along with a minute concentration of selenium produced significantly greater levels of glutathione than untreated cells. Biosynthesis is increased in a dose-dependent manner, plateauing at 5 mM concentration. FFAAP treatment at a concentration of 10 mM was toxic to the cells. The effectiveness of FFAAP was analyzed in ZIKV (PRVABC59) infected *Vero* cells by plaque assay showing a reduced number of plaques of 42% at 3mM and 85% at 6 mM. To evaluate whether intracellular glutathione was responsible for the reduction in ZIKV titers the virus reduction was quantified, while inhibiting the biosynthesis of glutathione by buthionine sulfoximine (BSO). They observed up to 90% reduction of ZIKV yields with FFAAP (2.5–5 mM) treatment of *JEG-3* and *Vero* cells. The reduction of virus titer is independent of the cell's utilization and concentrations of the intracellular glutathione (Vasireddi et al. 2019).

3.6.2 Bone Morphogenic Protein (BMP)

The expression of the iron hormone hepcidin is decreased upon chronic hepatitis C virus (HCV) infection, which is regulated by hepatic bone morphogenic protein (BMP)/SMAD signaling (Ryan et al. 2010; Babitt et al. 2006). Eddowes et al. (2019) observed that HCV infection and activation of the BMP/SMAD pathway are mutually antagonistic. Therefore, they explored the effects of several different BMPs and the related activin proteins on type I interferon (IFN) signaling, cellular antiviral responses, and replication of HCV. Upon the observation that BMP signaling controls expression of genes that regulate a broadly acting antiviral response, the group considered whether BMPs and activins may have activity against other viruses besides HCV, namely the hepatitis B virus (HBV), and ZIKV. They found that BMP6, as well as activin A, restrict replication of the DNA virus HBV and that activin A inhibits ZIKV replication. INF-α and activin A inhibited viral replication when used alone. However, the combination of both had synergistic effects (Eddowes et al. 2019).

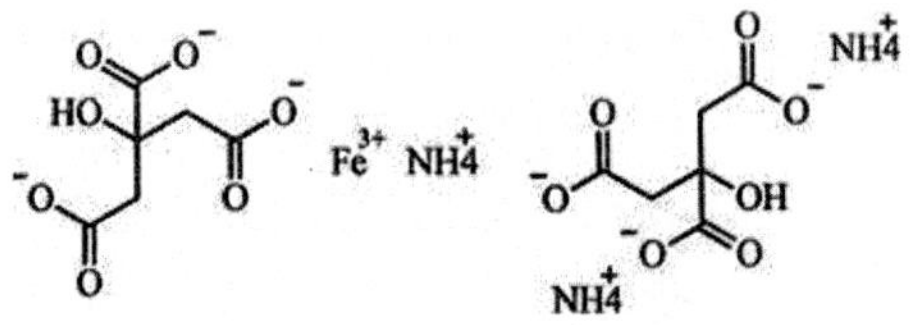

Fig. 59 Chemical structure of ferric ammonium citrate

3.6.3 Ferric Ammonium Citrate (FAC)

Ferric ammonium citrate (FAC; 100 μM, Fig. 59) blocks influenza A virus (IAV) PR8 replication and virus-induced expression of innate immune factors without being toxic even at high concentrations (10 mM). In cell-based assays, it was shown that FAC inhibits the endosomal release of IAV. In vivo assays showed that FAC protected mice from IAV PR8 virus infection reflected by weight loss and survival rate. Using a mouse ZIKV infection model, Wang et al. (2018) found that co-administration of FAC (15 mM) and ZIKV inhibited weight loss and increased survival rate compared to used controls. Neither with citrate nor with iron alone the anti-PR8 effect could be achieved, suggesting that specific ferric citrate complexes are required for antiviral activity. Testing the time window for application of FAC in vivo for prevention (FAC 30 min pre-treatment, then performing PR8 infection) or therapeutic (Pr8 infection, then FAC 30 min treatment) purposes, FAC had no inhibitory effect on the virus (Wang et al. 2018).

3.6.4 Benzoxazine/Carbon Dots (BZM-CD)

Carbon dots (CD) derived from hydrothermal treatment of benzoxazine (BZM) have been reported to possess antiviral activity against DNA and RNA viruses e.g. human herpes simplex virus type 1 (HSV-1), coronavirus, and norovirus (Ting et al. 2018; Barras et al. 2016; Dong et al. 2017). Huang et al. (2019) combined antimicrobial properties of benzoxazine (BZM) derivates (Sun et al. 2016; Iloni et al. 2018) with carbon dots (BZM-CD). To examine the effect of BZM-CDs on flaviviral infections they measured infectivity of JEV by plaque reduction assay in the presence and absence of BZM-CDs. Treatment with BZM-CDs prevented the viral infectivity in a dose-dependent manner (9.375–75 μg/mL) without being toxic. Regarding time dependency the authors found out that virions lost their infectivity after incubation with BZM-CDs for 10 min and no plaque was observed upon 60 min treatment. Similar to the results for JEV, the treatment of BZ-CDs reduced the ability of ZIKV to infect *Vero* cells. The EC_{50} value of BZ-CDs against ZIKV was determined to be 3.715 μg/mL. As shown by electron microscopy, BZM-CDs could directly bind to the surface of virions and block the first step of virus-cell interaction (Huang et al. 2019).

3.6.5 Guanylate-Binding Proteins (GBPs)

Guanylate-binding proteins (GBPs) have antiviral, antibacterial, and antiprotozoal activities (Kim et al. 2012; Vestal and Jeyaratnam 2011), nevertheless, through a yet unclear mechanism. Braun et al. (2019) showed that GBP2 and GBP5 target the cellular host protease furin and suppress the maturation and priming of several glycoproteins including those of different viruses. In this work, the research group focused on the suppression of the maturation of HIV and murine leukemia virus, somewhat it was identified similar effects for IAV, ZIKV, measles virus, and MARV. Finally, in some experiments they found, that not only furin activity but also its expression was reduced (Braun et al. 2019).

3.6.6 Protein–Protein Inhibitors (PPIs)

Kazmirchuk et al. (2017) designed synthetic competitive peptide inhibitors of host-ZIKV protein–protein interactions (PPIs). They initiated producing a comprehensive map of predicted human-ZIKV PPIs. They used two complementary sequence-based PPI prediction tools (PIPE and De Novo prediction) to identify sequences that served for the design of several peptides aiming at disrupting the corresponding PPI. From a list of 25 high priority human proteins that interact with ZIKV proteins, the research team focused mainly on four: RNASET2, ENO2, TRAF4, and CEP63. The results could be used to develop new therapeutics for the ZIKV (Kazmirchuk et al. 2017).

3.6.7 Vivo-Morpholino

Morpholino oligomers are antisense uncharged molecules that bind to RNA and can be applied to interrupt translation or to interfere with RNA processing. Further, *Vivo*-morpholino is comprised of a morpholino oligo covalently bonded to an octaguanidine dendrimer, a delivery portion that facilitates the cellular penetration of the morpholino and facilitates its delivery to the cytosol (Popik et al. 2018; Morcos et al. 2008; Moulton and Shan 2009). In this framework, *Vivo*-morpholino DWK-1 (Fig. 60) designed to be complementary to the 25-mer-nucleotide sequence within the ZIKV 5′-untranslated region (5′-UTR) suppressed the transcription of ZIKV in podocytes (>96%) and primary human retinal endothelial cells (98%). Also, it reduced ZIKV protein E expression (>98%) and suppressed the pro-inflammatory gene expression from IFN-β, RANTES, MIP-1α, TNF-α, and IL-1α in vitro. Lastly, DWK-1 (Fig. 60) demonstrates a potential antiviral effect against ZIKV, demanding additional in vivo investigations (Popik et al. 2018).

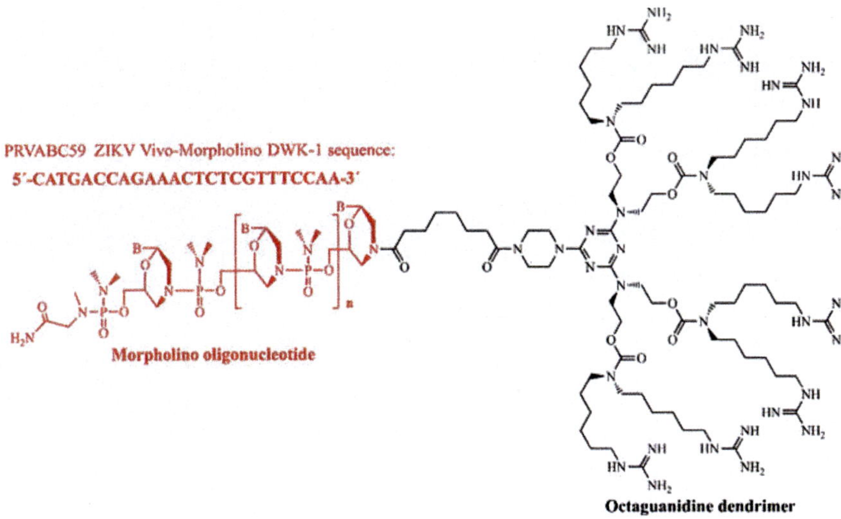

Fig. 60 Structure of Vivo-morpholino DWK-1. The sequence of DWK-1 complementary to the 25-mer of ZIKV 5′-UTR is shown (in red)

3.6.8 Cyclin-Dependent Kinase Inhibitors (CDKi)

A virtual screening study including about 6,000 compounds, counting approved and clinical trial drugs candidates, and also pharmacologically active compounds, was capable of identifying cyclin-dependent kinase inhibitors (CDKi) as potential in vitro inhibitors of ZIKV replication (Xu et al. 2016). Hence, PHA-690509 (Fig. 61) inhibited the replication of three different ZIKV strains, and it demonstrated an IC_{50} value of 1.72 μM in an assay that determined intracellular levels of ZIKV RNA. Furthermore, seliciclib, SNS-032, PHA-793887, AT7519, BS-194, RGB-286147, dinaciclib, and alvocidib (Fig. 61) inhibited ZIKV infection in SNB-19 cells, with IC_{50} values in a submicromolar range (Xu et al. 2016).

3.6.9 Non-competitive Inhibitor of Inosine-5′-Monophosphate Dehydrogenase

Merimepodib (Fig. 62), a potent non-competitive inhibitor of inosine-5′-monophosphate dehydrogenase (IMPDH), which is an enzyme involved in the De Novo synthesis of guanine nucleotides (Sintchak and Nimmesgern 2000), exhibited an in vitro antiviral effect by inhibiting ZIKV RNA replication, with an EC_{50} value of 0.6 μM. Moreover, the authors provided evidences that combinations of merimepodib with ribavirin (see Fig. 42) or favipiravir (Fig. 62) increased the antiviral effects. Somewhat, merimepodib was capable of inhibiting other meaningful

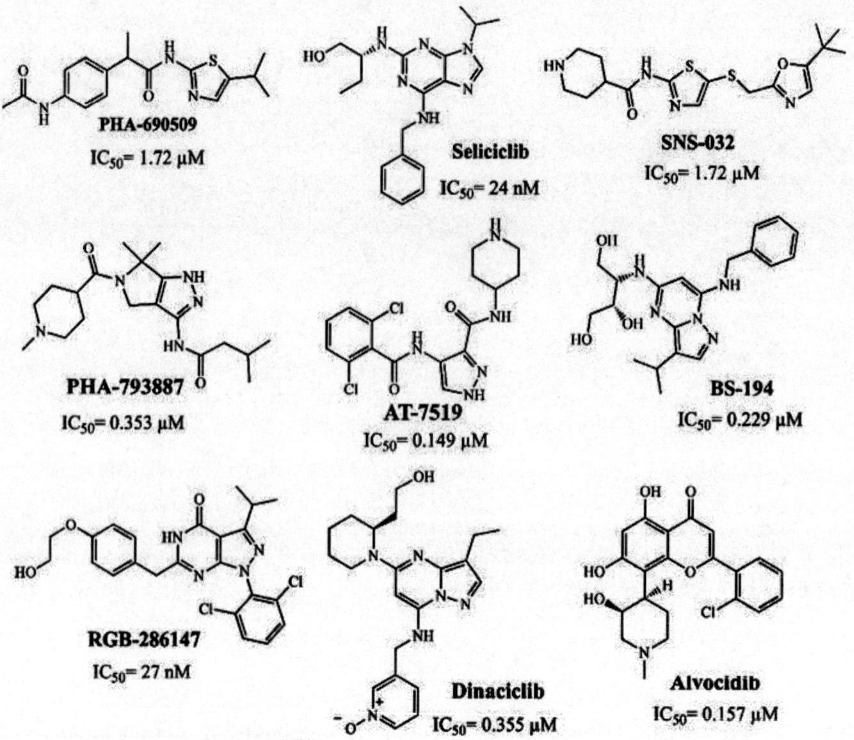

Fig. 61 Cyclin-dependent kinase inhibitors (CDKi) active against the Zika virus

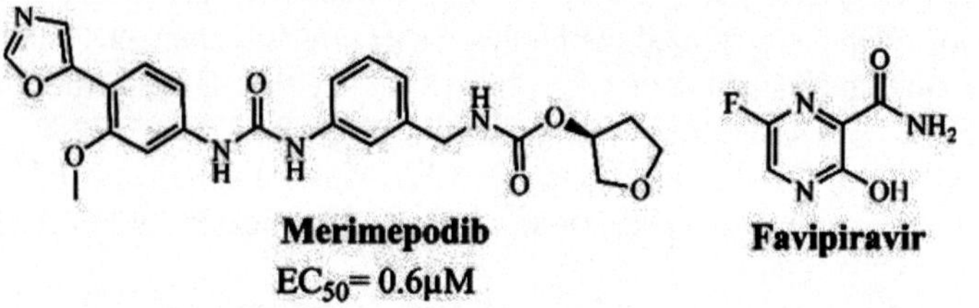

Fig. 62 Inosine-5′-monophosphate dehydrogenase inhibitor and antiviral drugs with activity against the Zika virus

viruses, such as Chikungunya (CHIKV), Junin (JUV), Lassa (LASV), and Ebola (EBOV) (Tong et al. 2018).

3.6.10 Pan-Caspase Inhibitors

Emricasan (Fig. 63), a pan-caspase inhibitor, is capable of inhibiting the increment in caspase-3 activity, thus protecting neural cells from ZIKV-induced death,

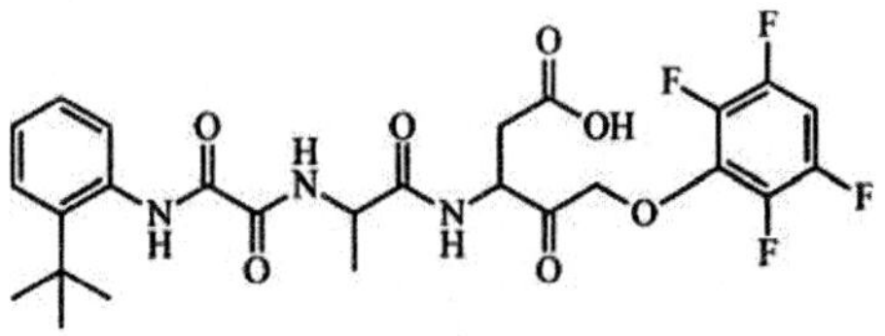

Fig. 63 Emricasan, a caspase-3 inhibitor with in vitro neuroprotective activity against Zika-induced death of neural cells

evidencing a neuroprotective activity but without inhibiting ZIKV replication (Xu et al. 2016).

3.6.11 Innate Immunity Targets

Components of innate immunity have been explored for antiviral effects (Makhluf et al. 2016). It was seen that primary human corneal epithelial cells (*Pr. HCEC*) were permissive to ZIKV infection and signals of the innate immune response were elicited, such as expression of pathogen recognition receptors (PRRs), induction of cytokines and chemokines, and upregulation of interferon-stimulated antiviral response genes. Finally, recombinant interferon-stimulated gene 15 (ISG15) inhibited ZIKV replication by direct inactivation and reduction of ZIKV attachment to Vero cells (Singh et al. 2019).

4 Final Considerations and Future Outlook

Deeming this scenario, the development of new alternatives medicines for the treatment of ZIKV is an unmet need since it is a Neglected Tropical Disease (NTD) and has not received considerable attention. Wherefore, there are no licensed vaccines or approved drugs to specifically treat this infectious disease. In contrast, various research groups worldwide have developed new strategies to discover efficient alternative drugs against this viral disease. These researchers have focused on searching for natural, nature-based, and synthetic compounds, as well as, repurposing of drugs. Somewhat, different strategies involving other potential inhibitors have also emerged.

It is remarkable how much medicinal chemistry in recent years has contributed to the development of new inhibitors against ZIKV. Often, the structure-based drug design or virtual screening campaigns have led to the identification of promising natural or synthetic derivatives with anti-ZIKV activity in the low- or sub-micromolar ranges. Considering the recent advances presented in this chapter, certain points should be highlighted. From the natural origin, patentiflorin A, isolated from *Justicia gendarussa* has demonstrated the most promising agent, with IC_{50} values ranging from 10 to 70 nM against several different viruses, such as DENV-1, JEV, TBEV, WNV, and EBOV. Additionally, nature-based tetrakistriazole analogs have exhibited best anti-ZIKV activities, with IC_{50} values varying from 0.16 to 0.24 μM. Still, quinolyloxyl analogs obtained semi-synthetically have been found to be promising against ZIKV. Amongst the synthetic compounds presented in this chapter, an AR-12 analog has shown the best results, with an IC_{50} value of 1.37 nM on *A549* infected cells. Notably, the drug repurposing strategy has been much explored to investigate a large chemical library of active ligands and FDA-approved drugs. Surprisingly, cerivastatin (a synthetic statin used to lower cholesterol and prevent cardiovascular disease) was found as a promissing drug against ZIKV, with an EC_{50} value of 0.02 μM against ZIKV-infected *Vero* cells. Finally, other inhibitors have exhibited meaningful activities against ZIKV, such as cyclin-kinase inhibitors and *Vivo*-morpholine.

ZIKV cases has been reported every year, asking for the development of new specific and safe antiviral agents. In this context, medicinal chemistry emerges as a powerful tool for overcoming difficulties associated with infectious diseases. Remarkable achievements have been highlighted in this chapter, showing the advances made by different research groups working on the ZIKV topic. This strategy has increased along with the advent of the emergence of new virtual protocols, more powerful machines, updated drug banks, crystallographic structures at high-resolution, and other factors. Lastly, it is expected that in the future, safer and active drugs will be discovered or synthesized as the number and types of viruses, including the mutant strains, will emerge.

Acknowledgements The authors thank the Coordenação de Aperfeiçoamento de Pessoal de Nível Superior—Brazil (CAPES), Fundação de Amparo à Pesquisa do Estado de Alagoas (FAPEAL), and the National Council for Scientific and Technological Development (CNPq)—Brazil for their support to the Brazilian Post-Graduate Programs. The authors also acknowledge the website https://smart.servier.com, which was used to generate the Fig. 2 in this chapter.

Funding None.

Declaration of Interest The authors have no relevant affiliations or financial involvement with any organization or entity with a financial interest in or financial conflict with the subject matter or materials discussed in the manuscript. This includes employment, consultancies, honoraria, stock ownership or options, expert testimony, grants or patents received or pending, or royalties.

References

Aagaard KM, Lahon A, Suter MA, Arya RP, Seferovic MD, Vogt MB, Hu M, et al. Primary human placental trophoblasts are permissive for Zika virus (ZIKV) replication. Sci Rep. 2017. https://doi.org/10.1038/srep41389.

Abdel-Hady M, Bansal S, Davison SM, Brown M, Tizzard SA, Mulla S, Barnes E, Davies P, Mieli-Vergani G, Kelly DA. Treatment of chronic viral hepatitis C in children and adolescents: UK experience. Arch Dis Child. 2014;99(6):505–10. https://doi.org/10.1136/archdischild-2013-304601.

Adcock RS, Chu Y-K, Golden JE, Chung D-H. Evaluation of anti-Zika virus activities of broad-spectrum antivirals and NIH clinical collection compounds using a cell-based, high-throughput screen assay. Antiviral Res. 2017;138:47–56.

Akaberi D, Chinthakindi PK, Båhlström A, Palanisamy N, Sandström A, Lundkvist Å, Lennerstrand J. Identification of a C2-symmetric diol based human immunodeficiency virus protease inhibitor targeting Zika virus NS2B-NS3 protease. J Biomol Struct Dyn. 2019;1102. https://doi.org/10.1080/07391102.2019.1704882.

Andral L, Brès P, Sérié C, Casals J, Panthier R. Studies on yellow fever in ethiopia. 3. Serological and virological studies of the woodland fauna. Bull World Health Organ. 1968.

Anggakusuma CC, Colpitts LM, Schang HR, Frentzen A, Pfaender S, Behrendt P, et al. Turmeric curcumin inhibits entry of all hepatitis C virus genotypes into human liver cells. Hepatology. 2014;63:1137–49. https://doi.org/10.1136/gutjnl-2012-304299.

de Araújo TVB, Rodrigues LC, de Alencar Ximenes RA, de Barros Miranda-Filho D, Montarroyos UR, de Melo APL, Valongueiro S, et al. Association between Zika virus infection and microcephaly in Brazil, January to May, 2016: preliminary report of a case-control study. Lancet Infect Dis. 2016;16(12):1356–63. https://doi.org/10.1016/S1473-3099(16)30318-8.

Ashwal S, Michelson D, Plawner L, Dobyns WB. Practice parameter: evaluation of the child with microcephaly (an evidence-based review): report of the quality standards subcommittee of the American academy of neurology and the practice committee of the child neurology society. Neurology. 2009. https://doi.org/10.1212/WNL.0b013e3181b783f7.

Assuncao-Miranda I, Cruz-Oliveira C, Neris RLS, Figueiredo CM, Pereira LPS, Rodrigues D, Araujo DFF, Da Poian AT, Bozza MT. Inactivation of dengue and yellow fever viruses by heme, cobalt-protoporphyrin IX and tin-protoporphyrin IX. Appl Microbiol. 2016;120:780–804. https://doi.org/10.1111/jam.13038.

Babitt JL, Huang FW, Wrighting DM, Xia Y, Sidis Y, Samad TA, Campagna JA, et al. Bone morphogenetic protein signaling by hemojuvelin regulates hepcidin expression. Nat Genet. 2006;38(5):531–9. https://doi.org/10.1038/ng1777.

Balasubramanian A, Teramoto T, Kulkarni AA, Bhattacharjee AK, Padmanabhan R. Antiviral activities of selected antimalarials against dengue virus type 2 and Zika virus. Antivir Res. 2017;137. Elsevier Ltd. https://doi.org/10.1016/j.antiviral.2016.11.015.

Barbés C, Sanchez J, Yebra MJ, Robert-Géro M, Hardisson C. Effects of sinefungin and S-adenosylhomocysteine on DNA and protein methyltransferases from streptomyces and other bacteria. FEMS Microbiol Lett. 1990;69:239–44.

Barbosa-Lima G, da Silveira LS, Pinto CR, Kaiser JL, Wardell CS, Freitas De, Vieira YR, Marttorelli A, et al. N-(2-(arylmethylimino)ethyl)-7-chloroquinolin-4-amine derivatives, synthesized by thermal and ultrasonic means, are endowed with anti-Zika virus activity. Eur J Med Chem. 2017;127:434–41. https://doi.org/10.1016/j.ejmech.2017.01.007.

Barras A, Pagneux Q, Sane F, Wang Qi, Boukherroub R, Hober D, Szunerits S. High efficiency of functional carbon nanodots as entry inhibitors of herpes simplex virus type 1. ACS Appl Mater Interfaces. 2016;8(14):9004–13. https://doi.org/10.1021/acsami.6b01681.

Barrows NJ, Campos RK, Steven Powell K, Prasanth R, Schott-lerner G, Soto-acosta R, Galarza-muñoz G, et al. A screen of FDA-approved drugs for inhibitors of Zika virus infection. Cell Host Microbe. 2017;20(2):259–70. https://doi.org/10.1016/j.chom.2016.07.004.A.

Barrows NJ, Campos RK, Powell ST, Routh A, Bradrick SS, Garcia-blanco MA, Barrows NJ, et al. Resource a screen of FDA-approved drugs for inhibitors of Zika virus infection resource a screen of FDA-approved drugs for inhibitors of Zika virus infection. 2016;259–70.

Basic M, Elgner F, Bender D, Sabino C, Herrlein M-L, Roth H, Glitscher M, et al. A synthetic derivative of houttuynoid B prevents cell entry of Zika virus. Antivir Res. 2019;172 (June):104644.

Bhargava S, Patel T, Gaikwad R, Patil UK, Gayen S. Identification of structural requirements and prediction of inhibitory activity of natural flavonoids against Zika virus through molecular docking and Monte Carlo based QSAR simulation. Nat Prod Res. 2019;33(6):851–7. https://doi.org/10.1080/14786419.2017.1413574.

Blázquez A, Escribano-romero E, Merino-ramos T, Saiz J-C, Cosby SL. Stress responses in flavivirus-infected cells: activation of unfolded protein response and autophagy. Front Microbiol. 2014;5(266):1–7. https://doi.org/10.3389/fmicb.2014.00266.

Boonyasuppayakorn S, Reichert ED, Manzano M, Padmanabhan R. Amodiaquine, an antimalarial drug, inhibits dengue virus type 2 replication and infectivity siwaporn. Antivir Res. 2015;106:125–34. https://doi.org/10.1016/j.antiviral.2014.03.014.

Braun E, Hotter D, Koepke L, Fackler OT, Kirchhoff F, Sauter D, Sparrer KMJ. Guanylate-binding proteins 2 and 5 exert broad antiviral activity by inhibiting furin-mediated processing of viral envelope proteins. Cell Rep. 2019;27:2092–104.

Braz-De-Melo HA, Pasquarelli-do-Nascimento G, Corrêa R, das Neves Almeida R, de Oliveira Santos I, Prado PS, Picolo V, de Bem AF, Pizato N, Magalhães KG. Potential neuroprotective and anti-inflammatory effects provided by Omega-3 (DHA) against Zika virus infection in human SH-SY5Y cells. Sci Rep. 2019;9(1):1–12. https://doi.org/10.1038/s41598-019-56556-y.

Bres P, Lacan A, Diop I, Michel R, Peretti P, Vidal C. Arboviruses in Senegal. A virological survey. [French]. Bull Soc Pathol Exotique. 1963.

Briant L, Desprès P, Choumet V, Missé D. Role of skin immune cells on the host susceptibility to mosquito-borne viruses. Virology. 2014. https://doi.org/10.1016/j.virol.2014.06.023.

Brown WC, Akey DL, Konwerski JR, Tarrasch JT, Skiniotis G, Kuhn RJ, Smith JL. Extended surface for membrane association in Zika virus NS1 structure. Nat Struct Mol Biol. 2016;23 (9):865–7. https://doi.org/10.1038/nsmb.3268.

Browning DJ. Hydroxychloroquine and chloroquine retinopathy. In: Hydroxychloroquine and chloroquine retinopathy; 2014. p. 35–63. https://doi.org/10.1007/978-1-4939-0597-3.

Bullard-Feibelman KM, Govero J, Zhu Z, Salazar V, Veselinovic M, Diamond MS, Geiss BJ. The FDA-approved drug sofosbuvir inhibits Zika virus infection. Antivir Res. 2017;137 (January):134–40. https://doi.org/10.1016/j.antiviral.2016.11.023.

Burns TM. Guillain-Barré syndrome. Semin Neurol. 2008. https://doi.org/10.1055/s-2008-1062261.

Calvet G, Aguiarv RS, Melo AS, Sampaio SA, de Filippis I, Fabri A, Araujo ES, et al. Case report of detection of Zika virus genome in amniotic fluid of affected fetuses: association with microcephaly outbreak in Brazil. Lancet Infect Dis. 2016. https://doi.org/10.1017/CBO9781107415324.004.

Carette JE, Raaben M, Wong AC, Herbert AS, Obernosterer G, Mulherkar N, Kuehne AI, et al. Ebola virus entry requires the cholesterol transporter niemann-pick C1. Nature. 2012;477 (7364):340–3. https://doi.org/10.1038/nature10348.

Carneiro BM, Batista MN, Braga ACS, Nogueira ML, Rahal P. The green tea molecule EGCG inhibits Zika virus entry. Virology. 2016;496:215–18. https://doi.org/10.1016/j.virol.2016.06.012.

de Carvalho OV, Félix DM, de Mendonça LR, de Araújo CMCS, de Oliveira Franca RF, Cordeiro MT, Júnior AS, Pena LJ. The thiopurine nucleoside analogue 6-methylmercaptopurine riboside (6MMPr) effectively blocks zika virus replication. Int J Antimicrob Agents. 2017;50(6):718–25. https://doi.org/10.1016/j.ijantimicag.2017.08.016.

Cataneo AH, Depieri DK, Koishi AC, Zanluca C, Silveira GF, Bonato T, de Arruda A, Suzukawa A, et al. The citrus flavonoid naringenin impairs the in vitro infection of human cells by Zika virus. Sci Rep. 2019;9(1):16348. https://doi.org/10.1038/s41598-019-52626-3.

Cauchemez S, Besnard M, Bompard P, Dub T, Guillemette-Artur P, Eyrolle-Guignot D, Salje H, et al. Association between Zika virus and microcephaly in French polynesia, 2013–15: a retrospective study. Lancet. 2016;387(10033):2125–32. https://doi.org/10.1016/S0140-6736(16)00651-6.

Cells H, Narayanan A, Kehn-hall K, Senina S, Lundberg L, Duyne RV, Guendel I, et al. Curcumin inhibits rift valley fever virus replication in human cells. J Biol Chem. 2012;287(40):33198–214. https://doi.org/10.1074/jbc.M112.356535.

Chan J-W, Zhu Z, Chu H, Yuan S, Chik K-H, Chan C-S, Poon V-M, et al. The celecoxib derivative kinase inhibitor AR-12 (OSU-03012) inhibits Zika virus via down-regulation of the PI3K/Akt pathway and protects Zika virus-infected A129 mice: a host-targeting treatment strategy. Antivir Res. 2018;160(December):38–47. https://doi.org/10.1016/j.antiviral.2018.10.007.

Chen JC, Wang Z, Huang H, Weitz SH, Wang A, Qiu X, Baumeister MA, Uzgiris A. Infection of human uterine fibroblasts by Zika virus in vitro: implications for viral transmission in women. Int J Infect Dis. 2016. https://doi.org/10.1016/j.ijid.2016.07.015.

Chen HH, Chen CC, Lin YS, Chang PC, Zi Yi Lu, Lin CF, Chen CL, Chang CP. AR-12 suppresses dengue virus replication by down-regulation of PI3K/AKT and GRP78. Antivir Res. 2017b;142:158–168. https://doi.org/10.1016/j.antiviral.2017.02.015.

Chen L, Liu Y, Wang S, Sun J, Wang P, Xin X, Qilin LZ, Xiao G, Wang W. Antiviral activity of peptide inhibitors derived from the protein E stem against Japanese encephalitis and Zika viruses. Antivir Res. 2017a;141:140–149. https://doi.org/10.1016/j.antiviral.2017.02.009.

Chen T-Y, Chen D-Y, Wen H-W, Jun-lin Ou, Chiou S-S, Chen J-M, Wong M-L, Hsu W-l. Inhibition of enveloped viruses infectivity by curcumin. PLoS ONE. 2013;8(5):e62482. https://doi.org/10.1371/journal.pone.0062482.

Chen W-C, Simanjuntak Y, Chu L-W, Ping Y-H, Lee Y-L, Lin Y-L, Li W-S. Benzenesulfonamide derivatives as calcium/calmodulin-dependent protein kinase inhibitors and antiviral agents against dengue and Zika virus infections. J Med Chem. 2020;63:1313–27.

Chen Y, Maguire T, Hileman RE, Fromm JR, Esko JD, Linhardt RJ, Marks RM. Dengue virus infectivity depends on envelope protein binding to target cell heparan sulfate. Nat Med. 1997;3(8):866–71. https://doi.org/10.1038/nm0897-866.

Chouin-Carneiro T, Vega-Rua A, Vazeille M, Yebakima A, Girod R, Goindin D, Dupont-Rouzeyrol M, Lourenço-de-Oliveira R, Failloux AB. Differential susceptibilities of aedes aegypti and aedes albopictus from the Americas to Zika virus. PLoS Negl Trop Dis. 2016. https://doi.org/10.1371/journal.pntd.0004543.

Chuang C-H, Chiou S-J, Cheng T-l, Wang Y-T. A molecular dynamics simulation study decodes the Zika virus NS5 methyltransferase bound to SAH and RNA analogue. Sci Rep. 2018;8:6336. https://doi.org/10.1038/s41598-018-24775-4.

Clark MJ, Miduturu C, Schmidt AG, Zhu X, Pitts JD, Wang J, Potisopon S, et al. GNF-2 inhibits dengue virus by targeting Abl Kinases and the viral e protein. Cell Chem Biol. 2016;23(4):443–52. https://doi.org/10.1016/j.chembiol.2016.03.010.

Coutard B, Barral K, Selisko B, Martin B, Aouadi W, Lombardia MO, Vasseur J-J, Guillemot JC, Canard B, Decroly E. Zika virus methyltransferase: structure and functions for drug design perspectives. J Virol. 2017;91(5):1–15.

Cumberworth SL, Barrie JA, Cunningham ME, Gomes DP, de Figueiredo V, Schultz AJ, Wilder-Smith BB, et al. Zika virus tropism and interactions in myelinating neural cell cultures: CNS cells and myelin are preferentially affected. Acta Neuropathol Commun. 2017. https://doi.org/10.1186/s40478-017-0450-8.

Dai L, Song J, Xishan Lu, Deng YQ, Musyoki AM, Cheng H, Zhang Y, et al. Structures of the Zika virus envelope protein and its complex with a flavivirus broadly protective antibody. Cell Host Microbe. 2016;19(5):696–704. https://doi.org/10.1016/j.chom.2016.04.013.

Dang Y, Yonggan Wu, Jia G, Anaya E, Abraham S, Choi J, Shi G, et al. A CRISPR-based screen identifies genes essential for West Nile virus-induced cell death. Cell Rep. 2015;12(4):673–83. https://doi.org/10.1016/j.celrep.2015.06.049.A.

Delvecchio R, Higa LM, Pezzuto P, Valadão AL, Garcez PP, Monteiro FL, Loiola EC, et al. Chloroquine, an endocytosis blocking agent, inhibits Zika virus infection in different cell models. Viruses. 2016;8(12):1–15. https://doi.org/10.3390/v8120322.

Dick GWA. Zika virus (I). Isolations and serological specificity. Trans Royal Soc Trop Med Hyg. 1952. https://doi.org/10.1016/0035-9203(52)90042-4.

Dong X, Moyer MM, Yang F, Sun YP, Yang L. Carbon dots' antiviral functions against noroviruses. Sci Rep. 2017;7(1):1–10. https://doi.org/10.1038/s41598-017-00675-x.

Duffy MR, Tai-Ho Chen W, Hancock T, Powers AM, Kool JL, Lanciotti RS, Pretrick M, et al. Zika virus outbreak on Yap Island, federated states of micronesia. N Engl J Med. 2009;360 (24):2536–43. https://doi.org/10.1056/NEJMoa0805715.

Dupont-Rouzeyrol M, Biron A, O'Connor O, Huguon E, Descloux E. Infectious Zika viral particles in breastmilk. The Lancet. 2016;387(10023):1051. https://doi.org/10.1016/S0140-6736(16)00624-3.

Dutta K, Ghosh D, Basu A. Curcumin protects neuronal cells from Japanese encephalitis virus-mediated cell death and also inhibits infective viral particle formation by dysregulation of ubiquitin—proteasome system. J Neuroimmune Pharmacol. 2009;4:328–37. https://doi.org/10.1007/s11481-009-9158-2.

ECDC, Rapid Risk Assessment. Rapid risk assessment Zika virus epidemic in the Americas: potential association with microcephaly and Guillain-Barré syndrome main conclusions disease background information Zika virus disease. ECDC. 2015.

Eddowes LA, Al-hourani K, Ramamurthy N, Frankish J, Baddock HT, Sandor C, Ryan JD, et al. Antiviral activity of bone morphogenetic proteins and activins. Nat Microbiol. 2019;4:339–51. https://doi.org/10.1038/s41564-018-0301-9.

Eijk AA, Der V, Van Genderen PJ, Verdijk RM, Reusken CB, Mögling R, Van Kampen JJA, Widagdo W, et al. Miscarriage associated with Zika virus infection. N Engl J Med. 2016. https://doi.org/10.1056/NEJMc1605898.

Elfiky AA, Ismail AM. Molecular docking revealed the binding of nucleotide/side inhibitors to Zika viral polymerase solved structures. SAR QSAR Environ Res. 2018;29(5):409–18. https://doi.org/10.1080/1062936X.2018.1454981.

Españo E, Nam JH, Song EJ, Song D, Lee CK, Kim JK. Lipophilic statins inhibit Zika virus production in vero cells. Sci Rep. 2019;9(1):1–11. https://doi.org/10.1038/s41598-019-47956-1.

Fangueiro JF, Calpena AC, Clares B, Andreani T, Egea MA, Veiga FJ, Garcia ML, Silva AM, Souto EB. Biopharmaceutical evaluation of epigallocatechin gallate-loaded cationic lipid nanoparticles (EGCG-LNs). In vivo, in vitro and ex vivo studies. Int J Pharm. 2016;502:161–9. https://doi.org/10.1016/j.ijpharm.2016.02.039.

Farias KJ, Silva PR, Machado L, Fonseca BALD. Chloroquine inhibits dengue virus type 2 replication in vero cells but not in C6/36 cells. Sci World J. 2013;2013. https://doi.org/10.1155/2013/282734.

Ferrero DS, Ruiz-Arroyo VM, Soler N, Usón I, Guarné A, Verdaguer N. Supramolecular arrangement of the full-length Zika virus NS5. Rey FA, editor. PLOS Pathog. 2019;15(4): e1007656. https://doi.org/10.1371/journal.ppat.1007656.

Fink SL, Vojtech L, Wagoner J, Slivinski NSJ, Jackson KJ, Wang R, Khadka S, Luthra P, Basler CF, Polyak SJ. The antiviral drug arbidol inhibits Zika virus. Sci Rep. 2018;8(1):8989. https://doi.org/10.1038/s41598-018-27224-4.

Foo SS, Chen W, Chan Y, Bowman JW, Chang LC, Choi Y, Yoo JS, et al. Asian Zika virus strains target CD14+ blood monocytes and induce M2-skewed immunosuppression during pregnancy. Nat Microbiol. 2017. https://doi.org/10.1038/s41564-017-0016-3.

Ganjhu KR, Mudgal PP, Maity H, Dowarha D, Devadiga S, Nag S, Arunkumar G. Herbal plants and plant preparations as remedial approach for viral diseases. Virus Dis. 2015;26(4):225–36. https://doi.org/10.1007/s13337-015-0276-6.

Ghezzi S, Cooper L, Rubio A, Pagani I, Rosaria M, Ippolito G, Pelletier J, et al. Heparin prevents Zika virus induced-cytopathic effects in human neural progenitor cells. Antivir Res. 2017;140:13–7. https://doi.org/10.1016/j.antiviral.2016.12.023.

Giulietti M, Righetti A, Cianfruglia L, Šabanović B, Armeni T, Principato G, Piva F. To accelerate the Zika beat: candidate design for RNA interference-based therapy. Virus Res. 2018;255 (August):133–40. https://doi.org/10.1016/j.virusres.2018.07.010.

Godoy AS, Lima GMA, Oliveira KIZ, Torres NU, Maluf FV, Guido RVC, Oliva G. Crystal structure of Zika virus NS5 RNA-dependent. Nat Commun. 2017;8:1–6. https://doi.org/10.1038/ncomms14764.

Gorshkov K, Shiryaev SA, Fertel S, Lin Y, Huang C. Zika virus: origins, pathological action, and treatment strategies. Front Microbiol. 2019;9 (3252). https://doi.org/10.3389/fmicb.2018.03252.

Grard G, Caron M, Mombo IM, Nkoghe D, Ondo SM, Jiolle D, Fontenille D, Paupy C, Leroy EM. Zika virus in gabon (Central Africa) - 2007: a new threat from aedes albopictus? PLoS Negl Trop Dis. 2014. https://doi.org/10.1371/journal.pntd.0002681.

Guerbois M, Fernandez-Salas I, Azar SR, Danis-Lozano R, Alpuche-Aranda CM, Leal G, Garcia-Malo IR, et al. Outbreak of Zika virus infection, Chiapas State, Mexico, 2015, and first confirmed transmission by aedes aegypti mosquitoes in the Americas. J Infect Dis. 2016;214 (9):1349–56. https://doi.org/10.1093/infdis/jiw302.

Ha SD, Ham B, Mogridge J, Saftig P, Lin S, Kim SO. Cathepsin B-mediated autophagy flux facilitates the anthrax toxin receptor 2-mediated delivery of anthrax lethal factor into the cytoplasm. J Biol Chem. 2010;285(3):2120–9. https://doi.org/10.1074/jbc.M109.065813.

Hamel R, Dejarnac O, Wichit S, Ekchariyawat P, Neyret A, Luplertlop N, Perera-Lecoin M, et al. Biology of Zika virus infection in human skin cells. J Virol. 2015;89(17):8880–96. https://doi.org/10.1128/jvi.00354-15.

Hasan SS, Sevvana M, Kuhn RJ, Rossmann MG. Structural biology of Zika virus and other flaviviruses. Nat Struct Mol Biol. 2018;25(January):13–20. https://doi.org/10.1038/s41594-017-0010-8.

Hastings AK, Hastings K, Uraki R, Hwang J, Gaitsch H, Dhaliwal K, Williamson E, Fikrig E. Loss of the TAM receptor Axl ameliorates severe Zika virus pathogenesis and reduces apoptosis in microglia. IScience. 2019;13:339–50. https://doi.org/10.1016/j.isci.2019.03.003.

Hayes EB. Zika virus outside Africa. 2009;15(9):2007–10. https://doi.org/10.3201/eid1509.090442.

Hengphasatporn K, Kungwan N, Rungrotmongkol T. Binding pattern and susceptibility of epigallocatechin gallate against envelope protein homodimer of Zika virus: a molecular dynamics study. J Mol Liq. 2019;274:140–7. https://doi.org/10.1016/j.molliq.2018.10.111.

Hilgenfeld R. Zika virus NS 1, a pathogenicity factor with many faces. EMBO J. 2016;35 (24):2631–33. https://doi.org/10.15252/embj.201695871.

Hitakarun A, Khongwichit S, Wikan N, Roytrakul S, Yoksan S, Rajakam S, Davidson AD, Smith DR. Evaluation of the antiviral activity of orlistat (tetrahydrolipstatin) against Dengue virus, Japanese encephalitis virus, Zika virus and Chikungunya virus. Sci Rep. 2020;10(1499). https://doi.org/10.1038/s41598-020-58468-8.

Holbrook M. Historical perspectives on flavivirus research. Viruses. 2017;9(5):97. https://doi.org/10.3390/v9050097.

Huang S, Jiangjiang Gu, Ye J, Fang B, Wan S, Wang C. Benzoxazine monomer derived carbon dots as a broad-spectrum agent to block viral infectivity. J Colloid Interface Sci. 2019;542:198–206. https://doi.org/10.1016/j.jcis.2019.02.0100021-9797.

Huang Y, Li Y, Zhang H, Zhao R, Jing R, Yinghua Xu, He M, et al. Zika virus propagation and release in human fetal astrocytes can be suppressed by neutral sphingomyelinase-2 inhibitor GW4869. Cell Discov. 2018;4(1):19. https://doi.org/10.1038/s41421-018-0017-2.

Ianevski A, Zusinaite E, Kuivanen S, Strand M, Lysvand H, Teppor M, Kakkola L, et al. Novel activities of safe-in-human broad-spectrum antiviral agents. 2018;154(April):174–82.

Iloni G, Vasam S, Guguloth V, Vadde R. One-pot multi-component synthesis of [1, 4] benzoxazine-isoxazole hybrids and their antibacterial activity. Int J Pharm Chem Biol Sci. 2018;8(1):118–24.

Irwin JJ, Shoichet BK. ZINC—a free database of commercially available compounds for virtual screening. J Chem Inf Model. 2005;45(1):177–82.

Julander JG, Siddharthan V, Evans J, Taylor R, Tolbert K, Apuli C, Stewart J, et al. Efficacy of the broad-spectrum antiviral compound BCX4430 against Zika virus in cell culture and in a mouse model. Antivir Res. 2017;137:14–22. https://doi.org/10.1016/j.antiviral.2016.11.003.

Kamiyama N, Soma R, Hidano S, Watanabe K, Umekita H, Fukuda C, Noguchi K, et al. Ribavirin inhibits Zika virus (ZIKV) replication in vitro and suppresses viremia in ZIKV-infected STAT1-deficient mice. Antivir Res. 2017;146:1–11. https://doi.org/10.1016/j.antiviral.2017.08.007.

Kang CB, Keller TH, Luo D. Zika virus protease: an antiviral drug target. Trends Microbiol. 2017;25(10):797–808. https://doi.org/10.1016/j.tim.2017.07.001.

Kazmirchuk T, Dick K, Burnside DJ, Barnes B, Moteshareie H, Hajikarimlou M, Omidi K, et al. Designing anti-Zika virus peptides derived from predicted human-Zika virus protein-protein interactions. Comput Biol Chem. 2017;71:180–7. https://doi.org/10.1016/j.compbiolchem.2017.10.011.

Ketkar H, Yang L, Wormser GP, Wang P. Lack of efficacy of ivermectin for prevention of a lethal Zika virus infection in a murine system. Diagn Microbiol Infect Dis. 2019;95(1):38–40. https://doi.org/10.1016/j.diagmicrobio.2019.03.012.

Khandelwal N, Chander Y, Dutt K, Riyesh T, Nishanth C, Sharma S, Jindal N, Tripathi BN, Barua S, Kumar N. Emetine inhibits replication of RNA and DNA viruses without generating drug-resistant virus variants. Antivir Res. 2017;144:196–204. https://doi.org/10.1016/j.antiviral.2017.06.006.

Kim B-H, Shenoy AR, Pradeep K, Bradfield CJ, MacMicking JD. IFN-inducible GTPases in host defense. Cell Host Microbe. 2012;12(4):432–44. https://doi.org/10.1016/j.chom.2012.09.007.

Kim K, Hwan K, Young H, Kook H, Sakamoto N, Cheong J. Curcumin inhibits hepatitis C virus replication via suppressing the Akt-SREBP-1 pathway. FEBS Lett. 2010;584:707–12. https://doi.org/10.1016/j.febslet.2009.12.019.

Kim SY, Koetzner CA, Payne AF, Nierode GJ, Yanlei Yu, Wang R, Barr E, et al. Glycosaminoglycan compositional analysis of relevant tissues in Zika virus pathogenesis and in vitro evaluation of heparin as an antiviral against Zika virus infection. Biochemistry. 2019;58:1155–66. https://doi.org/10.1021/acs.biochem.8b01267.

Kim SY, Zjao J, Liu X, Fraser K, Lin L, Zhang X, Zhang F, Dordick JS, Linhardt R. Interaction of Zika virus envelope protein with glycosaminoglycans. Biochemistry. 2017;56:1151–62. https://doi.org/10.1021/acs.biochem.6b01056.

King AMQ, Lefkowitz E, Adams MJ, Carstens EB. Flaviviridae-Ninth report of the international committee on taxonomy of viruses. Virus Taxon. 2012. https://doi.org/10.1016/B978-0-12-384684-6.00086-0.

Kiss R, Sandor M, Szalai FA. Http://Mcule.Com: a public web service for drug discovery. J Cheminform. 2012;4(S1):P17. https://doi.org/10.1186/1758-2946-4-s1-p17.

Kostyuchenko VA, Lim EXY, Zhang S, Fibriansah G, Ng T-S, Ooi JSG, Shi J, Lok S-M. Structure of the thermally stable Zika virus. Nature. 2016;533(7603):425–8. https://doi.org/10.1038/nature17994.

Krishnan MN, Ng A, Sukumaran B, Gilfoy FD, Uchil PD, Sultana H, Brass AL, et al. RNA interference screen for human genes associated with West Nile virus infection. Nature. 2008;11:242–5. https://doi.org/10.1016/S0213-005X(11)70040-4.

Kuivanen S, Bespalov MM, Nandania J, Ianevski A, Velagapudi V, De Brabander JK, Kainov DE, Vapalahti O. Obatoclax, saliphenylhalamide and gemcitabine inhibit Zika virus infection in vitro and differentially affect cellular signaling, transcription and metabolism. Antivir Res. 2017;139(2017):117–28. https://doi.org/10.1016/j.antiviral.2016.12.022.

Kumar D, Aarthy M, Kumar P, Singh SK, Uversky VN, Giri R. Targeting the NTPase site of Zika virus NS3 helicase for inhibitor discovery. J Biomol Struct Dyn. 2019;1102. https://doi.org/10.1080/07391102.2019.1689851.

Kusaka T, Yamamoto H, Shibata M, Muroi M, Kishi T, Mizuno K. Streptomyces citricolor Nov. SP. and a new antibiotic, aristeromycin. J Antibiot. 1968;21(4):255–63. https://doi.org/10.7164/antibiotics.21.255.

Lambert JD, Sang S, Hong J, Kwon S-J, Lee M-J, Ho C-T, Yang CS, Al LET. Peracetylation as a means of enhancing in vitro bioactivity and bioavailability of epigallocatechin-3-gallate ABSTRACT. Drug Metab Dispos. 2006;34:2111–6. https://doi.org/10.1124/dmd.106.011460.small.

Le Lee J, Loe MWC, Lee RCH, Chu JJH. Antiviral activity of pinocembrin against Zika virus replication. Antivir Res. 2019;167(March):13–24. https://doi.org/10.1016/j.antiviral.2019.04.003.

Leyssen P, De Clercq E, Neyts J. Perspectives for the treatment of infections with flaviviridae. Clin Microbiol Rev. 2000;13(1):67–82. https://doi.org/10.1128/CMR.13.1.67-82.2000.

Li C, Deng Y. 25-hydroxycholesterol protects host against Zika virus infection and its associated microcephaly in a article 25-hydroxycholesterol protects host against Zika virus infection and its associated microcephaly in a mouse model. Immunity. 2017;46:446–56. https://doi.org/10.1016/j.immuni.2017.02.012.

Li C, Zhu X, Ji X, Quanquin N, Deng YQ, Tian M, Aliyari R, et al. Chloroquine, a FDA-approved drug, prevents Zika virus infection and its associated congenital microcephaly in mice. EBio Med. 2017;24:189–94. https://doi.org/10.1016/j.ebiom.2017.09.034.

Li F, Lee EM, Sun X, Wang D, Tang H, Zhou G. Design, synthesis and discovery of andrographolide derivatives against Zika virus infection. Eur J Med Chem. 2020;187:111925. https://doi.org/10.1016/j.ejmech.2019.111925.

Li P-C, Jang J, Hsia C-Y, Groomes PV, Lian W, de Wispelaere M, Pitts JD, et al. Small molecules targeting the flavivirus E protein with broad-spectrum activity and antiviral efficacy in vivo. ACS Infect Dis. 2019;5(3):460–72. https://doi.org/10.1021/acsinfecdis.8b00322.

Lian W, Jang J, Potisopon S, Li P-C, Rahmeh A, Wang J, Kwiatkowski NP, Gray NS, Yang PL. Discovery of immunologically inspired small molecules that target the viral envelope protein. ACS Infect Dis. 2018;4:1395–406. https://doi.org/10.1021/acsinfecdis.8b00127.

Lim SP, Noble CG, Shi P-Y. The dengue virus NS5 protein as a target for drug discovery. Antivir Res. 2015;119:57–67. https://doi.org/10.1016/j.antiviral.2015.04.010.

Liu SY, Sanchez DJ, Aliyari R, Lu S, Cheng G. Systematic identification of type I and type II interferon-induced antiviral factors. PNAS. 2012;109(11):4239–44. https://doi.org/10.1073/pnas.1114981109.

Loe MW, Choy RC, Lee H, Chu JJH. Antiviral activity of the FDA-approved drug candesartan cilexetil against Zika virus infection. Antivir Res. 2019;172(June):104637. https://doi.org/10.1016/j.antiviral.2019.104637.

MacNamara FN. Zika virus: a report on three cases of human infection during an epidemic of jaundice in Nigeria. Trans R Soc Trop Med Hyg. 1954;48(2):139–45. https://doi.org/10.1016/0035-9203(54)90006-1.

Mairiang D, Zhang H, Sodja A, Murali T, Suriyaphol P, Malasit P, Limjindaporn T, Finley RL. Identification of new protein interactions between dengue fever virus and its hosts, human and mosquito. PLoS ONE. 2013;8(1):e53535. https://doi.org/10.1371/journal.pone.0053535.

Makhluf H, Kim K, Shresta S. Novel strategies for discovering inhibitors of dengue and Zika fever. Expert Opin Drug Discov. 2016;11(10):921–3. https://doi.org/10.1080/17460441.2016.1212013.

Marchette NJ, Garcia R, Rudnick A. Isolation of Zika virus from aedes aegypti mosquitoes in Malaysia. Am J Trop Med Hyg. 1969. https://doi.org/10.4269/ajtmh.1969.18.411.

Martín-Acebes MA, Vázquez-Calvo Á, Saiz J-C. Lipids and flaviviruses, present and future perspectives for the control of Dengue, Zika, and West Nile viruses. Prog Lipid Res. 2016;64:123–37. https://doi.org/10.1016/j.plipres.2016.09.005.

Martinez-Lopez A, Persaud M, Puray M, Zhang H, Rong L, Liu S, Wang TT, Sara SG, Diaz-griffero F. Glycosylated diphyllin as a broad-spectrum antiviral agent against Zika virus. EBio Med. 2019;47:269–83. https://doi.org/10.1016/j.ebiom.2019.08.060.

Mazumder A, Raghavan K, Weinstein J, Kohn KW, Pommier Y. Inhibition of human immunodeficiency virus type-1 integrase by curcumin. Biochem Pharmacol. 1995;49(8):1165–70.

Meertens L, Labeau A, Dejarnac O, Cipriani S, Sinigaglia L, Bonnet-Madin L, Le Charpentier T, et al. Axl mediates ZIKA virus entry in human glial cells and modulates innate immune responses. Cell Rep. 2017;18(2):324–33. https://doi.org/10.1016/j.celrep.2016.12.045.

Miner JJ, Diamond MS. Zika virus pathogenesis and tissue tropism. Cell Host Microbe. 2017;21(2):134–42. https://doi.org/10.1016/j.chom.2017.01.004.

Mlakar J, Korva M, Tul N, Popović M, Poljšak-Prijatelj M, Mraz J, Kolenc M, et al. Zika virus associated with microcephaly. N Engl J Med. 2016;374(10):951–8. https://doi.org/10.1056/NEJMoa1600651.

Moghadasian MH. Clinical pharmacology of 3-hydroxy-3-methylglutaryl coenzyme a reductase inhibitors. Life Sci. 1999;65(13):1329–37. https://doi.org/10.1016/S0024-3205(99)00199-X.

Mohr EL, McMullan LK, Lo MK, Spengler JR, Bergeron É, Albariño CG, Shrivastava-Ranjan P, et al. Inhibitors of cellular kinases with broad-spectrum antiviral activity for hemorrhagic fever viruses. Antivir Res. 2015;120:40–7. https://doi.org/10.1016/j.antiviral.2015.05.003.

Montes-Grajales D, Puerta-Guardo H, Espinosa DA, Harris E, Caicedo-Torres W, Olivero-Verbel J, Martínez-Romero E. In silico drug repurposing for the identification of potential candidate molecules against arboviruses infection. Antivir Res. 2020;173(November 2019):104668. https://doi.org/10.1016/j.antiviral.2019.104668.

Morcos PA, Li Y, Jiang S. Vivo-Morpholinos: a non-peptide transporter delivers morpholinos into a wide array of mouse tissues short technical reports. Biotechniques. 2008;45:613–23. https://doi.org/10.2144/000113005.

Moreira J, Peixoto TM, Siqueira AM, Lamas CC. Sexually acquired Zika virus: a systematic review. Clin Microbiol Infect. 2017;23(5):296–305. https://doi.org/10.1016/j.cmi.2016.12.027.

Motta IJF, Spencer BR, Suely G, da Silva C, Arruda MB, Dobbin JA, Gonzaga YBM, Arcuri IP, et al. Evidence for transmission of Zika virus by platelet transfusion. N Engl J Med. 2016;375 (11):1101–3. https://doi.org/10.1056/NEJMc1607262.

Moulton JD, Shan J. Gene knockdowns in adult animals: PPMOs and vivo-morpholinos. Molecules. 2009;14(3):1304–23. https://doi.org/10.3390/molecules14031304.

Mounce BC, Cesaro T, Carrau L, Vallet T, Vignuzzi M. Curcumin inhibits Zika and Chikungunya virus infection by inhibiting cell binding. Antivir Res. 2017;142:148–57. https://doi.org/10.1016/j.antiviral.2017.03.014.

Müller JA, Harms M, Krüger F, Groß R, Joas S, Hayn M, Dietz AN, et al. Semen inhibits Zika virus infection of cells and tissues from the anogenital region. Nat Commun. 2018;9(1):2207. https://doi.org/10.1038/s41467-018-04442-y.

Musso D, Roche C, Robin E, Nhan T, Teissier A, Cao-Lormeau V-M. Potential sexual transmission of Zika virus. Emerg Infect Dis. 2015;21(2):359–61. https://doi.org/10.3201/eid2102.141363.

Neris RLS, Figueiredo CM, Higa LM, Araujo DF, Carvalho CAM, Verçoza BRF, Silva MOL, et al. Co-protoporphyrin IX and Sn-protoporphyrin IX inactivate Zika, Chikungunya and other arboviruses by targeting the viral envelope. Sci Rep. 2018;8:9805. https://doi.org/10.1038/s41598-018-27855-7.

Nitsche C. Proteases from Dengue, West Nile and Zika viruses as drug targets. Biophys Rev. 2019;11(2):157–65. https://doi.org/10.1007/s12551-019-00508-3.

Oliveira Melo AS, Malinger G, Ximenes R, Szejnfeld PO, Alves Sampaio S, Bispo de Filippis AM. Zika virus intrauterine infection causes fetal brain abnormality and microcephaly: tip of the iceberg? Ultrasound Obstet Gynecol. 2016;47(1):6–7. https://doi.org/10.1002/uog.15831.

Olson JG, Ksiazek TG, Suhandiman G, Triwibowo V. Zika virus, a cause of fever in central Java, Indonesia. Trans R Soc Trop Med Hyg. 1981. https://doi.org/10.1016/0035-9203(81)90100-0.

Onawole AT, Sulaiman KO, Adegoke RO, Kolapo TU. Identification of potential inhibitors against the Zika virus using consensus scoring. J Mol Graph Modell. 2017;73:54–61. https://doi.org/10.1016/j.jmgm.2017.01.018.

Palanichamy K, Joshi A, Tugba Mehmetoglu-gurbuz M, Bravo F, Shlain MA, Schiro F, Naeem Y, Garg H, Braunschweig AB. Anti-Zika activity of a library of synthetic carbohydrate receptors. J Med Chem. 2019;62:4110–9. https://doi.org/10.1021/acs.jmedchem.9b00142.

Pan American Health Organisation. Neurological syndrome, congenital malformations, and Zika virus infection. Implications for Public Health in the Americas. Pan American Health Organization. 2015.

Pattnaik A, Palermo N, Sahoo BR, Yuan Z, Duoyi Hu, Annamalai AS, Vu HLX, et al. Discovery of a non-nucleoside RNA polymerase inhibitor for blocking Zika virus replication through in silico screening. Antivir Res. 2018;151:78–86. https://doi.org/10.1016/j.antiviral.2017.12.016.

Persaud M, Martinez-lopez A, Cindy Bu, Porcelli SA, Diaz-gri F. Infection by Zika viruses requires the transmembrane protein AXL, endocytosis and low PH. Virology. 2018;518:301–12. https://doi.org/10.1016/j.virol.2018.03.009.

Phoo WW, Zhang Z, Wirawan M, Chew EJC, Chew ABL, Kouretova J, Steinmetzer T, Luo D. Structures of Zika virus NS2B-NS3 protease in complex with peptidomimetic inhibitors. Antivir Res. 2018;160(August):17–24. https://doi.org/10.1016/j.antiviral.2018.10.006.

Pitts JD, Li P-C, De Wispelaere M, Yang PL. Antiviral activity of N-(4-hydroxyphenyl) retinamide (4-HPR) against Zika virus. Antivir Res. 2017;147:124–30. https://doi.org/10.1016/j.antiviral.2017.10.014.

Popik W, Khatua A, Hildreth JEK, Lee B, Alcendor DJ. Phosphorodiamidate morpholino targeting the 5′ untranslated region of the ZIKV RNA inhibits virus replication. Virology. 2018;519(April):77–85. https://doi.org/10.1016/j.virol.2018.04.001.

Prasad VM, Miller AS, Klose T, Sirohi D, Buda G, Jiang W, Kuhn RJ, Rossmann MG. Structure of the immature Zika virus at 9 Å resolution. Nat Struct Mol Biol. 2017;24(2):184–6. https://doi.org/10.1038/nsmb.3352.

Priya S, Senthil KN, Hemalatha S. Antiviral phytocompounds target envelop protein to control Zika virus. Comput Biol Chem. 2018;77:402–12. https://doi.org/10.1016/j.compbiolchem.2018.08.008.

Quek JP, Liu S, Zhang Z, Li Y, Yihui E, Ying Ru, Hung AW, Luo D, Kang C. Identification and structural characterization of small molecule fragments targeting Zika virus NS2B-NS3 protease. Antivir Res. 2020;175:104707. https://doi.org/10.1016/j.antiviral.2020.104707.

Quintana VM, Selisko B, Brunetti JE, Eydoux C, Guillemot JC, Canard B, Damonte EB, Julander JG, Castilla V. Antiviral activity of the natural alkaloid anisomycin against Dengue and Zika viruses. Antivir Res. 2020;176(January):104749. https://doi.org/10.1016/j.antiviral.2020.104749.

Ramharack P, Soliman MES. Zika virus NS5 protein potential inhibitors: an enhanced in silico approach in drug discovery. J Biomol Struct Dyn. 2018;36(5):1118–33. https://doi.org/10.1080/07391102.2017.1313175.

Rattanaburee T, Junking M, Panya A, Sawasdee N, Songprakhon P, Suttitheptumrong A, Limjindaporn T, Haegeman G, Yenchitsomanus PT. Inhibition of Dengue virus production and cytokine/chemokine expression by Ribavirin and compound A. Antivir Res. 2015;124:83–92. https://doi.org/10.1016/j.antiviral.2015.10.005.

Rausch K, Hackett BA, Weinbren NL, Reeder SM, Sadovsky Y, Hunter CA, Schultz DC, Coyne CB, Cherry S. Screening bioactives reveals nanchangmycin as a broad spectrum antiviral active against Zika virus. Cell Rep. 2017;18(3):804–15. https://doi.org/10.1016/j.celrep.2016.12.068.

Reis RAG, Calil FA, Feliciano PR, Pinheiro MP, Cristina Nonato M. The dihydroorotate dehydrogenases: past and present. Arch Biochem Biophys. 2017;632:175–91. https://doi.org/10.1016/j.abb.2017.06.019.

Renaudet J, Jan C, Ridet J, Adam C, Robin Y. Enquête Sérologique Pour Les Arbovirus Dans La Population Humaine Du Sénégal. A serological survey of arboviruses in the human population of senegal. Bull Soc Pathol Exotique Ses Filiales. 1978.

Richard AS, Shim B-S, Kwon Y-C, Zhang R, Otsuka Y, Schmitt K, Berri F, Diamond MS, Choe H. AXL-dependent infection of human fetal endothelial cells distinguishes Zika virus from other pathogenic flaviviruses. Proc Natl Acad Sci. 2017;114(8):2024–9. https://doi.org/10.1073/pnas.1620558114.

Rieth S, Miner MR, Chang CM, Hurlocker B, Braunschweig AB. Saccharide receptor achieves concentration dependent mannoside selectivity through two distinct cooperative binding pathways. Chem Sci. 2013;4:357–67. https://doi.org/10.1039/c2sc20873c.

Robin Y, Mouchet J. A serologic and entomologic survey on yellow fever in Sierra Leone. Bull Soc Pathol Exotique Ses Filiales. 1975.

Röcker AE, Müller JA, Dietzel E, Harms M, Krüger F, Heid C, Sowislok A, et al. The molecular Tweezer CLR01 inhibits Ebola and Zika virus infection. Antivir Res. 2018;152(April):26–35. https://doi.org/10.1016/j.antiviral.2018.02.003.

Rothan HA, Zhong Y, Sanborn MA, Teoh TC, Ruan J, Yusof R, Hang J, Henderson MJ, Fang S. Small molecule Grp94 inhibitors block Dengue and Zika virus replication. Antivir Res. 2019;171(November):104590. https://doi.org/10.1016/j.antiviral.2019.104590.

Ruan J, Rothan HA, Zhong Y, Yan W, Henderson MJ, Chen F, Fang S. A small molecule inhibitor of ER-to-cytosol protein dislocation exhibits anti-Dengue and anti-Zika virus activity. 2019; July:1–12.

Ryan JD, Ryan E, Fabre A, Lawless MW, Crowe J. Defective bone morphogenic protein signaling underlies hepcidin deficiency in HFE hereditary hemochromatosis. Hepatology. 2010;52 (4):1266–73. https://doi.org/10.1002/hep.23814.

Saavedra LM, Romanelli GP, Rozo CE, Duchowicz PR. The quantitative structure-insecticidal activity relationships from plant derived compounds against Chikungunya and Zika aedes aegypti (diptera: culicidae) vector. Sci Total Environ. 2018;610–611(January):937–43. https://doi.org/10.1016/j.scitotenv.2017.08.119.

Sacramento CQ, de Melo GR, de Freitas CS, Rocha N, Hoelz LVB, Miranda M, Fintelman-Rodrigues N, et al. The clinically approved antiviral drug sofosbuvir inhibits Zika virus replication. Sci Rep. 2017;7(1):40920. https://doi.org/10.1038/srep40920.

Samsa MM, Mondotte JA, Iglesias NG, Assunção-Miranda I, Barbosa-Lima G, Da Poian AT, Bozza PT, Gamarnik AV. Dengue virus capsid protein usurps lipid droplets for viral particle formation. PLoS Pathog. 2009;5(10). https://doi.org/10.1371/journal.ppat.1000632.

Santos FRS, Nunes DAF, Lima WG, Davyt D, Santos LL, Taranto AG, Jaqueline Ferreira MS. Identification of Zika virus NS2B-NS3 protease inhibitors by structure-based virtual screening and drug repurposing approaches. J Chem Inf Model. 2020b;60(2):731–737. https://doi.org/10.1021/acs.jcim.9b00933.

Santos FRS, Lima WG, Maia EHB, Assis C, Davyt D, Taranto AG, Ferreira JMS. Identification of a potential Zika virus inhibitor targeting NS5 methyltransferase using virtual screening and molecular dynamics simulations. J Chem Inform Model. 2020a. https://doi.org/10.1021/acs.jcim.9b00809.

Sarukhanyan E, Shityakov S, Dandekar T. In silico designed Axl receptor blocking drug candidates against Zika virus infection. ACS Omega. 2018;3(5):5281–90. https://doi.org/10.1021/acsomega.8b00223.

Scaturro P, Stukalov A, Haas DA, Cortese M, Draganova K, Płaszczyca A, Bartenschlager R, Götz M, Pichlmair A. An orthogonal proteomic survey uncovers novel Zika virus host factors. Nature. 2018;561(7722):253–7. https://doi.org/10.1038/s41586-018-0484-5.

Schmaljohn AL, McClain D. Alphaviruses (togaviridae) and flaviviruses (flaviviridae). Med Microbiol. 1996.

Schmidt AG, Lee K, Yang PL, Harrison SC. Small-molecule inhibitors of dengue-virus entry. PLoS Pathog. 2012;8(4). https://doi.org/10.1371/journal.ppat.1002627.

Šebera J, Dubankova A, Sychrovský V, Ruzek D, Boura E, Nencka R. The structural model of Zika virus RNA-dependent RNA polymerase in complex with RNA for rational design of novel nucleotide inhibitors. Sci Rep. 2018;8(1):11132. https://doi.org/10.1038/s41598-018-29459-7.

Sevvana M, Long F, Miller AS, Klose T, Buda G, Sun L, Kuhn RJ, Rossmann MG. Refinement and analysis of the mature Zika virus cryo-EM structure at 3.1 Å resolution. Structure. 2018;26 (9):1169–1177.e3. https://doi.org/10.1016/j.str.2018.05.006.

Shang Z, Song H, Shi Yi, Qi J, Gao GF. Crystal structure of the capsid protein from Zika virus. J Mol Biol. 2018;430(7):948–62. https://doi.org/10.1016/j.jmb.2018.02.006.

Sharma N, Murali A, Singh SK, Giri R. Epigallocatechin gallate, an active green tea compound inhibits the Zika virus entry into host cells via binding the envelope protein. Int J Biol Macromol. 2017;104:1046–54. https://doi.org/10.1016/j.ijbiomac.2017.06.105.

Shi Y, Gao GF. Structural biology of the Zika virus. Trends Biochem Sci. 2017;42(6):443–56. https://doi.org/10.1016/j.tibs.2017.02.009.

Shiryaev SA, Mesci P, Pinto A, Fernandes I, Sheets N, Shresta S, Farhy C, et al. Repurposing of the anti-Malaria drug chloroquine for Zika virus treatment and prophylaxis. Sci Rep. 2017;7 (15771). https://doi.org/10.1038/s41598-017-15467-6.

Si X, Wang Y, Wong J, Zhang J, Mcmanus BM, Luo H. Dysregulation of the ubiquitin-proteasome system by curcumin suppresses coxsackievirus B3 replication. J Virol. 2007;81(7):3142–50. https://doi.org/10.1128/JVI.02028-06.

da Silva-Júnior EF, Leoncini GO, Rodrigues ÉES, Aquino TM, Araújo-Júnior JX. The medicinal chemistry of Chikungunya virus. Bioorg Med Chem. 2017;25(16):4219–44. https://doi.org/10.1016/j.bmc.2017.06.049.

da Silva-Júnior EF, de Araújo-Júnior JX. Peptide derivatives as inhibitors of NS2B-NS3 protease from Dengue, West Nile, and Zika flaviviruses. Bioorg Med Chem. 2019;27(18):3963–78. https://doi.org/10.1016/j.bmc.2019.07.038.

da Silva-Júnior EF, Schirmeister T, Araújo-Júnior JX. Recent advances in inhibitors of flavivirus NS2B-NS3 protease from Dengue, Zika, and West Nile viruses. In: Vector-borne diseases and treatment; 2018. p. 1–25.

Simanjuntak Y, Liang JJ, Lee YL, Lin YL. Japanese encephalitis virus exploits dopamine D2 receptor-phospholipase C to target dopaminergic human neuronal cells. Front Microbiol. 2017;8(April):1–12. https://doi.org/10.3389/fmicb.2017.00651.

Singh A, Jana NK. Discovery of potential Zika virus RNA polymerase inhibitors by docking-based virtual screening. Comput Biol Chem. 2017;71:144–51. https://doi.org/10.1016/j.compbiolchem.2017.10.007.

Singh PK, Singh S, Farr D, Kumar A. Interferon-stimulated gene 15 (ISG15) restricts Zika virus replication in primary human corneal epithelial cells. Ocular Surf. 2019;17(3):551–9. https://doi.org/10.1016/j.jtos.2019.03.006.

Sinigaglia A, Riccetti S, Trevisan M, Barzon L. In silico approaches to Zika virus drug discovery. Expert Opin Drug Discov. 2018;13(9):825–35. https://doi.org/10.1080/17460441.2018.1515909.

Sintchak MD, Nimmesgern E. The structure of inosine 5'-monophosphate dehydrogenase and the design of novel inhibitors. Immunopharmacology. 2000;47(2–3):163–84. https://doi.org/10.1016/S0162-3109(00)00193-4.

Sirohi D, Chen Z, Sun L, Tomas Klose TC, Pierson MG, Rossmann, R. J. Kuhn. The 3.8 Å resolution cryo-EM structure of Zika virus. Science. 2016;352(6284):467–70. https://doi.org/10.1126/science.aaf5316.

Sirohi D, Kuhn RJ. Zika virus structure, maturation, and receptors. J Infect Dis. 2017;216 (Suppl_10):S935–44. https://doi.org/10.1093/infdis/jix515.

Sørensen MG, Henriksen K, Neutzsky-wulff AV, Dziegiel MH, Karsdal MA. Diphyllin, a novel and naturally potent V-ATPase inhibitor, abrogates acidification of the osteoclastic resorption lacunae and bone resorption. J Bone Miner Res. 2007;22(10):1640–8. https://doi.org/10.1359/JBMR.070613.

Stadler K, Allison SL, Schalich J, Heinz FX. Proteolytic activation of tick-borne encephalitis virus by furin. J Virol. 1997;71(11):8475–81. https://doi.org/10.1128/JVI.71.11.8475-8481.1997.

Steinmann J, Buer J, Pietschmann T, Steinmann E. Anti-infective properties of (EGCG), a component of green tea. Br J Pharmacol. 2013;168:1059–73. https://doi.org/10.1111/bph.12009.

Low SY, June, Chen KC, Wu KX, Ng MM-L, Chu JJH. Antiviral activity of emetine dihydrochloride against dengue virus infection. J Antivir Antiretrovir. 2009;01(01):062–071. https://doi.org/10.4172/jaa.1000009.

Sun J, Niu Y, Wang C, Zhang H, Xie B, Fengrong Xu, Jin H, Peng Y, Liang L, Ping Xu. Discovery of 3-benzyl-1,3-benzoxazine-2,4-dione analogues as allosteric mitogen-activated kinase kinase (MEK) inhibitors and anti-enterovirus 71 (EV71) agents. Bioorg Med Chem. 2016;24(16):3472–82. https://doi.org/10.1016/j.bmc.2016.05.055.

Surnar B, Kamran MZ, Shah AS, Basu U, Kolishetti N, Deo S, Jayaweera DT, Daunert S, Dhar S. Orally administrable therapeutic synthetic nanoparticle for Zika virus. ACS Nano. 2019;13 (10):11034–48. https://doi.org/10.1021/acsnano.9b02807.

Suroengrit A, Yuttithamnon W, Srivarangkul P, Pankaew S. Halogenated chrysins inhibit Dengue and Zika virus infectivity. Sci Rep. 2017;7(13696). https://doi.org/10.1038/s41598-017-14121-5.

Tan TY, Fibriansah G, Kostyuchenko VA, Ng TS, Lim XX, Zhang S, Lim XN, et al. Capsid protein structure in Zika virus reveals the flavivirus assembly process. Nat Commun. 2020;11(1):1–13. https://doi.org/10.1038/s41467-020-14647-9.

Tao Z, Cao R, Yan Y, Huang G, Lv K, Li W, Geng Y, et al. Design, synthesis and in vitro anti-Zika virus evaluation of novel sinefungin derivatives. Eur J Med Chem. 2018;157:994–1004. https://doi.org/10.1016/j.ejmech.2018.08.057.

Tian H, Ji X, Yang X, Xie W, Yang K, Chen C, Chen Wu, et al. The crystal structure of Zika virus helicase: basis for antiviral drug design. Protein Cell. 2016;7(6):450–4. https://doi.org/10.1007/s13238-016-0275-4.

Ting Du, Dong N, Fang L, Jian Lu, Bi J, Xiao S, Han H. Multisite inhibitors for enteric coronavirus: antiviral cationic carbon dots based on curcumin. ACS Appl Nano Mater. 2018;1(10):5451–9. https://doi.org/10.1021/acsanm.8b00779.

Tognarelli J, Ulloa S, Villagra E, Lagos J, Aguayo C, Fasce R, Parra B, et al. A report on the outbreak of Zika virus on Easter Island, South Pacific, 2014. Adv Virol. 2016. https://doi.org/10.1007/s00705-015-2695-5.

Tong X, Smith J, Bukreyeva N, Koma T, Manning JT, Kalkeri R, Kwong AD, Paessler S. Merimepodib, an IMPDH inhibitor, suppresses replication of Zika virus and other emerging viral pathogens. Antivir Res. 2018;149:34–40. https://doi.org/10.1016/j.antiviral.2017.11.004.

Upadhyay AK, Cyr M, Longenecker K, Tripathi R, Sun C, Kempf DJ. Crystal structure of full-length Zika virus NS5 protein reveals a conformation similar to Japanese encephalitis virus NS5. Acta Crystallogr Sect F Struct Biol Commun. 2017;73:116–22. https://doi.org/10.1107/S2053230X17001601.

Vasireddi M, Crum A, May H, Katz D, Hilliard J. A novel antiviral inhibits Zika virus infection while increasing intracellular glutathione biosynthesis in distinct cell culture models. Antivir Res. 2019;161:46–52. https://doi.org/10.1016/j.antiviral.2018.09.004.

Vassiliou AG, Maniatis NA, Orfanos SE, Mastora Z, Jahaj E, Paparountas T, Armaganidis A, Roussos C, Aidinis V, Kotanidou A. Induced expression and functional effects of aquaporin-1 in human leukocytes in sepsis. Crit Care. 2013;17(R199). https://doi.org/10.1186/cc12893.

Vestal DJ, Jeyaratnam JA. The guanylate-binding proteins: emerging insights into the biochemical properties and functions of this family of large interferon-induced guanosine triphosphatase. J Interferon Cytokine Res. 2011;31(1):89–97. https://doi.org/10.1089/jir.2010.0102.

Voss S, Nitsche C. Inhibitors of the Zika virus protease NS2B-NS3. Bioorg Med Chem Lett. 2020;30(5):126965. https://doi.org/10.1016/j.bmcl.2020.126965.

Wang H, Li Z, Niu J, Xu Y, Ma L, Lu A, Wang X, et al. Antiviral effects of ferric ammonium citrate. Cell Discov. 2018;4(14). https://doi.org/10.1038/s41421-018-0013-6.

Warren TK, Wells J, Panchal RG, Stuthman KS, Garza NL, Van Tongeren SA, Dong L, et al. Protection against filovirus diseases by a novel broad-spectrum nucleoside analogue BCX4430. Nature. 2014;508:402–5. https://doi.org/10.1038/nature13027.

de Wispelaere M, Lian W, Potisopon S, Li PC, Jang J, Ficarro SB, Clark MJ, et al. Inhibition of flaviviruses by targeting a conserved pocket on the viral envelope protein. Cell Chem Biol. 2018;25(8):1006-1016.e8. https://doi.org/10.1016/j.chembiol.2018.05.011.

Xu M, Lee EM, Wen Z, Cheng Y, Huang WK, Qian X, Tcw J, et al. Identification of small-molecule inhibitors of Zika virus infection and induced neural cell death via a drug repurposing screen. Nat Med. 2016;22(10):1101–7. https://doi.org/10.1038/nm.4184.

Yadav R, Selvaraj C, Aarthy M, Kumar P, Kumar A, Singh SK, Giri R. Investigating into the molecular interactions of flavonoids targeting NS2B-NS3 protease from ZIKA virus through in-silico approaches. J Biomol Struct Dyn. 2020:1–13. https://doi.org/10.1080/07391102.2019.1709546.

Yamaguchi K, Honda M, Ikigai H, Hara Y. Inhibitory effects of (−)-epigallocatechin gallate on the life cycle of human immunodeficiency virus type 1 (HIV-1). Antivir Res. 2002;53:19–34.

Yang CF, Gopula B, Liang JJ, Li JK, Chen SY, Lee YL, Chen CS, Lin YL. Novel AR-12 derivatives, P12–23 and P12–34, inhibit flavivirus replication by blocking host de novo pyrimidine biosynthesis. Emerg Microbes Infect. 2018a;7(1):187. https://doi.org/10.1038/s41426-018-0191-1.

Yang S, Xu M, Lee EM, Gorshkov K, Shiryaev SA, He S, Sun W, et al. Emetine inhibits Zika and Ebola virus infections through two molecular mechanisms: inhibiting viral replication and decreasing viral entry. Cell Discov. 2018b;4(31). https://doi.org/10.1038/s41421-018-0034-1.

Yang J, Xu Y, Yan Y, Li W, Zhao L, Dai Q, Li Y, et al. Small molecule inhibitor of ATPase activity of HSP70 as a broad-spectrum inhibitor against flavivirus infections. ACS Infect Dis. 2020. acsinfecdis.9b00376. https://doi.org/10.1021/acsinfecdis.9b00376.

Yang Y, Cao L, Gao H, Yue Wu, Wang Y, Fang F, Lan T, Lou Z, Rao Yu. Discovery, optimization, and target identification of novel potent broad-spectrum antiviral inhibitors. J Med Chem. 2019;62:4056–73. https://doi.org/10.1021/acs.jmedchem.9b00091.

Yang Y, Yu Y, Li X, Li J, Wu Y, Yu J, Ge J, et al. Target elucidation by cocrystal structures of NADH-ubiquinone oxidoreductase of plasmodium falciparum (Pf NDH2) with small molecule to eliminate drug-resistant Malaria. 2017. https://doi.org/10.1021/acs.jmedchem.6b01733.

Yebra MJ, Sanchez J, Martin CG, Hardisson C, Barbes C. The effect of sinefungin and synthetic analogues on RNA and DNA methyltransferases from streptomyces. J Antibiot. 1991;44 (10):1141–7. https://doi.org/10.7164/antibiotics.44.1141.

Yin Z, Chen YL, Schul W, Wang QY, Feng Gu, Duraiswamy J, Kondreddi RR, et al. An adenosine nucleoside inhibitor of Dengue virus. PNAS. 2009;106(48):20435–9. https://doi.org/10.1073/pnas.0907010106.

Yoon J-S, Kim G, Jarhad DB, Kim H-R, Shin Y-S, Shuhao Qu, Sahu PK, et al. Design, synthesis, and anti-RNA virus activity of 6′-fluorinated-aristeromycin analogues. J Med Chem. 2019;63:6346–62. https://doi.org/10.1021/acs.jmedchem.9b00781.

Yuan S, Cao S, Jiang R, Liu R, Bai J, Hou Qi. FLLL31, a derivative of curcumin, attenuates airway in Fl ammation in a multi-allergen challenged mouse model. Int Immunopharmacol. 2014;21:128–36. https://doi.org/10.1016/j.intimp.2014.04.020.

Zanluca C, Campos V, De Melo A, Luiza A, Mosimann P, Igor G, Nunes C, Luz K. First report of autochthonous transmission of Zika virus in Brazil. 2015;110(June):569–72. https://doi.org/10.1590/0074-02760150192.

Zhang S, Yi C, Li C, Zhang F, Peng J, Wang Q, Liu X, et al. Chloroquine inhibits endosomal viral RNA release and autophagy-dependent viral replication and e Ff ectively prevents maternal to fetal transmission of Zika virus. Antivir Res. 2019;169:104547. https://doi.org/10.1016/j.antiviral.2019.104547.

Zhang Z, Li Y, Loh YR, Phoo WW, Hung AW, Kang CongBao, Luo D. Crystal structure of unlinked NS2B-NS3 protease from Zika virus. Science. 2016;354(6319):1597–600. https://doi.org/10.1126/science.aai9309.

Zhong Y, Fang S. Live cell imaging of protein dislocation from the endoplasmatic reticulum. J Biol Chem. 2012;287(33):28057–66. https://doi.org/10.1074/jbc.M112.381798.

Zhu S, Zhang C, Huang LS, Zhang XQ, Xu Y, Fang X, Zhou J, et al. Discovery and computational analyses of novel small molecule Zika virus inhibitors. Molecules. 2019;24(8). https://doi.org/10.3390/molecules24081465.

Zhu Y-Z, Xu Q-Q, Wu D-G, Ren H, Zhao P, Lao W-G, Wang Y, et al. Japanese encephalitis virus enters rat neuroblastoma cells via a PH-dependent, dynamin and caveola-mediated endocytosis pathway. J Virol. 2012;86(24):13407–22. https://doi.org/10.1128/JVI.00903-12.

Zilbermintz L, Leonardi W, Jeong S, Sjodt M, Mccomb R, Ho CC, Retterer C, Gharaibeh D, Zamani R, Soloveva V. Identification of agents effective against multiple toxins and viruses by host-oriented cell targeting. Sci. Rep. 2015;5(13476). https://doi.org/10.1038/srep13476.

Human Papillomavirus

Human Papillomavirus and Its Role in the Development of Cancer

Sevic Ina, Romano Lucia, Cristina Carolina, and Alaniz Laura

Abstract Tumor-associated immune microenvironment triggered by the infection with high risk Human Papillomavirus (HPV) is well accepted today as one of the most important determinants for the development of HPV-related cervical cancer. HPV has developed different immune-evasion strategies that can escape the immune system and establish chronic infection leading to chronic inflammation. The immune response, even though it may abolish tumor, it may also further assist carcinogenesis by different signalling pathways. Even so, this is not enough for the development of the tumor. Viral oncoproteins E6 and E7 are actively involved in this process. They are known to be able to abolish cell cycle regulation, apoptosis and DNA repair systems and as a result aid the establishment of cancer. We discuss here the HPV life cycle and its involvement in the carcinogenesis process.

Keywords Human Papillomavirus · Tumor microenvironment · Carcinogenesis · Oncoproteins · Cervical cancer · E6 protein · E7 protein · Persistent infection · Inflammatory environment · High risk HPV

S. Ina (✉) · A. Laura
Laboratorio de Microambiente Tumoral, Centro de Investigaciones Básicas Y Aplicadas (CIBA), CIT NOBA, Universidad Nacional del Noroeste de La Pcia. de Bs. As. Consejo Nacional de Investigaciones Científicas Y Técnicas (UNNOBA-CONICET), Pcia. Bs. As, Junín, Argentina
e-mail: isevic@comunidad.unnoba.edu.ar

A. Laura
e-mail: ldalaniz@comunidad.unnoba.edu.ar

R. Lucia · C. Carolina
Laboratorio de Fisiopatología de La Hipófisis, Centro de Investigaciones Básicas Y Aplicadas (CIBA), CIT NOBA, Universidad Nacional del Noroeste de La Pcia. de Bs. As. Consejo Nacional de Investigaciones Científicas Y Técnicas (UNNOBA-CONICET), Pcia. Bs. As, Junín, Argentina
e-mail: mluciaromano8@gmail.com

C. Carolina
e-mail: carolina.cristina@nexo.unnoba.edu.ar

S. I. Ahmad (ed.), *Human Viruses: Diseases, Treatments and Vaccines*, https://doi.org/10.1007/978-3-030-71165-8_14

1 Introduction

It has been well accepted that the development of different types of cancer from the inflammatory environment might be a process driven by inflammatory cells on one side and a variety of mediators, including cytokines, chemokines and enzymes on the other side (Mangino et al. 2016). Chronic inflammation plays a significant role in the process of carcinogenesis and it can be caused by variety of different sources, like alcohol abuse or chronic infection with hepatitis B virus (HBV) or hepatitis C virus (HCV) in hepatocarcinogenesis and Human papillomavirus (HPV) infection in cervical cancer (Sevic et al. 2019; Graham 2017). Nevertheless, even though persistent infection is an important event, it may not be sufficient to progress into tumorigenesis fully. In the case of some viruses, several viral proteins are proven to be vital for the development of cancer. In HPV infection, the crucial change necessary for progression to cancer is the increase in expression of viral oncoproteins E6 and E7 in dividing, infected cells (Graham 2017). This chapter will provide an overview of the HPV replication and its involvement in carcinogenesis process.

2 Human Papillomavirus (HPV) Classification

Human papillomavirus (HPV) is a small, double-stranded DNA virus that has a tropism for epithelial cells and propagates itself in a specific epithelial niche. HPVs constitute a large family of viruses that have been divided into groups in various manners. According to the tissue of origin they have been divided into cutaneous and mucosal types. The mucosal HPVs belong to types defined as high risk for the development of cancer whereas the association of cutaneous HPV types with carcinogenesis is still under investigation (Flores and Lambert 1997; Mangino et al. 2016).

More than 200 HPV types have been described by now and are classified into five groups based on DNA sequences and tropism (Alpha, Beta, Gamma and Nu/Mu, papillomavirus). Each group shows some differences in the life cycle of the virus and its tropism (Doorbar et al. 2012; de Sanjose et al. 2018). De Villiers et al. described the advances in HPV classification and showed that in 2013 there had been 66 types described in genus Alpha papillomavirus, 45 in the genus Beta Papilloma virus and 54 in genus Gamma Papilloma virus (de Villiers 2013).With the arrival of new techniques more types are being described every year and these numbers are getting much higher, reaching 64 types reported in 2017 for alpha Papilloma virus. The Alpha genus includes genotypes that have been mostly associated with cancer, while Beta and Gamma genus are more associated with infections that are generally a asymptomatic, but in the state of immune-suppression (like HIV infection) these types can lead to the production of cutaneous Papilloma. Persistent infection with some Beta types is associated with the development of non-melanoma skin cancers, particularly in immune-suppressed patients. The Gamma and Nu/Mu types infect the cutaneous epithelium and are typically associated with the formation of benign papillomas (de Sanjose et al. 2018; Moody 2017).

On the other hand, HPV has been divided into two groups, low and high risk based on the risk they present for the development of cancer. The low risk group (like HPV6 or HPV11) produces benign lesions or warts, while the high risk group is associated with 90% of cervical cancers (Flores and Lambert 1997; de Sanjose et al. 2018). Among these, the HPV-16 and HPV-18 subtypes are the most prevalent in cancers and are considered to be responsible for approximately 70% of cervical cancers. At the same time, the rest is being associated with the other subtypes of the high risk group (like HPV-31, -33, -35, -39, -45, -51, -52, -56, -58, -59, -68, -73, and -82). Additionally, HPV-16 is associated with approximately 95% of HPV-positive oropharyngeal cancers (Berman and Schiller 2017; Stanley 2019). Nevertheless, most of HPV infections do not cause symptoms and are cleared in a period of 12–24 months post infection. Only a small portion of those infections persists and can progress to a preneoplastic lesion (de Sanjose et al. 2018).

3 HPV Infection Epidemiology and Diagnostics

As we previously mentioned, most cervical HPV infections are asymptomatic and more than 90% of infections clear within 24 months (Berman and Schiller 2017). However, cervical cancer is still a considerable burden worldwide being the fourth most common cancer in women. It was estimated that globally there is approximately 500,000 new diagnoses each year and amongst them, squamous cell carcinoma (SCC) is the most common histological type. Nearly all cervical cancers of this type are HPV-related. Adenocarcinomas account for approximately 25% of all cervical cancers and have much more heterogeneous origin, with around 15% being unrelated to HPV infection (Park 2020; Berman and Schiller 2017; Wakeham et al. 2017). Also, high risk HPV is now accepted as the principal cause of oropharyngeal squamous cell carcinoma in some parts of the world (Taberna et al. 2017).

HPV, unlike some other viruses, cannot easily be diagnosed by serology. This is due to the adaptive immune response being much slower than in other viruses, the antibody titters that are produced are low, and it has been reported that between 30 and 50% of infected women seem never to acquire antibodies (Frazer 2009; Pattyn et al. 2019). One of the possible explanations for this is that there is practically no viremia phase. Infection occurs entirely in the epithelium of the cervical mucosa. Women who develop antibodies do so between 8 and 12 months after the initial infection, but this depends on the individual and the HPV type. The majority of the developed antibodies are to the L1 protein, which does not necessarily protect against successive infections (Pattyn et al. 2019; World-Health-Organization 2009).

Antibodies can be determined by ELISA, using virus particles as a substrate; or by neutralization assays using virions and cells susceptible to infection (Frazer 2009; Pattyn et al. 2019; World-Health-Organization 2009). Regardless, as stated above, these tests can be used for population exposure evaluation and for the study of immunogenicity of vaccines (World-Health-Organization 2010). For the diagnosis of individual infection, it should be considered that only molecular methods directly detect HPV. The rest of the methods are dedicated to the detection of precancerous or cancerous lesions (Centers-for-Disease-Control-and-Prevention 2019).

There are three initial screening methods:
Molecular method that detects HPV DNA: The HPV screening test is extremely sensitive for detecting HPV infection in women. It is not recommended to apply these methods to women under 30 years of age because many young women are infected with these types of HPV, but most of these infections will clear spontaneously before they reach 30 years of age. It is recommended to perform these tests in 30–49 years old women (World-Health-Organization 2014).

Visual inspection method with two methodologies: Visual inspection with acetic acid and visual inspection with iodine from Lugol. In the first one a solution of acetic acid of 3–5% is used and the cervix is illuminated with a light source. The purpose is to identify the areas bleached by acetic acid, which may indicate that the tissue is undergoing precancerous changes. The second method involves the observation of the cervix after applying Lugol iodine to detect lesions. Both techniques require experienced medical personnel (World-Health-Organization 2014; Sankaranarayanan Wesley et al. 1998).

Cytological screening method: It is performed with the conventional Pap smear. This method uses a sample of cells taken from the cervix during a gynecological examination with a speculum. For a Pap smear, the sample is spread on a slide, fixed and stained, and then examined under a microscope. If abnormal epithelial cells are found in the cytological study, a positive result is reported, which does not necessarily indicate precancerous lesions (World-Health-Organization 2014; Berman and Schiller 2017).

On the other hand, as confirmatory tests for the diagnosis of precancerous lesions, methods like colposcopy, biopsy and endocervical curettage are currently performed (World-Health-Organization 2014; Centers-for-Disease-Control-and-Prevention 2019).

Colposcopy: It makes use of a colposcope to examine cervix, vagina and vulva for signs of disease. It allows the examination of the specific characteristics of the epithelial layer and the surrounding blood vessels. More recently, this test has also been done using specially designed video or digital cameras. This test is usually combined with targeted biopsies of all lesions where neoplasia is suspected (Torres García et al. 2007).

Biopsy: is the removal of small samples of abnormal tissue for microscopic examination. Special tweezers are required for biopsy and training is needed. The biopsy is used to determine the degree of abnormality of cervical cell changes and to rule out cancer. After the examination, a biopsy is classified as normal, as cervical intraepithelial neoplasia, or as an invasive carcinoma (Torres García et al. 2007; World-Health-Organization 2014).

Endocervical curettage. It is only recommended in some cases in which there is a strong suspicion of the lesion, but it is not observed in colposcopy or biopsies.

A spatula or curette, adaptable to the dimension of the neck, is used to scrape it and obtain a mucous material that contains cells. Staining and observation are then performed by trained personnel to find the lesion (Torres García et al. 2007).

To be noted that in different countries different diagnostic algorithms is applied depending on socioeconomic realities. However, they are all based on the previously mentioned methodologies.

4 HPV Structure

HPV is a small non-enveloped virus that has a circular double-stranded DNA genome, approximately 8 kb long. The genome organization of HPV16 is typical of the high-risk Alpha papillomavirus. It is divided into three main regions (Fig. 1). The early region (E) encodes genes that are necessary in different stages of the virus life cycle (E1–E7). These proteins are crucial for viral genome replication, virion synthesis and release, but, they also play a fundamental role in cell transformation and development of cancer. Second region is the late region (L), which encodes

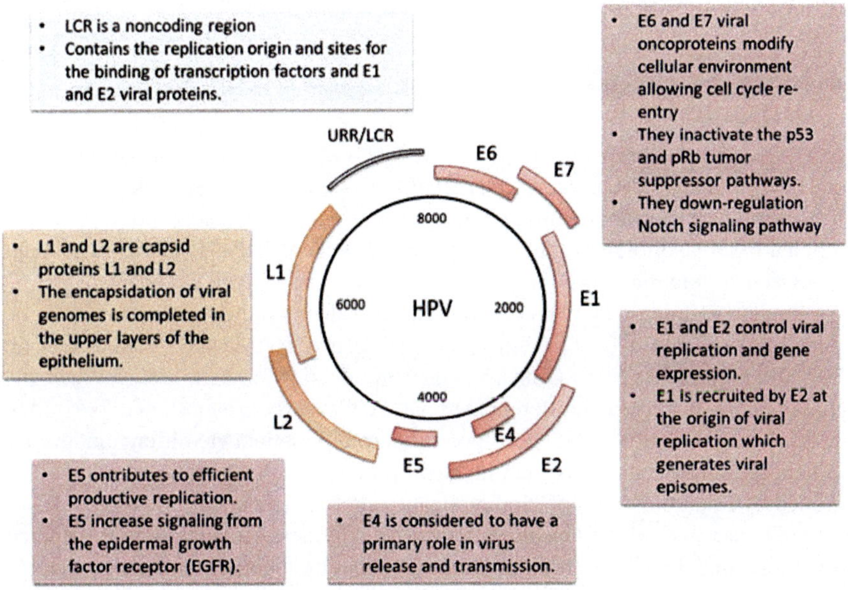

Fig. 1 Genome organisation and protein functions of HPV16. HPV16 is a typical model for the genome organization of high risk HPV. Genome is divided into three main regions. The early region (E) encodes genes that are necessary in different stages of the virus life cycle while late region (L) encodes capsid proteins. The third region is a non-coding region called upstream regulatory region (URR) or the long control region (LCR) which contributes to the regulation of DNA replication by controlling the viral gene transcription

capsid proteins L1 and L2 (de Sanjose et al. 2018; Doorbar et al. 2012). The most immunogenic protein of pathogenic HPVs is L1 protein, which forms the virion. The genome is encapsidated with 360 L1 protein units organized as 72 pentamers which probably have in the center one unit of L2 (Frazer 2009; Graham 2017). The third region is the upstream regulatory region (URR), also known as the long control region (LCR). This is a noncoding region that contains the replication origin and sites for the binding of transcription factors (like SP1, AP1, Oct1) and also for the E1 and E2 viral proteins that control replication and gene expression. This region contributes to the regulation of DNA replication by controlling the viral gene transcription and has post-transcriptional control via the LRE (late regulatory element) (de Sanjose et al. 2018; Graham 2017).

High-risk HPV genomes have two main promoters which are activated at different stages in the viral life cycle. In undifferentiated epithelial cells the early promoter which is situated adjacent to the E6 ORF in the URR is active and regulates viral gene expression. After epithelial differentiation the late promoter which is located in the E7 ORF is activated. This promoter leads the expression of the L1 and L2 genes, which allows the encapsidation of viral genomes to be completed in the upper layers of the epithelium. In HPV16, these two promoter elements are named PE (early promoter or p97) and PL (late promoter or p670), which regulate the expression of differentially-spliced mRNAs during epithelial differentiation (Doorbar et al. 2012; Moody 2017).

5 HPV Life Cycle

The HPV replicative cycle starts when HPV enters the basal layer of the epithelium where viral particles arrive through the wounds and infect keratinocytes of the epithelial basal layer. The productive replicative cycle of HPVs is highly dependent to host cell differentiation. HPV enters cell by an endocytosis mechanism that is most similar to micropinocytosis. After the initial entry the virus starts genome replication to create a pool of episomal viral genomes (Martinez-Ramirez et al. 2018; Woodby et al. 2016; Graham 2017; Flores and Lambert 1997). When the initial infection occurs in undifferentiated basal cells, early proteins are expressed in low levels (E1, E2, E6, and E7) postponing normal keratinocyte differentiation. E1 is a helicase/ATPase and it is the only viral protein with enzymatic activity. E1 is recruited by E2 at the origin of viral replication and this generates 50–100 copies of viral episomes per cell. When the infected cell divides, viral DNA is distributed between the daughter cells and it continues replicating together with cellular DNA. In this manner viral genomes are maintained in undifferentiated basal cells in a stable copy number. (Martinez-Ramirez et al. 2018; Stubenrauch and Laimins 1999).

During the epithelium differentiation, when the productive phase of the viral life cycle is activated, these numbers increase to thousands of DNA copies. In the middle and the upper differentiating epithelial layers, the high amounts of E6 and

E7 are being expressed. The expression of these two proteins is important for the viral life cycle primarily because they modify the cellular environment to help genome amplification by allowing cell cycle re-entry. These two proteins are very important for the high risk types as they seem also to drive cell proliferation in the basal and parabasal layers. On the other hand for the low risk HPV types (like HPV6 and HPV11), the exact role of these two proteins in the infected basal cells is not clear (Moody 2017; Egawa et al. 2015). High risk HPVs also express E5 protein that contributes to efficient productive replication. E5 protein is a small membrane protein, predominantly found in Golgi complex. One of the most described molecular activities of this protein is to increase signalling from the epidermal growth factor receptor (EGFR). The E2 protein is a transcription factor which regulates viral gene expression and has also been considered to be able to control some cellular promoters (Moody 2017; Woodby et al. 2016).

The viral life cycle is completed with the expression of L1 and L2 proteins in the upper layer of the epithelium, after which the viral genome is encapsidated and the mature virions are being released (Martinez-Ramirez et al. 2018; Doorbar et al. 2012).

Genome encapsidation includes the recruitment of L2 protein to the replication regions through E2 protein, before the expression of L1. Virus maturation takes place in the most superficial, dying keratinocytes, where the oxidizing environment allows the accumulation of disulphide bonds between L1 proteins. This leads to the production of very stable infectious virions. The E4 protein accumulates in very high levels in infected cells and is considered to have a primary role in virus release and transmission. In high risk types, this is considered to be driven possibly by E4 protein assembly into amyloid fibrils that disrupt keratin structure and compromise the normal assembly of the cornified envelope (Egawa et al. 2015; Doorbar et al. 2012).

When a complete viral life cycle occurs, it still does not imply a high risk for cancer development as most of these infections, even if produced by high risk types, is abolished within two years. The persistent infection that leads to the inflammatory environment, genomic instability and in many cases to viral genome integration into the host DNA is considered the main determining factor for cervical cancer development. It is well accepted for this cancer type that HPV integration allows larger expression of oncogenes E6 and E7 which, consecutively, aids oncogenesis (Taberna et al. 2017). Integration also can lead to oncogenesis by the inactivation of the E2 protein expression and the disruption of host genes due to the viral sequence insertion. If E2 protein is no longer expressed it cannot repress the P97 promoter and as a consequence it cannot restrict E6 and E7 expression (de Sanjose et al. 2018; Graham 2017). Tandem integration of the viral genome into the host DNA can also provoke the formation of super-enhancer-like elements which can additionally activate E6 and E7 protein expression. However, integration is not considered to be a part of the normal HPV replication cycle (Martinez-Ramirez et al. 2018; Graham 2017).

6 HPV-Induced Carcinogenesis

Microenvironment plays a very important role in the development of tumor. As we previously mentioned, HPV infection on its own in not enough for carcinogenesis. Some reports show that estrogen signaling in the stromal tumor microenvironment is associated with cervical cancer maintenance and progression. Spurgeon et al. have shown that the HPV oncogenes and estrogen together strongly alter host gene expression in the cervical stroma and propose that the pithelial-stromal crosstalk in cervical carcinogenesis is bidirectional (Spurgeon et al. 2017). Stromal cells as well as cytokines and chemokines they produce could be considered as factors that contribute to HPV-associated carcinogenesis because of the cross-talk between these cells and epithelial, tumor and immune cells that coexist the tumor microenvironment (Barros et al. 2018). One of the most important roles of stromal cells in carcinogenesis is to induce a proinflammatory milieu which in turn is a suitable condition for establishment of HPV transformed cells (Barros et al. 2018).

The interplay between chronic inflammation and oxidative stress induces different tissue alterations enabling HPV integration, which, as was previously mentioned, is one of the key events in the malignant transformation (Georgescu et al. 2018). The inflammation process involves reactive oxygen species, cytokines, chemokines, growth factors and enzymes (like metalloproteinases), which can induce changes in the proliferation processes as well as senescence and cell death. These factors can also lead to the generation of mutations and methylation of DNA and can stimulate angiogenesis. Westrich et al. described that E7 protein interacts with the DNA methyltransferase and stimulates its activity which could partially explain the changes in the host methylome that they have observed in HPV-positive keratinocytes (Westrich et al. 2017; Sadri Nahand et al. 2020).

Macrophages are one the most common immune cell in the skin and, like fibroblasts, can be activated in the presence of tumor cells forming tumor associated macrophages (TAMs). TAMs are known to be able to promote tumor growth and angiogenesis by secreting growth factors (like EGF and VEGF) and by releasing anti-inflammatory cytokines (like IL10 and TGFβ) (Woodby et al. 2016), with a phenotype like macrophages type 2 (M2). There is evidence which suggest that a high number of TAMs intumor microenvironment is strongly associated with viral infection. Chen et al. suggest that there is a critical role for CD163+ and CD68+ macrophages in HPV infection and show a strong association between the malignant transformation of cervical tissue and the increase in the number of these two types of macrophages (Chen et al. 2017). IL-6 which is necessary for the pro-tumorigenic response in monocytes is suppressed by HPV during productive infection while it is greatly up-regulated during later stages of cervical carcinogenesis. It has been shown that M2 macrophage differentiation is driven by IL-6 derived from cervical cancer together with prostagl and in E2 (Smola 2017). Also, the virus is able to escape the immune surveillance cells, mostly Langerhans cells, dermal myeloid dendritic cells and memory T cells, through the uncoordinated production of pro-inflammatory cytokines with the synthesis of IL-6 in combination with the inhibition of IL-1 release (Mangino et al. 2016).

HPVs show genotype-specific tropisms, infection type and pathologies. Different HPVs have become well adapted to their own epithelial niches and some of them, like those from Alpha papillomavirus genus, developed sophisticated immune evasion strategies allowing them to cause persistent infection, even in immune competent individuals (Egawa et al. 2015). Some studies proposed as one of the immune evasion strategies the decreased activity of keratinocytes. This might be due to viral oncoproteins being able to disarrange the expression of some genes and transcription factors (like NF-κB) in keratinocytes, like those involved in antigen presentation or cell communication (Barros et al. 2018).

E6 and E7 viral oncogenes are also implicated in down-regulation of Notch signaling pathway that in turn is involved in the control of stemness. In fact, it has been reported that inhibition of expression of E6 leads to loss of stemness of cancer stem cells in cervical cancer, possibly through down-regulation of downstream gene of Notch pathway, Hes1 (Gupta et al. 2018).

One of the most important factors for HPV-associated carcinogenesis is the ability of E6 and E7 proteins to inactivate the p53 and pRb (retinoblastoma) tumor suppressor pathways. The E6 of high risk HPV associates with the p53 protein and recruits ubiquitin-protein ligase which works as a ubiquitin ligase for a complex containing p53 and as a consequence p53 undergoes rapid degradation. Similarly, E7 binds to pRb resulting in the degradation and functional inactivation of pRb which leads to uncontrolled cell cycle, proliferation and progression (Stubenrauch and Laimins 1999; Hoppe-Seyler et al. 2018) (Fig. 2).

7 HPV Vaccines and Treatment

There is a prophylactic HPV-VLP (virus like particle) vaccine that is made by the synthesis and self-assembly of the major virus capsid protein L1 in vitro whose structure is geometrically and antigenically almost identical to virion. This particles are non-infectious and don't contain viral DNA. Currently available vaccines of this type are a bivalent HPV16/18, a quadrivalent HPV6/11/16/18 and a noncovalent HPV 6, 11, 16, 18, 31, 33, 45, 52, 58. These vaccines are highly effective and are mainly recommended for 9–14 year-old girls (Stanley 2019; Bednarczyk 2019).

In the case of the early stage cervical cancer the recommended treatment is surgical and often includes hysterectomy. Most cervical cancers are radiosensitive and can be treated with primary radiation therapy. Simultaneous chemotherapy with either cisplatin alone or with 5-fluorouracil is frequently administered with external-beam radiotherapy. Brachytherapy is a treatment that is frequently used for patients with advanced cervical cancer and is usually used in combination with external beam radiotherapy. Significantly advanced stages of cervical cancer are

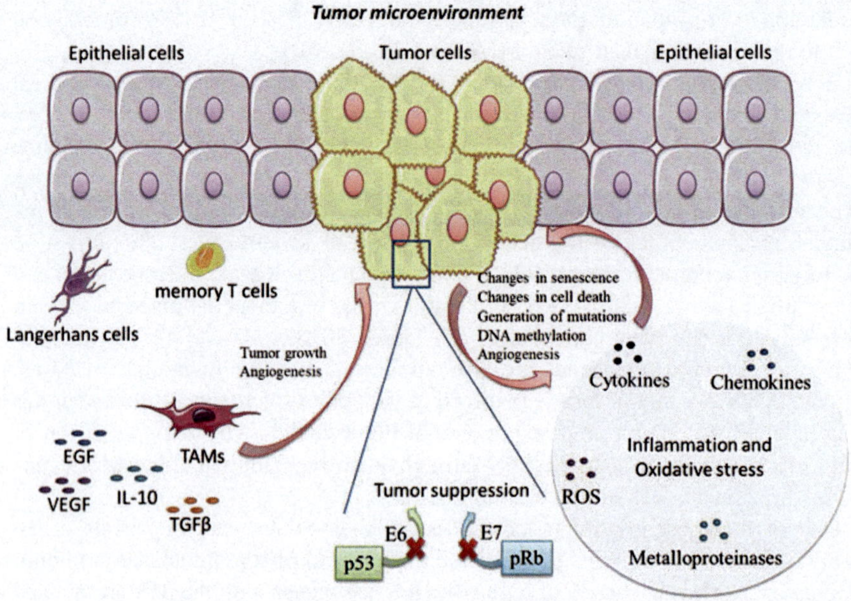

Fig. 2 Tumor microenvironment in HPV infection. HPV is able to escape the immune surveillance cells (like Langerhans cells and memory T cells) and establish infection. Chronic infection often leads to chronic inflammation. The inflammation process that includes ROS, cytokines, chemokines, growth factors and enzymes (like metalloproteinases) can induce changes in the proliferation processes, senescence, cell death, generation of mutations and methylation of DNA and can stimulate angiogenesis. Additionally, viral proteins E6 and E7 inactivate p53 and pRb tumor suppressor pathways. On the other hand TAMs are able to promote tumor growth and angiogenesis by secreting growth factors (like EGF and VEGF) and anti-inflammatory cytokines (like IL-10 and TGFβ). This complex interplay can lead to establishment and progression of tumor. ROS: Reactive oxygen species; TAMs: Tumor associated macrophages; EGF: Epidermal growth factor; VEGF: Vascular endothelial growth factor; IL-10: Interleukin 10; TGFβ: Transforming growth factor beta; pRb: Retinoblastoma protein

characterized by early treatment failure and a median survival of 7–12 months (Berman and Schiller 2017; Small et al. 2017). Taking into account that these standard therapies are not very efficient for advanced stages of cervical cancer, there have been many efforts for the development of immunotherapies that could provide better results. One of the promising treatments are DNA vaccines based on anti-tumor T cells which are the key players in host immune response against tumor. Multiple vaccines based on this principle are currently being designed and are on clinical trials (Chauhan and Bharadwaj 2018).

8 Conclusion

Despite the development of effective prophylactic vaccines cervical cancer is still 4th most common cancer in women. Immune-evasion strategies developed by HPV together with chronic inflammation unchains a great number of events that lead to deregulation of the cell cycle and apoptosis and disorganization of DNA repair systems and as a result aid the establishment of cancer. Further studies about these mechanisms will lead to a better understanding of the disease development and treatment failure.

References

Barros MR, JR, De Melo CML, Barros M, De Cassia Pereira De Lima R, De Freitas AC, Venuti A. Activities of stromal and immune cells in HPV-related cancers. J Exp Clin Cancer Res. 2018;37:137.

Bednarczyk RA. Addressing HPV vaccine myths: practical information for healthcare providers. Hum Vaccin Immunother. 2019;15:1628–38.

Berman TA, Schiller JT. Human papillomavirus in cervical cancer and oropharyngeal cancer: one cause, two diseases. Cancer. 2017;123:2219–29.

Centers-For-Disease-Control-And-Prevention. HPV-Associated Cancer Statistics; 2019. https://www.cdc.gov/cancer/hpv/statistics/.

Chauhan SR, Bharadwaj M. Gearing up T-cell immunotherapy in cervical cancer. Curr Probl Cancer. 2018;42:175–88.

Chen XJ, Han LF, Wu XG, Wei WF, Wu LF, Yi HY, … Wang W. Clinical significance of CD163 + and CD68+ tumor-associated macrophages in high-risk HPV-related cervical cancer. J Cancer. 2017;8:3868–75.

De Sanjose S, Brotons M, Pavon MA. The natural history of human papillomavirus infection. Best Pract Res Clin Obstet Gynaecol. 2018;47:2–13.

De Villiers EM. Cross-roads in the classification of papillomaviruses. Virology. 2013;445:2–10.

Doorbar J, Quint W, Banks L, Bravo IG, Stoler M, Broker TR, Stanley MA. The biology and life-cycle of human papillomaviruses. Vaccine. 2012;30(Suppl 5):F55-70.

Egawa N, Egawa K, Griffin H, Doorbar J. Human papillomaviruses; epithelial tropisms, and the development of neoplasia. Viruses. 2015;7:3863–90.

Flores ER, Lambert PF. Evidence for a switch in the mode of human papillomavirus type 16 DNA replication during the viral life cycle. J Virol. 1997;71:7167–79.

Frazer IH. Interaction of human papillomaviruses with the host immune system: a well evolved relationship. Virology. 2009;384:410–4.

Georgescu SR, Mitran CI, Mitran MI, Caruntu C, Sarbu MI, Matei C, … Tampa M. New Insights in the Pathogenesis of HPV Infection and the Associated Carcinogenic Processes: The Role of Chronic Inflammation and Oxidative Stress. J Immunol Res. 2018; 5315816.

Graham SV. The human papillomavirus replication cycle, and its links to cancer progression: a comprehensive review. Clin Sci (Lond). 2017;131:2201–21.

Gupta S, Kumar P, Das BC. HPV: Molecular pathways and targets. Curr Probl Cancer. 2018;42:161–74.

Hoppe-Seyler K, Bossler F, Braun JA, Herrmann AL, Hoppe-Seyler F. The HPV E6/E7 oncogenes: key factors for viral carcinogenesis and therapeutic targets. Trends Microbiol. 2018;26:158–68.

Mangino G, Chiantore MV, Iuliano M, Fiorucci G, Romeo G. Inflammatory microenvironment and human papillomavirus-induced carcinogenesis. Cytokine Growth Factor Rev. 2016;30:103–11.

Martinez-Ramirez I, Carrillo-Garcia A, Contreras-Paredes A, Ortiz-Sanchez E, Cruz-Gregorio A, Lizano M. Regulation of cellular metabolism by high-risk human papillomaviruses. Int J Mol Sci. 2018;19.

Moody C. Mechanisms by which HPV induces a replication competent environment in differentiating keratinocytes. Viruses. 2017;9.

Park KJ. Cervical adenocarcinoma: integration of HPV status, pattern of invasion, morphology and molecular markers into classification. Histopathology. 2020;76:112–27.

Pattyn J, Van Keer S, Tjalma W, Matheeussen V, Van Damme P., Vorsters A. Infection and vaccine-induced HPV-specific antibodies in cervicovaginal secretions. A review of the literature. Papillomavirus Res. 2019;8:100185.

Sadri Nahand J, Moghoofei M, Salmaninejad A, Bahmanpour Z, Karimzadeh M, Nasiri M, … Hamblin MR. Pathogenic role of exosomes and microRNAs in HPV-mediated inflammation and cervical cancer: a review. Int J Cancer. 2020;146:305–20.

Sankaranarayanan R, Wesley R, Somanathan T, Dhakad N, Shyamalakumary B, Amma NS, … Nair MK. Visual inspection of the uterine cervix after the application of acetic acid in the detection of cervical carcinoma and its precursors. Cancer. 1998;83:2150–6.

Sevic I, Spinelli FM, Cantero MJ, Reszegi A, Kovalszky I, Garcia MG, Alaniz L. The role of the tumor microenvironment in the development and progression of hepatocellular carcinoma. In: Tirnitz-Parker JEE (ed.) Hepatocellular Carcinoma. Brisbane (AU); 2019.

Small W, Jr, Bacon MA, Bajaj A, Chuang LT, Fisher BJ, Harkenrider MM, …, Gaffney DK. Cervical cancer: A global health crisis. Cancer. 2017;123:2404–12.

Smola S. Immunopathogenesis of HPV-associated cancers and prospects for immunotherapy. Viruses. 2017;9.

Spurgeon ME, Boon JAD, Horswill M, Barthakur S, Forouzan O, Rader JS, … Lambert PF. Human papillomavirus oncogenes reprogram the cervical cancer microenvironment independently of and synergistically with estrogen. PNAS. 2017; E9076–85.

Stanley M. HPV vaccines: alternative dosage schedules. Expert Rev Vaccines. 2019;18:1309–16.

Stubenrauch F, Laimins LA. Human papillomavirus life cycle: active and latent phases. Semin Cancer Biol. 1999;9:379–86.

Taberna M, Mena M, Pavon MA, Alemany L, Gillison ML, Mesia R. Human papillomavirus-related oropharyngeal cancer. Ann Oncol. 2017;28:2386–98.

Torres García LM, Concha López Á, González Paredes A. Valoración del endocérvix por legrado: correlación con citología, cambios colposcópicos y biopsias exocervicales. Progresos De Obstetricia Y Ginecología. 2007;50:577–83.

Wakeham K, Kavanagh K, Cuschieri K, Millan D, Pollock KG, Bell S, … Graham SV. HPV status and favourable outcome in vulvar squamous cancer. Int J Cancer. 2017;140:1134–46.

Westrich JA, Warren CJ, Pyeon D. Evasion of host immune defenses by human papillomavirus. Virus Res. 2017;231:21–33.

Woodby B, Scott M, Bodily J. The interaction between human papillomaviruses and the stromal microenvironment. Prog Mol Biol Transl Sci. 2016;144:169–238.

World-Health-Organization. Weekly epidemiological record. Human papillomavirus vaccines WHO position paper. Wkly Epidemiol Rec. 2009;15:117–32.

World-Health-Organization. Human papillomavirus laboratory manual, Switzerland. Department of Immunization, Vaccines and Biologicals: World Health Organization; 2010.

World-Health-Organization. Comprehensive cervical cancer control: a guide to essential practice. 2014.

Hepatitis B Virus

Characterization of Hepatitis B Virus with Complex Structural Variations

Kei Fujiwara

Abstract Chronic hepatitis B virus (HBV) infection is a major global public health issue. Fundamental research on HBV is necessary to eradicate and completely cure HBV infection. Conventionally, genome alterations have been recognized as nucleotide substitutions and canonical forms of structural variations (SVs) such as insertion, deletions, and duplication. However, novel forms of SVs, referred to as complex SVs, have been discovered. Complex SVs were initially reported in the human and mouse genomes. Subsequent reports have shown that complex SVs also occur in HBV, which can be detected using bioinformatical tools. Typically, complex SVs show low sequence similarity and sequence gaps in pair wise/multiple alignment. In addition, unique insertional motifs are observed, and are useful for detecting complex SVs. HBV strains with complex SVs can change the virological characteristics of HBV and the clinical course of HBV-infected patients, depending on the position in viral genome, type of insertional motifs, and combination of SVs. Notably, complex SVs that cause up-regulation of viral genome replication and excessive expression of viral protein may worsen the clinical course.

Keywords Hepatitis B virus · Genome · Complex structural variation · Genetic rearrangement · Bioinformatics · Insertional motif · Hepatocyte nuclear factor 1 binding site · Classification · Pair-wise/multiple alignment · Mutation

1 Introduction

Hepatitis B virus (HBV) research began with the discovery of Australian antigen in 1965 (Blumberg et al. 1965). Initially, immunological analyses were performed to characterize the virus. Serotyping was introduced as a method to classify the virus, and has been applied in epidemiological studies (Nordenfelt 1975; Holland 1975).

K. Fujiwara (✉)
Department of Gastroenterology and Metabolism, Nagoya City University Graduate School of Medical Sciences, 1 Kawasumi, Nagoya 467-8601, Mizuho, MizuhoAichi, Japan
e-mail: keifuji@med.nagoya-cu.ac.jp

S. I. Ahmad (ed.), *Human Viruses: Diseases, Treatments and Vaccines*, https://doi.org/10.1007/978-3-030-71165-8_15

In 1979, viral genome sequences of HBV were reported (Pasek et al. 1979; Galibert et al. 1979; Valenzuela et al. 1979), and research on the HBV genome was initiated.

In 1988, HBV genotypes were first defined by nucleotide differences of more than 8%, and 4 genotypes (genotypes A to D) were reported (Okamoto et al. 1988). Subsequent phylogenetic analyses revealed results identical to those obtained HBV genotyping, and 2 additional genotypes (genotypes E and F) were described (Norder et al. 1992; Norder et al. 1994; Magnius et al. 1995). A unique HBV strain containing 36 nucleotide (nt) insertion in the core open reading frame (ORF) was reported by Tran and colleagues (Tran et al. 1991), which was designated as novel HBV genotype G (Stuyver et al. 2000). HBV genotype H was subsequently reported in South American patients (Arauz-Ruiz et al. 2002). Recently 2 other tentative genotypes (I and J) have been reported (Tran et al. 2008; Tatematsu et al. 2009). Each HBV genotype has different geographical distribution (Orito et al. 2001a, b; Nakano et al. 2001; Ding et al. 2001; Sugauchi et al. 2001, 2002a, b; Ding et al. 2002; Arauz-Ruiz et al. 2002; Kato et al. 2002; Suzuki et al. 2003; Kramvis and Kew 2007; Raimondi et al. 2010).

In 1989, an important viral genome mutation, pre-core mutation G1896A, was reported (Carman et al. 1989), and further research revealed the relationship between fulminant hepatitis and pre-core mutation (Kosaka et al. 1991; Omata et al. 1991; Liang et al. 1991; Hasegawa et al. 1991). In 1994, core promoter (CP) mutations (A1762T/G1764A) were reported (Okamoto et al. 1994), and the mutations were found to be related with seroconversion from HBeAg positivity to anti-HBe antibody positivity (Lindh et al. 1998). Further research analyzing HBV genotypes and viral mutations was then initiated. HBV genotypes were found to be related to virological and clinical characteristics such as genotype specific mutation patterns (Orito et al. 2001a, b), as well as to mutation patterns and hepatocellular carcinoma (HCC) risk (Kao et al. 2003).

Following the studies on genotypes and mutations, intergenotypic recombination in HBV were studied and reported (Bollyky et al. 1996; Bowyer and Sim 2000; Morozov et al. 2000; Sugauchi et al. 2002b; Simmonds and Midgley 2005; Suwannkarn et al. 2005; Kurbanov et al. 2005; Chauhan et al. 2008; Ghosh et al. 2013; Araujo 2015). Sugauchi and colleagues (Sugauchi et al. 2002b) suggested that recombination could explain the differences between HBV genotype B in Japan and Taiwan. Genotype B in Taiwan was composed of genotype B and C recombination, and thus, the genetic differences between HBV genotype B in Japan and Taiwan reflected differences in clinical presentation. To date, the genetic differences in HBV have been explained according to genotypes, point mutations such as CP mutations and pre-core mutation, and recombination.

2 Complex Structural Variation

Structural variation (SV) in the genome is defined by canonical forms such as insertion, deletion, duplication, and inversion. Conversely, studies have elucidated novel variants that cannot be categorized by those canonical forms (Hastings et al. 2009; Zhang et al. 2009). The non-canonical forms of genetic variations are called complex SVs (Quinlan et al. 2010; Quinlan and Hall 2012; Yalcin et al. 2012; Guo et al. 2016; Collins et al. 2017). Quinlan and Hall (2012) reported that complex SVs comprise multiple breakpoints whose origin cannot be explained by a single end-joining or DNA exchange events. Yalcin and colleagues (Yalcin et al. 2012) characterized complex SVs as 2 or more SVs co-occurring at the same locus. In comparison with known canonical variants, complex SVs have a greater biological and clinical impact; for example, complex SVs can change regulatory genetic regions such as enhancer, promoter, or repressor sites into a novel form, and could disrupt normal regulation of the genome (Quinlan and Hall 2012).

3 Complex SVs in HBV

In 2005, an analysis of HBV genotype E led to the identification of a strange genetic alteration in HBV (Fujiwara et al. 2005). The genetic alteration was composed of an insertion, a deletion, and a duplication. In addition, those SVs co-occurred at the same locus. It was a novel rearrangement, and was distinct from SVs or mutations previously reported in the HBV genome. Nevertheless, a defined characterization system for genomic SVs in HBV was not established at the time. The concept of complex SVs has been developed to explain genome variations of model species. The current characterization systems to describe the complex SVs were adapted from previous studies of the human and mouse genomes. The HBV strain with the novel genomic alteration was confirmed to harbour complex SVs. A total of 70 HBV strains with complex SVs were later discovered and detailed data on those were published (Fujiwara et al. 2018).

4 Method of Analysis

Investigators defined complex SVs as SVs with multiple breakpoints and consisting of complicated combinations of insertions, inversions, deletions, and copy number gains such as duplications (Quinlan and Hall 2012; Yalcin et al. 2012). Compared with canonical SVs, the genetic sequence of complex SVs shows more nucleotide differences as compared with reference sequences. In these studies, pair-wise or multiple-alignment analysis was first performed using the CLUSTAL W program (Thompson et al. 1994), MAFFT (Katoh and Standley 2013), or T-coffee

(Notredame et al. 2000) program. Candidate nucleotide sequences were then aligned with reference genome sequences. When partial sequences with low sequence identities with and without sequence gaps were detected by visual inspection, a similarity search for the unique sequence fragments was performed using NCBI BLAST (Altschul et al. 1997). Our studies to date indicate that 80% of complex SVs in HBV contain sequence gaps (Fujiwara et al. 2018). Then additional analysis by visual inspection is conducted by examining the architecture of complex SVs in previous reports (Quinlan and Hall 2012; Yalcin et al. 2012) as references. In the initial report of complex SVs in HBV (Fujiwara et al. 2017), unique insertional motif sequences were discovered. The first such sequence identified was for hepatocyte nuclear factor 1 (HNF1) binding site. HNF1 binding site is located in the pre-S1 promoter region. The second sequence (GAAGAGCTCAAGCTTTGC) was an insertion of unknown origin. When new insertional motif sequences were found, they were searched in NCBI BLAST (Altschul et al. 1997) with HBV (taxid: 10,407) imputed in the "Organism" field of the "Choose Search Set" option. Based on the search results, candidate strains containing complex SVs were checked in a pair-wise analysis, and additional strains with complex SVs were identified. The NCBI-BLAST search was quite useful for detecting strains with complex SVs when pair-wise alignment did not contain sequence gaps. In the previous studies (Fujiwara et al. 2017, 2018), most of the complex SVs that did not show sequence gaps in pair-wise/multiple alignments were detected using the insertional motif and BLAST search method.

5 Architecture of Complex SVs in HBV

As an example, the architecture of complex SVs observed in CAR001 (AM494791, Bekondi et al. 2007) is shown in Fig. 1a. The sequence enlarged under the alignment shows the 32 nucleotides that are specific to CAR001 which were analyzed by NCBI BLAST (Altschul et al. 1997). The normal core upstream regulatory region (CURS) to basic core promoter (BCP) sequences were deleted, the pre-S1 promoter, including the HNF1 binding site, was externally inserted, and duplication occurred. Therefore, the complex SVs in CAR001 are composed of a deletion, an insertion, and a duplication. As a second example, the architecture of complex SVs observed in bne200 (FN594754, Abdou Chekaraou et al. 2010) is shown (Fig. 1b). The rearrangement in bne200 consisted of a deletion and an insertion of the pre-S1 promoter, including the HNF1 binding site. To date, 70 HBV strains with complex SVs have been reported, and the architectures of those strains have been described in recent reports (Fujiwara et al. 2017, 2018).

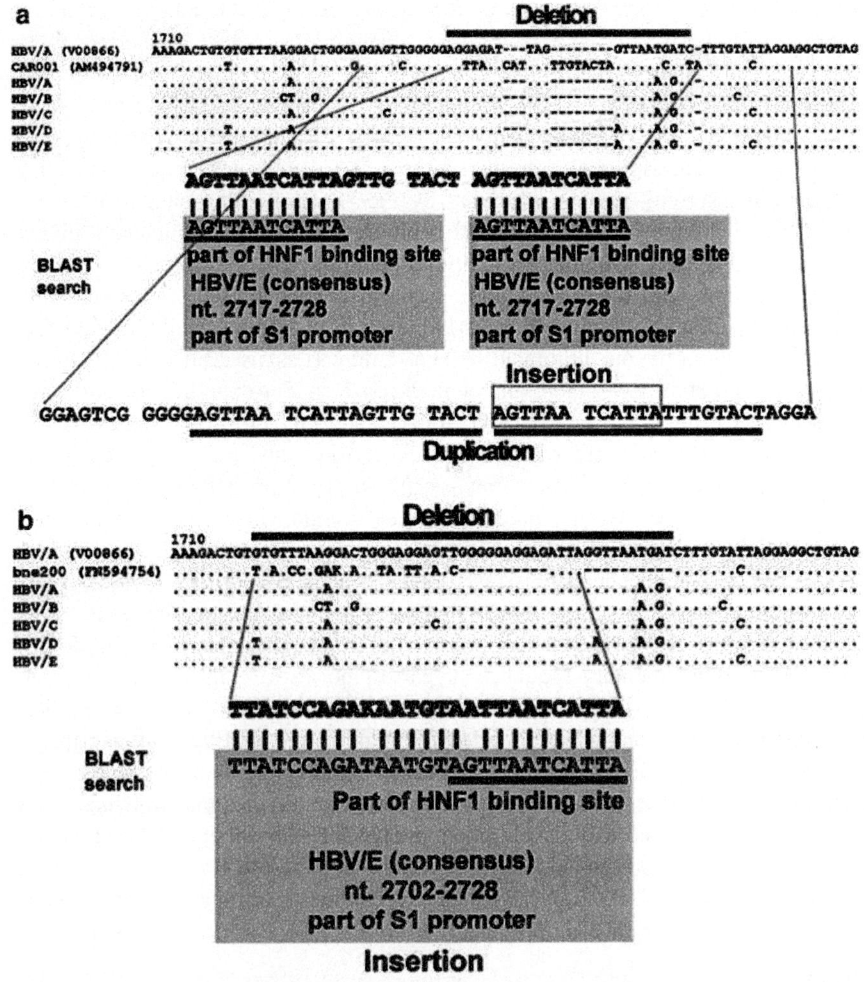

Fig. 1 Insertion patterns are shown in order of discovery. Some HBV strains with 2 SVs of having an identical pattern and 1 other SV are included in group I, III, and IV. SV, structural variation; HNF1, hepatocyte nuclear factor 1. Reproduced with permission of BMC Microbiology (Fujiwara et al. 2018)

6 Insertional Motifs

As mentioned above, unique insertional motif sequences in complex SVs were discovered and the first sequence coded for the HNF1 binding site is located in the pre-S1 promoter region. The consensus nucleotide sequence is "AGTTAATCATTAC". HNF1 is a transcriptional factor that binds to specific DNA sequences as homo- or hetero-dimers, and its binding sites are critical in regulating the genes in liver (Courtois et al. 1987, 1988; Cereghini et al. 1990;

Rey-Campos et al. 1991; Locker et al. 2002). For example, the albumin promoter contains an HNF1 binding site, which is indispensable for direct promoter activity (Tronche et al. 1990). Transcription of pre-S mRNA is regulated by the pre-S1 promoter, which contains an HNF1 binding site (Chang et al. 1989; Raney et al. 1990; Zhou and Yen. 1991; Nishizono et al. 1991). Raney and colleagues (1994) examined the transcriptional activity of the pre-S1 promoter and clarified that transcriptional activity was up-regulated by the HNF1 binding site. In addition, simple canonical insertion of the HNF1 binding site into BCP caused higher transcriptional activity (Gunther et al. 1996; Pult et al. 1997). Researchers reported that HBV strains with the HNF1 binding site inserted into BCP were frequently observed in patients following liver or kidney transplantation, and some patients were diagnosed as having fulminant hepatic failure (Laskus et al. 1994; Gunther et al. 1996; Pult et al. 1997; Gerolami et al. 2005). Previous experimental data have shown that BCP mutations (A1762T/G1764A) lead to the creation of a new HNF1 binding site in the BCP (Li et al. 1999). Therefore, introduction of the HNF1 binding site in BCP by point mutations, simple insertion, or complex SVs may be selected in chronic HBV infection. The second insertional motif was "GAAGAGCTCAAGCTTTCC (X-1)". The origin of X-1 is unknown. In addition, a complementary sequence of this insertional motif is also observed as an insertional motif (Fujiwara et al. 2017, 2018). The third insertional motif was "GGGCCGAACCAGA (X-2)". The origin of X-2 was also unknown. The fourth motif was a complementary sequence of part of box α in enhancer II. Box α in enhancer II affects transcription of the HBV genome. Yuh and colleagues (1992) noted that box α was one of several elements modulating BCP activity and that it can enhance BCP activity by more than 100-fold. Duplication of box α was reported in a previous study (Gunther et al. 1996). Box α mutation has been reported to be associated with development of HCC (Takahashi et al. 1998; Ito et al. 2007; Shinkai et al. 2007; Yuen et al. 2008; Kim et al. 2009; Bai et al. 2011; Lyu et al. 2013; Wang et al. 2016). These data suggest that insertional motif sequences have virological and clinical importance.

7 Position of Complex SVs

Since a number of HBV strains with complex SVs have been detected, the question arose as to whether these novel genetic rearrangements are equally distributed in HBV genome and therefore, the positions of complex SVs in HBV genome were investigated. The locations of complex SVs in each HBV strain were plotted, as shown in Fig. 2. Complex SVs accumulated in certain areas, that is, 94.3% (66/70) of complex SV were observed in nt 1500–2000. This area contains the X ORF, pre-core/core ORF. In addition, this area contains CURS/BCP, enhancer II, and DR1/DR2, which are regions that regulate the transcription of the HBV genome.

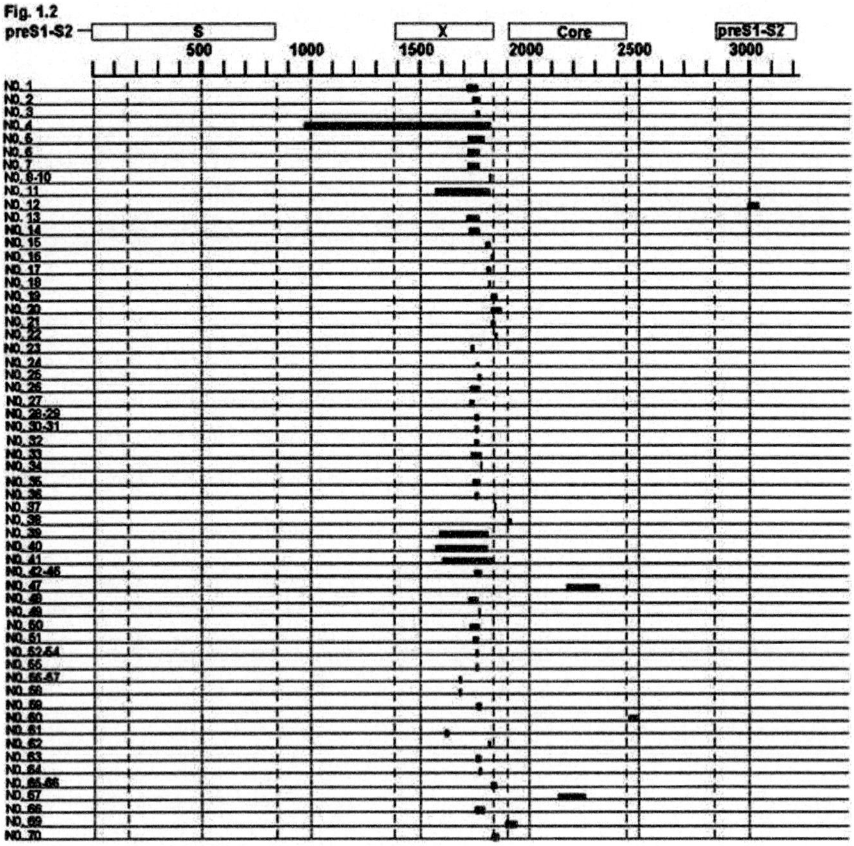

Fig. 2 Location of complex SVs in the HBV genome sequence. Positions where complex SVs were detected in the HBV genome sequence were investigated. The position and length of complex SVs were shown as each black rectangle. Reproduced with permission of BMC Microbiology (Fujiwara et al. 2018)

8 Classification of Complex SVs in HBV

As shown in Fig. 1, complex SVs have high variability compared to known genetic alterations reported in HBV. A previous study examined the composition of SVs and found that some classes were more common than others (Quinlan et al. 2010). Therefore, HBV strains with complex SVs were assorted provisionally according to their SV patterns in order to clarify the combination of highest frequency and elucidate the number of patterns of complex SVs. They were categorized into 6 groups, insertion (+)/deletion (+) (class I), insertion (+)/deletion (+)/duplication (+) (class II), insertion (+)/duplication (+) (class III), deletion (+)/duplication (+) (class

Table 1 Classification of HBV strains with complex SVs (n = 70)

I. Insertion (+), Deletion (+) 49 (70.0%)
Types of insertion
A. HNF1 binding site 24 (34.3%)
B. Insertion of unknown origin (X-1) 7 (10.0%)
C. Insertion of unknown origin (X-2) 3 (4.3%)
D. Sequence complementary to part of box α in enhancer II 6 (8.6%)
E. Miscellaneous 9 (12.9%)
II. Insertion (+), Deletion (+), Duplication (+) 6 (8.6%)
Types of insertion
A. HNF1 binding site 4 (5.7%)
B. Insertion of unknown origin (X-1) 1 (1.4%)
C. Insertion of unknown origin (X-2) 0 (0.0%)
D. Sequence complementary to part of box α in enhancer II 0 (0.0%)
E. Miscellaneous 1 (1.4%)
III. Insertion (+), Duplication (+) 5 (7.1%)
Types of insertion
A. HNF1 binding site 2 (2.9%)
B. Insertion of unknown origin (X-1) 3 (4.3%)
C. Insertion of unknown origin (X-2) 0 (0.0%)
D. Sequence complementary to part of box α in enhancer II 0 (0.0%)
E. Miscellaneous 0 (0.0%)
IV. Deletion (+), Duplication (+) 4 (5.7%)
V. Duplications (+) 1 (1.4%)
VI. Highly complicated (four or more SVs) 5 (7.1%)

IV), multiple duplications (class V), and highly complicated (class VI), which are composed of 4 or more SVs, as shown in Table 1. In addition, insertions were divided into 5 patterns.

Regarding frequency, the class I pattern (insertion (+)/deletion (+)) was more common (70.0%) than other patterns. More detailed analysis showed that HNF1 binding site insertion was the most frequent insertional motif sequence in the class I pattern. The Class I pattern with HNF1 binding site insertion were observed in 24/ 70 (34.3 %) of the HBV strains with complex SVs. The strain CAR001 was categorized as class II (insertion (+)/ deletion (+)/ duplication (+)), and class II showed the second highest frequency. For the insertional motif of the class II pattern, HNF1 binding site insertion showed the highest frequency.

9 Clinical Implications and Future Perspective

In 70 strains with complex SVs, at least 11, 2, and 2 strains were observed in patients with HCC, fulminant hepatitis, and severe chronic liver disease, respectively. The virologic properties and clinical manifestations of HBV strains with complex SVs heavily depend on the locations of the complex SVs and types of insertional motifs. Experimental data have shown that complex SVs with HNF1 insertion up-regulated transcription of the viral genome, and an excessive amount of viral protein was produced and localized in certain area of hepatocyte (Fujiwara et al. 2005), thus, the potential risk of severe or fulminant hepatitis is a concern. Further studies are required to clarify their importance in clinical medicine.

10 Conclusion

Novel genetic alterations known as complex SVs in HBV have been identified and shown to be completely different from known mutations or rearrangements such as CP mutations (A1762T/G1764A), pre-core mutation, or intergenotypic recombination. A total of 70 HBV strains with complex SVs were analyzed, and the characteristics of complex SVs in HBV such as frequent location in HBV genome and insertional motif patterns were clarified. In addition, complex SVs have now been thoroughly classified. Further studies are needed to elucidate the virological and clinical role of the rearrangements.

Acknowledgements This work was supported by a grant-in-aid from the JSPS (Grant Number 19K08427).

References

Abdou Chekaraou M, Brichler S, Mansour W, Le Gal F, Garba A, Deny P, et al. A novel hepatitis B virus (HBV) subgenotype D (D8) strain, resulting from recombination between genotypes D and E, is circulating in Niger along with HBV/E strains. J Gen Virol. 2010;91(Pt 6):1609–20.

Altschul SF, Madden TL, Schaffer AA, Zhang J, Zhang Z, Miller W, et al. Gapped BLAST and PSI-BLAST: a new generation of protein database search programs. Nucleic Acids Res. 1997;25(17):3389–402.

Araujo NM. Hepatitis B virus intergenotypic recombinants worldwide: an overview. Infect Genet Evol. 2015;36:500–10.

Arauz-Ruiz P, Norder H, Robertson BH, Magnius LO. Genotype H: a new Amerindian genotype of hepatitis B virus revealed in Central America. J Gen Virol. 2002;83(Pt 8):2059–73.

Bai X, Zhu Y, Jin Y, Guo X, Qian G, Chen T, et al. Temporal acquisition of sequential mutations in the enhancer II and basal core promoter of HBV in individuals at high risk for hepatocellular carcinoma. Carcinogenesis. 2011;32(1):63–8.

Bekondi C, Olinger CM, Boua N, Talarmin A, Muller CP, Le Faou A, et al. Central African Republic is part of the West-African hepatitis B virus genotype E crescent. J Clin Virol. 2007;40(1):31–7.

Blumberg BS, Alter HJ, Visnich S. A "New" Antigen in Leukemia Sera. JAMA. 1965;15 (191):541–6.

Bollyky PL, Rambaut A, Harvey PH, Holmes EC. Recombination between sequences of hepatitis B virus from different genotypes. J Mol Evol. 1996;42(2):97–102.

Bowyer SM, Sim JG. Relationships within and between genotypes of hepatitis B virus at points across the genome: footprints of recombination in certain isolates. J Gen Virol. 2000;81(Pt 2):379–92.

Carman WF, Jacyna MR, Hadziyannis S, Karayiannis P, McGarvey MJ, Makris A, et al. Mutation preventing formation of hepatitis B e antigen in patients with chronic hepatitis B infection. Lancet. 1989;2(8663):588–91.

Cereghini S, Yaniv M, Cortese R. Hepatocyte dedifferentiation and extinction is accompanied by a block in the synthesis of mRNA coding for the transcription factor HNF1/LFB1. EMBO J. 1990;9(7):2257–63.

Chang HK, Wang BY, Yuh CH, Wei CL, Ting LP. A liver-specific nuclear factor interacts with the promoter region of the large surface protein gene of human hepatitis B virus. Mol Cell Biol. 1989;9(11):5189–97.

Chauhan R, Kazim SN, Kumar M, Bhattacharjee J, Krishnamoorthy N, Sarin SK. Identification and characterization of genotype A and D recombinant hepatitis B virus from Indian chronic HBV isolates. World J Gastroenterol. 2008;14(40):6228–36.

Collins RL, Brand H, Redin CE, Hanscom C, Antolik C, Stone MR, et al. Defining the diverse spectrum of inversions, complex structural variation, and chromothripsis in the morbid human genome. Genome Biol. 2017;18(1):36.

Courtois G, Morgan JG, Campbell LA, Fourel G, Crabtree GR. Interaction of a liver-specific nuclear factor with the fibrinogen and alpha 1-antitrypsin promoters. Science. 1987;238 (4827):688–92.

Courtois G, Baumhueter S, Crabtree GR. Purified hepatocyte nuclear factor 1 interacts with a family of hepatocyte-specific promoters. Proc Natl Acad Sci U S A. 1988;85(21):7937–41.

Ding X, Mizokami M, Yao G, Xu B, Orito E, Ueda R, et al. Hepatitis B virus genotype distribution among chronic hepatitis B virus carriers in Shanghai, China. Intervirology. 2001;44(1):43–7.

Ding X, Mizokami M, Ge X, Orito E, Iino S, Ueda R, et al. Different hepatitis B virus genotype distributions among asymptomatic carriers and patients with liver diseases in Nanning, southern China. Hepatol Res. 2002;22(1):37–44.

Fujiwara K, Tanaka Y, Paulon E, Orito E, Sugiyama M, Ito K, et al. Novel type of hepatitis B virus mutation: replacement mutation involving a hepatocyte nuclear factor 1 binding site tandem repeat in chronic hepatitis B virus genotype E. J Virol. 2005;79(22):14404–10.

Fujiwara K, Matsunami K, Iio E, Nojiri S, Joh T. Novel non-canonical genetic rearrangements termed "complex structural variations" in HBV genome. Virus Res. 2017;15(238):84–93.

Fujiwara K, Matsuura K, Matsunami K, Iio E, Nojiri S. Characterization of hepatitis B virus with complex structural variations. BMC Microbiol. 2018;18(1):202.

Galibert F, Mandart E, Fitoussi F, Tiollais P, Charnay P. Nucleotide sequence of the hepatitis B virus genome (subtype ayw) cloned in E. coli. Nature. 1979;281(5733):646–50.

Gerolami R, Henry M, Borentain P, Colson P, Botta D, Tamalet C. Fulminant hepatitis B associated with a specific insertion in the basal core promoter region of hepatitis B virus DNA after immunosuppressive treatment. Clin Infect Dis. 2005;40(4):e24–7.

Ghosh S, Banerjee P, Deny P, Mondal RK, Nandi M, Roychoudhury A, et al. New HBV subgenotype D9, a novel D/C recombinant, identified in patients with chronic HBeAg-negative infection in Eastern India. J Viral Hepat. 2013;20(3):209–18.

Gunther S, Piwon N, Iwanska A, Schilling R, Meisel H, Will H. Type, prevalence, and significance of core promoter/enhancer II mutations in hepatitis B viruses from immunosuppressed patients with severe liver disease. J Virol. 1996;70(12):8318–31.

Guo Y, Gu X, Sheng Z, Wang Y, Luo C, Liu R, et al. A complex structural variation on chromosome 27 leads to the ectopic expression of HOXB8 and the muffs and beard phenotype in chickens. PLoS Genet. 2016;12(6):e1006071.

Hasegawa K, Huang JK, Wands JR, Obata H, Liang TJ. Association of hepatitis B viral precore mutations with fulminant hepatitis B in Japan. Virology. 1991;185(1):460–3.

Hastings PJ, Ira G, Lupski JR. A microhomology-mediated break-induced replication model for the origin of human copy number variation. PLoS Genet. 2009;5(1):e1000327.

Holland PV. Hepatitis B antigen subtypes-history, significance and immunogenicity. Am J Med Sci. 1975;270(1):161–4.

Ito K, Tanaka Y, Kato M, Fujiwara K, Sugauchi F, Sakamoto T, et al. Comparison of complete sequences of hepatitis B virus genotype C between inactive carriers and hepatocellular carcinoma patients before and after seroconversion. J Gastroenterol. 2007;42(10):837–44.

Kao JH, Chen PJ, Lai MY, Chen DS. Basal core promoter mutations of hepatitis B virus increase the risk of hepatocellular carcinoma in hepatitis B carriers. Gastroenterology. 2003;124 (2):327–34.

Katoh K, Standley DM. MAFFT multiple sequence alignment software version 7: improvements in performance and usability. Mol Biol Evol. 2013;30(4):772–80.

Kato H, Ruzibakiev R, Yuldasheva N, Hegay T, Kurbanov F, Achundjanov B, et al. Hepatitis B virus genotypes in Uzbekistan and validity of two different systems for genotyping. J Med Virol. 2002;67(4):477–83.

Kim JK, Chang HY, Lee JM, Baatarkhuu O, Yoon YJ, Park JY, et al. Specific mutations in the enhancer II/core promoter/precore regions of hepatitis B virus subgenotype C2 in Korean patients with hepatocellular carcinoma. J Med Virol. 2009;81(6):1002–8.

Kosaka Y, Takase K, Kojima M, Shimizu M, Inoue K, Yoshiba M, et al. Fulminant hepatitis B: induction by hepatitis B virus mutants defective in the precore region and incapable of encoding e antigen. Gastroenterology. 1991;100(4):1087–94.

Kramvis A, Kew MC. Epidemiology of hepatitis B virus in Africa, its genotypes and clinical associations of genotypes. Hepatol Res. 2007;37(s1):S9–19.

Kurbanov F, Tanaka Y, Fujiwara K, Sugauchi F, Mbanya D, Zekeng L, et al. A new subtype (subgenotype) Ac (A3) of hepatitis B virus and recombination between genotypes A and E in Cameroon. J Gen Virol. 2005;86(Pt 7):2047–56.

Laskus T, Rakela J, Steers JL, Wiesner RH, Persing DH. Precore and contiguous regions of hepatitis B virus in liver transplantation for end-stage hepatitis B. Gastroenterology. 1994;107 (6):1774–80.

Liang TJ, Hasegawa K, Rimon N, Wands JR, Ben-Porath E. A hepatitis B virus mutant associated with an epidemic of fulminant hepatitis. N Engl J Med. 1991;324(24):1705–9.

Lindh M, Gustavson C, Mardberg K, Norkrans G, Dhillon AP, Horal P. Mutation of nucleotide 1,762 in the core promoter region during hepatitis B e seroconversion and its relation to liver damage in hepatitis B e antigen carriers. J Med Virol. 1998;55(3):185–90.

Li J, Buckwold VE, Hon MW, Ou JH. Mechanism of suppression of hepatitis B virus precore RNA transcription by a frequent double mutation. J Virol. 1999;73(2):1239–44.

Locker J, Ghosh D, Luc PV, Zheng J. Definition and prediction of the full range of transcription factor binding sites–the hepatocyte nuclear factor 1 dimeric site. Nucleic Acids Res. 2002;30 (17):3809–17.

Lyu H, Lee D, Chung YH, Kim JA, Lee JH, Jin YJ, et al. Synergistic effects of A1896, T1653 and T1762/A1764 mutations in genotype c2 hepatitis B virus on development of hepatocellular carcinoma. J Viral Hepat. 2013;20(3):219–24.

Magnius LO, Norder H. Subtypes, genotypes and molecular epidemiology of the hepatitis B virus as reflected by sequence variability of the S-gene. Intervirology. 1995;38(1–2):24–34.

Morozov V, Pisareva M, Groudinin M. Homologous recombination between different genotypes of hepatitis B virus. Gene. 2000;260(1–2):55–65.

Nakano T, Lu L, Hu X, Mizokami M, Orito E, Shapiro C, et al. Characterization of hepatitis B virus genotypes among Yucpa Indians in Venezuela. J Gen Virol. 2001;82(Pt 2):359–65.

Nishizono A, Maeno M, Hiraga M, Hirai H, Esumi M, Shikata T. In vitro transcription of the hepatitis B virus gene by nuclear extracts of human hepatoma cells. Virology. 1991;182 (2):545–52.

Nordenfelt E. Some epidemiological and clinical aspects of hepatitis B antigen and its subtypes. Scand J Infect Dis. 1975;7(2):147–52.

Norder H, Hammas B, Lofdahl S, Courouce AM, Magnius LO. Comparison of the amino acid sequences of nine different serotypes of hepatitis B surface antigen and genomic classification of the corresponding hepatitis B virus strains. J Gen Virol. 1992;73(Pt 5):1201–8.

Norder H, Courouce AM, Magnius LO. Complete genomes, phylogenetic relatedness, and structural proteins of six strains of the hepatitis B virus, four of which represent two new genotypes. Virology. 1994;198(2):489–503.

Notredame C, Higgins DG, Heringa J. T-coffee: a novel method for fast and accurate multiple sequence alignment. J Mol Biol. 2000;302(1):205–17.

Okamoto H, Tsuda F, Sakugawa H, Sastrosoewignjo RI, Imai M, Miyakawa Y, et al. Typing hepatitis B virus by homology in nucleotide sequence: comparison of surface antigen subtypes. J Gen Virol. 1988;69(Pt 10):2575–83.

Okamoto H, Tsuda F, Akahane Y, Sugai Y, Yoshiba M, Moriyama K, et al. Hepatitis B virus with mutations in the core promoter for an e antigen-negative phenotype in carriers with antibody to e antigen. J Virol. 1994;68(12):8102–10.

Omata M, Ehata T, Yokosuka O, Hosoda K, Ohto M. Mutations in the precore region of hepatitis B virus DNA in patients with fulminant and severe hepatitis. N Engl J Med. 1991;324 (24):1699–704.

Orito E, Mizokami M, Sakugawa H, Michitaka K, Ishikawa K, Ichida T, et al. A case-control study for clinical and molecular biological differences between hepatitis B viruses of genotypes B and C. Japan HBV Genotype Research Group Hepatology. 2001a;33(1):218–23.

Orito E, Ichida T, Sakugawa H, Sata M, Horiike N, Hino K, et al. Geographic distribution of hepatitis B virus (HBV) genotype in patients with chronic HBV infection in Japan. Hepatology. 2001b;34(3):590–4.

Pasek M, Goto T, Gilbert W, Zink B, Schaller H, MacKay P, et al. Hepatitis B virus genes and their expression in E. coli. Nature. 1979;282(5739):575–9.

Pult I, Chouard T, Wieland S, Klemenz R, Yaniv M, Blum HE. A hepatitis B virus mutant with a new hepatocyte nuclear factor 1 binding site emerging in transplant-transmitted fulminant hepatitis B. Hepatology. 1997;25(6):1507–15.

Quinlan AR, Hall IM. Characterizing complex structural variation in germline and somatic genomes. Trends Genet. 2012;28(1):43–53.

Quinlan AR, Clark RA, Sokolova S, Leibowitz ML, Zhang Y, Hurles ME, et al. Genome-wide mapping and assembly of structural variant breakpoints in the mouse genome. Genome Res. 2010;20(5):623–35.

Raimondi S, Maisonneuve P, Bruno S, Mondelli MU. Is response to antiviral treatment influenced by hepatitis B virus genotype? J Hepatol. 2010;52(3):441–9.

Raney AK, Milich DR, Easton AJ, McLachlan A. Differentiation-specific transcriptional regulation of the hepatitis B virus large surface antigen gene in human hepatoma cell lines. J Virol. 1990;64(5):2360–8.

Raney AK, Easton AJ, McLachlan A. Characterization of the minimal elements of the hepatitis B virus large surface antigen promoter. J Gen Virol. 1994;75(Pt 10):2671–9.

Rey-Campos J, Chouard T, Yaniv M, Cereghini S. vHNF1 is a homeoprotein that activates transcription and forms heterodimers with HNF1. EMBO J. 1991;10(6):1445–57.

Shinkai N, Tanaka Y, Ito K, Mukaide M, Hasegawa I, Asahina Y, et al. Influence of hepatitis B virus X and core promoter mutations on hepatocellular carcinoma among patients infected with subgenotype C2. J Clin Microbiol. 2007;45(10):3191–7.

Simmonds P, Midgley S. Recombination in the genesis and evolution of hepatitis B virus genotypes. J Virol. 2005;79(24):15467–76.

Stuyver L, De Gendt S, Van Geyt C, Zoulim F, Fried M, Schinazi RF, et al. A new genotype of hepatitis B virus: complete genome and phylogenetic relatedness. J Gen Virol. 2000;81(Pt 1):67–74.

Sugauchi F, Mizokami M, Orito E, Ohno T, Kato H, Suzuki S, et al. A novel variant genotype C of hepatitis B virus identified in isolates from Australian Aborigines: complete genome sequence and phylogenetic relatedness. J Gen Virol. 2001;82(Pt 4):883–92.
Sugauchi F, Chutaputti A, Orito E, Kato H, Suzuki S, Ueda R, et al. Hepatitis B virus genotypes and clinical manifestation among hepatitis B carriers in Thailand. J Gastroenterol Hepatol. 2002a;17(6):671–6.
Sugauchi F, Orito E, Ichida T, Kato H, Sakugawa H, Kakumu S, et al. Hepatitis B virus of genotype B with or without recombination with genotype C over the precore region plus the core gene. J Virol. 2002b;76(12):5985–92.
Suwannakarn K, Tangkijvanich P, Theamboonlers A, Abe K, Poovorawan Y. A novel recombinant of Hepatitis B virus genotypes G and C isolated from a Thai patient with hepatocellular carcinoma. J Gen Virol. 2005;86(Pt 11):3027–30.
Suzuki S, Sugauchi F, Orito E, Kato H, Usuda S, Siransy L, et al. Distribution of hepatitis B virus (HBV) genotypes among HBV carriers in the Cote d'Ivoire: complete genome sequence and phylogenetic relatedness of HBV genotype E. J Med Virol. 2003;69(4):459–65.
Takahashi K, Akahane Y, Hino K, Ohta Y, Mishiro S. Hepatitis B virus genomic sequence in the circulation of hepatocellular carcinoma patients: comparative analysis of 40 full-length isolates. Arch Virol. 1998;143(12):2313–26.
Tatematsu K, Tanaka Y, Kurbanov F, Sugauchi F, Mano S, Maeshiro T, et al. A genetic variant of hepatitis B virus divergent from known human and ape genotypes isolated from a Japanese patient and provisionally assigned to new genotype J. J Virol. 2009;83(20):10538–47.
Thompson JD, Higgins DG, Gibson TJ. CLUSTAL W: improving the sensitivity of progressive multiple sequence alignment through sequence weighting, position-specific gap penalties and weight matrix choice. Nucleic Acids Res. 1994;22(22):4673–80.
Tran A, Kremsdorf D, Capel F, Housset C, Dauguet C, Petit MA, et al. Emergence of and takeover by hepatitis B virus (HBV) with rearrangements in the pre-S/S and pre-C/C genes during chronic HBV infection. J Virol. 1991;65(7):3566–74.
Tran TT, Trinh TN, Abe K. New complex recombinant genotype of hepatitis B virus identified in Vietnam. J Virol. 2008;82(11):5657–63.
Tronche F, Rollier A, Herbomel P, Bach I, Cereghini S, Weiss M, et al. Anatomy of the rat albumin promoter. Mol Biol Med. 1990;7(2):173–85.
Valenzuela P, Gray P, Quiroga M, Zaldivar J, Goodman HM, Rutter WJ. Nucleotide sequence of the gene coding for the major protein of hepatitis B virus surface antigen. Nature. 1979;280 (5725):815–9.
Wang Y, Zeng LI, Chen W. HBV X gene point mutations are associated with the risk of hepatocellular carcinoma: a systematic review and meta-analysis. Mol Clin Oncol. 2016;4 (6):1045–51.
Yalcin B, Wong K, Bhomra A, Goodson M, Keane TM, Adams DJ, et al. The fine scale architecture of structural variants in 17 mouse genomes. Genome Biol. 2012;13(3):R18.
Yuen MF, Tanaka Y, Shinkai N, Poon RT, But DY, Fong DY, et al. Risk for hepatocellular carcinoma with respect to hepatitis B virus genotypes B/C, specific mutations of enhancer II/ core promoter/precore regions and HBV DNA levels. Gut. 2008;57(1):98–102.
Yuh CH, Chang YL, Ting LP. Transcriptional regulation of precore and pregenomic RNAs of hepatitis B virus. J Virol. 1992;66(7):4073–84.
Zhang F, Khajavi M, Connolly AM, Towne CF, Batish SD, Lupski JR. The DNA replication FoSTeS/MMBIR mechanism can generate genomic, genic and exonic complex rearrangements in humans. Nat Genet. 2009;41(7):849–53.
Zhou DX, Yen TS. The ubiquitous transcription factor Oct-1 and the liver-specific factor HNF-1 are both required to activate transcription of a hepatitis B virus promoter. Mol Cell Biol. 1991;11(3):1353–9.

Enteroviruses

Impact of Genetic Changes in the Enterovirus 71 Genome on Virulence

Chit Laa Poh, Madiiha Bibi Mandary, and Seng-Kai Ong

Abstract Enterovirus 71 (EV-A71) is one of the main etiological agents of hand, foot and mouth disease (HFMD). EV-A71 mainly infects infants and young children below six years of age. Clinical manifestations of HFMD include fever, rashes with the presence of vesicles on the hand, feet and mouth. However, the onset of severe HFMD can lead to neurological complications such as acute flaccid paralysis, brainstem encephalitis and cardiac pulmonary failure that can be fatal. This chapter addresses how genetic events such as recombination and spontaneous mutations could change the genomic organization of EV-A71, leading to strains with higher virulence. An understanding of the recombination mechanism of the poliovirus and non-polio enteroviruses provides evidence for the emergence of novel EV-A71 strains responsible for fatal HFMD outbreaks. Currently, it is unknown if the virulence of EV-A71 is contributed by the events of recombination between EV-A71 and other enteroviruses or it is due to the presence of spontaneous mutations that effects its virulence.

Keywords EV-A71 · Enterovirus 71 · RNA · Epidemiology · Recombination · Spontaneous mutations · Virulence · Viral fitness · Pathogenicity · Antigenicity

1 Introduction

Enterovirus 71 (EV-A71) belongs to the genus Enterovirus within the Picornaviridae family and is classified as a human enterovirus species A (HEV-A) based on its genome sequence. Phylogenetic analysis of the VP1 region of EV-A71

C. L. Poh (✉) · M. B. Mandary
Centre for Virus and Vaccine Research, School of Medical and Life Sciences,
Sunway University, Kuala Lumpur, Selangor 47500, Malaysia
e-mail: pohcl@sunway.edu.my

S.-K. Ong
Department of Biological Sciences, School of Medical and Life Sciences,
Sunway University, Kuala Lumpur, Selangor 47500, Malaysia

S. I. Ahmad (ed.), *Human Viruses: Diseases, Treatments and Vaccines*, https://doi.org/10.1007/978-3-030-71165-8_16

has shown that the virus can be grouped into three main genotypes: A, B and C. Genotype A is represented by the prototype BrCr strain whereas genotypes B and C can be further segregated into five sub-genotypes, viz, B1 to B5 and C1 to C5, respectively (Brown et al. 1999). Newly discovered isolates were classified as genotypes D (isolated in India), E and F (isolated in Africa) (Bessaud et al. 2014).

EV-A71 was first isolated and characterized from a child with severe neurological disease in California, 1969 and since then, many hand, foot and mouth disease (HFMD) outbreaks have been reported around the world (Schmidt et al. 1974). EV-A71 together with CV-A16 (Coxsackievirus-16), CV-A6 and CV-A8 are the main causative agents of HFMD and these pathogens have caused millions of infections in China and South East Asian countries from 2008 to 2016. In 2015, a total of 2,014,999 cases of HFMD including 124 deaths were reported in China. Less than 12 months later in August 2016, 1,792,251 cases of HFMD were recorded by WHO, including 172 fatal cases in China (WHO 2017). Complications of HFMD infections include aseptic meningitis, brainstem encephalitis, poliovirus-like paralysis, shock and cardiac dysfunction (Huang et al. 1999; Solomon et al. 2010). Early reports of severe HFMD occurring in countries such as Bulgaria (Chumakov et al. 1979) and Hungary (Nagy et al. 1982) have been documented since 1975 to the early 1980s. However, the last decade has witnessed a significant increase in epidemic activities of EV-A71 in Asia. Countries in the Asia Pacific region such as China, Malaysia, Taiwan, Singapore, Vietnam and Thailand have witnessed recurrent outbreaks of EV-A71 epidemics originating from different sub-genotypes since the late 1990s (Table 1). A large number of HFMD-associated fatal cases were recorded in China, Malaysia and Taiwan. In China, enteroviruses such as EV-A71, CV-A16 and other enteroviruses have caused 7,200,092 HFMD cases from 2008 to 2012 and the mortality was highest among children aged 1 to 3 years of age. It was further reported that 82,486 patients developed neurological complications, and 1617 deaths were confirmed to be caused by EV-A71 (Xing et al. 2014).

EV-A71 has a single-stranded positive RNA (+ssRNA) genome which is contained within the non-enveloped capsid of the virus (Andino et al. 1999). The RNA genomic sequence is made up of 7411 base pairs and has an open reading frame (ORF) flanked on both ends by the 5′ non-translatable region (5′-NTR) and the 3′ non-transable region (3′-NTR) (Brown and Pallansch 1995). The 5′-NTR has an internal ribosome entry site (IRES) which controls cap-independent translation. The ORF comprising 6579 nucleotides can be classified into three polyprotein regions, namely, P1, P2 and P3. They encode four structural proteins (VP1 to VP4) in the P1 region and seven non-structural proteins in the P2 (2A-2C) and P3 regions (3A-3D) following proteolytic cleavage (Fig. 1) (Solomon et al. 2010). The viral capsid proteins VP1, VP2 and VP3 are displayed on the external surface of the EV-A71 viral particle whereas VP4 is found within the internal structures of the capsid. The 2A and 3C proteins are of particular interest as they are crucial for viral-host interactions, viral replication and cytopathic effects. As for the 3D polymerase of a +ssRNA virus, it is an RNA-dependent RNA polymerase (RdRp) with low fidelity, having 10^{-3} to 10^{-5} mutations per nucleotide copied in each replication cycle, and this could result in mis-incorporation of 1–2 new nucleotides (Domingo and Holland 1997; Domingo et al. 1978).

Table 1 EV-A71 subgenotypes circulating in Asia pacific region between 1997–2010

	1997	1998	1999	2000	2001	2002	2003	2004	2005	2006	2007	2008	2009	2010
Malaysia	C1, C2, B3, B4	C1	C1	B4, C1	–	C1	B5, C1	–	B5	–	–	–	–	–
Singapore	B3, B4	B3, C1	B3	B4	B4	B4, C1	B4	–	–	B5	–	B5	–	–
Taiwan	–	B4, C4, C2	B4	B4	B4	B4, C1	B4, B5	C4	C4	C5	B5, C5	B5	B5	C4
Japan	C2, B3, B4	C2	C2	C2, B4	C2	B4, C4, C2	C4, B5	–	–	C4	C4	–	–	–
China	–	C4	–	C4	C4	C4	C4	C4	C4	–	C4	A, C4	C4	–
Vietnam	–	–	–	–	–	–	–	–	C1, C4, C5	–	–	–	–	–
Australia	–	C2	B3, C2	B4, C1	B4, C1	C1	C1	C4	–	–	–	–	–	–
Korea	–	–	–	C3	–	–	C4	–	–	–	–	–	–	–
Thailand	C1,C2	–	C2	C2	C1	C1	C1	C1	–	B5, C1, C2, C4	B5, C1, C2, C4, C5	B5, C1, C2, C4	–	–

Schematic illustration is adapted and modified from Solomon et al. (2010) and Chang et al. (2016). Underlined groups refer to the predominant subgenotypes

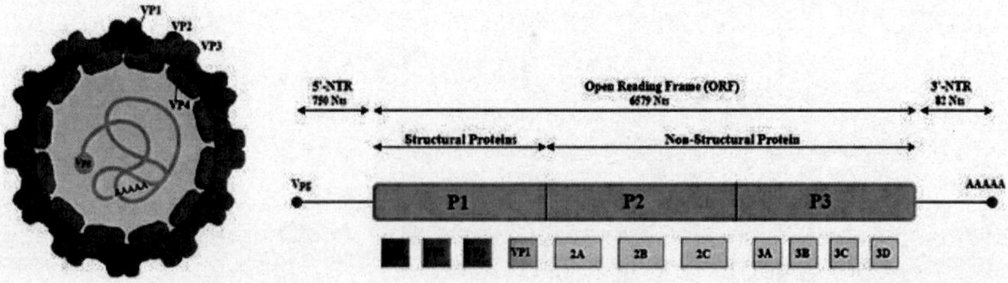

Fig. 1 **a** The EV71 capsid consists of four different structural proteins (VP1-VP4). **b** The EV-A71 genome (7.4 Kb). The position of the VPg primer is shown at the 5′-NTR end of the genome; the Open Reading Frame (ORF) encodes the structural viral protein P1 which is cleaved to yield VP1, VP2, VP3 and VP4, and non-structural viral proteins P2 (cleaved to yield 2A, 2B and 2C) and P3 (cleaved to yield 3A, 3B, 3C and 3D). The 3′-NTR end of the genome contains the poly (A) tail

The rate of mutation in RNA genome was reported to be in the order of 10^{-4} to 10^{-6} fold higher than the mutation rate of a stable DNA genome of similar complexity (Holland et al. 1982). Such high mutation rates were attributed to the absence of proofreading-repair activity of the viral RdRp (Ferrer-Orta et al. 2006). Mutations enable viruses to benefit from the advantages of having a few beneficial mutations without being affected by many deleterious mutations.

2 Recombination

The first discovery of RNA recombination in poliovirus was documented by Hirst (1962) and since then, numerous studies have demonstrated that significant levels of recombination frequently occur in enteroviruses such as EV-A71 (Hirst 1962; Ledinko 1963). It was further revealed that recombination might occur within a specific serotype or between different serotypes of enteroviruses. Here, we look into the different types of recombination and how they affected the evolution and virulence of EV-A71.

3 Recombination Between EV-A71 and Other Enteroviruses

The exchanges of genomic domains between poliovirus and other co-circulating enterovirus species affected the plasticity of the EV-A71 genome. It is intriguing that most of the outbreaks documented in the Asia–Pacific region in the last few decades arose from previously circulating genotypes/subgenotypes of EV-A71 and this raises the question concerning their origin and genetic complexity (Tee et al.

2010). A ladder-like tree structure was observed for B1 to B4, C1, C2 and C3 subgenotypes of EV-A71 and the patterns of the phylogenetic structure put forth the idea that a temporal strain replacement occurred through time on a global scale. This suggestion stems from the fact that viral sequences isolated at early time points are located closer to the tree root while sequences collected at a later time point are found further away from the tree root accompanied by limited genetic diversity observed at any one time. However, homologous recombination in the VP1 gene was detected less frequently than recombination outside the VP1 region as functional constraints tend to limit recombination in the capsid genes (Simmonds and Welch 2006). It is important also to note that recent outbreaks of HFMD in different countries caused by the same EV-A71 subgenotype might have arisen from different lineages rather than through the sequential transmission of a single lineage (Shimizu et al. 1999).

Furthermore, EV-A71, as an RNA virus, could undergo copy-choice recombination in order to generate intertypic and intratypic recombinant EV-A71 viruses. Copy choice recombination could occur when the RNA-dependent RNA polymerase dissociated from the genome of a virus and continued to replicate by binding to a second genome, which then generated a new mosaic-like genome with regions originating from different parental strains (Simon-Loriere and Holmes 2011; Worobey and Holmes 1999). In this connection, it was previously demonstrated that a mutation (D79H) in the RdRp of PV was able to reduce the rate of recombination without changing the fidelity of replication (Xiao et al. 2016).

In case of poliovirus, it was established that homologous recombination was mediated by template switching during replication (copy-choice recombination) (Xiao et al. 2016). Recombination in other enteroviruses such as EV-A71 could also follow the same mechanism. Recombination events were observed between human enterovirus species A, B and C. Intertypic recombination was observed between EV-A71 and different human enteroviruses (HEV-A) such as CV-A16, CV-A8 and CV-A10. Intertypic recombination between EV-A71 genotypes B and C could account for the emergence of EV-A7 1 strains causing severe HFMD in large outbreaks in Asia in recent years (Table 2).

The evolutionary rate and the time of emergence of different subgenotypes of EV-A71 revealed that the evolutionary rate of the VP1 gene of EV-A71 was approximately 4.6×10^{-3} and 4.2×10^{-3} substitutions/site/year for genotypes B and C, respectively. The common ancestors of subgenotype B1 (identified in 1972) have been circulating since 1967 before EV-A71 was classified as a human pathogen. Since then, other subgenotypes such as B2 (in the late 1970′s), B3, B4 (in the mid 1990′s) and B5 (in early 2000) were discovered. By comparing the estimates of origin and the date of first detection of each of the subgenotypes, it could be implied that different subgenotypes of EV-A71 genotype B have been circulating for approximately 2 to 5 years before a recombinant virus emerged to cause a large HFMD outbreak (Tee et al. 2010). The genetic diversity and population dynamics addressed whether a rise in EV-A71 cases in China reflected a real increase in the viral spread of a major sub-genotype or it was due to different sub-genotypes. Analysis of 257 eV-A71 subgenotype C4 strains from China between 1998 to 2010

Table 2 Recombination events in the genome of EV-A71

Recombinants of EV-A71	Genes of the EV-A71 involved in recombination	Significance of the recombination events	References
Recombination between species A and B human enteroviruses	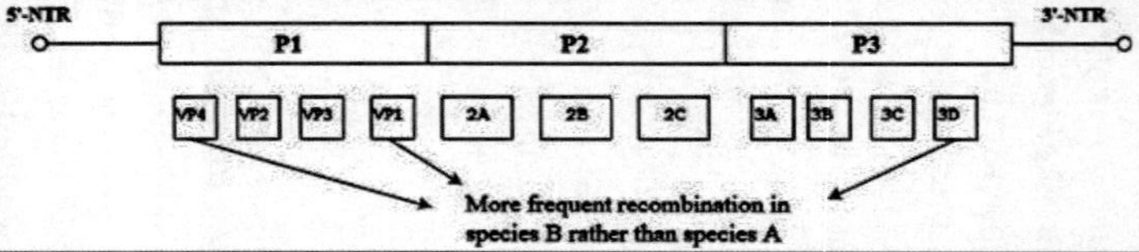	Enterovirus species B showed more recombination events between VP1 and $3D^{pol}$ as well as between VP1 and VP4	(Wang et al. 2012)
EV-A71 subgenotypes A, B (B2, B3 and B4) and C (C2 and C4)	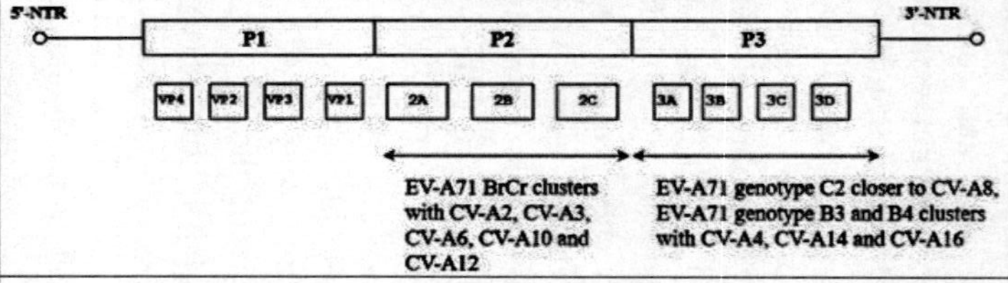	These similarity plots support the likelihood of intertypic recombination between EV-A71 and different Human Enteroviruses A (HEV-A)	(Huang et al. 2008)
EV-A71 strains SZ/HK08-5 and SZ/KH08-6	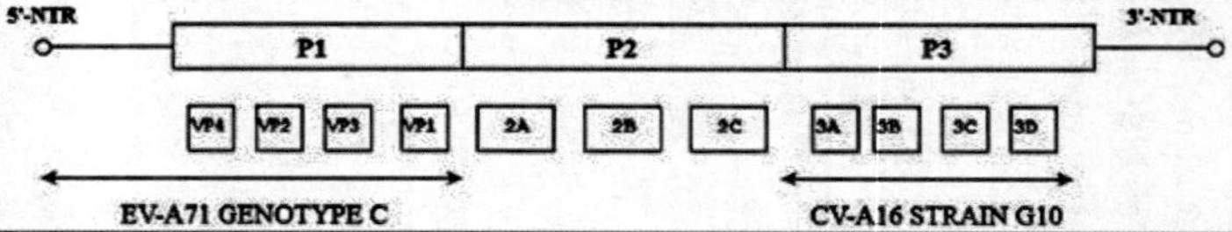	Both EV-A71 strains have > 80% similarity to the EV-A71 genotype C strain (Tainan/4643/98). Both EV-A71 strains showed similarity of ≥ 80% to the G-10 prototype strain of CV-A16	(Yip et al. 2013)
Seven full length EV-A71 C4 sequences from HFMD patients who had severe or mild diseases	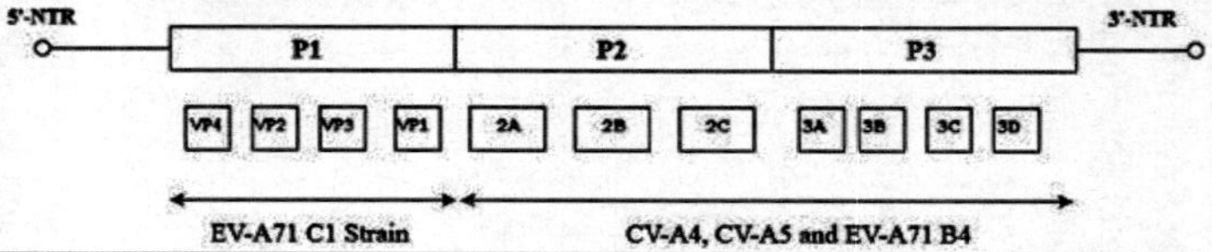	All seven strains might have originated from the same ancestor as they were found in the same cluster after phylogenetic analysis	(Chan and AbuBakar 2006)

(continued)

Table 2 (continued)

Recombinants of EV-A71	Genes of the EV-A71 involved in recombination	Significance of the recombination events	References
EV-A71 subgenotype C2 which was observed to be circulating in Taiwan in 1998	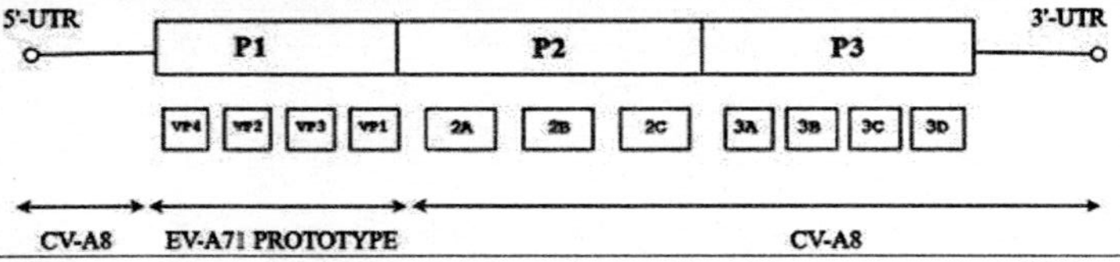	Recombination between CV-A8 and EV-A71 shows evidence of intertypic recombination	(Shimizu et al. 1999)
C4 isolates which circulated in China from 2004 to 2005	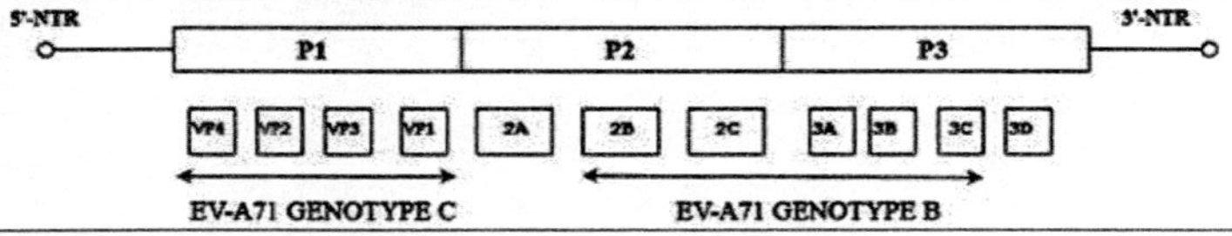	Proof of intratypic recombination was observed between EV-A71 subgenotype C and B	(Guan et al. 2012)

Recombination was observed between EV-A71 and other enteroviruses such as CV-A16, CV-A8 as well as other subgenotypes of EV-A71 revealed an increase in viral spread and continuous replacement of viral lineages by mutations in the VP1 through time (Guan et al. 2012).

Understanding the recombination processes that existed between non-polioviruses is very important. It can give an insight into the evolutionary patterns of EV-A71 by looking into its recombination history throughout the years and might lead to the prediction of new patterns of recombination. Analysis of the EV-A71 genomes showed that recombination events between EV-A71 and Coxsackie viruses are key factors in the evolution of EV-A71 genotypes/subgenotypes (Table 2). Understanding the intertypic and intratypic relationship is an opportunity to discover any possible epidemic strain that might arise to cause large outbreaks in the future. Genome similarities between Coxsackie viruses and EV-A71 reported are core to the idea whereby new EV-A71 strains could arise from recombination of part of the EV-A71 genome with Coxsackie viruses such as CV-A16, CV-A5, CV-A8, CV-A10 and CV-A12. The new recombinants could have acquired virulence determinants to cause major fatal outbreaks as observed in China in recent years.

4 Recombination with Poliovirus and its Implications for EV-A71 Associated HFMD

Poliovirus was nearly eradicated due to worldwide immunization efforts. With the absence of effective vaccines against EV-A71, the latter may become the next important neurotropic pathogen to replace poliovirus and pose a significant health threat to humans (Wang et al. 2012). There are marked similarities between the genome of poliovirus and EV-A71. This similarity is postulated to be the reason why EV-A71 is able to give rise to clinical complications that resemble poliomyelitis such as acute flaccid paralysis and other neurological complications that could impact the cognitive ability of young children (Brown and Pallansch 1995). Even though the nucleotide similarity between poliovirus and EV-A71 indicate that they are not highly genetically related, it is worthwhile noting that the significant neurovirulence determinants present in the 5′-NTR of the wild type Poliovirus are similar to some nucleotides in the 5′-NTR of EV-A71 subgenotype B4 strain 41 (GenBank: AF316321) (Fig. 2) (Brown and Pallansch 1995).

This observation is supported by the hypothesis that EV-A71 could have evolved from recombination between part of the genome of CV-A16 and 5′-NTR of poliovirus. It was revealed that the nucleotide identity of the EV-A71 (EV71/MS/7423) strain with CV-A16 and EV-A71 prototype BrCr strains is high at 77% and 81%, respectively. However, the overall nucleotide identity to Poliovirus is only 58%. The amino acid identity of the EV71/MS/7423 to both EV-A71 BrCr strain and CV-A16 is high at 95% and 89%, respectively. However, the overall amino

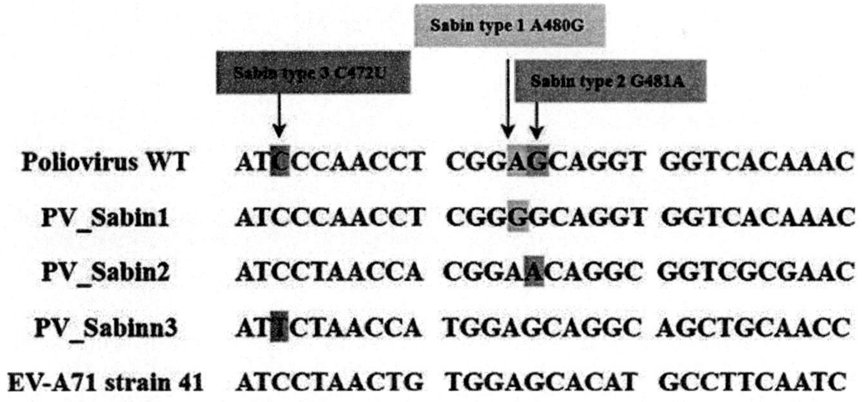

Fig. 2 Comparison of the nucleotide sequence between the wild type (WT) poliovirus, poliovirus (PV) Sabin Strain 1, 2 and 3 and EV-A71 subgenotype B4 strain 41 (GenBank: AF316321). PV Sabin 1, PV Sabin 2 and PV Sabin 3 contain altered nucleotides in the 5′-NTR of ssRNA. Shaded areas indicate the locations where the altered nucleotides are in each of the three Sabin strains

Table 3 Comparison of nucleotide and amino acid identities between the prototype strain of EV-A71 (BrCr), Coxsackie A16 (CV-A16) and wild type poliovirus (PV)

Nucleotide identity of strain EV71/MS/7423 with other enteroviruses												
	Whole Genome		5′-NTR		P1		P2		P3		3′ NTR	
EV-A71/BrCr	81		85		82		77		80		92	
CV-A16	77		86		68		82		79		79	
PV	58		71		53		58		61		41	
Amino acid identity of strain EV71/MS/7423 with other enteroviruses												
	Whole Genome	P1		P2	P3	VP1		VP2		VP3		VP4
EV-A71/BrCr	95	97		94	94	93		99		99		100
CV-A16	89	79		95	95	71		84		84		78
PV	55	46		59	62	36		55		45		58

The comparison among the different enteroviruses reported throughout specific regions of the genomes from 5′-NTR, P1, P2, P3 and 3′-NTR (Brown and Pallansch 1995)

acid identity to poliovirus is only 55% (Table 3) (Brown and Pallansch 1995). The greater identity between EV-A71 and CV-A16 could support the hypothesis that EV-A71 evolved from CV-A16 and acquired the virulence determinants from the poliovirus wild-type (Cardosa et al. 2003).

5 Mutation and its impact on pathogenesis of EV-A71

Another important aspect which plays a role in driving the virulence of EV-A71 could be the fidelity of the RNA polymerase and its interactions with host molecules. In an attempt to understand the role of the 3D polymerase in the virulence of EV-A71, it was discovered that the $3D^{pol}$ entered the nucleus and targeted the pre-mRNA processing factor, prp8, to block pre-mRNA splicing and mRNA synthesis of the cell. Specifically, the finger domain of the $3D^{pol}$ was found to associate with the C-terminal region of the prp8 which contained the Jab1/MPN domain. This interaction interfered with the second catalytic step, inhibiting mRNA production and inducing the build-up of its lariat form in the nucleus. This novel mechanism of viral attack contributed to the pathogenesis of the EV-A71 viral infection. However, this novel mechanism was only observed in certain viruses such as EV-A71 and PV. Other viruses such as CVB-3 and HRV-16 did not inhibit pre-mRNA splicing as it could not associate with the C-terminal of the cellular prp8 (Liu et al. 2014). Since recombination is a fairly common occurrence in EV-A71, this particular study raises the question whether recombination between EV-A71 and another enterovirus such as Coxsackie virus could affect the virulence of the newly evolved recombinant virus. Following copy-choice mechanism of recombination, the recombinant virus could possess a $3D^{pol}$ which lacked the ability to stop pre-mRNA splicing. This would ultimately affect the virulence of the recombinant virus whereby it would have restricted access to the host machinery to make more viral RNA (Liu et al. 2014).

In order to investigate whether mutations with the ability to decrease recombination could affect the fidelity of replication, the mutation D79H was observed to have reduced recombination but had no impact on fidelity, and recombinant polioviruses derived from the D79H mutation displayed a reduction in the accumulation of advantageous alleles and an increase in detrimental mutations. Using CirSeq analysis, the frequency of beneficial mutations was found to increase 100-fold over 7 passages for the wild type virus. However, no accumulation of beneficial mutations was observed for single mutants D79H and H273R. It is important to note that combining the mutations affecting fidelity and recombination resulted in major defects in the accumulation of beneficial mutations (Xiao et al. 2016). More in-depth analysis aimed to understand whether double mutations affecting recombination and fidelity inhibited viral adaptation was conducted in mice. Interestingly, the wild type virus was observed to cause fatality within 6 days while the recombination-deficient strain carrying the mutations D79H was as lethal as the WT. As for the low-fidelity mutant H273R, a milder phenotype was reported. Combining either high-fidelity or low-fidelity mutation with recombination-impairing mutations would dramatically reduce the virulence. It would appear that decreasing the rate of recombination as well as changing the replication fidelity could lower the ability of the virus to adapt within the infected host, resulting in a strong attenuation phenotype (Xiao et al. 2016). From the lethal dose (LD_{50}) observed, the lack of viral adaptation led to the alteration in tissue tropism which stopped the onset of infection in the CNS while allowing wild type levels of replication in the non-neuronal tissues (Xiao et al. 2016). Akin to these findings, it would be interesting to explore whether the RNA polymerase in EV-A71 could bear mutations which would decrease the recombination rate and affect viral fitness and adaptation. Findings from such investigations would help to shed some light on the successive emergence and disappearance of EV-A71 lineages. Ultimately, a clearer picture on the evolution strategies of EV-A71 would have a greater bearing on rational vaccine design.

6 Spontaneous Mutation in the 5′-NTR of EV-A71

When the nucleotide at position 158, cytosine, was substituted with uridine, the conformation of the RNA secondary structure of stem loop II within the 5′-NTR of isolate 4643 (subgenotype B1) obtained in 1998 was changed, leading to a decrease in viral translation and reduced virulence in mice (Yeh et al. 2011). Further evidence of spontaneous mutations affecting virulence of EV-A71 C4 and C4a subgenotypes was reported at three different positions: Val^{814}/Ile^{814} in the VP1, Val^{1148}/Ile^{1148} in the 3A and Ala^{1728}/Cys^{1728} and Val^{1728} in the 3C genes. These amino acids were highly conserved in the neuro-virulent strains of the subgenotype C4a. The implication from the study is that these amino acids are potential molecular determinants of virulence of EV-A71 subgenotype C4a. Variations of the secondary structure of the 5′-NTR were also observed in the fatal strain at three positions (C241T, A571T and C579T) and one position in the 3′-NTR (T7335C) and these amino acids might confer significant virulence determinants (Wen et al. 2013). When the sequences of EV-A71 responsible for severe and mild HFMD cases (25 SC-EV-A71 and 31 MC-EV-A71) across different genotypes and subgenotypes (A, B1-B5 and C1-C4) were compared, 4 amino acid residues ($Gly^{710}/Gln^{710}/Arg^{710}/Glu^{29}$) at 2 positions in the VP1, one residue (Lys^{930}) in the protease 2A and four nucleotides at three positions (G272, U488 and A700U) in the 5′-NTR were implied to be associated with the EV-A71 virulent phenotype (Li et al. 2011).

7 Spontaneous Mutations in the VP1 of EV-A71

Nine amino acid substitutions at different positions in the EV-A71 genome, namely: H22Q, P27S, N31S/D, E98K, E145G/Q, D164E, T240A/S, V249I, and A289T obtained from the alignment of VP1 sequences of the SC-EV-A71 (strains causing severe HFMD) versus the MC-EV-A71 (strains causing mild HFMD) were reported to have arisen spontaneously. The changes in these amino acids might be playing a role in conferring virulence to the EV-A71 subgenotype C4 strain (Liu et al. 2014). The authors hypothesized from these data that the VP1 of the EV-A71 genome could act as a sandwich switch for the stabilization of viral particles and cellular receptor attachments. When EV-A71 carries either G or Q at VP1-145 (VP1-145G/Q), it has the ability to interact with the residues of the cellular receptor, PSGL-1 N-terminus and as a molecular switch to modulate viral binding to the cell receptor by controlling the exposure of the amino acid (VP1-244 K) on the VP1 surface. Nonbinding viruses were observed to carry VP1-145E (Nishimura et al. 2013). Moreover, the critical substitution of E-Q at amino acid position 145 within the VP1 occurred multiple times in different virulent strains responsible for severe cases of HFMD (Chang et al. 2012). Therefore, specific mutations such as the E145G/Q detected in the VP1 of SC-EV-A71 have the ability to convert EV-A71 strains associated with mild infections to strains capable of causing severe HFMD (Liu et al. 2014). Other spontaneous mutations such as the K215A located at the VP1 GH loop was observed to increase the thermal stability of the virus, and thermostable mutants could be produced by changing the positively charged residues to alanine by alanine scanning of the VP1 (Yuan et al. 2015).

It was reported previously that an important contributor to virulence is a group of positively charged amino acids near the fivefold vertices of the EV-A71 capsid, and many of these positively charged amino acids are conserved across EV-A71 genotypes, suggesting their importance to the virus. Identification of single mutations in the conserved, positively charged fivefold region of the VP1 capsid protein such as G145E, E145Q, or K244E might therefore affect virulence in a significant way. Caine et al. (2016) discovered a mutation at nt. 244 (K244E) in the VP1 which is crucial for the virulence of the mouse-adapted EV-A71 (mEV-A71) strain as it led to the expansion of tissue tropism and viral spread in adult interferon-deficient AG129 mice. Another spontaneous VP1 mutation (H37R) arose when mEV-A71 was passaged in primate cells, and H37R was found to be important for the recovery of mEV-A71 in rhabdomyosarcoma cells (RD cells). It was postulated that H37E and K244E interactions were important for replication in primate cells but K244E alone was able to confer the ability of the virus to replicate in mice (Caine et al. 2016).

In a more recent study, molecular modelling indicated that mutations in the 5′-NTR and the VP1 influenced the EV-A71 engagement with human receptors SCARB2 and PSGL-1 and the virulence in vivo. During EV-A71 propagation in Vero (EV-V) and RD cells (EV-R), it was discovered that EV-V was more virulent while EV-R displayed altered virulence due to its acquired nucleotide substitution (T494C) in the 5′-NTR and mutations E145G, V146I and S241L in the VP1. The in vivo study revealed that 100% of mice receiving lethal dose of EV-V died while those which received EV-R survived. The authors speculated that mutations in the VP1 affected the structural conformation of the capsid which in turn altered the ability of the virus to bind to the cellular receptor. To investigate this, interactions of EV-V with SCARB2 were carried out and results indicated that EV-V had more interactions with SCARB2 when compared to EV-R. In contrast, the EV-A71 propagated in RD cells (EV-R) had a higher affinity to PSGL-1; hence, propagation in different host cells drove the virus to adapt to the physiological environment via mutations in the VP1 (Chang et al. 2018). Consequently, when a virus undergoes mutations which led to a change in the surface proteins, the antibodies elicited by an inactivated vaccine strain from previous vaccinations would not be effective to neutralize any mutated viruses that might arise (Huang et al. 2015). Thus, there is a real need to study the potential mutations which could compromise the efficacy of an inactivated EV-A71 vaccine.

8 Spontaneous Mutations in the 2A and 3C of EV-A71

Certain spontaneous mutations such as the non-synonymous mutation (K149I) within the VP2 of a mouse-adapted strain of EV-A71 were established to be responsible for an increase in virulence in mice. Moreover, the identification of the K216R within the 2C protein demonstrated an improved growth of the strain in vitro but did not lead to an increase in virulence in mice (Chua et al. 2008). A critical mutation characterized by Asn^{1617} within the 3C protease gene was present in highly virulent strains which were different from Asp^{1617} present in strains with low virulence. This highlights the position of this specific amino acid which led to a conformational change at the active-centre of the 3C proteinase ($3C^{pro}$) and could serve as a potential molecular determinant of virulence in the EV-A71 subgenotype C4a (Li et al. 2016). These findings establish the significance of single mutations which arose spontaneously within the viral genome and that have an impact on the growth in vitro and in vivo (Table 4). When the strains of EV-A71 subgenotype C4, isolated from severe HFMD were compared to those isolated from mild HFMD, the majority of mutations were located in the 5′-NTR and the VP1 regions. For instance, 10 mutations were observed in the VP1 alone (Liu et al. 2014; Nishimura et al. 2013; Yeh et al. 2011; Yuan et al. 2015). However, when 2 strains of EV-A71 subgenotype C4a were compared with the C4 subgenotype, the genes affected by the spontaneous mutations appeared to span across the entire length of the genome, ranging from the 5′-NTR to the 3′-NTR (Li et al. 2011; Wen et al. 2013). The question as to whether these mutations are genotype/subgenotype specific or they act as a universal indicator of virulence remains to be fully elucidated. Hence, it would be compelling to investigate the virulence of these strains by quantifying the degree of virulence conferred by each of the mutated genes so as to gain a better understanding of how exactly they could affect virulence (Table 4).

Interestingly, when the fatal EV-A71 strain isolated from the severe HFMD outbreak in Singapore in 2000 was analyzed, it was found to differ from the non-fatal strain C10 (5666/Sin/002,209) only at the nucleotide position 5262. The fatal strain carried an A nucleotide at position 5262 of the genome (Fig. 3) and this led to the replacement of alanine in the non-fatal strain with threonine in the fatal strain (Singh et al. 2002). To check whether threonine at this position could have contributed to the fatality of the strain, Yee et al. (2016) quantified the virulence of the fatal strain (C41) versus the non-fatal strain (C10) by site directed mutagenesis of the A $\rightarrow$ G nucleotide. The mutation at position 5262 (A5262G) led to a 75%

Table 4 Spontaneous mutations in different EV-A71 subgenotypes (BrCr, B1-B5, C1-C5)

Mutant strains of EV-A71	Position of amino acid(s) on the EV-A71 genomes of mutants	Significance of the mutations In the EV-A71 genome	References
Analysis of EV-A71 subgenotype B1 showed changes in the 5′-NTR	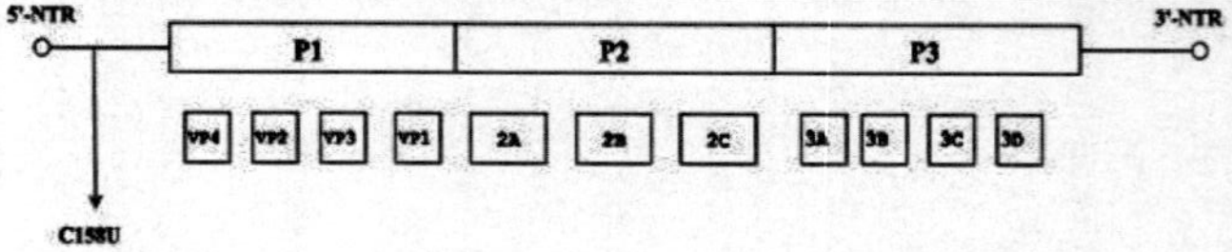	When nucleotide cytosine was substituted with uridine at position 158, the conformation of the RNA secondary structure of stem loop II in the 5′-NTR changed, leading to a decrease in viral translation and virulence in mice	(Yeh et al. 2011)
Nucleotide and amino acid changes in neuro-virulent strains of EV-A71 subgenotype C4a	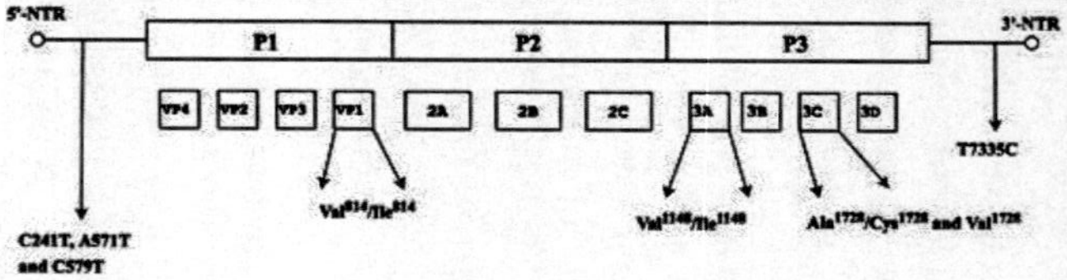	These amino acids are potential molecular determinants of virulence. Variations on the secondary structure of the 5′-NTR at three positions (C^{241}/T^{241}, A^{571}/T^{571} and C^{579}/T^{579}) and one position in the 3′-NTR (T^{7335}/C^{7335}) might confer fatality	(Wen et al. 2013)
Comparisons of EV-A71 across different genotypes (BrCr, B1-B5 and C1-C5)	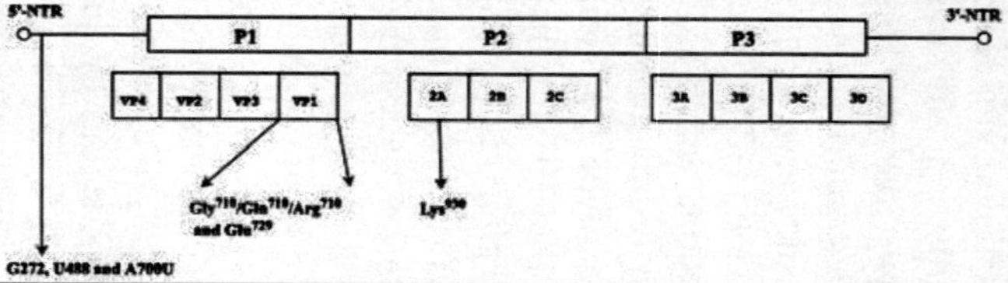	These amino acid residues might be associated with the EV-A71 virulent phenotype	(Li et al. 2011)

(continued)

Table 4 (continued)

Mutant strains of EV-A71	Position of amino acid(s) on the EV-A71 genomes of mutants	Significance of the mutations In the EV-A71 genome	References
Changes in VP1 sequences of EV-A71 subgenotype C4 causing severe HFMD	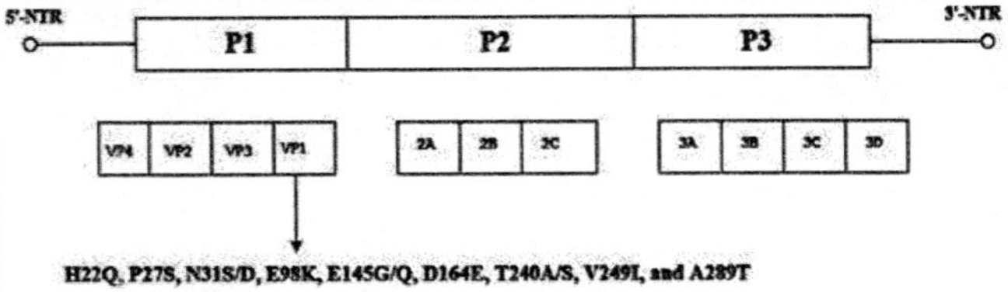	E145G/Q interacted with residues of PSGL-1 N-terminus and acted as a molecular switch to modulate binding to the cell receptor by controlling the exposure of the amino acid (VP1-244 K) on the VP1 surface	(Liu et al. 2014; Nishimura et al. 2013)
Analysis of EV-A71 subgenotype C4 showed changes in the VP1	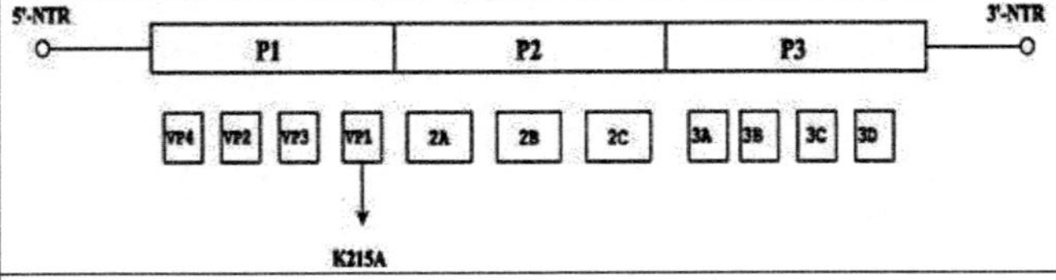	K215A located at the VP1 GH loop increased the thermal stability of the virus	(Yuan et al. 2015)
Role of K244E and H37R were investigated by reverse engineering in the EV71-B2 isolate, MS/7423/87	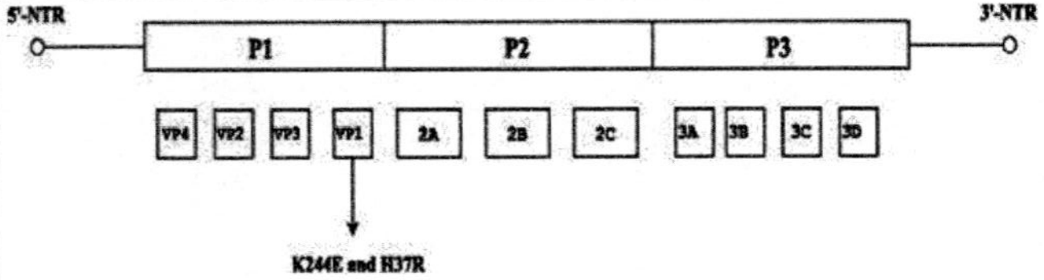	It was postulated that H37R and K244E interactions were important for replication in primate cells but K244E alone was able to confer the ability of the virus to replicate alone in a murine model	(Caine et al. 2016)

(continued)

Table 4 (continued)

Mutant strains of EV-A71	Position of amino acid(s) on the EV-A71 genomes of mutants	Significance of the mutations In the EV-A71 genome	References
Role of K216R, G145E and K149I in the mouse adapted strain of EV-A71 26 M/AUS/4/99	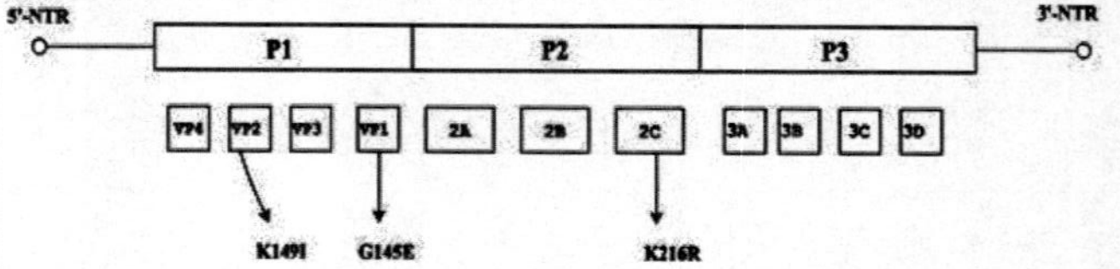	G145E mutation was solely responsible for an increase in virulence in mice whilst K149I led to an improved growth of the strain in vitro but did not lead to increased virulence in mice	(Chua et al. 2008)
Analysis of the genomes of six EV-A71 strains of subgenotype C4a identified the only change of amino acid Asn[1617] in the 3C gene	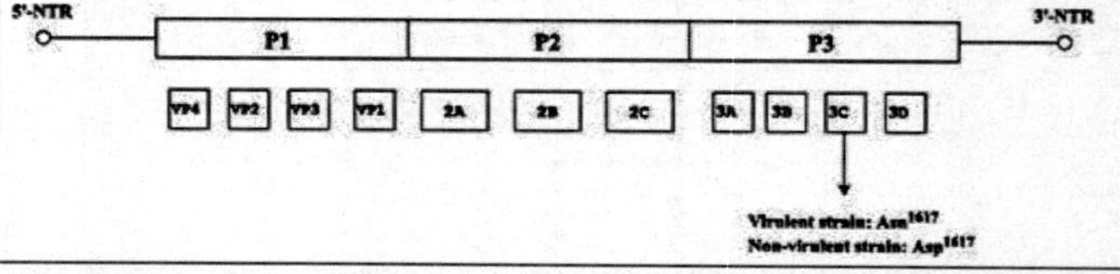	This specific amino acid which led to conformational change at the active centre of the 3C proteinase ($3C^{pro}$) and this could be a potential molecular determinant for the EV-A71	(Li et al. 2016)

Analysis of different EV-A71 strains showing the position of the spontaneous mutation s present in the genomes of EV-A71 and their significant impact on virulence

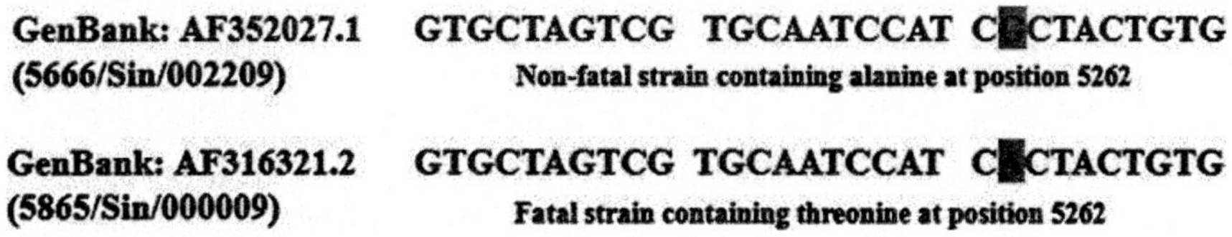

Fig. 3 Comparison of a fatal (AF316321.2) and a non-fatal (AF 352,027.1) strains isolated from the HFMD outbreak in Singapore (2000). The difference of a single nucleotide (A) at position 5262 (amino acid residue 1506) in the 3A non-structural region of the fatal strain is highlighted in grey (Singh et al. 2002)

reduction in viral RNA copy number and 90% reduction in plaque forming ability in the non-fatal strain when compared to the fatal strain (Yee et al. 2016).

Yee et al. (2016) further evaluated the effects of specific mutations of the EV-A71 subgenotype B4 strain 41 genome on virulence by mutating at multiple positions (158, 475, 486 and 487) and through partial deletion of 11 base pairs from nucleotides 475 to 486 in the 5′-NTR region. This strain was shown to encode for several significant molecular determinants. The mutant 475 (C475T) and the partial deletant (PD) (Δ11bps, nucleotides 475–486 in the 5′-NTR) showed significantly lower cytopathic effects and very low viral RNA copy numbers. However, analysis of RNA copy number showed that mutant A486G still produced high levels of viral RNA while mutant G487A showed minimal levels of RNA copy numbers in the RD cells. The data provided quantitative comparisons of the impact of molecular determinants of virulence in the EV-A71 subgenotype B4 strain 41 (Yee et al. 2016). Published data have shown that there were amino acid differences between fatal and non-fatal strains isolated from China (Chang et al. 2012; Li et al. 2016, 2011; Liu et al. 2014). Does this imply that some EV-A71 strains with higher virulence were causing the large-scale outbreaks in China? From the data presented, both intratypic and intertypic recombination could generate such highly virulent strains. The strong likelihood that multiple errors arise each time the EV-A71 virus replicates also raises the question as to whether a single mutation such as that observed in the VP1-145 residue is really the root cause of fatality or rather, a community of mutants carrying multiple amino acid substitutions increases the viral fitness of the viral population, thereby conferring higher pathogenecity. Under normal circumstances, a viral genome replicates and creates hundreds of progeny viruses which could differ by an amino acid substitution at a single position. In the following rounds of replication, more complex mutants could be generated which could differ significantly from the original master sequence and this ensemble of mutants form a quasispecies (Lauring and Andino 2010). It would be pertinent to question whether a single EV-A71 fatal strain arising from recombination and/or harbouring spontaneous mutations is the only rationale behind sudden outbreaks of HFMD in various countries or there could be the cooperation of diverse quasispecies with varying virulence. It remains to be confirmed whether this claim stands true.

9 Conclusion

EV-A71 is an enterovirus that is currently responsible for many large-scale outbreaks of HFMD with high fatality in the Asia Pacific region. In this chapter, we investigated whether recombination and spontaneous mutations are the key parameters that drive the diversity of a viral population. It was revealed that recombination occurred extensively throughout the genome of EV-A71 and the data suggests that this might be a key factor in determining the increased virulence in some strains. On the other hand, the effects of spontaneous mutations in the EV-A71 genome affecting virulence was discussed. Through an extensive study of recombination events and spontaneous mutations affecting virulence, more insights into the evolution of EV-A71 could elucidate its pathogenesis. Ultimately, more investigations of molecular determinants of pathogenesis of EV-A71 should be carried out as little is known whether a single universal or multiple molecular determinants of virulence are at play in the pathogenesis of EV-A71. Providing an answer to these questions could pave the way for the rational design of an effective and genetically stable live-attenuated vaccine (LAV) against EV-A71.

Acknowledgments Part of this work was presented at Vaccines R&D 2019 (Vaccines Research & Development), 18–20 November 2019 at Boston, MA, USA.

Funding This work was funded by the Sunway University Research Centre Grant to support research projects in the Centre for Virus and Vaccine Research, School of Medical and Life Sciences.

Conflicts of Interest The authors declare no conflict of interest.

References

Andino R, Böddeker N, Silvera D, Gamarnik AV. Intracellular determinants of picornavirus replication. Trends Microbiol. 1999;7(2):76–82.

Bessaud M, Razafindratsimandresy R, Nougairede A, Joffret ML, Deshpande JM, Dubot-Peres A, Heraud JM, de Lamballerie X, Delpeyroux F, Bailly JL. Molecular comparison and evolutionary analyses of VP1 nucleotide sequences of new African human enterovirus 71 isolates reveal a wide genetic diversity. PLoS ONE. 2014;9(3):e90624. https://doi.org/10.1371/journal.pone.0090624.

Brown BA, Oberste MS, Alexander JP Jr, Kennett ML, Pallansch MA. Molecular epidemiology and evolution of enterovirus 71 strains isolated from 1970 to 1998. J Virol. 1999;73(12):9969–75.

Brown BA, Pallansch MA. Complete nucleotide sequence of Enterovirus 71 is distinct from poliovirus. Virus Res. 1995;39(2–3):195–205.

Caine EA, Moncla LH, Ronderos MD, Friedrich TC, Osorio JE. A single mutation in the VP1 of enterovirus 71 is responsible for increased virulence and neurotropism in adult interferon-deficient mice. J Virol. 2016;90(19):8592–604. https://doi.org/10.1128/JVI.01370-16.

Cardosa MJ, Perera D, Brown BA, Cheon D, Chan HM, Chan KP, Cho H, McMinn P. Molecular epidemiology of human enterovirus 71 strains and recent outbreaks in the Asia-Pacific region: comparative analysis of the VP1 and VP4 genes. Emerg Infect Dis. 2003;9(4):462.

Chan YF, AbuBakar S. Phylogenetic evidence for inter-typic recombination in the emergence of human enterovirus 71 subgenotypes. BMC Microbiol. 2006;6:74. https://doi.org/10.1186/1471-2180-6-74.

Chang CK, Wu SR, Chen YC, Lee KJ, Chung NH, Lu YJ, Yu SL, Liu CC, Chow YH. Mutations in VP1 and 5'-UTR affect enterovirus 71 virulence. Sci Reports. 2018;8(1):6688. https://doi.org/10.1038/s41598-018-25091-7.

Chang PC, Chen SC, Chen KT. The current status of the disease caused by Enterovirus 71 infections: epidemiology, pathogenesis, molecular epidemiology, and vaccine development. Int J Environ Res Public Health. 2016; 13(9). doi:https://doi.org/10.3390/ijerph13090890.

Chang SC, Li WC, Chen GW, Tsao KC, Huang CG, Huang YC, Chiu CH, Kuo CY, Tsai KN, Shih SR, Lin TY. Genetic characterization of enterovirus 71 isolated from patients with severe disease by comparative analysis of complete genomes. J Med Virol. 2012;84(6):931–9. https://doi.org/10.1002/jmv.23287.

Chua BH, Phuektes P, Sanders SA, Nicholls PK, McMinn PC. The molecular basis of mouse adaptation by human enterovirus 71. J Gen Virol. 2008;89(Pt 7):1622–32. https://doi.org/10.1099/vir.0.83676-0.

Chumakov M, Voroshilova M, Shindarov L, Lavrova I, Gracheva L, Koroleva G, Vasilenko S, Brodvarova I, Nikolova M, Gyurova S, Gacheva M, Mitov G, Ninov N, Tsylka E, Robinson I, Frolova M, Bashkirtsev V, Martiyanova L, Rodin V. Enterovirus 71 isolated from cases of epidemic poliomyelitis-like disease in Bulgaria. Adv Virol. 1979;60(3–4):329–40.

Domingo E, Holland JJ. RNA virus mutations and fitness for survival. Annu Rev Microbiol. 1997;51(1):151–78.

Domingo E, Sabo D, Taniguchi T, Weissmann C. Nucleotide sequence heterogeneity of an RNA phage population. Cell. 1978;13(4):735–44.

Ferrer-Orta C, Arias A, Escarmis C, Verdaguer N. A comparison of viral RNA-dependent RNA polymerases. Curr Opin Struct Biol. 2006;16(1):27–34. https://doi.org/10.1016/j.sbi.2005.12.002.

Guan D, van der Sanden S, Zeng H, Li W, Zheng H, Ma C, Su J, Liu Z, Guo X, Zhang X, Liu L, Koopmans M, Ke C. Population dynamics and genetic diversity of C4 strains of human enterovirus 71 in Mainland China, 1998–2010. PLoS ONE. 2012;7(9):e44386. https://doi.org/10.1371/journal.pone.0044386.

Hirst GK. Genetic recombination with Newcastle disease virus, polioviruses, and influenza. Paper presented at the Cold Spring Harbor Symposia on Quantitative Biology. 1962.

Holland J, Spindler K, Horodyski F, Grabau E, Nichol S, VandePol S. Rapid evolution of RNA genomes. Science. 1982;215(4540):1577–85.

Huang CC, Liu CC, Chang YC, Chen CY, Wang ST, Yeh TF. Neurologic complications in children with enterovirus 71 infection. N Engl J Med. 1999;341(13):936–42. https://doi.org/10.1056/nejm199909233411302.

Huang SC, Hsu YW, Wang HC, Huang SW, Kiang D, Tsai HP, Wang SM, Liu CC, Lin KH, Su IJ, Wang JR. Appearance of intratypic recombination of enterovirus 71 in Taiwan from 2002 to 2005. Virus Res. 2008;131(2):250–9. https://doi.org/10.1016/j.virusres.2007.10.002.

Huang SW, Tai CH, Fonville JM, Lin CH, Wang SM, Liu CC, Su IJ, Smith DJ, Wang JR. Mapping enterovirus A71 antigenic determinants from viral evolution. J Virol. 2015;89(22):11500–6. https://doi.org/10.1128/jvi.02035-15.

Lauring AS, Andino R. Quasispecies theory and the behavior of RNA viruses. PLoS Pathog. 2010;6(7):e1001005. https://doi.org/10.1371/journal.ppat.1001005.

Ledinko N. Genetic recombination with poliovirus type 1: studies of crosses between a normal horse serum-resistant mutant and several guanidine-resistant mutants of the same strain. Virology. 1963;20(1):107–19.

Li P, Yue Y, Song N, Li B, Meng H, Yang G, Li Z, An L, Qin L. Genome analysis of enterovirus 71 strains differing in mouse pathogenicity. Virus Genes. 2016;52(2):161–71. https://doi.org/10.1007/s11262-015-1271-0.

Li R, Zou Q, Chen L, Zhang H, Wang Y. Molecular analysis of virulent determinants of enterovirus 71. PLoS ONE. 2011;6(10):e26237. https://doi.org/10.1371/journal.pone.0026237.

Liu Y, Fu C, Wu S, Chen X, Shi Y, Zhou B, Zhang L, Zhang F, Wang Z, Zhang Y, Fan C, Han S, Yin J, Peng B, Liu W, He X. A novel finding for enterovirus virulence from the capsid protein VP1 of EV71 circulating in mainland China. Virus Genes. 2014;48(2):260–72. https://doi.org/10.1007/s11262-014-1035-2.

Liu YC, Kuo RL, Lin JY, Huang PN, Huang Y, Liu H, Arnold JJ, Chen SJ, Wang RY, Cameron CE, Shih SR. Cytoplasmic viral RNA-dependent RNA polymerase disrupts the intracellular splicing machinery by entering the nucleus and interfering with Prp8. PLoS Pathog. 2014;10(6):e1004199. https://doi.org/10.1371/journal.ppat.1004199.

Nagy G, Takatsy S, Kukan E, Mihaly I, Domok I. Virological diagnosis of enterovirus type 71 infections: experiences gained during an epidemic of acute CNS diseases in Hungary in 1978. Adv Virol. 1982;71(3):217–27. https://doi.org/10.1007/bf01314873.

Nishimura Y, Lee H, Hafenstein S, Kataoka C, Wakita T, Bergelson JM, Shimizu H. Enterovirus 71 binding to PSGL-1 on leukocytes: VP1-145 acts as a molecular switch to control receptor interaction. PLoS Pathog. 2013;9(7):e1003511. https://doi.org/10.1371/journal.ppat.1003511.

Schmidt NJ, Lennette EH, Ho HH. An apparently new enterovirus isolated from patients with disease of the central nervous system. J Infect Dis. 1974;129(3):304–9.

Shimizu H, Utama A, Yoshii K, Yoshida H, Yoneyama T, Sinniah M, Yusof MA, Okuno Y, Okabe N, Shih SR, Chen HY, Wang GR, Kao CL, Chang KS, Miyamura T, Hagiwara A. Enterovirus 71 from fatal and nonfatal cases of hand, foot and mouth disease epidemics in Malaysia, Japan and Taiwan in 1997–1998. Jpn J Infect Dis. 1999;52(1):12–5.

Simmonds P, Welch J. Frequency and dynamics of recombination within different species of human enteroviruses. J Virol. 2006;80(1):483–93. https://doi.org/10.1128/jvi.80.1.483-493.2006.

Simon-Loriere E, Holmes EC. Why do RNA viruses recombine? Nat Rev Microbiol. 2011;9(8):617–26. https://doi.org/10.1038/nrmicro2614.

Singh S, Poh CL, Chow VT. Complete sequence analyses of enterovirus 71 strains from fatal and non-fatal cases of the hand, foot and mouth disease outbreak in Singapore (2000). Microbiol Immunol. 2002;46(11):801–8.

Solomon T, Lewthwaite P, Perera D, Cardosa MJ, McMinn P, Ooi MH. Virology, epidemiology, pathogenesis, and control of Enterovirus 71. Lancet Infect Dis. 2010;10(11):778–90.

Tee KK, Lam TT, Chan YF, Bible JM, Kamarulzaman A, Tong CY, Takebe Y, Pybus OG. Evolutionary genetics of human enterovirus 71: origin, population dynamics, natural selection, and seasonal periodicity of the VP1 gene. J Virol. 2010;84(7):3339–50. https://doi.org/10.1128/JVI.01019-09.

Wang X, Zhu C, Bao W, Zhao K, Niu J, Yu XF, Zhang W. Characterization of full-length enterovirus 71 strains from severe and mild disease patients in northeastern China. PLoS ONE. 2012;7(3):e32405. https://doi.org/10.1371/journal.pone.0032405.

Wen HL, Si LY, Yuan XJ, Hao SB, Gao F, Chu FL, Sun CX, Wang ZY. Complete genome sequencing and analysis of six enterovirus 71 strains with different clinical phenotypes. Virol J. 2013;10:115. https://doi.org/10.1186/1743-422X-10-115.

WHO. Hand, foot and mouth disease situation update number 511; 2017. Retrieved from https://www.wpro.who.int/emerging_diseases/HFMD.situation.updates.2017/en/.

Worobey M, Holmes EC. Evolutionary aspects of recombination in RNA viruses. J Gen Virol. 1999;80:2535–43. https://doi.org/10.1099/0022-1317-80-10-2535.

Xiao Y, Rouzine IM, Bianco S, Acevedo A, Goldstein EF, Farkov M, Brodsky L, Andino R. RNA recombination enhances adaptability and is required for virus spread and virulence. Cell Host Microbe. 2016;19(4):493–503. https://doi.org/10.1016/j.chom.2016.03.009.

Xing W, Liao Q, Viboud C, Zhang J, Sun J, Wu JT, Chang Z, Liu F, Fang VJ, Zheng Y, Cowling BJ, Varma JK, Farrar JJ, Leung GM, Yu H. Hand, foot, and mouth disease in China, 2008–12: an epidemiological study. Lancet Infect Dis. 2014;14(4):308–18. https://doi.org/10.1016/s1473-3099(13)70342-6.

Yee PT, Tan KO, Othman I, Poh CL. Identification of molecular determinants of cell culture growth characteristics of Enterovirus 71. Virol J. 2016;13(1):194. https://doi.org/10.1186/s12985-016-0645-9.

Yeh MT, Wang SW, Yu CK, Lin KH, Lei HY, Su IJ, Wang JR. A single nucleotide in stem loop II of 5'-untranslated region contributes to virulence of enterovirus 71 in mice. PLoS ONE. 2011;6 (11):e27082. https://doi.org/10.1371/journal.pone.0027082.

Yip CC, Lau SK, Woo PC, Yuen KY. Human enterovirus 71 epidemics: what's next? Emerg Health Threats J. 2013;6:19780. https://doi.org/10.3402/ehtj.v6i0.19780.

Yuan S, Li G, Wang Y, Gao Q, Wang Y, Cui R, Altmeyer R, Zou G. Identification of positively charged residues in enterovirus 71 capsid protein VP1 essential for production of infectious particles. J Virol. 2015;90(2):741–52. https://doi.org/10.1128/JVI.02482-15.

Merkel Cell Polyomavirus

Ubiquitous Merkel Cell Polyomavirus: Causative Agent of the Rare Merkel Cell Carcinoma

Naveed Shahzad, Usman Shah Gilani, Menahil Mahmood, Fareeda Tasneem, Muhammad Farhan Ul Haque, and Iqra Hussain

Abstract Merkel Cell polyomavirus (MCPyV) is the only member of the *Polyomaviridae* family that is directly linked to a type of human cancer, Merkel Cell Carcinoma (MCC). Predominantly, the clinical significance of MCPyV is due to the aggressive nature of MCC with a low survival rate and, presently, unavailability of a comprehensive and effective treatment regime. Secondly, the molecular mechanisms of MCPyV infection and oncogenic potential are currently poorly understood. Despite MCPyV ubiquitously infects humans, studies suggest that MCC is caused only after prolonged infection and integration of MCPyV DNA in the host genome. Mechanistically, transformation of normal cells into cancerous by MCPyV is mainly driven by truncated LT and sT antigens and the abnormal molecular modulation they carry out. Presently, conventional treatment options against MCC like surgery, radiotherapy and chemotherapy are on board, but, with limitations that pose them inadequate. However, modern treatment options are emerging which, in preliminary investigation, show great promise. However, extensive exploration needs to be carried out before their large scale acceptance and application as therapies against MCC.

1 Introduction

Polyoma viridae is a family of ubiquitous, icosahedral and non-enveloped viruses that contain small circular ds DNA genomes (DeCaprio and Garcea 2013). Till now, 14 different species of human polyoma viruses have been identified. Some of these polyomaviruses such as BKPyV, JCPyV, WUPyV, HPyV6, HPyV7 and TSPyV have been associated with human malignancies (Table 1). However, Merkel cell polyomavirus (MCPyV) is the only known human polyomavirus that is

N. Shahzad (✉) · U. S. Gilani · M. Mahmood · M. F. U. Haque · I. Hussain
School of Biological Sciences, University of the Punjab, Lahore 54590, Pakistan

F. Tasneem
Department of Biology, Virtual University of Pakistan, Lahore, Pakistan

S. I. Ahmad (ed.), *Human Viruses: Diseases, Treatments and Vaccines*, https://doi.org/10.1007/978-3-030-71165-8_17

Table 1 List of human polyomaviruses and their associated diseases

Sr. no.	Virus (Abbreviations)	Discovery (Year)	Sample	Associated diseases	References
1.	BK polyomavirus (BKPyV)	1971	Urine	Polyomavirus-associated nephropathy (PVAN), haemorrhagic cystitis	Gardner et al. (1971)
2.	JC polyomavirus (JCPyV)	1971	Urine, Brain	Progressive multifocal leukoencephalopathy	Padgett et al. (1971)
3.	Karolinska Institute polyomavirus (KIPyV)	2007	Nasopharyngeal tissue	Unknown	Allander et al. (2007)
4.	Washington University polyomavirus (WUPyV)	2007	Nasopharyngeal tissue	WU-PyV-associated bronchitis,	Gaynor et al. (2007)
5.	Merkel cell polyomavirus (MCPyV)	2008	Lesion	Merkel cell carcinoma	Feng et al. (2008)
6.	Human polyomavirus 6 (HPyV6)	2010	Skin	Unknown	Schowalter et al. (2010)
7.	Human polyomavirus 7 (HPyV7)	2010	Skin	HPyV7-associated keratosis	Schowalter et al. (2010)
8.	Trichodysplasia spinulosa polyomavirus (TSPyV)	2010	Lesion	Trichodysplasia spinulosa	van der Meijden et al. (2010)
9.	Human polyomavirus 9 (HPyV9)	2011	Skin, Blood, Urine	Unknown	Scuda et al. (2011)
10.	Malawi polyomavirus (MWPyV)	2012	Stool, Wart	WHIM syndrome	Siebrasse et al. (2012)
11.	Human polyomavirus 12 (HPyV12)	2013	Stool	Unknown	Korup et al. (2013)
12.	St Louis polyomavirus (STLPyV)	2013	Stool	Unknown	Lim et al. (2013)
13.	New Jersey polyomavirus (NJPyV)	2014	Muscle Biopsy	Unknown	Mishra et al. (2014)
14	Lyon IARC polyomavirus (LIPyV)	2017	Skin	Unknown	Gheit et al. (2017)

convincingly linked with the development of any human cancer, Merkel cell carcinoma (MCC), a relatively rare but aggressive neuroendocrine skin cancer (Feng et al. 2008; Wollebo et al. 2015; Harms et al. 2018). The MCPyV sequences in humans were first reported in 2008 by Moore and Chang while they were identifying non-human stretches of DNA in human cancerous tissue using a technique named as Digital Transcriptome Subtraction (DTS) (Feng et al. 2008). They associated MCPyV infection with MCC on observing clonal integration of MCPyV DNA in cancerous Merkel cells in more than 80% of cases. The MCPyV infection and integration of MCPyV DNA was reported to occur before the clonal expansion of MCC cells which further strengthens the etiological role of MCPyV in MCC. Following that, MCPyV was classified as a 2A carcinogen (Bouvard et al. 2012), and numerous studies described various mechanisms by which MCPyV transforms normal cells cancerous. Interestingly, the presence of MCPyV DNA is not limited to Merkel cells alone, but is also detected in a variety of other tumor types. These include non-melanoma skin cancer (Kassem et al. 2009), chronic lymphocytic leukemia (Pantulu et al. 2010; Teman et al. 2011), cutaneous squamous cell carcinoma (Murakami et al. 2011), cervical cancer (Imajoh et al. 2012), non-small-cell lung cancer (Hashida et al. 2013), CNS tumors (Sadeghi et al. 2015), and breast cancer (Reza et al. 2015). Although these studies suggest the presence of MCPyV, however, its causal and mechanistic role in these cancers is still unclear. This chapter focuses on molecular features of MCPyV, its ubiquitous presence in human populations across the world, and most importantly the mechanisms utilized by the virus to induce MCC. It also tends to describe available therapeutic modalities against MCC.

2 Genomic Organization and Molecular Virology of Merkel Cell Polyomavirus

The genome of MCPyV is circular and measures approximately 5.4 kb for the prototype MCV350 strain and the majority of other sequenced strains. Like other polyomaviruses, the MCPyV genome is divided into early and late gene cassettes which encode T antigens (LT, sT, 57kT and ALTO), structural proteins (VP1, VP2), and a miRNA, respectively (Fig. 1a). The early and late coding regions are separated by a regulatory non-coding region known as non-coding control region (NCCR). It includes the origin of replication, clustered binding sites for LT which are required for DNA replication (Kwun et al. 2009), and bi-directional promoter elements which control transcription of early and late genes and allows temporal regulation of gene expression (Feng et al. 2011).

Structurally, MCPyV early region is transcribed into four mRNAs (T1–T4) that are alternatively spliced and encode LT (T1), sT (T2 and T3), and 57kT (T4) proteins. As a consequence of alternative splicing, all of these antigens have common 78 amino acid sequence at their N-terminal that contains epitopes of B

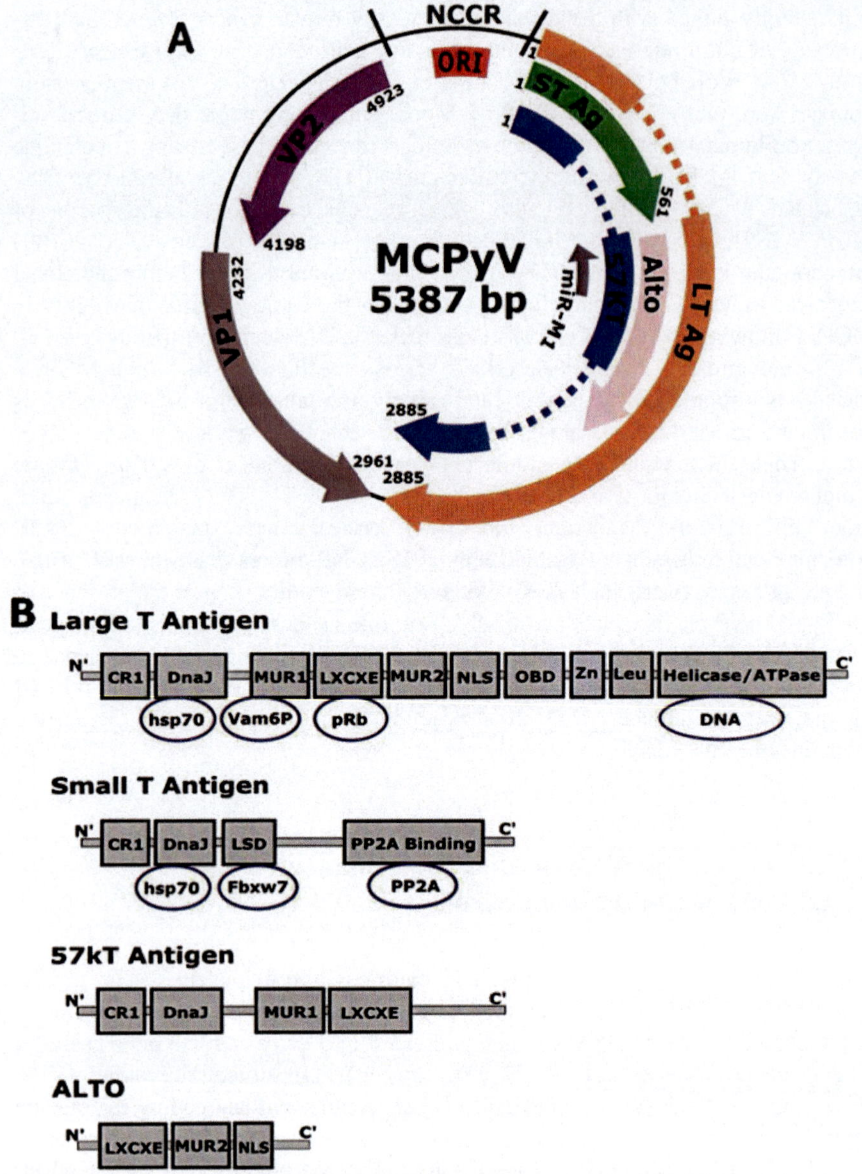

Fig. 1 Organization of MCPyV Genome. **a** The MCPyV early region encodes LT, sT and 57 kT whereas late region encodes VP1, VP2 and miRNA. **b** Interaction of MCPyV early gene products with various cellular host proteins

cells (Shuda et al. 2008). Through leaking scanning of early transcripts, an additional gene product of unknown function termed ALTO (Alternative Open Reading Frame) can be produced (Carter et al. 2013). Unlike early mRNAs, the late transcripts have not been described structurally. Although, it is known that the major and minor capsid proteins, VP1 and VP2, are encoded by open reading frames in this region. Moreover, in this open reading frame, despite the presence of an AUG codon that could initiate transcription of potential VP3, the Kozak consensus sequence is lacking, suggesting that MCPyV does not express a functional VP3 and it is also found to be absent in native MCPyV virions (Schowalter and Buck 2013). Additionally, mcv-miR-M1 is a miRNA encoded by MCPyV genome (Seo et al. 2009). It is located in antisense orientation to the early gene cassette and thus has the ability to direct cleavage of early transcripts (Seo et al. 2009; Grundhoff and Sullivan 2011; Kincaid and Sullivan 2012).

Functionally, immediately after infection, the early coding region, also called the T antigen locus, expresses its gene products which are mainly involved in the replication of viral genome. It has been described that MCPyV early gene products are expressed in an orderly manner where LT and 57kT are expressed first, followed by the expression of sT (Feng et al. 2011). After DNA replication, the late coding region becomes transcriptionally active and expresses viral structural components which assemble and result in the formation of progeny virions during the later stage of infection (Cole 2001).

Products of polyomavirus early regions are known to target cellular proteins involved in tumor suppression and cell cycle regulation (Fig. 1b). These gene products are called tumor antigens or T antigens and are important to initiate the synthesis of viral DNA. The LT antigen of MCPyV has certain conserved regions including the conserved region 1 (CR1) (LXXLL) and the heat shock protein binding DnaJ (HPDKGG) domains, an Origin Binding Domain (OBD), a pRb binding domain (LXCXE) and a helicase domain (Stakaitytė et al. 2014). In between the pRb binding domain and the OBD, there is a Nuclear Localization Signal (NLS) (Nakamura et al. 2010). Domains of the LT that are involved in viral DNA replication are mostly located at the C-terminal such as the OBD, helicase and ATPase domains. The C-terminal of LT is also where most of the tumor specific mutations occur (Feng et al. 2008). A mechanism unique for MCPyV LT is redistribution of LT to the cell's nucleus by binding with Vam6p through the MCPyV Unique Region (MUR) which is of about 200 amino acids and located at the beginning of the second exon close to the pRb binding (LXCXE) domain. The pRb binding domain of the MCPyV LT also contains a unique spacer region which separates the LXCXE motif from the 'Psycho' domain and also includes the viral micro RNA complementary sequence (Johnson 2010). This spacer region is present in the MUR and is unique to MCPyV, but its effect on the function of LT has not yet been elucidated.

The small T antigen (ST), also encoded by the same T antigen locus, has the same N-terminal as that of LT and 57kT, and thus possesses the CR1 and DnaJ domains. However, transcription beyond the first exon splice site results in the formation of a smaller protein of 186 amino acids which includes a protein phosphatase 2A (PP2A) binding site in its C-terminal. The ability of sT to bind to PP2A has been conserved in various polyomaviruses and is thought to be important in cellular transformation induced by polyomaviruses (Pallas et al. 1990). PP2A is an enzyme that removes phosphate group from the eukaryotic initiation factor 4E binding protein 1 (4E-BP1) and allows it to bind to the eukaryotic initiation factor 4E (eIF4E) to inhibit translation. Recent studies indicate that the polyomavirus sT binds to the PP2A Aα subunit and does not allow it to perform its phosphatase activity. Thus 4E-BP1 remains phosphorylated and is unable to sequester eIF4E, and translation continues to allow cells to proliferate (Shuda et al. 2011). Hence MCPyV sT is important for both viral DNA replication and cellular transformation.

Another T antigen expressed by the early coding region is the 57kT protein which is formed by the alternative splicing of a transcript to combine three exons that code for a 432 amino acid protein sharing many features with the LT and sT antigens. The 57kT N-terminal is the same as the other two T antigens with CR1 and DnaJ domains, and it also includes the pRb binding motif, MUR, and most of the C-terminal amino acids present in LT. Information about the role of 57kT in viral life cycle or virus induced tumorigenesis is still insufficient to depict a clear mechanistic picture. Certain studies consider 57kT as an analog of the SV40 17kT antigen which functions independently or with other T antigen to control cellular proliferation in vivo (Comerford et al. 2012).

The non-coding regulatory region (NCRR) includes a minimum 71 bps long origin of replication. Similar to other polyomaviruses, the origin of replication of MCPyV has an AT rich region, a large T binding domain and an early enhancer region. The central portion of the LT protein contains an Origin Binding Domain (OBD) that binds to the pentanucleotide consensus sequences G(A/G)GGC in the origin of replication. This binding then allows LT to perform its activity as a helicase and initiate the replication of viral DNA (Cole 2001). The origin of replication of MCPyV, however, includes more of these pentanucleotide sequences which also lie in closer proximity as compared to in other polyomaviruses (Johnson 2010). This arrangement allows OBD-OBD interactions between different LT proteins bound at the origin (Harrison et al. 2011). However, binding of LT to origin does not require these OBD-OBD interactions and these intermolecular interactions also do not influence the structure of viral DNA. No such interactions have been observed in SV40 as its pentanucleotide sequences are more spatially arranged, thus it indicates that MCPyV replication and LT seeding is more complex than other polyomaviruses (Meinke et al. 2007).

Major capsid protein named the viral protein 1 (VP1) and the minor viral capsid protein 2 (VP2), encoded by the late region, have the function to form the capsid of the viral particle. These proteins also have the ability to self-assemble into 45–55 nm diameter virus-like particles (VLPs) when expressed in mammalian or insect cells. Majority of the serological assays being developed for MCPyV are based on these VLPs (Tolstov et al. 2009, 2011; Touzé et al. 2010, 2011; Viscidi et al. 2011). The size of MCPyV virion is comparable to the size of other polyomavirus particles (Neumann et al. 2011).

The MCPyV also express a 22-nucleotide long miRNA called mcv-miR-M1 that negatively regulates the early gene expression during late phases of virion encapsidation. It also causes cleavage of early mRNA leading to the reduced expression of LT. In addition, it interacts with several cellular targets including PIK3CD and PSME3. This could potentially mediate the host immune response against MCPyV through down-regulation of PSME3-dependent antigen presentation by the host cell. The mcv-miR-M1 also plays an important role in viral replication as its expression level correlates with the viral genome copy number in MCC tumor. The MCPyV miRNA may decrease viral replication to increase viral persistence, as observed in case of BKPyV (Broekema and Imperiale 2013).

3 Frequency of MCPyV Infection in Human

The prevalence of MCPyV in healthy individuals is usually evaluated at two levels; seroprevalence and the presence of MCPyV DNA in host cells. Regarding seroprevalence, it is interesting to note that in healthy individuals, MCPyV seroprevalence is very high. This is mainly attributed to the primary, asymptomatic MCPyV infection thought to occur in early childhood (Carter et al. 2009; Kean et al. 2009; Pastrana et al. 2009; Tolstov et al. 2009, 2011; Sadeghi et al. 2010; Touzé et al. 2010; Chen et al. 2011; Faust et al. 2011; Touzé et al. 2011; Viscidi et al. 2011). The idea of early age infection is acquired by observing the seroprevalence of MCPyV to be age-specific. Among children of the age group 1–5 years, 20% are positive (Kean et al. 2009), and of the age group <10–15 years, 35–50% are positive (Tolstov et al. 2009; Chen et al. 2011; Viscidi et al. 2011). In adults, seroprevalence of 46–87.5% has been reported (Carter et al. 2009; Kean et al. 2009; Pastrana et al. 2009; Tolstov et al. 2009, 2011; Touzé et al. 2010, 2011; Viscidi et al. 2011). This huge variation in prevalence is found in adults because MCPyV antibodies are found to increase with age (Viscidi et al. 2011). In case of MCC, higher antibody titers are observed in patients suffering from MCPyV-positive MCC than those with tumors negative for the virus (Pastrana et al. 2012). Moreover, unlike VP1 specific antibodies that are commonly used for detection and are prevalent in general population, seroreactivity against early gene products is only detected in rare cases and at very low levels. Whereas, in patients with MCPyV-positive MCC, high T-Antigen titers are detected (Paulson et al. 2010). This suggests that during asymptomatic MCPyV infection LT levels are tightly regulated to evade immune recognition. Despite MCPyV sero-positivity is found to be associated with MCC, till now, no association is found with other chronic viral infections (Tolstov et al. 2011).

Healthy human skin is believed to harbors transient MCPyV infections. MCPyV DNA has been detected in 0–100% samples taken from different skin samples through PCR or nested PCR, qPCR or rolling circle amplification (Feng et al. 2008; Dworkin et al. 2009; Andres et al. 2010a, b; Mangana et al. 2010; Mertz et al. 2010a, b; Mogha et al. 2010; Wieland et al. 2011). However, DNA based detection greatly depends on the sensitivity of the technique, the sampling methods, and preparatory steps (Garneski et al. 2009; Foulongne et al. 2010; Schowalter et al. 2010; Faust et al. 2011; Wieland et al. 2011). Apart from skin, MCPyV DNA has been detected in various other anatomical sites; for instance, in a study MCPyV DNA has also been detected in anal and penile swabs with frequency 30% and 50% respectively (Wieland et al. 2009). Additionally, MCPyV DNA is detected at high level in the oral cavity, with detection rates ranging from 8.3 to 39–60% (Dworkin et al. 2009; Wieland et al. 2009; Loyo et al. 2010). MCPyV is also detected in whole blood (Shahzad et al. 2019) as well as other blood products i.e. peripheral blood mononuclear cells (PBMCs), serum, plasma, and buffy coats of healthy individuals with varying frequencies (Pastrana et al. 2009; Mertz et al. 2010a, b; Pancaldi et al. 2011). Interestingly, being able to amplify low levels of MCPyV through PCR based techniques from various human tissues (Feng et al. 2008; Kantola et al. 2009; Bergallo et al. 2010; Loyo et al. 2010), it is inferred that that MCPyV is distributed systemically, but in most tissues where it may undergo low-level replication, persistence, or latency, it is not pathogenic. Taken together, MCPyV infection is believed to be ubiquitous with frequent detection in different anatomical sites of healthy subjects.

4 Merkel Cells and Merkel Cell Carcinoma

In 1875, Friedrich Sigmund Merkel described Merkel cells as ‘touch cells’ considering their presence in the touch sensitive areas of the skin. These cells are identified to be present around hair follicles, some mucosal tissues and various skin sites, especially in area that is involved in the sensation of touch such as finger pads. In the skin, Merkel cells are located at the basal layer of epidermis connected with the ending of sensory nerves (Fig. 2a). Structurally, Merkel cells are round or oval shaped cells of around 10 μm diameter, with lobulated nucleus and certain cytoplasmic projections, mainly formed by microfilaments. Merkel cells also contain granules similar to those of packages hormones which are used to transmit information (Fig. 2b). The cytoplasmic projections anchor Merkel cells to neighboring cells by desmosome formation which enables them to communicate with surrounding cells by transferring and gathering information (Fig. 2c). Merkel cells express both neuroendocrine such as chromogranin-A, synaptophysin and epithelial, for instance, cytokeratins 8, 18, 19, and 20 markers. Among these markers, CK20 is considered as most sensitive and reliable (Lucarz and Brand 2007).

Despite their omnipresence in the skin, exact origin and function of Merkel cells is still unclear. However, it is established that they possess both epithelial and neuroendocrine phenotypes. Merkel cells possess the ability to synthesize and secrete neuropeptides located in the neurosecretory granules within their cytoplasm. These neuropeptides can be released into the intercellular space in response to specific stimuli. Once released, the neuropeptides may act as neuromodulators, mediators or neurotransmitters that may act on sensory nerve endings and perform

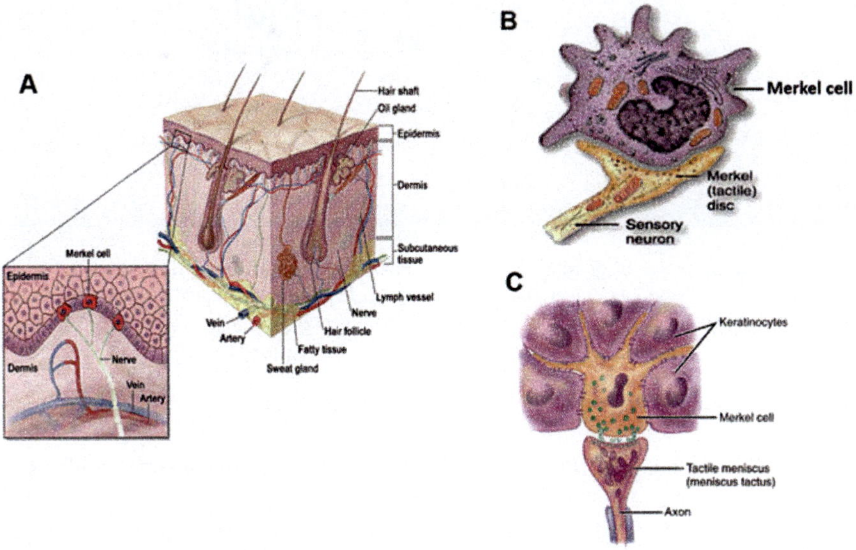

Fig. 2 Morphology and localization of Merkel Cells. **a** Location of Merkel cells in the skin (Image courtesy of the NIH, National Cancer Institute, USA). **b** Figure shows typical morphology of a Merkel cell having lobulated nucleus and certain cytoplasmic projections. **c** Desmosomes (shown as paired purple lines below) anchoring the Merkel cell to its neighboring keratinocytes

trophic roles on keratinocytes or some other endocrine function (Tachibana 1995). On the other hand, their presence in the epidermis and expression of epithelial cytokeratins (CKs) show that they might have an epidermal origin (Moll et al. 1996). Localization of these cells in the touch sensitive areas and the formation of Merkel cell neurites suggest that their role as mechanoreceptors (Ogawa 1996). Considering their well noted and characterized association with termini of somatosensory afferent nerve fibers in the epidermal basal layer adjacent to the dermis, their role of transducing mechanical stimuli from the skin to the central nervous system is hypothesized. The specialized junctions between unmyelinated dermal nerve fibers and Merkel cells are categorized as desmosome-like and synapse-like structures on their electron microscopic appearance. Innervation of Merkel cell clusters by slow adapting nerve fibers to form bulges known as the 'touch domes' in the epidermis is also noticed (Nakafusa et al. 2006).

Merkel cell carcinoma (MCC) was first observed as unusual tumors in the skin and named as trabecular carcinoma of the skin (Toker 1972). The MCC is typically represented as a rapidly growing, dome-shaped red or bluish nodule (Hodgson et al. 2005) (Fig. 3). However, sometimes it may have manifestation of a plaque-like appearance with small satellite lesions (Pectasides et al. 2006). The common primary sites of MCC are head and neck (40%), and then the upper extremities (19%) (Voog et al. 1999). Less than 10% of the cases affect the trunk region. Certain unusual primary lesion presentations of Merkel cell carcinoma have also been observed such as subcutaneous nodules in the inguinal region (Balaton et al. 1989),

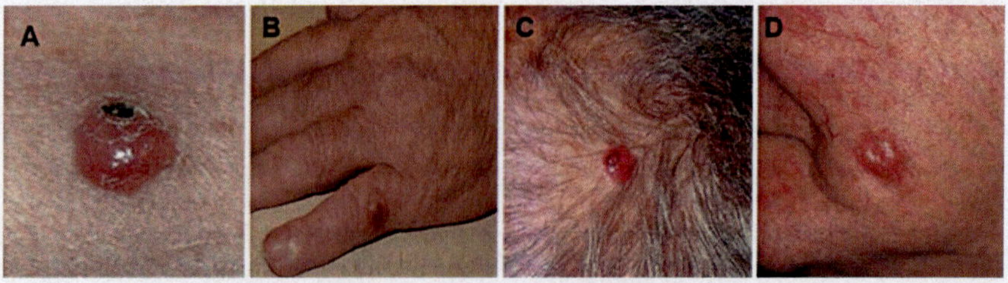

Fig. 3 Clinical presentation of Markel cell carcinoma. Rapidly growing flesh-colored or bluish red nodules on forehead (**a**), finger (**b**), head (**c**) and face (**d**). Figure opted from Mayo clinic (https://www.mayoclinic.org/) and merkelcell.org (https://merkelcell.org/)

an enlarged painless mass in the calvarium or ulceration of the vulva (Bottles et al. 1984).

The initial diagnosis of MCC is mostly based on histopathology and confirmed diagnosis can be made on the basis of both histological features and immunological markers' expression profiles of the lesions. Histologically, MCC shows an asymmetric dermal growth having irregular margins formed of tumor cells arranged in the form of strands (Plaza and Suster 2006). Atypical mitosis is usually observed in the tumors and the mitotic index remains high (Fig. 4). Usually the tumor occupies the entire thickness of the dermis and also commonly extends into the adjacent

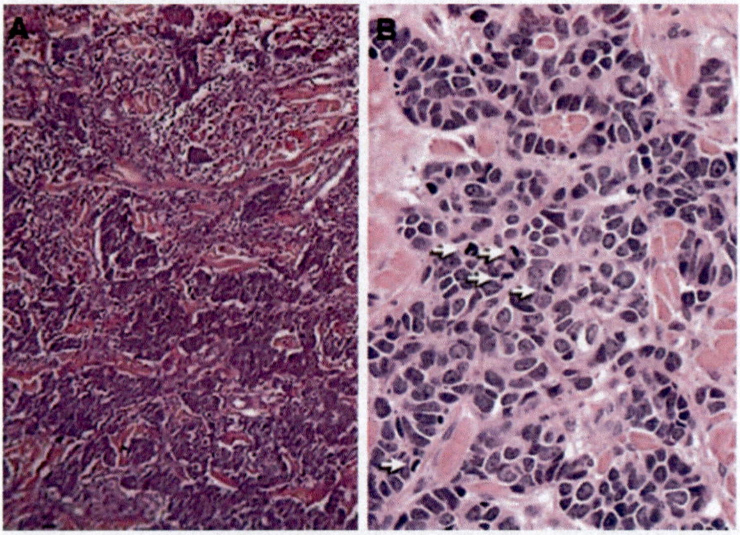

Fig. 4 Histological appearance of MCC. Hematoxylin and eosin (H&E) stained MCC tissues (A&B) contain scanty cytoplasm, round nucleus, dusty chromatin and high mitotic activity. **a** Figure shows sheets of small monotonous round cells extending throughout the dermis (Figure courtesy, Rafael et al. 2018) **b** 400X magnification of H&E stained tissue. Arrows show mitotic figures. Figure opted from Harms et al. (2018)

skeletal muscles and the subcutaneous fat layer (Silva et al. 1984). The tumor and the epidermis are generally separated by a dermal Grenz zone but epidermal changes such as ulceration have also been observed. The spread of tumor might also stimulate melanoma and rarely the tumor is seen to be exclusively intraepidermal.

Currently, three histological patterns have been recognized on the basis of arrangement and appearance of tumor cells in MCC: the trabecular, the intermediate and the small cell type. In between these three types, mixed and transitional forms are also common. The trabecular type of MCC is the least common. Mostly, the cells appear as diffused sheets having irregular borders with infiltrating growth pattern that intersect through the collagen fibers of the adjacent dermis. The trabecular type of MCC has tumor cells growing as narrow strands or ribbons. These ribbons are one to two cells-thick, and the cells have scanty amphophilic cytoplasm with oval or round nuclei. These nuclei stain pale, are vesicular and have delicate chromatin. In the small cell type, the nuclei may be hyperchromatic and spindle formed. Nucleoli are also commonly present but are not very prominent. The cells undergo numerous mitoses and thus result in atypical forms. Apoptosis is often marked and geographical areas of necrosis are seen. Small squamous foci and/or ductal differentiation may also be observed occasionally (Gould et al. 1988). Invasion of the lymphatic vessels is commonly present; rarely perineural infiltration is also seen. Usually the tumor infiltrate occurs together with lymphocytic infiltrate and sometimes the plasma cells are also present. The epidermis occasionally has in situ squamous cell carcinoma coexisting in it, but the merging of the two populations is rarely observed. The intermediate variant of MCC, namely the primary cutaneous neuroendocrine carcinoma, might at times coexist with the invasive squamous cell carcinoma with the two populations often blending together. However, they are distinguishable through immunocytochemistry (Iacocco et al. 1998). It is yet unclear if this coexistence implies origin from a common stem cell. Coexistence of neuroendocrine carcinoma and basal cell carcinoma is rarely seen (Cerroni and Kerl 1997).

A Characteristic immunohistological profile is usually observed for MCC. The cells of MCC express cytoskeletal keratins of both type I and II including CK8, CK18, CK19, and CK20 most importantly which is observed in the form of a paranuclear dot. Furthermore, neuroendocrine specific markers such as enolase, chromogranin, and synaptophysin are also expressed by MCC cells (Schmidt et al. 1998; Scott and Helm 1999). Major fraction of MCC (80%) express MCPyV T antigens and are denoted as MCPyV +ve MCC; whereas, a small subset of MCC cannot express viral oncoproteins and are named as MCPyV-ve MCC. It must be noted that two types of MCC (MCPyV +ve or MCPyV −ve) cannot be differentially diagnosed on the basis of immunohistochemistry alone since no immunological marker has been associated specifically with any type. However, both types can be distinguished by some characteristic features. For instance, MCPyV-ve MCCs have high mutational burden and do not express thyroid transcription factor 1 (TTF1), mammalian achaete-scute homologue 1 (ASH1), vimentin, S100B and CK7. In addition to that, a small subset of these MCCs (<10%) are negative for CK20 (Becker et al. 2017) (Fig. 5).

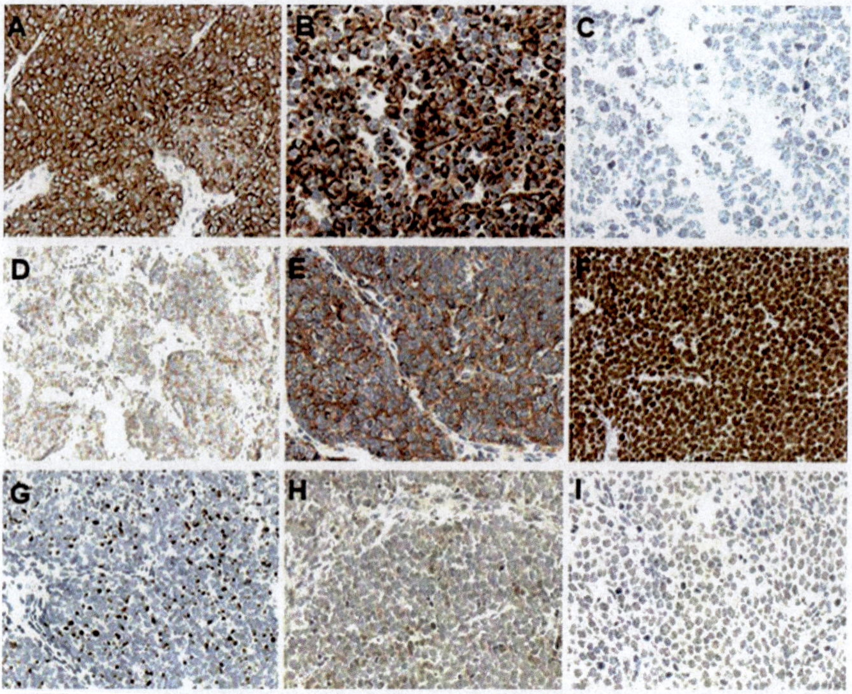

Fig. 5 Immunohistochemical features of MCC. **a** Chromogranin A cytoplasmic positivity, **b** cytokeratin 20 expression with paranuclear dot-pattern, **c** thyroid transcription factor-1 negativity, **d** membranous synaptophysin expression, **e** Membranous CD56 expression, **f** special AT-rich sequence-binding protein 2 (SATB2) nuclear expression, **g** neurofilament expression with a dot-pattern, **h** Terminal deoxy nucleotidyl transferase weak/moderate expression, **i** paired box 5 weak expression in tumor cells in comparison with intratumor lymphocytes. Images are taken from Kervarrec et al. (2019)

Several risk factors have been linked with the occurrence of MCC. These include prolonged exposure to UV and sunlight, old age and immune suppression. The individuals suffering from hematological neoplasms or having history of cutaneous tumors are also at higher risk of MCC (Reviewed in Becker et al. 2017). The disease is more prevalent among white population than the non-white (94.9% vs. 4.1%), and mean diagnosis in males is 73.6 and in females is 76.2 years (Albores-Saavedra et al. 2010). Although MCC has also been diagnosed in younger patients but in these cases it is most often related to immunosuppression due to organ transplantation (Lanoy et al. 2010). It has been indicated that patients undergoing organ transplantation have a 23.8-fold higher risk of MCC development than the immunocompetent patients (Clarke et al. 2015). If following transplantation, the immunosuppression is maintained for long periods of time the risk increases even further (Lanoy et al. 2010).

5 MCPyV: An Etiological Agent of MCC

Higher incidence of MCC in HIV-1 AIDS patients and immunocompromised (medically induced or in autoimmune diseases like rheumatoid arthritis) individuals, instigated the researchers to look for the pathogenic cause of MCC (Gooptu et al. 1997; Engels et al. 2002; Lanoy et al. 2010). Feng et al. identified some non-human stretches of DNA in human MCC tissue using a technique named as Digital Transcriptome Subtraction (DTS) (Feng et al. 2008). They detected MCPyV in 8 out of 10 studied MCC tumors by PCR and Southern blotting, indicating that a large subset of MCC is caused by MCPyV. Upon studying the primary tumor and a metastatic lymph node, they described that MCPyV is clonally integrated into the genome of MCC tumor cells. Later, several studies based on histopathology (Shuda et al. 2009; Busam et al. 2009; Bhatia et al. 2010; Shuda et al. 2011; Arora et al. 2019) and various molecular techniques (Feng et al. 2008; Busam et al. 2009; Duncavage et al. 2009; Fischer et al. 2010) confirmed the causal role of MCPyV in MCC. However, the percentage of MCPyV positive and negative MCC was found to be variable in different studies with an overall 80% MCPyV positivity.

6 How Does MCPyV Cause MCC?

MCPyV mediated cellular transformation and tumor development is exclusively linked with the activities of virus encoded oncoproteins, LT and sT. Both of these proteins have been immensely implicated in the initiation and progression of MCPyV-related carcinogenesis. This fact was first described by the finding of Houben et al. (2010) who observed decrease in proliferation and survival of MCC cells after knocking out T antigen locus, showing that continuous expression of LT and sT is required for the existence of MCPyV positive MCC cells (Houben et al. 2010). The expression of T antigens is found only restricted to the tumor cells lacking expression in surrounding healthy cells (Rodig et al. 2012). Despite the expression of both LT and sT was observed in majority of MCPyV induced MCC tumors suggesting the dependence of MCC cells on MCPyV T antigen proteins (Becker et al. 2017; Harms et al. 2018), their independent role in cellular transformation remained obscure. The knock down of sT in MCPyV positive MCC cell line by Shuda et al. (2011) resulted in halting cell growth but did not prime cell death, indicating other proteins might play a role in MCC survival. Whereas, sT alone was able to induce transformation in rodent fibroblasts by increasing phosphorylation and inactivating the eukaryotic translation initiation factor 4E-binding protein 1 (4E-BP1), ultimately resulting in the deregulation of cap-dependent translation that enhance proliferation and malignant transformation of cell (Shuda et al. 2011). The ability of sT to enhance phsopho-4EBP1 was linked to the Large T stabilization domain (LSD) of sT interaction with cell division cycle protein 20

(CDC20), rather than binding with protein phosphatase, PP2A (Shuda et al. 2015). The transforming activity of MCPyV sT was also confirmed in transgenic mice (Verhaegen et al. 2015). Furthermore, sT was described to play role in viral DNA replication, protection of MCPyV LT proteosomal degradation and of other cell-cycle regulators including c-Myc and cyclin E by targeting cellular ubiquitin ligase SCFFbw7 (Kwun et al. 2013). In fact, MCPyV sT has been shown to increase the level of LT protein via the activity of its LSD (Dye et al. 2019). Recently, the role of sT in inducing DNA damage response (DDR) pathway has been indicated in human MCC cells by observing sT over expression causing hyperphosphorylation of histone H2AX, a marker for DNA damage, along with the activation of ataxia telangiectasia mutant (ATM), an upstream kinase important for H2AX phosphorylation. Moreover, sT expression was also observed to induce hyperphosphorylation of other ATM downstream molecules (including 53BP1 and CHK2) as well as the hypermethylation of histone 3 and histone 4, indicating a novel link between sT and the DDR pathway in MCC (Wu et al. 2019). All these observations reinforce the role of MCPyV sT as a main driver of MCPyV mediated cellular transformation and oncogenesis. However, predominant role of sT is negated in study by Angermeyer et al. (2013) who showed that LT is more relevant in maintaining the proliferation and survival of MCC cell lines. Nevertheless, both LT and sT have been found essential for MCPyV medicated carcinogenesis.

Shuda et al. (2008) described that MCPyV genome is clonally integrated into the genome of MCC cells and this integration leads to mutations that generate premature stop codons resulting in the truncation of LT. The truncated LT retains N-terminal J-domain and RB binding motif, but lose the C-terminal regions. On the other hand, MCPyV strains isolated from healthy individuals were reported not to possess these signature mutations and encode full length LT. These observations suggested that truncation of LT plays an essential role in MCPyV mediated carcinogenesis. Owing to antitumor activities, there is a strong selective pressure to remove C-terminal during development of MCC. It was demonstrated that deletion of LT helicase domain eliminates the viral replication ability which is crucial in preventing the MCPyV cell death (Shuda et al. 2008). The truncated LT has been shown more efficient in promoting the growth and transformation of human and mouse fibroblasts when compared with full length LT and 57kT (Cheng et al. 2013). Similarly, truncated LT showed stronger binding affinity for pRb than full length LT (Borchert et al. 2014). It was found that the C-terminal of LT causes DNA damage stimulating host DNA damage responses which ultimately activate p53 and stop proliferation of cells that is not favorable for their malignant transformation (Li et al. 2013). Moreover, the C-terminal of full length LT was reported to halt the growth of several cell types (Cheng et al. 2013). It was also demonstrated that ATM cellular kinase may induce phosphorylation of MCPyV LT C-terminal, induce apoptosis and halt cell proliferation (Li et al. 2015). These studies altogether elucidate that deletion of the MCPyV LT C-terminus is not only necessary for the disruption of MCPyV replication but also important to overcome the growth-inhibitory properties of full length LT, henceforth, very crucial for oncogenic progression of MCPyV-associated carcinogenesis. Notably, the integration of

MCPyV does not induce any mutation in the sT leaving its expression in MCC in the native form.

Uncontrolled cell proliferation of cells is generally inhibited via one of two fundamental tumor suppression mechanisms mediated by retinoblastoma (pRb) and p53 proteins. The pRb restricts the progression of cell cycle by binding and sequestering the function of E2F transcription factor which transactivates the transcription of genes essential for G1 to S phase transition (Fisher et al. 2016). The MCPyV oncogenic potential is largely attributed to the binding and inactivation pRB by LT (Borchert et al. 2014). The LT binding to pRb is achieved via LXCXE motif of LT that is also retained in truncated LT. The truncated LT also preserves the ability to bind and inactivate the pRb as it retains the Dna J region and pRb binding pocket. The LT has been shown to bind with all proteins of pRb family (pRb, p107, p130). The inactivation of pRB by LT is crucial for the sustained growth of MCPyV positive MCC cells. In fact, it was described that pRb-binding motif is required by LT for promoting the growth of MCC cell (Houben et al. 2012). The LT also contain another 200 amino acids long domain named as Merkel unique region (MUR) which extends to LXCD domain, binds to cytoplasmic vacuolar sorting protein Vam6p and antagonizes its role in lysosomal clustering (Liu et al. 2011). However, this interaction does not seem to be critical in MCPyV mediated cellular transformation. The p53 is expressed under different stress conditions such as DNA damage and acts as a transcription factor which induces the expression of genes involved in DNA damage repair, apoptosis and senescence. Notably, no direct interaction between MCPyV LT and p53 has been described so far. However, full length LT was shown to significantly reduce p53-dependent transcription (Borchert et al. 2014). In most of the cancers, p53 pathway is inactivated through mutations in the p53 gene. However, p53 mutation seems to occur only occasionally in MCPyV positive MCC. It was suggested that p53 expression is negatively correlated with MCPyV DNA copy number and p53 mutations are only detected in MCPyV negative MCC (Sihto et al. 2011). These data suggest that p53 is most likely not a major driver of MCPyV-associated tumorigenesis. Recently, in a study, single-cell RNA-seq revealed that knockdown of T antigen directly inhibits the expression of Atoh1 by LT by up-regulating Sox2 through its retinoblastoma protein-inhibition domain (Harold et al. 2019). ATOH1 is a transcription factor that is a master regulator of Merkel cell development, with a controversial role in Merkel cell carcinoma (MCC) (Fan et al. 2020a, b).

Understanding the alteration in cellular protein network induced by MCPyV oncoproteins have also significantly contributed to our understanding of virus mediated cellular transformation and oncogenesis. In order to grab deep understanding of MCPyV mediated carcinogenesis, Gupta et al. 2019 performed expression profiling of NIKS cells expressing early genes of MCPyV. The MCPyV early genes down regulated the expression of tumor suppressor gene N-myc downstream-regulated gene 1 (NDRG1) which was found to exert its function in Merkel cell lines by regulating the expression of the cyclin-dependent kinase 2 (CDK2) and cyclin D1 proteins (Gupta et al. 2020). Likewise, the MCPyV small T antigen was described to activate the expression of LSD1 which was shown to antagonize the tumor suppressor non-canonical BAF (ncBAF) (Park et al. 2020).

In terms of the various cellular pathways, the expression of T antigens in MCC can lead to oncogenesis by altering various mitogenic pathways. One of important pathway downstream of the growth factor receptors that is commonly deregulated in cancers is the MAPK (mitogen activated protein kinase) pathway. This pathway carries growth signals from the plasma membrane to the nucleus and involves three consecutive kinases i.e. RAF, MEK and ERK. However, no activating mutations of these kinases have been observed in MCC samples. In addition, immunohistochemical analysis has revealed that the ERK kinase is present in the inactive non-phosphorylated form in MCC and thus the MAPK pathway is considered inactive (Houben et al. 2006). The cell survival pathway also involves another branch of signaling from the growth factor receptors which relies on the generation of a second messenger, phosphatidylinositol 3,4,5-triphosphate (PIP3). PIP3 is dephosphorylated to PIP2 by phosphatase and tensin homologue (PTEN), a tumor suppressor. PTEN thus inactivates growth signaling through PIP3. PTEN encoding gene is located on the long arm of chromosome 10, part of which is found to be frequently lost in case of MCC in a heterozygous pattern (van Gele et al. 1998). However, the inactivation of other alleles of PTEN through deletion or mutation occurs rarely in MCC, and other tumor suppressor genes might be the actual targets of loss of 10q (van Gele et al. 2001). Microarray analysis of MCC tissue, however, recently showed that the PTEN expression is very low in MCC tissues which hint toward epigenetic silencing of the second allele of PTEN (Fernandez-Figueras et al. 2007).

The comprehensive understanding of MCPyV mediated tumorigenesis is not possible without deciphering its interaction with the immune system of host. The evasion of the immune response is an important factor in determining the persistence and outcome of oncogenic viral infection. Owing to LT and sT oncoproteins, MCPyV has devised various strategies to escape the immune attack. One of these mechanisms is down- regulation of major histocompatibility complex class 1 (MHC-1) which prevents MCPyV from being recognized from the host immune system. MHC-1 has been found down regulated in 84% of MCC and this down-regulation was more pronounced in MCPyV MCC than in MCPyV negative tumors (Paulson et al. 2014). Another evasion mechanism was described by Shahzad et al. (2013) where MCPyV T antigen locus has been demonstrated to downregulate the expression of Toll-like Receptor 9 (TLR9) which are meant for the sensing of viral and bacterial dsDNA (Shahzad et al. 2013). It was also shown that TLR9 targeting plays a vital role in viral persistence, which is a prerequisite for cellular transformation in MCPyV mediated MCC. In another study, MCPyV sT has been shown to inhibit the transcription of NF-κB pathway associated genes, ultimately enabling MCPyV to undermine the host immune response in infected cells (Griffiths et al. 2013). MCPyV also encodes micro RNA (MCV-miR-M1) which facilitates the virus in establishing persistent infection and evasion of immune response. In fact, MCV-miR-M1 induces expression of several immune evasion related genes ultimately resulting in the attenuation of neutrophil chemotaxis toward MCPyV harboring Merkel cells (Akhbari et al. 2018). The MCPyV T-antigens also revealed to induce expression of several host micro RNAs including miR-375 which suppresses autophagy related genes ultimately protecting MCC cells from death (Kumar et al. 2020). Due to these mechanisms, several MCC patients having normal immune system probably fail to clear the MCPyV and ultimately become victim of MCC (Heath et al. 2008). However, better prognosis and survival rate has been shown in MCC patients being capable of generating vigorous immune system (Sihto and Joensuu 2012; Paulson et al. 2011).

Inhibition of apoptosis is another critical step in the development of tumorigenesis. Majority of cancers usually have elevated levels of anti-apoptotic members of the Bcl-2 family in order to control mitochondrial apoptosis by binding pro-apoptotic proteins (Bender et al. 2013). Bcl-2 family profiling and functional studies showed that these proteins are overexpressed in 94% of the MCC tumors and all 11 MCC cell lines (Brunner et al. 2008; Verhaegen et al. 2014). Furthermore, downregulation of Bcl-2 family proteins induced cell death in majority of MCC cell lines (Verhaegen et al. 2014). However, no correlation has been observed between Bcl-2 expression and aggressive behavior of MCC (Feinmesser et al. 2004). Many cancers also show high expression of another anti-apoptotic protein named survivin which is undetectable in differentiated human tissues. In a study all MCC cases have shown strong survivin staining which was localized either in the nucleus or the cytoplasm. Nuclear localization corresponded to more aggressive clinical outcomes with patients developing distant metastasis (Kim and McNiff 2008).

Another important driver of MCC particularly in case of MCPyV negative MCC is exposure to UV radiation. Striking differences have been observed in the genome of MCPyV positive and negative MCC tumor by recent studies. Notably, MCPyV containing MCC cells harbor very little somatic mutations, whereas, MCPyV lacking MCC cell contain high number of mutations (25–90-folds higher) most probably because of UV radiation (Wong et al. 2015; Starrett et al. 2017; González-Vela et al. 2017). These UV induced mutations in MCC cells are very similar to mutations in other sun exposed tumors including basal and squamous cell carcinoma (South et al. 2015; Martincorena et al. 2014). The mutation caused by UV radiation in MCPyV negative MCC cells impairs the functionality of *pRb, p53, NOTCH1, FAT1* and *HRAS* genes (Wong et al. 2015). The UV radiation is known to play a crucial role in MCPyV mediated MCC as both MCPyV negative and positive MCC occur in sun exposed area (Arora et al. 2019). The UV radiation is also known to modulate the expression of inflammatory mediators and alter the functionality of antigen-presenting dendritic cells, ultimately causing induction of immunosuppression, which is a prerequisite for MCPyV mediated carcinogenesis (Prasad and Katiyar 2017). The MCPyV positive MCC was also elaborated to harbor additional mutations that are known to activate PI3K pathway (Reviewed in DeCaprio et al. 2017). The UV exposure is linked with the increase in the activity of MCPyV non-coding regulatory region and transcription level of sT (Mogha et al. 2010). The presence of MCPyV also negatively impact the DNA damage repair mechanism in cells exposed to UV (Demetriou et al. 2012). It was shown that UV radiation induces certain growth factors (EGF and FGF) and activates WNT/β-catenin signaling pathway ultimately inducing expression of MMPs. These UV mediated activations may stimulate MCPyV infection and its potential to cause MCC, ultimately driving it towards MCC (Liu et al. 2016).

Epigenetic alterations in addition to genetic changes are also known to play critical role in the development of tumor. In fact, silencing of tumor suppressor genes (TSGs) by promoter hypermethylation has been established as an important mechanism in tumorigenesis (James et al. 2003). Like other tumors, several tumor suppressor genes have been described to be epigenetically silenced in MCC with varying frequencies. For instance, p16INK4a and p14ARFTSGs, which are important for cell cycle control, are found frequently hypermethylated in MCC (Horn et al. 2007). The epigenetic silencing ofp14ARF was observed in about half of the MCC cases, however, hypermethylation of the p16INK4a has not been significantly observed in MCC (Lassacher et al.,2008). Since p14ARF represses the permanent degradation of p53, it is possible that the p53 pathway is inactivated in case of MCC through the suppression of expression of p53 inducer p14ARF. Likewise, methylation of other cell cycle regulator such as CDKN2A was reported in MCC (Helmbold et al. 2009). Certain members of RASSFs tumor suppressors' family are also found aberrantly hypermethylated in MCC (Richter et al. 2013). In another study, the promoter of RB1 gene was found methylated in MCPyV positive as well negative MCC tumor (Sahi et al. 2014).

7 Treatment of MCC

Devising a comprehensive and effective therapy for MCC is currently a challenge. However, various treatment options are available which are recommended on the basis of different characteristics of MCC in a patient including location of the tumor, stage, regional lymph node involvement, co-morbidities and performance status of the patient (National Comprehensive Cancer Network 2016; Lebbe et al. 2015). An overview of therapeutics modalities for MCC is present in Fig. 6.

The preferred treatment option for primary MCC is surgery; however, this mode of treatment is limited to those patients who suffer only from locoregional primary MCC. Usually, it involves resection of a wide area (1–2 cm) of clinically free margins (Tai 2013). There are cases where surgery results in a positive margin which is then followed by re-excision (National Comprehensive Cancer Network 2016). Considering a wider area is resected, in cases where tissue sparing is crucial, a technique named as Mohs microsurgery is used (National Comprehensive Cancer Network 2016). However, its utility is debated as in few cases development of in-transit metastases is observed (Tai 2013; Hughes et al. 2014). In node-positive disease cases, complete lymph node dissection are conducted along with radiotherapy in a few cases (National Comprehensive Cancer Network 2016). Whereas, in patients that clinically appear node-negative, sentinel lymph node biopsy (SLNB) is mostly advised along with excision of primary MCC, which allows regional

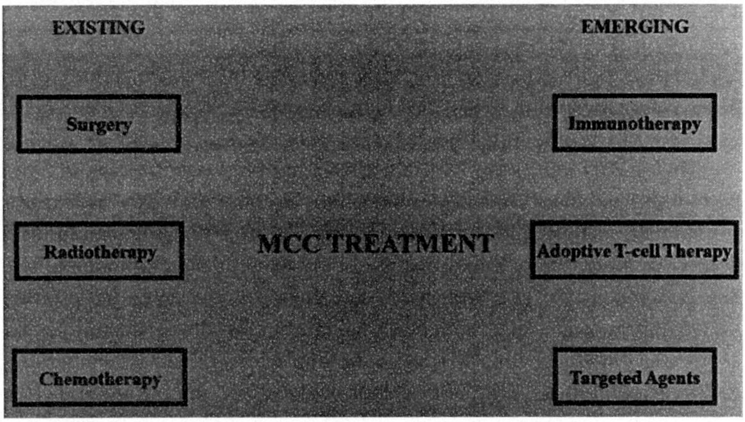

Fig. 6 Schematic representation of existing and emerging therapies against MCC

lymph node status at microscopic level (Paulson et al. 2013; Servy et al. 2016). If results of SLNB are positive, a complete lymph node dissection is conducted with or without radiotherapy to the nodal basin depending on the case (Cassler et al. 2016). However, further studies are required to determine the better suitability of both these surgical options; complete node dissection or radiotherapy, in patients with a positive SLNB result.

Radiotherapy could be considered at different disease stages whether as palliative treatment for cases of MCC where surgery is not possible or as an adjuvant treatment to surgery (Mortier et al. 2003; Veness et al. 2010; Poulsen et al. 2003). In numerous cases, adjuvant radiotherapy is observed to decrease recurrence in comparison to surgery alone (Lewis et al. 2006; Jabbour et al. 2007; Jouary et al. 2012; Chen et al. 2015; Hasan et al. 2013; Mojica et al. 2007; Veness et al. 2005). In patients where surgery in not possible due to poor performance status of the patient, radiotherapy can lead to long-term tumor control (Mortier et al. 2003). Despite surgery and/or radiotherapy may cure patients that suffer from local and regional MCC, recurrences are observed even if both are combined (Allen et al. 2005; Santamaria-Barria et al. 2013; Bichakjian et al. 2007). The benefit of adjuvant radiotherapy is also associated with the stage of MCC, as in an analysis of almost 7000 patients, an improved overall survival was observed in patients with stage I/II MCC as compared to surgery alone, but not in the case of stage III MCC (Bhatia et al. 2016).

Chemotherapy is recommended as standard treatment when MCC is at advanced-stage or becomes metastatic. However, responses are not durable and adjuvant radiotherapy is associated with better outcomes in comparison to adjuvant chemotherapy (Garneski et al. 2007). Survival benefit of chemotherapy is showed by only a limited number of studies, whereas, largely recurrences are seen to develop within 4–15 months after chemotherapy (Allen et al. 2005; Bichakjian et al. 2007; Poulsen et al. 2003, 2006; Saini et al. 2015). The complications associated with chemotherapy include increased morbidity, resistance to chemotherapy on recurrence, decreased quality of life and immune suppression. On the other hand, benefits are insignificant and it is not recommended in clinical guidelines (National Comprehensive Cancer Network 2016; Cassler et al. 2016; Bhatia et al. 2016; Desch and Kunstfeld 2013; Iyer et al. 2016), except for stage IV metastatic MCC in which it is now gradually being believed to palliate symptoms (National Comprehensive Cancer Network 2016; Lebbe et al. 2015). Various natural compounds and derivatives including etoposide, taxanes and anthracyclines, either alone or in various combinations are being used as chemotherapeutic agents for MCC. Very recently, Liu et al. (2020) screened a library of natural compounds and found that a natural product, glaucarubin, reduces viability of the MCPyV-positive MCC cell line MKL-1. The anti MCC effect of glaucarubin was enhanced when combined with a FDA-approved BCL-2 inhibitor, ABT-199 (Liu et al. 2020).

In this view, there is a pressing need to develop treatments which are able to induce durable responses in patients with metastatic or recurrent disease and simultaneously possess a good safety and tolerability profile. Among the other emerging treatments, immunotherapy has recently been implicated in the case of MCC, especially immune checkpoint inhibitors against members of the CD28 family such as CTLA-4, PD-1 and PD-L1 that is up-regulated in the tumor microenvironment and correlate with poor prognosis (Topalian et al. 2015; Postow et al. 2015). In a study, in patients who were treatment-naive with stage IIIB/IV MCC, pembrolizumab, an anti-PD-1 antibody, showed significant clinical activity. Similarly, in another study, durable responses with the use of avelumab which is a PD-L1 antibody were seen in 88 patients suffering from stage IV MCC and had failed first-line chemotherapy (Kaufman et al. 2016). These responses were observed in both MCPyV positive and negative patients. Despite the above mentioned drugs are currently under clinical trials and not available for treatment, immunotherapy is a promising treatment option for treating advanced stage MCC.

Adoptive T-cell therapy is another novel avenue currently being explored with regard to MCC treatment. In preclinical studies, effective killing of MCPyV positive MCC was observed by MCPyV-specific T-cells (Lyngaa et al. 2014). In this view, a phase 1/2 study is ongoing (NCT02584829) which is exploring the efficacy and safety of the use of avelumab combined with MHC-Iup regulation (via IFN-b administration) and autologous T-cell transfer. Moreover, phase 1 studies are being carried out to gauge the feasibility of glucopyranosyl lipid adjuvant-stable emulsion (GLA-SE). This adjuvant is an agonist of TLR-4 which plays a role in inducing a T helper 1 immune response through the stimulation of dendritic cells, in patients with MCC (NCT02035657). Presently, obtained results of these trials show that G100 (GLA-SE delivered intratumorally) has the ability to facilitate inflammatory changes in the tumor microenvironment, thereby activating T cells (Bhatia et al. 2016).

Another novel approach is to apply targeted agents, which in the case of MCC are mainly kinase inhibitors, considering that in MCC oncogenes and tumor suppressors are rarely mutated, whereas numerous receptor kinases and/or ligands are expressed, such as PI3K/Akt, c-kit, VEGFA, VEGFC, VEGFR-2, PDGF-a and PDGF-b (Nardi et al. 2012; Brunner et al. 2008; Hafner et al. 2012). Treatment with imatinib, a tyrosine kinase inhibitor, showed a complete response in a patient (Loader et al. 2013), however, due to lack of its efficacy observed in further studies, a phase 2 clinical trial was prematurely discontinued (Samlowski et al. 2010). Cabozantinib is a c-met and VEGFR-2 inhibitor and its clinical trials against recurrent and/or metastatic MCC is ongoing (NCT02036476). In comparison, pazopanib which is another multi targeted kinase inhibitor, a limited clinical activity in advanced MCC was observed (Davids et al. 2009; Nathan et al. 2016). Keeping in view that PI3K activating mutations are detected in MCC, idelalisib provided a complete clinical response in a single patient with stage IV MCC (Nardi et al. 2012; Shiver et al. 2015). As well as, mammalian target of rapamycin (mTOR) inhibitors MLN0128 and everolimus are currently being investigated in patients with advanced MCC as it is a regulator of the PI3K and MAPK pathways (NCT02514824 and NCT00655655) (Lin et al. 2014; Kannan et al. 2016). Similarly, somatostatin receptor type 2 (sst2) is expressed in MCC tumors and is therefore a potential imaging and treatment target (Buder et al. 2014). Treatment with lanreotide has shown response in MCC tumor expressing sst2 (Fakiha et al. 2010). Moreover, to evaluate the activity of lanreotide in patients with locally advanced or metastatic MCC, a phase 2 trial is currently being carried out (NCT02351128).

8 Conclusion

Merkel Cell polyomavirus (MCPyV) is the 7th known oncogenic virus in human, being responsible for Merkel Cell Carcinoma (MCC), a rare but aggressive neuroendocrine cancer of skin having low survival rate. The MCPyV causes ubiquitous infection in human with higher incidence in elderly individuals. The MCC is caused only after prolonged infection and integration of MCPyV in host genome which is followed by mutation at 3′ end of MCPyV LT gene leading to the truncation of its carboxyl terminus having helicase domain. Oncogenic potential of MCPyV is largely attributed to truncated LT and sT antigens which drive the transformation of normal cells into cancerous cells and induce cell proliferation by various mechanisms. Several treatment options like surgery, radiotherapy and chemotherapy are available for MCC but have some limitations like lower efficacy, relapse and dangerous side effects. Recently, other treatment options like immunotherapy, adoptive T-cell therapy and the application of targeted agents have been extensively explored.

References

Akhbari P, Tobin D, Poterlowicz K, Roberts W, Boyne JR. MCV-miR-M1 targets the host-cell immune response resulting in the attenuation of neutrophil chemotaxis. J Invest Dermatol. 2018;138(11):2343–54.

Albores-Saavedra J, Batich K, Chable-Montero F, Sagy N, Schwartz AM, Henson DE. Merkel cell carcinoma demographics, morphology, and survival based on 3870 cases: a population based study. J Cutan Pathol. 2010;37(1):20–7.

Allander T, Jartti T, Gupta S, Niesters HG, Lehtinen P, Osterback R, Vuorinen T, Waris M, Bjerkner A, Tiveljung-Lindell A, van den Hoogen BG, Hyypia T, Ruuskanen O. Human bocavirus and acute wheezing in children. Clin Infect Dis. 2007;44:904–10.

Allen PJ, Bowne WB, Jaques DP, Brennan MF, Busam K, Coit DG. Merkel cell carcinoma: prognosis and treatment of patients from a single institution. J Clin Oncol. 2005;23 (10):2300e9.

Andres C, Belloni B, Puchta U, Sander CA, Flaig MJ. Prevalence of MCPyV in Merkel cell carcinoma and non-MCC tumors. J Cutan Pathol. 2010a;37(1):28–34.

Andres C, Puchta U, Sander CA, Ruzicka T, Flaig MJ. Prevalence of Merkel cell polyomavirus DNA in cutaneous lymphomas, pseudolymphomas, and inflammatory skin diseases. Am J Dermatopathol. 2010b;32(6):593–8.

Angermeyer S, Hesbacher S, Becker JC, Schrama D, Houben R. Merkel cell polyomavirus-positive Merkel cell carcinoma cells do not require expression of the viral small T antigen. J Invest Dermatol. 2013;133(8):2059–64.

Arora R, Rekhi B, Chandrani P, Krishna S, Dutt A. Merkel cell polyomavirus is implicated in a subset of Merkel cell carcinomas, in the Indian subcontinent. Microb Pathog. 2019;137:103778.

Balaton AJ, Frederique C, Baviera EE. Neuroendocrine carcinoma (Merkel cell tumor?) presenting as a subcutaneous tumor: an ultrastructural and immunohistochemical study of three cases. Pathol Res Pract. 1989;184:211–6.

Becker JC, Stang A, DeCaprio JA, Cerroni L, Lebbe C, Veness M, Nghiem P. Merkel cell carcinoma. Nat Rev Dis Prim. 2017;3:17077.

Bender T, Martinou JC. Where killers meet-permeabilization of the outer mitochondrial membrane during apoptosis. Cold Spring Harb Perspect Biol. 2013;5:a011106.

Bergallo M, Costa C, Terlizzi ME, Astegiano S, Curtoni A, Solidoro P, Delsedime L, Cavallo R. Quantitative detection of the new polyomaviruses KI, WU and Merkel cell virus in transbronchial biopsies from lung transplant recipients. J Clin Pathol. 2010;63:722–5.

Bhatia S, Storer BE, Iyer JG, Moshiri A, Parvathaneni U, Byrd D, et al. Adjuvant radiation therapy and chemotherapy in Merkel cell carcinoma: survival analyses of 6908 cases from the National Cancer Data Base. J Natl Cancer Inst. 2016;108(9):djw042.

Bhatia K, Goedert JJ, Modali R, Preiss L, Ayers LW. Immunological detection of viral large T antigen identifies a subset of Merkel cell carcinoma tumors with higher viral abundance and better clinical outcome. Int J Cancer. 2010;127:1493–6.

Bichakjian CK, Lowe L, Lao CD, Sandler HM, Bradford CR, Johnson TM, et al. Merkel cell carcinoma: critical review with guidelines for multidisciplinary management. Cancer. 2007;110(1):1e12.

Borchert S, Czech-Sioli M, Neumann F, Schmidt C, Wimmer P, Dobner T, Grundhoff A, Fischer N. High-affinity Rb binding, p53 inhibition, subcellular localization, and transformation by wild type or tumor-derived shortened Merkel cell polyomavirus large T antigens. J Virol. 2014;88:3144–60.

Bottles K, Lancey CG, Goldberg J, et al. Merkel cell carcinoma of the vulva. Obstet Gynecol. 1984;63:61–5.

Bouvard V, Baan RA, Grosse Y, Lauby-Secretan B, El Ghissassi F, Benbrahim-Tallaa L, et al. Carcinogenicity of malaria and of some polyomaviruses. Lancet Oncol. 2012;13:339–40.

Broekema NM, Imperiale MJ. MiRNA regulation of BK polyomavirus replication during early infection. Proc Natl Acad Sci U S A. 2013;110:8200–5.

Brunner M, Thurnher D, Pammer J, et al. Expression of VEGF-A/C, VEGF-R2, PDGF-alpha/beta, c-kit, EGFR, Her-2/Neu, Mcl-1 and Bmi-1 in Merkel cell carcinoma. Mod Pathol. 2008;21:876–84.

Buder K, Lapa C, Kreissl MC, Schirbel A, Herrmann K, Schnack A, et al. Somatostatin receptor expression in Merkel cell carcinoma as target for molecular imaging. BMC Cancer. 2014;14 (268):2407e14–268.

Busam KJ, Jungbluth AA, Rekthman N, Coit D, Pulitzer M, Bini J, Arora R, Hanson NC, Tassello JA, Frosina D, Moore P, Chang Y. Merkel cell polyomavirus expression in Merkel cell carcinomas and its absence in combined tumors and pulmonary neuroendocrine carcinomas. Am J Surg Pathol. 2009;33:1378–85.

Carter JJ, Paulson KG, Wipf GC, et al. Association of Merkel cell polyomavirus-specific antibodies with Merkel cell carcinoma. J Natl Cancer Inst. 2009;101:1510–22.

Carter JJ, Daugherty MD, Qi X, Bheda-Malge A, Wipf GC, Robinson K, Roman A, Malik HS, Galloway DA. Identification of an overprinting gene in Merkel cell polyomavirus provides evolutionary insight into the birth of viral genes. Proc Natl Acad Sci. 2013;110:12744–9.

Cassler NM, Merrill D, Bichakjian CK, Brownell I. Merkel cell carcinoma therapeutic update. Curr Treat Options Oncol. 2016;17(7):36.

Cerroni L, Kerl H. Primary cutaneous neuroendocrine (Merkel cell) carcinoma in association with squamous- and basal-cell carcinoma. Am J Dermato Pathol. 1997;19:610–3.

Chen MM, Roman SA, Sosa JA, Judson BL. The role of adjuvant therapy in the management of head and neck Merkel cell carcinoma: an analysis of 4815 patients. JAMA Otolaryngol Head Neck Surg. 2015;141(2):137e41.

Chen T, Hedman L, Mattila PS, Jartti T, Ruuskanen O, SoderlundVenermo M, Hedman K. Serological evidence of Merkel cell polyomavirus primary infections in childhood. J Clin Virol. 2011;50:125–9.

Cheng J, Rozenblatt-Rosen O, Paulson KG, Nghiem P, DeCaprio JA. Merkel cell polyomavirus large T antigen has growth-promoting and inhibitory activities. J Virol. 2013;87(11):6118–26.

Clarke CA, Robbins HA, Tatalovich Z, Lynch CF, Pawlish KS, Finch JL, et al. Risk of Merkel cell carcinoma after solid organ transplantation. J Natl Cancer Inst. 2015;107(2):dju382.

Cole CN. Polyomaviridae: the viruses and their replication. Fields Virol. 2001;2141–2174.

Comerford SA, Schultz N, Hinnant EA, Klapproth S, Hammer RE. Comparativeanalysis of SV40 17kT and LT function in vivo demonstrates that LT's C-terminus re-programs hepatic gene expression and is necessary for tumorigenesis in the liver. Oncogenesis. 2012;1(9):e28.

Davids MS, Charlton A, Ng SS, Chong ML, Laubscher K, Dar M, et al. Response to a novel multitargeted tyrosine kinase inhibitor pazopanib in metastatic Merkel cell carcinoma. J Clin Oncol. 2009;27(26):e97e100.

DeCaprio JA. Merkel cell polyomavirus and Merkel cell carcinoma. Philos Trans R Soc Lond B Biol Sci. 2017;372(1732):20160276.

DeCaprio JA, Garcea RL. A cornucopia of human polyomaviruses. Nat Rev Micro. 2013;11 (4):264–76.

Demetriou SK, Ona-Vu K, Sullivan EM, Dong TK, Hsu SW, Oh DH. Defective DNA repair and cell cycle arrest in cells expressing Merkel cell polyomavirus T antigen. Intl J Cancer. 2012;131:1818–27.

Desch L, Kunstfeld R. Merkel cell carcinoma: chemotherapy and emerging new therapeutic options. J Skin Cancer. 2013;2013:327150.

Duncavage EJ, Zehnbauer BA, Pfeifer JD. Prevalence of Merkel cell polyomavirus in Merkel cell carcinoma. Mod Pathol. 2009;22:516–21.

Dworkin AM, Tseng SY, Allain DC, et al. Merkelcell polyomavirus in cutaneous squamous cell carcinomaof immunocompetent individuals. J Invest Dermatol. 2009;129:2868–74.

Dye KN, Welcker M, Clurman BE, Roman A. Galloway DA Merkel cell polyomavirus Tumor antigens expressed in Merkel cell carcinoma function independently of the ubiquitin ligases Fbw7 and β-TrCP. PLoS Pathog. 2019;15(1):e1007543.

Engels EA, Frisch M, Goedert JJ, Biggar RJ, Miller RW. Merkel cell carcinoma and HIV infection. Lancet. 2002;359:497–8.

Fakiha M, Letertre P, Vuillez JP, Lebeau J. Remission of Merkelcell tumor after somatostatin analog treatment. J Cancer Res Ther. 2010;6(3):382e4.

Fan K, Gravemeyer J, Ritter C, Rasheed K, Gambichler T, Moens U, Shuda M, Schrama D, Becker JC. MCPyV large T antigen-induced atonal homolog 1 is a lineage-dependency oncogene in Merkel cell carcinoma. J Invest Dermatol. 2020a;140(1):56–65.

Fan K, Zebisch A, Horny K, Schrama D, Becker JC. Highly Expressed miR-375 is not an intracellular oncogene in Merkel cell polyomavirus-associated Merkel cell carcinoma. Cancers (Basel). 2020b;12(3):529.

Faust H, Pastrana DV, Buck CB, et al. Antibodiesto Merkel cell polyomavirus correlate to presence of viral DNA in the skin. J Infect Dis. 2011;203:1096–100.

Feinmesser M, Halpern M, Kaganovsky E, et al. c-Kit expression in primary and metastatic Merkel cell carcinoma. Am J Dermatopathol. 2004;26:458–62.

Feng H, Shuda M, Chang Y, Moore PS. Clonal integration of a polyomavirus in human Merkel cell carcinoma. Science. 2008;319(5866):1096–100.

Feng H, Kwun HJ, Liu X, Gjoerup O, Stolz DB, Chang Y, Moore PS. Cellular and viral factors regulating Merkel cell polyomavirus replication. PLoS ONE. 2011;6(7):e22468.

Fernandez-Figueras MT, Puig L, Musulen E, et al. Expression profiles associated with aggressive behavior in Merkel cell carcinoma. Mod Pathol. 2007;20:90–101.

Fischer M, Grossmann P, Padi M, DeCaprio JA. Integration of TP53, DREAM, MMB-FOXM1 and RB-E2F target gene analyses identifies cell cycle gene regulatory networks. Nucleic Acids Res. 2016;44(13):6070–86.

Fischer N, Brandner J, Fuchs F, Moll I, Grundhoff A. Detection of Merkel cell polyomavirus (MCPyV) in Merkel cell carcinoma cell lines: cell morphology and growth phenotype do not reflect presence of the virus. Intl J Cancer. 2010;126:2133–42.

Foulongne V, Dereure O, Kluger N, et al. Merkelcell polyomavirus DNA detection in lesional and nonlesional skin from patients with Merkel cell carcinoma or other skin diseases. Br J Dermatol. 2010;162:59–63.

Gardner SD, Field AM, Coleman DV, Hulme B. New human papovavirus (BK) isolated from urine after renal transplantation. Lancet. 1971;1:1253–7.

Garneski KM, Nghiem P. Merkel cell carcinoma adjuvant therapy: current data support radiation but not chemotherapy. J Am Acad Dermatol. 2007;57(1):166e9.

Garneski KM, Warcola AH, Feng Q, et al. Merkelcell polyomavirus is more frequently present in North American than Australian Merkel cell carcinomatumors. J Invest Dermatol. 2009;129:246–8.

Gaynor AM, Nissen MD, Whiley DM, Mackay IM, Lambert SB, Wu G, Brennan DC, Storch GA, Sloots TP, Wang D. Identification of a novel polyomavirus from patients with acute respiratory tract infections. PLoS Pathog. 2007;3(5):e64.

Gheit T, Dutta S, Oliver J, et al. Isolation and characterization of a novel putative human polyomavirus. Virology. 2017;506:45–54.

González-Vela MDC, Curiel-Olmo S, Derdak S, Beltran S, Santibañez M, Martínez N, Castillo-Trujillo A, Gut M, Sánchez-Pacheco R, Almaraz C, Cereceda L, Llombart B, Agraz-Doblas A, Revert-Arce J, López Guerrero JA, Mollejo M, Marrón PI, Ortiz-Romero P, Fernandez-Cuesta L, Varela I, Gut I, Cerroni L, Piris MÁ, Vaqué JP. Shared oncogenic pathways implicated in both virus-positive and UV-induced Merkel cell carcinomas. J Invest Dermatol. 2017;137(1):197–206.

Gooptu C, Woollons A, Ross J, Price M, Wojnarowska F, Morris PJ, Wall S, Bunker CB. Merkel cell carcinoma arising after therapeutic immunosuppression. Br J Dermatol. 1997;137:637–41.

Gould E, Albores-Saavedra J, Dubner B, et al. Eccrine and squamous differentiation in Merkel cell carcinoma. An immunohistochemical study. Am J Surg Pathol. 1988;12:768–72.

Griffiths DA, Abdul-Sada H, Knight LM, Jackson BR, Richards K, Prescott EL, Peach AH, Blair GE, Macdonald A, Whitehouse A. Merkel cell polyomavirus small T antigen targets the NEMO adaptor protein to disrupt inflammatory signaling. J Virol. 2013;87:13853–67.

Grundhoff A, Sullivan CS. Virus-encoded micrornas. Virology. 2011;411:325–43.
Gupta P, Shahzad N, Harold A, Shuda M, Venuti A, Romero-Medina MC, Pacini L, Brault L, Robitaille A, Taverniti V, Hernandez-Vargas H, Durand G, Le Calvez-Kelm F, Gheit T, Accardi R, Tommasino M. Merkel cell polyomavirus downregulates N-myc downstream-regulated gene 1, leading to cellular proliferation and migration. J Virol. 2020;94(3): e00899–19.
Hafner C, Houben R, Baeurle A, Ritter C, Schrama D, Landthaler M, et al. Activation of the PI3K/AKT pathway in Merkel cell carcinoma. PLoS ONE. 2012;7(2):e31255.
Harms PW, Harms KL, Moore PS, et al. The biology and treatment of Merkel cell carcinoma: current understanding and research priorities. Nat Rev Clin Oncol. 2018;15:763–76.
Harold A, Amako Y, Hachisuka J, Bai Y, Li MY, Kubat L, Gravemeyer J, Franks J, Gibbs JR, Park HJ, Ezhkova E, Becker JC, Shuda M. Conversion of Sox2-dependent Merkel cell carcinoma to a differentiated neuron-like phenotype by T antigen inhibition. Proc Natl Acad Sci. 2019;116(40):20104–14.
Harrison CJ, Meinke G, Kwun HJ, Rogalin H, Phelan PJ, Bullock PA, et al. Asymmetric assembly of Merkel cell polyomavirus large T-antigen origin binding domains at the viral origin. J Mol Bio. 2011;409(4):529–42.
Hasan S, Liu L, Triplet J, Li Z, Mansur D. The role of postoperative radiation and chemo radiation in Merkel cell carcinoma: a systematic review of the literature. Front Oncol. 2013;4(3):276.
Hashida Y, Imajoh M, Nemoto Y, et al. Detection of Merkel cell polyomavirus with a tumour-specific signature in non-small cell lung cancer. Br J Cancer. 2013;108(3):629–37.
Heath M, Jaimes N, Lemos B, Mostaghimi A, Wang LC, Peñas PF, Nghiem P. Clinical characteristics of Merkel cell carcinoma at diagnosis in 195 patients: the AEIOU features. J Ame Acad Dermatol. 2008;58:375–81.
Helmbold P, Lahtz C, Enk A, Trost PH, Marsch CW, Kutzner H, Dammann RH. Frequent occurrence of RASSF1A promoter hypermethylation and merkel cell polyomavirus in merkel cell carcinoma. Mol Carcinog. 2009:4–10.
Hodgson NC. Merkel cell carcinoma: changing incidence trends. J Surg Oncol. 2005;89:1–4.
Horn HF, Vousden KH. Coping with stress: multiple ways to activate p53. Oncogene. 2007;26:1306–16.
Houben R, Michel B, Vetter-Kauczok CS, et al. Absence of classical MAP kinase pathway signalling in Merkel cell carcinoma. J Invest Dermatol. 2006;126:1135–42.
Houben R, Shuda M, Weinkam R, et al. Merkel cell polyomavirus-infected Merkel cell carcinoma cells require expression of viral T antigens. J Virol. 2010;84(14):7064–72.
Houben R, Adam C, Baeurle A, Hesbacher S, Grimm J, Angermeyer S, Henzel K, Hauser S, Elling R, Brocker EB, et al. An intact retinoblastoma protein-binding site in Merkel cell polyomavirus large T antigen is required for promoting growth of Merkel cell carcinoma cells. Int J Cancer. 2012;130:847–56.
Hughes MP, Hardee ME, Cornelius LA, Hutchins LF, Becker JC, Gao L. Merkel cell carcinoma: epidemiology, target, and therapy. Curr Dermatol Rep. 2014;3:46e53.
Iacocca MV, Abernethy JL, Stefanato CM, et al. Mixed Merkel cell carcinoma and squamous cell carcinoma of the skin. J Am Acad Dermatol. 1998;39:882–7.
Imajoh M, Hashida Y, Nemoto Y, et al. Detection of Merkel cell polyomavirus in cervical squamous cell carcinomas and adenocarcinomas from Japanese patients. Virol J. 2012;9:154.
Iyer JG, Blom A, Doumani R, Lewis C, Tarabadkar ES, Anderson A, et al. Response rates and durability of chemotherapy among 62 patients with metastatic Merkel cell carcinoma. 2016;5 (9):2294e301.
Jabbour J, Cumming R, Scolyer RA, Hruby G, Thompson JF, Lee S. Merkel cell carcinoma: assessing the effect of wide local excision, lymph node dissection, and radiotherapy on recurrence and survival in early stage disease-results from a review of 82 consecutive cases diagnosed between 1992 and 2004. Ann Surg Oncol. 2007;14(6):1943e52.
James GH, Stephen BB. Gene silencing in cancer in association with promoter hypermethylation. N Engl J Med. 2003; 349:2042–2054.

Johnson EM. Structural evaluation of new human polyomaviruses provides clues to pathobiology. Trends Microbiol. 2010;18(5):215–23.

Jouary T, Leyral C, Dreno B, Doussau A, Sassolas B, Beylot Barry M, et al. Adjuvant prophylactic regional radiotherapy versus observation in stage I Merkel cell carcinoma: a multicentric prospective randomized study. Ann Oncol. 2012;23(4):1074e80.

Kannan A, Lin Z, Shao Q, Zhao S, Fang B, Moreno MA, et al. Dual mTOR inhibitor MLN0128 suppresses Merkel cell carcinoma (MCC) xenograft tumor growth. Oncotarget. 2016;7 (6):6576e92.

Kantola K, Sadeghi M, Lahtinen A, et al. Merkelcell polyomavirus DNA in tumor-free tonsillar tissues and upper respiratory tract samples: implications for respiratory transmission and latency. J Clin Virol. 2009;45:292–5.

Kassem A, Technau K, Kurz AK, et al. Merkel cell polyomavirus sequences are frequently detected in nonmelanoma skin cancer of immunosuppressed patients. Int J Cancer. 2009;125 (2):356–61.

Kaufman HL, Russell J, Hamid O, Bhatia S, Terheyden P, D'Angelo SP, et al. Avelumab in patients with chemotherapy refractory metastatic Merkel cell carcinoma: a multicentre, single-group, open-label, phase 2 trial. Lancet Oncol. 2016;17(10):1374e85.

Kean JM, Rao S, Wang M, Garcea RL. Seroepidemiology of human polyomaviruses. PLoS Pathog. 2009;5:e1000363.

Kervarrec T, Samimi N, Guyétant S, Sarma B, Chéret J, Blanchard E, Berthon P, Schrama D, Houben R, Antoine T. Histogenesis of Merkel cell carcinoma: a comprehensive review. Front Oncol. 2019;9451.

Kim J, McNiff JM. Nuclear expression of survivin portends a poor prognosis in Merkel cell carcinoma. Mod Pathol. 2008;21:764–9.

Kincaid RP, Sullivan CS. Virus-encoded micrornas: an overview and a look to the future. PLoS Pathog. 2012;8:e1003018.

Korup S, Rietscher J, Calvignac-Spencer S, Trusch F, Hofmann J, et al. Identification of a novel human polyomavirus in organs of the gastrointestinal tract. PLoS ONE. 2013;8:e58021.

Kumar S, Xie H, Shi H, Gao J, Juhlin CC, Björnhagen V, Höög A, Lee L, Larsson C, Lui WO. Merkel cell polyomavirus oncoproteins induce microRNAs that suppress multiple autophagy genes. Int J Cancer. 2020;146(6):1652–66.

Kwun HJ, Guastafierro A, Shuda M, Meinke G, Bohm A, Moore PS, Chang Y. The minimum replication origin of Merkel cell polyomavirus has a unique large t-antigen loading architecture and requires small t-antigen expression for optimal replication. J Virol. 2009;83:12118–28.

Kwun HJ, Shuda M, Feng H, Camacho CJ, Moore PS, Chang Y. Merkel cell polyomavirus small T antigen controls viral replication and oncoprotein expression by targeting the cellular ubiquitin ligase SCFFbw7. Cell Host Microbe. 2013;14:125–35.

Lanoy E, Costagliola D, Engels EA. Skin cancers associated with HIV infection and solid-organ transplantation among elderly adults. Int J Cancer. 2010;126(7):1724–31.

Lassacher A, Heitzer E, Kerl H, Wolf P. p14ARF hypermethylation is common but INK4a-ARF locus or p53 mutations are rare in Merkel cell carcinoma. J Invest Dermatol. 2008;128: 1788–96.

Lebbe C, Becker JC, Grob JJ, Malvehy J, Del Marmol V, Pehamberger H, et al. Diagnosis and treatment of Merkel cellcarcinoma. European consensus-based interdisciplinary guideline. Eur J Cancer. 2015;51(16):2396e403.

Lewis KG, Weinstock MA, Weaver AL, Otley CC. Adjuvant local irradiation for Merkel cell carcinoma. Arch Dermatol. 2006;142(6):693e700.

Li J, Wang X, Diaz J, Tsang SH, Buck CB, You J. Merkel cell polyomavirus large T antigen disrupts host genomic integrity and inhibits cellular proliferation. J Virol. 2013;87:9173–88.

Li J, Diaz J, Wang X, Tsang SH, You J. Phosphorylation of Merkel cell polyomavirus large tumor antigen at serine 816 by ATM kinase induces apoptosis in host cells. J Biol Chem. 2015;290:1874–84.

Lim ES, Reyes A, Antonio M, et al. Discovery of STL polyomavirus, a polyomavirus of ancestral recombinant origin that encodes a unique T antigen by alternative splicing. Virology. 2013;436 (2):295–303.

Liu W, Krump NA, Herlyn M, You J. Combining DNA damage induction with BCL-2 inhibition to enhance Merkel cell carcinoma cytotoxicity. Biology. 2020;9:35.

Lin Z, McDermott A, Shao L, Kannan A, Morgan M, Stack Jr BC, et al. Chronic mTOR activation promotes cell survival in Merkel cell carcinoma. Cancer Lett. 2014;344(2):272e81.

Liu JH, Richardson SC, Basse PH, Toptan T, Moore PS, Gjoerup OV, Chang Y. Merkel cell polyomavirus large T antigen disrupts lysosome clustering by translocating human Vam6p from the cytoplasm to the nucleus. J Biol Chem. 2011;286:17079–90.

Liu W, Yang R, Payne AS, Schowalter RM, Spurgeon ME, Lambert PF, Xu X, Buck CB, You J. Identifying the target cells and mechanisms of Merkel cell polyomavirus infection. Cell Host Microbe. 2016;19(6):775–87.

Loader DE, Feldmann R, Baumgartner M, Breier F, Schrama D, Becker JC, et al. Clinical remission of Merkel cellcarcinoma after treatment with imatinib. J Am Acad Dermatol. 2013;69(4):e181e3.

Loyo M, Guerrero-Preston R, Brait M, et al. Quantitative detection of Merkel cell virus in human tissues and possible mode of transmission. Int J Cancer. 2010;126:2991–6.

Lucarz A, Brand G. Current considerations about Merkel cells. Eur J Cell Biol. 2007;86(5): 243–51.

Lyngaa R, Pedersen NW, Schrama D, Thrue CA, Ibrani D, Met O, et al. T-cell responses to oncogenic Merkel cell polyomavirus proteins distinguish patients with Merkel cell carcinoma from healthy donors. Clin Cancer Res. 2014;20(7):1768e78.

Mangana J, Dziunycz P, Kerl K, et al. Prevalenceof Merkel cell polyomavirus among Swiss Merkelcell carcinoma patients. Dermatology. 2010;221:184–8.

Martincorena I, Roshan A, Gerstung M, Ellis P, Van Loo P, McLaren S, et al. NOTCH1 mutations occur early during cutaneous squamous cell carcinogenesis. J Invest Dermatol. 2014;134:2630–8.

Meinke G, Phelan P, Moine S, Bochkareva E, Bochkarev A, Bullock PA, Bohm A. The crystal structure of the SV40 T-antigen origin binding domain in complex with DNA. PLoS Biol. 2007;5(2):e23.

Mertz KD, Junt T, Schmid M, Pfaltz M, Kempf W. Inflammatory monocytes are a reservoir for Merkel cell polyomavirus. J Invest Dermatol. 2010a;130(4):1146–51.

Mertz KD, Pfaltz M, Junt T, et al. Merkel cellpolyomavirus is present in common warts and carcinomain situ of the skin. Hum Pathol. 2010b;41:1369–79.

Mishra N, Pereira M, Rhodes RH, et al. Identification of a novel polyomavirus in a pancreatic transplant recipient with retinal blindness and vasculitic myopathy. J Infect Dis. 2014;210 (10):1595–9.

Mogha A, Fautrel A, Mouchet N, Guo N, Corre S, Adamski H, Watier E, Misery L, Galibert MD. Merkel cell polyomavirus small T antigen mRNA level is increased following in vivo UVradiation. PLoS ONE. 2010;5:e11423.

Mojica P, Smith D, Ellenhorn JD. Adjuvant radiation therapy is associated with improved survival in Merkel cell carcinoma of the skin. J Clin Oncol. 2007;25(9):1043e7.

Moll I, Zieger W, Schmelz M. Proliferative Merkel cells were not detected in human skin. Arch Dermatol Res. 1996;288(4):184–7.

Mortier L, Mirabel X, Fournier C, Piette F, Lartigau E. Radiotherapy alone for primary Merkel cell carcinoma. Arch Dermatol. 2003;139(12):1587e90.

Murakami M, Imajoh M, Ikawa T, et al. Presence of Merkel cell polyomavirus in Japanese cutaneous squamous cell carcinoma. J Clin Virol. 2011;50(1):37–41.

Nakafusa J, Narisawa Y, Shinogi T, Taira K, Tanaka T, Inoue T, Misago N. Changes in the number of Merkel cells with the hair cycle in hair discs on rat back skin. Br J Dermatol. 2006;155(5):883–9.

Nakamura T, Sato YD, Watanabe H, Ito N, Shimonohara T, Tsuji N, Nakajima Y, Suzuki K, Matsuo H, Nakagawa T, Sata H, Katano. Nuclear localization of Merkel cell polyomavirus large T antigen in Merkel cell carcinoma. Virology. 2010;398: 273–2792.

Nardi V, Song Y, Santamaria-Barria JA, Cosper AK, Lam Q, Faber AC, et al. Activation of PI3K signaling in Merkel cell carcinoma. Clin Cancer Res. 2012;18(5):1227e36.

Nathan PD, Gaunt P, Wheatley K, Bowden SJ, Savage J, Faust G, et al. UKMCC-01: a phase II study of pazopanib(PAZ) in metastatic Merkel cell carcinoma. ASCO Meet Abstr. 2016;34 (Suppl. 15):9542.

National Comprehensive Cancer Network. NCCN Clinical Practice Guidelines in Oncology: Merkel Cell Carcinoma. 2016.

Neumann F, Borchert S, Schmidt C, Reimer R, Hohenberg H, Fischer N, Grundhoff A. Replication, gene expression and particle production by a consensus Merkel Cell Polyomavirus (MCPyV) genome. PLoS ONE. 2011;6:e29112.

Ogawa H. The Merkel cell as a possible mechanoreceptor cell. Prog Neurobiol. 1996;49(4): 317–34.

Padgett BL, Walker DL, ZuRhein GM, Eckroade RJ, Dessel BH. Cultivation of papova-like virus from human brain with progressive multifocal leucoencephalopathy. Lancet. 1971;1:1257–60.

Pallas DC, Shahrik LK, Martin BL, Miller TB, Brautigan DL, Roberts TM. Polyoma small and middle T antigens and SV40 small t antigen form stable complexes with protein phosphatase 2A. Cell. 1990;60:167–76.

Pancaldi C, Corazzari V, Maniero S, et al. Merkel cell polyomavirus DNA sequences in the buffy coats of healthy blood donors. Blood. 2011;117(26):7099–101.

Pantulu ND, Pallasch CP, Kurz AK, Kassem A, et al. Detection of a novel truncating Merkel cell polyomavirus large T antigen deletion in chronic lymphocytic leukemia cells. Blood. 2010;116 (24):5280–4.

Park DE, Cheng J, McGrath JP, Lim MY, Cushman C, Swanson SK, Tillgren ML, Paulo JA, Gokhale PC, Florens L, Washburn MP, Trojer P, DeCaprio JA. Merkel cell polyomavirus activates LSD1-mediated blockade of non-canonical BAF to regulate transformation and tumorigenesis. Nat Cell Biol. 2020;22(6):752.

Pastrana DV, Tolstov YL, Becker JC, Moore PS, Chang Y, Buck CB. Quantitation of human sero responsiveness to Merkel cell polyomavirus. PLoS Pathog. 2009;5(9):e1000578.

Pastrana DV, Wieland U, Silling S, Buck CB, Pfister H. Positivecorrelation between Merkel cell polyomavirus viral load and capsid-specific antibody titer. Med Microbiol Immunol. 2012;201:17–23.

Paulson KG, Iyer JG, Byrd DR, Nghiem P. Pathologic nodal evaluation is increasingly commonly performed for patients with Merkel cell carcinoma. J Am Acad Dermatol. 2013;69(4):653e4.

Paulson KG, Carter JJ, Johnson LG, Cahill KW, Iyer JG, Schrama D, Becker JC, Madeleine MM, Nghiem P, Galloway DA. Antibodies to Merkel cell polyomavirus t antigen oncoproteins reflect tumor burden in Merkel cell carcinoma patients. Cancer Res. 2010;70:8388–97.

Paulson KG, Iyer JG, Tegeder AR, Thibodeau R, Schelter J, Koba S, Schrama D, Simonson WT, Lemos BD, Byrd DR, et al. Transcriptome-wide studies of Merkel cell carcinoma and validation of intratumoral CD8+ lymphocyte invasion as an independent predictor of survival. J Clin Oncol. 2011;29:1539–46.

Paulson KG, Tegeder A, Willmes C, Iyer JG, Afanasiev OK, Schrama D, Koba S, Thibodeau R, Nagase K, Simonson WT, et al. Downregulation of MHC-I expression is prevalent but reversible in Merkel cell carcinoma. Cancer Immunol Res. 2014;2:1071–9.

Pectasides D, Pectasides M, Economopoulos T. Merkel cell cancer of the skin. Ann Oncol. 2006;17:1489–95.

Plaza JA, Suster S. The Toker tumor: spectrum of morphologic features in primary neuroendocrine carcinomas of the skin (Merkel cell carcinoma). Ann Diagn Pathol. 2006;10:376–85.

Postow MA, Callahan MK, Wolchok JD. Immune check point blockade in cancer therapy. J Clin Oncol. 2015;33(17):1974e82.

Poulsen M, Rischin D, Walpole E, Harvey J, Mackintosh J, Ainslie J, et al. High-risk Merkel cell carcinoma of the skin treated with synchronous carboplatin/etoposide and radiation: a Trans-Tasman Radiation Oncology Group StudyeTROG 96:07. Clin Oncol. 2003;21 (23):4371e6.

Poulsen MG, Rischin D, Porter I, Walpole E, Harvey J, Hamilton C, et al. Does chemotherapy improve survival in highrisk stage I and II Merkel cell carcinoma of the skin? Int J Radiat Oncol Biol Phys. 2006;64(1):114e9.

Prasad R, Katiyar SK. Crosstalk among UV-induced inflammatory mediators, DNA damage and epigenetic regulators facilitates suppression of the immune system. Photochem Photobiol. 2017;93:930–6.

Rafael AS, Jose GC, Gabriela C, Fabio E, Grosso O, Luiza EBP Kassuga, Jorge LMT. Burden of disease, early diagnosis, and treatment of Merkel cell carcinoma in Latin America. J, Glob Oncol. 2018:4, 1–11.

Reza MA, Reza MH, Mahdiyeh L, Mehdi F, Hamid ZN. Evaluation frequency of Merkel cell polyoma, Epstein-Barr and mouse mammary tumor viruses in patients with breast cancer in Kerman, Southeast of Iran. Asian Pac J Cancer Prev. 2015;16(16):7351–7.

Richter AM, Haag T, Walesch S, Herrmann P, Marsch WC, Kutzner H, Helmbold P, Dammann RH. Aberrant promoter hypermethylation of RASSF family members in Merkel cell carcinoma. Cancers. 2013;5(4):1566–76.

Rodig SJ, Cheng J, Wardzala J, DoRosario A, Scanlon JJ, Laga AC, Martinez-Fernandez A, Barletta JA, Bellizzi AM, Sadasivam S, et al. Improved detection suggests all Merkel cell carcinomas harbor Merkel polyomavirus. J Clin Invest. 2012;122:4645–53.

Sadeghi M, Riipinen A, Väisänen E, et al. Newlydiscovered KI, WU, and Merkel cell polyomaviruses:no evidence of mother-to-fetus transmission. Virol J. 2010;7:251.

Sadeghi F, Salehi-Vaziri M, Ghodsi SM, et al. Prevalence of JC polyomavirus large T antigen sequences among Iranian patients with central nervous system tumors. Arch Virol. 2015;160 (1):61–8.

Sahi H, Savola S, Sihto H, Koljonen V, Bohling T, Knuutila S. RB1 gene in Merkel cell carcinoma: hypermethylation in all tumors and concurrent heterozygous deletions in the polyomavirus-negative subgroup. APMIS. 2014;122(12):1157–66.

Saini AT, Miles BA. Merkel cell carcinoma of the head and neck: pathogenesis, current and emerging treatment options. Onco Targets Ther. 2015;8:2157e67.

Samlowski WE, Moon J, Tuthill RJ, Heinrich MC, Balzer-Haas NS, Merl SA, et al. A phase II trial of imatinib mesylate in Merkel cell carcinoma (neuroendocrine carcinoma of the skin): a Southwest Oncology Group study (S0331). Am J Clin Oncol. 2010;33(5):495e9.

Santamaria-Barria JA, Boland GM, Yeap BY, Nardi V, Dias-Santagata D, Cusack Jr JC. Merkel cell carcinoma: 30-year experience from a single institution. Ann Surg Oncol. 2013;20 (4):1365e73.

Schmidt U, Muller U, Metz KA, Leder LD. Cytokeratin and neurofilament protein staining in Merkel cell carcinoma of the small cell type and small cell carcinoma of the lung. Am J Dermatopathol. 1998;20:346–51.

Schowalter RM, Buck CB. The Merkel cell polyomavirus minor capsid protein. PLoS Pathog. 2013;9:e1003558.

Schowalter RM, Pastrana DV, Pumphrey KA, et al. Merkel cell polyomavirus and two previously unknown polyomaviruses are chronically shed from humanskin. Cell Host Microbe. 2010;7:509–15.

Scott MP, Helm KF. Cytokeratin 20: a marker for diagnosing Merkel cell carcinoma. Am J Dermatopathol. 1999;21:16–20.

Scuda N, Hofmann J, Calvignac-Spencer S, Ruprecht K, Liman P, et al. A novel human polyomavirus closely related to the African green monkey-derived lymphotropic polyomavirus. J Virol. 2011;85:4586–90.

Seo GJ, Chen CJ, Sullivan CS. Merkel cell polyomavirus encodes a microrna with the ability to autoregulate viral gene expression. Virology. 2009;383:183–7.

Servy A, Maubec E, Sugier PE, Grange F, Mansard S, Lesimple T, et al. Merkel cell carcinoma: value of sentinel lymph-node status and adjuvant radiation therapy. Ann Oncol. 2016;27 (5):914e9.

Shahzad N, Shuda M, Gheit T, Kwun HJ, Cornet I, Saidj D, Zannetti C, Hasan U, Chang Y, Moore PS, et al. The T antigen locus of Merkel cell polyomavirus downregulates human Toll-like receptor 9 expression. J Virol. 2013;87:13009–19.

Shahzad N, Hussain I, Gilani US, Tayyeb A, Aslam MA, Khurshid M, Hassan U, Tasneem F, Umer U, Rashid N. Merkel cell polyomavirus DNA sequences in the blood of healthy population of Pakistan. Future Microbiol. 2019;14:599–608.

Shiver MB, Mahmoud F, Gao L. Response to idelalisib in a patient with stage IV Merkel-cell carcinoma. N Engl J Med. 2015;373(16):1580e2.

Shuda M, Arora R, Kwun HJ, Feng H, Sarid R, Fernandez-Figueras MT, et al. Human Merkel cell polyomavirus infection I. MCV T antigen expression in Merkel cell carcinoma, lymphoid tissues and lymphoid tumors. Int J Cancer. 2009;125:1243–9.

Shuda M, Feng H, Kwun HJ, Rosen ST, Gjoerup O, Moore PS, Chang Y. T antigen mutations are a human tumor-specific signature for Merkel cell polyomavirus. Proc Natl Acad Sci U S A. 2008;105(42):16272–7.

Shuda M, Kwun HJ, Feng H, Chang Y, Moore PS. Human Merkel cell polyomavirus small T antigen is an oncoprotein targeting the 4E-BP1 translation regulator. J Clin Invest. 2011;121 (9):3623–34.

Shuda M, Velasquez C, Cheng E, Cordek DG, Kwun HJ, Chang Y, Moore PS. CDK1 substitutes for mTOR kinase to activate mitotic cap-dependent protein translation. Proc Natl Acad Sci USA. 2015;112:5875–82.

Siebrasse EA, Reyes A, Lim ES, Zhao G, Mkakosya RS, et al. Identification of MW polyomavirus, a novel polyomavirus in human stool. J Virol. 2012;86(19):10321–6.

Sihto H, Kukko H, Koljonen V, Sankila R, Böhling T, Joensuu H. Merkel cell polyomavirus infection, large T antigen, retinoblastoma protein and outcome in Merkel cell carcinoma. Clin Cancer Res. 2011;17(14):4806–13.

Sihto H, Joensuu H. Tumor-infiltrating lymphocytes and outcome in Merkel cell carcinoma, a virus-associated cancer. Oncoimmunology. 2012;1:1420–1.

Silva EG, Mackay B, Goepfert H, et al. Endocrine carcinoma of the skin (Merkel cell carcinoma). Pathol Annu. 1984;19:1–30.

South AP, Purdie KJ, Watt SA, Haldenby S, den Breems NY, Dimon M, et al. Tumor evolution. High burden and pervasive positive selection of somatic mutations in normal human skin. Science. 2015;348:880–6.

Stakaitytė G, Wood JJ, Knight LM, Abdul-Sada H, Adzahar NS, Nwogu N, Macdonald A, Whitehouse A. Merkel cell polyomavirus: molecular insights into the most recently discovered human tumour virus. Cancers (Basel). 2014;6(3):1267–97.

Starrett GJ, Marcelus C, Cantalupo PG, Katz JP, Cheng J, Akagi K, Thakuria M, Rabinowits G, Wang LC, Symer DE, Pipas JM, Harris RS, DeCaprio JA. Merkel Cell polyomavirus exhibits dominant control of the tumor genome and transcriptome in virus-associated Merkel cell carcinoma. mBio. 2017;8(1):e02079–16.

Tachibana T. The Merkel cell: recent findings and unresolved problems. Arch Histol Cytol. 1995;58(4):379–96.

Tai P. A practical update of surgical management of Merkel cell carcinoma of the skin. ISRN Surg. 2013:850797.

Teman CJ, Tripp SR, Perkins SL, Duncavage EJ. Merkel cell polyomavirus (MCPyV) in chronic lymphocytic leukemia/small lymphocytic lymphoma. Leuk Res. 2011;35(5):689–92.

Toker C. Trabecular carcinoma of the skin. Arch Dermatol. 1972;105(1):107–10.

Tolstov YL, Pastrana DV, Feng H et al. Human Merkel cell polyomavirus infection II. MCV is a common human infection that can be detected by conformational capsid epitope immunoassays. Int J Cancer. 2009;125:1250–6.

Tolstov YL, Knauer A, Chen JG, et al. Asymptomaticprimary Merkel cell polyomavirus infection amongadults. Emerg Infect Dis. 2011;17:1371–80.

Topalian SL, Drake CG, Pardoll DM. Immune check point blockade: a common denominator approach to cancer therapy. Cancer Cell. 2015;27(4):450e61.

Touzé A, Gaitan J, Arnold F, et al. Generation of Merkel cell polyomavirus (MCV)-like particles and their application to detection of MCV antibodies. J Clin Microbiol. 2010;48:1767–70.

Touzé A, Bidre E, Laude H, et al. High levels of antibodies against Merkel cell polyomavirus identify a subset of patients with Merkel cell carcinoma with better clinical outcome. J Clin Oncol. 2011;29:1612–9.

Van der Meijden ME, Janssens RWA, Lauber C, Bouwes Bavinck JN, Gorbalenya AE, Feltkamp MCW. Discovery of a new human polyomavirus associated with trichodysplasia spinulosa in an immunocompromized patient. PLoS Pathog. 2010;6(7):e1001024.

Van Gele M, Speleman F, Vandesompele J, Van Roy N, Leonard JH. Characteristic pattern of chromosomal gains and losses in Merkel cell carcinoma detected by comparative genomic hybridization. Cancer Res. 1998;58:1503–8.

Van Gele M, Leonard JH, Van Roy N, Cook AL, De Paepe A, Speleman F. Frequent allelic loss at 10q23 but low incidence of PTEN mutations in Merkel cell carcinoma. Int J Cancer. 2001;92:409–13.

Veness M, Foote M, Gebski V, Poulsen M. The role of radiotherapy alone in patients with Merkel cell carcinoma: reporting the Australian experience of 43 patients. Int J Radiat Oncol Biol Phys. 2010;78(3):703e9.

Veness MJ, Morgan GJ, Gebski V. Adjuvant locoregional radiotherapy as best practice in patients with Merkel cell carcinoma of the head and neck. Head Neck. 2005;27(3):208e16.

Verhaegen ME, Mangelberger D, Weick JW, Vozheiko TD, Harms PW, Nash KT, Quintana E, Baciu P, Johnson TM, Bichakjian CK, Dlugosz AA. Merkel cell carcinoma dependence on bcl-2 family members for survival. J Invest Dermatol. 2014;134(8):2241–50.

Verhaegen ME, Mangelberger D, Harms PW, Vozheiko TD, Weick JW, Wilbert DM, Saunders TL, Ermilov AN, Bichakjian CK, Johnson TM, et al. Merkel cell polyomavirus small T antigen is oncogenic in transgenic mice. J Invest Dermatol. 2015;135(5):1415–24.

Viscidi RP, Rollison DE, Sondak VK, et al. Age-specific seroprevalence of Merkel cell polyomavirus, BK virus, and JC virus. Clin Vaccine Immunol. 2011;18:1737–43.

Voog E, Biron P, Martin JP, Blay JY. Chemotherapy for patients with locally advanced or metastatic Merkel cell carcinoma. Cancer. 1999;85:2589–95.

Wieland U, Silling S, Scola N et al. Merkelcell polyomavirus infection in HIV-positive men. Arch Dermatol. 2011;147:401–6.

Wieland U, Mauch C, Kreuter A, et al. Merkelcell polyomavirus DNA in persons without Merkelcell carcinoma. Emerg Infect Dis. 2009;15:1496–8.

Wollebo HS, Bellizzi A, Kaminski R, Hu W, White MK, Khalili K. CRISPR/Cas9 system as an agent for eliminating polyomavirus JC infection. PLoS ONE. 2015;10:e0136046.

Wong SQ, Waldeck K, Vergara IA, Schröder J, Madore J, Wilmott JS, Colebatch AJ, De Paoli-Iseppi R, Li J, Lupat R, Semple T, Arnau GM, Fellowes A, Leonard JH, Hruby G, Mann GJ, Thompson JF, Cullinane C, Johnston M, Shackleton M, Sandhu S, Bowtell DD, Johnstone RW, Fox SB, McArthur GA, Papenfuss AT, Scolyer RA, Gill AJ, Hicks RJ, Tothill RW. UV-associated mutations underlie the etiology of MCV-Negative Merkel cell carcinomas. Cancer Res. 2015;75(24):5228–34.

Wu JH, Narayanan D, Limmer AL, Simonette RA, Rady PL, Tyring SK. Merkel cell polyomavirus small T antigen induces DNA damage response. Intervirology. 2019;62 (2):96–100.

Hantavirus

Hantaviruses—A Concise Review of a Neglected Virus

María Victoria Vadell

Abstract Hantaviruses are negative sense single stranded, RNA viruses belonging to the family *Hantaviridae* (order Bunyavirales). They are harbored by a wide range of vertebrate reservoirs, including bats, rodents, shrews, moles, and recently, fish and reptiles. *Orthohantavirus* is the most numerous and diverse genus within this family, and the only one known to be able to cause disease to humans. Hantavirus disease is classified according to the main affected organ as Hantavirus Pulmonary Syndrome (HPS) and Hemorrhagic Fever with Renal Syndrome (HFRS), together with a mild form of HFRS commonly described as Nephropathia Epidemica (NE). HPS occurs almost exclusively in the Americas, while HFRS and NE cases are mostly distributed throughout Europe and Asia. Transmission to humans occurs by inhalation of aerosolized virus particles, which are shed in saliva, urine or feces of infected reservoirs. Human to human transmission is rare, and has only been demonstrated to occur in Argentina and Chile. Risk of acquiring Hantavirus disease depends strongly on seasonal and multiannual fluctuations of reservoir populations, and it is also affected by human behavior. Hantavirus infection has been associated to exposure to rodents or rodent excreta in or around dwellings, and to both recreational and occupational activities in the wild. To date there is no effective treatment against Hantavirus disease and therapy of both HPS and HFRS patients is usually based on supportive care.

Keywords Hantavirus · *Hantaviridae* · *Orthohantavirus* · Hemorrhagic Fever with Renal syndrome (HFRS) · Nephropathia Epidemica · Sin Nombre virus · *Rattus norvegicus* · *Peromyscus maniculatus* · Hantavirus Pulmonary Syndrome (HPS)

M. V. Vadell (✉)
CONICET (Consejo Nacional de Investigaciones Científicas y Técnicas, Argentina), Instituto Nacional de Medicina Tropical - ANLIS "Dr. Carlos G. Malbrán", Puerto Iguazú, Misiones, Argentina

S. I. Ahmad (ed.), *Human Viruses: Diseases, Treatments and Vaccines*, https://doi.org/10.1007/978-3-030-71165-8_18

1 Introduction

Still colloquially known as Hantavirus, the recently re-named genus *Orthohantavirus* is a worldwide group of viruses belonging to the family *Hantaviridae*, order Bunyavirales. Members of the family *Hantaviridae* are negative-sense single stranded, RNA viruses with genomes arranged in three segments of different lengths (Holmes and Zhang 2015). They form a large but not fully understood family, with around 100 described genotypes, and new ones being discovered or reclassified every few years. They are harbored by a wide range of vertebrate reservoirs, including bats, rodents, shrews, moles, and recently, fish and reptiles (Shi et al. 2018; Abudurexiti et al. 2019; Fig. 1). Based on phylogenetic analysis of the small and medium genomic segments, all the members of the former genus *Hantavirus* have recently been reclassified into four newly defined genera within the subfamily *Mammantavirinae* (Fig. 1). The *Orthohantavirus* genus includes most of the species of the former *Hantavirus* genus (derived from Greek ὀρθός [orthós], meaning "straight," historical genus Hantavirus, and genus suffix—virus), most of them hosted by rodents and shrews (Laenen et al. 2019). The other three genera within this subfamily are viruses hosted by bats (*Loanvirus*), by moles and bats (*Mobatvirus*), and by shrews (*Thottimvirus*; Fig. 1). Finally, the very recently discovered members of the family, the ones found in fish and reptiles, have been grouped in other three sub-families, *Actantavirinae* (hosted by ray-finned fish), *Agantavirinae* (hosted by jawless fish) and *Repantavirinae* (hosted by reptiles; Laenen et al. 2019; Fig. 1).

Certain viruses within the *Orthohantavirus* genus can cause disease in humans. According to the main affected organ, Hantavirus disease has been broadly classified as Hantavirus Pulmonary Syndrome (HPS), Hemorrhagic Fever with Renal Syndrome (HFRS), and a mild form of HFRS commonly described as Nephropathia Epidemica (NE) (Jonsson et al. 2010). HPS occurs almost exclusively in the Americas, which include North, Central and South American countries, while HFRS and NE cases are mostly distributed throughout Europe and Asia. Transmission to humans occurs by inhalation of aerosolized virus particles, which are shed in saliva, urine or feces of infected reservoirs (Forbes et al. 2018). Because these viruses are sensible to UV-light, and they are easily diluted by the effect of wind, transmission to humans is more likely to occur in closed, non-ventilated, shady environments.

2 Discoveries of Hantaviruses

Tracing the first records of a disease is never easy, and Hantavirus infections are no exceptions. Several illnesses share symptoms with hantavirus-associated diseases, including leptospirosis, which is also a rodent-borne zoonosis. The first report of HFRS-like disease is from China and as old as 960 A.D (Lee et al. 2014). However,

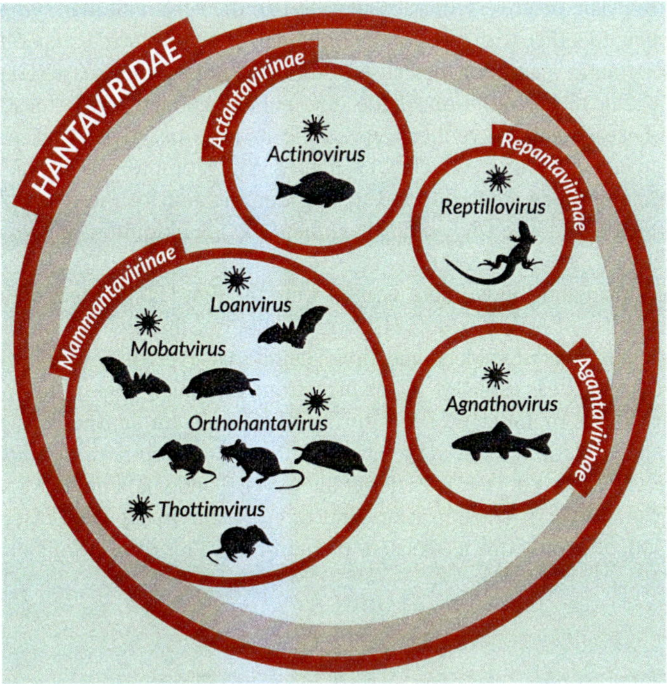

Fig. 1 Legend for the Fig. 1. Subfamilies, genera and known hosts of the family *Hantaviridae*. Known hosts include ray-finned fish, jawless fish, bats, moles, rodents, shrews and reptiles

it was not until the first half of the twentieth century that physicians all over Asia started acknowledging, describing and researching various forms of this disease, which often appeared as outbreaks associated to military operations during World War I and II (Escutenaire and Pastoret 2000; Lee et al. 2014). By the middle of the century, HFRS finally drew the attention of the Western world, when more than 3,000 United Nations soldiers contracted what was then known as Korean Hemorrhagic Fever (KHF) during the Korean War (1951–1953; Lee et al. 2014). Twenty five years later, and after much effort and cumulative research, Lee and colleagues finally identified the etiological agent of KHF, and its main reservoir, the striped field mouse, *Apodemus agrarius* (Lee et al. 1978; Johnson 2004). The newly discovered virus was named "Hantaan", after a small river located in an endemic area near the border between the two Koreas (Johnson 2004).

The identification of Hantaan virus marked the start of a whole new series of discoveries, including the identification of many different genotypes, both pathogenic and non-pathogenic, and many new reservoirs. However, its major turnover was the identification of a new kind of disease in the Americas, which caused acute-respiratory distress and was associated to an unknown Hantavirus, hosted by New World rodents. It all started in May 1993, when a previously healthy young man with shortness of breath died soon after being entered to a hospital in New

Mexico, USA. The discovery that his fiancée had died from similar symptoms the previous week, together with a series of similar cases that occurred soon after, led to the start of a frantic search for the cause of this new unknown disease (CDC 2020). Few weeks after the recognition of the outbreak, researchers from the CDC (US Center for Disease and Prevention) were able to link this mysterious pulmonary syndrome to a Hantavirus, which was named Sin Nombre virus. An intensive mammal trapping campaign soon led to the identification of the reservoir of this new virus, the deer mouse (*Peromyscus maniculatus*), a ubiquitous native rodent of North America (Yates et al. 2002).

While Hantaan and Sin Nombre viruses were the first Hantaviruses identified as causative agents of HFRS and HPS respectively, these were not the first Hantaviruses discovered. Indeed the first Hantavirus ever discovered was called Tottapalayam, and was isolated by chance from an Asian house shrew (*Suncus murinus*) in Southern India in 1964 during a research on Japanese Encephalitis virus, and later identified as a member of the then *Hantavirus* genus (family *Bunyaviridae*; Carey et al. 1971; Zeller et al. 1989). Tottapalayam virus remained a mystery for many years, since as more and more Hantaviruses were discovered, it was, until 2007, the only one associated to a non-rodent reservoir (Song et al. 2007; Laenen et al. 2019).

3 Ecology and Diversity of Hantaviruses

Since its recognition as etiologic agents of disease, the number of Hantaviruses discovered has been increasing at a great speed, eventually leading to the promotion of the genus *Hantavirus* into a new family and to the creation of new genera (Adams et al. 2017; Laenen et al. 2019). Though future studies could change this pattern, actually the genus *Orthohantavirus* is the most numerous and diverse, and its members are hosted by more than 20 genera in four families (Cricetidae, Muridae, Talpidae and Soricidae) within two orders (Rodentia and Eulipotyphla). Each virus is generally hosted by one or a few closely related species, though spillover and host switching events are frequently reported (Mull et al. 2020). Cross-species transmission, together with the recognition of more and more non-rodent Hantavirus reservoirs in the last decades led to the challenging of the long-thought idea that Hantavirus-host systems are the product of millions of years of co-evolution (Holmes and Zhang 2015).

The taxonomy and phylogeny of the genus *Orthohantavirus* is complex and not yet resolved, as is the case of so many other groups of viruses. The recent availability of molecular tools had led to an avalanche of viral genomic sequencing data, suggesting hundreds of potentially new viruses from different taxa (Shi et al. 2018; Laenen et al. 2019). However, genomic data is not sufficient to resolve the status of a taxa, and biological properties of the virus, such as host range, antigenic relatedness, and virion morphology are also needed, and not always available (Laenen et al. 2019). Because of this, from more than 90 *Orthohantavirus* genotypes that have been identified so far, the International Committee on Taxonomy of Viruses (ICTV) recognizes only 36 species and 58 viral names within these species (Abudurexiti et al. 2019; Laenen et al. 2019; Table 1). Particularly in South America, the distribution and diversity of the genus *Orthohantavirus* is highly complex, and its taxonomy seems far from being resolved (Carvalho de Oliveira et al. 2014; Rivera et al. 2015; Delfraro et al. 2017). Moreover, there are regions, such as Africa, where Hantavirids have recently been discovered and research has most probably been insufficient to capture their whole diversity (Witkowski et al. 2014; Meheretu et al. 2019). It is therefore very likely that the list of recognized Orthohantaviruses will largely increase in the future years.

4 Transmission of Hantaviruses

Orthohantaviruses are transmitted horizontally, either through social interactions or through the inhalation of infectious particles released into the environment (Forbes et al. 2018). It is generally agreed that infection in the rodent host is asymptomatic, though a slight effect on survival has been documented for certain species (Kallio et al. 2007; Luis et al. 2012). Shedding of the virus peaks shortly after infection but continues throughout the host′s life (Safronetz et al. 2008; Voutilainen et al. 2015). Aged males are often the most infected among host populations, and aggressive encounters among them have been proposed as the main route of viral transmission for several species (Hinson et al. 2004; Bagamian et al. 2012). For the indirect transmission, both within reservoir populations and from rodents to humans, the capability of Orthohantaviruses to survive outside the host is determinant. Studies on Puumala virus have revealed that viruses remain infective for about two weeks in contaminated beddings at room temperature (Kallio et al. 2006). Cool and damp conditions are generally thought to prolong survival outside the host, while high temperatures and ultraviolet radiations are expected to have detrimental effects (Forbes et al. 2018). However, it is likely that optimal conditions for survival vary among viral species, and future studies on different viruses around the world are needed to improve understanding on the environmental determinants of transmission.

5 Determinants of Prevalence and Distribution of Hantaviruses

Prevalence and distribution of Orthohantaviruses in reservoir populations vary in space and time and among the different virus-host systems. These changes are the product of the influence of a wide range of factors acting at different scales on these systems (Mills 2005). Major climatic events, for example, affect viral transmission through their effects on the conditions and resources which in turn affect host

Table 1 Orthohantaviruses officially accepted by the ICTV. The disease they cause (HPS, HFRS or UNK = unknown), the main reservoir and the known distribution of the virus are shown. Please note that the lack of capitalization in some viral names is correct (virus names should not be capitalized except when part of the name is a proper noun)

Species	Virus genotype	Disease	Main reservoir	Known virus distribution
Andes orthohantavirus	Andes virus (ANDV)	HPS	*Oligoryzomys longicaudatus*	Southwestern Argentina and Chile
	Castelo dos Sonhos virus (CASV)	HPS	*Oligoryzomys utiaritensis*	Central Brazil
	Lechiguanas virus (LECV = LECHV)	HPS	*Oligoryzomys flavescens*	East central Argentina
	Orán virus (ORNV)	HPS	*Oligoryzomys chacoensis*	Northwestern Argentina
Asama orthohantavirus	Asama virus (ASAV)	UNK	*Urotrichustalpoides*	Japan
Asikkala orthohantavirus	Asikkala virus (ASIV)	UNK	*Sorex minutus*	Czech Republic, Finland and Germany
Bayou orthohantavirus	bayou virus (BAYV)	HPS	*Oryzomys palustris*	Southeastern USA
	Catacamas virus (CATV)	UNK	*Oryzomys couesi*	Eastern Honduras
Black Creek Canal orthohantavirus	Black Creek Canal virus (BCCV)	HPS	*Sigmodon hispidus*	Southeastern USA, Venezuela and Peru
Bowe orthohantavirus	Bowé virus (BOWV)	UNK[a]	*Crocidura douceti*	Southwestern Guinea, Côte d'Ivoire
Bruges orthohantavirus	Bruges virus (BRGV)	UNK	*Talpa europaea*	Belgium, Germany and United Kingdom
Cano Delgadito orthohantavirus	Caño delgadito virus (CADV)	UNK	*Sigmodon alstoni*	Northwestern Venezuela

(continued)

Table 1 (continued)

Species	Virus genotype	Disease	Main reservoir	Known virus distribution
Cao Bang orthohantavirus	Cao Băng virus (CBNV)	UNK	*Anourosorex squamipes*	Vietnam
	Liánghé virus (LHEV)	UNK	*Anourosorex squamipes*	Southwestern China
Choclo orthohantavirus	Choclo virus (CHOV)	HPS	*Oligoryzomys fulvescens*	Panamá
Dabieshan orthohantavirus	Dàbiéshān virus (DBSV)	UNK	*Niviventer confucianus*	East and south China
Dobrava-Belgrade orthohantavirus	Dobrava virus (DOBV)	HFRS	*Apodemus flavicollis*	Most southeastern Europe countries
	Kurkino virus (KURV)	HFRS	*Apodemus agrarius*	Croatia, Estonia, Germany, Hungary, Russia, Slovenia and Slovakia
	Saaremaa virus (SAAV)	HFRS	*Apodemus agrarius*	Croatia, Denmark, Estonia, southeastern Finland, Germany, Hungary, Russia, Slovakia and Slovenia
	Sochi virus (SOCV)	HFRS	*Apodemus ponticus*	Western Russia
El Moro Canyon orthohantavirus	Carrizal virus (CARV)	UNK	*Reithrodontomys sumichrasti*	Mexico
	El Moro Canyon virus (ELMCV)	UNK	*Reithrodotomys megalotis*	Mexico, southern USA
	Huitzilac virus (HUIV)	UNK	*Reithrodontomys megalotis*	Mexico
Fugong orthohantavirus	Fúgòng virus (FUGV)	UNK	*Eothenomys eleusis*	Southern China
Fusong orthohantavirus	Fŭsōng virus (FUSV)	UNK	*Microtus fortis*	Eastern China

(continued)

Table 1 (continued)

Species	Virus genotype	Disease	Main reservoir	Known virus distribution
Hantaan orthohantavirus	Amur virus (AMRV)	HFRS	*Apodemus peninsulae*	China, Korea, Russia
	Hantaan virus (HTNV)	HFRS	*Apodemus agrarius*	China, Japan, Korea, Russia and Slovakia
	Soochong virus (SOOV)	HFRS	*Apodemus peninsulae*	Korea
Jeju orthohantavirus	Jeju virus (JJUV)	UNK	*Crocidura shantungensis*	Korea and China
Kenkeme orthohantavirus	Kenkeme virus (KKMV)	UNK	*Sorex roboratus*	Eastern Russia
Khabarovsk orthohantavirus	Khabarovsk virus (KHAV)	UNK	*Microtus maximowiczii*	Northeastern China and southeastern Russia
	Topografov virus (TOPV)	UNK	*Lemmus sibiricus*	Northern Russia
Laguna Negra orthohantavirus	Laguna Negra virus (LANV)	HPS	*Calomys laucha / C. callosus*	Northwestern Argentina, Bolivia, Brazil and Paraguay
	Maripa virus (MARV)	HPS	*Oligoryzomys delicatus / Zygodontomys brevicauda*	French Guyana
	Rio Mamoré virus (RIOMV)	HPS	*Oligoryzomys microtis*	Bolivia, northeastern Peru and northwestern Brazil
Luxi orthohantavirus	Lúxī virus (LUXV)	HFRS	*Eothenomys miletus*	China
Maporal orthohantavirus	Maporal virus (MAPV)	UNK	*Oligoryzomys delicatus*	Western Venezuela
Montano orthohantavirus	Montaño virus (MTNV)	UNK	*Peromyscus beatae*	Mexico
Necocli orthohantavirus	Necoclí virus (NECV)	UNK	*Zygodontomys cherriei*	Colombia

(continued)

Table 1 (continued)

Species	Virus genotype	Disease	Main reservoir	Known virus distribution
Oxbow orthohantavirus	Oxbow virus (OXBV)	UNK	*Neurotrichus gibbsii*	Western USA
Prospect Hill orthohantavirus	Prospect Hill virus (PHV)	UNK	*Microtus pennsylvanicus*	Canada, USA
Puumala orthohantavirus	Hokkaico virus (HOKV)	UNK	*Myodes rufocanus*	China, Japan, Russia
	Muju virus (MUJV)	UNK	*Myodes regulus*	Korea and China
	Puumala virus (PUUV)	NE	*Myodes glareolus*	All european countries except the British Isles, Cyprus, Spain, Portugal, Iceland, Italy, Lythuania and Malta
Rockport orthohantavirus	Rockport virus (RKPV)	UNK	*Scalopus aquaticus*	Central south USA
Sangassou orthohantavirus	Sangassou virus (SANGV)	HFRS	*Hylomyscus simus*	Gabon, Guinea
Seewis orthohantavirus	Seewis virus (SWSV)	UNK	*Sorex araneus*	Austria, Belgium, Czech Republic, Finland, Germany, Hungary, Slovakia, Slovenia, Switzerland and Russia
Seoul orthohantavirus	gōu virus (GOUV)	UNK[a]	*Rattus rattus*	China
	Seoul virus (SEOV)	HFRS	*Rattus norvegicus*	Nearly worldwide, though most prevalent in Asia
Sin Nombre orthohantavirus	New York virus (NYV)	HPS	*Peromyscus leucopus*	USA
	Sin Nombre virus (SNV)	HPS	*Peromyscus maniculatus*	Canada, Mexico and USA
Thailand orthohantavirus	Anjozorobe virus (ANJZV)	UNK[a]	*Rattus rattus*	Madagascar

(continued)

Table 1 (continued)

Species	Virus genotype	Disease	Main reservoir	Known virus distribution
	Serang virus (SERV)	UNK	*Rattus tanezumi*	Indonesia
	Thailand virus (THAIV)	UNK[a]	*Bandicota indica*	Thailand
Tigray orthohantavirus	Tigray virus (TIGV)	UNK	*Stenocephalemys albipes*	Ethiopia
Tula orthohantavirus	Adler virus (ADLV)	UNK	*Microtus majori*	South eastern Russia
	Tula virus (TULV)	UNK[a]	*Microtus arvalis*	Finland, France and most central and eastern European countries
Yakeshi orthohantavirus	Yákèshí virus (YKSV)	UNK	*Sorex isodon*	Northeastern China

[a]Though not conclusive, there is some evidence suggesting these viruses could be etiologic agents of disease in humans

abundance and distribution (Guterres and Lemos 2018). These effects have been clearly illustrated by studies in semiarid environments of the Southwestern United States, where increased precipitation during "El Niño" years causes a "trophic cascade", with increased primary production leading to increased rodent population sizes and increased numbers of infected rodents (Yates et al. 2002). Masting is also known to cause great eruptions of rodent abundance in many places, particularly in the Southern region of South America (Southern Cone) and in Central Europe. In western Patagonia, for example, the sudden availability of bamboo seeds after synchronic blooming leads to sometimes spectacular increases in abundance of many small mammal species, including some Hantavirus reservoir species, which increases the number of transmission events (Jaksic and Lima 2003).

Several behavioral and physiological factors, as the proposed aggression route of transmission, have also been observed to have an effect on the regulation of infection. In brown rats (*Rattus norvegicus*) and deer mice (*Peromyscus maniculatus*), agonistic encounters among males are thought to play a key role in viral transmission, while among bank voles (*Myodes glareolus*) aggressive encounters appear to be relevant to transmission only during the breeding season, as in winter grooming and communal nesting may be more relevant (Escutenaire et al. 2002; Hinson et al. 2004; Dizney and Dearing 2013). Behavior of animals infected with certain pathogens has long been observed to differ from the population at large, sometimes prior to infection and sometimes as a consequence of infection. Differences in behavior, either as cause or as a consequence of infection, can play an important role in transmission, as those individuals with certain behavior can be responsible of most transmission in the population. The frequency of transmission and infection has also been proposed to vary according to the degree of immunosuppression of the host, which in turn might be affected by stressors such as crowding (Mills 2005). These behavioral aspects have not yet been fully explored regarding Hantavirus infection and transmission, and therefore, deserve future studies.

Another characteristic that may affect transmission and that may help explain differences in viral prevalence across biomes and among hosts species is habitat selection. Host abundance is generally found to vary greatly across its spatial distribution, peaking in its optimal habitats and reaching minimum values at the edge of its distribution. Besides the effect of host abundance on the distribution and prevalence of the virus, habitat may affect viral transmission per se, since characteristics of the habitat such as temperature, humidity and vegetation cover may affect survival of the virus outside the host. How animals use their habitat, particularly their micro habitat, may also affect transmission rates. The vertical and horizontal use of space, the use of burrows and nests, and the use and sharing of tunnel-like runways may affect both direct and indirect viral transmission. In central Argentina, *Akodon azarae*, host of a non-pathogenic genotype (not yet recognized by the ICTV), overlaps its home range and shares runways with several other individuals, both conspecifics and from other species, in a clear example of a behavior that can promote transmission and spillover events (Maroli et al. 2015).

Anthropogenic activities may also affect transmission of the virus, mainly via the effects that habitat modification has on host demography. Through deforestation, agriculture and ranching human beings have been, particularly in the last two centuries, continuously advancing into natural habitats, causing great environmental changes. Tolerance to anthropogenic changes in the habitat varies greatly among species, particularly among rodents. Some rodent species, strict specialists for example, are very sensible to changes in their environment, and even a slight deterioration of their habitat could drastically lower their abundance and even drive them to local extinctions. Others are able to tolerate change, and can even take advantage of it, using buildings as refuge, and feeding on crops or left-overs by humans (Mills 2005). As the community becomes poorer, generalist and opportunistic species could even reach high numbers, as has been observed in numerous opportunities; extreme cases being those of commensal species such as several *Rattus* and *Mus* species (Aplin et al. 2003). As we have already seen, high densities tend to increase viral transmission, so those tolerant species that are also reservoirs of Orthohantaviruses are likely to have higher viral prevalences. Studies carried out in western United States have revealed that *Peromyscus maniculatus*, the main Hantavirus reservoir in the area, shows higher Hantavirus prevalences in peridomestic habitats than in natural habitats (Kuenzi et al. 2001). A very similar pattern has been observed for *Oligoryzomys longicaudatus* in southwestern Argentina, where peridomestic settings seem to favor the presence of this reservoir in comparison to other native species, and where higher seroprevalences of Hantavirus have been recorded (Piudo et al. 2011). It has been suggested that intraspecific encounters, and therefore direct transmission, could be favored by the restriction of movements and home range inside buildings in peridomestic environments. Buildings could also favor indirect transmission because the virus shed in urine and faeces is more concentrated than in outdoor environments, and is more protected from dissipation and from inactivation by ultraviolet light (Kuenzi et al. 2001).

Biodiversity has been proposed to affect Hantavirus infection, though the mechanisms involved and the generality of this phenomenon are not yet clear. Among one of the most plausible mechanisms, it has been suggested that the presence of other small rodent species in a community may reduce intraspecific encounters among Hantavirus hosts, resulting in fewer opportunities for pathogen transmission. Species diversity may also decrease prevalence by reducing host survival and persistence via predation or competition. However, some authors suggest that it is the abundance and the relative proportion of reservoirs which have an effect on Hantavirus infection, and not biodiversity per se (Yahnke et al. 2001; Vadell et al. 2020). If this holds true, a biodiversity effect would be observed in those systems where reduced diversity increases abundance and dominance of a reservoir species (as it happens with host species that are tolerant to anthropogenic changes in highly disturbed environments), but will not necessarily occur as a general rule. Given that the diversity and ecology of Orthohantaviruses are still not fully understood, it is not clear whether their reservoirs are more likely to be tolerant or non-tolerant to human activities. However, from a public health point of view, and taking into consideration the degree to which humans are modifying habitats all over the world, those reservoir species which are tolerant to anthropogenic changes, no matter whether they are the majority or not, are the ones which will thrive among us, and therefore, pose a risk to human health.

6 Epidemiology of Hantavirus Infections

Orthohantaviruses are considered to be emerging viruses due to the increasing recognition of their role as human pathogens and their cyclic reappearance during outbreaks. Tens of thousands of human cases are reported each year, at least 70% of which correspond to HFRS cases (Manigold and Vial 2014; Watson et al. 2014). Mortality rates are higher in HPS, reaching 40% for some viruses (for example, Andes virus) while fatality of HFRS and NE do not exceed 15% (Vaheri et al. 2013a; Delfraro et al. 2017). Though broadly considered distinct diseases, HPS and HFRS (together with NE) share several clinical signs, including overlap in the target organ in certain cases (i.e., kidney injury in severe cases of HPS and lung disease in cases of HFRS; Clement et al. 2014). HPS and HFRS both start with a febrile prodrome phase that persists several days, often accompanied by myalgia, abdominal discomfort and nausea, together with a series of laboratory's anomalies. The next phase of HPS is characterized by the onset of cough and hypoxia, caused by pulmonary edema, with severe cases progressing rapidly to respiratory failure and cardiogenic shock. HFRS patients develop proteinuria and elevated creatinine as a result of acute kidney injury. Pathogenesis and outcomes of both diseases depend strongly on the virus genotype, but they also seem to be affected by the age and immunologic characteristics of the patient (Jonsson et al. 2010; Vaheri et al. 2013b; Jiang et al. 2016). Several studies have reported presence of Hantavirus antibodies in human populations with no reports of symptoms compatible with Hantavirus disease, suggesting certain infections could be asymptomatic (Hukić et al. 2011; Souza et al. 2011; Amaral et al. 2018).

Human to human transmission is rare, and has only been demonstrated for Andes virus in Argentina and Chile (Martinez et al. 2005; Martinez-Valdebenito et al. 2014). It was first proposed as a possible route of transmission in 1996, shortly after the identification of Andes virus as etiological agent of HPS in the Southern Cone (Enria et al. 1996; Lopez et al. 1996). Since then, several outbreaks of infection clusters have been identified, from which the most outstanding was the one that occurred in 2018 in the locality of Epuyén in Argentine Patagonia. This outbreak was particular not only because of its magnitude, but because, unlike others, it needed drastic public health measures in order to be stopped (see box 1 for a case study of this outbreak).

Vertical transmission of Orthohantaviruses has not been described in humans, but some evidence suggests it cannot be ruled out. A recent study reported Andes virus infection in a baby born to an infected mother, suggesting possible novel routes of transmission, including trans-placental transmission and transmission via breast milk, though perinatal transmission through blood contact cannot be discarded (Bellomo et al. 2020).

Hantavirus disease is endemic in many regions throughout the world (except Australia and Antarctica), with seasonal and multi-annual fluctuations associated with host demographic variations. Risk of acquiring Hantavirus also depends on human behavior, and though associated behaviors vary from region to region, there are some general features that are correlated with increased risk in most human-hantavirus systems. For example, Hantavirus infection has been clearly associated to exposure to rodents or rodent excreta in or around dwellings, and to outdoor activities in the wild. In mountainous and cold regions, the use of summer cottages that are kept closed during the winter is considered a risk factor (Watson et al. 2014). These cottages are often non-rodent-proof and can be use as refuge by small mammals, including potentially infected rodents that can shed the virus inside the house. Humans typically enter the long-unused houses in spring or summer and inhale the aerosol particles with viruses, particularly when sweeping or dusting. Transmission has also been reported to occur inside tents in the wild, both during recreational camping and during occupational activities, as in the case of soldiers, animal trappers, wood cutters and fishermen. Rural workers living in precarious settlements are also at high risk in many endemic areas throughout the world (Watson et al. 2014).

HFRS is a major epidemic in Asia and Europe, though mainly in Asia where over 100,000 cases have been reported in peak years (Zhang et al. 2010; Vaheri et al. 2013a). Currently most cases occur in China and Russia, though South Korea also exhibits a relatively high incidence (almost 1 case every 100,000 inhabitants; Lee et al. 2013; Jiang et al. 2017). China leads the world ranking of accumulated HFRS cases with near 1,500,000 cases since 1950, though incidence have been decreasing in the last decades, following a comprehensive preventive strategy that includes public health education and promotion, rodent control and surveillance, and vaccination. Improvements in living conditions and increased urbanization are also believed to contribute to this decreasing trend (Zhang et al. 2010; Ke et al. 2016; Sun and Zou 2018). Hantaan virus and Seoul virus, hosted by *Apodemus agrarius* and *Rattus norvegicus*, respectively, are the main etiologic agents of Hantavirus disease in China, Hantaan being more pathogenic.

Russia has an annual incidence of almost 5 cases every 100,000 inhabitants. The overall fatality rate is around 0.4%, but rates vary greatly depending on the virus involved. Central regions of western Russia, where most cases occur and where Puumala virus is the most prevalent etiological agent, have case-fatality rates of around 0.3%, but the Black Sea coastal area, where highly pathogenic Sochi virus circulates, has a fatality rate of 14% (Tkachenko et al. 2019).

In Europe, Hantavirus infections are annually reported throughout most of the continent, but principally in Fennoscandia and Central Europe. The main etiologic agents of Hantavirus disease in Europe are Puumala, Dobrava and Saaremaa viruses. Tula virus is common in voles (mainly *Microtus arvalis*) throughout Central and Eastern Europe, but has only rarely been observed to cause disease in humans, and its infection is thought to be generally asymptomatic (Klempa et al.

2003; Mertens et al. 2011; Zelená et al. 2013). Puumala virus is the most widespread agent of Hantavirus disease in Europe, being found all over the continent, except in the southern Mediterranean areas, the British Isles and the northernmost tundra regions. Finland is the country with the highest number of cases in Europe, with 1,000 - 2,000 cases reported every year (ECDC 2020). Case fatality of Puumala virus infections is very low, ranging from 0.08 to 0.4% (Vaheri et al. 2013a). In the Balkans, the number of reported cases per year is around 100, but, unlike northern and central Europe, deaths are not rare due to the relatively high fatality rate of Dobrava virus (around 12%), the main etiologic agent in that region (Avšič Županc et al. 2014).

It is still not clear which are the relevance and dimensions of Hantavirus disease in Africa, as research and accurate diagnostic tools are scarce, health care coverage is not universal, and also, because the predominance of other diseases with similar etiology could be misleading diagnosis. The first reports of serological evidence of human hantavirus infections in Africa date from the 1980s, including an article reporting seroprevalences of up to 14% in humans in Cameroon, Central African Republic, Chad, Congo, Equatorial Guinea and Gabon (Coulaud et al. 1987; Gonzalez et al. 1989). More than 20 years later, and a few years after the identification of the first African Hantavirus, Sangassou virus, in 2006, Klempa and colleagues (2010) found the first conclusive proof that African Hantaviruses are able to cause disease in humans. According to some authors, at least three Orthohantaviruses, Sangassou, Anjozorobe and Bowe viruses are pathogenic to humans, and could be currently posing a risk to the inhabitants of several African countries (Klempa et al. 2010; Heinemann et al. 2016; Rabemananjara et al. 2020).

Compared to HFRS, HPS incidence is low, though its fatality rate is far larger. Since its first report in 1993, around 4,000 HPS cases have been reported in the Americas. Brazil, Argentina, Chile and the United States concentrate most cases, though almost every country has reports. Exceptions include Mexico, Ecuador, Suriname, Guyana and most countries in the Caribbean Sea. Given its proximity to endemic areas in US, the absence of reported cases in Mexico is quite puzzling (Vigueras-Galván et al. 2019). In fact, though Hantavirus research in this country is scarce, studies have reported seropositive individuals in both rodent and human populations, suggesting Hantavirus infections are being misdiagnosed or somehow less virulent than in the neighboring U.S (Milazzo et al. 2012; Vigueras-Galván et al. 2019).

Andes and *Sin Nombre* are the most prominent and most studied *Orthohantavirus* species in the Americas, and both are pathogenic. *Andes orthohantavirus* is responsible for most HPS cases in the Southern Cone, and its southernmost variant, Andes virus, is the one capable of human to human transmission (Martinez-Valdebenito et al. 2014). Viruses of this species are hosted by rodents of the genus *Oligoryzomys* (Sub-family *Sigmodontinae*), commonly known as "colilargos" (meaning "long-tailed"), with numerous species widespread throughout the continent. Several other viruses are reported as responsible of HPS in South America, including some highly pathogenic Brazilian strains such as Juquitiba and Araraquara, which, though quite mentioned in the literature, are not considered valid names by the ICTV (Delfraro et al. 2017). *Sin Nombre orthohantavirus* is the etiologic agent of most HPS cases in North America and it is carried by species of the genus *Peromyscus* (family Cricetidae), particularly *P. maniculatus* (host of Sin Nombre virus) and *P. leucopus* (host of New York virus), both fairly widespread across Canada, United States and Mexico.

As the geographic distribution of a Hantavirus is restricted to the distribution of its one or few hosts, most Hantaviruses are usually limited to certain habitats within a particular region. The exception to this is represented by those viruses harbored by commensal species, as is Seoul virus, hosted mainly by *Rattus norvegicus*. Though most Seoul virus infections still occur in Asia, Seoul virus antibodies have been detected in rats all over the world, and human cases associated to this virus are being increasingly reported worldwide. In France, the United Kingdom and the United States, human cases have even been associated to pet rat ownership (Clement et al. 2019), while in China, outbreaks associated to laboratory animals have been repeatedly reported (Zhang et al. 2010). Given the worldwide distribution of rats, and their tight relationship with humans, Seoul virus could eventually become a serious international public health concern.

To date there is no effective treatment against Hantavirus disease, and therapy of both HPS and HFRS patients is usually based on supportive care (Szabo 2017). Given the lack of effective treatment, and the relatively high morbidity and mortality of Hantavirus disease in some endemic areas, the development of safe and effective vaccines for their application in risk populations is of critical importance. But despite the efforts of the last decades, no vaccine has been yet approved in the Americas and Europe, though some are currently on trial (Brocato and Hooper). Vaccines based on inactivated virus particles are, however, of common used in Korea and China since the 1990's, where they have effectively reduced the annual number of HFRS cases (Jung et al. 2018; Brocato and Hooper 2019; Noh et al. 2019).

7 Box 1. 2018 Person to Person Outbreak in Argentine Patagonia—A Case Study

On November 3rd 2018, in Epuyén (Chubut province, Argentina), a small town located at the foot of the Andes, a man went to a 15th birthday party. He was not feeling well but he did not know why. He had been recollecting mushrooms in the wild in the previous days. Eleven days after the party he was diagnosed with HPS. By the end of November, five persons who had assisted to the party were diagnosed with HPS as well. They had not been exposed to rodents. To check for the possibility of shared exposure to rodents during the party, a thorough inspection and trapping was carried out in the place where the party was held, revealing no signs of rodent infestation. Out of these five cases representing the second chain of infection, two died. At least three more chains followed, with infected persons scattered among several localities, including one in Chile. Among other public health measures implemented by the health authorities, on December 30th selective respiratory isolation was imposed to every person who had had close contact with a confirmed case. On February 7th, the last two cases of the outbreak, both with epidemiological link to a case, were confirmed. The outbreak was over, resulting in a total of 36 infected persons, 12 of whom died. (Bologna et al. 2019). The sequencing of the complete viral genome revealed a 99.9% percent of genetic identity among cases, confirming person to person transmission as the route of infection in this outbreak (MSal 2019).

8 Conclusion

Until now, Hantavirus disease around the world has only been associated to rodent-borne Orthohantaviruses. However, it is worth keeping in mind the possibility that co-circulation of different viruses and insufficient research on the genotypes involved in human cases could be masking the pathogenicity of non-rodent-borne viruses. In fact, though not conclusive, there are a few studies that provide evidence in favor of Hantavirus infections caused by non-rodent Hantaviruses (Okumura et al. 2007; Heinemann et al. 2016). It is important to note that circumstances such as host-switching events, viral mutations, increased human exposure to wildlife pathogens, and changes in the geographic distribution of hosts and viruses could eventually help currently unimportant viruses become relevant human health threats. In addition, person to person transmission, though still rare and limited to only one virus in a relatively small geographic area, raises the alarm of the possibility of a pandemic.

References

Abudurexiti A, Adkins S, Alioto D, Alkhovsky SV, Avšič-Županc T, Ballinger MJ, et al. Taxonomy of the order Bunyavirales: Update 2019. Adv Virol. 2019;164:1949–65.

Adams MJ, Lefkowitz EJ, King AM, Harrach B, Harrison RL, Knowles NJ, et al. Changes to taxonomy and the international code of virus classification and nomenclature ratified by the international committee on taxonomy of viruses. Arch Virol. 2017;162:2505–38.

Amaral CD, Costa GB, de Souza WM, Alves PA, Borges IA, Tolardo AL, et al. Silent orthohantavirus circulation among humans and small mammals from central Minas Gerais, Brazil. *EcoHealth*; 2018:1–13.

Aplin KP, Chesser T, Ten Have J. Evolutionary biology of the genus *Rattus*: profile of an archetypal rodent pest. In: Singleton GR, Hynds LA, Krebs CJ, Spratt DM, editors. Rats, mice and people: rodent biology and management. Canberra: Australian Centre for International Agricultural Research; 2003. p. 487–98.

Avšič Županc T, Korva M, Markotić A. HFRS and hantaviruses in the Balkans/South-East Europe. Virus Res. 2014;187:27–33.

Bagamian KH, Towner JS, Kuenzi AJ, Douglass RJ, Rollin PE, Waller LA, et al. Transmission ecology of Sin Nombre Hantavirus in naturally infected North American deer mouse populations in outdoor enclosures. PLoS ONE. 2012;7:1–10.

Bellomo C, Alonso D, Coelho R, Iglesias A, Periolo N, Martinez VP. A newborn infected by Andes virus suggests novel routes of hantavirus transmission: a case report. Clin Microbiol Infect. 2020;26:130–1.

Bologna R, González S, and Ruvinsky S. Brote de hantavirus en Epuyen, Chubut, Argentina. *Medicina infantil*; 2019:53–6.

Brocato RL, Hooper JW. Progress on the prevention and treatment of hantavirus disease. Viruses. 2019;11:610.

Carey D, Reuben R, Panicker K, Shope R, Myers R. Thottapalayam virus: a presumptive arbovirus isolated from a shrew in India. Indian J Med Res. 1971;59:1758–60.

Carvalho de Oliveira R, Guterres A, Fernandes J, D'Andrea PS, Bonvicino CR, de Lemos ERS. Hantavirus reservoirs: current status with an emphasis on data from Brazil. Viruses. 2014;6:1929–73.

CDC. Tracking a mystery disease: the detailed story of hantavirus pulmonary syndrome (HPS); 2020. https://www.cdc.gov/hantavirus/outbreaks/history.html. *Accessed on* April 30, 2020.

Clement J, LeDuc JW, Lloyd G, Reynes J-M, McElhinney L, Van Ranst M, et al. Wild rats, laboratory rats, pet rats: Global Seoul hantavirus disease revisited. Viruses. 2019;11:652.

Clement J, Maes P, Van Ranst M. Hemorrhagic fever with renal syndrome in the new, and hantavirus pulmonary syndrome in the old world: paradi (se) gm lost or regained? Virus Res. 2014;187:55–8.

Coulaud X, Chouaib E, Georges A, Rollin P, Gonzalez J. First human case of haemorrhagic fever with renal syndrome in the Central African Republic. Trans R Soc Trop Med Hyg. 1987;81:686–96.

Delfraro A, Raboni SM, dos Santos CND. Hantavirus: general features and present situation in latin America, pp. 215–233 *in Human Virology in Latin America*. Springer; 2017.

Dizney L, Dearing MD. The role of behavioural heterogeneity on infection patterns: implications for pathogen transmission. Anim Behav. 2013;86:911–6.

ECDC. Hantavirus infection. Stockholm: European Centre for Disease Prevention and Control; 2020.

Enria D, Padula P, Segura EL, Pini N, Edelstein A, Posse CR, et al. Hantavirus pulmonary syndrome in Argentina. Possibility of person to person transmission. Medicina Buenos Aires. 1996;56:709–11.

Escutenaire S, Chalon P, De Jaegere F, Karelle-Bui L, Mees G, Brochier B, et al. Behavioral, physiologic, and habitat influences on the dynamics of Puumala virus infection in bank voles (*Clethrionomys glareolus*). Emerg Infect Dis. 2002;8:930–6.

Escutenaire S, Pastoret P-P. Hantavirus infections. Revue Scientifique Et Technique-Office International Des Epizooties. 2000;19:64–71.

Forbes KM, Sironen T, Plyusnin A. Hantavirus maintenance and transmission in reservoir host populations. Current Opinion in Virology. 2018;28:1–6.

Gonzalez J-P, Josse R, Johnson E, Merlin M, Georges A, Abandja J, et al. Antibody prevalence against haemorrhagic fever viruses in randomized representative Central African populations. Res Virol. 1989;140:319–31.

Guterres A, Lemos ERSd. Hantaviruses and a neglected environmental determinant. One Health. 2018;5:27–33.

Heinemann P, Tia M, Alabi A, Anon J-C, Auste B, Essbauer S, et al. Human infections by non-rodent-associated hantaviruses in Africa. J Infect Dis. 2016;214:1507–11.

Hinson ER, Shone SM, Zink MC, Glass GE, Klein SL. Wounding: the primary mode of Seoul virus transmission among male Norway rats. Am J Trop Med Hyg. 2004;70:310.

Holmes EC, Zhang Y-Z. The evolution and emergence of hantaviruses. Curr Opinion Virol. 2015;10:27–33.

Hukić M, Valjevac A, Tulumovic D, Numanovic F, Heyman P. Pathogenicity and virulence of the present hantaviruses in Bosnia and Herzegovina: the impact on renal function. Eur J Clin Microbiol Infect Dis. 2011;30:381–5.
Jaksic FM, Lima M. Myths and facts on ratadas: bamboo blooms, rainfall peaks and rodent outbreaks in South America. Austral Ecol. 2003;28:237–51.
Jiang H, Du H, Wang LM, Wang PZ, Bai XF. Hemorrhagic fever with renal syndrome: pathogenesis and clinical picture. Front Cellul Infect Microbiol. 2016;6:1.
Jiang H, Zheng X, Wang L, Du H, Wang P, Bai X. Hantavirus infection: a global zoonotic challenge. Virolog Sinica. 2017;32:32–43.
Johnson KM. The discovery of hantaan virus: comparative biology and serendipity in a world at war. J Infect Dis. 2004;190:1708–21.
Jonsson CB, Figueiredo LTM, Vapalahti O. A global perspective on hantavirus ecology, epidemiology, and disease. Clin Microbiol Rev. 2010;23:412–41.
Jung J, Ko S-J, Oh HS, Moon SM, Song J-W, Huh K. Protective effectiveness of inactivated hantavirus vaccine against hemorrhagic fever with renal syndrome. J Infect Dis. 2018;217:1417–20.
Kallio E, Klingstrom J, Gustafsson E, Manni T, Vaheri A, Henttonen H, et al. Prolonged survival of Puumala hantavirus outside the host: evidence for indirect transmission via the environment. J Gen Virol. 2006;87:2127–34.
Kallio ER, Voutilainen L, Vapalahti O, Vaheri A, Henttonen H, Koskela E, et al. Endemic hantavirus infection impairs the winter survival of its rodent host. Ecology. 2007;88:1911–6.
Ke G, Hu Y, Huang X, Peng X, Lei M, Huang C, et al. Epidemiological analysis of hemorrhagic fever with renal syndrome in China with the seasonal-trend decomposition method and the exponential smoothing model. Sci Rep. 2016;6:39350.
Klempa B, Koivogui L, Sylla O, Koulemou K, Auste B, Krüger DH, et al. Serological evidence of human hantavirus infections in Guinea, West Africa. J Infect Dis. 2010;201:1031–4.
Klempa B, Meisel H, Räth S, Bartel J, Ulrich R, Krüger DH. Occurrence of renal and pulmonary syndrome in a region of northeast Germany where Tula hantavirus circulates. J Clin Microbiol. 2003;41:4894–7.
Kuenzi AJ, Douglass RJ, White D, Bond CW, Mills JN. Antibody to Sin Nombre virus in rodents associated with peridomestic habitats in west central Montana. Am J Trop Med Hyg. 2001;64:137–46.
Laenen L, Vergote V, Calisher CH, Klempa B, Klingström J, Kuhn JH, et al. Hantaviridae: current classification and future perspectives. Viruses. 2019;11:1–17.
Lee HW, Lee PW, Johnson KM. Isolation of Hantaan virus, the etiologic agent of Korean Hemorrhagic fever. J Infect Dis. 1978;137:298–308.
Lee HW, Vaheri A, Schmaljohn CS. Discovery of hantaviruses and of the Hantavirus genus: personal and historical perspectives of the Presidents of the International Society of Hantaviruses. Virus Res. 2014;187:2–5.
Lee S-H, Chung B-H, Lee W-C, Choi I-S. Epidemiology of hemorrhagic fever with renal syndrome in Korea, 2001–2010. J Korean Med Sci. 2013;28:1552–4.
Lopez N, Padula P, Rossi C, Lázaro ME, Franze-Fernández MT. Genetic identification of a new hantavirus causing severe pulmonary syndrome in Argentina. Virology. 1996;220:223–6.
Luis AD, Douglass RJ, Hudson PJ, Mills JN, Bjørnstad ON. Sin Nombre hantavirus decreases survival of male deer mice. Oecologia. 2012;169:431–9.
Manigold T, Vial P. Human hantavirus infections: epidemiology, clinical features, pathogenesis and immunology. Swiss Med Wkly. 2014;144:1–10.
Maroli M, Vadell MV, Iglesias A, Padula PJ, and Villafañe IEG. Daily movements and microhabitat selection of hantavirus reservoirs and other sigmodontinae rodent species that inhabit a protected natural area of Argentina. *EcoHealth*; 2015:1–11.
Martinez-Valdebenito C, Calvo M, Vial C, Mansilla R, Marco C, Palma RE, et al. Person-to-person household and nosocomial transmission of Andes hantavirus, Southern Chile, 2011. Emerg Infect Dis. 2014;20:1629.

Martinez VP, Bellomo C, San Juan J, Pinna D, Forlenza R, Elder M, et al. Person-to-person transmission of Andes virus. Emerg Infect Dis. 2005;11:1848–53.

Meheretu Y, Stanley WT, Craig EW, Goüy de Bellocq J, Bryja J, Leirs H, et al. Tigray Orthohantavirus infects two related rodent species adapted to different elevations in Ethiopia. Vector-Borne Zoonotic Dis. 2019;19:950–3.

Mertens M, Hofmann J, Petraityte-Burneikiene R, Ziller M, Sasnauskas K, Friedrich R, et al. Seroprevalence study in forestry workers of a non-endemic region in eastern Germany reveals infections by Tula and Dobrava-Belgrade hantaviruses. Med Microbiol Immunol. 2011;200:263–8.

Milazzo ML, Cajimat MN, Romo HE, Estrada-Franco JG, Iñiguez-Dávalos LI, Bradley RD, et al. Geographic distribution of hantaviruses associated with neotomine and sigmodontine rodents, Mexico. Emerg Infect Dis. 2012;18:571–6.

Mills J. Regulation of rodent-borne viruses in the natural host: implications for human disease. Adv Virol. 2005;19:45–57.

MSal. ANLIS confirma la transmisión persona a persona en el brote de hantavirus de Epuyén; 2019. https://www.argentina.gob.ar/noticias/anlis-confirma-la-transmision-persona-persona-en-el-brote-de-hantavirus-de-epuyen. Accessed on June 10, 2020.

Mull N, Jackson R, Sironen T, Forbes KM. Ecology of Neglected Rodent-Borne American Orthohantaviruses. Pathogens. 2020;9:325.

Noh JY, Jung J, Song J-W. Hemorrhagic fever with renal syndrome. Infection & Chemotherapy. 2019;51:405–13.

Okumura M, Yoshimatsu K, Kumperasart S, Nakamura I, Ogino M, Taruishi M, et al. Development of serological assays for Thottapalayam virus, an insectivore-borne hantavirus. Clin Vaccine Immunol. 2007;14:173–81.

Piudo L, Monteverde MJ, Walker RS, Douglass RJ. Rodent community structure and Andes virus infection in sylvan and peridomestic habitats in Northwestern Patagonia, Argentina. Vector Borne Zoonotic Dis. 2011;11:315–24.

Rabemananjara HA, Raharinosy V, Razafimahefa RM, Ravalohery JP, Rafisandratantsoa JT, Andriamandimby SF, et al. Human Exposure to Hantaviruses Associated with Rodents of the Murinae Subfamily, Madagascar. Emerg Infect Dis. 2020;26:587.

Rivera PC, Gonzalez-Ittig RE, Gardenal CN. Preferential host switching and its relation with Hantavirus diversification in South America. J Gen Virol. 2015;96:2531–42.

Safronetz D, Drebot MA, Artsob H, Cote T, Makowski K, Lindsay LR. Sin Nombre virus shedding patterns in naturally infected deer mice (Peromyscus maniculatus) in relation to duration of infection. Vector-Borne and Zoonotic Diseases. 2008;8:97–100.

Shi M, Lin X-D, Chen X, Tian J-H, Chen L-J, Li K, et al. The evolutionary history of vertebrate RNA viruses. Nature. 2018;556:197–202.

Song J-W, Baek LJ, Schmaljohn CS, Yanagihara R. Thottapalayam virus, a prototype shrewborne hantavirus. Emerg Infect Dis. 2007;13:980.

Souza WMd, Machado AM, Figueiredo LTM, Boff E. Serosurvey of hantavirus infection in humans in the border region between Brazil and Argentina. Rev Soc Bras Med Trop. 2011;44:131–5.

Sun L, Zou L-X. Spatiotemporal analysis and forecasting model of hemorrhagic fever with renal syndrome in mainland China. Epidemiol Infect. 2018;146:1680–8.

Szabo R. Antiviral therapy and prevention against hantavirus infections. Acta Virol. 2017;61:3–12.

Tkachenko EA, Ishmukhametov AA, Dzagurova TK, Bernshtein AD, Morozov VG, Siniugina AA, et al. Hemorrhagic Fever with Renal Syndrome, Russia. Emerg Infect Dis. 2019;25:2325.

Vadell MV, Villafañe IEG, Carbajo AE. Hantavirus infection and biodiversity in the Americas. Oecologia. 2020;192:169–77.

Vaheri A, Henttonen H, Voutilainen L, Mustonen J, Sironen T, Vapalahti O. Hantavirus infections in Europe and their impact on public health. Rev Med Virol. 2013a;23:35–49.

Vaheri A, Strandin T, Hepojoki J, Sironen T, Henttonen H, Mäkelä S, et al. Uncovering the mysteries of hantavirus infections. Nat Rev Microbiol. 2013b;11:539–50.

Vigueras-Galván AL, López-Pérez AM, García-Peña GE, Rico-Chávez O, Sarmiento-Silva RE, Suzán G. Current situation and perspectives on hantaviruses in Mexico. Viruses. 2019;11:642.

Voutilainen L, Sironen T, Tonteri E, Bäck AT, Razzauti M, Karlsson M, et al. Life-long shedding of Puumala hantavirus in wild bank voles (Myodes glareolus). J Gen Virol. 2015;96:1238–47.

Watson DC, Sargianou M, Papa A, Chra P, Starakis I, Panos G. Epidemiology of Hantavirus infections in humans: A comprehensive, global overview. Crit Rev Microbiol. 2014;40:261–72.

Witkowski PT, Klempa B, Ithete NL, Auste B, Mfune JK, Hoveka J, et al. Hantaviruses in Africa. Virus Res. 2014;187:34–42.

Yahnke CJ, Meserve PL, Ksiazek TG, Mills JN. Patterns of infection with Laguna Negra virus in wild populations of *Calomys laucha* in the central Paraguayan chaco. Am J Trop Med Hyg. 2001;65:768–76.

Yates TL, Mills JN, Parmenter CA, Ksiazek TG, Parmenter RR, Vande Castle JR, et al. The ecology and evolutionary history of an emergent disease: hantavirus pulmonary syndrome. Bioscience. 2002;52:989–98.

Zelená H, Mrázek J, Kuhn T. Tula hantavirus infection in immunocompromised host, Czech Republic. Emerg Infect Dis. 2013;19:1873.

Zeller H, Karabatsos N, Calisher CH, Digoutte J-P, Cropp C, Murphy F, et al. Electron microscopic and antigenic studies of uncharacterized viruses. II. Evidence suggesting the placement of viruses in the family Bunyaviridae. Adv Virol. 1989;108:211–27.

Zhang Y-Z, Zou Y, Fu ZF, Plyusnin A. Hantavirus infections in humans and animals, China. Emerg Infect Dis. 2010;16:1195.

Flaviviruses

Introduction to Flaviviruses and Their Global Prevalence

Mansi Verma, Rajendra Phartyal, and Amit Bhatt

Abstract Family *Flaviviridae*, which encompasses several small dreadful viruses under its canopy has been a big challenge for human health globally.. With four genera and 89 species the family has drawn attention due to its genus *Flavivirus* known for its significant pathogenic members such as yellow fever virus (YFV), Dengue Virus (DENV), Japanese encephalitis Virus (JEV), West Nile Virus (WNV), Zika virus (ZIKV), tick borne encephalitis virus (TBEV).These viruses are lipid enveloped and have small (+) ssRNA genome of approximately 11 Kb with a single Open Reading Frame (ORF), coding for three structural and seven non-structural proteins. The members of this genus exhibit receptor mediated endocytosis for accessing the host cells and upon successful entry, they undergo replication using RNA-dependent RNA polymerase, a trait possessed by all ribo-viruses. Their high mutation rate leads to significant genetic diversity within the species; therefore designing specific antiviral drug(s) remains a challenge. Complexity also arises due to the spread of these viruses using vectors in their transmission cycles; arthropods being on the top notch among the vectors. Majority of them exploit mosquitos as their carriers and humans as their hosts, therefore global warming and human travel across the continents have aggravated their spread to other parts of the world, shifting their status from epidemic to pandemic. Although effective vaccine development is the need of the hour for the control of this group of viruses, but vector control seems to be the only plausible solution in current scenario so that their human transmissions can be controlled.

Keywords Flavivirus · Epidemic · Dengue · *Aedes* spp. · *Culex* spp. · Genome · Vectors · Vaccine · Infection · Pathogenesis · Structural proteins · Antivirals

M. Verma (✉) · R. Phartyal · A. Bhatt
Department of Zoology, Sri Venkateswara College, University of Delhi (South Campus), New Delhi 110021, India
e-mail: mansiverma@svc.ac.in

S. I. Ahmad (ed.), *Human Viruses: Diseases, Treatments and Vaccines*, https://doi.org/10.1007/978-3-030-71165-8_19

Abbreviations

(+) ssRNA	Positive single strand Ribosomal nucleic acid
ALF	Alfuy
C protein	Capsid protein
CDC	Centers for Disease Control and Prevention
DENV	Dengue virus
E protein	Envelope protein
HEP C	Hepatitis C virus
ISV	Insect-specific virus
JEV	Japanese Encephalitis virus
KV	Kokobera virus
MBFV	Mosquito-borne Flavivirus
MVEV	Murray Valley encephalitis virus
NCBI	National Centre for Biotechnology Information
NKV	No known vector
NS	Non-structural
ORF	Open Reading Frame
PrM protein	Pre-membrane virus
RdRp	RNA dependent RNA polymerase
SFV	Swine Fever Virus
SLEV	St. Louis Encephalitis virus
SV	Stratford virus
TBEV	Tick-borne encephalitis virus
TBFV	Tick-borne flavivirus
UTR	Untranslated region
WNV	West Nile virus
YFV	Yellow fever virus
ZIKAV	Zika virus

1 Introduction

Flaviviruses, the members of family *Flaviviridae* and a cause of significant economic loss and mortality among humans, tend to be a big challenge for researchers and health providers (Kok 2016). Mostly transmitted by infected mosquitoes and ticks, these are enveloped viruses with (+) single stranded RNA. Flaviviruses derive their name from Yellow fever virus (from Latin *Flavus* meaning yellow), the type virus for this genus as it causes yellow jaundice in its hosts (Waggoner et al. 2018). Yellow fever virus is said to have originated in Africa, from where it travelled to the Americas with the Atlantic slave trade (Chippaux and Chippaux 2018). Over the next two centuries, tropical America, the North American coastal cities, and Europe faced outbreaks of this disease (Strode et al. 1951). In 1881, Carlos Finlay suggested that the vector behind this widely spread disease was *Culex cubensis* (now known as *Aedes aegypti*) (Chaves-Carballo 2005), and this was the first evidence of vector involvement in the YFV transmission.

With 89 species, till date as per the International Committee on Taxonomy of Viruses (ICTV), the family *Flaviviridae* is divided into four genera (Shi et al. 2016) as follows:

i. *Flavivirus* (From Latin *Flavus* meaning yellow of yellow fever): More than 50 species are present comprising of important human pathogens like YFV, DENV, JEV, WNV and TBEV (Alkan et al. 2015).
ii. *Pestivirus* (from Latin *Pestis*, which means Plague) contains virus species that infect economically important farm animals like pigs, sheep goats and cattle as well as wild ruminants. The members of this genus include Bovine viral diarrhoea virus, Classical swine fever virus (SFV) (Smith et al. 2017).
iii. *Hepacivirus* (from Greek *hepar, hepatos* which means liver) which includes Hepatitis C virus (HEP C) that infects humans and causes damage to the liver. It also includes many other viruses that cause liver damage in mammals. Presently, there are 14 species belonging to *Hepacivirus* viz, *Hepacivirus A–N (*Smith et al. 2016*).*
iv. A fourth genus, *Pegivirus* is the latest addition to segregate the previously unclassified GB virus A (GBV-A), GBV-C, and GBV-D from other members in the family (Xiang et al. 2012; Pfaender et al. 2014). There are 11 species belonging to this genus marked as *Pegivirus A–K* (Smith et al. 2016).

In this chapter, our primary focus will be on the members of the genus Flavivirus as among *Flaviviridae* they are the cause of major human fatalities, sufferings, and economic losses (Table 1).

Flaviviruses are the most common arthropod borne viruses around the world with almost half of the members known to be human pathogens (Diosa-Toro et al. 2013). The importance of this genus is reflected by the number of complete genome sequences of its members in the public databases, with highest number of sequences belonging to DENV (serotype 1–4), followed by WNV, ZIKAV and JEV (Fig. 1).

With the advent of rapid and mass transportation of humans across the continents, flaviviruses have shown a significant advancement in their distribution across the globe. Flaviviruses are present with a varied distribution in tropical and subtropical regions across the world with some viruses like the WNV extending into the temperate regions (Daep et al. 2014). The Sylvatic cycle of flaviviruses among a variety of avian and mammalian hosts make containment efforts very difficult and serve as a source of repeated outbreaks in humans. These RNA viruses are emerging and evolving very fast due to their high rates of mutation, thus allowing them to coevolve with their vectors and adapt to a wide range of environments (Thomas et al. 2014). Control of Flaviviruses so far is achieved mainly through limiting the populations of their vectors (Kraemer et al. 2015) or preventing their

Table 1 Flaviviruses that significantly affect humans

Virus	Specific vector	Reservoir/ Host	Symptoms	Year of isolation	Global prevalence (no. of countries affected)
Yellow Fever Virus	*Aedes* spp., *Haemagogus* sp.	Monkeys, Humans	Fever, hemorrhage jaundice	1927	Tropical regions of Africa and South America (47 countries)
Dengue Virus (1–4)	*Aedes aegyptii*, *Aedes albopictus*	Monkey, Humans	Fever, aches and pain	1944	Tropical and subtropical areas worldwide (154 countries)
Zika Virus	*Aedes aegyptii*, *Aedes albopictus*	Monkeys, Humans	Fever, rash, conjunctivitis, muscle and joint pain	1947	Africa, the Americas, Asia and the Pacific (91 countries)
Japanese encephalitis	*Culex tritaeniorhynchus*	Pigs and Birds	Fever, encephalitis	1935	Asia (32 countries)
West Nile Virus	*Culex* spp. in particular *Cx. pipiens* (vertical transmission)	Birds	fever, headache, tiredness, and body aches and swollen lymph glands	1937	Africa, Europe, the Middle East, North America and West Asia (92 countries)
Tick-borne encephalitis virus	*Ixodes* Spp	Rodents	febrile illness with headache, myalgia, encephalitis	1937	Europe, Asia (36 countries)
Kyasanur Forest disease Virus	Haemaphysalis spinigera ticks	Rodents, shrews, and monkeys	Fever, hemorrhage, encephalitis	1957	Southwest India (India)

Data taken from CDC (www.cdc.gov/) and Gideon (www.gideononline.com)

contact with humans. Symptomatic treatment, development of new vaccines and effective implementation of authorized vaccines across the affected areas by concerned governments is a promising way forward.

2 Vectors of Flaviviruses

Each species of Flavivirus has a distinct geographic distribution based on its ecological habitat, climate and the invertebrate and vertebrate hosts in which it multiplies (Fig. 2). Zoonotic diseases are transmitted from vertebrate hosts to humans either by direct or indirect contact. For direct transmission, no vector is required but

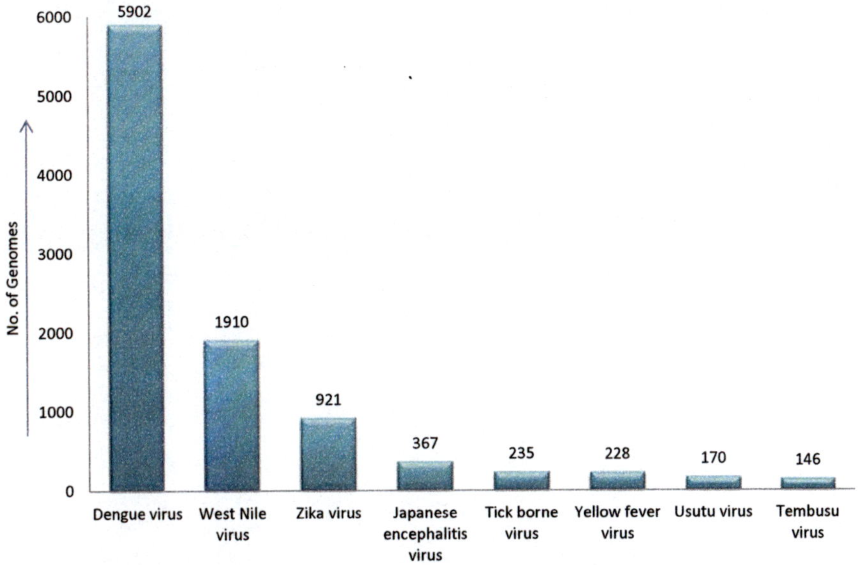

Fig. 1 Significant number of Flavivirus genome sequences available at NCBI. The highest number of sequenced genomes is of the members of four serotypes of Dengue virus (DENV), followed by WNV (1910 sequenced genomes), ZIKV (921), JEV (367) etc

for indirect transmission, vectors play a crucial role in completing the life cycle of the respective viruses (Valderrama et al. 2017; Gaunt et al. 2001). The vector transmission of flaviviruses is more common, with the arthropods being at the top notch among them (Table 2). When arthropods vectors are involved in the viral transmission, viruses belonging to various families are categorized as arboviruses (arthropod-borne) (Conway et al. 2014; Huang et al. 2019; Sukhralia et al. 2019).

Box 1: Genome based phylogenetic tree of flaviviruses
The tree shown in Fig. 2 represents genome sequences of genus *Flavivirus* (note that for some species of this genus, sequences were not available). The sequences were retrieved from NCBI (https://www.ncbi.nlm.nih.gov/) and aligned using Bio Edit (Tippmann 2004). Tree was constructed using ML method of MEGA X (Tamura et al. 2013) and edited using iTOL (Letunic and Bork 2016). Earlier phylogenetic trees constructed for flaviviruses were based on conserved proteins (Kuno et al. 1998), but the present tree is constructed using whole genome sequences and viral groups were found to clade as per their vectors.

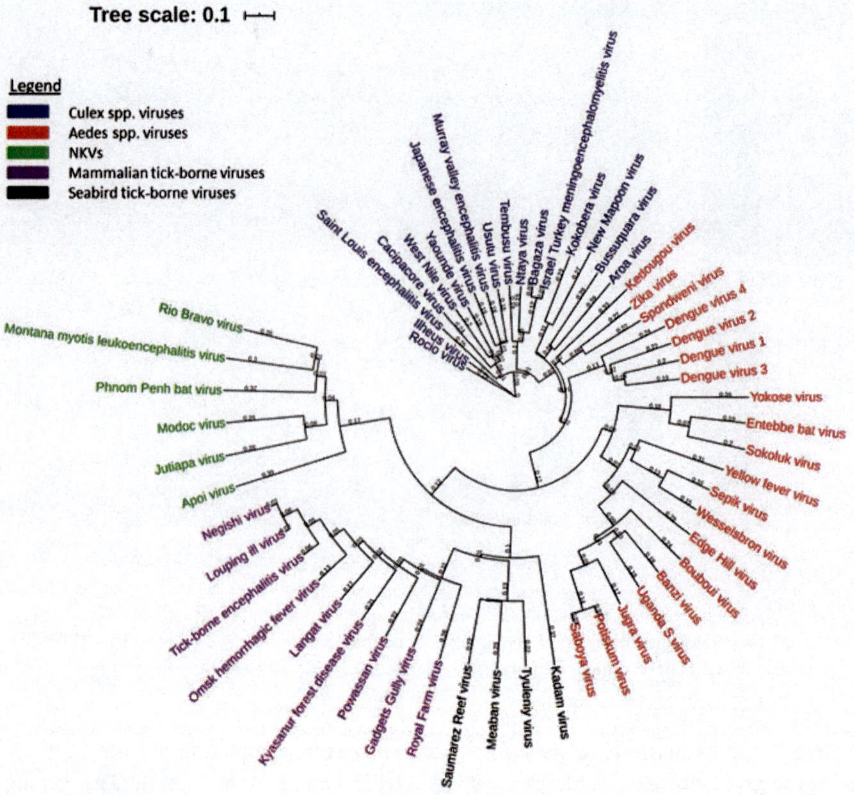

Fig. 2 Phylogenetic tree based on 53 whole genome sequences of flaviviruses available at NCBI as of April 2020. The tree shows distinct grouping of *Flavivirus* species according to their host specificity

2.1 *Flaviviruses as Arboviruses*

As stated above, the arboviruses involve arthropods as their vectors for transmission. Arthropod borne flaviviruses can be divided broadly into mosquito-borne flaviviruses (MBFV) and tick-borne flaviviruses (TBFVs) (Weaver and Barrett 2004; Valarcher et al. 2015; Nunez-Castilla et al. 2020). Other than mosquitoes and ticks, transmission using sandflies is recently reported in genus *Flavivirus* (Alkan et al. 2015). Hence, involvement of these vectors in viral transmission adds to the complexity in their life cycle, leading to failure in controlling the spread of these viruses.

The geographical distribution of mosquito-borne flaviviruses is dependent on the natural habitats of their vectors and the respective vertebrate hosts (Table 1). Further grouping of mosquito-borne arboviruses yield two major categories: one associated with *Aedes spp.* and the other with *Culex spp.* as the primary vector.

Table 2 Vectors of flaviviruses and the viruses transmitted through them (Data taken from ICTV)

Vector	Viral complexes	Species	Subtype
Mosquito-borne flaviviruses			
Aedes spp	Yellow fever Virus group	Sepik virus	Sepik virus
		Wesselsbron virus	Wesselsbron virus
		Yellow fever virus	Yellow fever virus
	Dengue virus group	Dengue virus	Dengue virus 1–4
Culex spp	Aroa virus group	Aroa virus	Aroa virus, Bussuquara virus, Iguape virus, Naranjal virus
	Ntaya virus group	Bagaza virus	Bagaza virus
		Ilheus virus	Ilhéus virus, Rocio virus
		Israel Turkey meningoencephalitis virus	Israel Turkey meningoencephalitis virus
		Ntaya virus	Ntaya virus
		Tembusu virus	Tembusu virus
		Zika virus	Zika virus
	Japanese encephalitis virus group	Cacipacore virus	Cacipacoré virus
		Japanese encephalitis virus	Japanese encephalitis virus
		Koutango virus	Koutango virus
		Murray Valley encephalitis virus	Alfuy virus, Murray Valley encephalitis virus
		St. Louis encephalitis virus	St. Louis encephalitis virus
		Usutu virus	Usutu virus
		West Nile virus	Kunjin virus, West Nile virus
		Yaounde virus	Yaoundé virus
	Kokobera virus Group	Kokobera virus	Kokobera virus, Stratford virus
Probably mosquito-borne	Kedougou virus group	Kedougou virus	Kédougou virus
	Edge Hill virus group	Banzi virus	Banzi virus
		Bouboui virus	Bouboui virus
		Edge Hill virus	Edge Hill virus
		Jugra virus	Jugra virus
		Saboya virus	Potiskum virus, Saboya virus
		Uganda S virus	Uganda S virus

(continued)

Table 2 (continued)

Vector	Viral complexes	Species	Subtype
Tick-borne flaviviruses			
Mammalian ticks	Mammalian tick-borne virus group	Gadgets Gully virus	Gadgets Gully virus
		Kyasanur Forest disease virus	Kyasanur Forest disease virus, Alkhumra hemorrhagic fever virus
		Langat virus	Langat virus
		Louping ill virus	Louping ill virus, British subtype, Irish subtype, Spanish subtype, Turkish sheep encephalitis virus subtype, Greek goat encephalitis virus subtype
		Omsk hemorrhagic fever virus	Omsk hemorrhagic fever virus
		Powassan virus	Powassan virus, Deer tick virus
		Royal Farm virus	Royal Farm virus
		Tick-borne encephalitis virus	European subtype, Far Eastern subtype, Siberian subtype
Seabird ticks	Sea-bird tick borne virus group	Meaban virus	Meaban virus
		Saumarez Reef virus	Saumarez Reef virus
		Tyuleniy virus	Tyuleniy virus
		Kadam virus	Kadam virus
No known vectors			
No Known Vector (NKVs)	Entebbe bat virus group	Entebbe bat virus	Entebbe bat virus, Sokuluk virus
		Yokose virus	Yokose virus
	Modoc virus group	Apoi virus	Apoi virus
		Cowbone Ridge virus	Cowbone Ridge virus
		Jutiapa virus	Jutiapa virus
		Modoc virus	Modoc virus
		Sal Vieja virus	Sal Vieja virus
		San Perlita virus	San Perlita virus
	Rio Bravo virus group	Bukalasa bat virus	Bukalasa bat virus
		Carey Island virus	Carey Island virus
		Dakar bat virus	Dakar bat virus
		Montana myotisleukoencephalitis virus	Montana myotisleukoencephalitis virus
		Phnom Penh bat virus	Batu Cave virus, Phnom Penh bat virus
		Rio Bravo virus	Rio Bravo virus

Arboviruses that exploit (but necessarily not restricted to) *Aedes* spp. include members of DENV and YFV groups (Nunez-Castilla et al. 2020). As *Aedes spp.* are widespread in the Old World, therefore their related viruses are also usually restricted to these regions. Exception to this is the prevalence of YFV and DENV in the New World as well (Gaunt et al. 2001; Weissenbock et al. 2010). Contrary to this, viruses associated with *Culex spp.* were known to be either restricted in New World (e.g., St Louis encephalitis virus (SLEV), Cacipacore virus) or Old World (e.g., WNV, JEV), depending upon the species of *Culex*. Some viruses associated with *Culex* spp. are also restricted to Australia and the regions of South-east Asia (Murray Valley encephalitis virus (MVE), Alfuy virus (ALF), Kokobera virus (KV), Stratford virus (SV)). Arboviruses exploiting *Culex spp.* as the vector are broadly categorized as Aroa virus (AV), Ntaya virus (NV), JEV and KV (Gaunt et al. 2001; Weissenbock et al. 2010). In general, the overlaps in the geographical distribution and the expansion of the territories of mosquito-borne arboviruses are attributed to the increase in travel across the continents, trade, urbanization and global warming (Petersen and Marfin 2005).

The viruses with ticks as their vectors are known as tick-borne viruses (TBVs). In case of flaviviruses, TBVs are also categorized as Mammalian tick-borne *Flavivirus* group {also known as Tick-borne encephalitis viruses (TBEVs) with rodents, humans and other forest animals as their hosts} and another small group of Tick-borne viruses with seabirds as their hosts (Gould et al. 2001; Shi et al. 2018). Tick-borne encephalitis viruses (TBEVs) are usually known to be restricted geographically (Ternovoi et al. 2019), and have three known genotypes, namely European (TBEV-Eu), Siberian (TBEV-Sib), and Far-eastern (TBEV-FE) (Shi et al. 2018). Some members of TBVs are known to have high mortality rate of domestic farm animals, and therefore, are of socio-economic importance.

2.2 Directly Transmitted Flaviviruses

Arboviruses and Tick borne viruses are spread by defined vectors between hosts and show horizontal transmission. Flaviviruses can be categorized as arthropod-restricted Flaviviruses or vertebrate-restricted Flaviviruses, depending on their direct transmission from either arthropod or vertebrate hosts, respectively (Blitvich and Firth 2015). Arthropod-restricted viruses which are presumed to be precursors of arboviruses, have arthropods as their hosts, exhibit a restricted range of transmission (without vector) and usually categorized as insect-specific viruses (ISVs) e.g. Palm Creek virus (PCV), Kamiti River virus (KRV) (Alkan et al. 2015; Blitvich and Firth 2017). A significant variety of flaviviruses are known to exist exclusively in the vertebrates like bats, rodents etc. and are labeled as the no known vector (NKV) flaviviruses (Billoir et al. 2000) which is in fact a non-taxonomic category. These viruses hold importance for comparing why some Flaviviruses require non-specific vectors for their transmission cycles, while some are restricted to their vertebrate or invertebrate hosts.

3 Flaviviral Genome Organisation

RNA viruses are known for their small sized genomes that sustain high mutation rates and lack proofreading (Belshaw et al. 2007). Therefore, the occupancy of large genome is not feasible. Another feature common to riboviruses is the presence of RNA-dependent-RNA polymerase (RdRp), required for the replication of (+) single strand RNA to make (−) single strand RNA.

Flaviviruses are lipid-enveloped, icosahedral RNA viruses that have single-stranded (+) RNA genome having a size of 11 kb approximately. Their GC content ranges from ~43% (NKV flaviviruses) to ~54% (of TBFV) (Schubert and Putonti 2010). The unsegmented RNA genome is capped but bears no polyA tail and comprises of a single Open Reading Frame (ORF) flanked by 5′ and 3′ untranslated regions (UTRs) (Table 3). The ORF of around 10 kb is translated into a polyprotein of ~3400 amino acids (Barrows et al. 2018). The polyprotein is further processed by viral and host proteases to yield 10 viral proteins, comprising of three structural proteins viz. capsid (C), membrane (M), and envelope (E); and seven non-structural (NS) proteins (NS1, NS2A, NS2B, NS3, NS4A, NS4B and NS5) (Fig. 3) (Holden et al. 2006; Sukhralia et al. 2019; Verma et al. 2019).

3.1 Structural Proteins

As mentioned above, there are three structural proteins in flavivirus genome, namely C, M and E (Table 3). The C protein required for the encapsidation of viral genome, is the first in the polyprotein and comprises of ~105 residues with a mass of ~12 kDa. Usually a virion has one viral RNA and several hundred copies of Capsid protein (Barrows et al. 2018). C protein is known to interact with a number of host proteins, and then interfere in the translation of host proteins; and can even interfere with ribosomal biogenesis by entering inside the nucleus (Sotcheff and Routh 2020). The E glycoprotein is about 53–60 kDa and plays crucial role in endocytosis of virus by interacting with the host cell membrane. E protein possesses receptor binding and fusogenic properties (Harrison 2008) and at low pH, it executes receptor mediated endocytosis for the entry of viral particles into the host. In fact, the glycolipid envelope of virus is formed by both envelope proteins as well as the membrane proteins, and it is the role of M proteins to avoid premature fusion of the envelope proteins with the endosomes. Membrane protein is synthesized as a precursor protein (PrM) and forms trimeric spikes with envelope protein; but during the release of virion from cell, the PrM is cleaved by furin proteases to form M protein, that makes a mature virion (Pierson and Diamond 2012). It is interesting to note that the E protein has conserved epitope regions that generate specific neutralizing antibodies (Kaufmann and Rossmann 2011; Zhang et al. 2017; Sukhralia et al. 2019).

Table 3 The details of the proteins encoded by the viral genome of the flaviviruses

Protein type	Protein	Function
Structural proteins	Capsid	Genome packaging
	Pre-Membrane/ Membrane	Formation and assembly of viral components and interacts with the host proteins
	Envelope	Viral entry via host cell membrane
Non-structural proteins	NS1	Immune recognition and acts as a cofactor for DNA complex
	NS2A	Viral assembly, RNA synthesis and replication
	NS2B	Acts as a co-factor for NS3
	NS3	Removes a phosphate from the 5′ terminus of the vRNA
	NS4A	NS4B translocation through ER via host membrane rearrangement
	NS4B	Inhibition of signal transduction cascade of interferon
	NS5	Capping of the neosynthetized vRNA. RdRp and methyltransferase activity

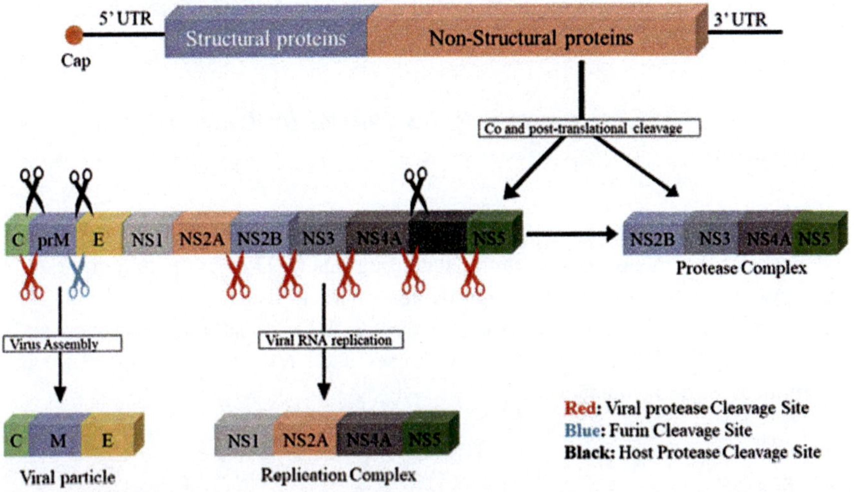

Fig. 3 Flaviviral genome is organised as 5′CAP(I)-5′UTR-C-prm-E-NS1-NS2A-NS2B-NS3-NS4A-NS4B-NS5-3′UTR. All coding region is translated as a polyprotein which is then acted upon by viral and host proteases to yield ten proteins. Structural proteins (C-M-E) are involved in viral particle formation whereas non-structural proteins form replication and protease complexes

3.2 Non-Structural Proteins

NS1 is the highly conserved protein among flaviviruses with molecular weight of ~46–55 kDa and plays an important role in viral replication as well as pathogenesis. In fact, the secreted form of NS1 is the first protein to appear in the blood circulation in large quantities during the viremic phase, and therefore targeted for designing detection kits (Muller and Young 2013; Rastogi et al. 2016).

NS3 protein, the second largest protein of flaviviruses, is a multifunctional protein with activities like serine protease, nucleoside triphosphatase and helicase. Together with NS2B as its cofactor (forming NS2B/NS3 heterodimer), it cleaves the viral polyprotein at specific sites. Due to helicase activity, it aids in RNA replication by melting the secondary structures of viral genomic RNA and unwinding it (Li et al. 2014; Yap et al. 2019).

NS5 is the largest protein in the genome and possesses most important enzymatic activities like methyl transferase at its N-terminal and RdRp at C-terminal. The structure of this protein is conserved across flaviviral genomes with 50–70% identities in the sequences. In fact, riboviruses are marked by the presence of RdRp, which is a crucial enzyme for replication. In flaviviruses, the methyl transferase domain helps in capping of the viral genome (El Sahili and Lescar 2017; Barrows et al. 2018).

NS2A, NS2B, NS4A and NS4B are integral membrane proteins. NS2A interacts with 3′UTR, NS3 and NS5 of replication complex and helps in viral assembly (Xie et al. 2013); whereas NS4A and NS4B are important for ER membrane-associated replication complex and are required for viral-host interactions (Zou et al. 2015).

3.3 5′ and 3′ UTRs: Crucial Roles Played by Untranslated Regions

As stated above, although the average genome size of flaviviruses is 11 Kb, these small sized RNA viruses possess untranslated regions (UTRs) at the extremities of their genomes. The non-coding regions are marked as 5′ UTR and 3′UTR, respectively (Rice et al. 1985; Lindenbach et al. 2007); and their presence and conservancy across the genomes in Flaviviruses itself reflects their importance in RNA replication and translation.

3.3.1 5′ UTR

Spanning for around 100 nucleotides at the 5′end of genome, this region is known to play crucial role in viral RNA synthesis. 5′ UTR comprises of type I cap structure (i.e., a m^7GpppAmpN1 cap) (Cleaves and Dubin 1979; Ng et al. 2017), two conserved stem-loop regions, named as Stem loop-A (SLA) and Stem Loop-B (SLB) (Brinton and Dispoto 1988) and one hairpin (HP; cHP). 5′-cap structure is formed by methylation aided by methyltranferase activities present in NS5 (Ray et al. 2006). Presence of cap at 5′ structure is known to play critical role in translation (except for DENV, which can show cap-independent mechanism for translation if required) (Edgil et al. 2006; Ng et al. 2017). SLA is 70 nucleotides long, and acts as the promoter for RNA synthesis, thus interacting with RdRp (Filomatori et al. 2006). SLA and SLB are separated by a poly (U) sequence. SLB is 30 nucleotides long and contains the 5′UAR (upstream AUG region) nucleotides that interacts with their complementary pair i.e., 3′UAR present near the 3′ terminus (Alvarez et al. 2008; Wang et al. 2017). Selection of the first AUG codon is enhanced by the cHP hairpin required for viral polyprotein synthesis. SLB is followed by 5′ cyclization sequence (dCS). Although this sequence resides in Capsid gene, nevertheless it is not a part of capsid protein and therefore is an untranslated region (Friebe et al. 2012).

3.3.2 3′UTR

This region is ~400 –600 bases long and rich in conserved regions across the flavivirus genomes (Brinton et al. 1986; Alvarez et al. 2008; Ng et al. 2017). The 3′ UTR comprises a variable region, two dumbbell structures, a short Hairpin (sHP), and the 3′ terminal Stem Loop (3′SL) (Wang et al. 2017). These structures are divided in three domains, out of which Domain 3 is the most conserved region, and contains the CS1, small hairpin (sHP), and a large terminal 3′SL structure (Shi et al. 1996; Gebhard et al. 2011; Ng et al. 2017; Wang et al. 2017). It is speculated that the absence of polyA tail in flaviviruses is compensated by the presence of these secondary structures at 3′ end (Alvarez et al. 2005; Brinton et al. 1986; Men et al. 1996; Shurtleff et al. 2001). CS1 contains the cyclization sequence (CS) complementary to dCS at the 5′ end of the genome (Hahn et al. 1987) as mentioned above. 3′ UAR and 3′ CS are for long-range RNA–RNA interactions and RNA cyclization (Alvarez et al. 2005). Overall, 3′ UTR is responsible for circularization and replication of a negative template from positive sense ssRNA genome.

4 Process of Viral Entry and Replication

The life cycle of flaviviruses is entirely dependent on the fate of genomic viral RNA (vRNA), the replication of which entirely occurs in the cytoplasm of host cells. The vRNA contains all the genetic information prerequisite for viral replication (Mazeaud et al. 2018). In order to complete its infectious life cycle, the virus must exploit the host machinery (Fig. 4).

4.1 Pathogenesis of Flaviviruses

Infection: The infection occurs when an infected vector introduces the virus into the host during blood meal. The introduced virus enters the host cells by

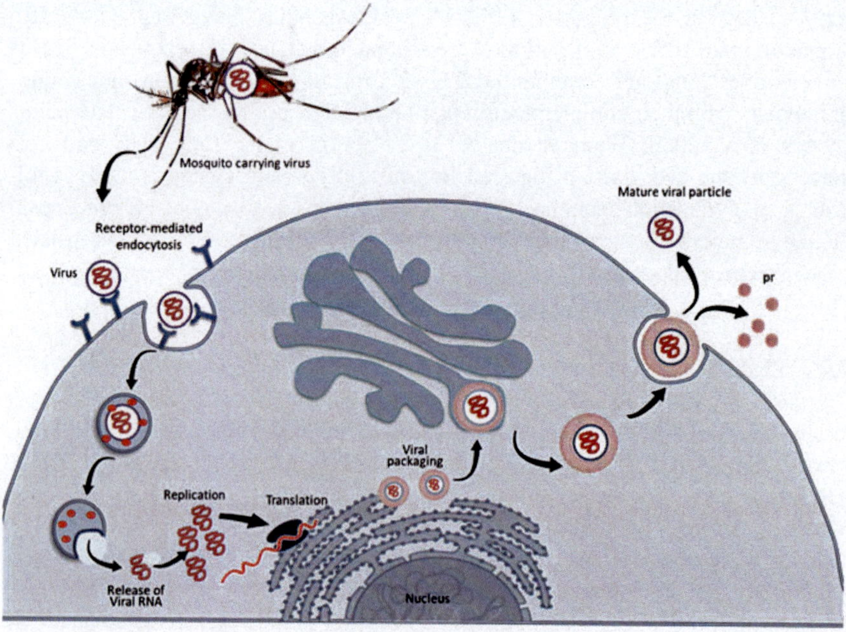

Fig. 4 Schematic diagram of viral entry and packaging of Flavivirus

receptor-mediated endocytosis; however, the receptor on the host cell is still unknown. Glycoprotein (E) undergoes conformational changes in the endocytic vesicle due to low pH environment (Bressanelli et al. 2004). This leads to the fusion of the viral envelope to the endocytic vesicle membrane. The glycoprotein capsid is disintegrated and the (+) vRNA is exposed (Selisko et al. 2014).

Viral translation: The exposed vRNA directly encodes for a 3,400 amino acid long polyprotein precursor, which is processed co- and posttranslationally (Ivanyi-Nagy and Darlix 2012). The capsid (C) protein remains on the cytoplasmic side of the RER and the Envelope (E) and pre-membrane (prM) enters the luminal side (Stadler et al. 1997). The NS2B/NS3 then form the remaining proteins i.e. three structural proteins and seven non-structural (NS) proteins (Lindenbach et al. 2007). The polyprotein is then activated by the viral (NS2B, NS3, NS4A and NS5), host and furin proteases (Fig. 3). A viral replication complex (VRC) is formed by the fusion of NS1, NS2A, NS4A and NS5 (Assenberg et al. 2009).

vRNA replication: vRNA replication is the most crucial step of the viral proliferation. Interestingly, it is de novo i.e. it occurs without initiation from a pre-existing primer. The circularisation of the vRNA occurs, followed by its attachment to the replication complex (Alvarez et al. 2005). Using the (+) RNA as the template, a (−) RNA copy is synthesised. Another round of replication occurs that uses (−) RNA as

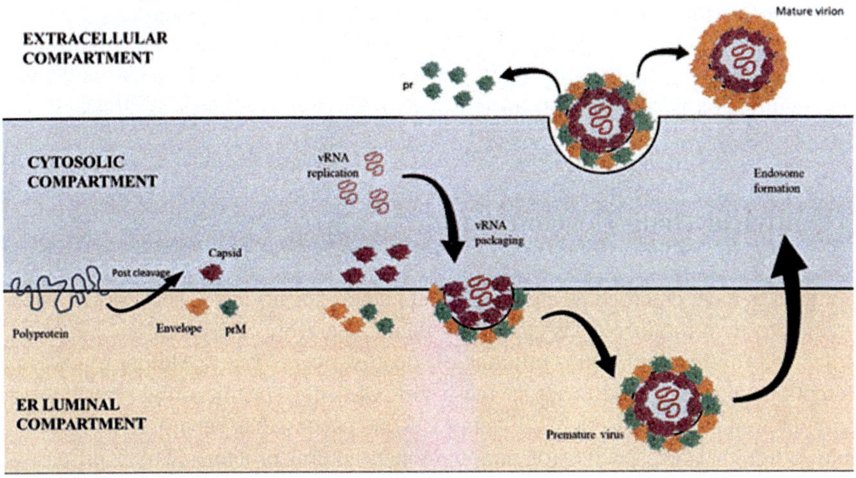

Fig. 5 Cellular compartments involved in assembly of viral structural proteins with the vRNA to form mature virion

the template and enough (+) RNA copies are made (Brinton 2002; Assenberg et al. 2009).

Viral assembly: The raw material required for the synthesis of new viral particles are synthesised and the preparations are made for the viral assembly. The vRNA binds to the capsid (C) proteins aggregated at the cytoplasmic site and gets packaged into the new viral proteins as it binds to (E) and (prM) and enters the ER (Fig. 5) (Apte-Sengupta et al. 2014).

Viral maturation: The viral maturation cannot occur in the ER, so the vesicle moves towards the trans Golgi network (TGN). Pre-Membrane protein (prM) covers the envelope to prevent premature fusion to the host membrane. The mature viral particles then leave the cell to infect the new cells (Gerold et al. 2017).

4.2 Role/Importance of Cyclization in Replication

For cyclization of genome, long-range RNA-RNA interaction is required to bring 5′ and 3′ ends together. Therefore, certain complementary structures are present in 5′ and 3′ UTR regions which contributes to the cyclization of viral genome (Alvarez et al. 2008; Villordo and Gamarnik 2009). These include 5′-3′ CS sequences and 5′-3′UAR in mosquito-borne flaviviruses (Alvarez et al. 2008; Villordo and Gamarnik 2009) and 5′-3′ CSA and 5′-3′CSB in case of tick-borne flaviviruses (Mandl et al. 1993). In addition, cyclization assists in binding of RdRp of NS5 to bind to SLA (promoter of flaviviruses), which is ~11 kb away from 3′ initiation site. By this, 3′

end is used as the template for initiation of minus strand RNA synthesis from an otherwise positive strand RNA genome (Filomatori et al. 2006; Villordo and Gamarnik 2009; Ng et al. 2017). Cyclization is also known to aid in detecting 5′ cap to be methylated by the methytransferases of NS5 gene (Egloff et al. 2002).

5 Antiviral Drug Development

Flaviviruses have emerged as a global threat and so far, no major success has been obtained to treat the infections caused by them. Therefore, these viruses are still a burden on humankind with millions of deaths every year. Although several attempts have been made to target these viruses using specific drugs, but due to high mutation rate of these RNA Flaviviruses, no success has been obtained so far. Many drugs that are specific for various non-structural proteins (which primarily includes RdRp, protease, methyl transferase, and helicase) have been targeted (Brecher et al. 2015; Zakaria et al. 2018). For example, temoporfin, niclosamide, and nitazoxanide, as flavivirus NS2B-NS3 interaction inhibitors drugs have been tested for NS3 proteases of flaviviruses, but due to the limitation of the structural active site formed by NS2B/NS3, protease inhibitors were found to work inefficiently (Xu et al. 2020; Li et al. 2017; Boldescu et al. 2017). Nevertheless, structural proteins of flaviviruses have also been targeted for antiviral therapy. For instance, E protein has been targeted by 2,4-diamino and 4,6-disubstituted pyrimidines that inhibit DENV entry at the very first step (Clark et al. 2016; de Wispelaere et al. 2018), but these drugs are still under clinical trials.

Other drugs used for symptomatic treatment for different flaviviruses are discussed below. Many of them have broad spectrum activity, and therefore, are not too specific. Umifenovir (also known as Arbidol) is an antiviral with broad spectrum activity (Panisheva et al. 1989). The mechanism of action of this drug depends upon its interference with the fusion of viral particle with the host cell membranes (Pecheur et al. 2016). This drug has recently been tested on ZIKV, WNV and TBEV and found to be promising for in vitro studies (Haviernik et al. 2018). Likewise, erythrosine B has recently been found to be effective on NS2B/NS3 proteases of several economically important flaviviruses including DENV2, ZIKV, WNV, YFV and JEV (Li et al. 2018).

Another broad spectrum drug, Chloroquine, is known to inhibit the viral replication cycle by increasing the pH of the endosome, thus blocking endocytosis of viral particles. This antiviral drug is known to be effective against several flaviviruses including ZIKAV, DENV, JEV and WNV infection (Farias et al. 2013; Delvecchio et al. 2016). Yet, alternative broad spectrum antivirals like favipiravir, novel benzavir-2 and ribavirin have been employed to treat infections caused by several Flaviviruses such as WNV, YFV, the TBEV, JEV, DENV and ZIKV (Gwon et al. 2020; Crance et al. 2003; Furuta et al. 2017).

Likewise, many more broad-spectrum drugs have been targeted for flaviviruses which were known to be effective on Hepatitis C virus (member of family

Flaviviridae), but flaviviruses were found to be almost insensitive to these drugs (Bollati et al. 2010). The present scenario indicates that there are no approved antiviral therapies for Flaviviruses (Wong et al. 2019), for the reason that they exhibit exceptional mutation rates, and thereby escape the available antiviral agents that target specifically viral proteins (Mateo et al. 2015).

Thus host based approaches are being implemented in the current pursuits of developing antiviral drugs against flaviviruses (Zakaria et al. 2018). This strategy is backed by the hypothesis that for any virus to complete its life cycle, the cellular machinery of the host is indispensable (Ahammad et al. 2019). The basis of the antiviral drug development via this strategy lies in the understanding of the mechanism of interaction of viral particles with the host factors that play a crucial role in viral infection and replication like the metabolic pathways, receptors and attachment factors, host proteins/enzymes, host immune factors, and anti-inflammatory pathways etc. (Ahammad et al. 2019; Zakaria et al. 2018). As no therapeutic agent has been known to work efficiently, therefore more emphasis is laid on designing vaccines to eliminate the viral existence in humans and certain livestock.

6 Vaccine Development

Though antiviral drugs are used as first line of defense to cure infections, but the higher mutability in viral genomes has resulted in appearance of drug resistant strains, which not only interrupt with treatment, but also remains a concern for immunocompromised patients (Strasfeld and Chou 2010). Contrary to this, vaccines are always preferred for preventing the devastation that can be caused by viral diseases, and have seldom contribution to induce resistant strains (Andreano et al. 2019).

The availability of licensed vaccines for Yellow Fever, Japanese Encephalitis and Tick-borne encephalitis has helped to release the worldwide load of the infections and deaths caused by the respective viruses (Barzon and Palu 2018; Frierson 2010; Barrett et al. 2003). In fact, the first vaccine designed for flaviviruses in 1930s was for YFV using its live attenuated form and was named as YFV-17D. This vaccine is known to be the safest and successful even after decades (Collins and Barrett 2017). JEV's live-attenuated vaccine was developed in China in 1988 under the CD.JEVAX™ trade name (Chengdu Institute of Biological Products, China) (Gromowski et al. 2015; Barzon and Palu 2018). After that, a wide range of vaccines have been developed for JEV and are available worldwide which are either inactivated, recombinant, or live attenuated vaccines (Guirakhoo et al. 1999) and are listed in Table 4.

So far, CYD-TDV is the only known licensed vaccine against DENV infections. Developed in 2015, by Sanofi-Pasteur this vaccine is available across 20 countries under the brand name Dengvaxia (Khetarpal and Khanna 2016). However, it exhibits low efficacy against serotypes 1 and 2. Other dengue vaccine candidates

Table 4 Vaccines developed or under development for socio-economically important flaviviruses

Flavivirus	Name of the vaccine	Type of the vaccine	Stage of development	Developer	Year	References
Dengue Virus (DENV)	rDEN2/4delta30(ME)	Live Attenuated Vaccine (LAV)	Phase 1 Clinical Trial	NIAID/Johns Hopkins Bloomberg School of Public Health	2008	Durbin et al. (2011)
	DENVax		Phase 3 Clinical Trial	Takeda	2011	Osorio et al. (2014)
	Chimerivax		Phase 2 Clinical Trial	Sanofi	2014	Dayan et al. (2013)
	CYD (Dengvaxia)		Liscensed		2018	Guy et al. (2011); Guy and Jackson (2016)
	D1ME100	DNA vaccine	Phase 1 Clinical Trial	U.S. Army Medical Research and Development Command and United States Army Medical Materiel Development Activity	2009	Zheng et al. (2017)
	TDENV-PIV	Purified Inactivated Vaccine	Phase 2 Clinical Trial		2015	Fernandez et al. (2015)
	DEN1–80E	Sub unit vaccine	Phase 1 Clinical Trial	Hawaii Biotech, Inc	2010	Swaminathan et al. (2016)
	V180		Phase 1 Clinical Trial	Merck	2013	Manoff et al. (2019)

(continued)

Table 4 (continued)

Flavivirus	Name of the vaccine	Type of the vaccine	Stage of development	Developer	Year	References
Japanese Encephalitis Virus (JEV)	JE-CV 12	Inactivated Vero cell culture vaccine	Licensed		2013	Kim et al. (2014)
	IXIARO-IC51	Purified, inactivated, whole virus vaccine	Licensed	Walter Reed Army Institute of Research	2009	Dubischar-Kastner et al. (2010)
	SA14-14–2	Live Attenuated Vaccine (LAV)	Licensed	Chengdu Institute of Biological Products Co., Ltd. (CDIBP)	2001	Jia et al. (2003)
Zika Virus (ZIKV)	GLS 5700	DNA Vaccine	Phase 1 Clinical Trial	Inovio Pharmaceuticals and GeneOne Life Science	2018	Makhluf and Shresta (2018)
	VRC-ZKADNA090-00-VP			National Institutes of Health Clinical Center (CC) (National Institute of Allergy and Infectious Diseases (NIAID))	2018	
	ZPIV	Purified Inactivated Virus		Walter Reed Army Institute of Research and NIAID	2019	Modjarrad et al. (2018)
	VLA1601			Valneva	2018	Garg et al. (2018)
	MR 766			Bharat Biotech		Frumence et al. (2019)
	MV-ZIKA	Live attenuated recombinant vaccine		Thermis Bioscience		Nurnberger et al. (2019)
	mRNA-1325	mRNA vaccine	Phase 2 Clinical Trial	Moderna Therapeutics		Barrett (2018)

(continued)

Table 4 (continued)

Flavivirus	Name of the vaccine	Type of the vaccine	Stage of development	Developer	Year	References
West Nile Virus	Hydrovax-001	Inactivated	Phase 1 Clinical Trial	National Institute of Allergy and Infectious Diseases (NIAID)	2016	Woods et al. (2019)
	ChimeriVax-WN02	Recombinant	Phase 2 Clinical Trial	Sanofi Pasteur	2009	Biedenbender et al. (2011)
	rWN/DEN4Δ30	Recombinant attenuated	Phase 1 Clinical Trial	National Institute of Allergy and Infectious Diseases (NIAID)	2011	Durbin et al. (2011)
	VRC WNV	DNA plasmid	Phase 1 Clinical Trial		2007	Gaudinski et al. (2018;) Ledgerwood et al. (2011)
	HBV-002	Recombinant	Phase 1 Clinical Trial	Hawaii Biotech	2016	Ulbert (2019)

have also been in the pipeline, with TDV and TDV 003/005 being in stage II of clinical trials and TDENV-PIV, a purified inactivated vaccine being in phase I clinical trial (Guy et al. 2011; Guy and Jackson 2016; Prompetchara et al. 2019; Redoni et al. 2020).

For other Flaviviruses, the situation remains the same although a number of potential vaccines are being formulated (vaccines for ZIKV being at the top of the list), and are under various stages of clinical trial. Among these are formulations like inactivated virus vaccines, whole-virus vaccines, subunit vaccines, and messenger RNA (mRNA), DNA, protein, and vector-based formulations (Poland et al. 2019; Pielnaa et al. 2020). The most advanced and the most promising vaccine candidate for ZIKV is a plasmid-based DNA vaccine VRC-ZKADNA090-00-VP undergoing phase 2 clinical trial (Gaudinski et al. 2018).

The situation is no different for West Nile Fever as no vaccine is currently available for the WNV infection. However, a good number of them are undergoing development and clinical trials are listed in Table 4.

7 Conclusion

Viruses are well known for their notorious behavior of using host's machinery to their own benefit, and harming them in the process. And flaviviruses are no different in this context. In fact, flaviviruses have exploited almost all means to make their existence successful, be it the involvement of vectors, no vectors, only vertebrates as hosts or solely residing in insects. They fool their hosts by impersonating through mutations, thus maintaining their existence from over 100,000 years (Pierson and Diamond 2020). Hence, the evolution of flaviviruses, their emergence and re-emergence, their global expansion due to climate change and unintentional spread due to traveling have made research in this area imperative. With no foolproof antivirals and unavailability of vaccines for majority of the flaviviral diseases, it seems to be unmanageable task to eradicate or control the spread of these viruses.

Although researchers are targeting the specific receptors involved in the viral entry, looking for the conservancy in structural and non-structural proteins, exploiting the viral biology using comparative genomics and proteomics, supportive care is the only mean to cure viral symptoms so far. Another way, although not completely infallible, is to control the spread of the vectors. Several ways are being employed to control vectors, for example, use of *Wolbachia*-based biocontrol (Jeffries and Walker 2015), or by genetically engineering mosquito species using RNAi (Pierson and Diamond 2020). But the success of flavivirus control lies on cumulative efforts of viral targeting using "omics", control of vectors, development of vaccines and human awareness about their spread.

Acknowledgements The authors greatly acknowledge Prof. C. Sheela Reddy, Principal, Sri Venkateswara College, University of Delhi, for her constant support.

References

Ahammad F,. Tengku Abd Rashid TR, Mohamed M, Tanbin S, Ahmad Fuad FA. Contemporary strategies and current trends in designing antiviral drugs against dengue fever via targeting host-based approaches. Microorganisms. 2019; 7(9). https://doi.org/10.3390/microorganisms7090296.

Alkan C, Zapata S, Bichaud L, Moureau G, Lemey P, Firth AE, Gritsun TS, Gould EA, de Lamballerie X, Depaquit J, Charrel RN. Ecuador Paraiso Escondido virus, a new flavivirus isolated from new world sand flies in Ecuador, is the first representative of a novel clade in the Genus Flavivirus. J Virol. 2015;89(23):11773–85. https://doi.org/10.1128/JVI.01543-15.

Alvarez DE, Ezcurra ALDL, Fucito S, Gamarnik AV. Role of RNA structures present at the 3′ UTR of dengue virus on translation, RNA synthesis, and viral replication. Virology. 2005;339 (2):200–12. https://doi.org/10.1016/j.virol.2005.06.009.

Alvarez DE, Filomatori CV, Gamarnik AV. Functional analysis of dengue virus cyclization sequences located at the 5' and 3'UTRs. Virology. 2008;375(1):223–35. https://doi.org/10.1016/j.virol.2008.01.014.

Andreano E, D'Oro U, Rappuoli R, Finco O. Vaccine evolution and its application to fight modern threats. Front Immunol. 2019;10:1722–1722. https://doi.org/10.3389/fimmu.2019.01722.

Apte-Sengupta S, Sirohi D, Kuhn RJ. Coupling of replication and assembly in flaviviruses. Curr Opin Virol. 2014;9:134–42. https://doi.org/10.1016/j.coviro.2014.09.020.

Assenberg R, Mastrangelo E, Walter TS, Verma A, Milani M, Owens RJ, Stuart DI, Grimes JM, Mancini EJ. Crystal structure of a novel conformational state of the flavivirus NS3 protein: implications for polyprotein processing and viral replication. (1098–5514 (Electronic)) 2009.

Barrett ADT. Current status of Zika vaccine development: Zika vaccines advance into clinical evaluation. NPJ Vaccines. 2018;3:24. https://doi.org/10.1038/s41541-018-0061-9.

Barrett PN, Schober-Bendixen S, Ehrlich HJ. History of TBE vaccines. Vaccine. 2003;21(Suppl 1):S41–9. https://doi.org/10.1016/s0264-410x(02)00814-9.

Barrows NJ, Campos RK, Liao KC, Prasanth KR, Soto-Acosta R, Yeh SC, Schott-Lerner G, Pompon J, Sessions OM, Bradrick SS, Garcia-Blanco MA. Biochemistry and molecular biology of flaviviruses. Chem Rev. 2018;118(8):4448–82. https://doi.org/10.1021/acs.chemrev.7b00719.

Barzon L, Palu G. Recent developments in vaccines and biological therapies against Japanese encephalitis virus. Expert Opin Biol Ther. 2018;18(8):851–64. https://doi.org/10.1080/14712598.2018.1499721.

Belshaw R, Pybus OG, Rambaut A. The evolution of genome compression and genomic novelty in RNA viruses. Genome Res. 2007;17(10):1496–504. https://doi.org/10.1101/gr.6305707.

Biedenbender R, Bevilacqua J, Gregg AM, Watson M, Dayan G. Phase II, randomized, double-blind, placebo-controlled, multicenter study to investigate the immunogenicity and safety of a West Nile virus vaccine in healthy adults. J Infect Dis. 2011;203(1):75–84. https://doi.org/10.1093/infdis/jiq003.

Billoir F, de Chesse R, Tolou H, de Micco P, Gould EA, de Lamballerie X. Phylogeny of the genus flavivirus using complete coding sequences of arthropod-borne viruses and viruses with no known vector. J Gen Virol. 2000;81(Pt 9):2339. https://doi.org/10.1099/0022-1317-81-9-2339.

Blitvich BJ, Firth AE. Insect-specific flaviviruses: a systematic review of their discovery, host range, mode of transmission, superinfection exclusion potential and genomic organization. Viruses. 2015;7(4):1927–59. https://doi.org/10.3390/v7041927.

Blitvich BJ, Firth AE. A review of flaviviruses that have no known arthropod vector. Viruses. 2017; 9(6). https://doi.org/10.3390/v9060154.

Boldescu V, Behnam MAM, Vasilakis N, Klein CD. Broad-spectrum agents for flaviviral infections: dengue, Zika and beyond. Nat Rev Drug Discov. 2017;16(8):565–86. https://doi.org/10.1038/nrd.2017.33.

Bollati M, Alvarez K, Assenberg R, Baronti C, Canard B, Cook S, Coutard B, Decroly E, de Lamballerie X, Gould EA, Grard G, Grimes JM, Hilgenfeld R, Jansson AM, Malet H, Mancini EJ, Mastrangelo E, Mattevi A, Milani M, Moureau G, Neyts J, Owens RJ, Ren J, Selisko B, Speroni S, Steuber H, Stuart DI, Unge T, Bolognesi M. Structure and functionality in flavivirus NS-proteins: perspectives for drug design. Antiviral Res. 2010;87(2):125–48. https://doi.org/10.1016/j.antiviral.2009.11.009.

Brecher M, Chen H, Liu B, Banavali NK, Jones SA, Zhang J, Li Z, Kramer LD, Li H. Novel broad spectrum inhibitors targeting the flavivirus methyltransferase. PLoS ONE. 2015;10(6): e0130062–e0130062. https://doi.org/10.1371/journal.pone.0130062.

Bressanelli S, Stiasny K, Allison SL, Stura EA, Duquerroy S, Lescar J, Heinz FX, Rey FA. Structure of a flavivirus envelope glycoprotein in its low-pH-induced membrane fusion conformation. EMBO J. 2004;23(4):728–38. https://doi.org/10.1038/sj.emboj.7600064.

Brinton MA. The molecular biology of West Nile Virus: a new invader of the western hemisphere. Annu Rev Microbiol. 2002;56:371–402. https://doi.org/10.1146/annurev.micro.56.012302.160654.

Brinton MA, Fernandez AV, Dispoto JH. The 3'-nucleotides of flavivirus genomic RNA form a conserved secondary structure. Virology. 1986;153(1):113–21. https://doi.org/10.1016/0042-6822(86)90012-7.

Brinton MA, Dispoto JH. Sequence and secondary structure analysis of the 5′-terminal region of flavivirus genome RNA. Virology. 1988;162(2):290–9. https://doi.org/10.1016/0042-6822(88)90468-0.

Chaves-Carballo E. Carlos Finlay and yellow fever: triumph over adversity. Mil Med. 2005;170(10):881–5. https://doi.org/10.7205/MILMED.170.10.881%JMilitaryMedicine.

Chippaux JP, Chippaux A. Yellow fever in Africa and the Americas: a historical and epidemiological perspective. J Venom Anim Toxins Incl Trop Dis. 2018;24:20. https://doi.org/10.1186/s40409-018-0162-y.

Clark MJ, Miduturu C, Schmidt AG, Zhu X, Pitts JD, Wang J, Potisopon S, Zhang J, Wojciechowski A, Hann Chu JJ, Gray NS, Yang PL. GNF-2 inhibits dengue virus by targeting Abl kinases and the viral E protein. Cell Chem Biol. 2016;23(4):443–52. https://doi.org/10.1016/j.chembiol.2016.03.010.

Cleaves GR, Dubin DT. Methylation status of intracellular dengue type 2 40 S RNA. Virology. 1979;96(1):159–65. https://doi.org/10.1016/0042-6822(79)90181-8.

Collins ND, Barrett AD. Live attenuated yellow fever 17D vaccine: a legacy vaccine still controlling outbreaks in modern day. Curr Infect Dis Rep. 2017;19(3):14. https://doi.org/10.1007/s11908-017-0566-9.

Conway MJ, Colpitts TM, Fikrig E. Role of the vector in arbovirus transmission. Annu Rev Virol. 2014;1(1):71–88. https://doi.org/10.1146/annurev-virology-031413-085513.

Crance JM, Scaramozzino N, Jouan A, Garin D. Interferon, ribavirin, 6-azauridine and glycyrrhizin: antiviral compounds active against pathogenic flaviviruses. Antiviral Res. 2003;58(1):73–9. https://doi.org/10.1016/s0166-3542(02)00185-7.

Daep CA, Muñoz-Jordán JL, Eugenin EA. Flaviviruses, an expanding threat in public health: focus on dengue, West Nile, and Japanese encephalitis virus. J Neurovirol. 2014;20(6):539–60. https://doi.org/10.1007/s13365-014-0285-z.

Dayan GH, Pugachev K, Bevilacqua J, Lang J, Monath TP. Preclinical and clinical development of a YFV 17 D-based chimeric vaccine against West Nile virus. Viruses. 2013;5(12):3048–70. https://doi.org/10.3390/v5123048.

de Wispelaere M, Lian W, Potisopon S, Li P-C, Jang J, Ficarro S, Clark M, Zhu X, Kaplan J, Pitts J, Wales T, Wang J, Engen J, Marto J, Gray N, Yang P. Inhibition of flaviviruses by targeting a conserved pocket on the viral envelope protein. Cell Chem Biol. 2018; 25. https://doi.org/10.1016/j.chembiol.2018.05.011.

Delvecchio R, Higa LM, Pezzuto P, Valadão AL, Garcez PP, Monteiro FL, Loiola EC, Dias AA, Silva FJM, Aliota MT, Caine EA, Osorio JE, Bellio M, O'Connor DH, Rehen S, Santana R, de Aguiar A, Savarino LC, Tanuri A. Chloroquine, an endocytosis blocking agent, inhibits Zika virus infection in different cell models. Viruses. 2016;8(12):322. https://doi.org/10.3390/v8120322.

Diosa-Toro M, Urcuqui-Inchima S, Smit JM. Arthropod-borne flaviviruses and RNA interference: seeking new approaches for antiviral therapy. Adv Virus Res. 2013;85:91–111. https://doi.org/10.1016/b978-0-12-408116-1.00004-5.

Dubischar-Kastner K, Eder S, Buerger V, Gartner-Woelfl G, Kaltenboeck A, Schuller E, Tauber E, Klade C. Long-term immunity and immune response to a booster dose following vaccination with the inactivated Japanese encephalitis vaccine IXIARO, IC51. Vaccine. 2010;28(32):5197–202. https://doi.org/10.1016/j.vaccine.2010.05.069.

Durbin AP, Whitehead SS, Shaffer D, Elwood D, Wanionek K, Thumar B, Blaney JE, Murphy BR, Schmidt AC. A single dose of the DENV-1 candidate vaccine rDEN1Delta30 is strongly immunogenic and induces resistance to a second dose in a randomized trial. PLoS Negl Trop Dis. 2011;5(8):e1267. https://doi.org/10.1371/journal.pntd.0001267.

Edgil D, Polacek C, Harris E. Dengue virus utilizes a novel strategy for translation initiation when cap-dependent translation is inhibited. J Virol. 2006;80(6):2976–86. https://doi.org/10.1128/JVI.80.6.2976-2986.2006.

Egloff MP, Benarroch D, Selisko B, Romette JL, Canard B. An RNA cap (nucleoside-2'-O-)-methyltransferase in the flavivirus RNA polymerase NS5: crystal structure and functional characterization. EMBO J. 2002;21(11):2757–68. https://doi.org/10.1093/emboj/21.11.2757.

El Sahili A, Lescar J. Dengue virus non-structural protein 5. Viruses. 2017; 9(4). https://doi.org/10.3390/v9040091.

Farias KJ, Machado PR, da Fonseca BA. Chloroquine inhibits dengue virus type 2 replication in Vero cells but not in C6/36 cells. Sci World J. 2013;2013:282734. https://doi.org/10.1155/2013/282734.

Fernandez S, Thomas SJ, De La Barrera R, Im-Erbsin R, Jarman RG, Baras B, Toussaint JF, Mossman S, Innis BL, Schmidt A, Malice MP, Festraets P, Warter L, Putnak JR, Eckels KH. An adjuvanted, tetravalent dengue virus purified inactivated vaccine candidate induces long-lasting and protective antibody responses against dengue challenge in rhesus macaques. Am J Trop Med Hyg. 2015;92(4):698–708. https://doi.org/10.4269/ajtmh.14-0268.

Filomatori CV, Lodeiro MF, Alvarez DE, Samsa MM, Pietrasanta L, Gamarnik AV. A 5' RNA element promotes dengue virus RNA synthesis on a circular genome. Genes Dev. 2006;20(16):2238–49. https://doi.org/10.1101/gad.1444206.

Friebe P, Peña J, Pohl MOF, Harris E. Composition of the sequence downstream of the dengue virus 5' cyclization sequence (dCS) affects viral RNA replication. Virology. 2012;422(2):346–56. https://doi.org/10.1016/j.virol.2011.10.025.

Frierson JG. The yellow fever vaccine: a history. Yale J Biol Med. 2010;83(2):77–85.

Frumence E, Viranaicken W, Bos S, Alvarez-Martinez MT, Roche M, Arnaud JD, Gadea G, Despres P. A chimeric Zika virus between viral strains MR766 and BeH819015 highlights a role for E-glycan loop in antibody-mediated virus neutralization. Vaccines (Basel). 2019; 7(2). https://doi.org/10.3390/vaccines7020055.

Furuta Y, Komeno T, Nakamura T. Favipiravir (T-705), a broad spectrum inhibitor of viral RNA polymerase. Proc Jpn Acad Ser B Phys Biol Sci. 2017;93(7):449–63. https://doi.org/10.2183/pjab.93.027.

Garg H, Mehmetoglu-Gurbuz T, Joshi A. Recent advances in Zika virus vaccines. Viruses. 2018; 10(11). https://doi.org/10.3390/v10110631.

Gaudinski MR, Houser KV, Morabito KM, Hu Z, Yamshchikov G, Rothwell RS, Berkowitz N, Mendoza F, Saunders JG, Novik L, Hendel CS, Holman LA, Gordon IJ, Cox JH, Edupuganti S, McArthur MA, Rouphael NG, Lyke KE, Cummings GE, Sitar S, Bailer RT, Foreman BM, Burgomaster K, Pelc RS, Gordon DN, DeMaso CR, Dowd KA, Laurencot C, Schwartz RM, Mascola JR, Graham BS, Pierson TC, Ledgerwood JE, Chen GL, Vrc, and V. R. C. study teams. Safety, tolerability, and immunogenicity of two Zika virus DNA vaccine candidates in healthy adults: randomised, open-label, phase 1 clinical trials. Lancet. 2018;391(10120):552–62. https://doi.org/10.1016/S0140-6736(17)33105-7.

Gaunt MW, Sall AA, Lamballerie X, Falconar AKI, Dzhivanian TI, Gould EA. Phylogenetic relationships of flaviviruses correlate with their epidemiology, disease association and biogeography. J Gen Virol. 2001;82(Pt 8):1867–76. https://doi.org/10.1099/0022-1317-82-8-1867.

Gebhard LG, Filomatori CV, Gamarnik AV. Functional RNA elements in the dengue virus genome. Viruses. 2011;3(9):1739–56. https://doi.org/10.3390/v3091739.

Gerold G, Bruening J, Weigel B, Pietschmann T. Protein interactions during the flavivirus and hepacivirus life cycle. Mol Cell Proteomics. 2017;16(4 suppl 1):S75. https://doi.org/10.1074/mcp.R116.065649.

Gould EA, Gaunt MW, Sall AA, Lamballerie X, Falconar AKI, Dzhivanian TI. Phylogenetic relationships of flaviviruses correlate with their epidemiology, disease association and biogeography. J Gen Virol. 2001;82(Pt 8):1867–76. https://doi.org/10.1099/0022-1317-82-8-1867.

Gromowski GD, Firestone CY, Bustos-Arriaga J, Whitehead SS. Genetic and phenotypic properties of vero cell-adapted Japanese encephalitis virus SA14-14-2 vaccine strain variants and a recombinant clone, which demonstrates attenuation and immunogenicity in mice. Am J Trop Med Hyg. 2015;92(1):98–107. https://doi.org/10.4269/ajtmh.14-0427.

Guirakhoo F, Zhang ZX, Chambers TJ, Delagrave S, Arroyo J, Barrett AD, Monath TP. Immunogenicity, genetic stability, and protective efficacy of a recombinant, chimeric yellow fever-Japanese encephalitis virus (ChimeriVax-JE) as a live, attenuated vaccine candidate against Japanese encephalitis. Virology. 1999;257(2):363–72. https://doi.org/10.1006/viro.1999.9695.

Guy B, Jackson N. Dengue vaccine: hypotheses to understand CYD-TDV-induced protection. Nat Rev Microbiol. 2016;14(1):45–54. https://doi.org/10.1038/nrmicro.2015.2.

Guy B, Barrere B, Malinowski C, Saville M, Teyssou R, Lang J. From research to phase III: preclinical, industrial and clinical development of the Sanofi Pasteur tetravalent dengue vaccine. Vaccine. 2011; 29(42):7229-41.https://doi.org/10.1016/j.vaccine.2011.06.094.

Gwon YD, Strand M, Lindqvist R, Nilsson E, Saleeb M, Elofsson M, Overby AK, Evander M. Antiviral activity of Benzavir-2 against emerging flaviviruses. Viruses. 2020; 12(3). https://doi.org/10.3390/v12030351.

Hahn CS, Hahn YS, Rice CM, Lee E, Dalgarno L, Strauss EG, Strauss JH. Conserved elements in the 3' untranslated region of flavivirus RNAs and potential cyclization sequences. J Mol Biol. 1987;198(1):33–41. https://doi.org/10.1016/0022-2836(87)90455-4.

Harrison SC. Viral membrane fusion. Nat Struct Mol Biol. 2008;15(7):690–8. https://doi.org/10.1038/nsmb.1456.

Haviernik J, Stefanik M, Fojtikova M, Kali S, Tordo N, Rudolf I, Hubalek Z, Eyer L, Ruzek D. Arbidol (Umifenovir): a broad-spectrum antiviral drug that inhibits medically important arthropod-borne flaviviruses. Viruses. 2018; 10(4). https://doi.org/10.3390/v10040184.

Holden KL, Stein DA, Pierson TC, Ahmed AA, Clyde K, Iversen PL, Harris E. Inhibition of dengue virus translation and RNA synthesis by a morpholino oligomer targeted to the top of the terminal 3' stem-loop structure. Virology. 2006;344(2):439–52. https://doi.org/10.1016/j.virol.2005.08.034.

Huang YS, Higgs S, Vanlandingham DL. Arbovirus-mosquito vector-host interactions and the impact on transmission and disease pathogenesis of arboviruses. Front Microbiol. 2019;10:22. https://doi.org/10.3389/fmicb.2019.00022.

Ivanyi-Nagy R, Darlix J-L. Reprint of: Core protein-mediated 5′–3′ annealing of the West Nile virus genomic RNA in vitro. Virus Res. 2012;169(2):448–57. https://doi.org/10.1016/j.virusres.2012.09.009.

Jeffries CL, Walker T. The potential use of Wolbachia-based mosquito biocontrol strategies for Japanese encephalitis. PLoS Negl Trop Dis. 2015;9(6):e0003576. https://doi.org/10.1371/journal.pntd.0003576.

Jia L, Wang Z, Yu Y. Protection of SA14-14-2 live attenuated Japanese encephalitis vaccine against the wild-type JE viruses. Chin Med J (Engl). 2003;116(6):941–3.

Kaufmann B, Rossmann MG. Molecular mechanisms involved in the early steps of flavivirus cell entry. Microbes Infect. 2011;13(1):1–9. https://doi.org/10.1016/j.micinf.2010.09.005.

Khetarpal N, Khanna I. Dengue fever: causes, complications, and vaccine strategies. J Immunol Res. 2016;2016:6803098. https://doi.org/10.1155/2016/6803098.

Kim DS, Houillon G, Jang GC, Cha SH, Choi SH, Lee J, Kim HM, Kim JH, Kang JH, Kim JH, Kim KH, Kim HS, Bang J, Naimi Z, Bosch-Castells V, Boaz M, Bouckenooghe A. A randomized study of the immunogenicity and safety of Japanese encephalitis chimeric virus vaccine (JE-CV) in comparison with SA14-14-2 vaccine in children in the Republic of Korea. Hum Vaccin Immunother. 2014;10(9):2656–63. https://doi.org/10.4161/hv.29743.

Kok WM. New developments in flavivirus drug discovery. Expert Opin Drug Discov. 2016;11 (5):433–45. https://doi.org/10.1517/17460441.2016.1160887.

Kraemer MU, Sinka ME, Duda KA, Mylne AQ, Shearer FM, Barker CM, Moore CG, Carvalho RG, Coelho GE, Van Bortel W, Hendrickx G, Schaffner F, Elyazar IR, Teng HJ, Brady OJ, Messina JP, Pigott DM, Scott TW, Smith DL, Wint GR, Golding N, Hay SI. The global distribution of the arbovirus vectors Aedes aegypti and Ae. albopictus. Elife. 2015; 4: e08347. https://doi.org/10.7554/eLife.08347.

Kuno G, Chang GJ, Tsuchiya KR, Karabatsos N, Cropp CB. Phylogeny of the genus Flavivirus. J Virol. 1998;72(1):73–83.

Letunic I, Bork P. Interactive tree of life (iTOL) v3: an online tool for the display and annotation of phylogenetic and other trees. Nucleic Acids Res. 2016;44(W1):W242–5. https://doi.org/10.1093/nar/gkw290.

Li K, Phoo WW, Luo D. Functional interplay among the flavivirus NS3 protease, helicase, and cofactors. Virol Sin. 2014;29(2):74–85. https://doi.org/10.1007/s12250-014-3438-6.

Li Z, Sakamuru S, Huang R, Brecher M, Koetzner CA, Zhang J, Chen H, Qin CF, Zhang QY, Zhou J, Kramer LD, Xia M, Li H. Erythrosin B is a potent and broad-spectrum orthosteric inhibitor of the flavivirus NS2B-NS3 protease. Antiviral Res. 2018;150:217–25. https://doi.org/10.1016/j.antiviral.2017.12.018.

Li Z, Zhang J, Li H. lavivirus NS2B/NS3 protease: structure, function, and inhibition. 2017; 163–88.

Lindenbach B, Thiel HJ, Rice CM. Flaviviridae: the viruses and their replication. Fields Virol. 2007; 1101–51.

Makhluf H, Shresta S. Development of Zika virus vaccines. Vaccines (Basel). 2018; 6(1). https://doi.org/10.3390/vaccines6010007.

Mandl CW, Holzmann H, Kunz C, Heinz FX. Complete genomic sequence of Powassan virus: evaluation of genetic elements in tick-borne versus mosquito-borne flaviviruses. Virology. 1993;194(1):173–84. https://doi.org/10.1006/viro.1993.1247.

Manoff SB, Sausser M, Falk Russell A, Martin J, Radley D, Hyatt D, Roberts CC, Lickliter J, Krishnarajah J, Bett A, Dubey S, Finn T, Coller BA. Immunogenicity and safety of an investigational tetravalent recombinant subunit vaccine for dengue: results of a Phase I randomized clinical trial in flavivirus-naive adults. Hum Vaccin Immunother. 2019;15 (9):2195–204. https://doi.org/10.1080/21645515.2018.1546523.

Mateo R, Nagamine CM, Kirkegaard K. Suppression of drug resistance in dengue virus. mBio. 2015; 6(6):e01960–15. https://doi.org/10.1128/mBio.01960-15.

Mazeaud C, Freppel W, Chatel-Chaix L. The multiples fates of the flavivirus RNA genome during pathogenesis. Front Genet. 2018;9:595. https://doi.org/10.3389/fgene.2018.00595.

Men R, Bray M, Clark D, Chanock RM, Lai CJ. Dengue type 4 virus mutants containing deletions in the 3' noncoding region of the RNA genome: analysis of growth restriction in cell culture and altered viremia pattern and immunogenicity in rhesus monkeys. J Virol. 1996;70(6):3930–7.

Modjarrad K, Lin L, George SL, Stephenson KE, Eckels KH, De La Barrera RA, Jarman RG, Sondergaard E, Tennant J, Ansel JL, Mills K, Koren M, Robb ML, Barrett J, Thompson J, Kosel AE, Dawson P, Hale A, Tan CS, Walsh SR, Meyer KE, Brien J, Crowell TA, Blazevic A, Mosby K, Larocca RA, Abbink P, Boyd M, Bricault CA, Seaman MS, Basil A, Walsh M, Tonwe V, Hoft DF, Thomas SJ, Barouch DH, Michael NL. Preliminary aggregate safety and immunogenicity results from three trials of a purified inactivated Zika virus vaccine candidate: phase 1, randomised, double-blind, placebo-controlled clinical trials. Lancet. 2018;391(10120):563–71. https://doi.org/10.1016/S0140-6736(17)33106-9.

Muller DA, Young PR. The flavivirus NS1 protein: molecular and structural biology, immunology, role in pathogenesis and application as a diagnostic biomarker. Antiviral Res. 2013;98(2):192–208. https://doi.org/10.1016/j.antiviral.2013.03.008.

Ng WC, Soto-Acosta R, Bradrick SS, Garcia-Blanco MA, Ooi EE. The 5' and 3' untranslated regions of the flaviviral genome. Viruses. 2017; 9(6). https://doi.org/10.3390/v9060137.

Nunez-Castilla J, Rahaman J, Ahrens JB, Balbin CA, Siltberg-Liberles J. Exploring evolutionary constraints in the proteomes of Zika, Dengue, and other Flaviviruses to find fitness-critical sites. J Mol Evol. 2020;88(4):399–414. https://doi.org/10.1007/s00239-020-09941-5.

Nurnberger C, Bodmer BS, Fiedler AH, Gabriel G, Muhlebach MD. A measles virus-based vaccine candidate mediates protection against Zika virus in an allogeneic mouse pregnancy model. J Virol. 2019; 93(3). https://doi.org/10.1128/JVI.01485-18.

Osorio JE, Velez ID, Thomson C, Lopez L, Jimenez A, Haller AA, Silengo S, Scott J, Boroughs KL, Stovall JL, Luy BE, Arguello J, Beatty ME, Santangelo J, Gordon GS, Huang CY, Stinchcomb DT. Safety and immunogenicity of a recombinant live attenuated tetravalent dengue vaccine (DENVax) in flavivirus-naive healthy adults in Colombia: a randomised, placebo-controlled, phase 1 study. Lancet Infect Dis. 2014;14(9):830–8. https://doi.org/10.1016/S1473-3099(14)70811-4.

Panisheva EK, Mikerova NI, Nikolaeva IS, Fomina AN, Cherkasova AA, Golovanova EA, Yu L, ChemInform Krylova J. Synthesis and antiviral activity of 5-hydroxyindole derivatives. 1989; 20(29):no-no.

Pecheur EI, Borisevich V, Halfmann P, Morrey JD, Smee DF, Prichard M, Mire CE, Kawaoka Y, Geisbert TW, Polyak SJ. The synthetic antiviral drug arbidol inhibits globally prevalent pathogenic viruses. J Virol. 2016;90(6):3086–92. https://doi.org/10.1128/JVI.02077-15.

Petersen LR, Marfin AA. Shifting epidemiology of flaviviridae. J Travel Med. 2005;12(Suppl 1): S3-11. https://doi.org/10.2310/7060.2005.12052.

Pfaender S, Brown RJP, Pietschmann T, Steinmann E. Natural reservoirs for homologs of hepatitis C virus. Emerg Microbes Infect. 2014;3(1):1–9. https://doi.org/10.1038/emi.2014.19.

Pielnaa P, Al-Saadawe M, Saro A, Dama MF, Zhou M, Huang Y, Huang J, Xia Z. Zika virus-spread, epidemiology, genome, transmission cycle, clinical manifestation, associated challenges, vaccine and antiviral drug development. Virology. 2020;543:34–42. https://doi.org/10.1016/j.virol.2020.01.015.

Pierson TC, Diamond MS. Degrees of maturity: the complex structure and biology of flaviviruses. Curr Opin Virol. 2012;2(2):168–75. https://doi.org/10.1016/j.coviro.2012.02.011.

Pierson TC, Diamond MS. The continued threat of emerging flaviviruses. Nat Microbiol. 2020;5 (6):796–812. https://doi.org/10.1038/s41564-020-0714-0.

Poland GA, Ovsyannikova IG, Kennedy RB. Zika vaccine development: current status. Mayo Clin Proc. 2019;94(12):2572–86. https://doi.org/10.1016/j.mayocp.2019.05.016.

Prompetchara E, Ketloy C, Thomas SJ, Ruxrungtham K. Dengue vaccine: global development update. Asian Pac J Allergy Immunol. 2019. https://doi.org/10.12932/AP-100518-0309.

Rastogi M, Sharma N, Singh SK. Flavivirus NS1: a multifaceted enigmatic viral protein. Virol J. 2016;13:131. https://doi.org/10.1186/s12985-016-0590-7.

Ray D, Shah A, Tilgner M, Guo Y, Zhao Y, Dong H, Deas TS, Zhou Y, Li H, Shi PY. West Nile virus 5'-cap structure is formed by sequential guanine N-7 and ribose 2'-O methylations by nonstructural protein 5. J Virol. 2006;80(17):8362–70. https://doi.org/10.1128/jvi.00814-06.

Redoni M, Yacoub S, Rivino L, Giacobbe D, Luzzati R, Di Bella S. Dengue: status of current and under-development vaccines. Rev Med Virol. 2020; e2101. https://doi.org/10.1002/rmv.2101.

Rice CM, Lenches EM, Eddy SR, Shin SJ, Sheets RL, Strauss JH. Nucleotide sequence of yellow fever virus: implications for flavivirus gene expression and evolution. Science. 1985;229 (4715):726–33. https://doi.org/10.1126/science.4023707.

Schubert AM, Putonti C. Evolution of the sequence composition of Flaviviruses. Infect Genet Evol. 2010;10(1):129–36. https://doi.org/10.1016/j.meegid.2009.11.004.

Selisko B, Wang C, Harris E, Canard B. Regulation of Flavivirus RNA synthesis and replication. Curr Opin Virol. 2014;9:74–83. https://doi.org/10.1016/j.coviro.2014.09.011.

Shi J, Hu Z, Deng F, Shen S. Tick-borne viruses. Virol Sin. 2018;33(1):21–43. https://doi.org/10.1007/s12250-018-0019-0.

Shi M, Lin XD, Vasilakis N, Tian JH, Li CX, Chen LJ, Eastwood G, Diao XN, Chen MH, Chen X, Qin XC, Widen SG, Wood TG, Tesh RB, Xu J, Holmes EC, Zhang YZ. Divergent viruses discovered in arthropods and vertebrates revise the evolutionary history of the flaviviridae and related viruses. J Virol. 2016;90(2):659–69. https://doi.org/10.1128/jvi.02036-15.

Shi PY, Brinton MA, Veal JM, Zhong YY, Wilson WD. Evidence for the existence of a pseudoknot structure at the 3' terminus of the flavivirus genomic RNA. Biochemistry. 1996;35 (13):4222–30. https://doi.org/10.1021/bi952398v.

Shurtleff AC, Beasley DW, Chen JJ, Ni H, Suderman MT, Wang H, Xu R, Wang E, Weaver SC, Watts DM, Russell KL, Barrett AD. Genetic variation in the 3' non-coding region of dengue viruses. Virology. 2001;281(1):75–87. https://doi.org/10.1006/viro.2000.0748.

Smith DB, Becher P, Bukh J, Gould EA, Meyers G, Monath T, Muerhoff AS, Pletnev A, Rico-Hesse R, Stapleton JT, Simmonds P. Proposed update to the taxonomy of the genera Hepacivirus and Pegivirus within the Flaviviridae family. J Gen Virol. 2016;97(11):2894–907. https://doi.org/10.1099/jgv.0.000612.

Smith DB, Meyers G, Bukh J, Gould EA, Monath T, Scott Muerhoff A, Pletnev A, Rico-Hesse R, Stapleton JT, Simmonds P, Becher P. Proposed revision to the taxonomy of the genus Pestivirus, family Flaviviridae. J Gen Virol. 2017;98(8):2106–12. https://doi.org/10.1099/jgv.0.000873.

Sotcheff S, Routh A. Understanding flavivirus capsid protein functions: the tip of the iceberg. Pathogens. 2020; 9(1). https://doi.org/10.3390/pathogens9010042.

Stadler K, Allison SL, Schalich J, Heinz FX. Proteolytic activation of tick-borne encephalitis virus by furin. J Virol. 1997;71(11):8475–81.

Strasfeld L, Chou S. Antiviral drug resistance: mechanisms and clinical implications. Infect Dis Clin North Am. 2010;24(2):413–37. https://doi.org/10.1016/j.idc.2010.01.001.

Strode GK, et al. Yellow fever. New York, Toronto & London: McGraw-Hill Book Co. Inc.; 1951.

Sukhralia S, Verma M, Gopirajan S, Dhanaraj PS, Lal R, Mehla N, Kant CR. From dengue to Zika: the wide spread of mosquito-borne arboviruses. Eur J Clin Microbiol Infect Dis. 2019;38 (1):3–14. https://doi.org/10.1007/s10096-018-3375-7.

Swaminathan G, Thoryk EA, Cox KS, Smith JS, Wolf JJ, Gindy ME, Casimiro DR, Bett AJ. A Tetravalent sub-unit dengue vaccine formulated with ionizable cationic lipid nanoparticle induces significant immune responses in rodents and non-human primates. Sci Rep. 2016;6(1):34215. https://doi.org/10.1038/srep34215.

Tamura K, Stecher G, Peterson D, Filipski A, Kumar S. MEGA6: molecular evolutionary genetics analysis version 6.0. Mol Biol Evol. 2013;30(12):2725–9. https://doi.org/10.1093/molbev/mst197.

Ternovoi VA, Gladysheva AV, Ponomareva EP, Mikryukova TP, Protopopova EV, Shvalov AN, Konovalova SN, Chausov EV, Loktev VB. Variability in the 3′ untranslated regions of the genomes of the different tick-borne encephalitis virus subtypes. Virus Genes. 2019;55(4):448–57. https://doi.org/10.1007/s11262-019-01672-0.

Thomas SJ, Martinez LJ, Endy TP. Flaviviruses: yellow fever, Japanese B, West Nile, and Others. In: Kaslow RA, Stanberry LR, Le Duc JW, editors. Viral infections of humans: epidemiology and control. Boston, MA: Springer US;2014. p. 383–415.

Tippmann HF. Analysis for free: comparing programs for sequence analysis. Brief Bioinform. 2004;5(1):82–7. https://doi.org/10.1093/bib/5.1.82.

Ulbert S. West Nile virus vaccines—current situation and future directions. Hum Vaccin Immunother. 2019;15(10):2337–42. https://doi.org/10.1080/21645515.2019.1621149.

Valarcher J, Hägglund S, Juremalm M, Blomqvist G, Renström L, Zohari S, Leijon M, Chirico J. Tick-borne encephalitis. Revue Scientifique Et Technique (International Office of Epizootics). 2015;34:453–66. https://doi.org/10.20506/rst.34.2.2371.

Valderrama A, Díaz Y, López-Vergès S. Interaction of Flavivirus with their mosquito vectors and their impact on the human health in the Americas. (1090–2104 (Electronic)). 2017.

Verma M, Bhatnagar S, Kumari K, Mittal N, Sukhralia S, Gopirajan At S, Dhanaraj PS, Lal R. Highly conserved epitopes of DENV structural and non-structural proteins: candidates for universal vaccine targets. Gene. 2019;695:18–25. https://doi.org/10.1016/j.gene.2019.02.001.

Villordo SM, Gamarnik AV. Genome cyclization as strategy for flavivirus RNA replication. Virus Res. 2009;139(2):230–9. https://doi.org/10.1016/j.virusres.2008.07.016.

Waggoner JJ, Rojas A, Pinsky BA. Yellow fever virus: diagnostics for a persistent arboviral threat. J Clin Microbiol. 2018; 56(10):e00827–18. https://doi.org/10.1128/JCM.00827-18.

Wang CC, Hsu YC, Wu HC, Wu HN. Insights into the coordinated interplay of the sHP hairpin and its co-existing and mutually-exclusive dengue virus terminal RNA elements for viral replication. Virology. 2017;505:56–70. https://doi.org/10.1016/j.virol.2017.02.007.

Weaver SC, Barrett AD. Transmission cycles, host range, evolution and emergence of arboviral disease. Nat Rev Microbiol. 2004;2(10):789–801. https://doi.org/10.1038/nrmicro1006.

Weissenbock H, Hubalek Z, Bakonyi T, Nowotny N. Zoonotic mosquito-borne flaviviruses: worldwide presence of agents with proven pathogenicity and potential candidates of future emerging diseases. Vet Microbiol. 2010;140(3–4):271–80. https://doi.org/10.1016/j.vetmic.2009.08.025.

Wong SH, Jassey A, Wang JY, Wang W-C, Liu C-H, Lin L-T. Virus-like particle systems for vaccine development against viruses in the flaviviridae family. Vaccines. 2019; 7(4). https://doi.org/10.3390/vaccines7040123.

Woods CW, Sanchez AM, Swamy GK, McClain MT, Harrington L, Freeman D, Poore EA, Slifka DK, Poer DeRaad DE, Amanna IJ, Slifka MK, Cai S, Shahamatdar V, Wierzbicki MR, Amegashie C, Walter EB. An observer blinded, randomized, placebo-controlled, phase I dose escalation trial to evaluate the safety and immunogenicity of an inactivated West Nile virus vaccine, HydroVax-001, in healthy adults. Vaccine. 2019;37(30):4222–30. https://doi.org/10.1016/j.vaccine.2018.12.026.

Xiang J, McLinden JH, Kaufman TM, Mohr EL, Bhattarai N, Chang Q, Stapleton JT. Characterization of a peptide domain within the GB virus C envelope glycoprotein (E2) that inhibits HIV replication. Virology. 2012;430(1):53–62. https://doi.org/10.1016/j.virol.2012.04.019.

Xie X, Gayen S, Kang C, Yuan Z, Shi PY. Membrane topology and function of dengue virus NS2A protein. J Virol. 2013;87(8):4609–22. https://doi.org/10.1128/jvi.02424-12.

Xu J, Shi PY, Li H, Zhou J. Broad spectrum antiviral agent niclosamide and its therapeutic potential. ACS Infect Dis. 2020;6(5):909–15. https://doi.org/10.1021/acsinfecdis.0c00052.

Yap BK, Lee C-Y, Choi SB, Kamarulzaman EE, Hariono M, Wahab HA. Silico identification of novel inhibitors. In: Ranganathan S, Gribskov M, Nakai K, Schönbach C, editors. Encyclopedia of bioinformatics and computational biology. Oxford: Academic Press; 2019. p. 761–79.

Zakaria MK, Carletti T, Marcello A. Cellular targets for the treatment of flavivirus infections. Front Cell Infect Microbiol. 2018;8:398. https://doi.org/10.3389/fcimb.2018.00398.

Zhang X, Jia R, Shen H, Wang M, Yin Z, Cheng A. Structures and functions of the envelope glycoprotein in flavivirus infections. Viruses. 2017;9(11):338. https://doi.org/10.3390/v9110338.

Zheng X, Chen H, Wang R, Fan D, Feng K, Gao N, An J. Effective protection induced by a monovalent DNA vaccine against Dengue Virus (DV) serotype 1 and a bivalent DNA vaccine against DV1 and DV2 in mice. Front Cell Infect Microbiol. 2017;7:175. https://doi.org/10.3389/fcimb.2017.00175.

Zou J, Xie X, Wang QY, Dong H, Lee MY, Kang C, Yuan Z, Shi PY. Characterization of dengue virus NS4A and NS4B protein interaction. J Virol. 2015;89(7):3455–70. https://doi.org/10.1128/jvi.03453-14.

HIV-HBV Viruses

HIV-HBV Co-infection, Clinical Concerns

Ryan D. Heath, Ali Syed, Suha Abu Khalaf, and Veysel Tahan

Abstract Both Human Immunodeficiency Virus and Hepatitis B virus result in chronic disease states resulting in clinical concerns both acute and chronic. There is no cure for either virus and both require chronic medication which can present challenges in terms of disease surveillance and management of decompensation. Even in the presence of appropriate antiviral therapy, mortality remains high in the HIV-HBV co-infected population with accelerated hepatic fibrosis and significantly increased incidence of hepatocellular cancer. While multiple risk factors and possible etiologies have been uncovered, the relative contribution of each has yet to be adequately determined, though further elucidation of these risks may present novel approaches for therapy. That stated, as discussed, antiviral therapy comes with its inherent risks in terms of both side effects of medication as well as its interactions. This review focuses to review the prevalence of co-infection with these viruses and the resultant special clinical concerns in this population.

Keywords Human Immunodeficiency virus · Hepatitis B virus · Co-infection · Hepatitis · Clinic

R. D. Heath
Fellow, Division of Gastroenterology and Hepatology, Department of Internal Medicine, University of Missouri, Columbia, MO, USA

A. Syed · S. A. Khalaf
Department of Internal Medicine, University of Missouri-Columbia, Columbia, MO, USA

V. Tahan (✉)
Associate Professor of Clinical Medicine, Division of Gastroenterology and Hepatology, Department of Internal Medicine, University of Missouri, Columbia, MO, USA
e-mail: tahanv@health.missouri.edu

S. I. Ahmad (ed.), *Human Viruses: Diseases, Treatments and Vaccines*, https://doi.org/10.1007/978-3-030-71165-8_20

1 Introduction

Human Immunodeficiency Virus (HIV), once a poorly understood and largely fatal disease, has since transformed into a treatable chronic condition with a potential for a normal life expectancy (Marcus et al. 2016). Significant advancements in therapies have increased life expectancy and dramatically reduced HIV-related mortality and morbidity (Maartens et al. 2014). As a result there is an increased prevalence of comorbidities which are observed in this patient population (Freiberg et al. 2013; Smith et al. 2014a). A common and important clinical scenario is co-infection with hepatitis B virus (HBV). When antiretroviral therapy (ART) was first introduced, there was conflicting data concerning the impact of HBV on the course of HIV progression. Early studies demonstrated increased rates of HIV progression to acquired immune deficiency syndrome (AIDS) in HBV co-infection Eskild et al. 1992). Other studies did not show any change in HIV progression or mortality (Gilson et al. 1997). More recent studies did not find any significant correlation between HBV co-infection and progression of HIV to AIDS or HIV suppression or CD4 T lymphocyte (CD4) cell responses following initiation of antiretroviral therapy (ART) or HIV-related death (Konopnicki et al. 2005; Hoffmann et al. 2009; Law et al. 2004). However, it is important to consider the toxicities of antiretroviral drugs which include several liver-associated complications. These can lead to flare in HBV activity after initiation or discontinuation of therapy which can ultimately affect the treatment of HIV in these individuals (Bellini et al. 2009; Law et al. 2004; Wit et al. 2002).

One of the main challenges in the management of both HIV and HBV is that antiviral treatment must be continued lifelong as both viruses have forms that can persist lifelong despite antiviral therapy. Research to find a cure for both HIV and HBV is being actively pursued, but current treatment strategies do not allow for complete viral eradication for co infected individuals. In this review we will focus on the prevalence of HIV and HBV coinfection as well as common complications and their relative risk in co-infected patients.

2 Prevalence

Approximately, 37 million people are infected with HIV globally and of that 5–20% are also co-infected with HBV (). Co-infection is common in the United States with an observed 70–90% of patients with HIV also having evidence of HBV infection, either active or past (Rodríguez-Méndez et al. 2000; Scharschmidt et al. 1992). Prevalence of chronic HBV varies geographically and is affected by multiple factors including mode of acquisition and age at the time of infection. In Western Europe and the United States, HBV is often contracted via sexual contact or IV drug use in adolescence or adulthood. Men who have sex with men and injection drug users experience the highest rates of coinfection with HIV and HBV in the

United States. Although immunocompetent individuals who acquire HBV in adulthood show a spontaneous clearance rate greater than 90%; those who are infected with HIV are half as likely to clear it when compared to their uninfected counterparts. As a result, 5–10% of HIV-infected individuals experience chronic HBV infection when exposed to HBV, a rate 10 times greater than that of the general population (Bodsworth et al. 1989; Alter 2006). In Asia vertical transmission is the most common way HBV in contracted while in sub-Saharan Africa early childhood exposure is. Overall, these countries have a higher prevalence of HBV and as a result experience a higher rate of coinfection with HIV and HBV which is estimated to be around 20–30% (Hoffmann and Thio 2007; Uneke et al. 2005; Askari et al. 2014).

Please see Table 1 for a review of HIV-HBV co-infection percentages complied through literature search.

Table 1 HIV-HBV co-infection published data globally

Continent	Country	Patient population/ City	Combined Sample size of HIV + patients	Percent of HBV/HIV co-infection	References #
Europe	Combined	Pregnant women from Western and Central Europe	1050	30/1050 (2.9%)	Landes et al. (2008)
	France	Nancy	383	172/383 (44.9%)	Bloquel et al. (2014)
		Migrants from sub-Saharan Africa, French Indies and French Guiana	3	0/3 (0%)	Aparicio et al. (2012)
		Multicenter	111	6/111 (5.4%)	Piroth et al. (2008)
	Greece	Meta-analysis	1729	104/1729 (6%)	Nikolopoulos et al. (2009)
	Italy	Multicenter	1593	27/1593 (1.7%)	Morsica et al. (2009)
	Switzerland	Multicenter	9053	996/9053 (11%)	Weber et al. (2012)
Asia	China	Multicenter	1944	161/1944 (8.3%)	Xie et al. (2016)
	India	South India	500	45/500 (9%)	Saravanan et al. (2007)
		North India	837	61/837 (7.3%)	Gupta and Singh (2010)
		Eastern India	874	73/874 (8.4%)	Pal et al. (2012)
		Sexually acquired HIV	58	17/58 (29.3%)	Rai et al. (2007)
	Indonesia	Yogyakarta	126	4/126 (3.2%)	Anggorowati et al. (2012)
	Iran	Injecting drug users	96	70/96 (72.9%)	Rahimi-Movaghar et al. (2009)

(continued)

Table 1 (continued)

Continent	Country	Patient population/ City	Combined Sample size of HIV + patients	Percent of HBV/HIV co-infection	References #
		Multicenter	92	23/92 (25%)	Ramezani et al. (2009)
		Mazandaran province	188	20/188 (10.6%)	Babamahmoodi et al. (2012)
		Northeastern Iran	168	3/168 (1.8%)	Moradi et al. (2011)
	Japan	Japanese men having sex with men	394	31/394 (7.9%)	Fujisaki et al. (2011)
	Thailand	Government hospitals	755	83/755 (11%)	Tsuchiya et al. (2012)
Australia	Australia	Multicenter	2086	1605/2086 (76.9%)	Lincoln et al. (2003)
Americas	Brazil	Belém	406	207/406 (51%)	Monteiro et al. (2004)
		São Paulo	401	164/401 (40.9%)	Souza et al. (2004)
		Cuiabá	1000	400/1000 (40%)	de Almeida Pereira et al. (2006)
	Canada	Ontario	1223	128/1223 (10.5%)	Gillis et al. (2012)
	Cuba	Multicenter	325	99/325 (30.5%)	Marite et al. (2011)
	USA	Atlanta veterans affairs hospital	2818	1685/2818 (59.8%)	Osborn et al. (2007)
		New York City	5639	252/5638 (4.5%)	Kim et al. (2008)
Africa	Burkina Faso	Pregnant women of Ouagadougou	115	14/115 (12.2%)	Ilboudo et al. (2010)
	Cameron	Yaoundé	169	14/169 (8.3%)	Mbougua et al. (2010)
	Mali	Blood donors in Bamako	522	131/522 (25.1%)	Tounkara et al. (2009)
	Nigeria	Pregnant women in Enugu	401	26/401 (6.5%)	Okeke et al. (2012)
		Lagos	102	29/102 (28.4%)	Bologna et al. (2012)
		Ibadan	1779	212/1779 (11.9%)	Otegbayo et al. (2008)
		Port Harcourt	342	33/342 (9.6%)	Ejele et al. (2004)
	South Africa	Multicenter	200	20/200 (10%)	Mayaphi et al. (2012)
		Johannesburg	100	35/100 (35%)	Lodenyo et al. (2000)

3 Complications in HIV-HBV Co-Infection

It is well demonstrated that mortality amongst individuals with HIV-HBV co-infection is higher relative to either population infected with only one of the aforementioned viruses (Chun et al. 2014; Pinchoff et al. 2016; Rajbhandari et al. 2016; Crowell et al. 2015; Coffin et al. 2013; Morlat et al. 2014; Weber et al. 2012). Cirrhosis is noted to be both prevalent and responsible for increased mortality in HBV patients co-infected with HIV (Thio et al. 2013; Zoufaly et al. 2012; Huang et al. 2016). Likely consequently, rates of hepatocellular carcinoma (HCC) and need for hospital utilization is noted to be increased with co-infection (Pinchoff et al. 2016; Rajbhandari et al. 2016; Coffin et al. 2013; Morlat et al. 2014; Thio et al. 2013; Prussing et al. 2015; Chun et al. 2012; Vinikoor et al. 2017; Ruiz-Artacho et al. 2013; Audsley et al. 2016; Boyd et al. 2014). Increased mortality in HBV patients has been documented with correlate with lower T cell counts; however, T cell counts appear to recover in patients receiving appropriate HBV therapy (Audsley et al. 2016; Hamers et al. 2013; Walsh and Locarnini 2012). Multiple cells in the liver appear conducive to HIV replication, including Kupffer cells, stellate ells, and hepatocytes (Kong et al. 2012; Iser et al. 2010; Housset et al. 1993; Cao et al. 1992; Penton and Blackard 2014).

Further suggesting potentiating of liver disease in the setting of co-infection, while on appropriate antiviral medication, it has been demonstrated that $\sim 10\%$ of individuals in this group have detectable HBV DNA (Ryom et al. 2016). Remarkably, relative loss of HBsAg—inhibiting adaptive immunity and production of antiviral antibodies—has been associated with co-infection (Hamers et al. 2013; Walsh and Locarnini 2012; Calisti et al. 2015). The magnitude of this relative HBsAg loss appears greater in patients with lower CD4 T cell prior to initiation of dual antiviral therapy as well as in cases of significant increase in CD4 T cell counts prior to initiation of the same therapy (Walsh and Locarnini 2012; Maylin et al. 2012; Zoutendijk et al. 2012; Strassl et al. 2014). This trend has been documented to be present even when the HIV is being appropriately treated (Lucifora and Protzer 2016; Thio et al. 2002). This remains true even in the setting of appropriate antiviral therapy (Huang and Núñez 2015). Despite multiple studies sequencing for mutations in HBV which may contribute to increased resistance, none have been found to explain this phenomenon (Audsley et al. 2009; Seto et al. 2013; Yang et al. 2016; Tseng et al. 2012). The reason for this increased adaptive response has been hypothetically linked to elevated levels of IL-18; however, further research is required to this relationship (Boyd et al. 2015). Worth noting is that beyond low T cell counts, low HBsAg at baseline before treatment is also associated with higher incidence of antigen loss after initiation of dual antiviral therapy (Hamers et al. 2013; Maylin et al. 2012; Tan et al. 2016). On the subject of initiation of antiviral therapy, the patient should be monitored for evidence of immune reconstitution disease. In particular, in the HIV-HBV co-infected patient, reconstitution is associated with a significant flare of transaminases. A prospective cohort study demonstrated 22% of patients developed a hepatitis flare in the setting of dual antiviral therapy suggestive of immune reconstitution (Mitsumoto et al. 2015). A significant concern in this co-infected population is the tendency to experience frequent flares of hepatic transaminases which can occur with immune reconstitution inflammatory syndrome secondary to treatment with ART. This can lead to interruption of HIV/HBV treatment which in turn can lead to development of resistance to HIV/HBV treatment (Jansen et al. 2015).

Despite a possibly expected improvement in complications of chronic HBV given this relatively increased loss of HBsAg, a striking consequence of chronic infection in this co-infection group is an increased risk of HCC. This risk has been demonstrated to be five to six times greater than in individuals with HIV mono-infection, even when on antiviral therapy (Chun et al. 2012; Gjaerde et al. 2016; Papatheodoridis et al. 2010). While lower CD4 T cell counts and higher baseline HBV DNA are associated with a relatively increased risk within this cohort, the persistent relative elevated risk in the setting of appropriate ART suggests further research is warranted to evaluate this relationship (Levrero and Zucman-Rossi 2016; Li et al. 2016). Given that alcohol abuse and steatosis is noted to be relatively increased in this population, it may be that potentiation of chronic liver disease due to these factors is adequate to explain this phenomenon; however, this studies have not adequately quantified the effect of such factors (Liao et al. 2017). This may explain previous studies noting an increased profibrotic activity in patients with HIV (Lester et al. 2009; Debes et al. 2016). Animal studies have demonstrated persistent HIV within Kupffer cells on appropriate antiviral therapy (Salloum et al. 2016; Shi et al. 2016). Of interest, heretofore unrecognized mutations in the HBV genome associated with more rapid onset of HCC may be contributing to this increased risk. One well documented HBV basal core double mutation, T1762/A1764, in genotypes B and C is a documented risk factor HCC, however, studies do not consistently demonstrate an increased prevalence of this mutation in HIV-HBV co-infected cohorts (Revill et al. 2007; Yu et al. 2016; Pollicino et al. 2014; Lim et al. 2017). Another mutation associated with increased HCC risk, pre-S deletions, have been demonstrated to have a higher prevalence in this HIV-HBV co-infected populations (Revill et al. 2007; Pollicino et al. 2014). Even with this positive association, studies defining and assessing genotypes and mutations in relation to other lifestyle factors are necessary to elucidate the relative effect of each.

4 Treatment

There is currently no cure available for either virus; both also require lifelong medication for replication suppression and regular monitoring of viral loads as well as potential iatrogenic effects of antiviral medication. Fortunately, there are multiple antiviral medications with activity against both HIV and HBV. Lamivudine, emtricitabine, and tenofovir are commonly prescribed drugs with dual antiviral activity. Tenofovir in particular is popular choice given significant improvement in viral replication, low viral resistance, and improvement in hepatic fibrosis (Khamduang et al. 2012; Shiels and Engels 2017; Crowell et al. 2014; Smith et al. 2014b). As previously noted, however, given the ongoing relative increased risk of complications such as HCC, the current popular regimens appear inadequate.

Novel therapies are in development for both viruses; however, in many cases the effect of the medication on HIV is unknown. One strategy has been medication designed to decrease HBsAg production, either by inhibiting its release or inhibition of viral protein production (Callebaut et al. 2015; Gallant et al. 2016). While in stage 2 clinical trials with some promise regarding HBV infection, any effect on HIV is unknown and would require further investigation to determine any positive outcome as well as any potential medication interaction. An area of increased promise is targeting covalently closed circular DNA (cccDNA), located in the nucleus of HBV-infected hepatocytes and considered responsible for persistent viremia in appropriately treated patients (Audsley et al. 2009). Such mechanisms include sulfonamides designed to inhibit rcDNA to cccDNA conversion as well as endonucleases which function to cleave DNA (Baril et al. 2016; Revill et al. 2016; Petersen et al. 2015; Chen et al. 2016). These drugs are all in the preclinical trial phase, however, may have some effect on HIV replication as well.

5 Conclusion

In summary, even in the presence of appropriate antiviral therapy, mortality remains high in the HIV-HBV co-infected population with accelerated hepatic fibrosis and significantly increased incidence of HCC. While multiple risk factors and possible etiologies have been uncovered, the relative contribution of each has yet to be adequately determined, though further elucidation of these risks may present novel approaches for therapy. That stated, as discussed, antiviral therapy comes with its inherent risks in terms of both side effects of medication as well as its interactions. Novel therapy may present opportunities for improved control of HBV, however, the effect of this medication on HIV and any prescribed ART must be fully evaluated before committing an already at risk population to increased possible harm.

Author Contributions: All authors equally contributed to this paper with conception and design of the study, literature review and analysis, drafting and critical revision and editing, and final approval of the final version.**Conflict-of-interest statement:** No potential conflicts of interest. No financial support.

References

Alter MJ. Epidemiology of viral hepatitis and HIV co-infection. J Hepatol. 2006;44(1 Suppl):S6-9.

Anggorowati N, Yano Y, Heriyanto DS, Rinonce HT, Utsumi T, Mulya DP, et al. Clinical and virological characteristics of hepatitis B or C virus co-infection with HIV in Indonesian patients. J Med Virol. 2012;84(6):857–65.

Aparicio C, Mourez T, Simoneau G, Magnier JD, Galichon B, Plaisance P, et al. Proposal of HIV, HBV and HCV targeted screening: short period feasibility study in a freeaccess outpatient medical structure. Presse Med. 2012;41(10):e517-523.

Askari A, et al. Prevalence of hepatitis B co-infection among HIV positive patients: narrative review article. Iran J Public Health. 2014;43(6):705–12.

Audsley J, Arrifin N, Yuen LK, Ayres A, Crowe SM, Bartholomeusz A, et al. Prolonged use of tenofovir in HIV/hepatitis B virus (HBV)-coinfected individuals does not lead to HBV polymerase mutations and is associated with persistence of lamivudine HBV polymerase mutations. HIV Med. 2009;10(4):229–35.

Audsley J, Robson C, Aitchison S, Matthews GV, Iser D, Sasadeusz J, et al. Liver fibrosis regression measured by transient elastography in Human Immunodeficiency Virus (HIV)-Hepatitis B Virus (HBV)-coinfected individuals on long-term HBV-active combination antiretroviral therapy. Open Forum Infect Dis. 2016;3(1) ofw035.

Audsley J, Bent SJ, Littlejohn M, Avihingsanon A, Matthews G, Bowden S, et al. Effects of long-term tenofovir-based cART in HIV-HBV co-infection on persistent HBV viremia and the role of HBV quasispecies diversity. Aids. 2016.

Babamahmoodi F, HeidariGorji MA, Mahdi Nasehi M, Delavarian L. The prevalence rate of hepatitis B and hepatitis C co-infection in HIV positive patients in Mazandaran province, Iran. Med GlasLjekkomoreZenickodobojkantona, 2012;9(2):299–303.

Baril JG, Angel JB, Gill MJ, Gathe J, Cahn P, van Wyk J, et al. Dual therapy treatment strategies for the management of patients infected with HIV: a systematic review of current evidence in ARV-Naive or ARV-experienced, virologically suppressed patients. PLoS One. 2016;11(2): e0148231.

Bellini C, Keiser O, Chave JP, et al. Liver enzyme elevation after lamivudine withdrawal in HIV-hepatitis B virus co-infected patients: the Swiss HIV Cohort Study. HIV Med. 2009;10 (1):12–8. https://www.ncbi.nlm.nih.gov/pubmed/18795964.

Bloquel B, Jeulin H, Burty C, Letranchant L, Rabaud C, Venard V. Occult hepatitis B infection in patients infected with HIV: report of two cases of hepatitis B reactivation and prevalence in a hospital cohort. J Med Virol. 2014;82(2):206–12.

Bodsworth N, Donovan B, Nightingale BN. The effect of concurrent human immunodeficiency virus infection on chronic hepatitis B: a study of 150 homosexual men. J Infect Dis. 1989;160 (4):577–82.

Bologna TM, Emmanuel S, Ojerinde EF. HIV, Hepatitis B and C viruses' co-infection among patients in a Nigerian tertiary hospital. Pan Afr Med J. 2012;12(100):100.

Boyd A, Gozlan J, Maylin S, Delaugerre C, Peytavin G, Girard PM, et al. Persistent viremia in human immunodeficiency virus/hepatitis B coinfected patients undergoing long-term tenofovir: virological and clinical implications. Hepatology (Baltimore, MD). 2014;60(2):497–507.

Boyd A, Maylin S, Gozlan J, Delaugerre C, Simon F, Girard PM, et al. Use of hepatitis B surface and "e" antigen quantification during extensive treatment with tenofovir in patients co-infected with HIV-HBV. Liver Int. 2015;35(3):795–804.

Calisti G, Muhindo R, Boum Y, 2, Wilson LA, Foster GM, Geretti AM, et al. Epidemiology of HBV infection in a cohort of Ugandan HIV-infected patients and rate and pattern of lamivudine-resistant HBV infection in patients receiving antiretroviral therapy. Trans R Soc Trop Med Hyg. 2015;109(11):723–9.

Callebaut C, Stepan G, Tian Y, Miller MD. In vitro virology profile of tenofoviralafenamide, a novel oral prodrug of tenofovir with improved antiviral activity compared to that of tenofovirdisoproxil fumarate. Antimicrob Agents Chemother. 2015;59(10):5909–16.

Cao YZ, Dieterich D, Thomas PA, Huang YX, Mirabile M, Ho DD. Identification and quantitation of HIV-1 in the liver of patients with AIDS. AIDS. 1992;6(1):65–70.

Chen M, Wong WW, Law MG, Kiertiburanakul S, Yunihastuti E, Merati TP, et al. Hepatitis B and C co-infection in HIV patients from the TREAT Asia HIV observational database: analysis of risk factors and survival. PLoS One. 2016;11(3):e0150512.

Chun HM, Roediger MP, Hullsiek KH, Thio CL, Agan BK, Bradley WP, et al. Hepatitis B virus coinfection negatively impacts HIV outcomes in HIV seroconverters. J Infect Dis. 2012;205 (2):185–93.

Chun HM, Mesner O, Thio CL, Bebu I, Macalino G, Agan BK, et al. HIV outcomes in hepatitis B virus coinfected individuals on HAART. J Acquir Immune Defic Syndr. 2014;66(2):197–205.

Coffin CS, Osiowy C, Myers RP, Gill MJ. Virology and clinical sequelae of long-term antiviral therapy in a North American cohort of hepatitis B virus (HBV)/human immunodeficiency virus type 1 (HIV-1) co-infected patients. J Clin Virol : The Official Publ Pan Am Soc Clin Virol. 2013;57(2):103–8.

Crowell TA, Gebo KA, Balagopal A, Fleishman JA, Agwu AL, Berry SA, et al. Impact of hepatitis coinfection on hospitalization rates and causes in a multicenter cohort of persons living with HIV. J Acquir Immune Defic Syndr. 2014;65(4):429–37.

Crowell TA, Berry SA, Fleishman JA, LaRue RW, Korthuis PT, Nijhawan AE, et al. Impact of hepatitis coinfection on healthcare utilization among persons living with HIV. J Acquir Immune Defic Syndr. 2015;68(4):425–31.

de Almeida Pereira RA, Mussi AD, de Azevedo e Silva VC, Souto FJ, . Hepatitis B Virus infection in HIV-positive population in Brazil: results of a survey in the state of Mato Grosso and a comparative analysis with other regions of Brazil. BMC Infect Dis. 2006;6(34):34.

Debes JD, Bohjanen PR, Boonstra A. Mechanisms of accelerated liver fibrosis progression during HIV infection. J Clin Transl Hepatol. 2016;4(4):328–35.

Delfino CM, Berini C, Eirin ME, Malan R, Pedrozo W, Krupp R, et al. New natural variants of hepatitis B virus among Amerindians from Argentina with mainly occult infections. J ClinVirol. 2012;54(2):174–9.

Ejele OA, Nwauche CA, Erhabor O. The prevalence of hepatitis B surface antigenaemia in HIV positive patients in the Niger Delta Nigeria. Niger J Med. 2004;13(2):175–9.

Eskild A, Magnus P, Petersen G, et al. Hepatitis B antibodies in HIV-infected homosexual men are associated with more rapid progression to AIDS. AIDS. 1992;6(6):571–4.

Freiberg MS, Chang C-CH, Keller LH, et al. HIV infection and the risk of acute myocardial infarction. JAMA Intern Med. 2013;173:614.

Fujisaki S, Yokomaku Y, Shiino T, Koibuchi T, Hattori J, Ibe S, et al. Outbreak of infections by hepatitis B virus genotype A and transmission of genetic drug resistance in patients coinfected with HIV-1 in Japan. J ClinMicrobiol. 2011;49(3):1017–24.

Gallant J, Brunetta J, Crofoot G, Benson P, Mills A, Brinson C, et al. Brief report: efficacy and safety of switching to a single-tablet regimen of Elvitegravir/Cobicistat/Emtricitabine/ TenofovirAlafenamide in HIV-1/Hepatitis B-coinfected adults. J Acquir Immune Defic Syndr. 2016;73(3):294–8.

Gillis J, Cooper C, Rourke S, Rueda S, O'Brien K, Collins E, et al. Impact of hepatitis B and C co-infection on healt-hrelated quality of life in HIV positive individuals. Qual Life Res. 2012;17:17.

Gilson RJ, Hawkins AE, Beecham MR, et al. Interactions between HIV and hepatitis B virus in homosexual men: effects on the natural history of infection. AIDS. 1997;11(5):597–606.

Gjaerde LI, Shepherd L, Jablonowska E, Lazzarin A, Rougemont M, Darling K, et al. Trends in incidences and risk factors for hepatocellular carcinoma and other liver events in HIV and hepatitis C virus-coinfected individuals from 2001 to 2014: a multicohort study. Clin Infect Dis: Official Publ Infect Dis Soc Am. 2016;63(6):821–9.

Gouws E, Williams BG, Sheppard HW, Enge B, Karim SA. High incidence of HIV-1 in South Africa using a standardized algorithm for recent HIV seroconversion. J Acquir Immune Defic Syndr. 2002;29(5):531.

Gupta S, Singh S. Occult hepatitis B virus infection in ART-naive HIV-infected patients seen at a tertiary care centre in north India. BMC Infect Dis. 2010;10(53):53.

Hamers RL, Zaaijer HL, Wallis CL, Siwale M, Ive P, Botes ME, et al. HIV-HBV coinfection in Southern Africa and the effect of lamivudine- versus tenofovir-containing cART on HBV outcomes. J Acquir Immune Defic Syndr. 2013;64(2):174–82.

Hoffmann CJ, Thio CL. Clinical implications of HIV and hepatitis B co-infection in Asia and Africa. Lancet Infect Dis. 2007;7(6):402–9.

Hoffmann CJ, Seaberg EC, Young S, et al. Hepatitis B and long-term HIV outcomes in coinfected HAART recipients. AIDS. 2009;23(14):1881–9. https://www.ncbi.nlm.nih.gov/pubmed/19550291.

Housset C, Lamas E, Courgnaud V, Boucher O, Girard PM, Marche C, et al. Presence of HIV-1 in human parenchymal and non-parenchymal liver cells in vivo. J Hepatol. 1993;19(2):252–8.

Huang AJ, Núñez M. Outcomes in HIV/HBV-coinfected patients in the tenofovir era are greatly affected by immune suppression. J InternatAssocProv AIDS Care (JIAPAC). 2015;14(4):360–8.

Huang SM, Cai WP, Hu FY, Lan Y, Liao BL, Chen YP, et al. Epidemiological and clinical characteristics of hepatitis B virus in HIV-infected patients in Guangdong, China. Int J STD AIDS. 2016;27(10):890–7.

Ilboudo D, Simpore J, Ouermi D, Bisseye C, Sagna T, Odolini S, et al. Towards the complete eradication of mother-to-child HIV/HBV co-infection at Saint Camille Medical Centre in Burkina Faso, Africa. Braz J Infect Dis. 2010;14(3):219–24.

Iser DM, Warner N, Revill PA, Solomon A, Wightman F, Saleh S, et al. Coinfection of hepatic cell lines with human immunodeficiency virus and hepatitis B virus leads to an increase in intracellular hepatitis B surface antigen. J Virol. 2010;84(12):5860–7.

Jansen K, Thamm M, Bock CT, Scheufele R, Kucherer C, Muenstermann D, et al. High prevalence and high incidence of coinfection with hepatitis B, hepatitis C, and syphilis and low rate of effective vaccination against hepatitis B in HIV-positive men who have sex with men with known date of HIV seroconversion in Germany. PLoS One. 2015;10(11):e0142515.

Khamduang W, Gaudy-Graffin C, Ngo-Giang-Huong N, Jourdain G, Moreau A, Luekamlung N, et al. Long-term hepatitis B virus (HBV) response to lamivudine-containing highly active antiretroviral therapy in HIV-HBV co-infected patients in Thailand. PLoS One. 2012;7(7): e42184.

Kim JH, Psevdos G, Suh J, Sharp VL. Coinfection of hepatitis B and hepatitis C virus in human immunodeficiency virus-infected patients in New York City, United States. World J Gastroenterol. 2008;14(43):6689–93.

Kong L, Cardona Maya W, Moreno-Fernandez ME, Ma G, Shata MT, Sherman KE, et al. Low-level HIV infection of hepatocytes. Virol J. 2012;9:157.

Konopnicki D, Mocroft A, de Wit S, et al. Hepatitis B and HIV: prevalence, AIDS progression, response to highly active antiretroviral therapy and increased mortality in the EuroSIDA cohort. AIDS. 2005;19(6):593–601. https://www.ncbi.nlm.nih.gov/pubmed/15802978.

Landes M, Newell ML, Barlow P, Fiore S, Malyuta R, Martinelli P, et al. Hepatitis B or hepatitis C co-infection in HIV-infected pregnant women in Europe. HIV Med. 2008;9(7):526–34.

Law WP, Dore GJ, Duncombe CJ, et al. Risk of severe hepatotoxicity associated with antiretroviral therapy in the HIV-NAT Cohort, Thailand, 1996–2001. AIDS. 2003;17(15):2191–9. https://www.ncbi.nlm.nih.gov/pubmed/14523276.

Law WP, Duncombe CJ, Mahanontharit A, et al. Impact of viral hepatitis co-infection on response to antiretroviral therapy and HIV disease progression in the HIV-NAT cohort. AIDS. 2004;18(8):1169–77.

Lester RT, Yao XD, Ball TB, McKinnon LR, Omange WR, Kaul R, et al. HIV-1 RNA dysregulates the natural TLR response to subclinical endotoxemia in Kenyan female sex-workers. PLoS One. 2009;4(5):e5644.

Levrero M, Zucman-Rossi J. Mechanisms of HBV-induced hepatocellular carcinoma. J Hepatol. 2016;64(1 Suppl):S84-101.

Liao H, Li X, Liu Y, Xu Z, Huang P, Nian X, et al. Intergenotype recombinant analysis of full-length hepatitis B virus genomes from 516 Chinese patients with different illness categories. J Med Virol. 2017;89(1):139–45.

Lim YS, Lee YS, Gwak GY, Byun KS, Kim YJ, Choi J, et al. Monotherapy with TenofovirDisoproxil fumarate for multiple drug-resistant chronic hepatitis B: 3-year trial. Hepatology (Baltimore, MD) 2017.

Lincoln D, Petoumenos K, Dore GJ; Australian HIV observational database. HIV/HBV and HIV/HCV coinfection, and outcomes following highly active antiretroviral therapy. HIV Med. 2003;4(3):241–9.

Li KW, Kramvis A, Liang S, He X, Chen QY, Wang C, et al. Higher prevalence of cancer related mutations 1762T/1764A and PreS deletions in hepatitis B virus (HBV) isolated from HBV/HIV co-infected compared to HBV-mono-infected Chinese adults. Virus Res. 2016;227:88–95.

Lodenyo H, Schoub B, Ally R, Kairu S, Segal I. Hepatitis B and C virus infections and liver function in AIDS patients at Chris Hani Baragwanath Hospital, Johannesburg. East Afr Med J. 2000;77(1):13–5.

Lucifora J, Protzer U. Attacking hepatitis B virus cccDNA–the holy grail to hepatitis B cure. J Hepatol. 2016;64(1 Suppl):S41-48.

Maartens G, Celum C, Lewin SR. HIV infection: epidemiology, pathogenesis, treatment, and prevention. Lancet. 2014;384:258–71.

Marcus JL, Chao CR, Leyden WA, et al. Narrowing the gap in life expectancy between HIV-infected and HIV-uninfected individuals with access to care. JAIDS J Acquir Immune DeficSyndr. 2016;73:39–46.

Marite B, Montalvo MC, Rodriguez Lde L, Sariego S, Verdasquera D, Vincent M, et al. Occult hepatitis B in Cuban HIV patients. MEDICC Rev. 2011;13(2):32–7.

Mayaphi SH, Roussow TM, Masemola DP, Olorunju SA, Mphahlele MJ, Martin DJ. HBV/HIV co-infection: the dynamics of HBV in South African patients with AIDS. S Afr Med J. 2012;102(3 Pt 1):157–62.

Maylin S, Boyd A, Lavocat F, Gozlan J, Lascoux-Combe C, Miailhes P, et al. Kinetics of hepatitis B surface and envelope antigen and prediction of treatment response to tenofovir in antiretroviral-experienced HIV-hepatitis B virus-infected patients. AIDS. 2012;26(8):939–49.

Mbougua JB, Laurent C, Kouanfack C, Bourgeois A, Ciaffi L, Calmy A, et al. Hepatotoxicity and effectiveness of a Nevirapine-based antiretroviral therapy in HIV-infected patients with or without viral hepatitis B or C infection in Cameroon. BMC Public Health. 2010;10(105):105.

Mitsumoto F, Murata M, Ura K, Takayama K, Hiramine S, Shimizu M, et al. The kinetics of the hepatitis B surface antigen level after the initiation of antiretroviral therapy for hepatitis B virus and human immunodeficiency virus coinfected patients. J Infect Chemother: Official J Japan Soc Chemother. 2015;21(4):264–71.

Monteiro MR, do Nascimento MM, Passos AD, Figueiredo JF, . Soroepidemiological survey of hepatitis B virus among HIV/AIDS patients in Belem, Para–Brasil. Rev Soc Bras Med Trop. 2004;37(Suppl 2):27–32.

Moradi A, Khodabakhshi B, Sadeghipour M, Besharat S, Tabarraei A. Concurrent infections of hepatitis C and HIV in hepatitis B patients in the north-east of Iran. Trop Doct. 2011;41 (3):129–31.

Morlat P, Roussillon C, Henard S, Salmon D, Bonnet F, Cacoub P, et al. Causes of death among HIV-infected patients in France in 2010 (national survey): trends since 2000. AIDS. 2014;28 (8):1181–91.

Morsica G, Ancarani F, Bagaglio S, et al. Occult hepatitis B virus infection in a Cohort of HIV-positive patients: Correlation with hepatitis C virus coinfection, virological and immunological features. Infection. 2009;37:445. https://doi.org/10.1007/s15010-008-8194-9.

Nikolopoulos GK, Paraskevis D, Hatzitheodorou E, Moschidis Z, Sypsa V, Zavitsanos X, et al. Impact of hepatitis B virus infection on the progression of AIDS and mortality in HIV-infected individuals: a cohort study and meta-analysis. Clin Infect Dis. 2009;48(12):1763–71.

Nordenstedt H, White DL, El-Serag HB. The changing pattern of epidemiology in hepatocellular carcinoma. Dig Liver Dis. 2010;42:S206–14.

Okeke TC, Obi SN, Okezie OA, Ugwu EO, Akogu SP, Ocheni S, et al. Coinfection with hepatitis B and C viruses among HIV positive pregnant women in Enugu south east, Nigeria. Niger J Med. 2012;21(1):57–60.

Osborn MK, Guest JL, Rimland D. Hepatitis B virus and HIV co-infection: relationship of different serological patterns to survival and liver disease. HIV Med. 2007;8(5):271–9.

Otegbayo JA, Taiwo BO, Akingbola TS, Odaibo GN, Adedapo KS, Penugonda S, et al. Prevalence of hepatitis B and C seropositivity in a Nigerian cohort of HIV-infected patients. Ann Hepatol. 2008;7(2):152–6.

Pal A, Panigrahi R, Biswas A, Datta S, Sarkar N, Guha SK, et al. Influence of HIV-associated degree of immune suppression on molecular heterogeneity of hepatitis B virus among HIV coinfected patients. Virology. 2012;7(12):00550–8.

Papatheodoridis GV, Lampertico P, Manolakopoulos S, Lok A. Incidence of hepatocellular carcinoma in chronic hepatitis B patients receiving nucleos(t)ide therapy: a systematic review. J Hepatol. 2010;53(2):348–56.

Penton PK, Blackard JT. Analysis of HIV quasispecies suggests compartmentalization in the liver. AIDS Res Hum Retroviruses. 2014;30(4):394–402.

Petersen J, Heyne R, Mauss S, Schlaak J, Schiffelholz W, Eisenbach C, et al. Effectiveness and safety of tenofovirdisoproxil fumarate in chronic hepatitis B: A 3-year prospective field practice study in Germany. Dig Dis Sci. 2015.

Pinchoff J, Tran OC, Chen L, Bornschlegel K, Drobnik A, Kersanske L, et al. Impact of hepatitis B on mortality and specific causes of death in adults with and without HIV co-infection in NYC, 2000–2011. Epidemiol Infect. 2016:1–11.

Piroth L, Lafon ME, Binquet C, Bertillon P, Gervais A, Lootvoet E, et al. Occult hepatitis B in HIV-HCV coinfected patients. Scand J Infect Dis. 2008;40(10):835–9.

Pollicino T, Cacciola I, Saffioti F, Raimondo G. Hepatitis B virus PreS/S gene variants: pathobiology and clinical implications. J Hepatol. 2014;61(2):408–17.

Prussing C, Chan C, Pinchoff J, Kersanske L, Bornschlegel K, Balter S, et al. HIV and viral hepatitis co-infection in New York City, 2000–2010: prevalence and case characteristics. Epidemiol Infect. 2015;143(7):1408–16.

Rahimi-Movaghar A, Razaghi EM, Sahimi-Izadian E, Amin-Esmaeili M. HIV, hepatitis C virus, and hepatitis B virus co-infections among injecting drug users in Tehran, Iran. Int J Infect Dis. 2009;14(1):e28-33.

Rai RR, Mathur A, Mathur D, Udawat HP, Nepalia S, Nijhawan S. Prevalence of occult hepatitis B & C in HIV patients infected through sexual transmission. Trop Gastroenterol. 2007;28 (1):19–23.

Rajbhandari R, Jun T, Khalili H, Chung RT, Ananthakrishnan AN. HBV/HIV coinfection is associated with poorer outcomes in hospitalized patients with HBV or HIV. J Viral Hepatitis. 2016;23(10):820–9.

Ramezani A, Mohraz M, Aghakhani A, Banifazl M, Eslamifar A, Khadem-Sadegh A, et al. Frequency of isolated hepatitis B core antibody in HIV-hepatitis C virus coinfected individuals. Int J STD AIDS. 2009;20(5):336–8.

Revill PA, Littlejohn M, Ayres A, Yuen L, Colledge D, Bartholomeusz A, et al. Identification of a novel hepatitis B virus precore/core deletion mutant in HIV/hepatitis B virus co-infected individuals. AIDS. 2007;21(13):1701–10.

Revill P, Testoni B, Locarnini S, Zoulim F. Global strategies are required to cure and eliminate HBV infection. Nat Rev GastroenterolHepatol. 2016;13(4):239–48.

Rodríguez-Méndez ML, González-Quintela A, Aguilera A, et al. Prevalence, patterns, and course of past hepatitis B virus infection in intravenous drug users with HIV-1 infection. Am J Gastroenterol. 2000;95(5):1316–22.

Ruiz-Artacho P, Tellez-Molina MJ, Vergas-Garcia J, Altali-Alhames K, Estrada-Perez V, Fernandez-Cruz-Perez A. HIV/HBV coinfection: serological control and therapeutic strategies. Revistaclinicaespanola. 2013;213(6):285–90.

Ryom L, Lundgren JD, De Wit S, Kovari H, Reiss P, Law M, et al. Use of antiretroviral therapy and risk of end-stage liver disease and hepatocellular carcinoma in HIV-positive persons. AIDS. 2016;30(11):1731–43.

Salloum S, Holmes JA, Jindal R, Bale SS, Brisac C, Alatrakchi N, et al. Exposure to human immunodeficiency virus/hepatitis C virus in hepatic and stellate cell lines reveals cooperative profibrotic transcriptional activation between viruses and cell types. Hepatology (Baltimore, MD) 2016.

Saravanan S, Velu V, Kumarasamy N, Nandakumar S, Murugavel KG, Balakrishnan P, et al. Coinfection of hepatitis B and hepatitis C virus in HIV-infected patients in south India. World J Gastroenterol. 2007;13(37):5015–20.

Scharschmidt BF, Held MJ, Hollander HH, et al. Hepatitis B in patients with HIV infection: relationship to AIDS and patient survival. Ann Intern Med. 1992;117(10):837–8.

Seto WK, Wong DK, Fung J, Huang FY, Lai CL, Yuen MF. Reduction of hepatitis B surface antigen levels and hepatitis B surface antigen seroclearance in chronic hepatitis B patients receiving 10 years of nucleoside analogue therapy. Hepatology (Baltimore, MD). 2013;58 (3):923–31.

Shiels MS, Engels EA. Evolving epidemiology of HIV-associated malignancies. CurrOpin HIV AIDS. 2017;12(1):6–11.

Shi L, Qin E, Zhou J, Zhao J, Nie W, Jiang T, et al. HIV and HCV Co-culture promotes profibrogenic gene expression through an epimorphin-mediated ERK signaling pathway in hepatic stellate cells. PLoS One. 2016;11(6):e0158386.

Smith CJ, Ryom L, Weber R, et al. Trends in underlying causes of death in people with HIV from 1999 to 2011 (D:A:D): a multicohort collaboration. Lancet. 2014a;384:241–8.

Smith CJ, Ryom L, Weber R, Morlat P, Pradier C, Reiss P, et al. Trends in underlying causes of death in people with HIV from 1999 to 2011 (D:A:D): a multicohort collaboration. Lancet. 2014b;384(9939):241–8.

Souza MG, Passos AD, Machado AA, Figueiredo JF, Esmeraldino LE. HIV and hepatitis B virus co-infection: prevalence and risk factors. Rev Soc Bras Med Trop. 2004;37(5):391–5.

Strassl R, Reiberger T, Honsig C, Payer BA, Mandorfer M, Grabmeier-Pfistershammer K, et al. Viral determinants predicting hepatitis B surface antigen (HBsAg) seroclearance in HIV-/ HBV-coinfected patients. J Viral Hepatitis. 2014;21(7):508–16.

Tan HY, Yong YK, Shankar EM, Paukovics G, Ellegard R, Larsson M, et al. Aberrant inflammasome activation characterizes tuberculosis-associated immune reconstitution inflammatory syndrome. J Immunol. 2016;196(10):4052–63.

Thio CL, Seaberg EC, Skolasky R, Jr, Phair J, Visscher B, Munoz A, et al. HIV-1, hepatitis B virus, and risk of liver-related mortality in the Multicenter Cohort Study (MACS) Lancet. 2002;360(9349):1921–6.

Thio CL, Smeaton L, Saulynas M, Hwang H, Saravanan S, Kulkarni S, et al. Characterization of HIV-HBV coinfection in a multinational HIV-infected cohort. AIDS. 2013;27(2):191–201.

Tounkara A, Sarro YS, Kristensen S, Dao S, Diallo H, Diarra B, et al. Seroprevalence of HIV/ HBV co-infection in Malian blood donors. J IntAssoc Physicians AIDS Care (Chic). 2009;8 (1):47–51.

Tseng TC, Liu CJ, Yang HC, Su TH, Wang CC, Chen CL, et al. High levels of hepatitis B surface antigen increase risk of hepatocellular carcinoma in patients with low HBV load. Gastroenterology. 2012;142(5):1140–9. e1143. quiz e1113–1144.

Tsuchiya N, Pathipvanich P, Rojanawiwat A, Wichukchinda N, Koga I, Koga M, et al. Chronic hepatitis B and C co-infection increased all-cause mortality in HAART-naive HIV patients in northern Thailand. Epidemiol Infect. 2012;1:1–9.

Uneke CJ, Ogbu O, Inyama PU, et al. Prevalence of hepatitis-B surface antigen among blood donors and human immunodeficiency virus-infected patients in Jos, Nigeria. Mem Inst Oswaldo Cruz. 2005;100(1):13–6.

UNAIDS. Global AIDS update. Edited by HIV/AIDS JUNPo. Geneva: United Nations; 2016.

Vinikoor MJ, Sinkala E, Chilengi R, Mulenga LB, Chi BH, Zyambo Z, et al. Impact of antiretroviral therapy on liver fibrosis among human immunodeficiency virus-infected adults with and without HBV coinfection in Zambia. Clin Infect Dis: Official Publ Infect Dis Soc Am. 2017;64(10):1343–9.

Walsh R, Locarnini S. Hepatitis B precore protein: pathogenic potential and therapeutic promise. Yonsei Med J. 2012;53(5):875–85.

Weber R, Ruppik M, Rickenbach M, Spoerri A, Furrer H, Battegay M, et al. Decreasing mortality and changing patterns of causes of death in the Swiss HIV Cohort Study. HIV Medicine. 2012.

Weber R, Ruppik M, Rickenbach M, Spoerri A, Furrer H, Battegay M, et al. Decreasing mortality and changing patterns of causes of death in the Swiss HIV Cohort Study. HIV Med. 2012.

Wit FW, Weverling GJ, Weel J, Jurriaans S, Lange JM. Incidence of and risk factors for severe hepatotoxicity associated with antiretroviral combination therapy. J Infect Dis. 2002;186 (1):23–31. https://www.ncbi.nlm.nih.gov/pubmed/12089658.

Xie J, Han Y, Qiu Z, et al. Prevalence of hepatitis B and C viruses in HIV-positive patients in China: a cross-sectional study. J Int AIDS Soc. 2016;19(1):20659. Published 2016 Mar 14.

Yang Y, Gao J, Li HL, Zheng W, Yang G, Zhang W, et al. Dose-response association between hepatitis B surface antigen levels and liver cancer risk in Chinese men and women. Int J Cancer. 2016;139(2):355–62.

Yu L, Zhang BF, Cheng ML, Zhao XK, Zhang Q, Hu YX, et al. Quantitative assessment of mutations in hepatitis B virus genome with liver cirrhosis and hepatocellular carcinoma development. Oncotarget. 2016.

Zoufaly A, Onyoh EF, Tih PM, Awasom CN, Feldt T. High prevalence of hepatitis B and syphilis co-infections among HIV patients initiating antiretroviral therapy in the north-west region of Cameroon. Int J STD AIDS. 2012;23(6):435–8.

Zoutendijk R, Zaaijer HL, de Vries-Sluijs TE, Reijnders JG, Mulder JW, Kroon FP, et al. Hepatitis B surface antigen declines and clearance during long-term tenofovir therapy in patients coinfected with HBV and HIV. J Infect Dis. 2012;206(6):974–80.

Rabies Virus

Catalysis of mRNA Capping with GDP Polyribonucleotidyltransferase Activity of Rabies Virus L Protein

Tomoaki Ogino and Todd J. Green

Abstract Rabies virus (RABV) causes a lethal neurological disease in humans, posing threats to human health particularly in developing countries. RABV possesses a nonsegmented negative-strand (NNS) RNA genome, which is transcribed and replicated by a viral RNA-dependent RNA polymerase (RdRp) complex composed of a catalytic large (L) protein and its co-factor phospho-(P) protein in host cells. The L protein co-transcriptionally modifies viral pre-mRNAs into mature mRNAs to imitate higher eukaryotic mRNAs with the 5′-cap 1 structure and 3′-poly (A) tail. Our studies revealed that the mechanism of pre-mRNA capping by a GDP polyribonucleotidyltransferase (PRNTase) domain of the RABV L protein is fundamentally different from that by eukaryotic mRNA capping enzyme. Furthermore, a unique structural element, named the "priming-capping loop", in the PRNTase domain was recently found to be required for transcription initiation as well as pre-mRNA capping. Thus, the virus-specific PRNTase domain represents a potential target for developing anti-RABV agents. This chapter focuses on the catalytic and regulatory roles of the RABV PRNTase domain in viral RNA biosynthesis.

Keywords Rabies virus · Lyssavirus · Nonsegmented negative-strand RNA virus · Transcription · Replication · mRNA capping · L protein · RNA-dependent RNA polymerase · GDP polyribonucleotidyltransferase · Guanosine 5′-triphosphatase · De novo initiation · Priming · Lyssavirus · *Rhabdoviridae*

T. Ogino (✉)
Department of Medical Microbiology and Immunology, College of Medicine and Life Sciences, University of Toledo, Toledo, OH 43614, USA
e-mail: tomoaki.ogino@utoledo.edu

T. J. Green
Department of Microbiology, School of Medicine, University of Alabama at Birmingham, Birmingham, AL 35294, USA

S. I. Ahmad (ed.), *Human Viruses: Diseases, Treatments and Vaccines*, https://doi.org/10.1007/978-3-030-71165-8_21

1 Introduction

Rabies virus (RABV) is a nonsegmented negative-strand (NNS) RNA virus belonging to the *Lyssavirus* genus of the *Rhabdoviridae* family in the order *Mononegavirales*. RABV causes a fatal neurological disease, called rabies, in humans and animals [reviewed in (Albertini et al. 2011; Davis et al. 2015; Fisher et al. 2018)]. Primary reservoir hosts of RABV are believed to be bat species belonging to the order *Chiroptera*. However, RABV is known to be sporadically transmitted to various warm-blooded animals, mainly species belonging to the order *Carnivora* including domestic dogs. The major route of RABV transmission to humans is via bites from rabid dogs or direct exposure to their saliva. RABV travels from initially infected peripheral muscle tissues to the brain via its retrograde axonal transport in motor neurons of the central nervous system and subsequently causes acute inflammation of the brain. Rabies is almost 100% lethal once infected humans or animals exhibit neurological symptoms. Although effective vaccines and immunoglobulins against RABV for post-exposure prophylaxis are available, RABV still kills nearly 60,000 people each year worldwide mostly in developing countries (Fooks et al. 2014; Lankester et al. 2014). Furthermore, RABV-related viruses, such as Duvenhage virus, European bat lyssaviruses 1 and 2, Australian bat lyssavirus, Mokola virus, and Irkut virus were reported to cause rabies-like diseases in humans (Johnson et al. 2010; Fisher et al. 2018).

Since purified RABV particles show extremely low transcription activities in vitro (Villarreal and Holland 1974; Kawai 1977; Flamand et al. 1978), it has remained challenging to characterize RNA synthesis and processing activities associated with RABV particles. On the other hand, vesicular stomatitis virus (VSV), a member of the *Vesiculovirus* genus of the *Rhabdoviridae* family, has been used as a prototype rhabdovirus to study molecular mechanisms of transcription and replication due to its strong RNA synthesis activity [reviewed in (Ogino and Green 2019a)]. Based on the findings with VSV (Ogino and Green 2019a), it can be speculated that RABV enzymes also generate viral mRNAs with a methylated cap structure and poly(A) tail at their 5′- and 3′-termini, respectively. However, the lack of efficient in vitro transcription systems for RABV has hampered progress in defining the molecular mechanisms of RABV RNA synthesis and processing.

The recent development of in vitro RNA synthesis and capping systems for RABV has significantly enhanced our understanding of roles of the multi-functional RABV RNA-dependent RNA polymerase (RdRp) in viral RNA biogenesis. This chapter describes the mechanisms of RABV transcription and replication and focuses on our recent studies on an enzymatic domain responsible for RABV mRNA capping and transcription initiation.

2 Transcription and Replication

The negative-strand RABV genome of 11.9 kilo nucleotides (nt) in length begins and ends with the 3′-leader (*Le*, 58 nt) and 5′-trailer (*Tr*, 70 nt) sequences, respectively, and contains tandemly connected genes for nucleocapsid (N), phospho-(P), matrix (M), glyco-(G), and large (L) proteins (Tordo et al. 1986a, b; Tordo et al. 1988; Conzelmann et al. 1990) (Fig. 1, middle). The genome is encapsidated with the N proteins (450 amino acids) to form a helical nucleocapsid, called N–RNA complex (Iseni et al. 1998), which serves as a template for transcription as well as replication. The RdRp complex comprises the catalytic L (2127 amino acids) and non-catalytic P (297 amino acids, homodimer) proteins, and is associated with the N–RNA template via an interaction between the N and P proteins to form a ribonucleoprotein (RNP) complex (Chenik et al. 1998; Schoehn et al. 2001; Castel et al. 2009; Ribeiro Ede et al. 2009). In RABV-infected cells, transcription and replication may occur in cytoplasmic liquid-like replication organelles, similar to Negri Bodies (Lahaye et al. 2009; Nikolic et al. 2016, 2017).

As suggested for VSV (Abraham et al. 1975a; Abraham and Banerjee 1976; Ball and White 1976; Testa, Chanda, and Banerjee 1980; Emerson 1982), the RABV RdRp complex may enter from the 3′-terminal of the genome to initiate synthesis of

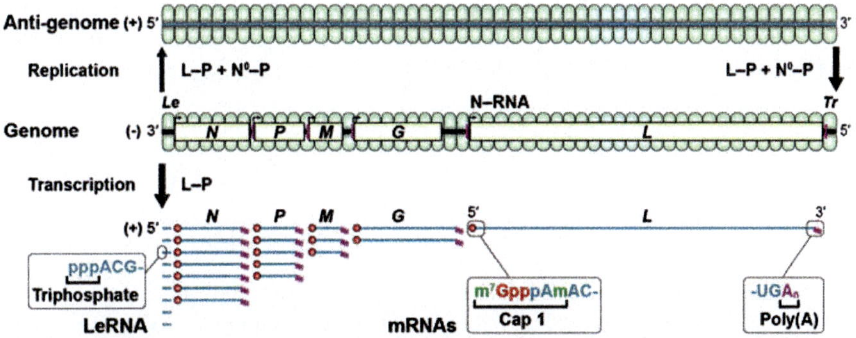

Fig. 1 Transcription and replication of RABV. The RABV genome with a negative polarity (middle) is encapsidated with the nucleocapsid (N) proteins and acts as a template (called N–RNA template) for transcription (lower) and replication (upper). The genome begins and ends with the 3′-leader (*Le*) and 5′-trailer (*Tr*) regions, respectively, and contains five tandemly arranged genes (*N*, *P*, *M*, *G*, and *L*). Each gene starts and ends with the conserved gene-start (3′-UUGURRNGA, bent arrows) and gene-end (3′-ACUUUUUUU, vertical magenta lines) sequences, respectively, which serve as transcription initiation and polyadenylation/termination signals, respectively. The RNA-dependent RNA polymerase (RdRp) complex composed of the large (L) and phospho-(P) proteins sequentially transcribes the *Le* region and internal genes into the non-coding leader RNA (LeRNA) and mRNAs, respectively, by the stop-start transcription mechanism (lower). The mRNAs possess a 5′-cap 1 structure (m^7GpppAm-) and 3′-poly(A) tail. For replication, the RdRp copies the genome into the positive-sense anti-genome, which is in turn used as a template for synthesis of progeny genomes. The complex of the RNA-free N protein with the P protein (N^0–P) is required for co-replicational encapsidation of the genome/antigenome with the N proteins

the leader RNA (LeRNA, 55–58 nt) (Kurilla et al. 1984), and then sequentially transcribes the internal genes into monocistronic mRNAs by the stop-start transcription mechanism (Abraham and Banerjee 1976; Ball and White 1976; Iverson and Rose 1981) (Fig. 1, lower). The conserved gene-start (3′-UUGURRNGA, R = G/A) and gene-end (3′-ACUUUUUUU) sequences of each gene may act as signals for transcription reinitiation and polyadenylation/termination, respectively (Ogino and Green 2019b), as reported for VSV (Barr et al. 1997; Stillman and Whitt 1999). RABV mRNAs produced in infected cells possess a 3′-poly(A) tail of 100–250 nt in length (Holloway and Obijeski 1980). Similar to VSV (Abraham and Banerjee 1976; Ball and White 1976; Iverson and Rose 1981), attenuation of transcription reinitiation at each RABV gene-start sequence results in generation of a gradient in mRNA abundance in the following order: $N > P > M > G > L$ (Flamand and Delagneau 1978; Finke et al. 2000). Transcription of a downstream gene is significantly more attenuated with a longer intergenic region (*G*/*L*, 24 nt > *P*/*M* and *M*/*G*, 5 nt > *N*/*P*, 2 nt) between upstream gene-end and downstream gene-start sequences (Finke et al. 2000).

For replication, the RdRp initiates RNA synthesis at the 3′-terminal of the genome, but ignores the internal termination signals for LeRNA and mRNAs to generate a full-length anti-genome (Fig. 1, upper). The anti-genome is in turn used as a template for synthesis of progeny genomes. The genome and anti-genome are encapsidated with the N proteins during replication. As reported for VSV (Peluso 1988; Peluso and Moyer 1988; Masters and Banerjee 1988; Gupta and Banerjee 1997; Blumberg et al. 1981, 1983), a heterodimeric complex between RNA-free N and P proteins (N^0–P) of RABV (Mavrakis et al. 2003) may be required for the co-replicational genome encapsidation.

3 The L Protein

As reported for other NNS RNA viral L proteins (Hunt and Hutchinson 1993; Grdzelishvili et al. 2005; Li et al. 2005; Ogino et al. 2005; Ogino and Banerjee 2007; Galloway and Wertz 2008; Rahmeh et al. 2009; Morin et al. 2012; Paesen et al. 2015; Jordan et al. 2018), the RABV L protein is most probably responsible for all enzymatic activities required for RNA synthesis, mRNA 5′-capping, cap methylation, and 3′-polyadenylation. However, to date, only its RNA synthesis and capping activities of the RABV L protein have been experimentally demonstrated (Ogino et al. 2016, 2019; Morin et al. 2017). The RNA synthesis activity of the RABV L protein is stimulated with the P protein at both the initiation (Ogino et al. 2019) and elongation (Morin et al. 2017) steps. The RNA capping activity of the L protein does not require the P protein (Ogino and Banerjee 2007; Ogino et al. 2016).

Recently, a three-dimensional structure of the RABV L protein (SADB19 strain) in complex with an N-terminal part of the P protein (residues 1–91) has been solved by cryo-electron microscopy (Horwitz et al. 2020). Similar to the VSV L protein (PDB id: 5A22) (Liang et al. 2015), the RABV L protein has an N-terminal ring-like structure composed of RdRp (residues 29–879) and capping (Cap, residues 880–1351) domains and three C-terminal globular domains [connector (CD, residues 1352–1627), methyltransferase (MTase, residues 1628–1926), and C-terminal (CTD, residues 1927–2127)]. The N-terminal part of the RABV P protein interacts with amino acid residues within the RdRp, CD, and CTD of the L protein. Horwitz et al. (2020) suggested that the N-terminal ring-like structure consists of the two domains, RdRp and Cap, although their domain boundaries have not been biochemically determined. On the other hand, we have recently suggested that the N-terminal ring-like structure of the RABV L protein, RC-HL strain, can be divided into four domains (Fig. 2a, b): N-terminal (NTD) (residues 35–375), RdRp (residues 376–946), bridge (residues 953–1084), and mRNA capping enzyme (GDP polyribonucleotidyltransferase, PRNTase, residues 1093–1349 (Fig. 2c) domains.

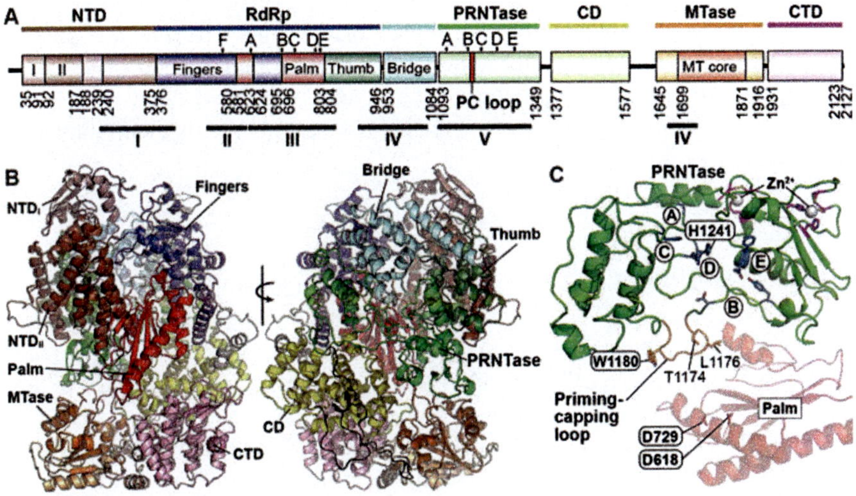

Fig. 2 The RABV L protein. **a** The RABV L protein (2127 amino acids) was predicted to be composed of the following domains (Ogino and Green 2019b): NTD, N-terminal domain (subdomains: I, light brown; II, brown); RdRp, RNA-dependent RNA polymerase (subdomains: fingers, blue; palm, red; thumb, dark green); Bridge (cyan); PRNTase, GDP polyribonucleotidyltransferase (green); CD, connecter domain (yellow); MTase, methyltransferase (MTase core, orange; other regions, light orange); CTD, C-terminal domain (pink). The positions of RdRp motifs (A–F), PRNTase motifs (A–E), and priming-capping (PC) loop are denoted (Ogino and Green 2019b). The six conserved regions (I–VI) (Poch et al. 1990) are indicated by black lines. **b** Two views of a predicted three-dimensional structure of the RABV L protein (RC-HL strain; www.modelarchive.org, id: ma-pe4nf) are shown as ribbon models (Ogino and Green 2019b). The predicted domains and subdomains are colored as in panel (**a**). **c** The predicted structure of the RABV PRNTase domain (green) is depicted as a ribbon model with essential amino acid residues (blue stick models) including a catalytic histidine residue (H1241) in PRNTase motifs A–E (circled capital letters). The priming-capping loop (orange) is shown with key amino acid residues. Amino acid residues required for binding to two Zn^{2+} ions (silver spheres) are represented as magenta stick models. The palm subdomain (pale red) of the RdRp domain possesses two catalytic aspartate residues (D618 and D729, red stick models)

The NTD is composed of two subdomains and displays a topological similarity to the C-terminal domain of the influenza virus RdRp PA subunit (PA_C) (He et al. 2008; Obayashi et al. 2008; Pflug et al. 2014; Reich et al. 2014; Hengrung et al. 2015), the PA_C-like domain of the LACV RdRp L (Gerlach et al. 2015), and the N-terminal domain of the reovirus RdRp λ3 (Tao et al. 2002). The NTD of the RABV L protein may play an essential role in transcription as reported for VSV (Qiu et al. 2016). The RdRp domain composed of fingers, palm, and thumb subdomains shares structural similarities with other viral RdRps with structural motifs A–F (Poch et al. 1990; O'Reilly and Kao 1998; Bruenn 2003; Lang et al. 2013; Ogino and Green 2019a). Two universally conserved aspartate residues, D618 and D729 are present in motifs A and C, respectively, and may play critical roles in two-metal dependent nucleotide polymerization as suggested for other polymerases (Steitz 1998; Florian et al. 2003, 2005; Castro et al. 2007, 2009; Gong and Peersen 2010; Genna et al. 2016). Mutations of D729 in the RABV L protein were shown to abolish its RNA synthesis activity (Schnell and Conzelmann 1995; Ogino et al. 2019). The bridge domain may form a template exit channel (Horwitz et al. 2020) and appears to be involved in transcription rather than capping (unpublished data), although it has been originally proposed to be a part of the Cap domain. Therefore, it will be necessary to determine exact domain boundaries based on their biological functions.

4 mRNA Capping

One of structural hallmarks of eukaryotic mRNA is the presence of the 5′-cap structure [m^7GpppN_1-(cap 0)], in which N^7-methylguanosine (m^7G) is linked to the first nucleoside (N_1) of mRNA via an inverted 5′-5′ triphosphate bridge (ppp) [reviewed in (Banerjee 1980; Furuichi and Shatkin 2000; Shuman 2001; Ghosh and Lima 2010; Ramanathan et al. 2016)]. In higher eukaryotes, the cap 0 structure is further methylated at the 2′-*O* position of the first nucleoside into the cap 1 (m^7GpppN_1m-) structure (Fig. 3a). The cap 0 structure is required for mRNA splicing, export, translation, and stability (Banerjee 1980; Furuichi and Shatkin 2000; Ramanathan et al. 2016). The formation of the cap 1 structure is critical to avoid triggering anti-viral innate immune responses [reviewed in (Hyde and Diamond 2015; Leung and Amarasinghe 2016)]. In the nucleus of eukaryotic cells, eukaryotic mRNA capping enzyme co-transcriptionally caps pre-mRNA, synthesized by RNA polymerase II, by RNA 5′-triphosphatase (RTPase) and mRNA guanylyltransferase (GTase) activities (Furuichi and Shatkin 2000; Shuman 2001; Ghosh and Lima 2010; Ramanathan et al. 2016). RTPase hydrolyzes a 5′-triphosphate group of pre-mRNA into a 5′-diphosphate group and inorganic phosphate (P_i) (Fig. 3b). GTase transfers a GMP moiety from GTP to the diphosphate end of pre-mRNA through a covalent enzyme-(lysyl-$N^ζ$)-GMP (E-pG) intermediate to form the cap core structure (GpppN-). The cap core structure is subsequently methylated by mRNA (guanine-N^7)-methyltransferase (G-N^7-MTase) followed by

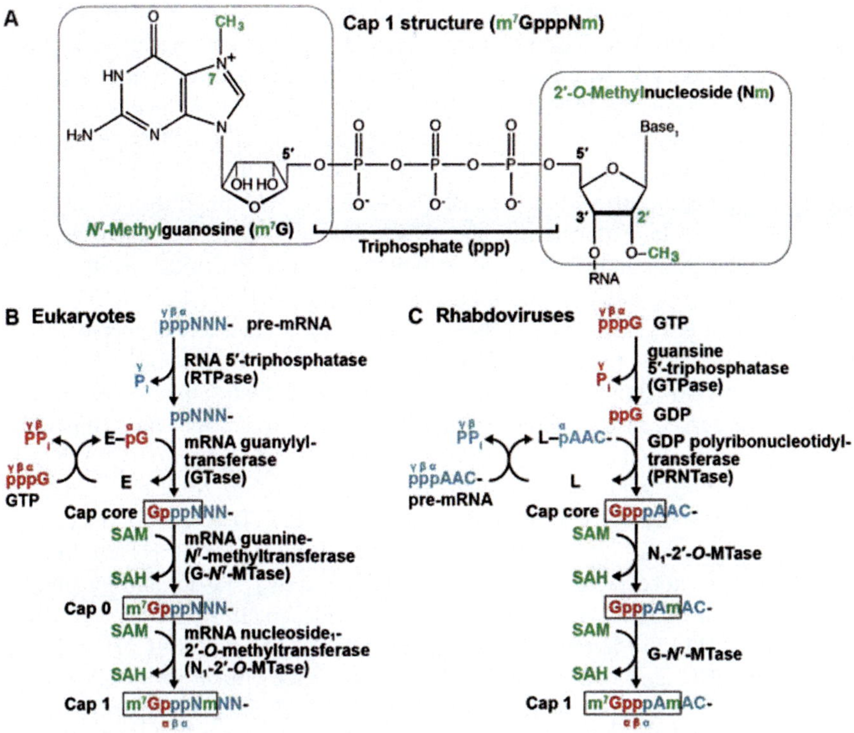

Fig. 3 Conventional and unconventional mechanisms of mRNA capping. **a** The 5′-terminal cap 1 structure ($m^7GpppNm$) of higher eukaryotic mRNA is composed of N^7-methylguanosine (m^7G) lined to 2′-*O*-methylnucleoside (Nm) via a 5′-5′-triphosphate bridge (ppp). **b** and **c** The pathway of the cap 1 formation in rhabdoviruses (**c**) is compared with that in eukaryotes (**b**). pre-mRNA, GTP, and *S*-adenosyl-l-methionine (SAM) are shown in light blue, red, and green, respectively. E and L indicate the GTase domain of eukaryotic mRNA capping enzyme and the PRNTase domain of the L protein, respectively. P_i, PP_i, and SAH indicate inorganic phosphate, inorganic pyrophosphate, and *S*-adenosyl-l-homocysteine, respectively

mRNA (nucleoside$_1$-2′-*O*-)-methyltransferase (N_1-2′-*O*-MTase) to generate the cap 1 structure (Banerjee 1980).

In contrast to the GMP transfer mechanism for eukaryotic mRNA capping, as described above, rhabdoviruses, such as VSV (Abraham et al. 1975a, b; Ogino and Banerjee 2007), Chandipura virus (Ogino and Banerjee 2010), and RABV (Ogino et al. 2016), employ a novel pRNA transfer mechanism for viral mRNA capping (Fig. 3c). First, a guanosine 5′-triphosphatase (GTPase) activity associated with the L protein hydrolyzes GTP into GDP (a pRNA acceptor) and P_i (Ogino and Banerjee 2007, 2008; Ogino et al. 2016). Then the PRNTase domain of the RABV L protein specifically caps 5′-triphosphorylated RABV pre-mRNAs (pppAAC-started RNAs), but not LeRNA, with GDP by the pRNA transfer activity (Ogino et al. 2016). The 5′-terminal AAC sequence conserved in lyssaviral pre-mRNAs is essential for its pRNA donor substrate activity (Ogino et al. 2016). As demonstrated for VSV (Ogino and Banerjee 2007; Ogino et al. 2010), it can be speculated that the pRNA transfer reaction with the RABV L protein proceeds through the formation of a covalent enzyme-(histidyl-$N^{\varepsilon 2}$)-pRNA (called L-pRNA) intermediate to yield the GpppA cap core structure. However, it remains difficult to detect the putative intermediate formation activity of the RABV L protein due to its low capping activity (Ogino et al. 2016). For VSV, the isolated L-pRNA intermediate has been shown to be capable of transferring the pRNA moiety to GDP, in which the guanine-C^2-amino group and ribose-2′ or 3′-hydroxyl group are critical for its pRNA accepter activity (Ogino et al. 2010; Ogino and Ogino 2017).

The RABV PRNTase domain possesses five collinear sequence elements, Rx(3)Wx(3–8)ΦxGxζx(P/A) (motif A), (Y/W)ΦGSxT (motif B), W (motif C), HR (motif D), and ζxxΦx(F/Y)QxxΦ (motif E) (Φ, hydrophobic; ζ, hydrophilic amino acids), which are conserved in L proteins of other NNS RNA viruses including important human pathogens, such as Ebola (*Filoviridae*), measles (*Paramyxoviridae*), Nipah (*Paramyxoviridae*), and respiratory syncytial (*Pneumoviridae*) viruses (Ogino et al. 2016; Ogino and Green 2019a, b). Similar to the VSV PRNTase domain (Neubauer et al. 2016; Ogino and Green 2019a), motifs B–E of the RABV PRNTase domain are localized in close proximity to form a unique active site for the enzymatic domain (Horwitz et al. 2020) and motif A may provide a platform for the active site organization. G1112 in motif A, T1170 in motif B, W1201 in motif C, H1241 and R1242 in motif D, and F1285 and Q1286 in motif E are critical for the PRNTase activity of the RABV L protein (Ogino et al. 2016). The VSV counterparts of these residues are essential for virus gene expression and propagation in host cells (Ogino 2014; Neubauer et al. 2016). In VSV, a lone pair of electrons at the $N^{\varepsilon 2}$ position of the H residue in motif D was suggested to nucleophilically attack the α-phosphorus in the 5′-triphosphate group of pre-mRNA to form the L–pRNA intermediate with concomitant release of inorganic pyrophosphate (PP_i) (Fig. 4) (Ogino et al. 2010). During the intermediate formation, the R residue in motif D may be required for an interaction with the 5′-triphosphate group of pre-mRNA and/or proton transfer to leaving PP_i. An oxyanion on the β-phosphate group of GDP subsequently attacks the α-phosphorus in the L–pRNA intermediate to release the GpppA-capped pre-mRNA from the L protein. Key residues in motifs B–E surrounding motif D together with other less-conserved residues form a crevice, which may serve as binding sites for the substrates and products during the pRNA transfer reaction as recently predicted (Ogino and Green 2019a).

The cap core structure (GpppA-) of RABV mRNAs is believed to be methylated into the cap 1 structure (m^7GpppAm-) by the putative MTase domain of the RABV L protein as reported for VSV mRNAs (Abraham, Rhodes, and Banerjee 1975a; Testa and Banerjee 1977). In VSV, the MTase domain of the L protein methylates the cap core structure at the N_1-2′-*O*-position followed by G-N^7-position during transcription (Testa and Banerjee 1977; Rahmeh et al. 2009). This order (GpppA- → GpppAm- → m^7GpppAm-) is opposite to the order of cap methylation by the eukaryotic enzymes as described above.

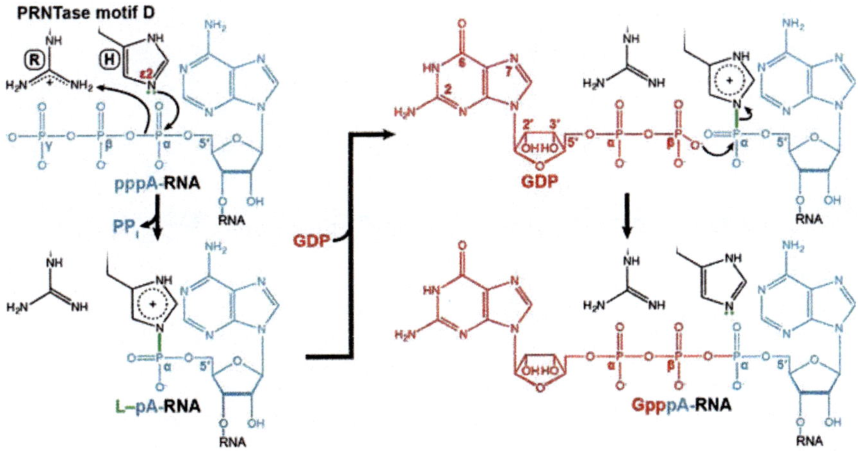

Fig. 4 Proposed roles of PRNTase motif D in NNS RNA viral mRNA capping. Partial structures of the catalytic histidine (H) and arginine (R) residues in PRNTase motif D of the L protein are shown. The 5′-terminal pppA residue of pre-mRNA and GDP are colored light blue and red, respectively. A pair of green dots indicates a lone pair of electrons on the $N^{\varepsilon 2}$ position of the H residue. Curly arrows illustrate the movement of electrons. For details, see text

5 Roles of the Priming-Capping Loop of the RABV PRNTase Domain in RNA Biosynthesis

The PRNTase domain of RABV (Horwitz et al. 2020) as well as VSV (Liang et al. 2015) has a large loop structure, which extends into the active site cavity of the RdRp domain. We have revealed that the loop, named the "priming-capping loop", plays dual roles in transcription initiation and capping (Fig. 5a) (Ogino et al. 2019). Similar to a tyrosine residue on a priming loop of the bacteriophage Φ6 RdRp (Butcher et al. 2001), an aromatic tryptophan residue (W1180 for RABV, W1167 for VSV) on the priming-capping loop is essential for terminal de novo initiation (first phosphodiester bond formation) at the 3′-terminal of the genome to synthesize LeRNA (Ogino et al. 2019). We have modeled a structure of a putative RABV initiation complex with the 3′-terminal of the RABV genome (3′-UGCG), initiator ATP, incoming CTP, and divalent cations (Fig. 5b, Model Archive id: ma-uqibu), and suggested that the indole group of the W1180 residue on the priming-capping loop π-stacks with the adenine base of ATP to stabilize the initiation complex formed on the 3′-terminal UG sequence (Ogino and Green 2019b). In contrast, W1180 is dispensable for internal de novo initiation at the gene-start sequence, elongation, or capping (Ogino et al. 2019), suggesting that the priming-capping loop is retracted from the active site cavity of the RdRp domain after terminal de novo initiation to open a path for elongating transcripts.

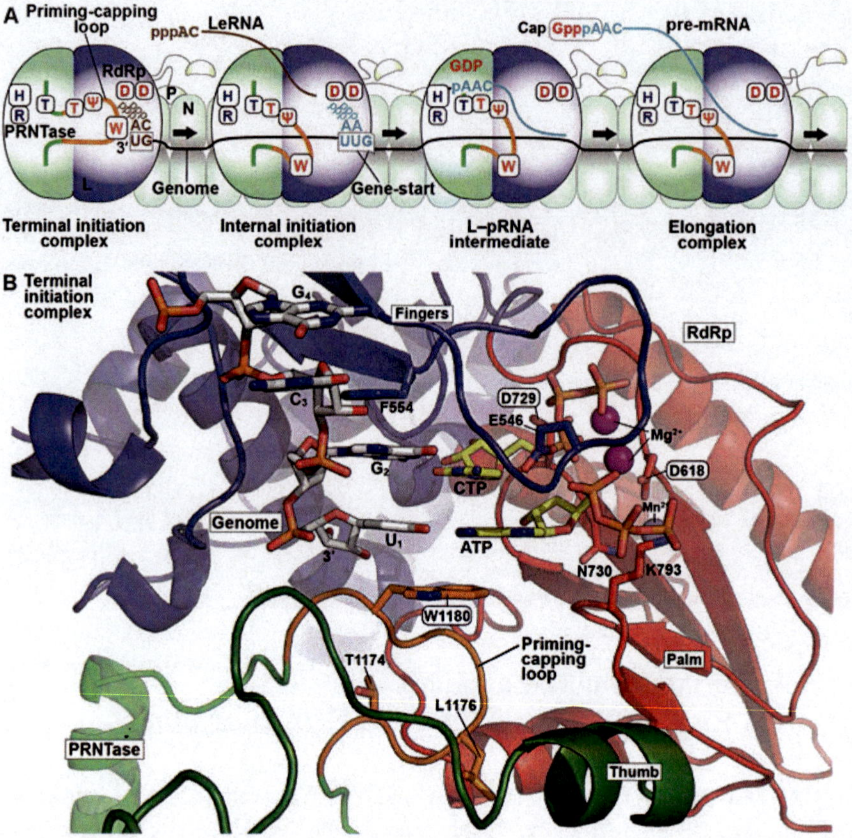

Fig. 5 The dual-functional priming-capping loop of rhabdoviral L proteins regulate terminal transcription initiation and pre-mRNA capping. **a** The RdRp (blue) and the PRNTase (green) domains of the L protein are illustrated with a flexible priming-capping loop (orange). Letters in round squares indicate key amino acid residues required for RNA biosynthesis: D residues (red) in RdRp motifs A and C (for RABV, D618 and D729); T residue (blue) in PRNTase motif B (T1170); H and R residues (blue) in PRNTase motif D (H1241 and R1242); T, Ψ (Ψ, aliphatic), and W residues (orange) on the priming-capping loop (T1174, L1176, and W1180). The RdRp complex composed of the L and P (pale yellow) proteins forms a terminal initiation complex at the 3′-terminal UG sequence of the genome [encapsulated with the N proteins (pale green)] to initiate synthesis of the leader RNA (LeRNA). After LeRNA synthesis, the RdRp forms an internal initiation complex at the *N* gene-start sequence (3′-UUG-) to initiate synthesis of *N* mRNA. During mRNA chain elongation, the L protein forms a covalent complex with the 5′-end of pre-mRNA (L-pRNA intermediate), and subsequently transfers it to GDP to generate the GpppA cap structure on pre-mRNA. The W residue and TxΨ motif on the priming-capping loop are critical for terminal de novo initiation and capping intermediate formation, respectively (Ogino et al. 2019). **b** The active site of the RdRp domain in a modeled RABV L terminal initiation complex (Model Archive, id: ma-uqibu) is shown as a ribbon model with key amino acid residues (stick models) (Ogino and Green 2019b). The fingers, palm, and thumb RdRp subdomains are colored as in Fig. 2. A model genome terminal (3′-UGCG-5′, white carbon backbone), initiator and incoming nucleotides (ATP and CTP, respectively; yellow carbon backbone), and divalent metal ions (Mg^{2+}, purple; Mn^{2+}, obscured) are indicated. The priming-capping loop extended from the PRNTase domain is colored orange

On the other hand, a TxΨ (Ψ, aliphatic amino acid) motif (T1174-x-L1176) in the priming-capping loop of the RABV PRNTase domain is critical for mRNA capping, but not for transcription initiation (Ogino et al. 2019). In VSV, the TxΨ motif has been demonstrated to be essential for the L-pRNA intermediate formation (Ogino et al. 2019). Therefore, the flexible loop may undergo a further structural rearrangement to form the capping intermediate during mRNA chain elongation (Fig. 5a). In contrast to the PRNTase motifs, the TxΨ motif and the priming W residue are conserved in L proteins of rhabdoviruses infecting animals and/or arthropods, but not in other NNS RNA viral L proteins (Ogino et al. 2019; Ogino and Green 2019a). Interestingly, mutations in the TxΨ motif as well as in the PRNTase motifs of the VSV L protein cause frequent termination and reinitiation of transcription using cryptic termination and initiation signals within the *N* gene, leading to production of short 5′-triphosphorylated *N* mRNA fragments (Ogino 2014; Neubauer et al. 2016; Ogino et al. 2019). Thus, pre-mRNA capping mediated by the priming-capping loop seems to be critical for synthesis of full-length mRNAs.

6 Conclusion

We believe that our biochemical studies on the rhabdovirus L proteins have contributed significantly to the understanding of their unique roles in viral RNA biosynthesis. Especially, the establishment of our in vitro RNA synthesis and capping systems with the recombinant L proteins has led to the discoveries of the novel enzymatic and regulatory roles of their PRNTase domain in capping and transcription (Ogino et al. 2019, 2016; Ogino and Banerjee 2007; Ogino et al. 2010). Since the PRNTase domain with the unique activities and structure is critical for rhabdovirus replication in host cells (Ogino 2014; Neubauer et al. 2016; Ogino et al. 2019), it may serve as a potential target for the future development of antiviral drugs. Importantly, because PRNTase active sites are highly conserved in L proteins of all NNS RNA viruses, including Ebola, measles, and respiratory syncytial viruses, belonging to the order *Mononegavirales* (Neubauer et al. 2016; Ogino and Green 2019a), the catalytic mechanism of mRNA capping discussed in this chapter is expected to be common among these viruses. Further biochemical and structural studies on the NNS RNA viral PRNTase domains are warranted to rationally design antiviral drugs against them.

Acknowledgements The authors' work described herein was supported by funding from the National Institutes of Health to TO (AI093569 and AI146172) and to TJG (AI116738).

References

Abraham G, Banerjee AK. Sequential transcription of the genes of vesicular stomatitis virus. Proc Natl Acad Sci USA. 1976;73(5):1504–8.

Abraham G, Rhodes DP, Banerjee AK. The 5′ terminal structure of the methylated mRNA synthesized in vitro by vesicular stomatitis virus. Cell. 1975;5(1):51–8.

Abraham G, Rhodes DP, Banerjee AK. Novel initiation of RNA synthesis in vitro by vesicular stomatitis virus. Nature. 1975;255(5503):37–40.

Albertini AA, Ruigrok RW, Blondel D. Rabies virus transcription and replication. Adv Virus Res. 2011;79:1–22.

Ball LA, White CN. Order of transcription of genes of vesicular stomatitis virus. Proc Natl Acad Sci USA. 1976;73(2):442–6.

Banerjee AK. 5′-terminal cap structure in eucaryotic messenger ribonucleic acids. Microbiol Rev. 1980;44(2):175–205.

Barr JN, Whelan SP, Wertz GW. cis-Acting signals involved in termination of vesicular stomatitis virus mRNA synthesis include the conserved AUAC and the U7 signal for polyadenylation. J Virol. 1997;71(11):8718–25.

Blumberg BM, Giorgi C, Kolakofsky D. N protein of vesicular stomatitis virus selectively encapsidates leader RNA in vitro. Cell. 1983;32(2):559–67.

Blumberg BM, Leppert M, Kolakofsky D. Interaction of VSV leader RNA and nucleocapsid protein may control VSV genome replication. Cell. 1981;23(3):837–45.

Bruenn JA. A structural and primary sequence comparison of the viral RNA-dependent RNA polymerases. Nucleic Acids Res. 2003;31(7):1821–9.

Butcher SJ, Grimes JM, Makeyev EV, Bamford DH, Stuart DI. A mechanism for initiating RNA-dependent RNA polymerization. Nature. 2001;410(6825):235–40.

Castel G, Chteoui M, Caignard G, et al. Peptides that mimic the amino-terminal end of the rabies virus phosphoprotein have antiviral activity. J Virol. 2009;83(20):10808–20.

Castro C, Smidansky ED, Arnold JJ, et al. Nucleic acid polymerases use a general acid for nucleotidyl transfer. Nat Struct Mol Biol. 2009;16(2):212–8.

Castro C, Smidansky E, Maksimchuk KR, et al. Two proton transfers in the transition state for nucleotidyl transfer catalyzed by RNA- and DNA-dependent RNA and DNA polymerases. Proc Natl Acad Sci USA. 2007;104(11):4267–72.

Chenik M, Schnell M, Conzelmann KK, Blondel D. Mapping the interacting domains between the rabies virus polymerase and phosphoprotein. J Virol. 1998;72(3):1925–30.

Conzelmann KK, Cox JH, Schneider LG, Thiel HJ. Molecular cloning and complete nucleotide sequence of the attenuated rabies virus SAD B19. Virology. 1990;175(2):485–99.

Davis BM, Rall GF, Schnell MJ. Everything You Always Wanted to Know About Rabies Virus (But Were Afraid to Ask). Annu Rev Virol. 2015;2(1):451–71.

Emerson SU. Reconstitution studies detect a single polymerase entry site on the vesicular stomatitis virus genome. Cell. 1982;31(3 Pt 2):635–42.

Finke S, Cox JH, Conzelmann KK. Differential transcription attenuation of rabies virus genes by intergenic regions: generation of recombinant viruses overexpressing the polymerase gene. J Virol. 2000;74(16):7261–9.

Fisher CR, Streicker DG, Schnell MJ. The spread and evolution of rabies virus: conquering new frontiers. Nat Rev Microbiol. 2018;16(4):241–55.

Flamand A, Delagneau JF. Transcriptional mapping of rabies virus in vivo. J Virol. 1978;28 (2):518–23.

Flamand A, Delagneau JF, Bussereau F. An RNA polymerase activity in purified rabies virions. J Gen Virol. 1978;40(1):233–8.

Florian J, Goodman MF, Warshel A. Computer simulation of the chemical catalysis of DNA polymerases: discriminating between alternative nucleotide insertion mechanisms for T7 DNA polymerase. J Am Chem Soc. 2003;125(27):8163–77.

Florian J, Goodman MF, Warshel A. Computer simulations of protein functions: searching for the molecular origin of the replication fidelity of DNA polymerases. Proc Natl Acad Sci USA. 2005;102(19):6819–24.

Fooks AR, Banyard AC, Horton DL, Johnson N, McElhinney LM, Jackson AC. Current status of rabies and prospects for elimination. Lancet. 2014;384(9951):1389–99.

Furuichi Y, Shatkin AJ. Viral and cellular mRNA capping: past and prospects. Adv Virus Res. 2000;55:135–84.

Galloway SE, Wertz GW. S-adenosyl homocysteine-induced hyperpolyadenylation of vesicular stomatitis virus mRNA requires the methyltransferase activity of L protein. J Virol. 2008;82 (24):12280–90.

Genna V, Vidossich P, Ippoliti E, Carloni P, De Vivo M. A self-activated mechanism for nucleic acid polymerization catalyzed by DNA/RNA polymerases. J Am Chem Soc. 2016;138 (44):14592–8.

Gerlach P, Malet H, Cusack S, Reguera J. Structural Insights into bunyavirus replication and its regulation by the vRNA promoter. Cell. 2015;161(6):1267–79.

Ghosh A, Lima CD. Enzymology of RNA cap synthesis. Wiley Interdiscip Rev RNA. 2010;1 (1):152–72.

Gong P, Peersen OB. Structural basis for active site closure by the poliovirus RNA-dependent RNA polymerase. Proc Natl Acad Sci USA. 2010;107(52):22505–10.

Grdzelishvili VZ, Smallwood S, Tower D, Hall RL, Hunt DM, Moyer SA. A single amino acid change in the L-polymerase protein of vesicular stomatitis virus completely abolishes viral mRNA cap methylation. J Virol. 2005;79(12):7327–37.

Gupta AK, Banerjee AK. Expression and purification of vesicular stomatitis virus N-P complex from *Escherichia coli*: role in genome RNA transcription and replication in vitro. J Virol. 1997;71(6):4264–71.

He X, Zhou J, Bartlam M, et al. Crystal structure of the polymerase PA(C)-PB1(N) complex from an avian influenza H5N1 virus. Nature. 2008;454(7208):1123–6.

Hengrung N, El Omari K, Serna Martin I, et al. Crystal structure of the RNA-dependent RNA polymerase from influenza C virus. Nature. 2015;527(7576):114–7.

Holloway BP, Obijeski JF. Rabies virus-induced RNA synthesis in BHK21 cells. J Gen Virol. 1980;49(1):181–95.

Horwitz JA, Jenni S, Harrison SC, Whelan SPJ. Structure of a rabies virus polymerase complex from electron cryo-microscopy. Proc Natl Acad Sci USA. 2020;117(4):2099–2107

Hunt DM, Hutchinson KL. Amino acid changes in the L polymerase protein of vesicular stomatitis virus which confer aberrant polyadenylation and temperature-sensitive phenotypes. Virology. 1993;193(2):786–93.

Hyde JL, Diamond MS. Innate immune restriction and antagonism of viral RNA lacking 2-O methylation. Virology. 2015;479–480:66–74.

Iseni F, Barge A, Baudin F, Blondel D, Ruigrok RW. Characterization of rabies virus nucleocapsids and recombinant nucleocapsid-like structures. J Gen Virol. 1998;79(Pt 12):2909–19.

Iverson LE, Rose JK. Localized attenuation and discontinuous synthesis during vesicular stomatitis virus transcription. Cell. 1981;23(2):477–84.

Johnson N, Vos A, Freuling C, Tordo N, Fooks AR, Muller T. Human rabies due to lyssavirus infection of bat origin. Vet Microbiol. 2010;142(3–4):151–9.

Jordan PC, Liu C, Raynaud P, et al. Initiation, extension, and termination of RNA synthesis by a paramyxovirus polymerase. PLoS Pathog. 2018;14(2):e1006889.

Kawai A. Transcriptase activity associated with rabies virion. J Virol. 1977;24(3):826–35.

Kurilla MG, Cabradilla CD, Holloway BP, Keene JD. Nucleotide sequence and host La protein interactions of rabies virus leader RNA. J Virol. 1984;50(3):773–8.

Lahaye X, Vidy A, Pomier C, et al. Functional characterization of Negri bodies (NBs) in rabies virus-infected cells: evidence that NBs are sites of viral transcription and replication. J Virol. 2009;83(16):7948–58.

Lang DM, Zemla AT, Zhou CL. Highly similar structural frames link the template tunnel and NTP entry tunnel to the exterior surface in RNA-dependent RNA polymerases. Nucleic Acids Res. 2013;41(3):1464–82.

Lankester F, Hampson K, Lembo T, Palmer G, Taylor L, Cleaveland S. Infectious Disease. Implementing Pasteur's vision for rabies elimination. Science. 2014;345(6204):1562–4.

Leung DW, Amarasinghe GK. When your cap matters: structural insights into self vs non-self recognition of 5′ RNA by immunomodulatory host proteins. Curr Opin Struct Biol. 2016;36:133–41.

Li J, Fontaine-Rodriguez EC, Whelan SP. Amino acid residues within conserved domain VI of the vesicular stomatitis virus large polymerase protein essential for mRNA cap methyltransferase activity. J Virol. 2005;79(21):13373–84.

Liang B, Li Z, Jenni S, et al. Structure of the L protein of vesicular stomatitis virus from electron cryomicroscopy. Cell. 2015;162(2):314–27.

Masters PS, Banerjee AK. Complex formation with vesicular stomatitis virus phosphoprotein NS prevents binding of nucleocapsid protein N to nonspecific RNA. J Virol. 1988;62(8):2658–64.

Mavrakis M, Iseni F, Mazza C, et al. Isolation and characterisation of the rabies virus N degrees-P complex produced in insect cells. Virology. 2003;305(2):406–14.

Morin B, Liang B, Gardner E, Ross RA, Whelan SP. An in vitro RNA synthesis assay for rabies virus defines ribonucleoprotein interactions critical for polymerase activity. J Virol. 2017;91 (1):e01508-e1516.

Morin B, Rahmeh AA, Whelan SP. Mechanism of RNA synthesis initiation by the vesicular stomatitis virus polymerase. EMBO J. 2012;31(5):1320–9.

Neubauer J, Ogino M, Green TJ, Ogino T. Signature motifs of GDP polyribonucleotidyltransferase, a non-segmented negative strand RNA viral mRNA capping enzyme, domain in the L protein are required for covalent enzyme-pRNA intermediate formation. Nucleic Acids Res. 2016;44(1):330–41.

Nikolic J, Civas A, Lama Z, Lagaudriere-Gesbert C, Blondel D. Rabies virus infection induces the formation of stress granules closely connected to the viral factories. PLoS Pathog. 2016;12(10): e1005942.

Nikolic J, Le Bars R, Lama Z, et al. Negri bodies are viral factories with properties of liquid organelles. Nat Commun. 2017;8(1):58.

O'Reilly EK, Kao CC. Analysis of RNA-dependent RNA polymerase structure and function as guided by known polymerase structures and computer predictions of secondary structure. Virology. 1998;252(2):287–303.

Obayashi E, Yoshida H, Kawai F, et al. The structural basis for an essential subunit interaction in influenza virus RNA polymerase. Nature. 2008;454(7208):1127–31.

Ogino M, Gupta N, Green TJ, Ogino T. A dual-functional priming-capping loop of rhabdoviral RNA polymerases directs terminal de novo initiation and capping intermediate formation. Nucleic Acids Res. 2019;47(1):299–309.

Ogino M, Ito N, Sugiyama M, Ogino T. The rabies virus L protein catalyzes mRNA capping with GDP polyribonucleotidyltransferase activity. Viruses. 2016;8(5):144.

Ogino M, Ogino T. 5′-Phospho-RNA acceptor specificity of GDP polyribonucleotidyltransferase of vesicular stomatitis virus in mRNA capping. J Virol. 2017;91(6):e02322-e2416.

Ogino T. Capping of vesicular stomatitis virus pre-mRNA is required for accurate selection of transcription stop-start sites and virus propagation. Nucleic Acids Res. 2014;42(19):12112–25.

Ogino T, Banerjee AK. Unconventional mechanism of mRNA capping by the RNA-dependent RNA polymerase of vesicular stomatitis virus. Mol Cell. 2007;25(1):85–97.

Ogino T, Banerjee AK. Formation of guanosine(5′)tetraphospho(5′)adenosine cap structure by an unconventional mRNA capping enzyme of vesicular stomatitis virus. J Virol. 2008;82 (15):7729–34.

Ogino T, Banerjee AK. The HR motif in the RNA-dependent RNA polymerase L protein of Chandipura virus is required for unconventional mRNA-capping activity. J Gen Virol. 2010;91 (Pt 5):1311–4.

Ogino T, Green TJ. RNA synthesis and capping by non-segmented negative strand RNA viral polymerases: lessons from a prototypic virus. Front Microbiol. 2019;10:1490.

Ogino T, Green TJ. Transcriptional control and mRNA capping by the GDP polyribonucleotidyltransferase domain of the rabies virus large protein. Viruses. 2019b;11(6).

Ogino T, Kobayashi M, Iwama M, Mizumoto K. Sendai virus RNA-dependent RNA polymerase L protein catalyzes cap methylation of virus-specific mRNA. J Biol Chem. 2005;280(6):4429–35 Epub 2004 Nov 30.

Ogino T, Yadav SP, Banerjee AK. Histidine-mediated RNA transfer to GDP for unique mRNA capping by vesicular stomatitis virus RNA polymerase. Proc Natl Acad Sci USA. 2010;107 (8):3463–8.

Paesen GC, Collet A, Sallamand C, et al. X-ray structure and activities of an essential Mononegavirales L-protein domain. Nat Commun. 2015;6:8749.

Peluso RW. Kinetic, quantitative, and functional analysis of multiple forms of the vesicular stomatitis virus nucleocapsid protein in infected cells. J Virol. 1988;62(8):2799–807.

Peluso RW, Moyer SA. Viral proteins required for the in vitro replication of vesicular stomatitis virus defective interfering particle genome RNA. Virology. 1988;162(2):369–76.

Pflug A, Guilligay D, Reich S, Cusack S. Structure of influenza A polymerase bound to the viral RNA promoter. Nature. 2014;516(7531):355–60.

Poch O, Blumberg BM, Bougueleret L, Tordo N. Sequence comparison of five polymerases (L proteins) of unsegmented negative-strand RNA viruses: theoretical assignment of functional domains. J Gen Virol. 1990;71(Pt 5):1153–62.

Qiu S, Ogino M, Luo M, Ogino T, Green TJ. Structure and function of the N-terminal domain of the vesicular stomatitis virus RNA polymerase. J Virol. 2016;90(2):715–24.

Rahmeh AA, Li J, Kranzusch PJ, Whelan SP. Ribose 2'-O methylation of the vesicular stomatitis virus mRNA cap precedes and facilitates subsequent guanine-N-7 methylation by the large polymerase protein. J Virol. 2009;83(21):11043–50.

Ramanathan A, Robb GB, Chan SH. mRNA capping: biological functions and applications. Nucleic Acids Res. 2016;44(16):7511–26.

Reich S, Guilligay D, Pflug A, et al. Structural insight into cap-snatching and RNA synthesis by influenza polymerase. Nature. 2014;516(7531):361–6.

Ribeiro Ede A, Jr., Leyrat C, Gerard FC, et al. Binding of rabies virus polymerase cofactor to recombinant circular nucleoprotein-RNA complexes. J Mol Biol. 2009;394(3):558–75.

Schnell MJ, Conzelmann KK. Polymerase activity of in vitro mutated rabies virus L protein. Virology. 1995;214(2):522–30.

Schoehn G, Iseni F, Mavrakis M, Blondel D, Ruigrok RW. Structure of recombinant rabies virus nucleoprotein-RNA complex and identification of the phosphoprotein binding site. J Virol. 2001;75(1):490–8.

Shuman S. Structure, mechanism, and evolution of the mRNA capping apparatus. Prog Nucleic Acid Res Mol Biol. 2001;66:1–40.

Steitz TA. A mechanism for all polymerases. Nature. 1998;391(6664):231–2.

Stillman EA, Whitt MA. Transcript initiation and 5'-end modifications are separable events during vesicular stomatitis virus transcription. J Virol. 1999;73(9):7199–209.

Tao Y, Farsetta DL, Nibert ML, Harrison SC. RNA synthesis in a cage–structural studies of reovirus polymerase lambda3. Cell. 2002;111(5):733–45.

Testa D, Banerjee AK. Two methyltransferase activities in the purified virions of vesicular stomatitis virus. J Virol. 1977;24(3):786–93.

Testa D, Chanda PK, Banerjee AK. Unique mode of transcription in vitro by Vesicular stomatitis virus. Cell. 1980;21(1):267–75.

Tordo N, Poch O, Ermine A, Keith G. Primary structure of leader RNA and nucleoprotein genes of the rabies genome: segmented homology with VSV. Nucleic Acids Res. 1986;14(6):2671–83.

Tordo N, Poch O, Ermine A, Keith G, Rougeon F. Walking along the rabies genome: is the large G-L intergenic region a remnant gene? Proc Natl Acad Sci U S a. 1986;83(11):3914–8.

Tordo N, Poch O, Ermine A, Keith G, Rougeon F. Completion of the rabies virus genome sequence determination: highly conserved domains among the L (polymerase) proteins of unsegmented negative-strand RNA viruses. Virology. 1988;165(2):565–76.

Villarreal LP, Holland JJ. Transcribing complexes in cells infected by vesicular stomatitis virus and rabies virus. J Virol. 1974;14(3):441–50.

Monkeypox Virus

Unveiling the Arcane of an Elusive Virus from the Heart of the African Continent: The Monkeypox

Bien-Aimé M. Mandja and Jean-Paul Gonzalez

Abstract Monkeypox is a zoonotic disease caused by the monkeypox virus (MPXV) of the same *Poxiviridae* family and the same genus *Orthopoxvirus* than the close related smallpox virus. MPXV affects humans and animals frequently living in the tropical rainforest mostly in Central and West Africa. Human monkeypox clinical signs are mimicking those of smallpox lesions and need a comprehensive differential diagnostic with others poxvirus infection. MPXV can infect a wide range of mammalian species from pan-geographical to local distribution. The frequency and geographic spread of human MPX cases have increased in recent years with occurrence of outbreaks outside of Africa (USA, UK, Israel, and Singapore). Human-to-human transmission of MPX is increasingly reported especially in immunocompromised people. In addition, there is a more and more active circulation of MPXV often encountered among travelers visiting known endemic areas. This set, and this low-noise extension of MPXV, favors an epidemic risk not only at regional but also international level. This could potentially lead to a pandemic threat as it happens by the past, with its very close parent of smallpox virus (i.e. Variola) and, in another context but also from a single zoonotic source, with the SARS-Cov-2 (i.e. COVID-19) today.

Keywords Monkeypox · Orthopoxvirus · Central and West Africa · Democratic Republic of Congo · Emerging infectious diseases · Zoonosis · Disease outbreaks

---

B.-A. M. Mandja (✉)
Faculty of Medicine, Microbiology Department, Kinshasa University, Kinshasa, Democratic Republic of Congo

B.-A. M. Mandja
Chrono-Environment Laboratory, UMR 6249 CNRS, Bourgogne Franche-Comté, University Besançon, Besançon, France

J.-P. Gonzalez
Department of Microbiology & Immunology, Division of Biomedical Graduate Research Organization, Georgetown University, School of Medicine, Washington, DC 20057, US

J.-P. Gonzalez
Commonwealth Trade Partner Inc., Entaurus Biotech LLC, Alexandria, VA 22314, US

S. I. Ahmad (ed.), *Human Viruses: Diseases, Treatments and Vaccines*, https://doi.org/10.1007/978-3-030-71165-8_22

1 Introduction

Monkeypox (MPX) is a zoonotic disease caused by the monkeypox virus (MPXV) a very close parent of the smallpox virus within the same family of poxviruses. MPXV affects humans and animals commonly living in the tropical regions of the rainforest areas of Central and West Africa, also, it can occasionally be disseminated to other territories (Arita et al. 1985; Breman 2000). MPXV was indeed discovered in Singapore in 1958, and first isolated from Cynomolgus spp. monkeys when outbreaks of a pox-like disease occurred in monkey colonies kept for research (Table 1), hence the name 'monkeypox. Also, a large pox-like outbreak was recorded in rhesus monkey from Bengal in India in 1936 and potentially attributed to MPXV. Von Magnus et al. (1959), then, others outbreaks of monkeypox were reported (Prier et al. 1960; Sauer et al. 1960; McConnell et al. 1962; Peters 1966; Arita and Henderson 1968) including *Cynomolgus spp.*, orangutans, gorilla and *Cercopithecus* spp., marmoset, *Macaca philippinensis*, *Macaca mulatta*, *Cercopithecus aethiops* var. *sabaeus,* from which primate species other type of poxvirus than MPXV were ruled out (Arita and Henderson 1968) (Table 1).

Table 1 Monkeypox virus emerging and resurgences in human and vertebrate hosts overtime and by country

Country (1)	Year	Human Confirmed Cases (2)	Animal	Reference (3)
France&	1767	0	NHP	*Barrier, quoted by Schmidt (1870)
Panama& (India)	1841	0	Rhesus NHP	*Anderson (1861)
France&	1842	0	NHP	*Rayer, quoted by Schmidt (1870)
Trinidad	1858	0	NHP	*Furlong, quoted by Schmidt (1870
Brazil	1922	0	*Mycetesseniculus, Cebuscapucinus*	*Bleyer (1922)
India	1936	0	*Macacamulatta*	*Rahman (pers.comm. 1967)
Indonesia	1949	0	Orang-utan	*Gispen (1949)
Denmark # (Singapore)	1958	0	*Macacacynomolgus*	*Von Magnus et al. (1959)
USA #	1959	0	*Macacaphilippinensis, M. mulatta, Cercopithecus aethiopsvar.sabaeus*	*Prier et al. (1960) and Sauer et al. (1960)
Washington D.C., USA	1962	0	*Cynomolgusspp*, Rhesus, African Green	*McConnell et al. (1962)
Netherlands #	1964	0	*Cynomolgusspp*	*Gispen and Kapsenberg (1968)

(continued)

Table 1 (continued)

Country (1)	Year	Human Confirmed Cases (2)	Animal	Reference (3)
Netherlands #	1965	0	*Cynomolgusspp*	*Gispen and Kapsenberg (1968)
Rotterdam Zoo #	1964	0	Giant anteater, Orangutan, Chimpanzee, Gorilla, Cercopithecus, marmoset, gibbon, squirrel monkey	*Peters (1966)
USA (India)*	1965	0	African green monkeys	*Bierly, perscomm (1967)
USA (Philippines – Malaysia) *	1965	0	*Macacairus*	*J.H.Vickers, per.comm. (1967)
UCLA USA (India)	1967	0	Rhesus	*K. E. Hamlin per.comm. (1967)
Zaire or current DRC	1970	1	NA (4)	Breman et al. (1980); Ježek and Fenner (1988)
Liberia	1970	4	NA	Bremanet al. (1980); Ježek and Fenner (1988)
Sierra Leone	1970	1	NA	Breman et al. (1980); Ježek and Fenner (1988)
Ivory Coast	1971	1	NA	Breman et al. (1980); Ježek and Fenner (1988)
Nigeria	1971	2	NA	Breman et al. (1980); Ježek and Fenner (1988)
Zaire or current DRC	1972	5	NA	Breman et al. (1980); Ježek and Fenner (1988)
Zaire or current DRC	1973	3	NA	Breman et al. (1980); Ježek and Fenner (1988)
Zaire or current DRC	1974	1	NA	Breman et al. (1980); Ježek and Fenner (1988)
Zaire or current DRC	1975	2	NA	Breman et al. (1980); Ježek and Fenner (1988)
Zaire or current DRC	1976	3	NA	Breman et al. (1980); Ježek and Fenner (1988)
Zaire or current DRC	1977	6	NA	Breman et al. (1980); Ježek and Fenner (1988)
Nigeria	1978	1	NA	Breman et al. (1980); Ježek and Fenner (1988)

(continued)

Table 1 (continued)

Country (1)	Year	Human Confirmed Cases (2)	Animal	Reference (3)
Zaire or current DRC	1978	11	NA	Breman et al. (1980); Ježek and Fenner (1988)
Zaire or current DRC	1979	4	NA	Breman et al. (1980); Ježek and Fenner (1988)
Cameroon	1979	2	NA	Ježek and Fenner (1988)
Zaire or current DRC	1980	4	NA	Ježek and Fenner (1988)
Ivory Coast	1981	1	NA	Ježek and Fenner (1988)
Zaire or current DRC	1981	7	NA	Ježek and Fenner (1988)
Zaire or current DRC	1982	40	NA	Ježek and Fenner (1988)
Zaire or current DRC	1983	84	NA	Ježek and Fenner (1988)
Central African Republic	1984	6	NA	Khodakevitch et al. (1985)
Zaire or current DRC	1984	86	NA	Khodakevitch et al. (1985)
Zaire or current DRC	1985	62	Rope squirrel (*Funisciuruscongicus*)	Ježek and Fenner (1988)
Sierra Leone	1986	1	NA	Ježek and Fenner (1988)
Zaire or current DRC	1986	59	NA	Ježek and Fenner (1988)
Gabon	1987	5	NA	Meyer et al. (1991)
Zaire or current DRC	1987	1	NA	Heymann et al. (1998)
Cameroon	1989	11	NA	Heymann et al. (1998); Tchokoteu et al. (1991)
Gabon	1991	5	NA	Heymann et al. (1998)
Zaire or current DRC	1992	1	NA	Heymann et al. (1998)
DRC	1996–1997	511	Squirrel species, Gambian rats, elephant shrew, domestic pig	CDC (1997); Hutin et al. (2011)
DRC	2001	6	NA	Heymann et al. (2002)
Central African Republic	2001	4	NA	Durski et al. (2008)

(continued)

Table 1 (continued)

Country (1)	Year	Human Confirmed Cases (2)	Animal	Reference (3)
Republic of Congo	2003	11	NA	Learned et al. (2003)
United States (Ghana)	2003	37	Prairie Dog (2); *Funiscuirusspp.* or rope squirrel, *(2) Cricetomysspp.*or Gambian-pouched rat *(1); Graphiurusspp.* or Africandormouse *(3)*	CDC (2003); Guarner et al. (2004)
Sudan (DRC)	2005	19	NA	Damon et al. (2006); Formenty et al. (2010)
DRC	2005–2007	760	NA	Rimoin et al. (2010)
Republic of Congo	2009	2	NA	Reynolds et al. (2013)
Central African Republic	2010	2	NA	Berthet et al. (2011)
Ivory Coast	2012	0	*CercocebusMangabey* (1)	Radonic et al. (2012)
Sierra Leone	2014	1	NA	Li et al. (2017)
Central African Republic	2015	12	NA	Kalthan et al. (2016); Nakoune et al. (2017)
Central African Republic	2016	3	NA	WHO (2016)
Central African Republic	2017	8	NA	WHO (2017b), candd
Liberia	2017	2	NA	WHO (2018a)
Republic of Congo	2017	7	Cricetomysemini	WHO (2017a), Doshi et al. (2019)
Sierra Leone	2017	1	NA	WHO (2017e)
Nigeria	2017	80	NA	Yinka-Ogunleye (2019 et 2020)
Nigeria	2018	38	NA	Yinka-Ogunleye (2019 et 2020)
Cameroon	2018	1	NA	WHO (2018c)
Central African Republic	2018	14	*Civettictiscivetta, Cricetomysemini, Funesciurusanerythrus*	Besombes et al. (2019)
Israel (Nigeria)	2018	1	NA	Erez et al. (2019)
United Kingdom (Nigeria)	2018	7	NA	Vaughan et al. (2018 et 2020)

(continued)

Table 1 (continued)

Country (1)	Year	Human Confirmed Cases (2)	Animal	Reference (3)
Nigeria	2019	46	NA	ProMED (2019)
Singapore (Nigeria)	2019	1	NA	ProMED (2019b)
Republic of the Congo	2019	152	NA	ProMED (2019c)
Republic of the Congo	2020	73	NA	ProMED (2020)
Cameroon	2020	1	NA	ProMED (2020a)

Legend (1) & = No virus isolation; In parenthesis (origin of imported case); # = NHP from a biological Institution; (2) Bold = suspected, very few biologically confirmed cases or data of biologically confirmed cases not yet available; (3) * = quoted by Arita et al. (1985); (4) NA- Not Applicable. The corresponding event are of human origin without animal associated with

Caption CDC, Center of Disease Control; DC, District Colombia, DRC, Democratic Republic of Congo; NASIM, No Animal Source of Infection Mentioned; NHP, No Human Primate; UCLA, University California of Los Angeles

This work used only official published sources (research articles, reviews, WHO reported)

Before 1958, the observed outbreak of poxvirus in Primates where attributed, with respect to the contagiousness of these viruses, to Monkey pox virus rather than smallpox, when no concomitant human case was observed

The first human case was a 9 years old infant reported in Zaire (alias Democratic Republic of Congo, DRC), Equateur province, in 1970, in the midst of the Global Smallpox Eradication Campaign (Ladnyj et al. 1972; Marennikova et al. 1972). In order to assess the importance of this emerging phenomenon, the World Health Organization (WHO) engaged an active surveillance program for MPX in the DRC between 1981 and 1986 (Ježek and Fenner 1988). The epidemiological data obtained by this surveillance made it possible to develop a model for predicting the human-to-human spread of MPXV. This inaugural work in collaboration with the CDC made it possible to conclude that there was a very low chance of maintaining MPX in human populations. In addition, these initial results showed that human infection with MPXV did not, at that time, considered as a major public health issue in the potential endemic areas studied (Beaux et al. 1988; Ježek and Fenner 1988). However, after 40 years of cessation of vaccination campaigns against smallpox, it generally was admitted that, there is a steady increase of MPX disease cases in several tropical regions of Africa, such an upsurge starting to arise a real public health concern (Fuller et al. 2011; Rimoin et al. 2010).

2 The Monkeypox Virus

Structure and taxonomy: MPXV is a double-stranded DNA (170–250 kbp), that belongs to the *Orthopoxvirus* genus of the *Poxviridae* family including the variola major (variola virus, VARV), cowpox (CPXX), and the Vaccinia (VACV) viruses among others 27 species and subspecies identified (Hendrickson et al. 2010). Virions are brick-shaped, of about 200 nm section by 250 nm length (Jahrling et al. 2007) (Figs. 1). Currently, there are two different genomic groups of MPXV that have been identified which include: (1) The Central African group; and, (2) The West African group (Chen et al. 2005). The Central African strains are characterized by their high virulence, efficient human-to-human transmission and by having a high level of lethality (1.5 to 10%) (Heymann et al. 1998; Ježek and Fenner 1988). The West African strains are less pathogenic and less virulent (Likos et al. 2005; Saijo et al. 2009). Although, MPXV is not a direct ancestor to VARV, it causes a disease clinically mirroring smallpox clinical pattern, but with a milder rash, and lower death rate of 0 to 10% as compared to 10 to 30% case-fatality rate for variola major.

Transmission: There are two modes of MPXV transmission: (1) An animal transmission or primary infection (from a natural reservoir to a natural permissive host i.e. animal or human) And, (2) a human-to-human transmission or, secondary infection. Primary as well as secondary cases are infected by direct contact with

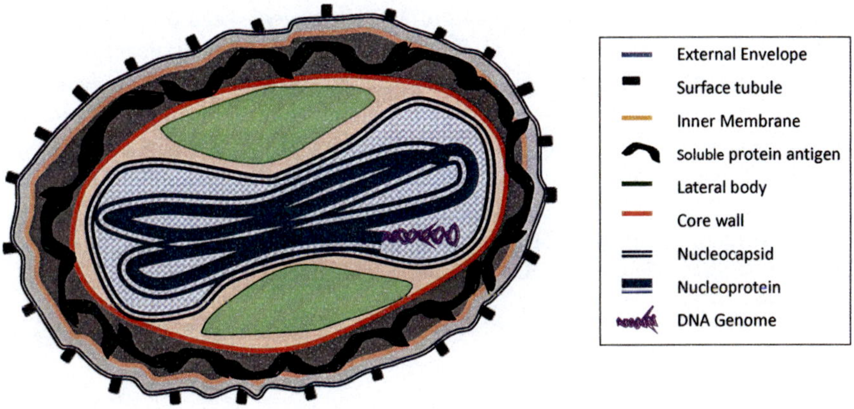

Fig. 1 Schematic structure of the Monkeypox virus. *Legend* The monkeypox virus is a brick-wrapped virion 250 nm long and 200 nm wide. The surface membrane has tubules or surface filaments. There are two types of infectious viral particles depending on the maturity of the replication of the virus: the mature intracellular virus and then the enveloped extracellular virus. The Genome is made of a linear double helix of 170–250 kb. Looks like a linear structure flanked by inverted terminal repeating sequences that form hairpin terminations at each end

biological products such as blood, lymph from skin vesicles, bites (saliva) from infected animals (Breman et al. 1980; Huhn et al. 2005; Ježek et al. 1988). Human-to-human transmission can result from respiratory infection during prolonged close or direct contact with bodily secretions of an infected person, or from any recently infected objects by the MPXV (clothing, bed sheets, etc.) (Arita et al. 1985; Fenner et al. 1988; Ježek et al. 1988). One can identify major risk factors of MPXV infection mainly associated with trapping and hunting of potential MPXV host animals, transport or butchering of such infected prey, children handling carcasses of infected animals or a by a bite of an injured animal, and also, in the preparation of wild infected animal or in its consumption (Jones et al. 2008).

3 Clinic

The clinical presentation of MPX mimics that of smallpox. The incubation period varies between 10 and 14 days. Then the invasion period (lasts between 0–5 days) disease begins with a prodromal syndrome characterized by fever, severe headache, back pain, myalgia, severe asthenia and lymphadenopathy. Then will appear the characteristic maculopapular rashes. These lesions generally follow a centrifugal distribution: they first appear at the extremities (face, forearm, hands, legs and feet) before spreading all over the body. These eruptions evolve in papules towards vesicles, pustules, umbilicated lesions and crusts, for approximately 14 to 21 days before flaking occurs, often leaving depigmented scars (Fig. 2a and b).

Unlike smallpox, a hemorrhagic sign of MPX has never been documented (Di Giulio and Eckburg 2004; Nalca et al. 2005). In some patients, significant sequelae may be observed such as smallpox-like scars, keloid scars, total or partial hair loss (alopecia), corneal opacities and cases of unilateral or bilateral blindness. Some patients may develop secondary bacterial infections of the skin and soft tissue, vomiting with dehydration, bronchopneumonia, encephalitis and sepsis (Di Giulio and Eckburg 2004; Nalca et al. 2005). Also, because the exacerbation of the multiple severe syndromes (dehydration, bronchopneumonia, encephalitis and concurrent sepsis), death may occur for one in ten subjects with the disease (Ježek and Fenner 1988).

4 Surveillance and Prevalence

One can distinguish several and original epidemiological patterns of MPX disease according to the environment and the origins of transmission. Human monkeypox infections have been documented four times outside of Africa; in the United States in 2003 (37 confirmed cases among 47 suspected) (Croft et al. 2007; Ligon 2004), and in both the United Kingdom (7 cases including 2 imported from Nigeria and the other form contact with UK) (Vaughan et al. 2018 and 2020) and Israel (1 case) in

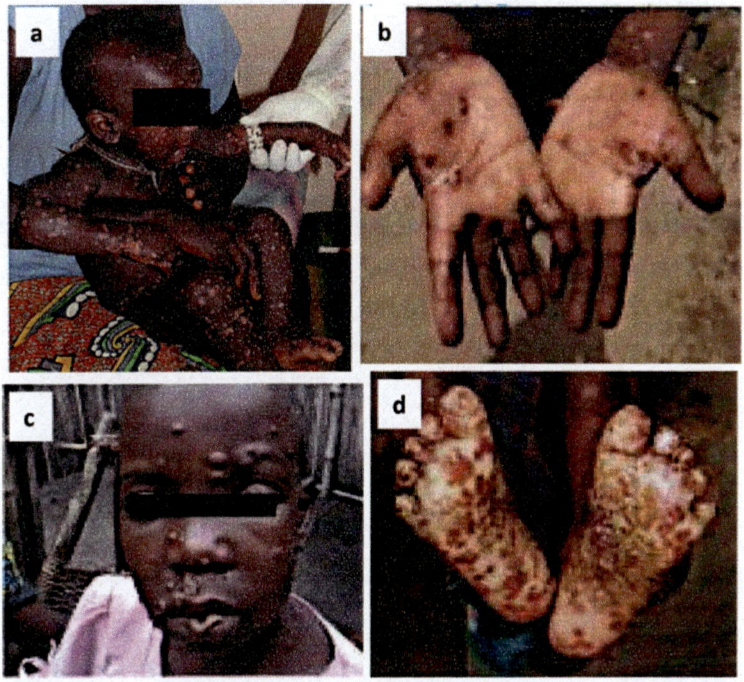

Fig. 2 Clinical presentation of Monkeypox disease among young children. *Legend* **a** Young baby from West Africa (2015 Sierra Leone); **b**, **c**, **d** monkeypox patients in young children from Central Africa (2009 Sankuru, Democratic Republic of Congo). *Credit* **a** Linda Mose and Nadia Wauquier; **b**, **c**, **d** Placide Mbala

2018 (Erez et al. 2019), and Singapore (1 case) in 2019 (Ng et al. 2019). Also, people living in or near the tropical rain forest may have indirect or low-level exposure to MPXV, eventually leading to subclinical infection (https://www.who.int/news-room/fact-sheets/detail/monkeypox).

MPX of West and Central Africa. Most human cases of MPX occurred in West and Central Africa (Breman 2000; Learned et al. 2005). In West Africa cases have been reported from Nigeria, Ivory Coast, Liberia and Sierra Leone (Breman 2000; Foster et al. 1972; Lourie et al. 1972), while in Central Africa, Cameroon, Central African Republic, DRC, Gabon and Republic of Congo (Berthet et al. 2011; Heymann et al. 1998; Learned et al. 2005; Meyer et al. 1991; Müller et al. 1988; WHO 2011; Reynolds et al. 2013; Mandja et al. 2019a) (Fig. 3).

From 1970 to 1979, 47 new cases of MPX were identified in Africa, including 38 cases notified in DRC with a case-fatality rate of 17% (Fig. 4). 83% of the cases were children under the age of ten years old. Four cases were attributed to secondary human-to-human transmission showing a secondary attack rate of 7.5% (Breman et al. 1980; WHO 1980).

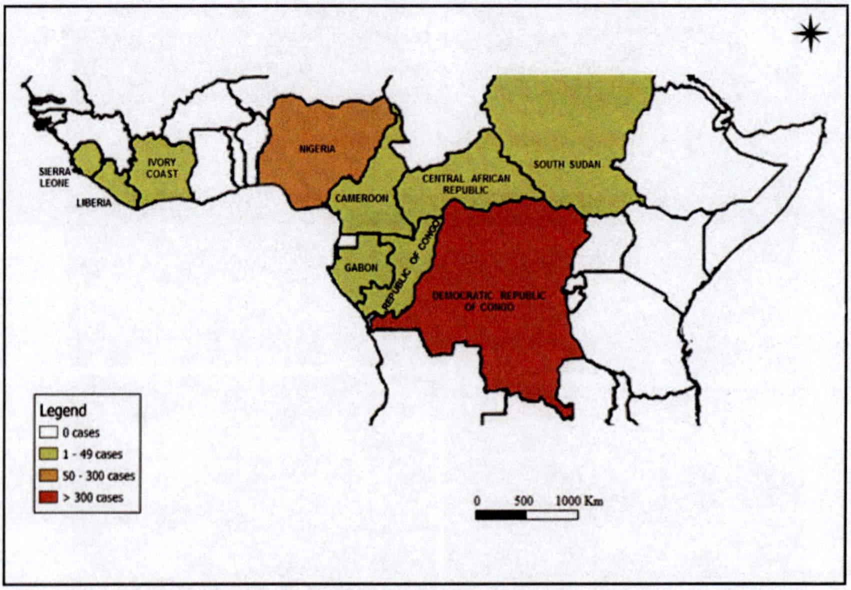

Fig. 3 Countries reporting human monkeypox cases in Africa 1970–2018

The WHO active MPX surveillance program (1981 to 1986) was able to record 404 cases of MPX with total of 386 out of 404 cases identified in DRC (Ježek and Fenner 1988) (Fig. 4). An animal source of infection was mentioned for 245 of the 338 cases in the DRC, as a 72% primary transmission ate. Most affected were children (86%) were under the age of 10 years old. An observed 10% fatality rate was lower than the 17% found before 1980. While at the end of the intensive WHO surveillance program (1986) and until 1992, 13 cases of MPX were reported in Central Africa from Gabon, Cameroon and the DRC. Then, no cases were reported from 1993 to 1995, linked to the dramatic decrease of active MPX surveillance (Heymann et al. 1998) (Fig. 4).

Then, from February 1996 to October 1997, occurred the largest ever reported epidemic of MPX that occurred in the Katako-Kombe and Lodja health zones (Sankuru health district, Kasai Oriental province, DRC). A total of 511 suspected cases were recorded (Fig. 4) with a case-fatality rate of 1.5 to 3%. Most affected patients were under the 16 years old. This epidemic was characterized by a very large human-to-human transmission estimated at 78% of the cases. However, the limitation of biologically confirmed cases and the low case-fatality rate suggested a concurrent circulation of MPX and Chicken pox viruses having a confusing clinical presentation (CDC 1997; Hutin et al. 2001).

In the DRC, from January 1998 to December 2002, 31 patients were identified (Meyer et al. 2002) while four other cases occurred in the Central African Republic in 2001 (Durski et al. 2008). Altogether, for the period (Kebela 2004) 1,265 suspected cases of MPX were recorded with 44.6% cases of the 10 to 24 class of age.

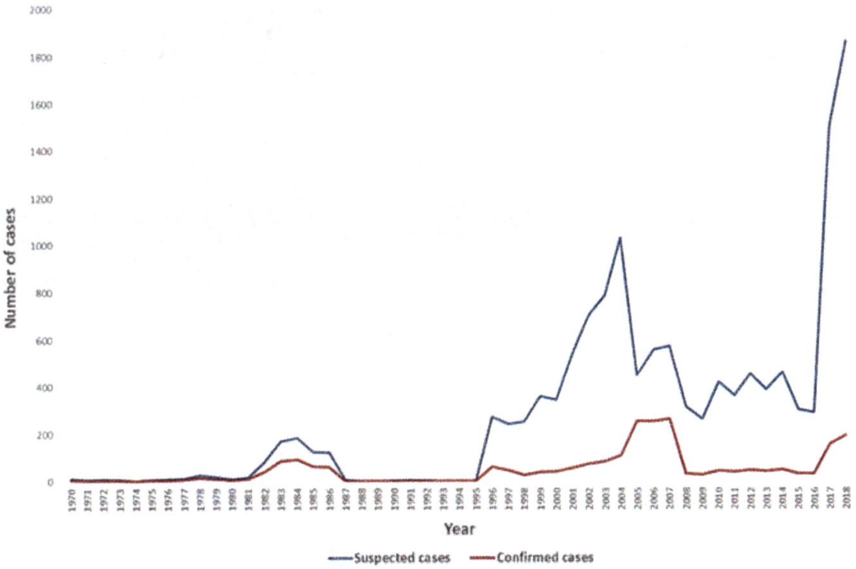

Fig. 4 Timeline of Monkeypox disease emergence in Africa, 1970–2018. *Caption* For the elaboration of this figure, this work used both published sources (research articles, reviews, WHO reported) and unpublished data mainly data of DRC National Passive Disease Surveillance System. Although, the validity of unofficial data is not guaranteed and is often missing, a study showed the importance of this source of information and demonstrated that they can be used for the research (Mandja et al. 2019b)

Among 215 samples tested, 41% were MPX laboratory confirmed. Another report from the DRC National Passive Disease Surveillance System indicated that 2,734 MPX suspected cases had been reported from January 1, 2001 to December 31, 2004 (Sklenovská and Van Ranst 2018) (Figs. 4 and 6).

The active surveillance program for MPX in the DRC between November 2005 and November 2007 had identified 760 biologically confirmed MPX cases in Sankuru district (Figs. 4, 5 and 6). The average age of the cases was 11.9 years old and 92.1% were born after the cessation (1980) of smallpox vaccination (Rimoin et al. 2010).

In 2010, 10 cases (2 confirmed, 8 suspected) were notified by the Republic of Congo in the Likouala Department (Reynolds et al. 2013) and two cases confirmed by the Central African Republic (Berthet et al. 2011). From 2010 to 2014, 2,000 suspected cases of MPX were reported each year mainly from the provinces of Equateur and Kasaï Oriental (Mwamba et al. 2014; Malekani et al. 2014; Nolen et al. 2016) (Figs. 4, 5 and 6).

In 2014, MPX re-emerge in Sierra Leone, after 40 years since last outbreak, while only one case was biologically confirmed (Li et al. 2017). In the Central African Republic from December 2015 to February 2016, 12 cases of MPX had been reported from the provinces of Bangassou and Mbomou (Kalthan et al. 2016; Nakouné et al; 2017), and then in August 2016, 26 suspected cases (3 confirmed)

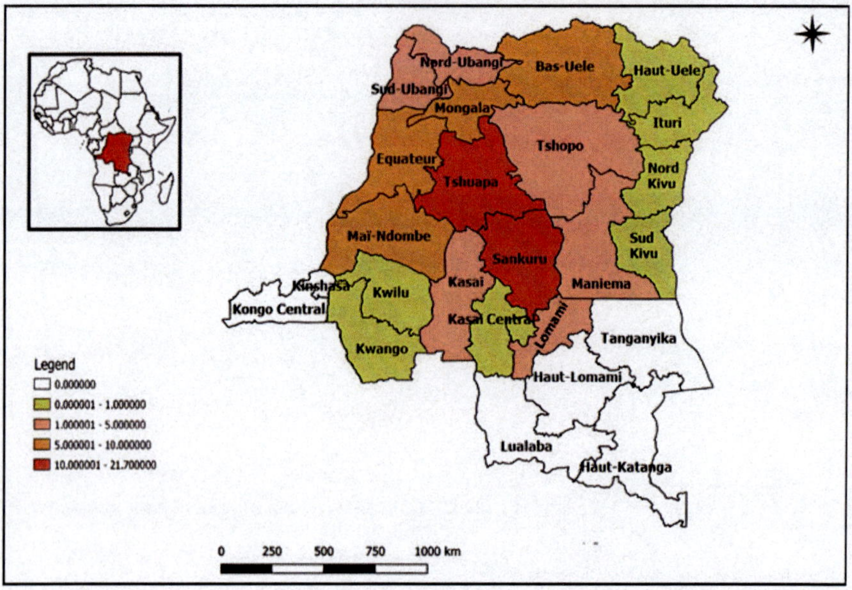

Fig. 5 Geography distribution of human Monkeypox among the Democratic Republic of Congo provinces, 2000–2018. *Caption* Data of DRC National Passive Disease Surveillance System

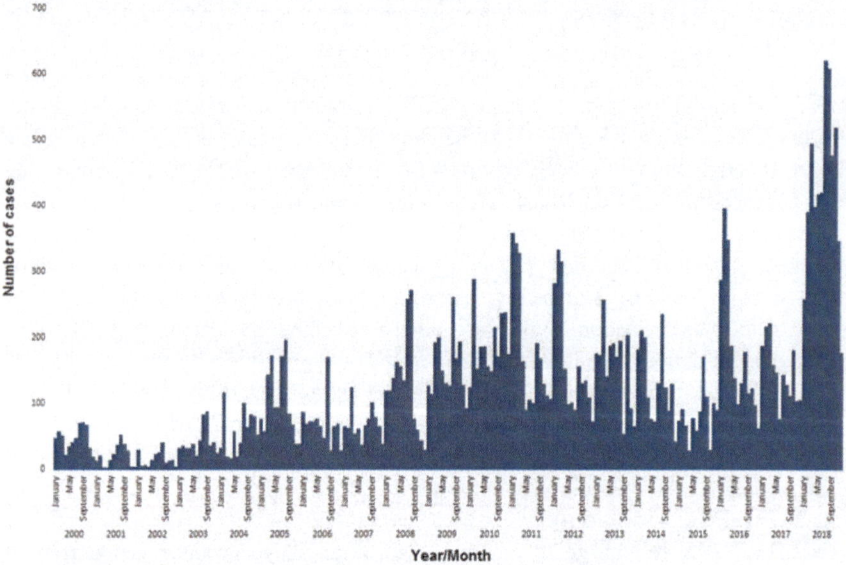

Fig. 6 Timeline of Monkeypox disease evolution in DRC, 2000–2018. *Caption* Data of DRC National Passive Disease Surveillance System

occurred in the provinces of Basse-Kotto and Haute-Kotto (WHO 2016). While from September 2014 to February 2016, 587 suspected cases were reported in DRC (Sklenovská 2017) (Figs. 4 and 6).

Between January and August 2017, the Likouala Department of the Republic of Congo reported an epidemic of 88 suspected MPX cases with 7 confirmed (WHO 2017a). In 2017, two outbreaks have been reported in the Central African Republic: one February in the province of Mbomou with 47 suspected cases (7 confirmed) (WHO 2017b and c) and the second in April in Mbaiki district (3 cases, 1 confirmed) (WHO 2017d). One isolated confirmed case was reported in district of Pujehan in Sierra Leone (WHO 2017e) (Fig. 4).

Between September 2017 and August 2018 in Nigeria, the largest epidemic of MPX in West Africa occurred: A total of 262 suspected cases (113 confirmed and Seven deaths) had years old (Yinka-Ogunleye et al. 2018 and 2019).

After 40 years of silence, from November to December 2017, a MPX outbreak was reported in Liberia with a total of 16 suspected cases (2 confirmed) (WHO 2018a) (Fig. 4).

Between the first and 24th week of 2018, 2,845 suspected cases had been reported by the DRC National Passive Disease Surveillance System (WHO 2018b) (Fig. 6). Between 17 March and 24 April 2018, 20 suspected MPX cases (9 confirmed) had been reported in the Central African Republic (WHO 2018c). In May 2018, 16 suspected cases (1 confirmed) were identified in Cameroon (WHO 2018c) (Fig. 4).

An MPX expending geographic epidemic. In recent years, it has been observed MPX spill over to three other continents (North American, European and Asian) following the movement of MPXV infected person and the transportation of naturally infected animal. Moreover, this unexpected extension, out of MPX endemic area, was also highly due to the mobility of the people and growing trade in and out of Africa.

MPX epidemic in the USA. In May 2003, an outbreak of MPX was discovered in the state of Wisconsin. The disease was introduced into the USA by a shipment from Ghana of MPXV infected exotic rodents (giant Gambian rat, African dormouse and ground squirrel) that contaminated local prairie dogs housed together in the pet store. From Wisconsin, the epidemic spread to the states of Indiana, Illinois, Missouri, Kansas and Ohio. A total of 47 cases were identified (37 confirmed, no deaths). The patients were aged from 3 to 43 years old, both sexes were affected equally (Croft et al. 2007; Ligon 2004).

MPX outbreak in Sudan From September to December 2005, there was an outbreak of MPX in the State of Unity in Sudan (alias Sud Sudan).Ten out of 19 patients (52%) were women, 80% were under the age of 20, no patient died. Although initially, a new virus derived from the Central African strain was suspected in this epidemic (Damon et al. 2006; Formenty et al. 2010) recently the Sudan MPXV isolates were shown close to those of northern DRC. Therefore, it is probable that MPXV originated from infected imported animals from the Congo Basin and/or due to the movement of infected population, with respect to the socio-political instability at the time of the epidemic (Nakazawa et al. 2013).

MPX epidemic in the UK. In September 2018, two separately imported confirmed cases were identified in the United Kingdom (UK) in Cornwall and in Blackpool. These patients were not related and returned from Abuja in Nigeria, where an MPX outbreak started in September 2017 to this date. For the first time MPX was notified in Europe (Vaughan et al. 2018). A healthcare worker was contaminated from one of these patients, while among 134 healthcare worker contacts, four became ill. No patient had died (Vaughan et al. 2020).

An imported MPX case in Israel. In October 2018, a 38-year-old Israel resident, who returned to Jerusalem from Rivers State in Nigeria, where MPX circulated, tested positive for MPX, and survived. For the first time MPX was reported in the Asian continent. No secondary cases were detected (Erez et al. 2019).

An imported MPX case in Singapore. In May 2019, an imported confirmed MPX case was identified in Singapore. The patient was a Nigerian male, who travelled in Singapore survived. No secondary cases were detected (Ng et al. 2019).

Virus Natural Reservoir(s) and Hosts

Although notable progress has been made on the knowledge of hosts and reservoirs of the MPXV, several gray areas still persist: Numerous serological surveys have shown that a large number of animals can be infected in nature. The most incriminated animals are monkeys (crab-eating macaque, *Macaca fascicularis*—the first original host, and potentially other macaques species and the chimpanzee, *Pan troglodytes*) (Breman et al. 1977; Mutombo et al. 1983; Arita et al. 1985), squirrels (Thomas's rope squirrel, *Funisciurus anerythrus*; Red-legged Squirrel, *Heliosciurus rubobrachium;* and, Congo rope squirrel, *Funisciurus congicus*) (Khodakevich et al. 1986, 1988) and, rats (Gambian pouched rat, *Cricetomys spp.*; African dormice, *Graphiurus spp.*; Tetradactyl horn rat, *Petrodromus tetradactylus*) (Hutin et al. 2001; Reynolds et al. 2010) (Table 1).

However, the primary reservoir of the MPXV remains unknown. From the original hypothesis of an exclusive simian reservoir (*Macaca fascicularis*) at the origin of the discovery of MPX (Arita et al. 1985; Mutombo et al. 1983), we have moved on to the hypothesis of a more diversified host-reservoir chain of transmission, through primates as secondary host of the MPXV transmitted from squirrels (*Funisciurus anerythrus*), whereas squirrels are the only mammals from which the MPXV has so far been isolated in nature (Khodakevich et al. 1986; Khodakevich and German 1988). Moreover, several monkey species are known to pray on squirrels (Parker and Buller 2013) (Table 1).

5 Diagnostic

Clinical and Differential Diagnosis

Typically, the MPX rash evolves sequentially from macules (lesions with a flat base) to papules (slightly raised firm lesions), vesicles (lesions filled with clear fluid), pustules (lesions filled with yellowish fluid), and crusts which dry up and fall off (Petersen et al. 2019).

Several diseases can be characterized by vesiculopustular rashes very similar to those present during MPX. These are, among others: from the poxvirus family, chickenpox, herpetic eczema, herpetiform dermatitis, tanapox, skunkpox, raccoonpox, taterapox; and caused by Rickettsia and molluscum contagiosum, allergic dermatitis, variola dermatitis (Breman 2000; Di Giulio and Eckburg 2004). The main difference between symptoms of smallpox and monkeypox is that monkeypox causes lymph nodes to swell (lymphadenopathy) while smallpox does not.

However, among all these pathologies, the most difficult challenge is to distinguish MPX from the severe and atypical form of chickenpox (Breman 2000; Fenner 2000; MacNeil et al. 2009). Clinically, the chickenpox skin lesions are smaller, much more superficial, localized preferentially on thoracic segment of the body with a centripetal distribution of the skin lesions than centrifugal for MPX and, then gradually transform into crusts after 3 or 5 days (Auvin et al. 2002; Jezek et al. 1988; MacNeil et al. 2009; Nalca et al. 2005). The occurrence of rashes on the palms of the hands and soles of the feet is more characteristic of MPX (75% of the patients) than chickenpox. In addition, there are almost no deaths from chickenpox (Table 2). Human tanapox, which is endemic to the DRC and other African countries, can be differentiated from MPX because it produces skin lesion types which can persist for several weeks, even months (Jezek et al. 1985). Also, some atypical rashes such as those caused by allergic reactions due to medication and insect bites may resemble MPX and chickenpox diseases for clinicians unfamiliar with these two diseases (Breman 2000).

Laboratory diagnosis. Several laboratory tests can be used to establish the specific diagnosis of human MPX: immunohistochemical tests, electron microscopy, virus culture, serological tests and molecular detection techniques (i.e. PCR, viral DNA sequencing and Diagnostic DNA microarrays or "DNAchip") (Chastel and Charmot 2004; Di Giulio and Eckburg 2004; Nalca et al. 2005).

For the sensitivity of these tests, it is important to take the following sample specimens: scabs, secretions from skin lesions, skin biopsies and blood. All samples must be handled according to level 2 biosecurity rules for their collection, packaging and transport (Damon and Esposito 2003; https://www.cdc.gov/poxvirus/monkeypox/lab-personnel/index.html).

The combined use of histological examinations and immunological tests (immunohistochemical analyzes) using polyclonal or anti-poxvirus reacting monoclonal antibodies, can make it possible positively identify the MPXV (Bayer-Garner 2005).

Electron microscopy: for skin biopsies has been one of the main viral diagnostic tests in the past (Gentile and Gelderblom 2005). If available, it can be used as

Table 2 Monkey pox, Smallpox and Chickenpox (Varicella) Diseases Differential diagnosis

Case history and clinical signs	MPX	Variola	Varicella
Anamnesis			
Recent contact with an exotic animal	Yes	No	Yes
Recent exposure to a patient with vesicular-pustular rashes	Possible	No	Yes
Smallpox vaccination	10–15%	Rare	Yes
Incubation period (day)	Oct-14	Oct-14	14–16
Prodromal period (day)	01-Mar	02-Apr	0–2
Main clinical signs			
Prodromal fever and malaise	Yes	Yes	Yes
Lymphadenopathy	Yes	Non	Non
Skin lesion Distribution	Centrifugal	Centrifugal	Centripetal
Skin lesion depth	Superficial	Intradermal	Intradermal
Skin lesion evolution	Uniform	Uniform	Polyform
Skin drying/peeling (day post rash)	22–24	14–21	Jun-14
Hand palm and feet sole Lesions	Specific	Specific	Rare
Secondary clinical signs			
Bronchitis pneumopathy	12%	Likely	3–16%
Secondary infection of the skin and soft tissue	19%	Rare	Likely
Ocular lesion	4–5%	5–9%	Non
Encephalitis	< 1%	< 1%	< 1%
Laboratory diagnostic			
Specific PCR for each virus	MPX	Variola	Varicella Zoster
Electron microscopy	Poxvirus like	Poxvirus like	Herpesvirus
Virus isolation on embryonated egg	yes	yes	No
Serology (reacting antibodies)	MPXV (1)	Variola (1)	Varicella (2)

Legend (1) = anti-orthopoxvirus IgG and IgM antibody test performed by ELISA; (2) anti Varicella-Zoster IgG and IgM tests, performed using ELISA. Modified from: Di Giulio and Eckburg (2004)

Caption ELISA, Enzyme Linked Immunosorbent Assay; Ig, immunoglobulin, MPX, Monkeypox; PCR, Poly Chain Reaction; MPXV; Monkeypox Virus

a first-line method in biological confirmations of poxvirus infections with respect to the original shape of the virus.

MPXV Isolation by culturing the virus on a cell line or in the embryonated egg and its characterization by PCR techniques are currently the most precise methods for identifying Orthopoxvirus species. PCR and viral DNA sequencing are more sensitive and specific than other conventional virological and immunological methods. These techniques allow better differentiation of species or strains (Damon and Esposito 2003).

The serological tests for MPXV are not very specific because of the close antigenic relationship between the surface antigens common to all Orthopoxviruses. Moreover, the sensitivity of these tests is limited (50 to 90%) and such tests are not used for the diagnosis of MPX in the acute phase of the disease (Damon and Esposito 2003). During the active phase, biological confirmation has to be done by PCR analysis of the vesicle's fluid or the crusts.

6 Treatment, Control and Prevention

There is no specific treatment for MPXV infection. However, smallpox vaccination has been shown to provide cross-protection against this zoonosis (Arita et al. 1985; Jezek et al. 1986; Ježek and Fenner 1988). This is why the CDC recommends smallpox vaccination before exposure for anyone investigating or managing any suspected MPX infected human or animal, or by handling suspected samples likely to contain MPXV (CDC (https://www.cdc.gov/ncidod/monkeypox/treatmentguidelines.htm). The effectiveness of vaccinia immunoglobulin, formerly used to treat smallpox, has not yet been proven to potentially treat MPX patients. However, such immunoglobulin could be eventually used as a prophylactic treatment for those exposed with severe immunodeficiency in whom smallpox vaccination is prohibited.

The treatment of MPX is mainly symptomatic with the use of antipyretics, analgesics and skin antiseptics. Licensed antiviral drugs effective for the treatment of MPXV infection do not yet exist on the market. In the curative treatment against MPX, the most promising antiviral drug remains Cidofovir®, marketed under the name of Vistide® (De Clercq 2002). However, Cidofovir has never been administered in the curative management of human MPX and has unsuitable nephrotoxic effects. Other drugs derived from Cidofovir (HPMPO-DApy and ST-246) have shown some efficacy in in vitro treatment against Orthopoxvirus infection and could be being administered orally (Sbrana et al. 2007; Stittelaar et al. 2006). However, there is yet no pharmaceutical marketing authorization of any of these products for the treatment of MPX.

7 Conclusion

MPX is a zoonotic disease which has recently appeared on the African continent and which has become the most important infection in humans, due to an orthopoxvirus after the eradication of the terrible pandemic smallpox in the last century. Because of the resemblance of the clinical picture of MPX to that of smallpox and other poxvirus infections, the challenge is to clinically distinguish MPX from many infections caused by poxviruses but also from other skin diseases.

Today MPX is a growing threat to the public health of people living in endemic regions of Africa but also jeopardizes global health security, as demonstrated by epidemics with local transmission outside the continent of Africa. The availability and speed of international transport, combined with the vulnerability due to a decreasing immunity against smallpox, increases the potential risk of seeing MPX spread on a regional scale but. Moreover, at an international level, as done in its time by the variola pandemic, and today the SARS-COV-2 (Covid-19) pandemic having also, like the MPXV, a zoonotic origin followed by an inter-human transmission.

Acknowledgements This manuscript has been developed with some unpublished material of BAM doctoral thesis done at the University of _Bourgogne Franche-Comté and Kinshasa University, therefore we would like to thanks all directors and supervisors of the thesis: Jean-Jacques Muyembe, Frédéric Mauny, Didier Bompangue and Pascal Handschumacher. As well as extending my warm acknowledgement to the members of the Research and Training Unit on the Ecology and Control of Infectious Diseases of the University of Kinshasa and, all the members of the Clinical Methodology Center of Teaching Hospital Center of Besançon, France, for their constant support. The authors thank also Dr Placide Mbala for providing the clinical photographs of monkeypox patients in Sankuru in DRC, and the patients who gave permission to have the photos taken.

References

Arita I, Henderson DA. Smallpox and monkeypox in non-human primates Bull. World Health Organ. 1968;39(2):277–83.

Arita I, Jezek Z, Ruti K, Khodakevich L. Human Monkeypox: a newly emerged orthopoxvirus zoonosis in the tropical rain forests of Africa. Am J Trop Med Hyg. 1985;34:781–9. https://doi.org/10.4269/ajtmh.1985.34.781.

Auvin S, Catteau B, Ganga-Zandzou PS, Ythier H. Atypical varicella with palm and sole involvement. Int J Dermatol. 2002;41:903–5.

Bayer-Garner IB. Monkeypox virus: histologic, immunohistochemical and electron-microscopic findings. J Cutan Pathol. 2005;32:28–34.

Beaux P, Jezek Z, Grab B, Dixon H. Le potentiel de transmission du virus de Monkeypox dans les populations humaines. Am J Epidemiol. 1988;17:643–50.

Berthet N, Nakouné E, Whist E, Selekon B, Burguière AM, Manuguerra JC, Gessain A, Kazanji M. Maculopapular lesions in the Central African Republic. The Lancet. 2011;378:1354.

Besombes C, Gonofio E, Konamna X, Selekon B, Grant R, Gessain A, Berthet N, Manuguerra JC, Fontanet A, Nakouné E. Intrafamily transmission of monkeypox virus, Central African Republic, 2018. Emerg Infect Dis. 2019;25(8):1602–4. https://doi.org/10.3201/eid2508.190112 Epub 2019 Aug 17 PMID: 31216261.

Breman JG, Bernadou J, Nakano JH. Poxvirus in West African nonhuman primates: serological survey results. Bull World Health Organ. 1977;55:605.

Breman JG, Nakano JH, Coffi E, Godfrey H, Gautun JC. Human poxvirus disease after smallpox eradication. Am J Trop Med Hyg. 1977;26(2):273–81.

Breman JG, Kalisa-Ruti MV, Zanotto E, Gromyko AI, Arita I. Human monkeypox, 1970–79. Bull World Health Organ. 1980;58:165.

Breman JG. Monkeypox: an emerging infection for humans? Emerg Infect. 2000;4:45–67. https://doi.org/10.1128/9781555816971.ch5.

Damon IK, Esposito JJ. Poxviruses that infect humans. Man Clin Microbiol Wash Am Soc Microbiol. 2003;1583–91.

Damon IK, Roth CE, Chowdhary V. Discovery of monkeypox in Sudan. N Engl J Med. Augu;355 (9):962–3. https://doi.org/10.1056/NEJMc060792.PMID16943415.

Chastel C, Charmot G. Epidémies bactériennes et virales d'origine zoonotique. Rôle de la chasse et du dépeçage d'animaux sauvages. Bull Soc Pathol Exot. 2004;97:207–12.

CDC. Human monkeypox–Kasai Oriental, Democratic Republic of Congo, February 1996-October 1997. MMWR Morb Mortal Wkly Rep. 1997;46:1168.

CDC. 2020. https://www.cdc.gov/poxvirus/monkeypox/about.html (visited on Feb 15, 2020).

CDC. 2020. https://www.cdc.gov/poxvirus/monkeypox//lab-personnel/index.html (visited on May 20, 2020).

CDC. 2020. https://www.cdc.gov/ncidod/monkeypox/treatmentguidelines.htm (visited on Feb 15, 2020).

CDC. Update: Multistate outbreak of monkeypox- Illinois, Indiana, Missouri, Ohio, and Wisconcin. Morbidity Mortality Weekly Rep. 2003;52:642–6. Available online at: https://www.cdc/gov/mmwr/preview/mmwrhtml/mm5227a5.htm.

Chen N, Li G, Liszewski MK, Atkinson JP, Jahrling PB, Feng Z, Schriewer J, et al. Virulence differences between monkeypox virus isolates from West Africa and the Congo basin. Virology. 2005;340:46–63.

Croft DR, Sotir MJ, Williams CJ, Kazmierczak JJ, Wegner MV, Rausch D, et al. Occupational risks during a monkeypox outbreak, Wisconsin, 2003. Emerg Infect Dis. 2007;13:1150–7. https://doi.org/10.3201/eid1308.061365.

De Clercq E. Cidofovir in the treatment of poxvirus infections. Antiviral Res. 2002;55:1–13.

Di Giulio DB, Eckburg PB. Human monkeypox: an emerging zoonosis. Lancet Infect Dis. 2004;4:15–25. https://doi.org/10.1016/S1473-3099(03)00856-9.

Doshi RH, Guagliardo SAJ, Doty JB, Babeaux AD, Matheny A, Burgado JB, Townsend MB, Morgan CN, Satheshkumar PS, Ndakala N, Kanjingankolo T, Kitembo L, Malekani J, Kalemba L, Pukuta E, N'kaya T, Kangoula F, Moses C, McCollum AM, Reynolds MG, Mombouli JV, Nakazawa Y, Petersen BW. Epidemiologic and ecologic investigations of monkeypox, likouala department, Republic of the Congo, 2017. (February 2019) Emerging Infectious Diseases. www.cdc.gov/eid. Vol. 25, No. 2, DOI: https://doi.org/https://doi.org/10.3201/eid2502.181222.

Durski KN, McCollum AM, Nakazawa Y, Petersen BW, Reynolds MG, Briand S, et al. Emergence of Monkeypox–West and Central Africa, 1970–2017. MMWR. 2018;67:306–10.

Erez N, Achdout H, Milrot E, Schwartz Y, Wiener-Well Y, Paran N, et al. Diagnosis of imported monkeypox, Israel, 2018. Emerg Infect Dis. 2019;25:980–3. https://doi.org/10.3201/eid2505.190076.

Fenner F. Adventures with poxviruses of vertebrates. FEMS Microbiol Rev. 2000;24:123–33.

Formenty P, Muntasir MO, Damon I, Chowdhary V, Opoka ML, Monimart C, Mutasim EM, et al. Human monkeypox outbreak caused by novel virus belonging to congo basin clade, sudan, 2005. Emerg Infect Dis. 2010;16:1539–45. https://doi.org/10.3201/eid1610.100713.

Foster SO, Brink EW, Hutchins DL, Pifer JM, Lourie B, Moser CR, Cummings ECK, et al. Human monkeypox. Bull World Health Organ. 1972;46:569–76.

Fuller T, Thomassen HA, Mulembakani PM, Johnston SC, Lloyd-Smith JO, Kisalu NK, et al. Using remote sensing to map the risk of human monkeypox virus in the congo basin. EcoHealth. 2010;8:14–25. https://doi.org/10.1007/s10393-010-0355-5.

Guarner J, Johnson BJ, Paddock CD, et al. Monkeypox transmission and pathogenesis in prairie dogs. Emerg Infect Dis. 2004;10(3):426–31. https://doi.org/10.3201/eid1003.030878.

Gentile M, Gelderblom HR. Rapid viral diagnosis: role of electron microscopy. New Microbiol. 2005;28:1–12.

Gispen R, Verlinde JD, Zwart P. Arch Ges Virusforsch. 1967;21:205–16.

Gispen R, Kapsenberg JG. Verslagen en medelingenbetseffende deVolksgezondheit. The Hague: Gezondheidsraad; 1968.

Hahon N, Wilson BJ. Amer J Hyg. 1960;71, 69–80.

Hahon N. J Infect Dis. 1961;109:294.

Hendrickson RC, Wang C, Hatcher EL, Lefkowitz EJ. Orthopoxvirus genome evolution: the role of gene loss. Viruses. 2010;2:1933–67.

Heymann DL, Szczeniowski M, Esteves K. Re-emergence of monkeypox in Africa: a review of the past six years. Br Med Bull. 1998;54:693–702. https://doi.org/10.1093/oxfordjournals.bmb.a011720.

Horgan ES, Mansour AH. J Hyg (Lond). 1939;39:615–37.

Horgan ES, Haseeb MA, Sati MH. 1948.

Huhn GD, Bauer AM, Yorita K, Graham MB, Sejvar J, Likos, A, Damon IK. Clinical characteristics of human monkeypox, and risk factors for severe disease. Clin Infect Dis. 2005;1742–1751.

Hutin YJF, Williams RJ, Malfait P, Pebody R, Loparev VN, Ropp SL, Rodriguez M, et al. Outbreak of human monkeypox, democratic Republic of Congo, 1996 to 1997. Emerg Infect Dis. 2001;7:434–8. https://doi.org/10.3201/eid0703.010311.

Jahrling PB, Huggins JW, Ibrahim MS, Lawler JV, Martin JW. Smallpox and related orthopoxviruses. Med Asp Biol Warf TMM Publ Off Surg Gen. 2007. Wash. DC 215–240.

Jezek Z, Arita I, Szczeniowski M, Paluku KM, Ruti K, Nakano JH. Human tanapox in Zaire: clinical and epidemiological observations on cases confirmed by laboratory studies. Bull World Health Organ. 1985;63:1027.

Ježek Z, Fenner F. Human monkeypox. Monog Virologie. S. Karger AG. 1988.

Ježek Z, Grab B, Szczeniowski M, Paluku KM, Mutombo M. Clinico-epidemiological features of monkeypox patients with an animal or human source of infection. Bull World Health Organ. 1988;66:459–64.

Jones KE, Patel NG, Levy MA, Storeygard A, Balk D, Gittleman JL, Daszak P. Global trends in emerging infectious diseases. Nature. 2008;451:990–3. https://doi.org/10.1038/nature06536.

Kalthan E, Dondo-Fongbia JP, Yambele S, Dieu-Creer LR, Zepio R, Pamatika CM. Twelve cases of monkeypox outbreak in Bangassou district (Central African Republic) in December 2015. Bull Soc Pathol Exot. 2016;109:358–63.

Kebela BI. Le profil épidémiologique de monkeypox en RDC 1998–2002. Bull Epidemiol Republ Democr Congo. 2004;29:2–6.

Khodakevich L, Widy-Wirski R, Arita I, Marennikova SS, Nakano J, Meunier D. Monkey pox virus infection in humans in the Central African Republic. Bull Soc Pathol Exot Filiales. 1985;78(3):311–20.

Khodakevich L, Jezek Z, Kinzanzka K. Isolation of monkeypox virus from wild squirrel infected in nature. The Lancet. 1986;327:98–9. https://doi.org/10.1016/S0140-6736(86)90748-8.

Khodakevich L, Ježek Z, Messinger D. Monkeypox virus: ecology and public health significance. Bull World Health Organ. 1988;66:747–52.

Ladnyj ID, Ziegler P, Kima E. A human infection caused by monkeypox virus in Basankusu Territory, Democratic Republic of the Congo. Bull World Health Organ. 1972;46 (5): 593–7. PMC 2480792. PMID 4340218.

Learned LA, Reynolds MG, Wassa DW, Li Y, Olson VA, Karem K, Stempora LL, Braden ZH, Kline R, Likos A. Extended interhuman transmission of monkeypox in a hospital community in the Republic of the Congo, 2003. Am J Trop Med Hyg. 2005;73:428–34.

Li D, Wilkins K, McCollum AM, Osadebe L, Kabamba J, Nguete B, et al. Evaluation of the genexpert for human monkeypox infection diagnosis. Am J Tropic Med Hygiene. 2017;96:405–10.

Ligon BL. Monkeypox: a review of the history and emergence in the Western hemisphere. Semin Pediatr Infect Dis. 2004;15:280–7.

Likos AM, Sammons SA, Olson VA, Frace AM, Li Y, Olsen-Rasmussen M, Davidson and al. A tale of two clades: monkeypox viruses. J Gen Virol. 2005;86: 2661–72.

Lourie B, Bingham PG, Evans HH, Foster SO, Nakano JH, Herrmann KL. Human infection with monkeypox virus: laboratory investigation of six cases in West Africa. Bull World Health Organ. 1972;46:633.

Maber HB. Smallpox and monkeypox in non-human primates. Brit J Exp Path. 1966; 47: 453–46.

MacNeil A, Reynolds MG, Carroll DS, Karem K, Braden Z, Lash R, Moundeli A, Mombouli JV, Jumaan AO, Schmid DS. Monkeypox or varicella? Lessons from a rash outbreak investigation in the Republic of the Congo. Am J Trop Med Hyg. 2009; 80:503–7.

Malekani MJ, McCollum A, Monroe BP, Malekani VD, Mulumba ML, Tshilenge CG, et al. Possible commune animal hosts for monkeypox and ebola viruses in the Tshuapa District, Province of Equateur. Ann African Med: Democratic Republic of Congo; 2014. p. 7.

Mandja BM, Brembilla A, Handschumacher P, Bompangue D, Gonzalez JP, Muyembe JJ, Mauny F. Temporal and spatial dynamics of monkeypox in democratic Republic of Congo, 2000–2015. Ecohealth. 2019a Aug 13;. Doi: https://doi.org/10.1007/s10393-019-01435-1. [Epub ahead of print] PubMed PMID: 31410720.

Mandja BM, Bompangue D, Handschumacher P, Gonzalez JP, Salem G, Muyembe JJ, Mauny F. The score of integrated disease surveillance and response adequacy (SIA): a pragmatic score for comparing weekly reported diseases based on a systematic review. BMC Public Health. 2019b;19(1):624. https://doi.org/10.1186/s12889-019-6954-3.PubMedPMID:31118016;PubMedCentralPMCID:PMC6532185.

Marennikova SS, Šeluhina EM, Mal'ceva NN, Čimiškjan KL, Macevič,GR. Isolation and properties of the causal agent of a new variola-like disease (monkeypox) in man. Bull World Health Organ. 1972; 46:599–611.

Mcconnell SJ, Herman YF, Mattson DE, Erickson L. Monkey pox disease in irradiated cynomologous monkeys. Nature. 1962;195:1128–9.

McConnell SJ, Spertzel RD, HuXsoll DL, Eliot L, Yager RH. J Bact. 1964; 87:238–9.

McConnell SJ, HermanYF, Mattson DE, Huxsoll DL, Lang CM, Yager RH. Amer J Vet Res. 1964; 25:192.

Meyer A, Esposito JJ, Gras F, Kolakowski T, Fatras M, Muller G. First appearance of monkey pox in human beings in Gabon. Med Trop Rev Corps Sante Colon. 1991;51:53–7.

Mika LA, Pirsch JB. J Bact. 1960;80:861.

Müller G, Meyer A, Gras F, Emmerich P, Kolakowski T, Esposito JJ. Monkeypox virus in liver and spleen of child in Gabon. The Lancet. 1988;331:769–70.

Meyer H, Perrichot M, Stemmler M, Emmerich P, Schmitz H, Varainne F, et al. Outbreaks of disesase suspected of being due to human Monkeypox virus infection in Democratic Republic of Congo in 2001. J Clin Microbiol. 2002; 40:2919–21.
DOI https://doi.org/10.1128/JCM.40.8.2919-2912.2002.

Mutombo M, Jezek Z, Arita I. Human monkeypox transmitted by a chimpanzee in a tropical rain-forest area of Zaire. The Lancet. 1983;321:735–7.

Mwamba DK, Kebela BI, Shongo RL, Pukuta E, Kayembe NJM. Profil épidémiologique du monkeypox en RDC, 2010–2014. Ann African Med. 2014;8:1855–60.

Nakazawa Y, Emerson GL, Carroll DS, Zhao H, Li Y, Reynolds MG, et al. Phylogenetic and ecologic perspectives of a monkeypox outbreak, southern Sudan, 2005. Emerg Inf Dis. 2013, February; 19(2):237–45. Doi:https://doi.org/10.3201/eid1902.121220. PMC 3559062. PMID 23347770.

Nakouné E, Lampaert E, Gervais Ndjapou S, Janssens C, Zuniga I, Van Herp M, et al. A nosocomial outbreak of human monkeypox in the Central African Republic. Open Forum Infect Dis. 2017;4:1–4.

Nalca A, Rimoin AW, Bavari S, Whitehouse CA. Reemergence of monkeypox: prevalence, diagnostics, and countermeasures. Clin Infect Dis. 2005;41:1765–71.

Ng OT, Lee V, Marimuthu K, Vasoo S, Chan G, Pin Lin RT, et al. A case of imported Monkeypox in Singapore. Lancet Infect Dis. 2019;19(11):1166. https://doi.org/10.1016/S1473-3099(19)30537-7.

Nolen LD, Osadebe L, Katomba J, Likofata J, Mukadi D, Monroe B, et al. Extended human-to-human transmission during a monkeypox outbreak in the Democratic Republic of the Congo. Emerg Infect Dis. 2016;22:1014–21.

Parker S, Buller RM. A review of experimental and natural infections of animals with monkeypox virus between 1958 and 2012. Future Virol. 2013;8(2):129–57. https://doi.org/10.2217/fvl.12.130.

Petersen E, Abubakar I, Ihekweazu C, et al. Monkeypox—Enhancing public health preparedness for an emerging lethal human zoonotic epidemic threat in the wake of the smallpox post-eradication era. Int J Infect Dis. 2019;78:78–84.

Peters JC. T Diergeneesk. 1966;91:387–91.

Prier JE, Sauer RM. A pox disease of monkeys. Ann NY Acad Sci. 1960; 85:951–9.

ProMED. Published Date: 2020–01–22 12:30:25 Subject: PRO/AH/EDR> Monkeypox Africa (03): Nigeria Archive Number: 20200122.6904535. 2019a.

ProMED. Published Date: 2019–05–11 12:00:16 Subject: PRO/AH/EDR> Monkeypox - Singapore ex Nigeria Archive Number: 20190511.6464598. 2019b.

ProMED. Published Date: 2019–04–09 21:28:41 Subject: PRO/AH/EDR> Monkeypox—Africa (03): Congo DR, Republic of the Congo Archive Number: 20190409.6413379. 2019c.

ProMED. Published Date: 2020–02–18 15:33:24 Subject: PRO/AH/EDR> Monkeypox—Africa (04): Congo DR, WHO Archive Number: 20200218.7001717. 2020a.

ProMED. Published Date: 2020–01–17 21:30:23 Subject: PRO/AH/EDR> Monkeypox - Africa (02): Cameroon Archive Number: 20200117.6894022. 2020b.

Radonic A, Metzger S, Dabrowski PW, et al. Fatal monkeypox in wild-living sooty mangabey, Cote d'Ivoire, 2012. Emerg Infect Dis. 2012;2014(20):1009–11.

Reynolds MG, Emerson GL, Pukuta E, Karhemere S, Muyembe JJ, Bikindou A, et al. Detection of human monkeypox in the Republic of the Congo following intensive community education. Am J Trop Med Hyg. 2013;88:982–5. https://doi.org/10.4269/ajtmh.12-0758.

Reynolds MG, Carroll DS, Olson VA, Hughes C, Galley J, Likos A, Montgomery JM, et al. A silent enzootic of an orthopoxvirus in Ghana, West Africa: evidence for multi-species involvement in the absence of widespread human disease. Am J Trop Med Hyg. 2010;82:746–54.

Reynolds MG, Wauquier N, Li Y, Satheshkumar PS, Kanneh LD, Monroe B, Maikere J, Saffa G, Gonzalez JP, Fair J, Carroll DS, Jambai A, Dafae F, Khan SH, Moses LM. Human Monkeypox in Sierra Leone after 44-Year Absence of Reported Cases. Emerg Infect Dis. 2019 May;25 (5):1023–1025. Doi: https://doi.org/10.3201/eid2505.180832. Epub 2019 May 17. PubMed PMID: 30753125; PubMed Central PMCID: PMC6478203.

Rimoin AW, Mulembakani PM, Johnston SC, Lloyd Smith JO, Kisalu NK, Kinkela TL, et al. Major increase in human monkeypox incidence 30 years after smallpox vaccination campaigns cease in the Democratic Republic of Congo. Proc Natl Acad Sci. 2010;107:16262–7. https://doi.org/10.1073/pnas.1005769107.

Saijo M, Ami Y, Suzaki Y, Nagata N, Iwata N, Hasegawa H, Iizuka I, et al. Virulence and pathophysiology of the Congo Basin and West African strains of monkeypox virus in non-human primates. J Gen Virol. 2009;90:2266–71.

Sauer RM, Prier JE, Buchanan RS, Creamer AA, Fegley HC. Studies on a pox disease of monkeys. I Pathology Am J Vet Res. 1960;21:377–80.

Sbrana E, Jordan R, Hruby DE, Mateo RI, Xiao SY, Siirin M, Newman PC, Da Rosa APT, Tesh RB. Efficacy of the antipoxvirus compound ST-246 for treatment of severe orthopoxvirus infection. Am J Trop Med Hyg. 2007;76:768–73.

Schmidt M. ZoologischeKlinik. Die Krankheiten der Affen, Berlin: A. Hirschwald; 1870. p. 97.

Stittelaar KJ, Neyts J, Naesens L, Van Amerongen G, Van Lavieren RF, Holỳ A, De Clercq E, Niesters HG, Fries E, Maas C. Antiviral treatment is more effective than smallpox vaccination upon lethal monkeypox virus infection. Nature. 2006;439:745.

Sklenovská N. Epidemiology and laboratory diagnosis of Monkeypox virus in Africa. Master's Thesis. Leuven: KU Leuven; 2017.

Sklenovská N, Van Ranst M. Emergence of monkeypox as the most important orthopoxvirus infection in humans. Front Public Health. Sept;6:241. https://doi.org/10.3389/fpubh.2018.00241.PMC6131633.PMID30234087.

Tchokoteu PF, Kago I, Tetanye E, Ndoumbe P, Pignon D, Mbede J. Variola or a severe case of varicella? A case of human variola due to monkeypox virus in a child from the Cameroon. Ann Soc Belg Med Trop. 1991;71(2):123–8 Epub 1991/06/01.

Vaughan A, Aarons E, Astbury J, Balasegaram S, Beadsworth M, Beck CR, et al. Two cases of monkeypox imported to the United Kingdom, September 2018. Euro Surveill. 2018;23. ES.2018.23.38.1800509.

Vaughan A, Aarons E, Astbury J, Brooks T, Chand M, Flegg P, Hardman A, et al. Human-to-human transmission of monkeypox virus, United Kingdom, October 2018. Emerg Infect Dis. 2020;26(4):782–5. https://doi.org/10.3201/eid2604.191164WHO,1980.

Von Magnus P, Andersen EK, Petersen KB, Birch-Andersen A. A pox-like disease in cynomolgus monkeys. Acta Path Micro Scand. 1959;46(2):156–176. Discovery of monkeypox virus.

Westwood JCN, Boulter EA, Bowen ETW, der Affen, Berlin A. Hirschwald, p. 97.

WHO. The global eradication of smallpox. Final Report of the Global Commission for the Certification of Smallpox Eradication. Geneva, Switzerland; 1980.

WHO. Monkeypox. Wkly Epidemiol Rec. 2011;41:448–51.

WHO. Monkeypox in Central African Republic; 2016. Available online at: https://www.who.int/csr/don/13-october-2016-monkeypox-caf/en/ (Accessed 14/02/2020).

WHO. Weekly Bulletin on Outbreaks and Other Emergencies, Week 18 2017; 2017a. Available online at: https://apps.who.int/iris/bitstream/handle/10665/255272/OEW18-294552017.pdf (Accessed 14/02/2020).

WHO. Weekly Bulletin on Outbreaks and Other Emergencies, Week 22 2017, 2017b. Available online at: https://apps.who.int/iris/bitstream/10665/255624/1/OEW22-270262017.pdf (Accessed 14/02/2020).

WHO. Weekly Bulletin on Outbreaks and Other Emergencies, Week 38 2017; 2017c. Available online at: https://apps.who.int/iris/bitstream/10665/258984/1/OEW38-162292017.pdf (Accessed 14/02/2020).

WHO. Weekly Bulletin on Outbreaks and Other Emergencies, Week 28 2017, 2017d. Available online at: https://apps.who.int/iris/bitstream/10665/255895/1/OEW28-81472017.pdf (Accessed 14/02/2020).

WHO. Weekly Bulletin on Outbreaks and Other Emergencies, Week 16 2017; 2017e. Available online at: https://apps.who.int/iris/bitstream/10665/255050/1/OEW16-152142017.pdf (Accessed 14/02/2020).

WHO. Weekly Bulletin on Outbreaks and Other Emergencies, Week 8 2018; 2018a. Available online at: https://apps.who.int/iris/bitstream/10665/260335/1/OEW8-1723022018.pdf (Accessed 14/02/2020).

WHO. Weekly Bulletin on Outbreaks and Other Emergencies, Week 14 2018; 2018b. Available online at: https://apps.who.int/iris/bitstream/10665/272343/1/OEW14-310306042018.pdf (Accessed 14/02/2020).

WHO. Weekly Bulletin on Outbreaks and Other Emergencies, Week 14 2018; 2018c. Available online at: https://apps.who.int/iris/bitstream/10665/272981/OEW26-2329062018.pdf (Accessed 14/02/2020).

Yinka-Ogunleye A, Aruna O, Ogoina D, Aworabhi N, Eteng W, Badaru S, et al. (2018) Reemergence of human monkeypox in Nigeria. Emerg Infect Dis. 2017;24(6):1149–51. https://doi.org/10.3201/eid2406.180017 PMID: 29619921.

Yinka-Ogunleye A, Aruna O, Dalhat M, Ogoina D, McCollum A, Disu Y, et al. CDC Monkeypox Outbreak Team. Outbreak of human monkeypox in Nigeria in 2017–18: a clinical and epidemiological report. Lancet Infect Dis. 2019;19:872–9. https://doi.org/10.1016/S1473-3099(19)30294-4.

Measles Virus

Brief Introduction of Measles Virus and Its Therapeutic Strategies

Igor José dos Santos Nascimento,
Paulo Fernando da Silva Santos-Júnior,
and Edeildo Ferreira da Silva-Júnior

Abstract Measles virus (MV) is an infectious disease related to thousands of deaths in previous outbreaks. Health agencies estimated that approximately 42% of countries would have the eradication of this pathogen. However, the virus remains as a challenging health problem for countries' economies. This pathogen is highly transmissible via the respiratory droplets of infected individuals, which remain in the air and surfaces for several hours, increasing its rate of transmission. Prophylactic vaccination is the most effective strategy in controlling this infection. In recent years, political-economic factors and the anti-vaccine movement have contributed to its reemergence. Despite apparently being a controlled disease, these factors contributed to the increase in the number of cases, including fatalities. Also, there is still no drug used in the disease's clinical management, justifying the investment in new alternatives. This chapter will be addressed to the main advances in new therapies against MV, highlighting *hit* compounds, focusing on non- and nucleoside, peptide, quinoline, retinoid analogs, and other derivatives. Besides the main advances in biological approaches such as glycosaminoglycan and interferon; also immunological approaches exist such as plant-and peptide-based vaccines. Nucleoside and non-nucleoside analogs are the most explored chemical agents, resulting in several promising molecules. Also, natural products could be effective in controlling MV. Finally, advances in immunological strategies, such as discovering new vaccines show that (despite a vaccine's existence) it is necessary to find another more effective and accepted by the population, leading to the definitive eradication of this disease.

Keywords Measles virus · Drug discovery · Viral disease · Vaccines · Rinderpest virus

---

I. J. dos S. Nascimento · P. F. da S. Santos-Júnior · E. F. da Silva-Júnior (✉)
Chemistry and Biotechnology Institute, Federal University of Alagoas, Maceió, AL, Brazil
e-mail: edeildo.junior@iqb.ufal.br

E. F. da Silva-Júnior
Laboratory of Medicinal Chemistry, Pharmaceutical Sciences Institute,
Federal University of Alagoas, Maceió, AL, Brazil

S. I. Ahmad (ed.), *Human Viruses: Diseases, Treatments and Vaccines*, https://doi.org/10.1007/978-3-030-71165-8_23

1 Introduction

Measles is a highly contagious viral disease responsible for approximately two million deaths in previous outbreaks (WHO 2019). However, efforts focused on immunization by using prophylactic vaccination resulted in decreased deaths between the years 2000–2018, preventing about 23 million deaths with a reduction of 73% (WHO 2019). Despite these encouraging data and perspectives worldwide for the virus eradication in approximately 42% of countries, in recent years, outbreaks of measles have been reported in countries that apparently are free of this disease, such as the European Union, United States, and China (CDC 2020; Brown et al. 2020; Ding et al. 2020; Cui et al. 2020).

The most effective strategy for eradicating the measles virus (MV) is the prophylactic vaccination (Bonneton et al. 2020). It has 95–99% effectiveness, being one dose administered at 12 months of age has 93% of immunity, while two doses have 97% effectiveness (Ashkenazi et al. 2020; Camilloni et al. 2020). The administration of two doses is responsible for interrupting transmission and possible eradication of MV (Ashkenazi et al. 2020). Despite the scientific data showing the effectiveness of the vaccine and the epidemiological data showing the decrease in the incidence of the disease, in recent years, the "anti-vaccine" movement has emerged and driven by celebrities, digital influencers, and certain high ranking politicians, leading to the re-emergence of this disease, previously considered as eradicated (Benecke and DeYoung 2019).

Multiple factors are associated with this MV reemergence, such as deficiency in health policies and false information propagated by social media, leading to under-vaccinated populations (Benecke and DeYoung 2019; Vink et al. 2020). Lack of investment by some countries' governments is also a significant challenge to control this disease (Sibeudu et al. 2020; De Broucker et al. 2020). Social media were also responsible for the decline in vaccination and the spread of false information regarding possible side-effects, such as induction of autism (Benecke and DeYoung 2019; Arendt and Scherr 2020; Schober 2020). In this context, such information generates fear in the children's legal guardians, directly interfering in the vaccination program (Griffith et al. 2020; Christianson et al. 2020).

Besides difficulties and immunization, associated with high MV transmissibility are intrinsically related to current outbreaks (Dunn et al. 2020; Xu et al. 2020). In this sense, the transmission is via respiratory droplets from infected individuals, since these droplets could remain in the air and on other surfaces for several hours, which increases the rate of transmission (Dunn et al. 2020). Another important fact is that an infected patient could transmit the virus up to four days before the onset of symptoms (skin rashes), with a high probability of transmission (up to 90%) to susceptible individuals (Dunn et al. 2020).

Although vaccination is the most effective strategy for controlling MV, the aforementioned facts show the need for new approaches to help in disease control, prevention, and treatment (Kobayashi et al. 2020; Harrison 2014). In this context, drug discovery in searching new alternatives to treat MV should be increased (Harrison 2014; Plemper 2020).

This chapter addresses to the main therapeutic approaches studied in recent years that may play effective roles in MV treatment. Among these are included promising molecules and their biological testing with results, in addition to natural products, new approaches to discover vaccines, and biological strategies against this disease.

2 Measles Virus: History and Re-Emergence of the Virus

Literature reviews show that the MV appeared more than 1,000 years ago, a period marked by the domestication of cattle and Rinderpest virus (RPV) emergence. The RPV and MV are closely related (genus *Morbillivirus* and *Paramyxoviridae* family for both), and some findings indicate that these agents appeared at the same period (Roeder et al. 2013). The prolonged human contact with these animals led to RPV mutations, allowing their transmission from cows to humans, initiating its infectious cycle in human hosts, and making the human as its the natural host. After emerging in the Old World, it became endemic in several countries, such as the United States, leading to several deaths (Goodson and Seward 2015). With increased travel and the beginning of globalization, the modern era was the landmark for the growth and spread of MV worldwide (Paterson et al. 2013). Lastly, this topic will provide a brief history of the emergence of the MV and its current re-emergence in the New World.

2.1 A Brief History of Measles Virus

The origin of the MV is not yet fully determined. In certain 9th-century finding written by a Persian doctor indicate the first signs of this disease. Later, in 1757, Francis Home identified MV as an infectious agent in patients' blood (WHO 2019). This pathogen was prevalent in Europe from 16th to 18th centuries as a non-fatal disease. However, in the late 18th century, this disease became more virulent without explanation, generating more epidemics, resulting in thousands of deaths from 18th to 20th centuries. Like other diseases, MV moved from Europe to the New World (American continent), generating serious outbreaks in Mexico, Canada, and United States, from 1531 to 1772 (Cunha 2004).

During 19th and 20th centuries, MV was a disease responsible for many deaths. The most significant increase in numbers of cases was registered during the First World and American Civil wars. However, decline in transmission rates was observed during Second World War, as result of the availability of vaccines and antibiotics (Shanks et al. 2014). Before vaccine discovery, there was a high rate of infection in children of under 15 years, with approximately 2 million deaths, before 1960s (Shanks et al. 2014; Lee et al. 2020). In 1980s, the vaccine was introduced in the European Union to eliminate the propagation of MV. In this context, World Health Organization (WHO) program "Health for All in the 21st century" estimated the eradication of MV until the year 2010 (Peña-Rey et al. 2009).Unfortunately, this objective could not be achieved, and MV remains WHO target for its elimination (Magurano et al. 2017; Wong et al. 2020; Ma et al. 2020).

2.2 Measles Virus Re-Emergence

The emergence of the internet has been a crucial system in making health-related decisions, given the availability of medical information online, encouraging more lay people to make controversial decisions in face of health agencies. Also, social media like *Facebook* and *Instagram* allowed lay people to share their opinions by promoting social activism, leading to the spread of false information to the general public. As a consequence of this "contrary revolution" was the "anti-vaccine" movement disseminating incorrect information on social media, leading to the MV's reappearance. However, certain projections indicate the eradication of this virus in 2020 (WHO 2019; Oehler 2020). Thus, the increase in the number of MV cases in recent years is associated with insufficient vaccination program (Fragkou et al. 2020). The 2019 data show that the number of infections has been the highest in the last 13 years, indicating the above factors have contributed to this unfavorable result (Hotez et al. 2020).

3 Promising Drug Targets

Knowledge of physiology, structure, function, and binding sites of the virus is essential for designing new compounds (Danishuddin and Khan 2015). Stages of the life cycle and certain proteins found in paramyxovirus (including MV) are unique, making them attractive subject for pharmacological intervention studies. In this context, promising strategies involving inhibition of viral entry and RNA-dependent RNA polymerase (RdRp) could be promising as new alternatives of treatment, resulting in inhibition of mRNA synthesis and viral genome (Plemper 2020).

3.1 Measles Virus Structure and Its Replication

Measles Virus belongs to the genus *Morbillivirus* from the *Paramyxoviridae* family, causing infection in both humans and non-human primates (Laksono et al. 2020). It is a single-stranded negative-sense RNA virus, with six tandemly linked genes, encoding from 3' to 5' regions, presenting two surface glycoproteins

(Hemagglutinin (H) and Fusion (F), four internal proteins matrix (M), phosphoprotein (P), large polymerase (L), nucleocapsid (N), and two non-regulatory structural proteins (C and V) (Fig. 1) (Griffin 2021; Düx et al. 1912; Takayama et al. 2012). The H protein from the viral surface along with a non-covalent association involving the fusion glycoprotein are responsible for viral entry, representing a target for antibodies (Griffin 2021). Also, H protein interaction with its cell adhesion receptor (CD46) can generate downregulation of its expression. The viral spreading is directly related to F and H proteins' expression in infected cells, allowing virus dissemination in the absence of any exposure to antivirals or even the immune system (Barnard 2004). Therefore, preventing H protein binding to CD46 could be a useful strategies in discovering new compounds against MV (Barnard 2004).

MV entry into host cells depends on neutral pH, starting from the interaction of H protein with receptors of the host cells, and F protein then triggers the fusion of the viral envelope with the membrane of host cells (Fig. 2) (Fukuhara et al. 2020; Navaratnarajah et al. 2012). Subsequently, H protein binds to CD46 and CD150 receptors (or SLAM receptors) (Moss and Griffin 2006). MV replication occurs mainly in lymphoid tissues, spreading rapidly through the respiratory tract and other organs prior the appearance of clinical symptoms (i.e. skin rashes). Despite replicating in macrophages, dendritic, epithelial, and endothelial cells, MV mainly affects lymphoid tissue, in $CD4^{+}$ and $CD8^{+}$ T cells, in addition to memory B cells that have the SLAM/CD150 receptor. Then, depletion of these cells of the lymphoid tissue occurs (Griffin 2020).

Interestingly, replication complex appears to be a promising target, although not much is known about it in scientific studies. This process comprises the N protein, which is responsible for encapsidating the viral genome, along with RdRp and P proteins acting as a cofactor. Therefore, N protein binds to six nucleotides that are part of the viral genome during the encapsidation process, forming helical nucleocapsid, and resulting in the protection of the viral RNA by the innate immune system, preventing its recognition and degradation (Guseva et al. 2020).

Fig. 1 Structure of Measles Virus (MV) and its proteins

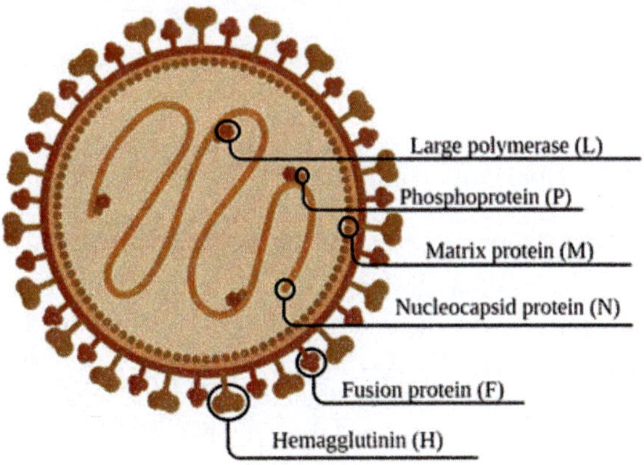

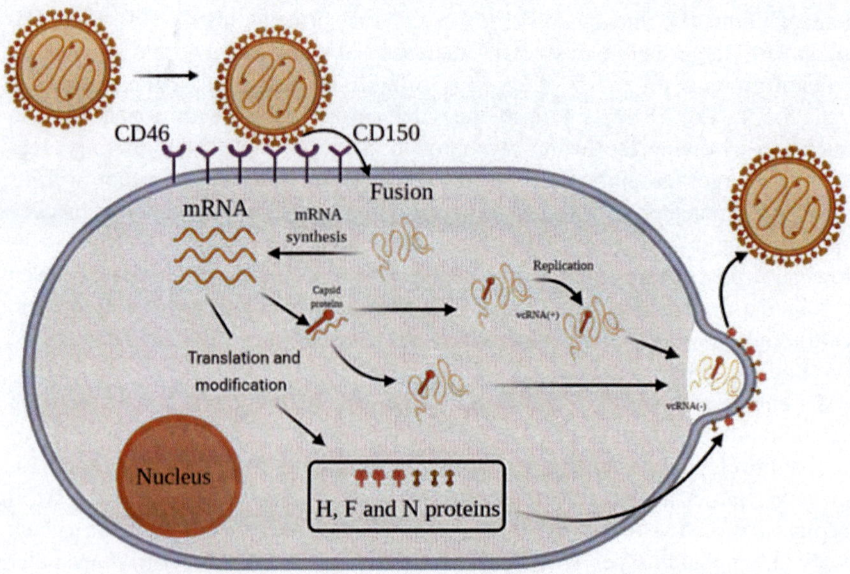

Fig. 2 Replication of the Measles virus

4 Therapeutic Strategies Against Measles Virus

As aforementioned there is no a specific drug treatment for MV; however, the prophylactic vaccination is the most effective measure in controlling this disease (Paules et al. 2019). Ribavirin, interferon-α, and other antivirals have been associated with some improvements in more severe cases. Besides, there is evidence that vitamin A may act as a preventive and antibiotic agent to relieve the symptoms (Caseris et al. 2015). Therefore, the development of new drugs that can induce effective treatment or even an innovative vaccine is challenging to control MV (WHO 2019; Caseris et al. 2015).

4.1 Synthetic Compounds

4.1.1 Nucleoside Inhibitors

The potential of nucleoside analogs against parasitic diseases inspired Ugarkar and colleagues (1984) to synthesize a series of these derivatives and evaluate their antiviral activity (Ugarkar et al. 1984). Compound **(1)** (Fig. 3) showed the best activity against MV (ED_{50} of 1.4×10^{-9}M), with low toxic effects.

Another study that evaluated nucleoside analogs was performed by Schneller et al. (1984), who reported the synthesis of adenine analogs as MV inhibitors. The

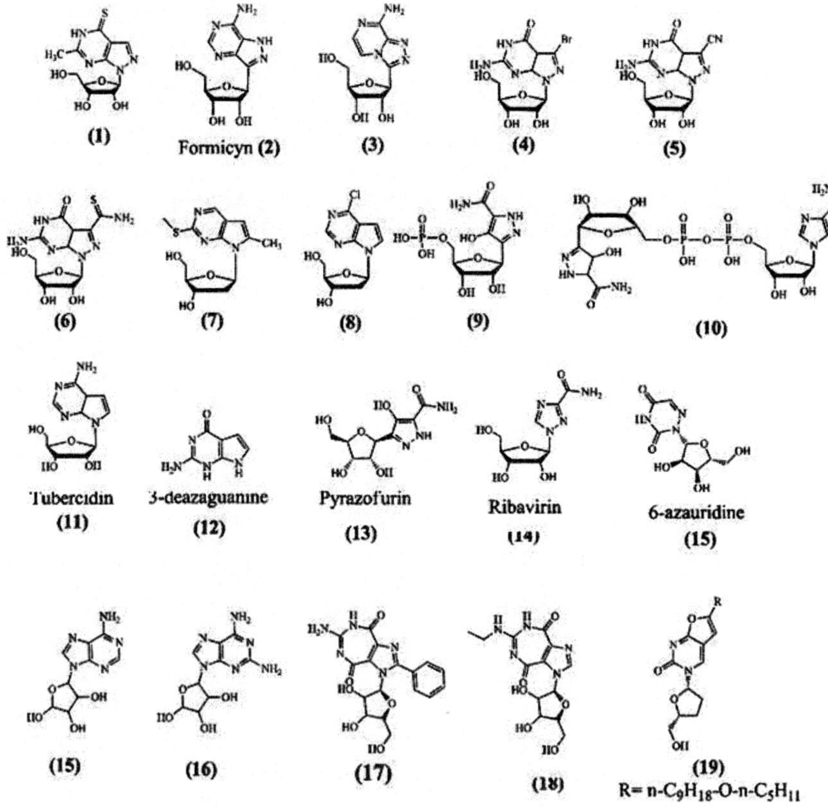

Fig. 3 Most promising nucleoside analogs found in the literature

authors designed structural modifications in the scaffold of formycin **(2)** (Fig. 3), notably studied as antitumor, antiviral, antifungal and antibacterial (Drach and Sidwell 1981; Shaban 1997; Erian 1982; Wang et al. 2019). Thus, five new analogs were synthesized and screened against S-adenosyl-homocysteine hydrolase (AdoHcyase) enzyme. This biomolecule is essential for MV replication since it catalyzes the reversible hydrolysis of S-adenosyl-homocysteine (AdoHcy) to homocysteine (Hcy) and adenosine (Ado) (Turner et al. 2000; Singha et al. 2015; Nakao et al. 2015). Compound **(3)** (Fig. 3) was found to be the most promising analog, exhibiting antiviral activity at 200 μg/mL concentration, although the broad-spectrum antiviral ribavirin (positive control), showed an antiviral activity at 20 μg/mL. Furthermore, compound **(3)** was characterized as an irreversible inhibitor against this target, displaying a *Ki* value of 78 μM.

The promising activity of pyrazolo[3,4-*d*]pyrimidine nucleoside derivatives inspired Petrie and colleagues (1984) to synthesize a series of these compounds and evaluate them against antitumor and diverse viruses, including MV (Petrie et al. 1985). Compounds **(4, 5, 6)** (Fig. 3) showed more significant activities against MV

(ED_{50}values of 2 × 10^{-5}, 5.4 × 10^{-5}, and 6.5 × 10^{-5} M, respectively), similar effects were observed by ribavirin (ED_{50}value of 3.3 × 10^{-5} M), and without toxic effects up to the tested concentrations. Simultaneously, Cottam et al. (1985) carried out the synthesis and investigation of antitumor and antiviral activities of pyrrolo [2,3-*d*]pyrimidine 2'-deoxyribonucleoside derivatives (Cottam et al. 1985). Among the synthesized compounds, analogs **(7, 8)** (Fig. 3) showed the best activities against MV (ED_{50} values of 9.5 × 10^{-5} and 8.9 ×10^{-3} M, respectively), with activity comparable to ribavirin (ED_{50}value of 3.3 × 10^{-5} M), presenting tolerated toxicity.

Petrie et al. (1986) investigated the activity of nucleoside and nucleotide derivatives of pyrazofurin (Petrie et al. 1986). Thus, compounds **(9, 10)** (Fig. 3) were found the most active against MV (ED_{50}values of 2 × 10^{-5} and 1.6× 10^{-5} M, respectively). Besides, these compounds showed less toxicity, indicating better potential among the tested compounds.

Posteriorly, adenine analogs were explored by Clercq et al. (1987), who investigated the antiviral activity of pyrrolo[2,3-*d*]pyrimidine nucleosides with changes in sugar moiety (Clercq et al. 1987). Among these, tubercidin **(11)** (Fig. 3) demonstrated the most potent activity against different viruses, including MV (MIC_{50}value of 0.3 μg/mL) with low cytotoxicity. Finally, the authors suggested that changes in sugar ring could be better explored to improve antiviral activity against this virus.

Furthermore, Hosoya (1989) evaluated the in vitro inhibitory antiviral effects of compounds against various Subacute Sclerosing Panencephalitis (SSPE) virus strains, a condition caused by effects of MV in the body. Then, 3-deazaguanine **(12)**, pyrazofurin **(13)**, ribavirin **(14)**, and 6-azauridine **(15)** (Fig. 3) (ED_{50}values of 0.9, 0.13, 8, 0.78 μg/mL, respectively) were presented, presenting inhibition at a lower concentration than the cytotoxic one. The authors pointed out that pyrazofurin, 3-deazaguanine, and 6-azauridine affect cell morphology by inhibiting protein synthesis. On the other hand, ribavirin acts by inhibiting the synthesis of DNA or RNA. Thus, these compounds could be considered as promising agents for treating SSPE.

Inspired by studies that indicate the possible activity of ribavirin **(14)** (Fig. 3) against MV, Gururangan et al. (1990) reported a case of a nine-years-old child infected with MV giant cell pneumonia together with Hodgkin's disease, and treated with anticancer therapy by using vinblastine, chlorambucil, procarbazine, and prednisolone drugs (Gururangan et al. 1990). Additionally, as the patient did not have Koplik's and was at risk of giant cell pneumonia. Therapy with ribavirin and nebulization for seven days showed effectiveness. From this the authors concluded that the early start of drug therapy could generate satisfactory results in treating measles in immunocompromised children.

Patil et al. (1992) synthesized four carbocyclic nucleosides and investigated their antiviral activities against a number of different viruses including MV (Patil et al. 1992). Against all those tested, compounds **(15, 16)** (Fig. 3) demonstrated the best results, including MV (MIC values of 0.4 and 20 μg/mL). Additionally, K_i assay revealed that their activities may be due to inhibition of *S*-adenosyl-*L*-homocysteine hydrolase enzyme, showing inhibitions at 0.0111±0.0048 and 0.85±0.15 μM concentrations, respectively. In another study, Zhanget et al. (2002) inspired by nucleoside analogs and their biological potentials, synthesized new five nucleosides fused to imidazodiazepine rings (Zhang et al. 2002). Thus, compounds **(17)** (ED_{50} CPE inhibition of 10 μM and Neutral red of 0.5 μM) and **(18)** (ED_{50} CPE inhibition of 2 μM and Neutral red of 1.2 μM) (Fig. 3) were identified useful. The authors stated that adding more hydrophobic groups at 2 and 6 positions could improve their molecular activities.

Dideoxybicyclic pyrimidine nucleoside analogs (ddBCNAs) with *D*-chirality is recognized as an important scaffold in drug design against human cytomegalovirus and vaccinia virus (VACV) (McGuigan et al. 2004). Deeming this, McGuigan et al. (2013) (McGuigan et al. 2004) designed new inhibitors using *L*-analogs and performed a screening against different viruses. Thus, 20 new analogs were screened against VACV, in which compound **(19)** (Fig. 3) was more active against it (IC_{50} value of 0.19 μM), compared to cidofovir (IC_{50} value of 3.4 μM). Later on, the authors screened this analog also against MV wild-type upon infected B95a cells, in which was obtained a $TCID_{50}$ value of 7.5 μM, exhibiting a cytotoxic safety profile (CC_{50} > 100 μM towards BSC-1 and HFFF cells).

4.1.2 Non-nucleoside Inhibitors

Potential of 3-deazaguanine against various diseases inspired Schneller et al. (1984) to synthesize the 3,7-dideazaguanine **(20)** analog (Fig. 4) and evaluate its biological potential (Schneller et al. 1984). Also, they showed that 3-deazaguanine **(12)** (Fig. 3) has better biological potential against MV (MIC value of 2 μg/mL) and other viruses (herpes, vaccinia, reovirus, parainfluenza, sindbis, coxsacki, VSV, and poliovirus). Thus, 3,7-dideazaguanine **(20)** did not act either as a substrate or as an inhibitor towards hypoxanthine-guanine phosphoribosyl-transferase (HGPRTase) and other enzymes, confirming the importance of nitrogen for purine derivatives.

In another study, Clercq et al. (1985) designed and synthesized 15 new adenosine analogs as broad-spectrum antivirals. Compound **(21)** (Fig. 4) showed more promising activity against MV, with 12 μg/mL, in addition to 400 μg/mL cytotoxicity against HeLa cells, constituting a promising scaffold to be used in the development of antivirals against MV.

Given the evidence of the possible activity of optical enantiomers of (±)-C-Ado and C-AMP, Herdewijn et al. (1985) performed the synthesis of these compounds and evaluated them against some viruses (e.g. reovirus, herpes, parainfluenza, vaccinia, sindbis, coxsacki, VSV, poliovirus, among others), including MV (Herdewijn et al. 1985). It was verified that the optical mixture of enantiomers **(22, 23)** (Fig. 4) showed the best results against MV (MIC values of 3 and 10 μg/mL, respectively), revealing that the racemic mixture had the best biological potential.

As reported above, the inhibition of the *S*-adenosylhomocysteine enzyme may be related to the control of viral infections, as well as for MV. Thus, Madhavan et al. (1988) designed a series of aristeromycin analogs as inhibitors containing a

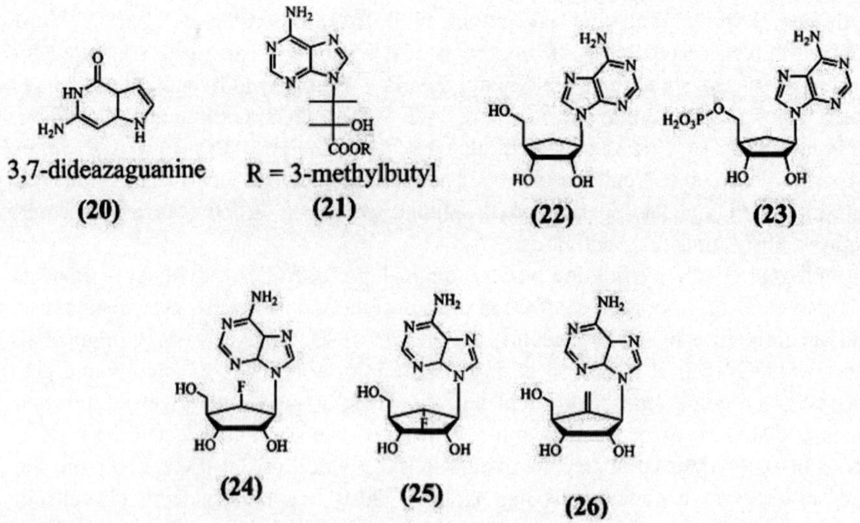

Fig. 4 Non-nucleoside inhibitors with activity against MV

fluoride atom at position 6, indicating possible irreversible inhibition of the enzyme (Hydrolase et al. 1988). Thus, compounds **24, 25** and **26** (Fig. 4) showed promising potential against the enzyme (IC_{50} of 8, 180, and 80 nM, respectively). The best activities of these compounds are linked to the substituent at position 6, which generates the formation of a reactive intermediate with the possibility of forming a covalent bond.

Subsequently, a study performed by Shuto et al. (1992) reported the synthesis of eight new analogs designed as broad-spectrum antivirals. The authors designed their analogs considering neplanocin A **(27)** (Fig. 5), to act as a broad-spectrum antiviral for RNA and DNA viruses. Then, modifications based on the monovalent bioisosterism approach, idealizing the insertion of a chlorine atom **(28)** (Fig. 5) instead of the hydroxyl group from neplanocin A **(27)**. This new compound presented an IC_{50}value of 0.8 μg/mL against MV (Vero cells), in comparison with its precursor (0.2 μg/mL). Additionally, compound **(28)** induced cytotoxicity at concentrations 50-fold more than the inhibitory activity against MV (40 μg/mL), which indicates a safety profile in the administration of this promising compound.

In 1996, the same research group continued to search for broad-spectrum inhibitors based on the structure of neplanocin A **(27)** (Fig. 5), resulting in eight new analogs. Among these, they highlighted compound **(29)** (Fig. 5), in which the insertion of a fluorine atom was designed by using monovalent bioisosterism approach since the insertion of a halogenated group at the 2 position of adenosine confers resistance to the action of the adenosine deaminase enzyme. In this context, compound **(29)** presented an IC_{50}value of 0.03 μM, besides being safe (CC_{50} value of 200 μM upon Vero cells). Among the adenosine analogs reported so far, the most promising analog is this, constituting a potential drug candidate to combat MV infections.

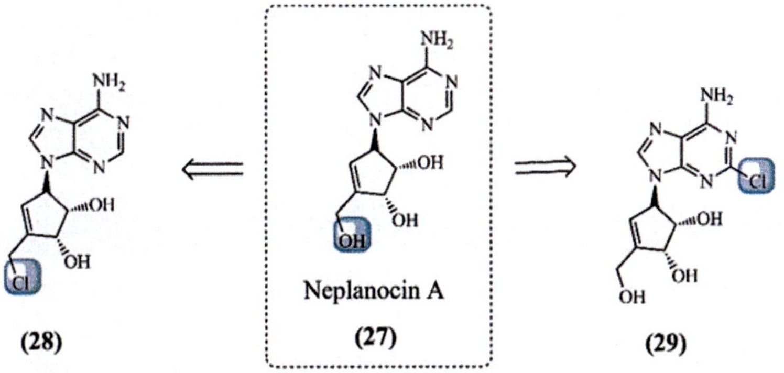

Fig. 5 Neplanocin A (27) and its analogs. In blue, structural bioisoster modification zone used for designing compounds (28) and (29)

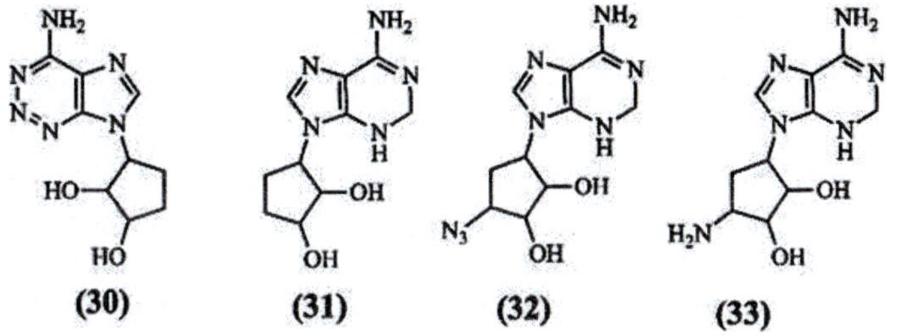

Fig. 6 Non-nucleoside analogs synthesized by Barnard et al. (2001)

Several studies suggest the promising activity of ribavirin **(14)** (Fig. 3) against MV. Thus, Barnard et al. (2001) evaluated analog compounds (non- and nucleosides) in order to identify new alternatives against MV (Barnard et al. 2001). Compounds **(30–33)** (Fig. 6) were found to be the most potent analogs from the CPE inhibition assay (EC_{90}values of 3, 0.4, 11, and 0.1 μg/mL, respectively). Also, these compounds were non-toxic up to 100 μg/mL concentration. Additionally, compounds **(31** and **33)** demonstrated synergism (additive effect) along with ribavirin, without virucidal activity at the same concentrations that inhibited the viral replication. Such results suggest that these compounds could be tested in more advanced biological assays.

4.1.3 Peptide and Non-peptide Analogs

RdRp constitutes an important drug able to targets present in several viruses, being involved in aspects of complete replication-transcription (Ganeshpurkar et al. 2019). Thus, an important study involving design of compounds targeting this

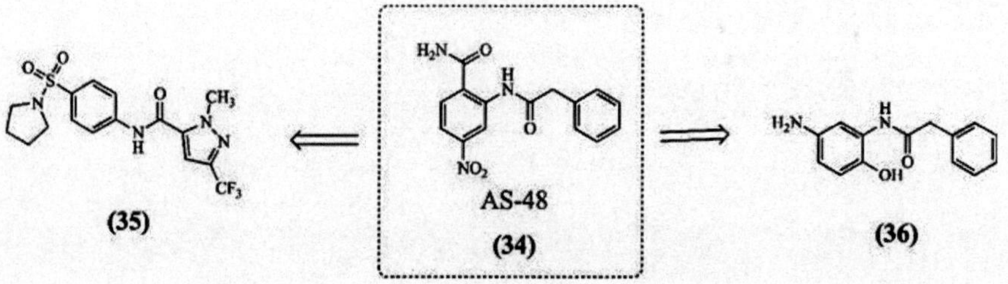

Fig. 7 First-in-class non-nucleoside inhibitor against RdRp-MV

macromolecule was developed by White et al. (2007). The authors analyzed by using HTS a library of compounds containing 34,000 ligands from Emory Chemical Biology Discovery Center, considering AS-48 **(34)** (Fig. 7) (IC_{50} values ranging from 0.6 to 3.0 μM for different isolated MV strains) (Sánchez-Barrena et al. 2003; Plemper et al. 2005; Hashiguchu et al. 2018). Thus, the authors identified four compounds containing a sulfonamide group, in which compound **(35)** (Fig. 7) presented an EC_{50} value of 0.24 μM, as well as good tolerance (CC_{50} value > 300 μM), in addition to its high selectivity index (SI > 1,250). Also, compound **(35)** inhibits the activity of the target in question, being a reversible inhibitor (Fig. 7). Their results suggested this new derivative as an important scaffold to be used in the development of new inhibitors against MV.

Nonpeptide-based entry inhibitors have been reported by Sun et al. (2006) from studies involving the MVF protein, based on the crystal structure of the Newcastle disease virus protein. Then, the authors designed analogs from theAS-48 **(34)** structure, applying some modifications and 2D-QSAR studies, resulting in the derivative **(36)** (Fig. 7). Finally, was demonstrated the activity of compound **(36)** with an IC_{50} value of 0.26 μM, by suppression of virus-induced cytopathic effects (CPE) upon the MV Edmonston strain. Furthermore, it showed cytotoxicity with a CC_{50}value of 17 μM and SI of 65.

Sun et al. (2007) reported their work carried out on specific structural changes in compound **(35)**, and a SAR study shown (Fig. 8). Then, 34 new compounds were designed and analyzed for their activity against MV-Alaska, resulting in compound **(37)**. It presented an EC_{50}value < 2.3 μM, showing some benefits in the expansion from 5 members pyrrolidine ring to 6 members (piperidine), with SI > 130 and CC_{50} > 300 μM upon Vero cells.

One year later, the same research group (Sun et al. 2008; Yoon et al. 2009) presented new modifications from HTS of the previously listed derivatives **(34, 35** and **37)**, using the same specific regions as shown in Fig. 8. So, 41 new compounds were designed and synthesized, in which analog **(38)** showed potential activity against MV-Alaska, with an EC_{50} value of 0.0005 μM, in addition to high SI (85000) and a CC_{50} value > 425 μM. Structurally, this analog was designed in a similar way as compound **(37)** previously described, with the expansion of the piperidine ring, resulting in a 7-membered ring, such as observed for compound **(38)**. In addition, the piperidine ring has also been expanded to 8 members **(39)**,

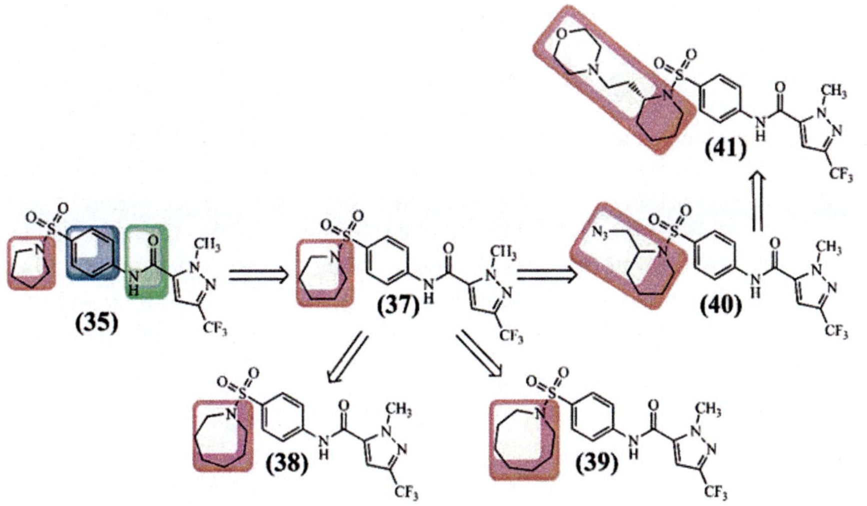

Fig. 8 Modifications proposed from the AS-48 inhibitor **(34)**. In red, modifications of the pyrrolidine ring; In blue, phenyl ring substitution and pyrazole ring replacement; In green, modification in the linker region

although even with promising activity (EC_{50}value of 0.045 μM), comparatively, the activity declined 90-fold, besides presenting considerable CC_{50} and low SI values, being 16 μM and > 7, respectively.

Following the studies, Ndungu et al. (2012) reported structural changes by SAR analysis, based on the aforementioned compounds **(37–39)**, and screened the new analogs against MV-Alaska, resulting in 27 new derivatives (Fig. 8). In this context, derivative **(40)** was found to be the most active analog, with an EC_{50} value of 1.5 μM, contributing to the development of new antivirals against MV. Finally, they described the allosteric inhibitors of MV-RdRp **(40** and **41)**. In contrast, the structural modification resulted in a 6.5-times more active analog **(41)**, with an EC_{50} value of 0.23 μM, and SI of 346 towards MV-Alaska strain, when compared to its precursor **(40).** Additionally, it was more active against the MV-Ibadan strain NIE/97/1 (B3-2), showing an EC_{50} value of 0.07 μM and SI of 1071.

Inspired by the recent elucidated antiviral activity of the peptide mucroporin, isolated from the venom of the scorpion *Lychas mucronatus*, Li et al. (2011) investigated its action and another optimized derivative (Mucroporin-M1) against SARS-CoV, H5N1 influenza, and MV(Li et al. 2011). Interestingly, only the peptide mucroporin did not show activity against any of the viruses. In contrast, the optimized peptide mucroporin M1 showed significant inhibition mainly in the MV infectivity assay, with an EC_{50} value of 7.15 μg/mL, with low toxicity (CC_{50} value of 70.46 μg/mL), and SI of 9.85. Thus, replacing Gly and Pro residues from mucroporin with Lys or Arg from mucroporin-M1 improves the peptide activity. Additionally, the authors suggested that the peptide does not interfere with viral replication but interacts directly with the viral particles, with virucidal activity being proposed.

4.1.4 Quinoline Derivatives

Isoquinolin-4-Ones

One of the first studied and described isoquinolin-4-one to act as MV inhibitor was developed by Santagati et al. (2003), since there existed compounds active against myxoviruses and paramyxovirus containing such heterocycle UK 2054 (famotine), UK 2371 (memotine), and DIQA, 2-(3,4-dihydroisoquinolin-1-yl) acetamide (Beare et al. 1968; Swallow 1978). Among 28 designed compounds, analog **(42)** (Fig. 9) presented the best activity against MV (IC_{50} value of 12.5 μM), besides cytotoxic concentration about 8-times more (CC_{50} value of 100 μM against Vero cells and SI of 8).

Isozaxoles

Niedermeier et al. (2009) demonstrated new compounds based on the scaffold of *N*-heterocycle, since such small molecules have been reported as inhibitors of *Paramyxoviriridae* F protein (Plemper et al. 2005). Thus, the authors screened 18 libraries of compounds against MV, resulting in derivative **(43)** (Fig. 10), being a small molecule from isoxazole class. This aforementioned compound presented an EC_{50} value of 0.4 μM and CC_{50} value of 8 μM, showing selectivity for infected cells with MV (SI = 20).

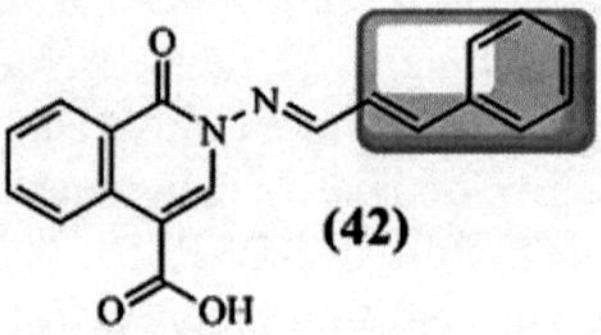

(42)

Fig. 9 Isoquinolin-4-one reported with MV activity. In blue, modifications by SAR promoted in the demarcated region, involving extension of the alkyl chain, as well as halogens, donors and acceptors of H-bonding interactions

Fig. 10 Isoxazole reported as anti-MV compound

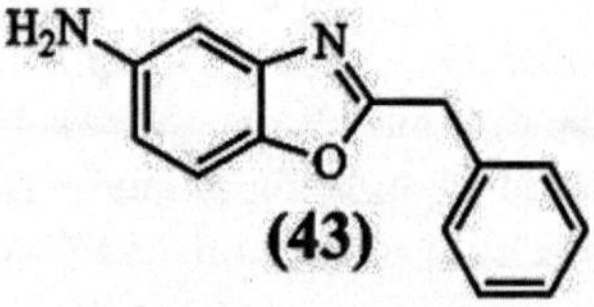

(43)

Fig. 11 Retinoid derivative presented as active against MV

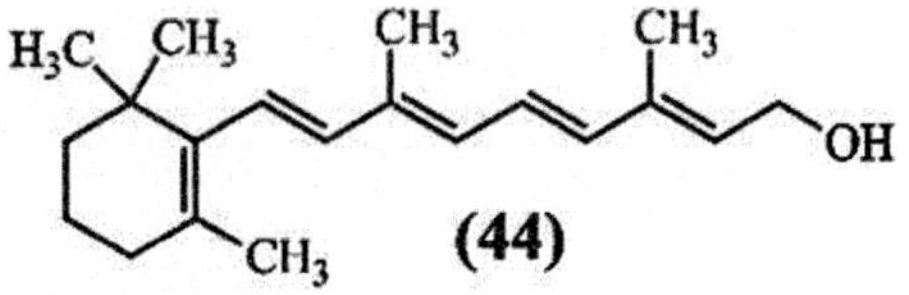

(44)

4.1.5 Retinoids

A study reporting the class of retinoid with activity against MV was described by Trottier et al. (2008), in which it was analyzed 3 compounds (ATRA, 9cRA, and ROH **(44)**) (Fig. 11), where the latter was most active against NB4 cells. Its activity was measured at EC_{50}, revealed a value of 2.03×10^{-10} μM, as well as CC_{50} of 5.09×10^{-5} μM and SI was quantified in terms of 2.51×10^{2}. Finally, the authors suggested that its mechanism of action could be related to the receptor signaling pathways.

4.1.6 Peptide-Conjugated Morpholino Oligomers

Peptide-conjugated phosphorodiamidate morpholino oligomers (PPMO) refer to antisense compounds, being active in blocking mRNA (Forand et al. 2020). In this sense, Sleeman et al. (2009) designed a series of PPMO, aiming to inhibit RNA sequences from MV, in order to inhibit viral replication. Then, PPMO 454 (sequence 5'-3': GGCCATCTCGGATATCCTTAATC) has been described as readily active against several MV genotypes, with action on the coding of mRNA for the N protein, constituting a promising approach in the development of anti-MV therapies.

4.1.7 Pyrimidines and Their Derivatives

In Munier-Lehmann et al. (2015) described the pyrimidine compounds with activity against MV, constituting one of the first studies, since this class is reported having antiviral activity. They described compound **(45)** as a potential inhibitor of viral replication, with a MIC_{50}value of 2.72 μM, then being modified by means of bioisosterism approach, obtaining compound **(46)** (Fig. 12). These changes potentially increased the activity of the new derivative, exhibiting aMIC_{50}value of 0.7 nM. Furthermore, its activity was idealized to be involved with the target dihydroorotate dehydrogenase (DHODH), which is the fourth enzyme in the pyrimidine biosynthetic pathway. Besides, studies with recombinant human DHODH have shown promising activity against the target, with an IC_{50}value of 13 nM, constituting a druggable target to be explored in drug design against MV.

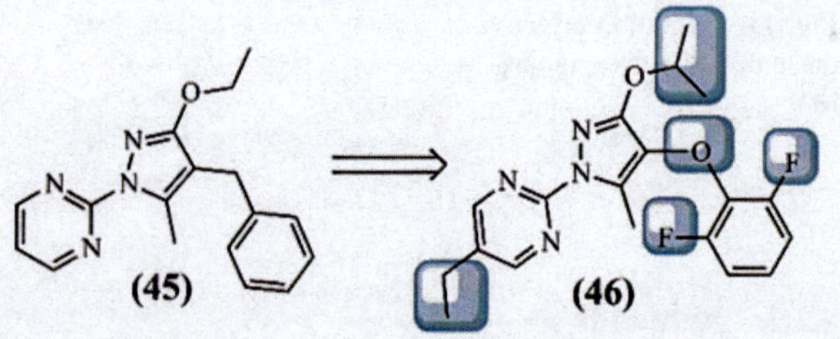

Fig. 12 Pyrimidine-based derivatives as a DHODH inhibitor for MV. The blue color, structural changes performed to obtain the best compound

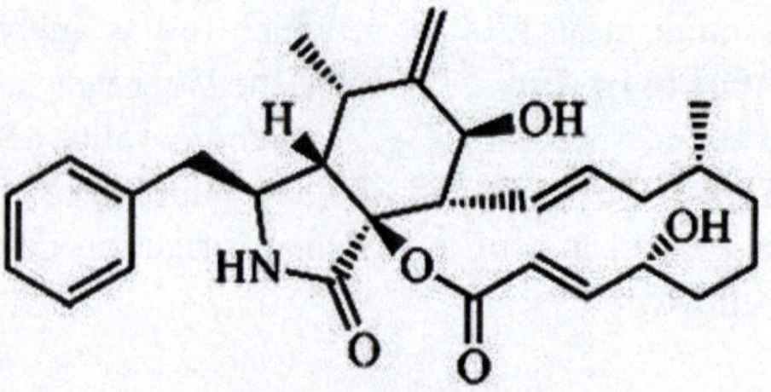

Fig. 13 Chemical structure of cytochalasin B

4.1.8 Cytochalasin B as MV Inhibitor

Given the evidence that cytochalasin B **(47)** (Fig. 13) could function as an inhibitor of protein glycosylation, Stallcup et al. (1983) investigated the potential of the compound against MV (Stallcup et al. 1983). MV is the composed of H and F glycoproteins, in which the H and F glycosylation inhibition may be related to the inhibition of viral none-effectiveness. Thus, cytochalasin B **(47)** was responsible for inhibiting viral, about 88–99%. Another important fact was that the compound did not inhibit viral synthesis. Besides, virus inhibition was more pronounced when added in the first third of the replication cycle. Finally, interaction with microfilament was also proposed, indicating a fundamental role in the release of virions.

4.2 Biological Approaches

4.2.1 Glycosaminoglycan

Terao-Muto et al. (2008) presented a study with the preventive activity of MV infection in signaling lymphocyte activation molecule (SLAM)-negative cells based on heparin-like glycosaminoglycans- these polysaccharides being ubiquitously present on cell surfaces. In this study, the authors constructed a recombinant MV model that expressed the enhanced green fluorescent protein (EGFP) (rMV-EGFP), infecting cells that do not express SLAM, which is known as a cellular receptor for morbilli viruses. Thus, soluble heparin was able to inhibit the infectivity of rMV-EGFP in SLAM-negative cells, in a dose-dependent manner, although the infectious process has not been completely blocked, requiring further studies to clarify this mechanism.

4.2.2 Interferon

Since interferons (IFN) are involved in the innate response from viral infectious processes, this protein produced by leukocytes and fibroblasts is studied as a promising key for inhibiting MV infections (Shaffer et al. 2003; Sarkar et al. 2006; Braun et al. 2019). In this context, paramyxoviruses have non-structural proteins that act to inhibit the IFN response, so a study reported by Shaffer et al. (2003) indicated that MV protein C is able to inhibit it, suggesting it is a promising indicator of inhibiting viral entry.

Subsequently, Berghäll et al. (2006) suggested that MV infection results from the expression of the mda-5, RIG-I, and TLR3 genes, followed by the expression of IFN-*b*, interleukin (IL)-28 and IL-29, CCL5, and CXCL10 genes, where these factors are involved in the entry of the virus into the human body, although not fully elucidated.

Braun et al. (2019) evaluated the interference of Guanylate-binding protein (GBP) 5 in the viral infection process associated with MV, since it is involved in the infectivity of HIV-I (also an RNA virus). Finally, the authors assessed that GBP2 and GBP5 are involved in reducing the infectivity of MV, Zika, and Influenza A, exercising broad-spectrum antiviral activity.

4.3 Immunological Approach

One of the goals of vaccine developers is discovering highly effective and safe vaccines against infectious diseases. However, several failures came on the way due to the variability of pathogens and hosts (Poland et al. 2011). However, one of the most successful was the MV vaccine, the most significant public health advances of the 20th century (Baker 2011). The discovery of this vaccine provided several advances in immunology and the organism's natural response against MV. Although, an estimation indicates high vaccine effectiveness in one or two doses over a lifetime, it is not yet known whether immunity could last throughout life (Naniche 2009). Thus, studies on the discovery of new vaccines are needed to fill this gap (Poland et al. 2011; Naniche 2009).

4.3.1 Peptide-Based Vaccine

Most vaccines consist of inactivated, attenuated, or genetically modified live pathogens. The goal is to stimulate an immune response in the body by presenting B and T cells' epitopes in a conformation that will help fight infection by improving pathogen recognition. On the other hand, peptide- or protein-based vaccines, although useful, face some limitations and may require repeated immunizations. Despite in recent years several technologies have been developed to improve this limitation, such as the development of nanoparticles, delivery systems, especially related to the delivery of the genetic material that encodes a specific protein resulting in the production of the immunogen by the host (Malonis et al. 2020).

In this context, given the potential of peptide-based vaccines could be beneficial against MV, Pütz et al. (2003) demonstrated how peptide-based vaccines could be helpful in association with the vaccine containing the virus (Pütz and Muller 2003). Thus, their work indicated that the peptide administration helps to fight infection without interfering with the response produced by immunization by the complete virus. Therefore, a great alternative would be early vaccination with peptide to complement subsequent immunization with the virus. Finally, the vaccine does not induce antibodies to the peptide, allowing a serological distinction between infected and vaccinated patients, leading to a diagnostic tool for epidemiological monitoring programs for eradication of MV.

4.3.2 Plant-Based Vaccine

The discovery of gene transfer mechanisms between microorganisms and plants and advances in genetic engineering have provided new perspectives in the development of drugs from natural products (Azegami et al. 2019; Singh et al. 2018). In this context, around 1990s, concepts of plants producing edible vaccines emerged (Singh et al. 2018). After that, genetic engineering, biotechnology, and pharmaceutical industries developed remarkable efforts to make plant vaccines (Singh et al. 2018; Rybicki 2009). The justification for investing in this technology is mainly related to the administration's convenience, as they are used orally, low-cost of production, accessible storage, and transport. Plants are used to make proteins of biological interest, reduce costs related to the process, and the possibility of oral administration, being a permissive technology against viral diseases, addressed against MV (Azegami et al. 2019; Singh et al. 2018; Rybicki 2009; Webster et al. 2005).

With the potential of a plant-based vaccine, Webster et al. (2005) discussed the advantages of using this approach against MV (Webster et al. 2005). Thus, given the evidence for the expression of MV H protein in tobacco, they showed that immunizations in mice with plant-derived MV H protein resulted in the production of antibodies against MV, as well as humoral and mucosal immune responses due to IgA production. Another interesting finding was that reinforcement of plant-derived H protein associated with the DNA primer led to a more effective response comparable to that of any vaccine. However, despite these encouraging findings, the authors argued that tobacco does not produce sufficient MV H protein, requiring discovering a more useful plant species in this regard. With this, it will be possible to improve this protein's production, allowing a more consistent and effective immune response, with oral immunizations enabling a reduction in the number of doses administered.

4.4 Natural Compounds

Humanity has used traditional medicines as medicinal herbs for thousands of years, responsible for relieving the symptoms and possible treatment of various diseases, including infectious diseases caused by viruses. Thus, oral administration of extracts of natural sources has been in common use as a prophylactic and even curative treatment for various viral infections, including MV (Lin et al. 2014; Kitazato et al. 2007). Also, herbal medicines purified from natural products offer privileged structures that could be used as scaffolds in drug discovery campaigns. Thus, researchers worldwide invest their researches in the discovery of new therapeutic alternatives through natural products (Lin et al. 2014).

Since the 1990s, evidences point to the promising activity of traditional herbal medicines as promising alternatives in face of antiviral activity against MV. Thus, to discover new therapies against this disease, Kurokawa et al. (1993) evaluated the potential in vitro and in vivo of 142 traditional medicines used to treat viral infections in China, Indonesia, and Japan against various viral diseases, including MV (Kurokawa et al. 1993). Thus, 30 extracts with effectiveness against MV were identified; in which 11 of them were effective at 100 µg/mL concentration and 19 other extracts at 300 or 500 µg/mL concentrations, emphasizing that extracts of *Alpinia officinarum*, *Geum japonicum*, *Machilus thunhergii*, *Polygomcm cuspidatum*, and *Zanthoxylum bungeanum* could be interesting subjects to be further explored.

In the 2000s, Parker et al. (2007) instigated plants to be inserted in the diets of children of the Maasai pastoralists of East Africa against MV (Parker et al. 2007). Initially, the study showed that the extracts did not produce toxicity at concentrations below 1 µg/mL concentration. In this way, they were tested against MV neutralization, in which extracts of *Rhus natalenses*, *Albizia amara*, *Olinia rochetiana,* and *Warburgia ugandenses* showed potent antiviral activity, with neutralization of MV above 50%. On the other hand, only *Olinia rochetiana* and *Rhus natalenses* extracts reduced viral replication in the PRN assay. The results indicate that Maasai diet plants could be useful for controlling MV infection.

In addition to using products from plants, compounds that makeup living organisms, such as constituents of mollusks, have been considered a vast biological potential including as antiviral agents. In this context, the biological potential of land slug *Phyllocaulis boraceiensis* was investigated by Toledo-Piza et al. (2016) in vitro with infected cells with MV (de Toledo-Piza et al. 2016). The results from qPCR and direct electron microscopy revealed that the fraction 39 was more active (80% inhibition), which was attributed to the presence of polyunsaturated fatty acids, such as hydroxytritriacontapentaenoic acid and hydroxypentatriacontapentaenoic acid, identified through HPLC-DAD-ESI-MS/MS and infrared spectroscopy. Besides, fraction 50 also showed antiviral activity, although it presented a lower percentage of this polyunsaturated fatty acids, being oxopentatetracontaheptenoic acid, the main constituent of it. The authors pointed out that their activities may be through the disruption of cell membranes and viral envelope.

5 Challenges and Opportunities

Despite the lack of an effective vaccine to control MV, there are still some challenges to be overcome in preventing infection. Perhaps the main one is the political-economic issues, as there is a lack of priority for this disease in countries with low infection rates (Cutts et al. 1999). Also, the estimated cost of MV control, including prophylaxis through vaccines, laboratory diagnostic tests, healthcare professionals for patient care, and promotion and prevention campaigns related to this disease, could reach approximately 2.7 to 5.3 million dollars. This high investment creating lack of priority on the part of certain governments (Sundaram et al. 2019).

Despite appearing around year 1830, time attributed to vaccination against smallpox, the anti-vaccine movement is returning and the re-emergence of diseases such as MV (Stern and Markel 2005; Holzmann et al. 2016). This is resulting the outbreak observed in several regions in the world, notably affecting unvaccinated patients (Cutts et al. 2013; Shrivastava et al. 2015). Thus, it is necessary to combine the efforts of health professionals, politicians, researchers, and educators to take this movement more seriously and end the *demonization* of vaccines, as this strategy remains the most effective in controlling MV to avoid human fatalities (WHO 2019; Lancella et al. 2017; Hussain et al. 2018; Spika et al. 2003).

Difficulties in developing a new medication against MV are also related to the lack of a well-understood target. In drug discovery, a target is only definitely elucidated when there is an approved drug that has the route as the primary biological target (Lage et al. 2018; De Rycker et al. 2018). Although studies indicate inhibition of RdRp or viral entry by interference by blocking H and F proteins as promising strategies, further advances in drug design are still needed to elucidate such targets as the main ones against MV (Ndungu et al. 2012; Prussia et al. 2008).

Besides, there are also difficulties in clinical trials for the discovery of a new drug. Such challenges include: (1) rapid viral replication and acute manifestations, which makes therapeutic interventions difficult; (2) clinical manifestations occur when the viral load is decreased; (3) the predominant population of pediatric patients, and (4) limited size of patient groups (Plemper 2020).

In addition to drug development, vaccine development is also a complex process that has undergone numerous advances over the past few years (Ovsyannikova and

Poland 2011). Besides, limiting factors such as incomplete understanding of developing the immune response, the genetic variability of pathogens and hosts, and even the population's skepticism regarding vaccines‘ effectiveness make the process more challenging. Thus, Omics technologies and the growth of bioinformatics have been moving towards progress in overcoming such difficulties and resulting in the discovery of new vaccines that could be useful against MV (Kennedy et al. 2020).

6 Conclusion and Future Outlooks

Despite being a disease that appeared to be none-offensive, MV has been collecting several fatal cases worldwide, creating a global public health problem. For many years MV was controlled, showing wispy incidence due to an effective vaccine to control the infection. However, the lack of drug treatment linked to the return to the anti-vaccine movement made the eradication of this pathogen a public health priority, which justifies the investment in new treatments against this disease.

Research on drug discovery against MV is mainly focused on nucleoside and non-nucleoside analogs. Over the years, in addition to ribavirin, several promising analogs have been presented in the scientific literature, such as quinoline derivatives, retinoids, and others that may represent new *lead* compounds. Natural products such as the use of plant extracts with reported MV activity also show versatility. Besides, the biological approaches, focused on discovering inhibitors against glycosaminoglycan, and the use of interferon and peptide derivatives, also represent advances in drug discovery. Also, vaccines' discovery is also a strategy that has made many advances in recent years, mainly using more current approaches, such as peptide- and plant-based vaccines.

Finally, we hope that this chapter will act as a guide for research groups worldwide searching for new therapeutic alternatives and thus could inspire researchers to discover a new drug or vaccines capable of saving several lives affected by this disease.

Acknowledgements The authors thank the Coordenação de Aperfeiçoamento de Pessoal de Nível Superior—Brazil (CAPES), Fundação de Amparo à Pesquisa do Estado de Alagoas (FAPEAL), and the National Council for Scientific and Technological Development (CNPq)—Brazil for their support to the Brazilian Post-Graduate Programs. Finally, we would like to thank the BioRender (https://biorender.com/) for providing figures to ellaborate our biological schemes.

Funding None.

Declaration of interest

The authors have no relevant affiliations or financial involvement with any organization or entity with a financial interest in or financial conflict with the subject matter or materials discussed in the manuscript. This includes employment, consultancies, honoraria, stock ownership or options, expert testimony, grants or patents received or pending, or royalties.

References

Arendt F, Scherr S. Social science & medicine news-stimulated public-attention dynamics and vaccination coverage during a measles outbreak : an observational study. Soc Sci Med. 2020:113495. https://doi.org/10.1016/j.socscimed.2020.113495.

Ashkenazi S, Livni G, Klein A, Kremer N, Havlin A, Berkowitz O. The relationship between parental source of information and knowledge about measles/measles vaccine and vaccine hesitancy. Vaccine. 2020;38:7292–8. https://doi.org/10.1016/j.vaccine.2020.09.044.

Azegami T, Yuki Y, Kiyono H. Plant-based mucosal vaccine delivery systems. Mucosal Vaccines Innov Prev Infect Dis. 2019:357–70. https://doi.org/10.1016/B978-0-12-811924-2.00020-1.

Baker JP. The first measles vaccine. Pediatrics. 2011;128:435–7. https://doi.org/10.1542/peds.2011-1430.

Barnard DL. Inhibitors of measles virus. Antivir Chem Chemother. 2004;15:111–9. https://doi.org/10.1177/095632020401500301.

Barnard DL, Stowell VD, Seley KL, Hegde VR, Das SR, Rajappan VP, et al. Inhibition of measles virus replication by 5′-Nor carbocyclic adenosine analogues. Antivir Chem Chemother. 2001;12:241–50. https://doi.org/10.1177/095632020101200405.

Beare A, Bynoe ML, Tyrrell DA. Prophylaxis of influenza with a synthetic isoquinoline. Lancet. 1968;20:843–5.

Benecke O, DeYoung SE. Anti-vaccine decision-making and measles resurgence in the United States. Glob Pediatr Heal. 2019;6:2333794X1986294. https://doi.org/10.1177/2333794x19862949.

Bonneton M, Antona D, Danis K, Aït-Belghiti F, Levy-Bruhl D. Are vaccinated measles cases protected against severe disease? Vaccine. 2020;38:4516–9. https://doi.org/10.1016/j.vaccine.2020.05.005.

Braun E, Hotter D, Koepke L, Fackler OT, Kirchhoff F, Sauter D, et al. Antiviral activity by inhibiting furin-mediated processing of viral envelope proteins guanylate-binding proteins 2 and 5 exert broad antiviral activity by inhibiting furin-mediated processing of viral envelope proteins 2019:2092–104.

Brown DW, Warrener L, Scobie HM, Donadel M, Waku-Kouomou D, Mulders MN, et al. Rapid diagnostic tests to address challenges for global measles surveillance. Curr Opin Virol. 2020;41:77–84. https://doi.org/10.1016/j.coviro.2020.05.007.

Camilloni B, Stracci F, De Lio MC, Mencacci A, Cenci E, Bozza S. Measles immunity in healthcare workers of an Italian hospital. J Infect Public Health. 2020;13:1123–5. https://doi.org/10.1016/j.jiph.2020.04.001.

Caseris M, Burdet C, Lepeule R, Houhou N, Yeni P, Yazdanpanah Y, et al. Actualité de la rougeole. Rev Med Interne. 2015;36:339–45. https://doi.org/10.1016/j.revmed.2014.10.362.

Christianson B, Sharif-Mohamed F, Heath J, Roddy M, Bahta L, Omar H, et al. Parental attitudes and decisions regarding MMR vaccination during an outbreak of measles among an undervaccinated Somali community in Minnesota. Vaccine. 2020;38:6979–84. https://doi.org/10.1016/j.vaccine.2020.09.022.

Clercq E, Balzarini J, Madej D, Hansske F, Robins MJ. Nucleic acid related compounds. 51: synthesis and biological properties of sugar-modified analogues of the nucleoside antibiotics tubercidin, toyocamycin, sangivamycin, and formycin. J Med Chem 1987;30:481–6. https://doi.org/10.1021/jm00386a007

Clercq E, Holy A. Alkyl esters of 3-Adenin-9-yl-2-hydroxypropanoic acid: a new broad-spectrum antiviral agents. J Med Chem 1985;28:282–287

Cottam HB, Kazimierczuk Z, Geary S, McKernan PA, Revankar GR, Robins RK. Synthesis and biological activity of certain 6-substituted and 2,6-disubstituted 2’-deoxytubercidins prepared via the stereospecific sodium salt glycosylation procedure. J Med Chem. 1985;28:1461–7. https://doi.org/10.1021/jm00148a015.

Cui A, Zhang Y, Zhu Z, Wang H, Mao N, Song J, et al. Classification of measles cases from 2014 to 2018: implications for progress towards measles elimination in China. Vaccine. 2020;38:3832–8. https://doi.org/10.1016/j.vaccine.2020.03.049.

Cunha BA. Smallpox and measles: historical aspects and clinical differentiation. Infect Dis Clin North Am. 2004;18:79–100. https://doi.org/10.1016/S0891-5520(03)00091-6.

Cutts FT, Henao-Restrepo AM, Olivé JM. Measles elimination: progress and challenges. Vaccine. 1999;17:47–52. https://doi.org/10.1016/S0264-410X(99)00309-6.

Cutts FT, Lessler J, Metcalf CJE. Measles elimination: progress, challenges and implications for rubella control. Expert Rev Vaccines. 2013;12:917–32. https://doi.org/10.1586/14760584.2013.814847.

Danishuddin M, Khan AU. Structure based virtual screening to discover putative drug candidates: necessary considerations and successful case studies. Methods. 2015;71:135–45. https://doi.org/10.1016/j.ymeth.2014.10.019.

De Broucker G, Ssebagereka A, Apolot RR, Aloysius M, Ekirapa Kiracho E, Patenaude B, et al. The economic burden of measles in children under five in Uganda. Vaccine X. 2020;6: https://doi.org/10.1016/j.jvacx.2020.100077.

De Rycker M, Baragaña B, Duce SL, Gilbert IH. Challenges and recent progress in drug discovery for tropical diseases. Nature. 2018;559:498–506. https://doi.org/10.1038/s41586-018-0327-4.

de Toledo-Piza AR, Figueiredo CA, de Oliveira MI, Negri G, Namiyama G, Tonelotto M, et al. The antiviral effect of mollusk mucus on measles virus. Antiviral Res. 2016;134:172–81. https://doi.org/10.1016/j.antiviral.2016.09.005.

Ding Y, Chen W, Lei Y, Mao N, Gao Z, Xu W, et al. Evaluating the population measles susceptibility in Tianjin China. Vaccine. 2020;38:4829–36. https://doi.org/10.1016/j.vaccine.2020.05.053.

Drach JC, Sidwell RW. Antiviral agents. Annu Rep Med Chem. 1981;16:149–59. https://doi.org/10.1016/S0065-7743(08)61282-8.

Dunn JJ, Baldanti F, Puchhammer E, Panning M, Perez O, Harvala H. Measles is back—considerations for laboratory diagnosis. J Clin Virol. 2020;128: https://doi.org/10.1016/j.jcv.2020.104430.

Düx A, Lequime S, Patrono LV, Vrancken B, Boral S, Gogarten J, et al. The history of measles: from a 1912 genome to an antique origin. 2019:1–18. https://doi.org/10.1101/2019.12.29.889667.

Elnagdi MH, Al-Awadi N, Erian AW. Bicyclic 5–6 systems: other four heteroatoms 2:2. Compr Heterocycl Chem II A Rev Lit 1982–1995, vol 7, Elsevier Ltd; 1996, p. 431–88. https://doi.org/10.1016/B978-008096518-5.00156-8.

Forand A, Muchir A, Mougenot N, Sevoz-Couche C, Peccate C, Lemaitre M, et al. Combined treatment with peptide-conjugated phosphorodiamidate morpholino oligomer-PPMO and AAV-U7 rescues the severe DMD phenotype in mice. Mol Ther Methods Clin Dev. 2020;17:695–708. https://doi.org/10.1016/j.omtm.2020.03.011.

Fragkou PC, Thomas K, Sympardi S, Liatsos GD, Pirounaki M, Sambatakou H, et al. Clinical characteristics and outcomes of measles outbreak in adults: a multicenter retrospective observational study of 93 hospitalized adults in Greece. J Clin Virol. 2020;131: https://doi.org/10.1016/j.jcv.2020.104608.

Fukuhara H, Mwaba MH, Maenaka K. Structural characteristics of measles virus entry. Curr Opin Virol. 2020;41:52–8. https://doi.org/10.1016/j.coviro.2020.04.002.

Ganeshpurkar A, Gutti G, Singh SK. Polymerases and their emerging roles in antiviral therapy. Viral Polym. 2019:1–42. https://doi.org/10.1016/B978-0-12-815422-9.00001-2.

Goodson JL, Seward JF. Measles 50 years after use of measles vaccine. Infect Dis Clin North Am. 2015;29:725–43. https://doi.org/10.1016/j.idc.2015.08.001.

Griffin DE. Measles virus persistence and its consequences. Curr Opin Virol. 2020;41:46–51. https://doi.org/10.1016/j.coviro.2020.03.003.

Griffin DE. Measles immunity and immunosuppression. Curr Opin Virol. 2021;46:9–14. https://doi.org/10.1016/j.coviro.2020.08.002.

Griffith BC, Ulrich AK, Becker AB, Nederhoff D, Koch B, Awan FA, et al. Does education about local vaccination rates and the importance of herd immunity change US parents' concern about measles? Vaccine. 2020. https://doi.org/10.1016/j.vaccine.2020.09.076.

Gururangan S, Stevens RF, Morris DJ. Ribavirin response in measles pneumonia. J Infect. 1990;20:219–21. https://doi.org/10.1016/0163-4453(90)91094-T.

Guseva S, Milles S, Jensen MR, Schoehn G, Ruigrok RW, Blackledge M. Structure, dynamics and phase separation of measles virus RNA replication machinery. Curr Opin Virol. 2020;41:59–67. https://doi.org/10.1016/j.coviro.2020.05.006.

Harrison C. Infectious disease: small molecule combats measles virus. Nat Rev Drug Discov. 2014;13:416–7. https://doi.org/10.1038/nrd4343.

Hashiguchu T, Fukuda Y, Matsuoda R, Kuroda D, Kubota M, Shirogane Y, et al. Structure of the prefusion form of measles virus fusion protein in complex with inhibitors. PNAS. 2018;115:2496–501. https://doi.org/10.1073/pnas.1718957115.

Herdewijn P, Balzarini J, De Clercq E, Vanderhaeghe H. Resolution of aristeromycin enantiomers. J Med Chem. 1985;28:1385–6. https://doi.org/10.1021/jm00148a002.

Holzmann H, Hengel H, Tenbusch M, Doerr HW. Eradication of measles: remaining challenges. Med Microbiol Immunol. 2016;205:201–8. https://doi.org/10.1007/s00430-016-0451-4.

Hotez PJ, Nuzhath T, Colwell B. Combating vaccine hesitancy and other 21st century social determinants in the global fight against measles. Curr Opin Virol. 2020;41:1–7. https://doi.org/10.1016/j.coviro.2020.01.001.

Hussain A, Ali S, Ahmed M, Hussain S. The anti-vaccination movement: a regression in modern medicine. Cureus. 2018;10. https://doi.org/10.7759/cureus.2919.

Hydrolase MIS, Prisbe EJ, Madhavan B, Mcgee DPC, Rydzewski RM, Boehme R, et al. Synthesis and antiviral evaluation of g' -substituted aristeromycins: potential. 1988:1798–804.

Kennedy RB, Ovsyannikova IG, Palese P, Poland GA. Current challenges in vaccinology. Front Immunol. 2020;11:1–18. https://doi.org/10.3389/fimmu.2020.01181.

Kitazato K, Wang Y, Kobayashi N. Viral infectious disease and natural products with antiviral activity. Drug Discov Ther. 2007;1:14–22.

Kobayashi A, Shimada T, Tanaka-Taya K, Kanai M, Okuno H, Kinoshita M, et al. Epidemiology of a workplace measles outbreak dominated by modified measles cases at Kansai international airport, Japan, during august–september 2016. Vaccine. 2020;38:4996–5001. https://doi.org/10.1016/j.vaccine.2020.05.067.

Kurokawa M, Ochiai H, Nagasaka K, Neki M, Xu H, Kadota S, et al. Antiviral traditional medicines against herpes simplex virus (HSV-1), poliovirus, and measles virus in vitro and their therapeutic efficacies for HSV-1 infection in mice. Antiviral Res. 1993;22:175–88. https://doi.org/10.1016/0166-3542(93)90094-Y.

Lage OM, Ramos MC, Calisto R, Almeida E, Vasconcelos V, Vicente F. Current screening methodologies in drug discovery for selected human diseases. Mar Drugs. 2018;16:1–31. https://doi.org/10.3390/md16080279.

Laksono BM, de Vries RD, Duprex WP, de Swart RL. Measles pathogenesis, immune suppression and animal models. Curr Opin Virol. 2020;41:31–7. https://doi.org/10.1016/j.coviro.2020.03.002.

Lancella L, Di Camillo C, Vittucci AC, Boccuzzi E, Bozzola E, Villani A. Measles lessons in an anti-vaccination era: Public health is a social duty, not a political option. Ital J Pediatr. 2017;43:1–4. https://doi.org/10.1186/s13052-017-0420-6.

Lee YC, Lee YH, Lu CW, Cheng SY, Yang KC, Huang KC. Measles immunity gaps in an era of high vaccination coverage: A serology study from Taiwan. Travel Med Infect Dis. 2020;36: https://doi.org/10.1016/j.tmaid.2020.101804.

Li Q, Zhao Z, Zhou D, Chen Y, Hong W, Cao L, et al. Virucidal activity of a scorpion venom peptide variant mucroporin-M1 against measles, SARS-CoV and influenza H5N1 viruses. Peptides. 2011;32:1518–25. https://doi.org/10.1016/j.peptides.2011.05.015.

Lin LT, Hsu WC, Lin CC. Antiviral natural products and herbal medicines. J Tradit Complement Med. 2014;4:24–35. https://doi.org/10.4103/2225-4110.124335.

Ma C, Hao L, Rodewald L, An Q, Wannemuehler KA, Su Q, et al. Risk factors for measles virus infection and susceptibility in persons aged 15 years and older in China: A multi-site case-control study, 2012–2013. Vaccine. 2020;38:3210–7. https://doi.org/10.1016/j.vaccine.2020.03.006.

Magurano F, Baggieri M, Filia A, Del Manso M, Lazzarotto T, Amendola A, et al. Towards measles elimination in Italy: Virological surveillance and genotypes trend (2013–2015). Virus Res. 2017;236:24–9. https://doi.org/10.1016/j.virusres.2017.05.009.

Malonis RJ, Lai JR, Vergnolle O. Peptide-based vaccines: current progress and future challenges. Chem Rev. 2020;120:3210–29. https://doi.org/10.1021/acs.chemrev.9b00472.

McGuigan C, Pathirana RN, Snoeck R, De Andrei G, Clercq E, Balzarini J. Discovery of a new family of inhibitors of human cytomegalovirus (HCMV) based upon lipophilic alkyl furano pyrimidine dideoxy nucleosides: action via a novel non-nucleosidic mechanism. J Med Chem. 2004;47:1847–51. https://doi.org/10.1021/jm030857h.

WHO 2019. https://www.who.int/news-room/fact-sheets/detail/measles. Accessed 13 Nov 2020.

CDC 2020. History of Measles. https://www.cdc.gov/measles/about/history.html. Accessed 13 Nov 2020.

Moss WJ, Griffin DE. Global measles elimination. Nat Rev Microbiol. 2006;4:900–8. https://doi.org/10.1038/nrmicro1550.

Munier-Lehmann H, Lucas-hourani M, Guillou S, Helynck O, Zanghi G, Noel A, et al. Original 2 – (3-Alkoxy – 1 H – pyrazol-1-yl)pyrimidine Derivatives as Inhibitors of Human Dihydroorotate Dehydrogenase (DHODH). J Med. 2015;58:860–77. https://doi.org/10.1021/jm501446r.

Nakao A, Suzuki H, Ueno H, Iwasaki H, Setsuta T, Kashima A, et al. Discovery and structural analyses of S-adenosyl-l-homocysteine hydrolase inhibitors based on non-adenosine analogs. Bioorganic Med Chem. 2015;23:4952–69. https://doi.org/10.1016/j.bmc.2015.05.018.

Naniche D. Human immunology of measles virus infection. Curr Top Microbiol Immunol. 2009;330:151–71. https://doi.org/10.1007/978-3-540-70617-5_8.

Navaratnarajah CK, Miest TS, Carfi A, Cattaneo R. Targeted entry of enveloped viruses: Measles and herpes simplex virus i. Curr Opin Virol. 2012;2:43–9. https://doi.org/10.1016/j.coviro.2011.12.002.

Ndungu JM, Krumm SA, Yan D, Arrendale RF, Reddy GP, Evers T, et al. Non-nucleoside inhibitors of the measles virus RNA-dependent RNA polymerase: Synthesis, structure-activity relationships, and pharmacokinetics. J Med Chem. 2012;55:4220–30. https://doi.org/10.1021/jm201699w.

Niedermeier S, Singethan K, Rohrer SG, Matz M, Kossner M, Diederich S, et al. A small-molecule inhibitor of Nipah virus envelope protein-mediated membrane fusion. J Med Chem. 2009;52:4257–65. https://doi.org/10.1021/jm900411s.

Oehler RL. On measles, vaccination, social media activism, and how to win back our role as our patients' best advocates. Clin Infect Dis. 2020;70:338–40. https://doi.org/10.1093/cid/ciz656.

Ovsyannikova IG, Poland GA. Vaccinomics: Current findings, challenges and novel approaches for vaccine development. AAPS J. 2011;13:438–44. https://doi.org/10.1208/s12248-011-9281-x.

Parker ME, Chabot S, Ward BJ, Johns T. Traditional dietary additives of the Maasai are antiviral against the measles virus. J Ethnopharmacol. 2007;114:146–52. https://doi.org/10.1016/j.jep.2007.06.011.

Paterson BJ, Kirk MD, Cameron AS, D'este C, Durrheim DN. Historical data and modern methods reveal insights in measles epidemiology: a retrospective closed cohort study. BMJ Open. 2013;3:1–10. https://doi.org/10.1136/bmjopen-2012-002033.

Patil SD, Schneller SW, Hosoya M, Snoeck R, Andrei G, Balzarini J, et al. The synthesis and antiviral properties of (.+-.)-5'-noraristeromycin and related purine carbocyclic nucleosides: a new lead for anti-human cytomegalovirus agent design. J Med Chem. 1992;35:3372–7. https://doi.org/10.1021/jm00096a012.

Paules CI, Marston HD, Fauci AS. Measles in 2019—going backward. N Engl J Med. 2019;380:2185–7. https://doi.org/10.1056/NEJMp1905099.

Peña-Rey I, Martínez de Aragón V, Mosquera M, de Ory F, Echevarría JE. Measles risk groups in Spain: implications for the European measles-elimination target. Vaccine. 2009;27:3927–34. https://doi.org/10.1016/j.vaccine.2009.04.024.

Petrie CR, Cottam HB, McKernan PA, Robins RK, Revankar GR. Synthesis and biological activity of 6-azacadeguomycin and certain 3,4,6-trisubstituted pyrazolo[3,4-d]pyrimidine ribonucleosides. J Med Chem. 1985;28:1010–6. https://doi.org/10.1021/jm00146a007.

Petrie CR, Revankar GR, Dailey NK, George RD, McKernan PA, Hamill RL, et al. Synthesis and biological activity of certain nucleoside and nucleotide derivatives of pyrazofurin. J Med Chem. 1986;29:268–78. https://doi.org/10.1021/jm00152a016.

Plemper RK. Measles resurgence and drug development. Curr Opin Virol. 2020;41:8–17. https://doi.org/10.1016/j.coviro.2020.02.007.

Plemper RK, Doyle J, Sun A, Prussia A, Cheng L, Rota PA, et al. Design of a small-molecule entry inhibitor with activity against primary measles virus strains. Antimicrob Agents Chemother. 2005;49:3755–61. https://doi.org/10.1128/AAC.49.9.3755-3761.2005.

Poland GA, Kennedy RB, Ovsyannikova IG. Vaccinomics and personalized vaccinology: Is science leading us toward a new path of directed vaccine development and discovery? PLoS Pathog. 2011;7. https://doi.org/10.1371/journal.ppat.1002344.

Prussia AJ, Plemper RK, Snyder JP. Measles virus entry inhibitors: a structural proposal for mechanism of action and the development of resistance. Biochemistry. 2008;47:13573–83. https://doi.org/10.1021/bi801513p.

Pütz MM, Muller CP. The rationale of a peptide-conjugate vaccine against measles. Vaccine. 2003;21:663–6. https://doi.org/10.1016/S0264-410X(02)00576-5.

Roeder P, Mariner J, Kock R. Rinderpest: the veterinary perspective on eradication. Philos Trans R Soc B Biol Sci. 2013;368:20120139. https://doi.org/10.1098/rstb.2012.0139.

Rybicki EP. Plant-produced vaccines: promise and reality. Drug Discov Today. 2009;14:16–24. https://doi.org/10.1016/j.drudis.2008.10.002.

Sánchez-Barrena MJ, Martínez-Ripoll M, Gálvez A, Valdivia E, Maqueda M, Cruz V, et al. Structure of bacteriocin AS-48: from soluble state to membrane bound state. J Mol Biol. 2003;334:541–9. https://doi.org/10.1016/j.mb.2003.09.060.

Santagati NA, Bousquet E, Garozzo A, Prezza O, Spadaro A. Synthesis and anti-measles virus activity of new isoquinolin-4-one derivatives. Farm. 2003;58:1217–25. https://doi.org/10.1016/j.farmac.2003.07.003.

Sarkar D, Julkunen I, Bergha H, Fisher PB, Vainionpa R. The interferon-inducible RNA helicase, mda-5, is involved in measles virus-induced expression of antiviral cytokines. 2006;8:2138–44. https://doi.org/10.1016/j.micinf.2006.04.005.

Schneller SW, Thompson RD, Cory JG, Olsson RA, De Clercq E, Kim I, et al. Biological activity and isomer of formycin. J Med Chem. 1984a;27:924–8.

Schneller SW, Luo JK, Hosmane RS, De Clercq E, Stoeckler JD, Agarwal KC, et al. Synthesis and biological evaluation of 6-amino-1H-pyrrolo[3,2-c]pyridin-4(5H)-one (3,7-dideazaguanine). J Med Chem. 1984b;27:1737–9. https://doi.org/10.1021/jm00378a037.

Schober T. Effects of a measles outbreak on vaccination uptake. Econ Hum Biol. 2020;38: https://doi.org/10.1016/j.ehb.2020.100871.

Shaban MAE. The chemistry of C-Nucleosides and their analogs II: C-Nucleosides of condensed heterocyclic bases. Adv Heterocycl Chem. 1997;70:163–337. https://doi.org/10.1016/s0065-2725(08)60931-9.

Shaffer JA, Bellini WJ, Rota PA. The C protein of measles virus inhibits the type I interferon response. Virology. 2003;315:389–97. https://doi.org/10.1016/S0042-6822(03)00537-3.

Shanks GD, Hu Z, Waller M, Lee SE, Terfa D, Howard A, et al. Measles epidemics of variable lethality in the early 20th century. Am J Epidemiol. 2014;179:413–22. https://doi.org/10.1093/aje/kwt282.

Shrivastava SRBL, Shrivastava PS, Ramasamy J. Measles in India: Challenges and recent developments. Afr J Disabil. 2015;5. https://doi.org/10.3402/iee.v5.27784.

Shuto S, Obara T, Toriya M, Hosoya M, Snoeck R, Andrei G. New neplanocin analogues 1: synthesis of B' -Modified neplanocin a derivatives as broad-spectrum antiviral agents. 1992:324–31.

Sibeudu FT, Onwujekwe OE, Okoronkwo IL. Cost analysis of supplemental immunization activities to deliver measles immunization to children in Anambra state, south-east Nigeria. Vaccine. 2020;38:5947–54. https://doi.org/10.1016/j.vaccine.2020.06.072.

Singh J, Sharma D, Kumar G, Sharma NR. Microbial bioprospecting for sustainable development. Microb Bioprospecting Sustain Dev 2018:1–397. https://doi.org/10.1007/978-981-13-0053-0.

Singha UK, Miller MB, Nes CR, Anyatonwu H, Lecordier L, Patkar P, et al. Discovery of an ergosterol-signaling factor that regulates Trypanosoma brucei growth. J Lipid Res. 2015;56. https://doi.org/10.1194/jlr.M054643.

Sleeman K, Stein DA, Tamin A, Reddish M, Iversen PL, Rota PA. Inhibition of measles virus infections in cell cultures by peptide-conjugated morpholino oligomers. Virus Res. 2009;140:49–56. https://doi.org/10.1016/j.virusres.2008.10.018.

Spika JS, Wassilak S, Pebody R, Lipskaya G, Deshevoi S, Güriş D, et al. Measles and rubella in the World Health Organization European region: diversity creates challenges. J Infect Dis. 2003;187:191–7. https://doi.org/10.1086/368336.

Stallcup KC, Raine CS, Fields BN. Cytochalasin B inhibits the maturation of measles virus. Virology. 1983;124:59–74. https://doi.org/10.1016/0042-6822(83)90290-8.

Stern AM, Markel H. The history of vaccines and immunization: Familiar patterns, new challenges if we could match the enormous scientific strides of the twentieth century with the political and economic investments of the nineteenth, the world's citizens might be much heal. Health Aff. 2005;24:611–21. https://doi.org/10.1377/hlthaff.24.3.611.

Sun A, Prussia A, Zhan W, Murray EE, Doyle J, Cheng L, et al. Nonpeptide inhibitors of measles virus entry. 2006:5080–92.

Sun A, Chandrakumar N, Yoon J, Plemper RK, Snyder JP. Non-nucleoside inhibitors of the measles virus RNA-dependent RNA polymerase complex activity: synthesis and in vitro evaluation. Bioorg Med Chem Lett. 2007;17:5199–203. https://doi.org/10.1016/j.bmcl.2007.06.084.

Sun A, Yoon JJ, Yin Y, Prussia A, Yang Y, Min J, et al. Potent non-nucleoside inhibitors of the measles virus RNA-dependent RNA polymerase complex. J Med Chem. 2008;51:3731–41. https://doi.org/10.1021/jm701239a.

Sundaram ME, Guterman LB, Omer SB. The true cost of measles outbreaks during the postelimination era. JAMA. 2019;321:1155. https://doi.org/10.1001/jama.2019.1506.

Swallow DL. Antiviral agents. Prog Drug Res. 1978;22:267–326.

Takayama I, Sato H, Watanabe A, Omi-Furutani M, Sugai A, Kanki K, et al. The nucleocapsid protein of measles virus blocks host interferon response. Virology. 2012;424:45–55. https://doi.org/10.1016/j.virol.2011.12.011.

Terao-muto Y, Yoneda M, Seki T, Watanabe A, Tsukiyama-kohara K, Fujita K, et al. Heparin-like glycosaminoglycans prevent the infection of measles virus in SLAM-negative cell lines. Antiviral Res. 2008;80:370–6. https://doi.org/10.1016/j.antiviral.2008.08.006.

Trottier C, Chabot S, Mann KK, Colombo M, Chatterjee A, Miller WH, et al. Retinoids inhibit measles virus in vitro via nuclear retinoid receptor signaling pathways. Antiviral Res. 2008;80:45–53. https://doi.org/10.1016/j.antiviral.2008.04.003.

Turner MA, Yang X, Yin D, Kuczera K, Borchardt RT, Howell PL. Structure and function of S-adenosylhomocysteine hydrolase. Cell Biochem Biophys. 2000;33:101–25. https://doi.org/10.1385/CBB:33:2:101.

Ugarkar BG, Cottam HB, Mckernan PA, Robins RK, Revankar GR. Synthesis and antiviral/antitumor activities of certain. 1984;1026–1030

Vink JP, Snell LB, Bernard K, Mitchell H, Heathcock RT, Cordery R, et al. Mapping a nosocomial outbreak of measles, coinciding with a period of sustained transmission in South London in 2018. J Hosp Infect. 2020;105:747–51. https://doi.org/10.1016/j.jhin.2020.06.004.

Wang SA, Ko Y, Zeng J, Geng Y, Ren D, Ogasawara Y, et al. Identification of the formycin A biosynthetic gene cluster from streptomyces kaniharaensis illustrates the interplay between biological pyrazolopyrimidine formation and de novo purine biosynthesis. J Am Chem Soc. 2019;141:6127–31. https://doi.org/10.1021/jacs.9b00241.

Webster DE, Thomas MC, Huang Z, Wesselingh SL. The development of a plant-based vaccine for measles. Vaccine. 2005;23:1859–65. https://doi.org/10.1016/j.vaccine.2004.11.018.

White LK, Yoon J, Lee JK, Sun A, Du Y, Fu H, et al. Nonnucleoside inhibitor of measles virus RNA-dependent RNA polymerase complex activity. 2007;51:2293–303. https://doi.org/10.1128/AAC.00289-07.

Wong AH, Ho YII, Tang KPS, Lai RWM. Comparison of CaptiaTM Measles IgG Assay with Vidas® Measles IgG Assay for determination of immune protection status against measles virus. J Clin Virol. 2020;127: https://doi.org/10.1016/j.jcv.2020.104342.

Xu L, Chen J, Zhan Z, Chen T, Liu W, Qian J, et al. Evaluating the effectiveness of national measles elimination action in mainland China during 2004–2016: A multi-site interrupted time-series study. Vaccine. 2020;38:4440–7. https://doi.org/10.1016/j.vaccine.2020.04.053.

Yoon J, Krumm SA, Ndungu JM, Hoffman V, Bankamp B, Rota PA, et al. Target analysis of the experimental measles therapeutic AS-136A. Antimicrob Agents Chemother. 2009;53:3860–70. https://doi.org/10.1128/AAC.00503-09.

Zhang N, Chen HM, Sood R, Kalicharran K, Fattom AI, Naso RB, et al. In vitro inhibition of the measles virus by novel ring-expanded ('fat') nucleoside analogues containing the imidazo [4,5-e][1,3]diazepine ring system. Bioorganic Med Chem Lett. 2002;12:3391–4. https://doi.org/10.1016/S0960-894X(02)00762-X.

Corona Virus and SARS-CoV-2

Lethal Human Coronavirus Infections and the Role of Vaccines in Their Prevention

Jeremy Baldwin and Nikolai Petrovsky

Abstract The Severe Acute Respiratory Syndrome coronavirus (SARS CoV) was first identified in 2002 when it caused an epidemic of fatal human pneumonia cases that from an epicenter in Hong Kong rapidly spread to multiple countries. SARS caused approximately 774 deaths before it was eradicated from the human population by quarantine control measures. Repeated outbreaks of Middle East respiratory syndrome coronavirus (MERS-CoV) have occurred since 2012 with 858 reported human deaths to date. Most recently, in late 2019 a new respiratory virus infection now known as COVID-19 due to infections by the SARS-CoV-2 virus started in China before spreading rapidly to the rest of the world, resulting in approximately 1.5 million deaths to date. All these coronaviruses (CoV) causing fatal human respiratory infections originally had zoonotic origins, with SARS-like and MERS-like coronaviruses circulating in bats. Immunity in humans induced by most CoV infections is rapidly waning; presenting the problem those convalescent patients can become reinfected. Vaccines present the best strategy to protect the human population against CoV infection but face several challenges. One concern is that CoV vaccines, particularly those containing Th2-biased alum adjuvants, may cause problems of disease enhancement such as eosinophilic lung immunopathology upon virus exposure. This problem was seen with SARS vaccines in animal testing. Another concern is that vaccine protection, like natural immunity to CoV infection, may be short-lived. There is also a concern that like influenza, SARS-CoV-2 might mutate its spike protein receptor to circumvent existing human immunity including that induced by vaccines. This could slow or prevent the development of human herd immunity. Ideally CoV vaccines should provide robust long-lived immunity providing broad protection against mutated strains. This chapter describes the current state of development of CoV vaccines including against COVID-19, the issue of CoV-associated lung immunopathology and the form those current and future CoV pandemic vaccines might take.

J. Baldwin · N. Petrovsky (✉)
Vaxine Pty Ltd, 11 Walkley Avenue, Warradale 5046, Australia
e-mail: nikolai.petrovsky@flinders.edu.au

N. Petrovsky
College of Medicine and Public Health, Flinders University, Adelaide 5042, Australia

S. I. Ahmad (ed.), *Human Viruses: Diseases, Treatments and Vaccines*, https://doi.org/10.1007/978-3-030-71165-8_24

Keywords Coronavirus · SARS-CoV-2 · SARS CoV · MERS · COVID-19 · Vaccine · Adjuvant · Lung pathology · Immunity · Infection

1 Introduction

Coronaviruses (CoV) are a family of positive-sense single-stranded RNA viruses approximately 26–32 kb in length that causes disease in mammalian and avian species. The first known CoV pathogens in humans were identified in the 1960s after virus samples collected from patients could not be cultivated using known techniques for common cold viruses (Kapikian 1975). Subsequent characterization under the electron microscope revealed that these virus samples shared a similar morphological structure, due to a crown-shaped (corona) spike protein projecting from the virus envelope,with this unique featuregiving these viruses their name (Almeida and Tyrrell 1967). Since their initial discovery over 20 new CoV types have been identified. Novel strains of CoV have led to three epidemic events in modern history, namely severe acute respiratory syndrome coronavirus (SARS) in 2002, Middle East respiratory syndrome-related coronavirus (MERS-CoV) in 2012 and severe acute respiratory syndrome coronavirus 2 (SARS-CoV-2) in 2019. The SARS outbreak infected over 8,000 people in 29 countries before it was brought under control by quarantine measures (Drosten et al. 2003; Rota et al. 2003). The MERS outbreak infected over 2,500 people in 27 countries and was again brought under control through quarantine measures, although there still continues to be ongoing sporadic clusters of human cases as recently as March 2020 (Mubarak et al. 2019). The SARS-CoV-2 virus has at the time of writing this chapter infected 67 million cases worldwide in over 213 countries and remains a significant ongoing health crisis. The failure of quarantine measures to bring the SARS-CoV-2 pandemic under control highlights the need for development of effective CoV vaccines. Given it has been only one year since the start of theSARS-CoV-2 outbreak and research into the virus is still ongoing, previous data on SARS and MERS CoV might provide useful insights into SAR-CoV-2 vaccine design. In this chapter we compare and contrast the features of the different types of CoVs, summarize lessons learned from SARS and MERS vaccines, and explore the different vaccine platform approaches for SARS-CoV-2 together with potential challenges.

1.1 Features of SARS, MERS and SARS-CoV-2

In order to examine past vaccine strategies against SARS and MERS and its applicability to SAR-CoV-2 vaccine design, it is first important to understand the differences between these viruses. Coronaviruses are zoonotic in origin with bats being a natural CoV reservoir, with the virus typically transmitted to human through intermediate hosts. The intermediate host for SARS is believed to have been palm civets (Guan et al. 2003) and for MERS dromedary camels (Chu et al. 2014) based on the high genomic homology of strains isolated from these animals with the original human isolates. The origins and any intermediate host for SARS-CoV-2 remains unknown with pangolins, bats and even snakes being proposed but largely discounted as potential intermediate hosts along with the possibility that the source might have been an accidental laboratory release of a lab-cultured bat virus (Ji et al. 2020; Zhang et al. 2020; Zhou et al. 2020). The SARS-CoV-2 genome shares a homology of almost 79% with SARS, and only 50% with MERS (Hoffmann et al. 2020). The closest reported relatives to SARS-CoV-2 is recently published Chinese bat CoV sequence known as RATG13 although no virus correlating to this sequence has ever been cloned.

Structurally coronaviruses consist primarily of the same structural proteins, namely (i) nucleocapsid, (ii) spike, (iii) envelope, and (iv) membrane proteins, plus additional accessory proteins (V'kovski et al. 2020). The distinctive S-protein mediates attachment and viral entry into host cells making it an attractive target for vaccines (Li 2016). SARS was shown to bind to angiotensin-converting enzyme-2 (ACE2) and use it as a receptor to enter cells, whilst MERS was shown to utilize dipeptidyl peptidase 4 (DPP4) for this purpose. Due to the homology of its spike protein with SARS (77%) SAR-CoV-2 was predicted and confirmed to also use ACE-2 for cell entry. SARS-CoV-2 has been shown to have an 10–20-fold higher binding affinity with ACE-2 than SARS CoV (Wrapp et al. 2020). In addition, Transmembrane Serine Protease 2 (TMPRSS2) has been shown to play a role in SARS, MERS and SARS-CoV-2 cell entry by priming the spike protein and allowing fusion of the viral and cellular membrane (Hoffmann et al. 2020). The expression levels and distribution of these different virus receptors varies between cells/tissues in the body and can influence disease pathology (Yin and Wunderink 2018).

1.2 Disease Transmission and Pathology of Coronaviruses

Coronaviruses spread primarily by aerosol in the form of respiratory droplets, or through close personal contact, being absorbed through mucous membranes. The incubation period between initial infection and development of symptoms is similar between coronaviruses at 4–6 days (Jiang et al. 2020). The serial interval is the time between the primary case symptom onsets to secondary case symptom onset and this varies between CoVs. SARS has a mean serial interval of 8–12 days (Lipsitch et al. 2003) and MERS has a mean serial interval of 7–12 days (Cowling et al. 2015). A number of systematic reviews have suggested SARS-CoV-2 has a lower serial interval of ~4–5 days (Nishiura et al. 2020).The shorter serial interval indicates that SARS-CoV-2 virus has a higher transmission rate compared to SARS and MERS, and this may explain in part why quarantine measures alone have failed to contain the SARS-CoV-2 outbreak. Moreover, viral shedding, defined as release of virus from host body into the environment, has been shown to peak early for SARS-CoV-2 when compared to SARS and MERS (He et al. 2020).

Coronavirus infection symptoms are common among the different CoVs with high fever, dry cough, shortness of breath, headache, muscle aches, sore throat, fatigue and diarrhea, typical (Zhu et al. 2020).Lung histology in fatal cases revealed a marked diffuse alveolar damage, inflammatory cell infiltrate, bronchial epithelial denudation, loss of cilia, and squamous metaplasia (Zhu et al. 2020). Deaths result from respiratory, heart, neurological and/or liver failure and the highest fatality rates have been observed in the elderly or individuals with co-morbidities such as diabetes, obesity or cardiovascular disease. Disease mortality differs significantly between CoVs with estimated fatality rates of 10% for SARS, 34% for MERS and 1–3% for SAR-CoV-2 (Abdelrahman et al. 2020). The difference in fatality rate of the different CoVs may be due to viral properties, or in the case of MERS the high apparent fatality rate may be derived from hospitalized patients as patients with mild or no symptoms may remain undetected.

Infected individuals recovering from coronavirus infections may become susceptible to re-infection due to rapidly waning immunity (Callow et al. 1990; Tang et al. 2011). In fact, those with waning immunity may be at risk of even more severe disease upon coronavirus re-infection (Wang et al. 2014). SARS re-infection studies showed that although immune animals cleared lung virus much faster than naïve animals, the incidence and severity of lung inflammation was not reduced (Clay et al. 2012). This suggests that illness severity is not just dictated by viral load but also may be adversely influenced by host factors including pre-existing immunity.

1.3 Lessons Learned from Vaccine Approaches to SARS and MERS CoVs

Although some SARS and MERS vaccine candidates showed good immunogenicity and virus protection in animal models, there are currently no human vaccines for SARS and MERS that have moved beyond Phase II clinical trials. One of the reasons why these vaccine candidates have failed to progress further in development has been ongoing concerns about the potential risk of vaccine-enhanced disease. The SARS and MERS vaccine literature still represents a valuable resource due to genetic similarity and pathogenesis of the viruses and the lack of available data on SAR-CoV-2, in particularon long-term vaccine protection and safety.

2 CoV Preclinical Models

One of the major hurdles in initial vaccine development in response to the SARS and MERS outbreaks was a lack of suitable animal models. An ideal animal model should closely replicate disease mechanisms/pathology to the virus and also match the likely human vaccine immune response. Mice are one of the most widely used animal models in immunology and vaccine research due to low cost and ease of manipulation; however a lack of homology between mouse and human ACE-2 in the case of SARS or DPP4 in the case of MERS limited their use. Day et al. were able to overcome this challenge by generating a mouse adapted strain of SARS through serial passage and accumulated mutations in the spike protein that enabled it to bind to mouse ACE-2 receptor (Day et al. 2009). Another approach was the development of transgenic mice expressing the human ACE-2 or DPP4 genes. Mice were also transfected with ACE-2 or DPP4 using viral vectors but these lacked physiological expression as all cells expressed ACE-2 or DPP4 (Agrawal et al. 2015), however subsequent advances in genetic engineering using CRISPR, enabled the insertion of human ACE-2 or DPP4 into the corresponding mouse gene loci (Cockrell et al. 2016). Given that SARS-CoV-2 also uses ACE-2 to enter cells, human ACE-2 transgenic mice have been shown to be useful for studying SARS-CoV-2, but have unfortunately been limited in supply.

Although hamster and ferret models do not support MERS infection (Raj et al. 2014; de Wit et al. 2013), these animals were useful for studying SARS and more recently SARS-CoV-2. Hamsters are good models for SARS as the virus is able to replicate to high viral titers in the respiratory tract, with good correlation to lung pathology although they do not display outward symptoms of the disease. (Roberts et al. 2005; Lamirande et al. 2008; Roberts et al. 2006). Ferrets were a useful SARS model, displaying symptoms including fever and nasal shedding making them beneficial for studying vaccine immunity but also transmission (Martina et al. 2003; Weingartl et al. 2004; Chu et al. 2008; Ter Meulen et al. 2004). There was a wide range of variability amongst non-human primates (NHP) for SARS and MERS infections, due to genetic diversity and differences in NHP sub-species (Gretebeck and Subbarao 2015). Rhesus macaques developed only moderate symptoms (De Wit et al. 2013; Yao et al. 2014) whilst severe pathology and even lethality was observed in Marmosets (Falzarano et al. 2014). Still NHPs replicated both SARS and MERS disease pathology closest to humans. Thus far, animal models used for SARS-CoV-2 are similar to previously described models for SARS and MERS, being human ACE2 transgenic mice, hamsters, ferrets and NHP. Hamster show high replication of SARS-CoV-2 and outward symptoms (weight-loss/respiratory distress) and age-dependent differences (Chan et al. 2019; Osterrieder et al. 2020). Viral replication and lung pathology are observed in the ferret models of SARS-CoV-2 infection, but they only exhibit moderate symptoms and no lethality (Kim et al. 2020). A review of different NHP models highlighted differences in sub-species susceptibility to SARS-CoV-2 infection, with macaques showing the greatest pathological featuresto SARS-CoV-2 infection (Lu et al. 2020).

3 Vaccine Approaches

MERS and SARS vaccines were primarily based around traditional vaccine technologies including inactivated virus, recombinant protein, and viral vectors, with less common examples of use of newer vaccines technology platforms, such as DNA and mRNA.

3.1 Inactivated CoV Vaccines

Inactivated viruses in which whole virus is rendered inert either physically (heat/irradiation) or chemically (formalin fixation) is one of the oldest forms of vaccine technology. Inactivated viruses can be rapidly developed as the inactivation process is relatively straightforward and doesn't require genetic manipulation of the antigen (Delrue et al. 2012). Nevertheless, generating live virus requires high level biosafety containment and specialized manufacturing facilities(Delrue et al. 2012). The use of whole virus as vaccine antigen means that a broad range of different epitopes are present to which the immune system can generate a response (Delrue et al. 2012). However, excess inactivation treatment can change the native conformation of the virus epitopes and can create high batch-to-batch product variability. A number of inactivated viruses were developed for SARS and MERS and stimulated an antibody response (Deng et al. 2018; Agrawal et al. 2016; Wirblich et al. 2017; Lin et al. 2007). However there were reports of lung hypersensitivity in animals vaccinated with whole inactivated SARS or MERS virus (Agrawal et al. 2016). Lung immunopathology was particularly evident when an inactivated SARS vaccine was formulated with Alum adjuvant, suggesting an excess Th2 immune response was the culprit.

Fortunately this lung immunopathology side effect was prevented when the inactivated antigen was formulated with Advax, an adjuvant providing a strong Th1 T cell response (Honda-Okubo et al. 2015). Although a Phase I clinical trial of an inactivated SARS vaccine did not report side effects, the immunized subjects were notably not exposed to the SARS virus itself, which is what triggered the lung immunopathology in the mice, so no conclusions can be drawn on the safety of inactivated SARS vaccines from this study (Zhang et al. 2020). Notably, several similar inactivated SARS-CoV-2 vaccine candidates are currently under development including the Chinese CoronaVac vaccine from Sinovac which is in Phase III trials after demonstrating immunogenicity in mice, rats, and NHP (Gao et al. 2020). No immunopathology was reported in NHP following SAR-CoV-2 viral challenge (Gao et al. 2020) or Phase I clinical trials (Zhang et al. 2020), although long-term safety in humans in still unknown.

3.2 Protein Subunit CoV Vaccines

Unlike inactivated virus vaccines that use the whole virus as an antigen, protein subunit vaccines take a targeted approach and use specific virus proteins as the antigen source. Multiple MERS and SARS proteins have been used for vaccine design (He et al. 2004; Kam et al. 2007; Du et al. 2007, 2013; Tai et al. 2017; Wang et al. 2015; Lan et al. 2015; Jiaming et al. 2017), but with the spike protein being most successful in inducing antibodies that neutralize the virus and block infection. An even more restricted approach that used just the receptor binding domain (RBD) of the SARS spike protein to induce neutralizing antibody showed protection in animals (Du et al. 2007), although the lack of proper protein folding and lack of adequate T cell epitopes in such a small antigen may be a limitation of this RBD approach. Protein subunit units typically require an adjuvant to generate sufficient immunogenicity, with the choice of adjuvant influencing the type and strength of immune response generated (Petrovsky and Aguilar 2004). SARS and MERS proteins were also shown to be more immunogenic when expressed as self-assembling virus-like particles (VLP) (Liu et al. 2011; Lan et al. 2018). A VLP vaccine without adjuvant protected mice from a lethal SARS challenge, whilst the equivalent protein subunit only protected when formulated with alum adjuvant (Liu et al. 2011). A two proline (2P) substitution in the MERS spike protein to stabilize it in a pre fusion conformation significantly increased neutralizing antibody titers (Pallesen et al. 2017), and this same 2P mutation has been incorporated into many SARS-CoV-2 vaccines under development. Mutations have also been introduced to disrupt the furin cleavage site of the SARS-CoV-2 spike protein, a critical cleavage site that does not exist in SARS CoV spike protein or any other closely related CoVs, raising questions as to where it might have come from. Hence, the antigen conformation and choice of adjuvant play a key role in CoV vaccine immunogenicity. NVX-CoV2373 (Novavax, USA) is a SARS-CoV-2 vaccine based on a recombinantfull-lengthSARS-CoV-2 spike protein with the 2P mutation and mutated furin cleavage site formulated with Matrix-M1 adjuvant based on the saponin, QS21. NVX-CoV2373 induced high neutralizing antibody levels and a Th1 T cell bias, but less than 0.5% of the CD4+T-cells on average were polyfunctional (IFN-γ, TNF-α and IL-2) with data on CD8+responses not shown (Keech et al. 2020).

Another SARS-CoV-2 vaccine currently in Phase I clinical trials is the authors own COVAX-19 vaccine (Vaxine, Australia), a protein subunit vaccine which comprises insect cell expressed SARS-CoV-2 spike protein complete extracellular domain with the furin cleavage site removed and formulated with Advax-SM adjuvant. The same vaccine platform as used for COVAX-19 was shown to induce a robust Th1 cellular immune response when used for previous SARS and MERS CoV vaccines (Honda-Okubo et al. 2015; Adney et al. 2019).The proven long-term safety and efficacy of protein subunit vaccines for infections such as hepatitis B and human papilloma virus make them the most promising candidates for a successful CoV vaccine approach. However, the major downside to protein subunit vaccines is the complexity of their design and manufacture as unlike the other vaccine platforms such as nucleic acid and viral vectors where the antigenic protein is produced and folded inside the body, in the case of recombinant protein vaccines the antigenic protein needs to be produced synthetically using cultured cell lines, then purified from the other proteins these cells make before it can be used. During this process, the recombinant protein is sensitive to degradation by cellular enzymes and also may be misfolded so that the final product does not expose the right antigenic structures to the immune system. To tackle these issues special assays, many of which use monoclonal antibodies need to be developed, adding considerable time to the development process. This explains why during the COVID-19 pandemic, vaccines based on nucleic acid or virus vectors were first to enter human trials, with the protein subunit vaccines taking considerably longer to reach this stage. Nevertheless, given their exemplary human safety and efficacy track record, protein subunit vaccines remain the leading candidates to become successful long-term COVID-19 vaccines.

3.3 *Viral Vectors*

Viral vector vaccines work by inducing intracellular antigen expression, rather than extracellular antigen presentation, shuttling genetic material encoding immunogenic components of the target virus into host cells that then express them (Ura et al. 2014). The two main types of virus vectors used in MERS and SARS vaccines in animals were Adenoviruses and modified vaccinia virus Ankara (MVA) (Guo et al. 2015; Kim et al. 2014; Munster et al. 2017; Alharbi et al. 2017; Fett et al. 2013; Volz et al. 2015). Multiple SARS-CoV-2 vaccines are currently under development that utilize adenovirus vectors and include a chimpanzee adenovirus (ChAdOx1) (Oxford–AstraZeneca) as well as Adeno5 and Adeno26 vectors. An issue with use of human adenoviruses and MVA vaccines is pre-existing immunity in the population against the viral vector impairs vaccine efficacy and may induce a harmful inflammatory response (Weingartl et al. 2004; Fausther-Bovendo and Kobinger 2014). Use of ChAdOx1 was intended as a way around this limitation due to the low human seroprevalence to chimpanzee adenovirus (Dicks et al. 2012) but showed poor immunogenicity with a single vaccine dose, but despite two doses being used for the Phase 3 human trialthis resulted in just a 62% per protocol protection rate against symptomatic infection. This compares to protection rates of over 90% reported for several mRNA vaccine Phase 3 trials (see below). ChAdOx1 has unknown human safety having never previously been given to large numbers of human subjects, and cases of multiple sclerosis and transverse myelitis during the Phase 3 trial triggered temporary clinical holds. Overall, the high rate of side effects, low efficacy and unproven human safety of the adenovirus vector approach make it a relatively weak choice for use as a CoV vaccine platform despite the speed with which such vaccines can be designed using existing virus backbones.

3.4 mRNA CoV Vaccines

Similar to viral vector vaccines, mRNA operate as gene therapies to induce intracellular protein expression. A key to the success of mRNA vaccines was the use of lipid based nanoparticles into which the mRNA could be loaded which then act as a delivery mechanism allowing for rapid uptake into the cell cytoplasm where the mRNA is translated into protein (Pardi et al. 2018). mRNA is new as a vaccine technology and therefore there is limited data on its use in SARS or MERS vaccines in animals and next to no previous human safety or efficacy data. A mRNA formulated in lipid nanoparticles (LNP) vaccine for MERS was under evaluationat the time of the SARS-CoV-2 outbreak (Corbett et al. 2020). The mRNA–LNP which utilized the 2P mutation at the apex of the central helix and heptad repeat 1 (Pallesen et al. 2017) elicited pseudovirus-neutralizing antibodies and protected against a lethal MERS-CoV challenge in transgenic hDPP4mice (Corbett et al. 2020). TheSARS-CoV-2 outbreak occurred during preclinical evaluation of this MERS vaccine and it was adapted to generate a SARS-CoV-2 vaccine (mRNA-1273) that bypassed the normal animal testing and entered Phase I clinical trials under 2 months from release of the viral sequence. This vaccine has shown over 60% efficacy in preventing symptomatic SARS-CoV-2 infection in Phase III trials although the durability of such protection is currently unknown. Another mRNA vaccine candidate (BNT162b1) by BioNTech/Pfizer that similarly encodes the SARS-CoV-2 spike protein and is formulated with LNP has also completed Phase 3 trials with short-term protection rates of over 90% (Mulligan et al. 2020). A potential benefit of mRNA apart from the speed with which a new vaccine can be designed, is that it can be chemically synthesized which should speed up manufacturing, although it still requires the cell based production of enzymes needed for its synthesis (Graham et al. 2018). Another key drawback of mRNA technology is its instability with most mRNA vaccines requiring storage temperatures of −20 to −80 °C, which may affect the rollout of the vaccine, in particular in developing countries that lack appropriate freezer in frastructure with the normal vaccine cold chain in these countries being based on vaccines that are stable under 2–8 °C refrigeration.

3.5 DNA CoV Vaccines

DNA vaccines act by transfecting cells with engineered DNA plasmids encoding protein sequences that can be then translated in the cell nucleus to RNA and then into protein in the cytoplasm. DNA vaccines can be delivered via a variety of way including intramuscular/intradermal needle, liposomal carrier, gene gun and electroporation devices. A major benefit of DNA vaccines over mRNA is thermal stability even being stable at room temperature (Stenler et al. 2014). DNA CoV vaccines have been tested in mice, macaques and camels (Muthumani et al. 2015;

Woo et al. 2005; Shi et al. 2006; Chi et al. 2017; Al-Amri et al. 2017) and a MERS DNA vaccine had progressed to human Phase II clinical trials at the time of SARS-CoV-2 outbreak. GLS-5300, a MERS DNA vaccine has completed phase I clinical trial in the US and has entered phase II trials inKorea. The Phase I trial showed a high rate of injection-site reactions (93%) due to the electroporation device used to deliver the DNA vaccine, with antibody responses detected in more than 85% of participants after two vaccinations (Modjarrad et al. 2019). Inovio Pharmaceuticals adapted the GLS-5300 vaccine and electroporation delivery system to SARS-CoV-2 and this SARS-CoV-2 vaccine (INO-4800) is currently in Phase 2/3 clinical trials (Smith et al. 2020).The advantage of DNA vaccines are their rapid design and synthesis, with major downsides being the need for a separate electroporator device making administration more complex and painful than traditional vaccines. Also, DNA vaccines typically require up to 3 vaccine doses even with electroporation to generate satisfactory immunogenicity. These issues and the lack of a previous example of a successful licensed human DNA vaccine, makes the role of DNA vaccines in COVID-19 highly uncertain, particularly until Phase 3 data showing efficacy becomes available.

4 Vaccine-Enhanced Disease

Many papers have reported immunopathologic reactions in mice vaccinated with either SARS or MERSCoV vaccines, based on recombinant spike protein, virus-like particles, inactivated virus, recombinant vaccinia virus and adenovirus-based vectors (Agrawal et al. 2016; Tseng et al. 2012; Deming et al. 2006; Yasui et al. 2008; See et al. 2008). The root cause behind this immunopathology remains to be fully elucidated. Demming et al. performed passive transfer of anti-N, anti-HA, or anti-S sera from a SARS vaccine into naive mice and suggested the increased inflammatory response was not due to vaccine-induced antibodies (Deming et al. 2006). Instead, the immunopathology seen by CoV vaccines may be due to an unbalanced T cell response triggered by some vaccines. CD8 T cells are responsible for virus clearance and adoptive transfer of immune splenocytes or SARS-specific T cells reduced lung virus titers and enhanced survival (Zhao et al. 2010). Autopsy lung samples of patients dying after SARS infection revealed downregulation of type 1interferon and CXCL10, a chemokine involved in T-cell recruitment and inhibition of the whole STAT1 pathway (Kong et al. 2009). CoVs have been shown to use multiple mechanisms to inhibit the host type 1 interferon response with open reading frame(ORF) 3b/6 inhibiting both interferon synthesis and STAT1 signaling and nucleocapsid proteins blocking interferon production (Kopecky-Bromberg et al. 2007). Prompetechara et al. speculated also that the delayed type 1 IFN caused by SARS and MERSs in early viral control and lead to influx of hyperinflammatory cells and manifesting in lung immunopathology (Prompetchara et al. 2020).

This suggests that CoV vaccines that prime for a Th-2 biased immune response might further exacerbate the Th-2 immunomodulatory effect of the CoVs themselves. A study by our group observed severe lung immunopathology when animals were immunizedwith a recombinant spike protein or inactivated SARS viruspar-ticularly when the latter was formulated with alum adjuvant, a known Th2-polarising agent (Honda-Okubo et al. 2015). However when the recombinant protein or inactivated virus antigens were formulated with Advax-CpG adjuvant no lung eosinophilic pathology was observed post-challenge in immunized animalsand this was associated with a strong Th1 immune response in the immunized animals (Honda-Okubo et al. 2015). Whilst the risk of vaccine enhanced disease in COVID-19 remains unknown, a sensible precaution would be to only formulate subunit protein vaccines with strongly Th1-inducing adjuvants to avoid any such risk of an aberrant immune response upon virus exposure. This would be consistent with the high levels of protection seen with the mRNA vaccines that also induce a strong Th1-biased immune response.

5 Important Considerations and Future Outlook for CoV Vaccines

The COVID-19 pandemic has been like no other outbreak in modern history with high global mortality and unprecedented shutdowns of businesses, economies and travel globally. The development of a COVID-19 vaccine has also been like no other drug development program in history following an accelerated timescale of just months instead of years. The compressed development of SARS-CoV-2 vaccines means that in most cases pre-clinical and clinical stages have been run in parallel. Furthermore, vaccines have had to be developed whilst simultaneously trying to decipher the fundamental biology, mechanisms, and modes of transmission of the SARS-CoV-2 virus. The speed with which this has been done has been unprecedented but nevertheless the question of when is fast, too fast still needs to be addressed when it comes to assessment of vaccine efficacy and safety prior to large scale roll out of pandemic vaccines. Is 2 months of human data even in tens of thousands of subjects really adequate to get sufficient efficacy and safety data to give a new vaccine to billions of people? This question is particularly pertinent to the new vaccine technologies like mRNA vaccines and chimpanzee adenovirus vectors which are effectively being used for the first time in humans during this actual pandemic. Due to genetic similarity and pathogenesis, data from earlier SARS and MERS vaccines has provided valuable insights into the types of vaccines that might work, but also provides cautions. While recently developed CoV vaccine technologies might accelerate vaccines of the future, are they ready for large scale implementation right now? Do we know enough about the long-term efficacy and safety implications of these new vaccine approaches? Are they really ready to replace the true and tried approaches based on recombinant proteins and inactivated virus?

A key takeaway message from this chapter is that given the ability of coronaviruses to manipulate the immune system, the type of immune response induced by vaccine candidates is likely to be critical to avoid potential immunopathology. Use of the right adjuvants with recombinant proteins or inactivated virus vaccines is likely to be a key to achieving this. Plus, further research is required to better understand the mechanism behind vaccine-induced immunopathology and whether the new vaccine technologies may lead to similar responses.

Another increasing trend in vaccine technology as exhibited by the COVID-19 vaccines is antigen engineering, such as their presentation as VLPs and introduction of mutations to stabilize proteins and remove cleavage sites as discussed in this chapter. Several studies have demonstrated that such modifications can increase capacity of the vaccines to induce neutralizing antibodies, however in some cases they can have detrimental effects and paradoxically destabilize the antigen. Hence computer modelling and in silico predictions of protein structure will be critical to assist in such engineering attempts to ensure creation of immune epitopes that induce the right type of neutralizing antibodies or cytotoxic T cells.

The COVID-19 pandemic has yet again highlighted the importance of availability of good animal models for vaccine testing. A number of models for SARS, such as transgenic ACE-2 mice, hamsters, and non-human primates were eventually found to be reasonable models for SARS-CoV-2, but this took time. The choice of species comes down to research question being asked and other factors such as the availability of species-specific reagents, i.e. assays and antibodies for mouse and non-human primates due to cross-reactivity with humans are more broadly available than reagents for species such as ferrets and hamsters. Moreover, there is a limited availability of BSL-3 animal facilities, constraining the access to challenge models for vaccine testing.

A key challenge moving forward is the current pandemic is the high global daily cases of SARS-CoV-2 infection. Each human case presents the opportunity for the virus to mutate, engineer immune escape and transform into a new strain. As of June 2020, over 15,000 SARS-CoV-2 samples had been genotyped and 8309 single mutations in the genome had been identified (Wang et al. 2020). Now the number of prevalent strains is steadily increasing with fourth and fifth generation clades now readily distinguishable in different parts of the world. If this trend continues then the new vaccines may rapidly lose efficacy and multivalent vaccines may become needed, just as is the case for influenza vaccines. This would require further redesign of vaccine candidates and could pose major problems for virus vector vaccines where multiple serotypes cannot be easily co-administered or for nucleic acid vaccines where maximal doses have already been reached at the price of a high rate of reactogenicity which would prohibit addition of further mRNA to cover additional virus variants. This again leaves protein subunit vaccines as potentially the only viable candidates as the amount of antigen variants and thereby protein in the vaccine can be increased markedly without reducing tolerability as demonstrated for quadrivalent influenza vaccines that can contain up to 180 micrograms of protein and yet remain well tolerated.

References

Abdelrahman Z, Li M, Wang X. Comparative review of SARS-CoV-2, SARS-CoV, MERS-CoV, and influenza a respiratory viruses. Front Immunol. 2020;11:2309.

Adney DR, et al. Efficacy of an adjuvanted Middle East respiratory syndrome coronavirus spike protein vaccine in dromedary camels and alpacas. Viruses. 2019;11(3):212.

Agrawal AS, et al. Generation of a transgenic mouse model of Middle East respiratory syndrome coronavirus infection and disease. J Virol. 2015;89(7):3659–70.

Agrawal AS, et al. Immunization with inactivated Middle East Respiratory Syndrome coronavirus vaccine leads to lung immunopathology on challenge with live virus. Human Vaccines Immunother. 2016;12(9):2351–6.

Al-Amri SS, et al. Immunogenicity of candidate MERS-CoV DNA vaccines based on the spike protein. Sci Rep. 2017;7:44875.

Alharbi NK, et al. ChAdOx1 and MVA based vaccine candidates against MERS-CoV elicit neutralising antibodies and cellular immune responses in mice. Vaccine. 2017;35(30):3780–8.

Almeida JD, Tyrrell DA. The morphology of three previously uncharacterized human respiratory viruses that grow in organ culture. J Gen Virol. 1967;1(2):175–8.

Callow KA, et al. The time course of the immune response to experimental coronavirus infection of man. Epidemiol Infect. 1990;105(2):435–46.

Chan JFW et al. Simulation of the clinical and pathological manifestations of Coronavirus Disease 2019 (COVID-19) in golden Syrian hamster model: implications for disease pathogenesis and transmissibility. Clin Infect Dis. 2020.

Chi H, et al. DNA vaccine encoding Middle East respiratory syndrome coronavirus S1 protein induces protective immune responses in mice. Vaccine. 2017;35(16):2069–75.

Chu YK, et al. The SARS-CoV ferret model in an infection–challenge study. Virology. 2008;374 (1):151–63.

Chu DK, et al. MERS coronaviruses in dromedary camels, Egypt. Emerging Inf Dis. 2014;20 (6):1049.

Clay C, et al. Primary severe acute respiratory syndrome coronavirus infection limits replication but not lung inflammation upon homologous rechallenge. J Virol. 2012;86(8):4234–44.

Cockrell AS, et al. A mouse model for MERS coronavirus-induced acute respiratory distress syndrome. Nat Microbiol. 2016;2(2):1–11.

Corbett KS, et al. SARS-CoV-2 mRNA vaccine design enabled by prototype pathogen preparedness. Nature. 2020;586(7830):567–71.

Cowling BJ, et al. Preliminary epidemiological assessment of MERS-CoV outbreak in South Korea, May to June 2015. Eurosurveillance. 2015;20(25):21163.

Day CW, et al. A new mouse-adapted strain of SARS-CoV as a lethal model for evaluating antiviral agents in vitro and in vivo. Virology. 2009;395(2):210–22.

de Wit E, et al. The Middle East respiratory syndrome coronavirus (MERS-CoV) does not replicate in Syrian hamsters. PLoS ONE. 2013a;8(7).

De Wit E, et al. Middle East respiratory syndrome coronavirus (MERS-CoV) causes transient lower respiratory tract infection in rhesus macaques. Proc Natl Acad Sci. 2013;110(41):16598–603.

Delrue I, et al. Inactivated virus vaccines from chemistry to prophylaxis: merits, risks and challenges. Expert Rev Vaccines. 2012;11(6):695–719.

Deming D, et al. Vaccine efficacy in senescent mice challenged with recombinant SARS-CoV bearing epidemic and zoonotic spike variants. PLoS Med. 2006;3(12).

Deng Y, et al. Enhanced protection in mice induced by immunization with inactivated whole viruses compare to spike protein of middle east respiratory syndrome coronavirus. Emerg Microbes Infect. 2018;7(1):1–10.

Dicks MD, et al. A novel chimpanzee adenovirus vector with low human seroprevalence: improved systems for vector derivation and comparative immunogenicity. PLoS ONE. 2012;7 (7).

Drosten C, et al. Identification of a novel coronavirus in patients with severe acute respiratory syndrome. N Engl J Med. 2003;348(20):1967–76.
Du L, et al. Receptor-binding domain of SARS-CoV spike protein induces long-term protective immunity in an animal model. Vaccine. 2007;25(15):2832–8.
Du L, et al. Identification of a receptor-binding domain in the S protein of the novel human coronavirus Middle East respiratory syndrome coronavirus as an essential target for vaccine development. J Virol. 2013;87(17):9939–42.
Falzarano D, et al. Infection with MERS-CoV causes lethal pneumonia in the common marmoset. PLoS Pathog. 2014;10(8).
Fausther-Bovendo H, Kobinger GP. Pre-existing immunity against Ad vectors: humoral, cellular, and innate response, what's important? Human Vaccines Immunother. 2014;10(10):2875–84.
Fett C, et al. Complete protection against severe acute respiratory syndrome coronavirus-mediated lethal respiratory disease in aged mice by immunization with a mouse-adapted virus lacking E protein. J Virol. 2013;87(12):6551–9.
Gao Q, et al. Development of an inactivated vaccine candidate for SARS-CoV-2. 2020.
Graham BS, Mascola JR, Fauci AS. Novel vaccine technologies: essential components of an adequate response to emerging viral diseases. JAMA. 2018;319(14):1431–2.
Gretebeck LM, Subbarao K. Animal models for SARS and MERS coronaviruses. Curr Opinion Virol. 2015;13:123–9.
Guan Y, et al. Isolation and characterization of viruses related to the SARS coronavirus from animals in southern China. Science. 2003;302(5643):276–8.
Guo X, et al. Systemic and mucosal immunity in mice elicited by a single immunization with human adenovirus type 5 or 41 vector-based vaccines carrying the spike protein of Middle East respiratory syndrome coronavirus. Immunology. 2015;145(4):476–84.
He Y, et al. Receptor-binding domain of SARS-CoV spike protein induces highly potent neutralizing antibodies: implication for developing subunit vaccine. Biochem Biophys Res Commun. 2004;324(2):773–81.
He X, et al. Temporal dynamics in viral shedding and transmissibility of COVID-19. Nat Med. 2020;26(5):672–5.
Hoffmann M, et al. SARS-CoV-2 cell entry depends on ACE2 and TMPRSS2 and is blocked by a clinically proven protease inhibitor. Cell. 2020;52(5):731–3.
Honda-Okubo Y, et al. Severe acute respiratory syndrome-associated coronavirus vaccines formulated with delta inulin adjuvants provide enhanced protection while ameliorating lung eosinophilic immunopathology. J Virol. 2015;89(6):2995–3007.
Ji W, et al. Cross-species transmission of the newly identified coronavirus 2019-nCoV. J Med Virol. 2020;92(4):433–40.
Jiaming L, et al. The recombinant N-terminal domain of spike proteins is a potential vaccine against Middle East respiratory syndrome coronavirus (MERS-CoV) infection. Vaccine. 2017;35(1):10–8.
Jiang X, Rayner S, Luo MH. Does SARS-CoV-2 has a longer incubation period than SARS and MERS? J Med Virol. 2020;92(5):476–8.
Kam YW, et al. Antibodies against trimeric S glycoprotein protect hamsters against SARS-CoV challenge despite their capacity to mediate FcγRII-dependent entry into B cells in vitro. Vaccine. 2007;25(4):729–40.
Kapikian AZ. The corona viruses. Dev Biol Stand. 1975;28:42–64.
Keech C et al. Phase 1–2 trial of a SARS-CoV-2 recombinant spike protein nanoparticle vaccine. N Engl J Med. 2020.
Kim YI, et al. Infection and rapid transmission of SARS-CoV-2 in ferrets. Cell Host Microbe. 2020.
Kim E, et al. Immunogenicity of an adenoviral-based Middle East Respiratory Syndrome coronavirus vaccine in BALB/c mice. Vaccine. 2014;32(45):5975–82.
Kong SL, et al. Elucidating the molecular physiopathology of acute respiratory distress syndrome in severe acute respiratory syndrome patients. Virus Res. 2009;145(2):260–9.

Kopecky-Bromberg SA, et al. Severe acute respiratory syndrome coronavirus open reading frame (ORF) 3b, ORF 6, and nucleocapsid proteins function as interferon antagonists. J Virol. 2007;81(2):548–57.

Lamirande EW, et al. A live attenuated severe acute respiratory syndrome coronavirus is immunogenic and efficacious in golden Syrian hamsters. J Virol. 2008;82(15):7721–4.

Lan J, et al. Recombinant receptor binding domain protein induces partial protective immunity in rhesus macaques against Middle East respiratory syndrome coronavirus challenge. EBioMedicine. 2015;2(10):1438–46.

Lan J, et al. Significant spike-specific IgG and neutralizing antibodies in mice induced by a novel chimeric virus-like particle vaccine candidate for Middle East respiratory syndrome coronavirus. Virol Sinica. 2018;33(5):453–5.

Li F. Structure, function, and evolution of coronavirus spike proteins. Ann Rev Virol. 2016;3:237–61.

Lin JT, et al. Safety and immunogenicity from a phase I trial of inactivated severe acute respiratory syndrome coronavirus vaccine. Antiviral Ther. 2007;12(7):1107–13.

Lipsitch M, et al. Transmission dynamics and control of severe acute respiratory syndrome. Science. 2003;300(5627):1966–70.

Liu YV, et al. Chimeric severe acute respiratory syndrome coronavirus (SARS-CoV) S glycoprotein and influenza matrix 1 efficiently form virus-like particles (VLPs) that protect mice against challenge with SARS-CoV. Vaccine. 2011;29(38):6606–13.

Lu S, et al. Comparison of nonhuman primates identified the suitable model for COVID-19. Signal Transduct Targeted Ther. 2020;5(1):1–9.

Martina BEE, et al. SARS virus infection of cats and ferrets. Nature. 2003;425(6961):915–6.

Modjarrad K, et al. Safety and immunogenicity of an anti-Middle East respiratory syndrome coronavirus DNA vaccine: a phase 1, open-label, single-arm, dose-escalation trial. Lancet Infect Dis. 2019;19(9):1013–22.

Mubarak A, Alturaiki W, Hemida MG. Middle east respiratory syndrome coronavirus (MERS-CoV): infection, immunological response, and vaccine development. J Immunol Res. 2019.

Mulligan MJ, et al. Phase I/II study of COVID-19 RNA vaccine BNT162b1 in adults. Nature. 2020;586(7830):589–93.

Munster VJ, et al. Protective efficacy of a novel simian adenovirus vaccine against lethal MERS-CoV challenge in a transgenic human DPP4 mouse model. npj Vaccines. 2017;2(1):1–4.

Muthumani K, et al. A synthetic consensus anti–spike protein DNA vaccine induces protective immunity against Middle East respiratory syndrome coronavirus in nonhuman primates. Sci Transl Med. 2015. 7(301): 301ra132–301ra132.

Nishiura H, Linton NM, Akhmetzhanov AR. Serial interval of novel coronavirus (COVID-19) infections. Int J Infect Dis. 2020;93:284–6.

Osterrieder N, et al. Age-dependent progression of SARS-CoV-2 infection in syrian hamsters. Viruses. 2020; 12(7).

Pallesen J, et al. Immunogenicity and structures of a rationally designed prefusion MERS-CoV spike antigen. PNAS. 2017;114(35):E7348–57.

Pardi N, et al. mRNA vaccines—a new era in vaccinology. 2018;17(4):261.

Petrovsky N, Aguilar JC. Vaccine adjuvants: current state and future trends. Immunol Cell Biol. 2004;82(5):488–96.

Prompetchara E, Ketloy C, Palaga T. Immune responses in COVID-19 and potential vaccines: Lessons learned from SARS and MERS epidemic. Asian Pac J Allergy Immunol. 2020;38 (1):1–9.

Raj V, Stalin et al. Adenosine deaminase acts as a natural antagonist for dipeptidyl peptidase 4-mediated entry of the Middle East respiratory syndrome coronavirus. J Virol 2014;88 (3):1834–1838.

Roberts A, et al. Severe acute respiratory syndrome coronavirus infection of golden Syrian hamsters. J Virol. 2005;79(1):503–11.

Roberts A, et al. Therapy with a severe acute respiratory syndrome–associated coronavirus–neutralizing human monoclonal antibody reduces disease severity and viral burden in golden Syrian Hamsters. J Infect Dis. 2006;193(5):685–92.

Rota PA, et al. Characterization of a novel coronavirus associated with severe acute respiratory syndrome. Science. 2003;300(5624):1394–9.

See RH, et al. Severe acute respiratory syndrome vaccine efficacy in ferrets: whole killed virus and adenovirus-vectored vaccines. J Gen Virol. 2008;89(9):2136–46.

Shi S-Q, et al. The expression of membrane protein augments the specific responses induced by SARS-CoV nucleocapsid DNA immunization. Mol Immunol. 2006;43(11):1791–8.

Smith TR, et al. Immunogenicity of a DNA vaccine candidate for COVID-19. Nat Commun. 2020;11(1):1–13.

Stenler S, Blomberg P, Smith CE. Safety and efficacy of DNA vaccines: Plasmids vs. minicircles. Human Vaccines Immunother. 2014;10(5):1306–1308.

Tai W, et al. Recombinant receptor-binding domains of multiple Middle East respiratory syndrome coronaviruses (MERS-CoVs) induce cross-neutralizing antibodies against divergent human and camel MERS-CoVs and antibody escape mutants. J Virol. 2017;91(1).

Tang F, et al. Lack of peripheral memory B cell responses in recovered patients with severe acute respiratory syndrome: a six-year follow-up study. J Immunol. 2011;186(12):7264–8.

Ter Meulen J, et al. Human monoclonal antibody as prophylaxis for SARS coronavirus infection in ferrets. The Lancet. 2004;363(9427):2139–41.

Tseng CT, et al. Immunization with SARS coronavirus vaccines leads to pulmonary immunopathology on challenge with the SARS virus. PLoS ONE. 2012;7(4).

Ura T, Okuda K, Shimada MJV. Developments in viral vector-based vaccines. Vaccines. 2014;2 (3):624–41.

V'kovski P, et al. Coronavirus biology and replication: implications for SARS-CoV-2. Nat Rev Microbiol. 2020:1–16.

Volz A, et al. Protective efficacy of recombinant modified vaccinia virus Ankara delivering Middle East respiratory syndrome coronavirus spike glycoprotein. J Virol. 2015;89(16):8651–6.

Wang R, et al. Decoding SARS-CoV-2 transmission, evolution and ramification on COVID-19 diagnosis, vaccine, and medicine. J Chem Inf Model. 2020.

Wang SF, et al. Antibody-dependent SARS coronavirus infection is mediated by antibodies against spike proteins. Biochem Biophys Res Commun. 2014;451(2):208–14.

Wang L, et al. Evaluation of candidate vaccine approaches for MERS-CoV. Nat Commun. 2015;6 (1):1–11.

Weingartl H, et al. Immunization with modified vaccinia virus Ankara-based recombinant vaccine against severe acute respiratory syndrome is associated with enhanced hepatitis in ferrets. J Virol. 2004;78(22):12672–6.

Wirblich C, et al. One-Health: a safe, efficient, dual-use vaccine for humans and animals against middle east respiratory syndrome coronavirus and rabies virus. J Virol. 2017;91(2).

Woo PC, et al. SARS coronavirus spike polypeptide DNA vaccine priming with recombinant spike polypeptide from Escherichia coli as booster induces high titer of neutralizing antibody against SARS coronavirus. Vaccine. 2005;23(42):4959–68.

Wrapp D, et al. Cryo-EM structure of the 2019-nCoV spike in the prefusion conformation. Science. 2020;367(6483):1260–3.

Yao Y, et al. An animal model of MERS produced by infection of rhesus macaques with MERS coronavirus. J Infect Dis. 2014;209(2):236–42.

Yasui F, et al. Prior immunization with severe acute respiratory syndrome (SARS)-associated coronavirus (SARS-CoV) nucleocapsid protein causes severe pneumonia in mice infected with SARS-CoV. J Immunol. 2008;181(9):6337–48.

Yin Y, Wunderink RG. MERS, SARS and other coronaviruses as causes of pneumonia. Respirology. 2018;23(2):130–7.

Zhang Y et al. Safety, tolerability, and immunogenicity of an inactivated SARS-CoV-2 vaccine in healthy adults aged 18–59 years: a randomised, double-blind, placebo-controlled, phase 1/2 clinical trial. Lancet Infect Dis. 2020.

Zhang T, Wu Q, Zhang Z. Probable pangolin origin of SARS-CoV-2 associated with the COVID-19 outbreak. Curr Biol. 2020;30(7):1346–51.

Zhao J, Zhao J, Perlman S. T cell responses are required for protection from clinical disease and for virus clearance in severe acute respiratory syndrome coronavirus-infected mice. J Virol. 2010;84(18):9318–25.

Zhou P, et al. A pneumonia outbreak associated with a new coronavirus of probable bat origin. Nature. 2020;579(7798):270–3.

Zhu Z, et al. From SARS and MERS to COVID-19: a brief summary and comparison of severe acute respiratory infections caused by three highly pathogenic human coronaviruses. Respir Res. 2020;21(1):1–14.

SARS-CoV-2 Mutations: An Insight

Rajendra Phartyal and Mansi Verma

Abstract A new terror to mankind, the novel coronavirus, known as SARS-CoV-2, previously known 2019-nCoV, is a recent member in RNA virus group and has shaken almost every country around the world. Since December 2019, world has witnessed global lockdown in different phases, and with every phase the severity of infection is increasing. Already many countries around the world are reeling under the second wave of novel coronavirus infection which is much more severe than the first wave. Although the average death rate across the world is ~2.2%, this rate varies from country to country and may be attributed to geo-distribution of different strains and rich genetic variations as the result of mutations. This chapter focuses on the hotspots of mutations across the genomes sequenced so far and the evolutionary forces leading to stabilizing these mutations. Although SARS-CoV-2 is mutating slow, nevertheless few mutations can be associated with increased infectivity.

Keywords SARS-CoV-2 · Betacoronavirus · Mutations · Evolution

Abbreviations

(+) ssRNA	Positive single stranded Ribosomal nucleic acid
ACE2	Angiotensin-converting enzyme2
BatCoV	Bat Coronavirus
GISAID	Global Initiative on Sharing all Influenza Data
MERS-CoV	Middle East respiratory syndrome coronavirus
NS	Non-structural
NSP	Nonstructural protein
ORF	Open Reading Frame
RdRp	RNA dependent RNA polymerase
SARS-CoV	Severe acute respiratory syndrome coronavirus
SARS-CoV-2	Severe Acute Respiratory Syndrome coronavirus 2

R. Phartyal · M. Verma (✉)
Department of Zoology, Sri Venkateswara College, University of Delhi (South Campus), New Delhi 110021, India
e-mail: mansiverma@svc.ac.in

S. I. Ahmad (ed.), *Human Viruses: Diseases, Treatments and Vaccines*, https://doi.org/10.1007/978-3-030-71165-8_25

SNP	Single nucleotide polymorphism
UTR	Untranslated Region
WHCV	Wuhan-Human 1 coronavirus
WHO	World Health Organization

1 Introduction

Within a year of getting familiar with the name SARS-CoV-2, the deadly virus has already caused more than 1.6 million deaths and the numbers are continuously increasing as the virus is tightening its grip in newer areas (https://www.worldometers.info/coronavirus/). Many pharmaceutical companies and research institutes across the globe are in a race against time to compete for designing the most effective vaccine. Reported first in December 2019 from Wuhan city in China (Li et al. 2020), SARS-CoV-2 stands for Severe Acute Respiratory Syndrome Coronavirus 2 and belongs to family Coronaviridae and shows pneumonia like symptoms. The disease caused by this virus is named as COVID-19 and was declared pandemic on March 11, 2020 by WHO (Astuti and Ysrafil 2020). The severity of this disease can be understood by the fact that within a very short period from the time first case was reported, 72,764,982 people have been infected worldwide as on 13th December, 2020. So, what is SARS-CoV-2?

2 SARS-CoV-2: A Betacoronavirus

As mentioned earlier, SARS-CoV-2 belongs to member of family Coronaviridae and genus betacoranavirus, the same genus to which other life-threatening viruses like severe acute respiratory syndrome coronavirus (SARS-CoV) and Middle East respiratory syndrome coronavirus (MERS-CoV) belong (Tang et al. 2015). Members of betacoronaviruses are (+) ssRNA viruses infecting wide variety of hosts and possess the large RNA genomes of around 30 Kb. Within Betacoronaviruses, four lineages are recognized namely A, B, C and D. SARS-CoV-2 belongs to B lineage of coronaviruses (Zhu et al. 2020; Llanes et al. 2020). Peculiar feature of all coronaviruses is the presence of spikes in their structure. Both SARS-CoV and SARS-CoV-2 exploit human receptor angiotensin-converting enzyme2 (ACE2) for their entry inside the host cells (Shang et al. 2020), but the affinity of Spike protein of SARS-CoV-2 is found out to be ten to 20-fold higher, making it more infectious (Wrapp et al. 2020).

3 Genomics of SARS-CoV-2

Release of first genome in GenBank on January 11, 2020 (accession no. MN908947.3) from an isolate designated as WH-Human 1 coronavirus (WHCV) or '2019-nCoV' obtained from 41-year-old man (Wu et al. 2020a; Rahimi et al. 2020) has marked the beginning of sequencing of SARS-CoV-2 isolates for comparative studies. With 12 open reading frames (ORFs) which are flanked by 5' and 3'UTRs, the genome of SARS-CoV-2 encodes for 27 proteins including structural, non-structural and accessory proteins (Abduljalil and Abduljalil 2020; Wu et al. 2020b) (Fig. 1).

Structural proteins are required for interaction with host cells for invasion and include spike (S), envelope (E), membrane (M), and nucleocapsid (N) proteins (Pyrc et al. 2007; Rahimi et al. 2020). Whereas there are 16 non-structural proteins required for replication and maintenance of viral particles.

4 Mutations Reported in SARS-CoV-2 Over the Time

Since the time when this virus started spreading like wildfire and became a serious health issue and threat to life, scientists are looking for its source of origin and the reason for host switching with better efficiency of SARS-CoV-2 Spike protein binding to human receptors. SARS-CoV-2 shares 96.2% homology with the Bat-CoV RaTG13, ~79% sequence identity with SARS-CoV, and 50% sequence identity with the Middle East respiratory syndrome (MERS) (Zhou et al. 2020).

Lately, it has been traced that receptor binding motif of SARS-CoV-2 could be an outcome of recombination between different strains of coronaviruses existing in pangolin (pangolin-CoV) and bat-CoV (RaTG13) (Wong et al. 2020), making it more proficient in establishing humans as its new host (Fig. 2).

With more and more genomes sequenced for SARS-CoV-2 isolates inhabiting different geographical locations, more light is being shed on the evolution of this virus. Global Initiative on Sharing All Influenza Data (GISAID) (https://www.gisaid.org/) presently hubs 2,36,400 genomes from all around the world, making the mutational analysis feasible. A recent research by van Dorp et al. (2020) reported 12,706 mutations in 48,454 SARS-CoV-2 assemblies. Comparative studies have showed that a variety of mutations including missense, synonymous, indels and non-coding mutations have contributed to the genetic diversity of SARS-CoV-2 (Plante et al. 2020; Kumar et al. 2020), with missense and synonymous mutations being more recurrent than others (Table 1). A recent work classified SARS-CoV-2 into four main groups which were further divided into 10 subgroups, among which group D was found to be rapidly expanding (Chen et al. 2020). Even hydrophobic mutations are being favored over hydrophilic mutations (van Dorp et al. 2020). Overall, transitions were found to be favored on transversions. Many studies have supported C to T transitions to be more prominent than

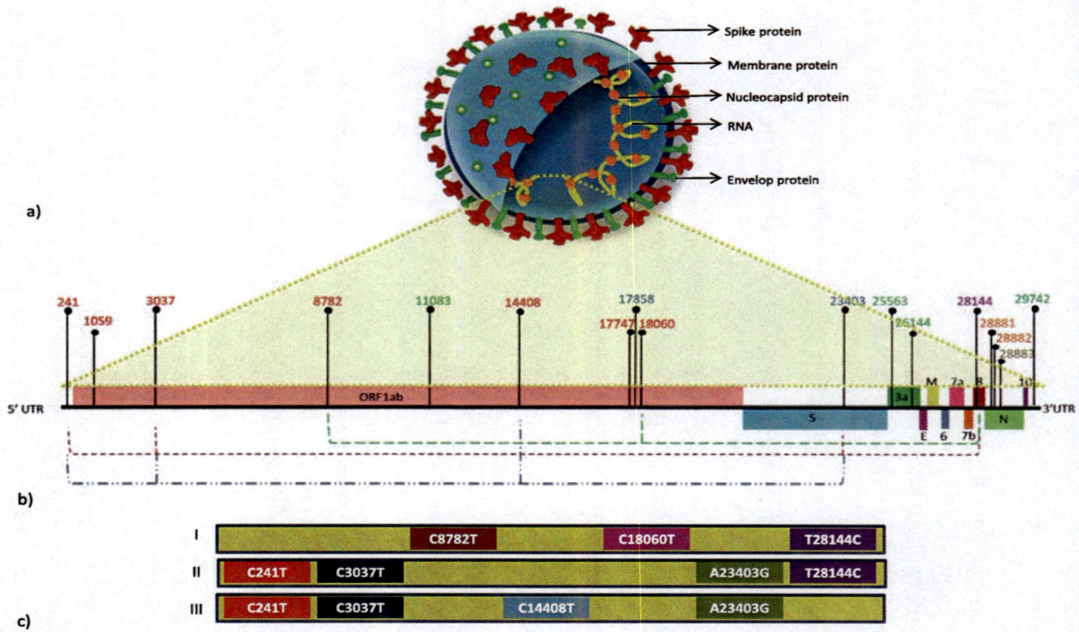

Fig. 1 Depiction of Structure and genomics of SARS-CoV-2. **a** The virion comprises of four structural proteins namely Spike, Membrane, Envelope and Nucleocapsid. Spike protein is involved in interaction with human receptors, whereas Nucleocapsid proteins are associated with RNA genome of viruses. **b** the RNA genome is ~29–30 kb with 12 Open Reading Frames (ORFs) and 27 proteins, flanked by 5'and 3'Untranslated Regions (UTRs). 17 major mutation hotspot positions are also represented; positions marked with red are C > T transitions, positions in green are G > T, blue positions mark A > G transitions, purple as T > C, orange as G > A, whereas brown color represents mutation as G > C at position 28883. Co-mutations are represented by three dotted lines. **c** Three major co-mutations found to exist together showing evolutionary conservancy

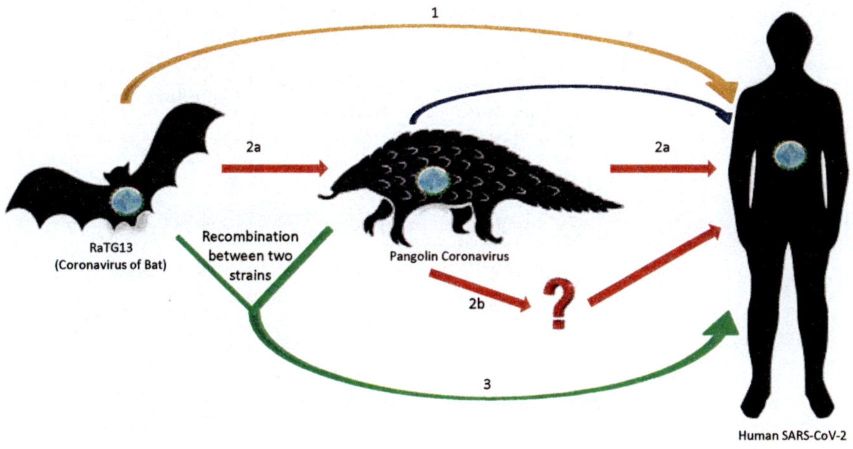

Fig. 2 Possible transmission path of SARS-CoV-2 to humans. Pathway 1: Bat source of SARS-CoV-2. Pathway 2a: Bat to pangolin to human transmission. Pathway 2b: Pangolin to unknown intermediate host to human transmission. Pathway 3: bat CoV and pangolin CoV recombination resulted in origin of SARS-CoV-2. Recently, pathway 3 is known to be more plausible

any other mutation, followed by A > G transition, then G > T transversions (Mercatelli and Giorgi 2020). Although coronaviruses have proofreading mechanism, but nsp14 cannot eliminate U resulted from cytosine deamination (Khrustalev et al. 2020; Hosseini and McLellan 2020), attributing to more pyrimidines (C and T) in codons.

Many of these mutations are continent specific, with A > G being more frequent in Africa, Europe and America; and G > T mutation dominating in Asia and Oceania. Recent study also supported that while Kazakhstan, Bangladesh, India and Congo showed higher mutation rates as compared to the world's average rate (being 7.63%); few countries like Germany, Italy, Greece, Kenya, Hong Kong and Japan showed relatively lesser mutation burden (Mercatelli and Giorgi 2020). It is also reported that there is a characteristic mutation patterns in European, North American and Asian strains (Pachetti et al. 2020).

The mutations were not found to be constant even within the genome and few regions were found to show higher mutations as compared to other genomic regions. Moreover, it was also seen that not all mutations effected the change in amino acids. The hotspots for mutations were found to be Spike Protein, ORF3a, ORF8, NSP12, NSP13, NSP14, NSP15, NSP16 and 5'& 3'UTRs (Kumar et al. 2020; Laamarti et al. 2020; Rehman et al. 2020). Several studies have reported 58–67% of SNPs resulting in changes at amino acid level, silent mutations (not leading to change in amino acid) to be ~28%, whereas mutations in intergenic regions (5' and 3' UTRs) to be 12.6% (Gupta et al. 2020; Mercatelli and Giorgi 2020). Note that the percentage varies depending on the number of genomes considered in respective studies.

Table 1 Various mutational hotspots categorized as missense, synonymous and non-coding (Data adapted from Rahimi et al. 2020)

Mutation type	Nucleotide change	Nucleotide position	Amino acid change	Gene/location
Missense mutation	A>G	17858	Y541C	ORF1ab (*nsp13*)
		23403	D614G	*S*
	C>T	1059	T85I	ORF1ab (*nsp2*)
		14408	P323L	ORF1ab (*nsp12*)
		17747	P504L	ORF1ab (*nsp13*)
	G>A	28881	R203K	*N*
	G>C	28883	G204R	*N*
	G>T	11083	L37F	ORF1ab (*nsp6*)
		25563	Q57H	*ORF3a*
		26144	G251V	*ORF3a*
	T>C	28144	L84S	*ORF8*
Synonymous mutation	C>T	3037	F105F	ORF1ab (*nsp3*)
		8782	S75S	ORF1ab (*nsp4*)
		18060	L6L	ORF1ab (*nsp14*)
	G>A	28882	R203R	*N*
Non-coding	C>T	241	NA	5′ UTR
	G>T	29742	NA	3′ UTR

Out of 46,723 genetically diverse genomes analyzed in one of the studies, total 12,706 variable positions were marked across the alignment (van Dorp et al. 2020). Till date, 17 mutations have been found to be highly occurring in all the genomes (Rahimi et al. 2020). These mutation hotspots were also found to be related to their geographical distribution, with mutations at positions 2891, 3036, 14408, 23403 and 28881 more predominant in Europe; whereas mutations at positions 17746, 17857 and 18060 are entirely present in North America (Pachetti et al. 2020). Among all the mutations, one mutation is most explored and is marked as D614G leading to change in 614th amino acid of Spike protein from Aspartate (D) to Glycine (G). This mutation has been associated with the increased transmissibility of SARS-CoV-2 by enhancing viral replication in human lung epithelial cells and is recurring at an alarming rate (Korber et al. 2020). This mutation is found to be associated with the fitness advantage rather than founder effect/genetic drift (Plante et al. 2020). The plausible reason for influencing the transmission may be that while the binding affinity of Spike protein with ACE-2 receptor of humans is decreased after this mutation, but it leads to enhanced binding of Spike protein with TMPRSS-2 receptor of humans (Gupta et al. 2020). It is already known that SARS-CoV-2 interacts with ACE-2 receptor for entry in host cell and TMPRSS-2 is a serine protease of host cells which is involved in Spike protein priming (Gupta et al. 2020; Zhang et al. 2020). Although a recent publication suggests that D614G mutation does not enhance transmissibility (van Dorp et al. 2020) leaving all the doors open for discussion.

Some co-mutations have also been traced in the genomes. For example, along with nucleotide mutation A23403G (=D614G amino acid mutation of Spike protein), other three mutations are also known to exist. These mutations are C14408T, C241T, C3037T and follow same evolutionary process (https://www.sciencedaily.com/releases/2020/10/201026114157.htm). Similarly, co-mutations in Nsp13 at locations 5768 (Nsp13_1) and 5731(Nsp13_2) in the isolates sequenced from USA were also found. In fact, these co-mutations were also reflected in the phylogenetic clading of the genome sequences (Gupta et al. 2020).

BOX 1: Mutations in SARS-CoV-2: a Matter of Concern

A recent study showed that mutations in ORF3a may elevate mortality rate for SARS-CoV-2 infection (Majumdar and Niyogi 2020). Furthermore, some variants of SARS-CoV-2 Spike protein (although occurring at low frequencies) are found to resist neutralizing antibodies, and therefore limiting the use of monoclonal antibodies for therapeutics (Weisblum et al. 2020). Yet, another study based on mathematical modeling evaluating all possible 3686 future mutations of Spike glycoprotein has traced that SARS-CoV-2 has tendency to become more infectious in future. Particularly, residues 452, 489, 500, 501, and 505 on receptor-binding motif of Spike protein may prove to be more infectious after mutation (Chen et al. 2020). To summarize, although mutations can occur throughout the genome, but some mutations can have more terrible effects in making the novel coronavirus highly infectious.

5 SARS-CoV-2: Mutating Fast or Slow?

Viruses emerge in three stages, starting from gaining ability to infect new host cells, to adapting to the new host environment for aiding transmission, and finally gaining ability to spread in host populations for spreading epidemically. Viruses have an inherent capability to reproduce very fast in host cells and evolve very fast endowing them with a potential to multiply and diversify rapidly (Saiz et al. 2013). Knowing the mutation rate of a virus helps in understanding its evolution and aids in practical applications like drug designing, vaccination strategies and risk assessment of emergent infectious disease (Sanjuán et al. 2010). The almost million times higher mutation rate of RNA viruses compared to their hosts correlates with increased virulence and adaptability (Duffy 2018). Apart from Coronaviruses, all viral RNA-dependent RNA polymerases (RdRp) lack 3'exonuclease proofreading activity and the same is true of all reverse transcriptases (RTs) leading to incorporation of mismatches of newly synthesized genetic material (Combe and Sanjuan 2014).The genome of each individual RNA virus differs from its parental genome in one nucleotide on average and this is due to the error prone replication caused by the low fidelity RNA polymerase. This leads to a highly variable RNA virus population a raw material for their rapid evolution. Available mutation rates for RNA viruses range between 10^{-6} and 10^{-4} s/n/c (nucleotide substitutions per nucleotide per round of cell infected) (Peck and Lauring 2018).

As can be seen in Table 2, some RNA viruses like Influenza and HIV are known to mutate at a very faster rate, while some groups have very slow mutation rate. Coronaviruses have largest genome sizes among RNA viruses (Pachetti et al. 2020) making them more susceptible to mutations, but their proofreading mechanism makes them less prone to error, contributing towards slower mutation rates over the time. For example, SARS-CoV and MERS-CoV have average mutation rate of $0.80–2.38 \times 10^{-3}$ nucleotide substitution per site per year (Zhao et al. 2004) and 1.12×10^{-3} nucleotide substitution per site per year (Cotten et al. 2014), respectively, which is far too slower than HIV. Although initially the rate of mutation of novel coronavirus as documented by WHO was 1.12×10^{-3}, but recently SARS-CoV-2 is estimated to have mutation rate of 9.8×10^{-4} (van Dorp et al. 2020), reflecting that the mutations are getting stabilized over time. It is also estimated that the rate of mutation of coronavirus is half of Influenza virus and one fourth of HIV mutation rates. Although this rate is slow when comparing with

Table 2 Rate of mutations of few RNA viruses

RNA virus	Mutation rate (substitutions/site/year)	References
(+) ss RNA viruses		
SARS-CoV-2	9.8×10^{-4}	van Dorp et al. (2020)
SARS-CoV	$0.80\text{-}2.38 \times 10^{-3}$	Zhao et al. (2004)
MERS-CoV	1.12×10^{-3}	Cotten et al. (2014)
Dengue Virus	7.6×10^{-4}	Costa et al. (2012)
HCV	1.92×10^{-3}	Ogata et al. (1991)
(-) ss RNA viruses		
Influenza A Virus	2.28×10^{-3}	Zhao et al. (2004)
Ebola Virus	4.7×10^{-4}	Whitfield et al. (2020)
Rabies Virus	2.32×10^{-4} to 1.38×10^{-3}	Hughes et al. (2005)
Retroviruses		
HIV-I	4.1×10^{-3}	Andrews and Rowland-Jones (2017)
Bovine Leukemia Virus	4.8×10^{-6}	Zhao et al. (2004)

a)
b)

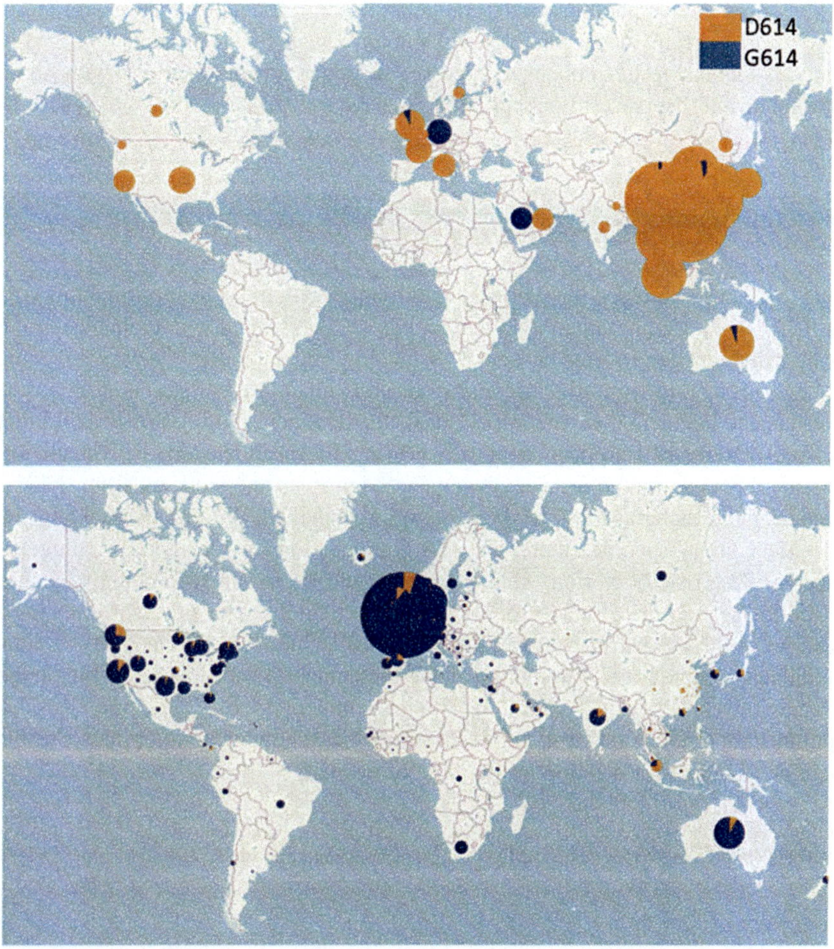

Fig. 3 Predominance of G614 mutation in spike protein over D614 D614/G614 existence as on **a** 13th February 2020 and **b** 12th December, 2020 (Data taken from GISAID Database (https://cov.lanl.gov/apps/covid-19/map/)

rapidly evolving viruses, few of the mutations are already getting fixed at alarming rates. For example, D614G mutation just took few months to predominate over the ancestral strain, with 70% of sequences showing this mutation (Zhang et al. 2020) (Fig. 3).

While most of the mutations are deleterious, higher mutation rates are beneficial with changing environments as there are better chances of favorable mutations to arise (Duffy 2018). The various mutations and recombinations changing the different regions of the virus have enabled the SARS-CoV-2 to become a highly infectious novel pathogen to humans (Rehman et al. 2020). A mutation may enhance transmissibility and hence increase the spread of the virus provided other population genetics parameters like range expansion, founder effects, and random genetic drift are in favor. One particular mutation in Spike protein, i.e., D614G which is increasing in frequency suggests a selective advantage along with higher viral loads and decrease in patient age (Volz et al. 2020).

6 Conclusion

Year 2020 has witnessed the world transformation with the spread of dreadful virus SARS-CoV-2, forcing mankind into lockdown, slowing down global economy along with the new norms of social distancing and mandatory masks in public places. Originating from the ancestral strain in China, different strains of this virus have created havoc in many countries across the continents with USA, Brazil, India, Mexico and many European countries bearing the brunt in terms of death toll. These strains are mutating at a slower pace, but even the slow mutation pace is uncontrollable as far as the spread of disease is concerned. With mutational hotspots that are country-specific, there may be complications in designing a universal vaccine. The rise of various mutant strains and their spread aided by Natural selection, genetic drift/founder effect in various regions with different climates is continuing the unobstructed march of this novel pathogen. While many vaccine candidates are under trial around the globe, the launch of vaccines against the novel coronavirus by Pfizer-BioNTech and other competitors like Moderna, Russia's SputnikV and University of Oxford/AstraZeneca showing extraordinary results there are finally rays of hope in this time of despair.

Acknowledgements The authors greatly acknowledge Prof. C. Sheela Reddy, Principal, Sri Venkateswara College, University of Delhi, for her constant support.

References

Abduljalil JM, Abduljalil BM. Epidemiology, genome, and clinical features of the pandemic SARS-CoV-2: a recent view. New Microbes New Infect. 2020;35:100672. https://doi.org/10.1016/j.nmni.2020.100672, PMID 32322400.

Andrews SM, Rowland-Jones S. Recent advances in understanding HIVevolution. F1000Res. 2017;6:597. https://doi.org/10.12688/f1000research.10876.1, PMID 28529718.

Astuti I, Ysrafil. Severe Acute Respiratory Syndrome Coronavirus 2 (SARS-CoV-2): An overview of viral structure and host response. Diabetes Metab Syndr 14. 2020;2:407–12. https://doi.org/10.1016/j.dsx.2020.04.020.

Chen J, Wang R, Wang M, Wei GW. Mutations strengthened SARS-CoV-2 infectivity. J Mol Biol. 2020;432(19):5212–26. https://doi.org/10.1016/j.jmb.2020.07.009, PMID 32710986.

Chen Z-W, Li Z, Li H, Ren H, Hu P. Global genetic diversity patterns and transmissions of SARS-CoV-2. medRxiv. 2020 2020.2005.2005.20091413.

Combe M, Sanjuán R. Variation in RNA virus mutation rates across host cells. PLOS Pathog. 2014;10(1):e1003855. 10.1371/journal.ppat.1003855, PMID 24465205.

Costa RL, Voloch CM, Schrago CG. Comparative evolutionary epidemiology of dengue virus serotypes. Infect Genet Evol. 2012;12(2):309–14. https://doi.org/10.1016/j.meegid.2011.12.011.

Cotten M, Watson SJ, Zumla AI, Makhdoom HQ, Palser AL, Ong SH, Al Rabeeah AA, Alhakeem RF, Assiri A, Al-Tawfiq JA, Albarrak A, Barry M, Shibl A, Alrabiah FA, Hajjar S, Balkhy HH, Flemban H, Rambaut A, Kellam P, Memish ZA. Spread, circulation, and evolution of the Middle East respiratory syndrome coronavirus. mBio. 2014;5(1):e01062–13. https://doi.org/10.1128/mbio.01062-13, PMID 24549846.

Duffy S. Why are RNA virus mutation rates so damn high? PLOS Biol. 2018;16(8):e3000003. https://doi.org/10.1371/journal.pbio.3000003, PMID 30102691.

Gupta V, Haider S, Verma M, Ponnusamy K, Malik MZ, Singhvi N, Verma H, Kumar R, Sood U, Hira P, et al. Multi-omics and integrated network approach to unveil evolutionary patterns, mutational hotspots, functional cross talk and regulatory interactions in SARS-CoV-2. bioRxiv. 2020. https://doi.org/10.1101/2020.06.20.162560.

Hosseini Rad Sm A, McLellan AD. Implications of SARS-CoV-2 mutations for genomic RNA structure and host microRNA targeting. Int J Mol Sci. 2020;21(13):4807. https://doi.org/10.3390/ijms21134807, PMID 32645951.

Hughes GJ, Orciari LA, Rupprecht CE. Evolutionary timescale of rabies virus adaptation to North American bats inferred from the substitution rate of the nucleoprotein gene. J Gen Virol. 2005;86(5):1467–74. https://doi.org/10.1099/vir.0.80710-0, PMID 15831959.

Khrustalev VV, Giri R, Khrustaleva TA, Kapuganti SK, Stojarov AN, Poboinev VV. Translation-associated mutational U-pressure in the first ORF of SARS-CoV-2 and other coronaviruses. bioRxiv. 2020. https://doi.org/10.1101/2020.05.05.078238.

Korber B, Fischer WM, Gnanakaran S, Yoon H, Theiler J, Abfalterer W, Hengartner N, Giorgi EE, Bhattacharya T, Foley B, Hastie KM, Parker MD, Partridge DG, Evans CM, Freeman TM, de Silva TI, Sheffield COVID-19 Genomics Group, McDanal C, Perez LG, Tang H, Moon-Walker A, Whelan SP, LaBranche CC, Saphire EO, Montefiori DC. Tracking changes in SARS-CoV-2 spike: evidence that D614G increases infectivity of the COVID-19 virus. Cell. 2020;182(4):812–827.e19. https://doi.org/10.1016/j.cell.2020.06.043, PMID 32697968.

Kumar R, Verma H, Singhvi N, Sood U, Gupta V, Singh M, Kumari R, Hira P, Nagar S, Talwar C, Nayyar N, Anand S, Rawat CD, Verma M, Negi RK, Singh Y, Lal R. Comparative genomic analysis of rapidly evolving SARS-CoV-2 reveals mosaic pattern of phylogeographical distribution. mSystems. 2020;5(4):e00505–20. https://doi.org/10.1128/msystems.00505-20, PMID 32723797.

Laamarti M, Alouane T, Kartti S, Chemao-Elfihri MW, Hakmi M, Essabbar A, Laamart M, Hlali H, Allam L, Hafidi NEL, Jaoudi REL, Allali I, Marchoudi N, Fekkak J, Benrahma H, Nejjari C, Amzazi S, Belyamani L, Ibrahimi A. Large scale genomic analysis of 3067 SARS-CoV-2 genomes reveals a clonal geo-distribution and a rich genetic variations of hotspots mutations. bioRxiv. 2020:2020.2005.2003.074567.

Li Q, Guan X, Wu P, Wang X, Zhou L, Tong Y, Ren R, Leung KSM, Lau EHY, Wong JY, Xing X, Xiang N, Wu Y, Li C, Chen Q, Li D, Liu T, Zhao J, Liu M, Tu W, Chen C, Jin L, Yang R, Wang Q, Zhou S, Wang R, Liu H, Luo Y, Liu Y, Shao G, Li H, Tao Z, Yang Y, Deng Z, Liu B, Ma Z, Zhang Y, Shi G, Lam TTY, Wu JT, Gao GF, Cowling BJ, Yang B, Leung GM, Feng Z. Early transmission dynamics in Wuhan, China, of novel coronavirus-infected pneumonia. N Engl J Med. 2020;382(13):1199–207. https://doi.org/10.1056/NEJMoa2001316, PMID 31995857.

Llanes A, Restrepo CM, Caballero Z, Rajeev S, Kennedy MA, Lleonart R. Betacoronavirus Genomes: how Genomic Information has been Used to Deal with Past Outbreaks and the COVID-19 Pandemic. Int J Mol Sci. 2020;21(12):4546. https://doi.org/10.3390/ijms21124546, PMID 32604724.

Majumdar P, Niyogi S. ORF3a mutation associated with higher mortality rate in SARS-CoV-2 infection. Epidemiol Infect. 2020;148:e262. Published 2020 Oct 26. https://doi.org/10.1017/s0950268820002599, PMID 33100263.

Mercatelli D, Giorgi FM. Geographic and genomic distribution of SARS-CoV-2 mutations. Front Microbiol. 2020;11:1800. doi:10.3389/fmicb.2020.01800, PMID 32793182.

Ogata N, Alter HJ, Miller RH, Purcell RH. Nucleotide sequence and mutation rate of the H strain of hepatitis C virus. Proc Natl Acad Sci U S A. 1991;88(8):3392–6. https://doi.org/10.1073/pnas.88.8.3392, PMID 1849654.

Pachetti M, Marini B, Benedetti F, Giudici F, Mauro E, Storici P, Masciovecchio C, Angeletti S, Ciccozzi M, Gallo RC, Zella D, Ippodrino R. Emerging SARS-CoV-2 mutation hot spots include a novel RNA-dependent-RNA polymerase variant. J Transl Med. 2020;18(1):179. https://doi.org/10.1186/s12967-020-02344-6, PMID 32321524.

Peck KM, Lauring AS. Complexities of viral mutation rates. J Virol. 2018;92(14):e01031–17. https://doi.org/10.1128/JVI.01031-17, PMID 29720522.

Plante JA, Liu Y, Liu J, Xia H, Johnson BA, Lokugamage KG, Zhang X, Muruato AE, Zou J, Fontes-Garfias CR, Mirchandani D, Scharton D, Bilello JP, Ku Z, An Z, Kalveram B, Freiberg AN, Menachery VD, Xie X, Plante KS, Weaver SC, Shi PY. Spike mutation D614G alters SARS-CoV-2 fitness. Nature. 2020. https://doi.org/10.1038/s41586-020-2895-3, PMID 33106671.

Pyrc K, Berkhout B, van der Hoek L. The novel human coronaviruses NL63 and HKU1. J Virol. 2007;81(7):3051–7. https://doi.org/10.1128/JVI.01466-06, PMID 17079323.

Rahimi A, Mirzazadeh A, Tavakolpour S. Genetics and genomics of SARS-CoV-2: a review of the literature with the special focus on genetic diversity and SARS-CoV-2 genome detection. Genomics. 2020;S0888–7543(20):30876-4. https://doi.org/10.1016/j.ygeno.2020.09.059, PMID 33007398.

Rehman SU, Shafique L, Ihsan A, Liu Q. Evolutionary trajectory for the emergence of novel coronavirus SARS-CoV-2. Pathogens. 2020;9(3):240. https://doi.org/10.3390/pathogens9030240, PMID 32210130.

Saiz JC, Sobrino F, Sevilla N, Martín V, Perales C, Domingo E. Molecular and evolutionary mechanisms of viral emergence. In: Singh SK, editor Viral infections and global change; 2013.

Sanjuán R, Nebot MR, Chirico N, Mansky LM, Belshaw R. Viral mutation rates. J Virol. 2010;84(19):9733–48. https://doi.org/10.1128/JVI.00694-10, PMID 20660197.

Shang J, Ye G, Shi K, Wan Y, Luo C, Aihara H, Geng Q, Auerbach A, Li F. Structural basis of receptor recognition by SARS-CoV-2. Nature. 2020;581(7807):221–4. https://doi.org/10.1038/s41586-020-2179-y, PMID 32225175.

Tang Q, Song Y, Shi M, Cheng Y, Zhang W, Xia XQ. Inferring the hosts of coronavirus using dual statistical models based on nucleotide composition. Sci Rep. 2015;5:17155. https://doi.org/10.1038/srep17155, PMID 26607834.

van Dorp L, Richard D, Tan CCS, Shaw LP, Acman M, Balloux F. No evidence for increased transmissibility from recurrent mutations in SARS-CoV-2. Nat Commun. 2020;11(1):5986. https://doi.org/10.1038/s41467-020-19818-2, PMID 33239633.

Volz E, Hill V, McCrone JT, Price A, Jorgensen D, O'Toole Á, Southgate J, Johnson R, Jackson B, Nascimento FF, Rey SM, Nicholls SM, Colquhoun RM, da Silva Filipe A, Shepherd J, Pascall DJ, Shah R, Jesudason N, Li K, Jarrett R, Pacchiarini N, Bull M, Geidelberg L, Siveroni I; COG-UK Consortium, Goodfellow I, Loman NJ, Pybus OG, Robertson DL, Thomson EC, Rambaut A, Connor TR, Connor TR. Evaluating the effects of SARS-CoV-2 spike mutation D614G on transmissibility and pathogenicity. Cell. 2020 Nov 19; S0092-8674(20):31537–3. https://doi.org/10.1016/j.cell.2020.11.020, PMID 33275900, PMCID PMC7674007.

Weisblum Y, Schmidt F, Zhang F, DaSilva J, Poston D, Lorenzi JC, Muecksch F, Rutkowska M, Hoffmann HH, Michailidis E, Gaebler C, Agudelo M, Cho A, Wang Z, Gazumyan A, Cipolla M, Luchsinger L, Hillyer CD, Caskey M, Robbiani DF, Rice CM, Nussenzweig MC, Hatziioannou T, Bieniasz PD. Escape from neutralizing antibodies by SARS-CoV-2 spike protein variants. eLife. 2020;9:e61312. Published 2020 Oct 28. https://doi.org/10.7554/elife.61312, PMID 33112236.

Whitfield ZJ, Prasad AN, Ronk AJ, Kuzmin IV, Ilinykh PA, Andino R, Bukreyev A. Species-specific evolution of Ebola virus during replication in human and bat cells. Cell Rep. 2020;32(7): https://doi.org/10.1016/j.celrep.2020.108028.

Wong MC, Javornik Cregeen SJ, Ajami NJ, Petrosino JF. Evidence of recombination in coronaviruses implicating pangolin origins of nCoV-2019. bioRxiv. 2020. https://doi.org/10.1101/2020.02.07.939207, PMID 32511310.

Wrapp D, Wang N, Corbett KS, Goldsmith JA, Hsieh CL, Abiona O, Graham BS, McLellan JS. Cryo-EM structure of the 2019-nCoV spike in the prefusion conformation. Science. 2020;367(6483):1260–3. https://doi.org/10.1126/science.abb2507, PMID 32075877.

Wu A, Peng Y, Huang B, Ding X, Wang X, Niu P, Meng J, Zhu Z, Zhang Z, Wang J, Sheng J, Quan L, Xia Z, Tan W, Cheng G, Jiang T. Genome composition and divergence of the novel coronavirus (2019-nCoV) originating in China. Cell Host Microbe. 2020a;27(3):325–8. https://doi.org/10.1016/j.chom.2020.02.001, PMID 32035028.

Wu F, Zhao S, Yu B, Chen YM, Wang W, Song ZG, Hu Y, Tao ZW, Tian JH, Pei YY, Yuan ML, Zhang YL, Dai FH, Liu Y, Wang QM, Zheng JJ, Xu L, Holmes EC, Zhang YZ. A new coronavirus associated with human respiratory disease in China. Nature. 2020b;579(7798):265–9. https://doi.org/10.1038/s41586-020-2008-3, PMID 32015508.

Zhang L, Jackson CB, Mou H, Ojha A, Rangarajan ES, Izard T, Farzan M, Choe H. The D614G mutation in the SARS-CoV-2 spike protein reduces S1 shedding and increases infectivity. bioRxiv. 2020. https://doi.org/10.1101/2020.06.12.148726.

Zhao Z, Li H, Wu X, Zhong Y, Zhang K, Zhang YP, Boerwinkle E, Fu YX. Moderate mutation rate in the SARS coronavirus genome and its implications. BMC Evol Biol. 2004 Jun 28;4:21. https://doi.org/10.1186/1471-2148-4-21, PMID 15222897, PMCID PMC446188.

Zhou P, Yang XL, Wang XG, Hu B, Zhang L, Zhang W, Si HR, Zhu Y, Li B, Huang CL, Chen HD, Chen J, Luo Y, Guo H, Jiang RD, Liu MQ, Chen Y, Shen XR, Wang X, Zheng XS, Zhao K, Chen QJ, Deng F, Liu LL, Yan B, Zhan FX, Wang YY, Xiao GF, Shi ZL. A pneumonia outbreak associated with a new coronavirus of probable bat origin. Nature. 2020;579(7798):270–3. https://doi.org/10.1038/s41586-020-2012-7, PMID 32015507.

Zhu N, Zhang D, Wang W, Li X, Yang B, Song J, Zhao X, Huang B, Shi W, Lu R, Niu P, Zhan F, Ma X, Wang D, Xu W, Wu G, Gao GF, Tan W, China Novel Coronavirus Investigating and Research Team. A Novel Coronavirus from Patients with Pneumonia in China, 2019. N Engl J Med. 2020;382(8):727–33. https://doi.org/10.1056/nejmoa2001017, PMID 31978945.

A Brief Introduction to Coronavirus Disease 2019 (COVID-19) and the Roles of Zoonotic Spillover

D. Katterine Bonilla-Aldana and Alfonso J. Rodriguez-Morales

Abstract Coronaviruses are eminently present in animals since they were discovered over half a century ago. In 2002 and 2012 two major zoonotic emerging epidemics occurred in Asia, the Severe Acute Respiratory Syndrome (SARS) in China, and the Middle East Respiratory Syndrome (MERS), in Saudi Arabia, both extended to other countries in the regions and beyond. In 2019 the SARS-CoV-2 and the Coronavirus disease 2019 (COVID-19) begun to be apparent due to the first epidemic in Wuhan, China. All of them had a zoonotic origin, in the first two, clearer, than the current pandemic virus. In the current chapter, we discuss and explore a different point of views regarding the COVID-19 and the roles of zoonotic spillover, based on the available evidence, still on developing (March 2021).

Keywords Coronavirus · SARS-CoV-2 · COVID-19 · Epidemiology · Public health · Spillover · One health · Pandemic · Animal health · Human health · Zoonotic

1 Introduction

Coronaviruses (CoVs) are single-stranded, positive-sense RNA viruses that are enveloped, highly diverse, and cause respiratory, digestive, neurological, and liver disorders (Vargas-Gandica et al. 2020). CoVs are grouped into four genuses:

D. K. Bonilla-Aldana
Semillero de Investigacion en Zoonosis (SIZOO), Biodiversity and Ecosystem Conservation Research Group (BIOECOS), Fundación Universitaria Autónoma de las Américas, Sede Pereira, Pereira, Risaralda, Colombia

D. K. Bonilla-Aldana · A. J. Rodriguez-Morales
Latin American Network of COVID-19 Research, LANCOVID, Pereira, Risaralda, Colombia

A. J. Rodriguez-Morales (✉)
Grupo de Investigación Biomedicina, Faculty of Medicine, Fundación Universitaria Autónoma de las Américas, Pereira, Risaralda, Colombia
e-mail: alfonso.rodriguez@uam.edu.co

S. I. Ahmad (ed.), *Human Viruses: Diseases, Treatments and Vaccines*, https://doi.org/10.1007/978-3-030-71165-8_26

Alphacoronavirus, Betacoronavirus, Gammacoronavirus, and Deltacoronavirus (Chen et al. 2020). These genuses include more than 40 different virus species (Dhama et al. 2020b).

Among the Betacoronaviruses, with the most significant interest, the Severe Acute Respiratory Syndrome Coronavirus (SARS-CoV-1) originated in Guangdong, China, in 2002, causing 8,096 reported cases with 774 deaths in 37 different countries. Ten years later, the Middle East respiratory syndrome (MERS-CoV) (Bonilla-Aldana et al. 2020a; Rabaan et al. 2020), which appeared in Saudi Arabia in 2012 caused 2,521 infections with 866 deaths (Rabaan et al. 2020; The Lancet 2020; Totura and Baric 2012).

In December 2019, an outbreak of pneumonia of unknown etiology was detected in the city of Wuhan China (Biscayart et al. 2019). A week later, a new coronavirus, initially designated as 2019-nCoV and later (Holshue et al. 2020; Li et al. 2020; Munster et al. 2020), SARS-CoV-2, causing the COVID-19 disease, a disease that has so far generated more than 2.6 million deaths in all the countries in the world (Zhan et al. 2020; Zhu et al. 2020). It is worthy of mentioning that although the virus was recognized in December 2019, there is growing evidence in Asia and Europe, that indicates that SARS-CoV-2 was previously circulating months earlier (Gerbaud et al. 2020; Miglietta and Levi 2020).

The SARS-CoV-2 virus belongs to the Orthocoronavirinae subfamily, genus Coronavirus and the Sarbecovirus subgenus (beta-coronavirus, beta-2b) and within them to clade or lineage 2. The SARS-CoV-2 genome is formed by a single-stranded positive strand RNA of about 30,000 nucleotides and 6 ORFs (open reading frames), identical to the rest of coronaviruses, and several additional genes. Most of these genes are 80% homologous to the ancient SARS-CoV virus (Lu et al. 2020).

SARS-CoV-2 has been shown to use the same receptor as SARS-CoV in its process of entering the cell; this is because it has a receptor-binding domain (RBD) that binds with greater affinity to angiotensin II converting enzyme (ACE2) generating the potential for transmission from animal to human (Zhou et al. 2020; Rodriguez-Morales et al. 2020c).

2 Zoonotic Spill-Over

Coronaviruses infect a wide range of animals such as camels, canine, birds, bats, among others (Ahmad et al. 2020). Current scientific evidence proposes that SARS-CoV-2 had a zoonotic origin, and everything points to its natural reservoir being bats, particularly of the species *Rhinolophus affinis* (horseshoe bat) (Bonilla-Aldana et al. 2020b; Tiwari et al. 2020).

Bats have coevolved with viruses for millions of years. Coevolution has led bats to constitutively express higher levels of type I and type III interferons in their body cells. Interferons induce a wide variety of antiviral genes, called interferon-stimulated genes (ISG), which limit viral replication (Ye et al. 2020; Long et al. 2020).

Coronaviruses have been observed to have the ability to jump from one species to another, as is the case with SARS-CoV-2 where it has been proposed that there is possibly an intermediate (or amplifier) host between bats and humans. Among the alleged intermediaries are palm civets, snakes or other wild animals traded in Chinese wet markets (Zhang and Holmes 2020; Dhama et al. 2020a), but this still requires more studies to confirm it. Recently a study mentions that pangolins, an endangered species of mammal that is traded in Wuhan, China, could be the intermediary animals. The study focused on a species of pangolin (*Manis javanica*), finding that the genome of the coronavirus identified in this species has between 85.5 and 92.4% sequence similarity with SARS-CoV-2 (Lam et al. 2020).

From the point of view of natural infection in animals, different infected species have been found, such as canines, small and big cats (Rodriguez-Morales et al. 2020b), minks, and ferrets mainly, as has been reported in many countries in the world. The susceptibility in the different species is currently being investigated. Additionally, it has been demonstrated experimentally in animals (Abdel-Moneim and Abdelwhab 2020). At the laboratory, multiple animal species would be infected, as different studies have showed their susceptibility to the experimental infection with SARS-CoV-2, by different routes (Trimpert et al. 2020).

3 Clinical Manifestations

Limited information is available on the clinical manifestations of SARS-CoV-2 in animals, but pathological findings in necropsy studies suggest some inflammation-mediated complications would be similar to human beings (Vasquez-Bonilla et al. 2020). Existing evidence suggests that clinical manifestations can range from asymptomatic infections to symptomatic illness with signs that may include sneezing, cough, runny nose, shortness of breath, eye discharge, diarrhoea or vomiting (Villamizar-Pena et al. 2020), lethargy and fever, among others (Munir et al. 2020), similar to the observed in human beings (Rodriguez-Morales et al. 2020a).

SARS-CoV-2 infects and replicates efficiently in pneumocytes, macrophages, and dendritic cells in the deepest parts of the lung parenchyma where the ACE-2 cell receptor resides, causing mild respiratory symptoms up to severe pneumonia in humans (Dhama et al. 2020b).

4 Conclusion

Due to deforestation, economic development, and climate change, an increase in human interactions with these zoonotic hosts is expected; thus, knowledge of the natural and intermediate reservoirs of this virus can help us control and prevent related infectious diseases in humans, avoiding possible epidemics in the future (Wong et al. 2020; Bonilla-Aldana et al. 2019). Besides, a collaborative, multi-sectoral, and transdisciplinary approach must be adopted globally to achieve the best possible health outcomes for people, animals, plants, and their shared environment (Bonilla-Aldana et al. 2020c).

In the specific case of the SARS-CoV-2/COVID-19, there is still a gap of knowledge about its zoonotic origin. There is growing evidence showing that bats are natural hosts for this emerging coronavirus, nevertheless few evidence is available regarding the intermediate hosts that would be involved. Beyond that, there is a lot of concern regarding the possible human-animal infection, especially for those that are in close contact with the man. The spillover has occurred and will continue to do so, then, more research is needed, and derived from, a One Health and holistic approach in order to prevent and mitigate the consequences of such transmission (Salajegheh Tazerji et al. 2020).

References

Abdel-Moneim AS, Abdelwhab EM. Evidence for SARS-CoV-2 infection of animal Hosts. Pathogens. 2020;9(7).

Ahmad T, Khan M, Haroon, Musa TH, Nasir S, Hui J, et al. COVID-19: zoonotic aspects. Travel Med Infect Dis. 2020;36:101607.

Biscayart C, Angeleri P, Lloveras S, Chaves T, Schlagenhauf P, Rodriguez-Morales AJ. The next big threat to global health? 2019 novel coronavirus (2019-nCoV): what advice can we give to travellers?—Interim recommendations January 2020, from the Latin-American society for travel medicine (SLAMVI). Travel Med Infect Dis. 2020;33:101567.

Bonilla-Aldana DK, Suarez JA, Franco-Paredes C, Vilcarromero S, Mattar S, Gomez-Marin JE, et al. Brazil burning! What is the potential impact of the Amazon wildfires on vector-borne and zoonotic emerging diseases?—A statement from an international experts meeting. Travel Med Infect Dis. 2019;31:101474.

Bonilla-Aldana DK, Cardona-Trujillo MC, Garcia-Barco A, Holguin-Rivera Y, Cortes-Bonilla I, Bedoya-Arias HA, et al. MERS-CoV and SARS-CoV infections in animals: a systematic review and meta-analysis of prevalence studies. Infez Med. 2020a;28(suppl 1):71–83.

Bonilla-Aldana DK, Jimenez-Diaz SD, Arango-Duque JS, Aguirre-Florez M, Balbin-Ramon GJ, Paniz-Mondolfi A, et al. Bats in ecosystems and their wide spectrum of viral infectious potential threats: SARS-CoV-2 and other emerging viruses. Int J Infect Dis. 2020b;102:87–96.

Bonilla-Aldana DK, Holguin-Rivera Y, Perez-Vargas S, Trejos-Mendoza AE, Balbin-Ramon GJ, Dhama K, et al. Importance of the one health approach to study the SARS-CoV-2 in Latin America. One Health. 2020c;10:100147.

Chen Y, Liu Q, Guo D. Emerging coronaviruses: Genome structure, replication, and pathogenesis. J Med Virol. 2020;92(10):2249.

Dhama K, Patel SK, Sharun K, Pathak M, Tiwari R, Yatoo MI, et al. SARS-CoV-2 jumping the species barrier: zoonotic lessons from SARS, MERS and recent advances to combat this pandemic virus. Travel Med Infect Dis. 2020a;37:101830.

Dhama K, Khan S, Tiwari R, Sircar S, Bhat S, Malik YS, et al. Coronavirus disease 2019-COVID-19. Clin Microbiol Rev. 2020b;33(4).

Gerbaud L, Guiguet-Auclair C, Breysse F, Odoul J, Ouchchane L, Peterschmitt J, et al. Hospital and population-based evidence for COVID-19 early circulation in the east of France. Int J Environ Res Public Health. 2020;17(19).

Holshue ML, DeBolt C, Lindquist S, Lofy KH, Wiesman J, Bruce H, et al. First case of 2019 novel coronavirus in the United States. N Engl J Med. 2020;382(10):929–36.

Lam TT, Jia N, Zhang YW, Shum MH, Jiang JF, Zhu HC, et al. Identifying SARS-CoV-2-related coronaviruses in Malayan pangolins. Nature. 2020;583(7815):282–5.

Li Q, Guan X, Wu P, Wang X, Zhou L, Tong Y, et al. Early transmission dynamics in Wuhan, China, of novel coronavirus-infected pneumonia. N Engl J Med. 2020;382(13):1199–207.

Long QX, Tang XJ, Shi QL, Li Q, Deng HJ, Yuan J, et al. Clinical and immunological assessment of asymptomatic SARS-CoV-2 infections. Nat Med. 2020;26(8):1200–4.

Lu R, Zhao X, Li J, Niu P, Yang B, Wu H, et al. Genomic characterisation and epidemiology of 2019 novel coronavirus: implications for virus origins and receptor binding. Lancet. 2020;395 (10224):565–74.

Miglietta A, Levi M. Documented early circulation of coronavirus disease 2019 (COVID-19) in Florida, USA, since February 2020. J Public Health (Oxf). 2020;42(3):517–8.

Munir K, Ashraf S, Munir I, Khalid H, Muneer MA, Mukhtar N, et al. Zoonotic and reverse zoonotic events of SARS-CoV-2 and their impact on global health. Emerg Microbes Infect. 2020;9(1):2222–35.

Munster VJ, Koopmans M, van Doremalen N, van Riel D, de Wit E. A novel coronavirus emerging in China—Key questions for impact assessment. N Engl J Med. 2020;382(8):692–4.

Rabaan AA, Al-Ahmed SH, Haque S, Sah R, Tiwari R, Malik YS, et al. SARS-CoV-2, SARS-CoV, and MERS-COV: a comparative overview. Infez Med. 2020;28(2):174–84.

Rodriguez-Morales AJ, Cardona-Ospina JA, Gutierrez-Ocampo E, Villamizar-Pena R, Holguin-Rivera Y, Escalera-Antezana JP, et al. Clinical, laboratory and imaging features of COVID-19: a systematic review and meta-analysis. Travel Med Infect Dis. 2020a;34:101623.

Rodriguez-Morales AJ, Dhama K, Sharun K, Tiwari R, Bonilla-Aldana DK. Susceptibility of felids to coronaviruses. Vet Rec. 2020b.

Rodriguez-Morales AJ, MacGregor K, Kanagarajah S, Patel D, Schlagenhauf P. Going global—Travel and the 2019 novel coronavirus. Travel Med Infect Dis. 2020c;33:101578.

Salajegheh Tazerji S, Magalhaes Duarte P, Rahimi P, Shahabinejad F, Dhakal S, Singh Malik Y, et al. Transmission of severe acute respiratory syndrome coronavirus 2 (SARS-CoV-2) to animals: an updated review. J Transl Med. 2020;18(1):358.

The Lancet. Emerging understandings of 2019-nCoV. The Lancet. 2020;395(10221):311.

Tiwari R, Dhama K, Sharun K, Iqbal Yatoo M, Malik YS, Singh R, et al. COVID-19: animals, veterinary and zoonotic links. Vet Q. 2020;40(1):169–82.

Totura AL, Baric RS. SARS coronavirus pathogenesis: host innate immune responses and viral antagonism of interferon. Curr Opin Virol. 2012;2(3):264–75.

Trimpert J, Vladimirova D, Dietert K, Abdelgawad A, Kunec D, Dokel S, et al. The Roborovski Dwarf Hamster is a highly susceptible model for a rapid and fatal course of SARS-CoV-2 infection. Cell Rep. 2020;33(10):108488.

Vargas-Gandica J, Winter D, Schnippe R, Rodriguez-Morales AG, Mondragon J, Escalera-Antezana JP, et al. Ageusia and anosmia, a common sign of COVID-19? A case series from four countries. J Neurovirol. 2020.

Vasquez-Bonilla WO, Orozco R, Argueta V, Sierra M, Zambrano LI, Munoz-Lara F, et al. A review of the main histopathological findings in coronavirus disease 2019. Hum Pathol. 2020;105:74–83.

Villamizar-Pena R, Gutierrez-Ocampo E, Rodriguez-Morales AJ. Pooled prevalence of Diarrhea among COVID-19 patients. Clin Gastroenterol Hepatol. 2020.

Wong G, Bi YH, Wang QH, Chen XW, Zhang ZG, Yao YG. Zoonotic origins of human coronavirus 2019 (HCoV-19/SARS-CoV-2): why is this work important? Zool Res. 2020;41 (3):213–9.

Ye ZW, Yuan S, Yuen KS, Fung SY, Chan CP, Jin DY. Zoonotic origins of human coronaviruses. Int J Biol Sci. 2020;16(10):1686–97.

Zhan J, Li H, Yu H, Liu X, Zeng X, Peng D, et al. 2019 novel coronavirus (COVID-19) pneumonia: CT manifestations and pattern of evolution in 110 patients in Jiangxi, China. Eur Radiol. 2020.

Zhang YZ, Holmes EC. A genomic perspective on the origin and emergence of SARS-CoV-2. Cell. 2020;181(2):223–7.

Zhou P, Yang XL, Wang XG, Hu B, Zhang L, Zhang W, et al. Addendum: a pneumonia outbreak associated with a new coronavirus of probable bat origin. Nature. 2020;588(7836):E6.

Zhu N, Zhang D, Wang W, Li X, Yang B, Song J, et al. A novel coronavirus from patients with pneumonia in China, 2019. N Engl J Med. 2020;382(8):727–33.

Treatment of COVID-19 by Combination Therapy with 5-fluorouracil, Ribonucleosides and Ribavirin—A Modified Strategy

Shamim I. Ahmad

Abstract The pandemic of corona virus, SARS-CoV-2, starting in 2019 from Wuhan, China, and now spread almost all over the countries of the world, and unknown as to how long it is going to stay with us in future, remains most worrying. It appears that almost every epidemics and pandemics prevailed in the world; perhaps this is the most virulent and most difficult human virus disease (COVID-19) to have a limited control over it. Besides a large number of human infections and fatalities, its other effects have been immense including economic turmoil, loss of employments, reduced industrial productivity, and agricultural destruction. Other human suffering includes psychological disturbances including severe depression, hunger, and increased domestic violence. Attempts have been going on to control the pandemic, but in this most advanced era in which scientists have achieved so much success in developing new medications, analysis of lives up to molecular levels, and gene technologies, have been failing to find either to eliminate or successfully medically treat patients suffering from this disease. A large number of antiviral drugs have been tried to cure the patients suffering from this disease (COVID-19) but none has been found that can be used with full satisfaction. Vaccines programmes have been on-going but it is too early to draw any conclusion especially their short and long-terms side effects. Here the author is proposing a modified strategy to control this virus, but before its human trial carried out, a stringent number of in vitro and in vivo laboratory in vitro testing, animal testing and dose adjustment must be carried out before applying for clinical trials.

Keywords 5-fluorouracil · Adenosine · Ribavirin · Remdesivir · Lethal mutagenesis · SARS-CoV-2 · Chloroquine phosphate and hydroxychloroquine sulphate · RNA-dependent RNA polymerase · Middle East Respiratory Syndrome · Chikungunya Virus · Crimean-Congo Hemorrhagic fever · Dengue virus · Ebola virus · Epstein-Barr virus · Enterovirus · Flavivirus · Hantavirus · Human cytomegalovirus · Human papillomavirus · Hepatitis B virus, Hepatitis E · Herpes

S. I. Ahmad (✉)
Former Senior Lecturer, School of Science and Technology, Nottingham Trent University, Clifton Lane, Nottingham NG11 8NS, England
e-mail: shamim.ahmad@ntu.ac.uk

S. I. Ahmad (ed.), *Human Viruses: Diseases, Treatments and Vaccines*, https://doi.org/10.1007/978-3-030-71165-8_27

viruses · Human adenovirus · Human Rhinovirus · J C polyomavirus · Lassa virus · Measles virus · Merkel cell polyomavirus · Monkey pox virus · Measles virus · Mumps virus · Newcastle disease virus · Nipah virus · Polio virus, Rabies virus · Rubella virus · Varicella-Zoster virus · Vesicular stomatitis virus · Yellow fever virus, West Nile fever · Zika virus

1 Introduction

Microbial infections mostly occur by the groups of microorganisms include viruses, bacteria and protozoa. Several hundred microorganisms are known to infect all kinds of lives from lower to higher groups of animals, algae, fungi, human even plants and trees. Viruses additionally are known to infect bacteria. For this reason, viruses so far discovered, their types remain almost uncountable and their classification difficult.

COVID-19, the human disease, started in 2019 and prevailing over the past one year, and as a pandemic generating a large number of calamity in which the world is currently experiencing not only a massive number of infections and deaths, but associated problems such as economic turmoil, loss of employments, reduced industrial productivity, and agricultural destruction; other human suffering includes psychological disturbance, hunger, and increased domestic violence. More worryingly is that it is unknown how long this pandemic will last and how much more catastrophic impacts it will have. Furthermore, the prediction is that even when the current wave of infection is over, the virus may re-appear either in mutated variant forms or by finding another suitable environment and host.

In this publication, the author proposes to employ a combination of three drugs which are already in use: a nucleic acid base analogue, 5-fluorouracil, a nucleoside, adenosine to provide the source of Ribose-1-Phosphate plus ribavirin. The chapter starts from the brief introduction of viruses.

1.1 Genetic Makeup of Viruses

In contrast to eukaryotes which have DNA as their genetic materials, viruses belonging to prokaryotes, have either DNA or RNA as their genetic materials. DNA viruses are sub-classified into single stranded DNA and the double stranded DNA viruses. RNA viruses are usually having single-stranded RNA (ssRNA) but double-stranded RNA (dsRNA) viruses are also known. The ssRNA viruses are further sub-classified into: (+)ssRNA viruses, (−)ssRNA, and ssRNA-RT. Human diseases which have drawn most attentions caused by RNA viruses include the common cold, influenza, SARS, MERS, COVID-19, hepatitis C, hepatitis E, West Nile fever, Ebola virus, rabies, polio and measles. As far the viral replication cycles and infectivity concern, significant differences have been noted between (−)

ssRNA and (+)ssRNA viruses. Here the focus is on SARS-CoV-2 which is a (+) ssRNA viruses.

1.2 Virus Replication Cycle

The replication process of SARS-CoV-2, in essence includes viral attachment using its spikes to its host cell, then through making a passage it enters inside. There it releases its nucleic acid and starts using host translation system to synthesise viral replicative enzymes, which generate new RNA genome and mRNA. During this process occurs proof reading of the RNA so that wrong bases are not retained and end terminal capping not affected. This is followed by the synthesis of viral structural components and assembly of new viral progenies. For this, two co-factors Nsp8 and Nsp10 play an important role to hold the entire process and enhance the efficiency of RNA replication (Romano et al. 2020). It appears that knowledge about the replication cycle of this virus has not yet been fully worked out and requires fast progression.

2 Viruses of Past Epidemics and Pandemics and Continuing Diseases

Major global epidemics and pandemics occurred due to viruses include small pox virus, Yellow fever virus, a variety of flu viruses including H1N1 Swine flu pandemic of 2009; some of these are still continuing although sporadically. However, during nineteenth and twentieth centuries the world has seen a large number of human viral diseases including West African Ebola pandemic started in 2014 which is currently on hold; the Zika virus epidemic of 2015 and continuing mostly in South East Asian countries, Autoimmune Deficiency Syndrome (AIDS) pandemic of 1981 and continuing globally; Severe Acute Respiratory Syndrome (SARS 2002) virus and Middle East Respiratory Syndrome (MERS, of 2012) virus etc.

Certain other viruses commonly prevailing and causing human diseases (presented in this book) are: Chikungunya virus, Crimean-Congo Hemorrhagic fever virus, Dengue virus, Ebola virus, Epstein-Barr virus, Hanta virus, Hepatitis virus, Enteroviruses, Human Papilloma virus, HIV-HVB viruses, Measles virus, Merkel cell polyomavirus, Flaviviruses, Rabies viruses, Monkey Pox virus, Poliovirus and Zika virus.

Other viruses which have drawn considerable attention yet more studies required include: Newcastle disease virus, Herpes viruses, Rabies virus, Rubella virus, Varicella-Zoster virus, Vesicular stomatitis virus, Yellow fever virus, Human adenovirus, Human cytomegalovirus, JC polyomavirus, Lassa virus, Human Rhinovirus, Mumps virus, Nipah virus and West Nile virus.

3 Present Pandemic of SARS-CoV-2

The SARS-CoV-2 infection and human illness leading to death was first identified in Wuhan, China in December 2019, was declared as a disaster in the third week of January 2020. The World Health Organization (WHO) declared this as a Public Health Emergency of International Concern (PHEIC) on 31st January 2020, and finally a pandemic on 11th March 2020. From Wuhan, infection spread first to other cities in China, and subsequently to neighbouring countries including South Korea and Japan, and then, according to worldometers.info, to 218 countries and Territories around the world. This has resulted until December 18, 2020 the confirmed infection number of 75,407,507 and the death number 1,671,135, which amounts to the death rate of around 2.2% amongst infected persons.

4 Consequence of the Pandemic

As mentioned above Covid-19 calamity is causing massive effects of all kinds on human lives including infections and deaths. In the beginning the virus was considered to be normal but now a variety of mutant strains have emerged, possibly with different phenotypes including infectivity and virulence. Associated problems include economic turmoil, loss of employments, reduced industrial productivity, agricultural destruction, travel and transport problems. On personal and psychological levels, the sufferings include panic, psychological disturbance due to death in the family, hunger due to loss of income, and increased domestic violence. More worryingly, it is not known how long this pandemic will last and how much more catastrophic impacts it will have on human. Furthermore, the prediction is that even when the current wave of infection is over, the virus may re-appear either in mutated forms or by finding another suitable environment and host. *For zoonotic virus readers are referred chapter 26.*

Emergence of this new strain of SARS-CoV-2 (reported on BBC TV) has been categorised to be highly transmissible and presumably can lead to higher infectivity. Although no other phenotypic alterations are apparent, nevertheless the question about the mutation in the gene, responsible to synthesise spike protein, remains questionable. Furthermore, the variants of this virus in different countries may be different. Most common reason for the development of variant or strains is the mutations (or alteration in the nucleic acid bases) in the genome of the virus. Certain mutations occur spontaneously (may be due to environmental stress) and others due to genetic recombination between two strains. Development of genetic changes due to mutation(s) is a common phenomenon and there seems no way this process can be altered or controlled by any means; development of vaccine and therapeutic findings remain saviour of this disease. *For details readers are referred to chapter 25 of this book.*

5 Finding Treatment for COVID-19

Because the SARS-CoV-2 pandemic had started in China, their researchers were the first to start their work on finding the treatment of COVID-19. For this, they carried out more than a hundred clinical studies on known antiviral drugs, anti-malarial drugs, glucocorticoids and certain Western drugs of this new Coronavirus infection. In this trial were also included 50% of the Chinese medicines and the results were inconclusive. The Chinese herbal medicines were those commonly used to treat respiratory diseases (Zhang et al. 2020). The results show that out of the natural compounds that were screened, only 13 were found to have potential for anti-2019-nCoV (SARS-CoV-2) activity (Zhang et al. 2020).

Other antiviral drugs have been put to clinical trials, to test their safety and efficacies, are those which are either suspected to have certain probability to work on COVID-19 or were at experimental phase; they were tested by administrating one or more of these drugs to patients suffering from Covid-19. These are Chloroquine phosphate and hydroxychloroquine sulphate (also known as anti-malarial drugs), arbidol, Remdesivir (GS-5734), Favipiravir and ribavirin. So far, some promising results have been seen (Al-Tawfiq et al. 2020; Dong et al. 2020; Gao et al. 2020) yet none of these antiviral drugs for managing coronavirus infection in humans has left the ground.

In another study has evaluated 48 antiviral drugs and selected 24 of them which showed potential activities against SARS-CoV-2 (Jeon et al. 2020). Studies also have been carried out employing convalescent plasma to treat COVID-19 (Roback and Guarner 2020), but this treatment appears to be un-popular possibility because it requires that the plasma ought to be taken from those patients have suffered from COVID-19 and recovered. Also the amount of plasma required for transfusing one patient; the blood is required from at least eight patients. Furthermore, the donor required to be free from all kinds of medical conditions which may adversely affect the recipients (personal communication).

Certain antiviral drugs which showed limited success includes remdesivir which was successfully used to treat the Ebola virus, showed some promising result with COVID-19 in that it was able to inhibit replication of SARS-CoV and MERS-CoV in tissue culture and in non-human animal models infection, nevertheless this could not be much promising (Martinez 2020). Lopinavir/Ritonavir and interferon β (LPV/RTV-INFb) form, another combination drug also was clinically tried on a humanized mice model have shown that remdesivir is a better drug to treat MERS-CoV (Martinez 2020; Gordon et al. 2020). In another study in South Korea, Lopinavir/Ritonavir was administered to patients suffering from Covid-19. Following treatment, viral loads were found to be significantly decreased, and none or little Coronavirus titres were observed (Lim et al. 2020). Other drugs included were hydroxychloroquine (antimalarial drug) and dexamethasone (anti-inflammatory drug) (Ortolani and Pastorello 2020). Global attempts are ongoing to procure drug(s) for treatment, and existing anti-viral drugs have proven to be non-effective or marginally effective to treat COVID-19.

Also a number of laboratories have already developed and have released for vaccination and others are continuing development against this virus (Thanh Le et al. 2020; Diamond and Pierson 2020); but uncertainty exists over what kind and when the most suitable as well free from any side effects vaccine will be available. Other unknown is as to how long the effect of each vaccine will stay in the body. Most specifically it is not know when these will successfully vaccinate the global population of over seven billions. *For detail about vaccines, readers are referred to chapter 24 of this book.*

6 Author's Proposal for the Treatment

6.1 Theory Behind the Treatment

In this section the author proposes to employ a combination of three drugs: a biological product, the adenosine, a base analogue, 5-fluorouracil, a commonly used anticancer drug and ribavirin, a chemically modified drug used as antiviral agent. The nucleoside is to provide the source of Ribose-1-Phosphate for the conversion of 5-FU to its nucleoside, 5-Fluorouridine. 5-Fluorouridine stepwise can convert to 5-Fluorouridinetriphosphate by a number of kinases and finally polymerised into the genome of SARS-CoV-2 (see Table 1). Our research on *Escherichia coli* has shown that exposure of cells to 2.5 μg ml^{-1} of 5-FU can allow selection of mutant resistant to this concentration of the agent. The resistant mutant then can be re-sensitised if adenosine, 200 μg ml^{-1} is added. This suggests that 5-FU on its own is not assimilated in the nucleic acid as effectively as in the presence of a source of Ribose-1-Phosphate such as adenosine (Pritchard and Ahmad 1971). Ribavirin is added to induce lethal mutagenesis in virus (see below).

Table 1 Metabolic pathway of 5-Fluorouracil

5-FU metabolism	Catalysed by
A-Rib or G-Rib + Pi → Rib-1-P + A or G	Purine nucleoside phosphorylase
*5-FU + Rib-1-P → 5-FUR + Pi	Uridine phosphorylase
5-FUR + Pi → 5-FUMP	Uridine kinase
5-FUMP + Pi → 5FUDP	UMP kinase
5-FUDP + Pi → 5-FUTP	UDP kinase
5-FUTP + Pi → RNA genome	RNA dependent RNA polymerase or RNA polymerase

A: adenine, **G:** guanine, **A-Rib:** Adenosine, **G-Rib:** guanosine, **5-FU:** 5-Fluorouracil, **5-FUR:** 5-fluoro uridine**, 5-FUMP:** 5-fluoro uridine monophospate; **5-FUDP:** 5-fluorouridine diphosphate, **5-FUUTP** 5-fluoro uridine triphosphate, **Rib-1-P:** Ribose-1-Phosphate, **Pi:** inorganic phosphate

6.2 5-Fluorouracil

5-Fluuorouracil is a pyrimidine analogue, one of the most researched and used to treat human cancers and virus infections. When used in therapy the compound is metabolised via the route through which pyrimidine is metabolised to be polymerised in DNA. In DNA, 5-FU is inserted in place of natural pyrimidine and this causes damage to DNA and the cellular lethality. Colorectal cancer and liver metastasis, digestive cancer, hepatocellular carcinoma, gastric, oesophageal and pancreatic cancers, breast cancer, colon cancer, non-melanoma skin cancer, head and neck cancer, squamous cell carcinoma, bladder cancer, small cell lung cancer are some of the examples to be treated by 5FU (two many to be cited). Out of 59,669 publications on 5-fluorouracil (PubMed 16th December 2020) here only a selected research references are presented used as antiviral agents: Zika and Usutu Virus (Bassi et al. 2018), Dengue virus (Liu and Askov 2018), Hepatitis C virus (Mihalik and Feigelstock 2013), Foot and Mouth disease virus (Pariente and Airaksinen 2003; Pariente et al. 2005, Agudo et al. 2009), Arena virus (De la Torre 2005), genital human papillomavirus (Krebs 1986; Davila et al. 1996; Holmes et al. 1999), Genital HPV (Niwa et al. 2003) Human Immunodeficiency Virus Type 1 (Panozzo et al. 1996), Rinderpest Virus (Ghosh et al. 1996), Influenza Virus (Lavrov 1968), Polyoma Virus (Albert et al. 1967), RNA containing bacteriophage R17 (Graham et al. 1965) and Rous Sarcoma Virus (Golde and Vigier 1963).

6.3 Ribavirin and Lethal Mutagenesis

An essential step in SARS-CoV-2 reproduction is the polymerization of its nucleic acid by the host polymerase, Nsp12 or RNA dependent RNA polymerase (RdRp). From the studies carried until late 2020, it appears that the most important target to inactivate RNA viruses is the RdRp. Ribavirin, an anticancer agent, is a complex synthetic guanine analogue, a FDA approved nucleoside, best viewed as a ribosyl purine analogue (full name: 1-β-d-ribofuranosyle-1-H-1,2,4-triazole-3-carboxamide) (Moreno et al. 2011, Casaos et al. 2019). Although the mechanism of action of ribavirin is not fully understood, the suggestion is that when introduced in biological system, during RNA replication this analogue pairs with cytosine or uracil. Also impairs the activity of RNA dependent RNA polymerase which can stops viral replication and viral mRNA capping, leading to lethal mutagenesis and inactivation (Crotty et al. 2002; Ortega-prieto Ana et al. 2013).

Lethal mutagenesis is explained as the induction of a heavy mutation rate, against which RNA viruses have low mutational tolerance, especially when used in combination with an antiviral drug. The lethal mutagenesis leads to a reduction in the number of infective progeny. Complete eradication of virus populations can be achieved when sufficient mutational burden is applied leading to lethal mutagenesis (Pauly and Lauring 2015). Other suggested activities of ribavirin are: (i) to act as a mutagen and reduce the infectivity of new virions (ii) reduce viremia by impairing viral production, (iii) modulate cell damage by reducing inflammation, (iv) enhance antiviral immunity (Carrillo-Bustamante et al. 2017)). Ribavirin applied to treat human viruses includes, 2019-nCoV (Khalili et al. 2020; Zhang et al. 2019; Du and Chen 2020; Li et al. 2020), SARS-CoV-2 (Elfiki 2020), Lassa virus (Carrillo-Bustamante 2017)), Arenaviruses (Moreno et al. 2011), Hepatitis C virus, Respiratory syncytial virus (Crotty et al. 2002), Parainfluenza and AIDS viruses (Fernandez et al. 1986).

In case of SARS-CoV-2 a bifunctional nsp14 contains 3–5 prime exoribonuclease (Exon N) and guanine –N7-methyle transferase (N7-Metase) whose roles are proofreading and RNA capping (Ogando et al. 2020). If 5-FU and ribavirin combination is used without the nucleoside it is possible that it may not induce lethal mutagenesis, but in the presence of a source of Ribose-1-phosphate to potentiate the activity of 5-FU, it may enhance the process.

7 Drug Delivery

Pathogenic RNA viruses, such as SARS-CoV-2, have specifically been proven difficult to be treated by conventional antiviral drugs because either of their unique features or that they mutate and become resistant to the drugs; however, lethal mutagenesis can be induced if ribavirin plus a known antiviral drug used (Agudo et al. 2009). PubMed search (December 22, 2020) to find if 5-FU used or even proposed to treat COVID-19, resulted in only one publication and this is by the author proposing this modified treatment (Ahmad 2020).

Although the final decision will be taken by researchers and clinicians to implement the proposed treatment, here the author is suggesting some preliminary ideas to carry out the program of work.

The project should be carried out in 3 stages. In stage one there would be an in vitro testing to determine as to which combination and the concentrations of the three agents and what is the most appropriate timing to be used to inactivate the virus. Also to be determined which in vitro experiment is causing the smallest damage to the human cell cultures or tissues used. For this, an appropriate human cell line (preferably lung epithelial cells) may be grown in monolayer, and exposed to attenuated virus plus the analogue and ribavirin. In stage two the most appropriate combination may be tested on laboratory animals. In stage three after the approval from the Drug Regulatory Authorities to be sent for clinical trials.

8 Conclusion

Although a large number of tests have been carried out to treat COVID-19, none so far has come out clear showing satisfactory treatment of this virus. As mentioned earlier the most important target to inactivate this RNA virus is RdRp. Also a detailed analysis is required of this enzyme to lead to development of new drugs or bring about some modifications in the already approved existing anti-viral drugs to target this enzyme hopefully, to overcome the difficulties to treat the emerging zoonotic RNA viruses.

The proposal presented in this paper is with the hope that if successful it can save millions of lives, especially as it is predicted that this pandemic may continue until safe, effective and reliable vaccines are produced as well treatment found. Another worry is the extent of the immune response induced by vaccines; if it is short lived then another cycle of this disease may return. Hence along with the development of vaccines it is essential that one or more promising treatment is available to eliminate human fatalities or at least minimise as much as possible. Due to their rapid mutagenesis and recombination activities and movement from one host to another, it is possible that coronaviruses will emerge again sometime in the near future and we should be prepared for that from now.

References

Agudo R, Arias A, Domingo E. 5-fluorouracil in lethal mutagenesis of Foot-And-Mouth Disease virus. Future Med Chem. 2009 Jun;1(3):529–39. https://doi.org/10.4155/fmc.09.26.

Ahmad SI. 5-fluorouracil in combination with deoxynucleosides and deoxyribose as possible therapeutic option for the Coronavirus, COVID-19 infection. Medical Hypotheses. 142 (2020) 109754. https://doi.org/10.1016/j.mehy2020.109754.

Albert TB T, Raymond SK, Tennant W. Effect of 5-fluorouracil on the multiplication of a virulent (pseudorabies) and an oncogenic virus (polyoma). Virology July 1967;32(3):445–456. https://doi.org/10.1016/0042-6822(67)90296-6 Get right and content.

Al-Tawfiq JA, Al-Homoud AA. Memish ZA. Remdesivir as a possible therapeutic option for the COVID-19. Travel Med Infect Dis. Mar-Apr 2020;34:101615. https://doi.org/10.1016/j.tmaid.2020.101615. [Epub 2020 Mar 5].

Bassi MR, Sempere RN, Meyn P. et al. Extinction of Zika virus and Usutu virus by lethal mutagenesis reveals different patterns of sensitivity to three mutagenic drugs. Antimicrob Agents Chemother. 2018 Aug 27;62(9):e00380–18. https://doi.org/10.1128/AAC.00380-18. Print 2018 Sep.

Carrillo-Bustamante P, Nguyen THT, Oestereich L, et al. Determining ribavirin's mechanism of action against lassa virus infection. 2017 Sep 15;7(1):11693. https://doi.org/10.1038/s41598-017-10198-0.

Casaos J, Gorelick NL, Huq S, et al. The use of ribavirin as an anticancer therapeutics: Will it go viral? Mol Cancer Ther. 2019 Jul;18(7):1185–94. https://doi.org/10.1158/1535-7163.MCT-18-0666.

Crotty S, Cameron C, Andino R. Ribavirin's antiviral mechanism of action: lethal mutagenesis. J Mol Med (Berl) 2002 Feb;80(2):86–95. https://doi.org/10.1007/s00109-001-0308-0. [Epub 2001 Dec 4].

Crotty S, Cameron C, Andino R. Ribavirin's antiviral mechanism of action: lethal mutagenesis. J Mol Med (Berl). 2002 Feb;80(2):86–95. https://doi.org/10.1007/s00109-001-0308-0. [Epub 2001 Dec 4].

Davila GW, Shroyer KR. Topical 5-fluorouracil in the treatment of cervical Human Papillomavirus infection. Gynecol Obstet Invest. 1996;41(4):275–7. https://doi.org/10.1159/000292283.

De la Torre J. Arena virus extinction through lethal mutagenesis. Virus Res. 2005 Feb;107(2):207–14. https://doi.org/10.1016/j.viruses.2004.11.010.

Diamond MS, Pierson TC. The challenges of vaccine development against a new virus during a pandemic. Cell Host and Microbe. May 13, 2020. Doi: https://doi.org/10.1016/j.chom.2020.04.021.

Dong L, Hu S, Gao J. Discovering drugs to treat coronavirus disease 2019 (COVID-19). DRUG Discov Ther. 2020;14(1):58–60. https://doi.org/10.5582/ddt.2020.01012.

Du Y, Chen X. Favipiravir: pharmacokinetics and concerns about clinical trials for 2019-nCoV infection. Clin Pharmacol Ther. 2020 Apr 4. https://doi.org/10.1002/cpt.1844. Online ahead of print.

Elfiki AA. Ribavirin, remdesivir, sofosbuvir, Galidesivir, and tenofovir against SARS-CoV-2 RNA dependent RNA polymerase (RdRp): a molecular docking study. Life Sci. 2020 Jul 15;253:117592. https://doi.org/10.1016/j.lfs.2020.117592. [Epub 2020 Mar 25].

Fernandez H, Banks G, Smith R. Ribavirin: a clinical overview. Eur J Epidemiol. 1986 Mar;2 (1):1–14. https://doi.org/10.1007/BF00152711.

Gao J, Tian Z, Yang X. Breakthrough:chloroqione phosphate has shown apparent efficacy in treatment of COVID-19 associate pseumonia in clinical studies. Biosci Trends 2020 Feb 19. https://doi.org/10.5582/bst.2020.01047. [Epub ahead of print].

Ghosh A, Nayak R, Shaila MS. Inhibition of replication of Rinderpest virus by 5-fluorouracil. Antiviral Res. 1996;31(1–2, June):35–44.

Golde A, Vigier P. Kinetic study of the inhibition of growth of Rous Sarcoma virus and cells by 5-fluorouracil. Viorology. 1963 Jul;20:420–32. https://doi.org/10.1016/0042-6822(63)90091-6.

Gordon CJ, Tchesnokov EP, Feng JY. The antiviral compound remdesivir potentially inhibits RNA-dependent RNA polymerase from Middle East respiratory syndrome coronavirus. J Biol Chem. 2020 Feb 24. https://doi.org/10.107/jbc.AC120.013056. [Epub ahead of print].

Graham AF, Kirk C. Effect of 5-fluorouracil on the growth of bacteriophage R17. J Bacteriol. 1965 Oct;90(4):928–35 PMID: 584.

Holmes MM, Weaver 2nd, Vermillion ST. A randomised, double-blind, placebo-controlled trial of 5-fluorouracil for the treatment of cervicovaginal Human Papillomavirus. Infect Dis Obstet Gynecol 1999;7(4):186–189. https://doi.org/10.1002/(SICI)1098-0997 (1999)7:4 186:: AID-IDOG4 3.0.C);2-Z.

Jeon S, Ko M, Lee J, et al. Identification of antiviral drug candidates against SARS-CoV-2 from FDA-approved drugs. Antimicrobes Agents Chemother. 2020 Jul;64(7):e00819-820. https://doi.org/10.1128/AAC.00819-20.

Khalili JS, Zhu H, Mak NSA et al. Novel coronavirus treatment with ribavirin: ground work for an evaluation. J Med Virol 2020 Jul;92(7):740746. https://doi.org/10.1002/jmv.25798. [Epub 2020 April 10].

Krebs HB. Prophylactic topical 5-fluorouracil following treatment of human papillomavirus-associated lesions of the vulva and vagina. Obstet Gynecol. 1986 Dec;68(6):837–41 PMID:3785797.

Lavrov SV. Inhibition of reproduction of Influenza virus in tissue culture with 5-fluorouracil. Vopr Virusol. 1968 Jan–Feb;13(1):13–18. PMID: 5693778.

Lim J, Jeon S, Shin HY et al. Case of the index patient who caused tertiary transmission of COVID-19 infection in Korea: the application of Lopinavir/Ritonavir for the treatment of COVID-19 infected pneumonia monitored by quantitative RT-PCR. J Korean Med Sci. 2020 Feb 17;35(6):e79. https://doi.org/10.3346/jkms.2020.35.e79.

Liu WJ, Askov JG. Fitness peaks of Dengue virus population. PLoS One. 2018 Jan 2;13(1): e0189554. https://doi.org/10.1371/journal.pone.0189554.eCollection2018.

Li H, Yang YM, Xu JY. Potential antiviral therapeutics for 2019 novel Coronavirus Zhonghua Jie He He Hu Xi Za Zhi 2020 Feb 5;43(0):E002. https://doi.org/10.3760/cma.j.issn.1001-0939.2020.0002. Online ahead of print.

Martinez MA. Compounds with therapeutic potential against novel respiratory 2019 coronavirus. Antimicrobe Agents Chemother 2020 Mar 9. https://doi.org/10.1128/AAC.00399-20. [Epub ahead of print].

Mihalik KB, Feigelstock DA. Sensitivity of a ribavirin resistant mutant of Hepatitis C virus to other antiviral drugs. PLoS One 2013 Sep 5;8(9):e74027. https://doi.org/10.1371/journal pone.0074027. eColletion 2013.

Moreno H, Gallego I, Seville N. et al. Ribavirin can be mutagenic for arenaviruses. J Virol.2011 Jul;85(14):7246–55. https://doi.org/10.1128/JVI.00614-11. [Epub 2011 May 11].

Niwa K, Tagami K, Lian Z, et al. Topical vidarabin or 5-fluorouraciltreatment against persistent HPV in genital (Pre)cancerous lesions. Oncol Rep Sep–Oct. 2003;10(5):1437–41 PMID: 12883720.

Ogando NS, Zevenhoven-Dobbe JC, van der Meer Y. The enzyme activity of the nsp14 exoribonuclease is critical for replication of MERS-CoV and SARS-CoV-2. J Virol. 2020 Nov 9;94(23):e01246-e1320. https://doi.org/10.1128/JVI.01246-20.Print2020Nov9.

Ortega-prieto Ana M, Sheldon J, Grande-Perez A et al. Extinction of Hepatitis C virus by ribavirin in hepatoma cells involved in lethal mutagenesis. PLoS One 2013 Aug 16;8(8):e71039. https://doi.org/10.1371/journal.pone.0071039. eCollection 2013.

Ortolani C, Pastorello EA. Hydroxychloroquine and dexamethasone in COVID-19: who won and who lost? Clin Mol Allergy. 2020 Sep 9;18:17. https://doi.org/10.1186/s12948-020-00132-7. eCollection2020.

Panozzo J, Akan E, Griffiths TD, et al. The effects of 5-fluorouracil and doxorubicin on expression of Human Immunodeficiency virus type 1 long terminal repeat. Cancer Lett. 1996 Aug 2;105 (2):217–23. https://doi.org/10.1016/0304-3835(96)04285-1.

Pariente N, Airaksinen A, Domingo E. Mutagenesis versus inhibition in the efficiency of extinction of Foot-and-Mouth Disease virus. J Virol 2003 Jun;77(12):7131–7138. https://doi.org/10.1128/jvi.77.12.7131-7138.2003

Pariente N, Sierra S, Airaksinen A. Action of mutagenic agents and antiviral inhibitors on Foot-and Mouth Disease virus. Virus Res. 2005 Feb;107(2):183–93. https://doi.org/10.1016/j.viruses.2004.11.008.

Pauly MD, Lauring AS. Effective lethal mutagenesis of influenza virus by three nucleoside analogs. J Virol, 2015 Apr;89(7):3584–3597. https://doi.org/10.1128/JVI.03483-14. [Epub 2015 Jan 14].

Pritchard RH, Ahmad SI. Fluorouracil and the isolation of mutants lacking uridine phosphorylase in *Escherichia coli*. Molec Gen Genet. 1971;111:84–8.

Roback JD, Guarner J. Convalescent plasma to treat COVID-19: possibilities and challenges. JAMA. 2020 April 28;323(16):1561–2. https://doi.org/10.1001/jama.2020.4940.

Romano M, Ruggiero A, Squeglia et al. A structural view of SARS-CoV-2 RNA replication machinery: RNA synthesis, proof reading and final capping. Cells 2020 May 20;9(5):1267. https://doi.org/10.3390/cells9051267.

Thanh Le T, Andreadakis Z, Kumar A. et al. The COVID-19 vaccine development landscape. Nature Reviews Drug Discovery 9 April 2020;19(5):305–306. https://doi.org/10.1038/d41573-020-00073-5.

Zhang Q, Wang Y, Qi C et al. Clinical trial analysis of 2019-nCoV therapy registered in China. J Med Virol. 2020 Feb 28. https://doi.org/10.1002/jmv.25733. [Epub ahead of print].

Zhang C, Huang S, Zheng F et al. Controversial treatments: an update understanding of the Coronavirus disease 2019. J Med Virol. 2020 Mar 26;https://doi.org/10.1002/jmv.25788. Online ahead of print.

Zhang D, Wu K, Zhang X. In silico screening of Chinese herbal medicines with the potential to directly inhibit 2019 novel coronavirus. J Integr Med 2020 Mar;18(2):152–158. https://doi.org/10.1016/j.joim.2020.02.005. [Epub 2020 Feb 20].

Antiviral Agents

HIV-1 and HBV RNase H as Metal-Chelating Inhibitors: Discovery and Medicinal Chemistry Strategies

Fenju Wei, Edeildo Ferreira da Silva-Júnior, Xinyong Liu, and Peng Zhan

Abstract Human immunodeficiency viruses (HIV) and hepatitis B viruses (HBV) are highly pathogenic reverse-transcribing viruses. Despite significant progress achieved so far in antiviral therapy, the currently available treatments are difficulty to eradicate latent viral reservoirs due to severe side effects and the rapid emergence of viral drug resistance. In this context, ribonuclease H (RNase H) is regarded as an attractive target for antiretroviral therapy and it would represent a mechanistically novel class of drugs that can be particularly valuable in counteracting the burden of drug resistance. In this chapter, we have reviewed medicinal chemistry strategies in the discovery of RNase H inhibitors, including anti-HIV/AIDS and anti-HBV infection.

Keywords Ribonuclease H · HIV-1 · HBV · Medicinal chemistry strategies · Metal-chelating inhibitors · RNAase · Neucleos(t)ide · Drug design

1 Introduction

Human immunodeficiency viruses type 1 and type 2 (HIV-1 and HIV-2, respectively) and hepatitis B viruses (HBV) are the etiological agents of the most threatening human diseases (Menéndez-Arias et al. 2017). Up to now, there are about 38 million people living with HIV (www.unaids.org), while more than 250 million people are chronically infected with the HBV. In addition, approximately 10% of the HIV-infected population worldwide is coinfected with HBV, although this figure can be higher in several regions of Southeast Asia (Singh et al. 2017).

F. Wei · P. Zhan (✉)
Department of Medicinal Chemistry, Key Laboratory of Chemical Biology, Ministry of Education, School of Pharmaceutical Sciences, Shandong University, Ji'nan 250012, China
e-mail: zhanpeng1982@sdu.edu.cn

E. F. da Silva-Júnior · X. Liu
Institute of Chemistry and Biotechnology, Federal University of Alagoas, Campus AC. Simões, Lourival Melo Mota Avenue, Maceió 57072-970, Brazil

S. I. Ahmad (ed.), *Human Viruses: Diseases, Treatments and Vaccines*, https://doi.org/10.1007/978-3-030-71165-8_28

Approved drugs by the U.S. Food and Drug Administration (FDA) so far are primarily RT reverse transcriptase (RT) inhibitors against HIV infection (Tachikawa 2012) and nucleos(t)ide-analogue therapies against HBV infection (Pei et al. 2017). In another aspect, combinational therapy (Greig and Deeks 2015) is widely used in the treatment and management of antiretroviral infection. However, emerging adverse effects, drug resistance as well as multiple drawbacks caused by life-long treatments limit their wide application (Hawkins 2010; Schneider et al. 2016) and therefore novel therapeutics that act on previously unexplored target are urgently needed to improve current treatments. Ribonuclease H (RNase H) (Cerritelli and Crouch 2009), which catalyzes the cleavage of RNA in RNA/DNA substrates, is still not targeted by any current drug and can be considered as an appealing target.

HIV RNase H is one of the functional domains of RT that selectively cleaves the RNA strand within the RNA: DNA heteroduplex replication intermediate along with processing primers (Sarafianos et al. 2009). The HIV RNase H active site shares with prokaryotic and eukaryotic RNases H, a highly conserved "DEDD" motif that in the case of HIV-1 contains the carboxylate residues Asp443 (D443), Glu478 (E478), Asp498 (D498) and Asp549 (D549) (Fig. 1a) (Schultz and Champoux 2008). Based on the mechanism of action, RNase H inhibitors can be divided in two classes: metal-chelating active site inhibitors and allosteric inhibitors (Tramontano et al. 2019). Among them, metal-chelating active site inhibitors coordinate the two catalytic Mg^{2+} ions in the RNase H active site, usually through a triad of basic atoms or electron donors (Nowotny et al. 2007).

The role of the HBV RNase H is to degrade the pregenomic viral RNA after being transcribed into DNA by the viral polymerase in order to facilitate synthesis of the positive strand of DNA. Experimental evidence indicates that RNase H inhibition has effects on the synthesis of both viral DNA strands (i.e. plus and minus polarity) (Edwards et al. 2019). The HBV RNase H has been successfully expressed as a fusion recombinant protein containing an N-terminal maltose-binding protein (MBP) and C-terminal hexahistidine (H6) tag. Catalytic

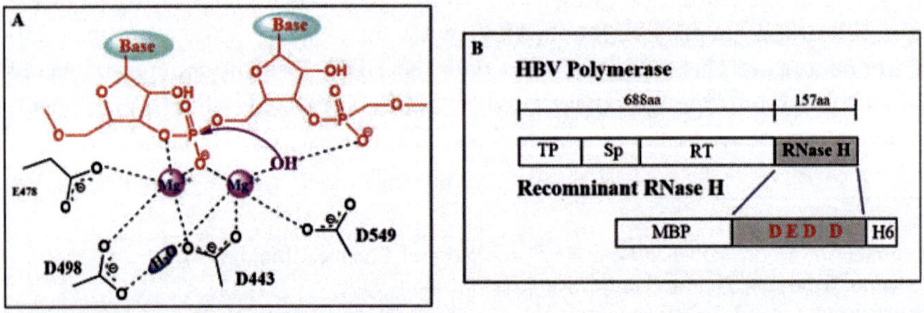

Fig. 1 **a** The two-metal ion mechanism of catalysis for HIV-1 RNase H activity, **b** The HBV RNase H. TP, terminal protein domain that primes DNA synthesis; Sp, spacer domain; RT, reverse transcriptase domain; RNase H, RNase H domain; MBP, maltose-binding protein, H6, hexahistidine tail

residues (Asp702, Glu731, Asp750 and Asp790) equivalent to those indicated above for HIV-1 RNase H have been identified by introducing non-conservative substitutions in the "DEDD" motif (Fig. 1b) (Chen and Marion 1996; Villa et al. 2016). The HBV RNase H is Mg^{2+}-dependent and all known HBV RNase H inhibitors have metal-coordinating moieties that could chelate the active site cations. Likewise, all HBV RNase H inhibitors identified to date feature a three-oxygen pharmacophore or an equivalent structure as a triad that binds two Mg^{2+} cations required for catalysis (Fig. 1).

From a mechanistic point of view, there are important differences in the process of reverse transcription in HBV and HIV (Menéndez-Arias et al. 2017). However, the RNase H activity of the viral polymerase is required in both cases to degrade viral genomic or pregenomic RNA as a necessary step to facilitate the formation of double-stranded DNA that will be translocated into the nucleus of the infected cell. The RNase H domain is located at the C-terminal region of the HIV RT and the HBV polymerase. HBV and HIV RNases H share about 23% identity and 33% similarity in the core catalytic domain, which both belong to the nucleotidyl transferase superfamily (Hu et al. 2013; Nowotny 2009). According to the reports, most HIV metal chelation inhibitors also have antiviral activity against HBV RNase H. In view of the strong structural similarities exerted in the active sites of HBV and HIV inhibitors, we summarized the HBV/HIV RNase H inhibitors in recent years through a plethora of pharmacochemical strategies.

2 Medicinal Chemistry Approaches in Lead Discovery

2.1 Anti-RNase H Agents from Natural Products

Natural extracts have been investigated while searching for innovative biochemical compounds valuable against many targets including HIV RNase H (Tatyana et al. 2009). In 2017, Sonar et al. found that extracts from leaves of *Ocimum sanctum* L contained molecules with anti-HIV RNase H activity. The compound with meaningful antiviral activity was a ferulaldehyde containing a catechol moiety that showed activity in the low micromolar range in enzymatic assays (IC_{50} = 2.4 μM) (Sonar et al. 2017). Derivatives of that compound included *N*-oleylcaffeamide (**1**) (Fig. 2) which showed remarkable inhibitory activity against HIV-1 RNase H and RNA-dependent DNA polymerase activities with IC_{50} values of 0.68 μM and 2.3 μM, respectively (Sanna et al. 2018). In another study prenylatedphloroglucinols (**2–4**), obtained from the angiosperm *Hypericumscruglii* (a plant related to the common Saint John's wort) was shown to inhibit HIV-1 replication with EC_{50} values of 3.5–8 μM, seven to fourteen times lower than their CC_{50} (Sanna et al. 2018). One of these compounds (**4**) inhibited HIV-1 IN, RNase H and DNA polymerase activities in the micromolar range (Fig. 2).

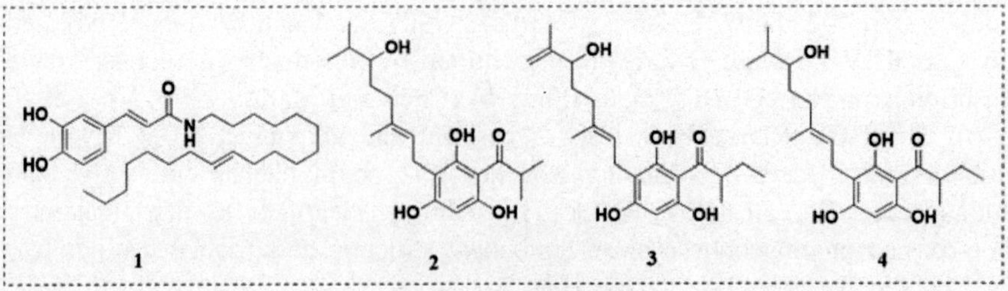

HIV RNase H inhibitors

Fig. 2 Chemical structures of HIV RNase H inhibitors **1–4**

2.2 High-Throughput Screening Approach

High-throughput screening (HTS) of large collections of molecules is a valuable technique to seek novel hits and remains as a major procedure at the beginning of a drug discovery campaign (Jadhav 2014; Sun et al. 2016). Furthermore, HTS has demonstrated its efficiency and flexibility in the identification of antiviral agents recently (Fig. 3).

In 2014, Himmel et al. identified the dihydroxycoumarin, F3284-8495 (**5**) by HTS,which showed low micromolar HIV-1 RT RNase H inhibitory activity (IC_{50} = 4.8 μM) but was unable to block DNA synthesis catalyzed by the RT at 10 μM concentration (Himmel et al. 2014). After further analysis of the co-crystal structure of **5** bound to the RNase H active site, researchers replaced the ethanoic acid moiety by a series of larger substituents, in order to introduce additional interactions within the RNase H domain and improve their inhibitory activity. A series of analogs of compound **5** was obtained (Himmel et al. 2014). As expected, compounds F3385-2581 (**6**) and F3385-2590 (**7**), containing a piperazine ring linked to one or multiple aromatic rings, instead of the ethanoic acid substituent, showed inhibitory activity against an engineered HIV RNase H domain known as p15-EC RNase H at submicromolar levels (IC_{50} = 0.1 μM and 0.2 μM for compounds **6** and **7**, respectively).

Apart from the inhibitors described above, several reports have demonstrated the HIV-1 RNase H inhibitory activity of tropolone derivatives (Budihas et al. 2005;

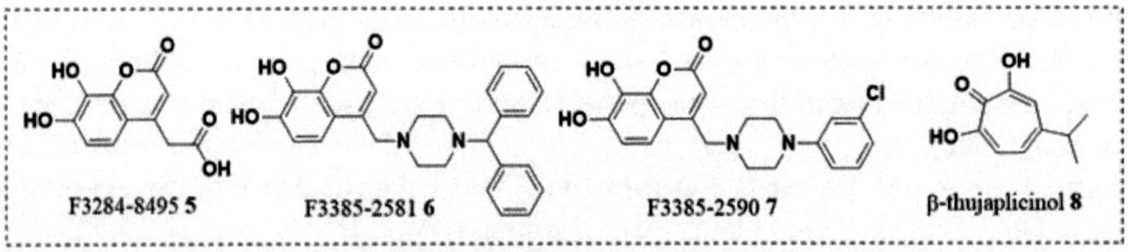

HIV RNase H inhibitors

Fig. 3 Structures of HIV RNase H inhibitors dihydroxycoumarin (**5–7**) and *β*-thujaplicinol (**8**)

Herman and Sluis-Cremer 2013; Semenova et al. 2006). Among them, a tropolone derivative with a 7-OH substitution discovered in a HTS campaign using National Cancer Institute libraries of pure natural products, was found to inhibit retroviral, bacterial and human RNases H (Herman and Sluis-Cremer 2013). The compound, known as β-thujaplicinol (**8**), is found in the heartwood of a western red cedar, a coniferous tree of the *Cupressaceae* family, and showed an IC_{50} value of 0.21 μM in HIV-1 RNase H inhibition assays (Budihas et al. 2005) (Fig. 3).

2.3 *Drug Repositioning and Phenotypic Screening*

Drug repurposing which consists of giving old inhibitors a new use (outside the scope of the original medical indication) by exploring new molecular pathways and targets for intervention has recently become very popular (Jones and Bunnage 2017; Strittmatter 2014; Wu et al. 2019).

Anti-influenza derivative **9** based on the 1,2,4-triazolo[1,5-*a*]-pyrimidine (TZP) scaffold disrupts the interaction of the polymerase acidic protein-basic protein 1 (PA-PB1) subunits of influenza virus (Flu) RNA-dependent RNA polymerase (IC_{50} = 1.1 μM) (Massari et al. 2015). In 2020, TZP scaffold was repurposing as a new chemotype to inhibit HIV RNase H in the low micromolar range without showing RT-associated polymerase inhibitory activity. The further structural modification entailed the most potent 5-methyl-7-phenyl derivative **10** and its C2 inverse amide analogue **11** showed anti-RNase H activity in the sub-micromolar range (IC_{50} values of 0.8 and 0.41 μM, respectively) (Fig. 4) (Desantis et al. 2020).

As an extension of drug repurposing, existing inhibitors with anti-HIV activity could confer significant antiviral effect of HBV by phenotypic screening. It has been demonstrated that many compounds selected for the ability to suppress HIV RNase H activity also inhibit HBV RNase H in enzymatic assays. Some can also block HBV replication in cell cultures via suppression of the viral RNase H activity (Cai et al. 2014; Hu et al. 2013). Such as *β*-thujaplicinol, it could suppress HIV RNase H while displaying antiviral activity of the HBV RNase H. Tavis et al.

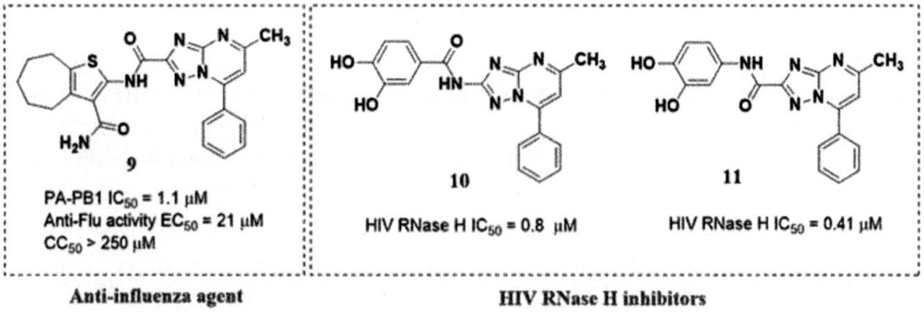

Fig. 4 Chemical structures of anti-influenza agent (**9**) and HIV RNase H inhibitors (**10–11**)

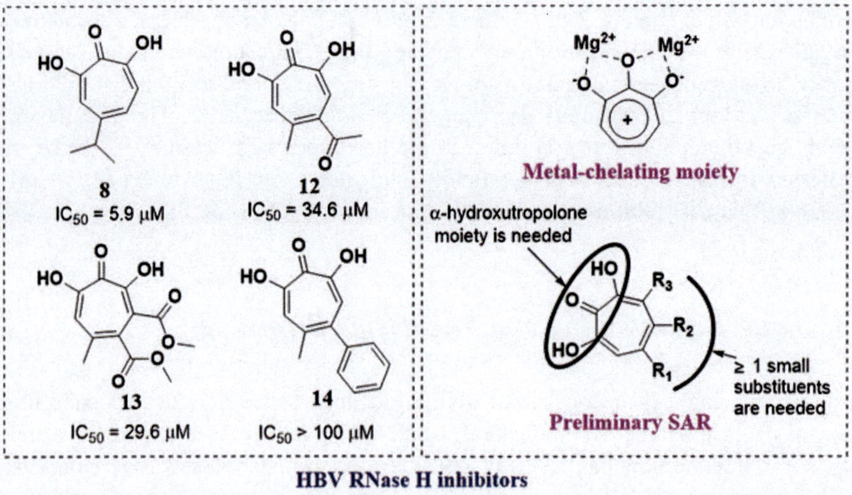

Fig. 5 Structures of HBV RNase H inhibitors (**8**, **12–14**) and their metal-chelating moiety and preliminary SAR

demonstrated its antiviral activity using a cell-based phenotypic screening assay. Major stabilizing interactions of the HBV RNase H inhibitor complexes are probably electrostatic and appear to be primarily established between the chelating moiety of the ligands and the polar catalytic center of the enzyme (Fig. 5) (Hu et al. 2013).

Preliminary structure–activity relationships (SAR) using hydroxytropolones indicated that the intact α-tropolone moiety is needed because deleting one of the three oxygen molecule on the tropolone ring ablates inhibition. This implies that hydroxytropolones inhibit the HBV RNase H by the metal-chelating mechanism employed against the HIV RNase H. The most active hydroxytropolone was the least-substituted molecule, compound **8**, which has a single isopropyl group with an IC_{50} value of 5.9 μM against HBV (Lu et al. 2015). The substituents at positions R1, R2 and R3 should be short for substantial activity (length equal or smaller than four atoms) (Hu et al. 2013). This size limitation suggests that the HBV RNase H active site is probably narrower than the HIV RNase H active site. A narrower active site is expected to facilitate the design of compounds with adequate specificity for the RNase H (Tavis and Lomonosova 2015).

Meanwhile, *N*-hydroxyimides exhibit anti-HBV RNase H activity while exerting potential anti-HIV activity. Representatives of this group of compounds include *N*-hydroxyisoquinolinediones (HID) (**15**), HPD (**16**) and *N*-hydroxynapthyridinones (HNO) (**17**). Compounds **15–17** have been characterized as HBV RNase H inhibitors in replicon assays, with EC_{50} values of 1.4, 0.69 and 3.4 μM. *N*-hydroxyimides cores provide different directionalities for branching, and preliminary SAR for different substituents and molecular cores have been empirically determined (Fig. 6) (Tavis et al. 2019).

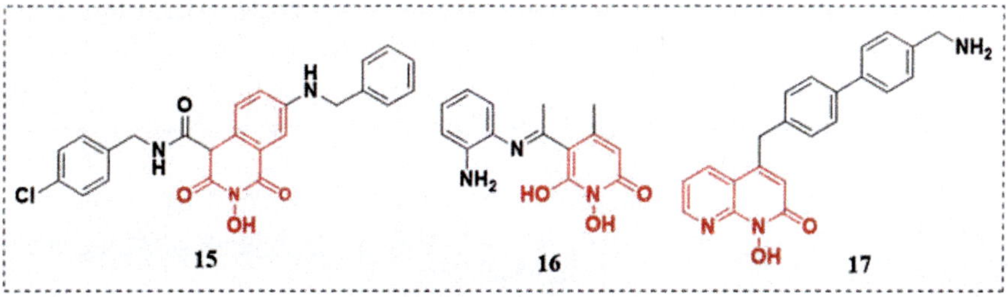

HBV RNase H inhibitors

Fig. 6 Chemical structures of representative *N*-hydroxyimides (**15–17**)

HID and HPD compounds inhibit the HIV RNase H activity in vitro through binding two Mg^{2+} cations in the RNase H active site. The inhibitory effects of HBV replication for HID and HPD derivatives has been measured by oligonucleotide-directed RNA cleavage assays as well as cell-based HBV replication assays (Billamboz et al. 2019). Both chemotypes (HID and HPD) have been identified as attractive scaffolds for antiviral development (Cai et al. 2014; Huber et al. 2017).

Recently, a preliminary SAR for HID and HPD derivatives was reported (Edwards et al. 2017). The six-membered nitrogenous HPD ring appears to be the minimal pharmacophore because it is shared by all active compounds (Fig. 7). Addition of a second six-membered ring bridging positions five and six of the HPD ring creates the HID scaffold. Both the HPD and HID scaffolds have an oxygen trident at positions C1, N2 and C3 which is essential for their activity. The loss of any one of these oxygens results in inactive compounds. For example, compound **18** lacks the oxygen at N2 and is inactive (Fig. 7). This is consistent with the known mechanism by which the HIDs inhibit the HIV RNase H (Suchaud et al. 2014).

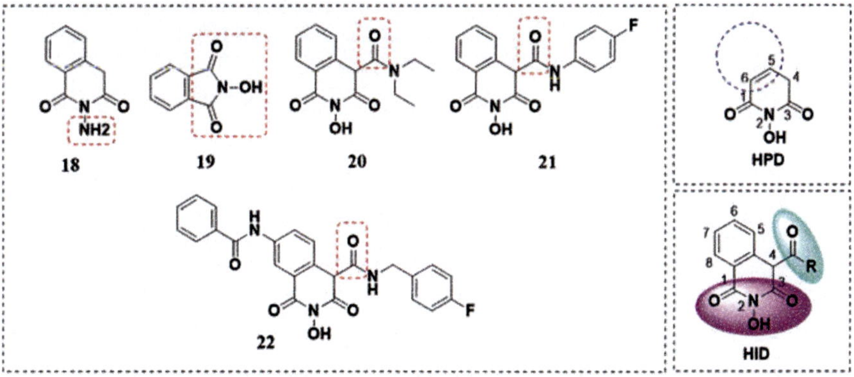

HBV RNase H inhibitors

Fig. 7 Chemical structures of compounds **18–22** and their preliminary SAR

The six-membered ring is the minimum ring necessary for activity. Compound **19** that contains a five-membered ring is also inactive, presumably due to an inability to coordinate the Mg^{2+} ions at the active site of the HBV RNase H. This failure could be due to inappropriate bond angles and/or lack of relative acidic hydrogen at R4 of the HID ring.

Substitutions by alkyl groups at C4 are not tolerated. However, significant anti-HBV activity can be achieved by introducing a carbonyl group (ester or amide function) at C4 (**22**, EC_{50} = 1.4 μM)), which increases the acidity of the hydrogen linked to C4 and therefore the ability to coordinate the Mg^{2+} ions (Billamboz et al. 2011), followed by large hydrophobic groups (aryl or alkaryl).

.

2.4 Novel RNase H Inhibitors Obtained by Using "Click Chemistry"

Click chemistry is a term that was first proposed by Kolb et al. (2001). Cornerstones of click chemistry are high yield, wide scope, less cytotoxic by products, high stereospecificity and a simple reaction. Click chemistry reactions can occur under physiological conditions and the resulting chemical bonds are irreversible. Therefore, click chemistry is widely used for the modification of biomolecules, such as nucleic acids, lipids, and proteins using different compounds (Takayama et al. 2019).

In 2016, a series of 4-[4-(aryl)-1H-1,2,3-triazol-1-yl] benzenesulfonamides were prepared using "click chemistry" reactions and identified as novel HIV-1 RNase H inhibitors through an in-house screening campaign. Among them, three compounds (**23–25**) showed selective potency against the HIV-1 RNase H at micromolar concentrations (Fig. 8) (Pala et al. 2016). Notably, their structures were different from those of previously reported for divalent metal-chelating RNase H active site

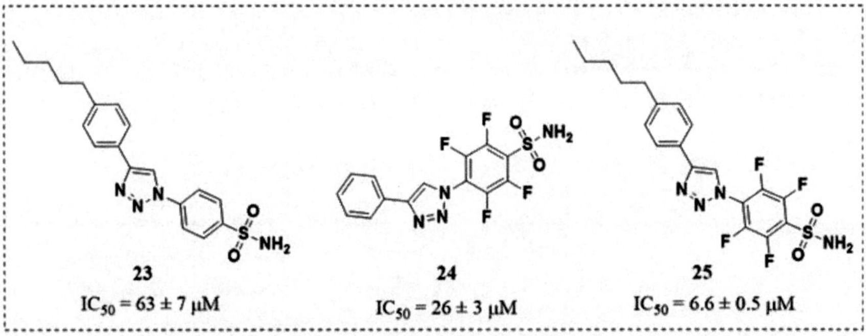

Fig. 8 Chemical structures ofHIV RNase H inhibitors (**23–25**)

inhibitors, and were considered as good lead compounds for the development of a new generation of anti-HIV agents. Compound **25** was the most potent inhibitor in enzymatic assays showing an IC_{50} of 6.6 μM.

3 Medicinal Chemistry Approaches for Structural Optimization of RNase H Inhibitors

3.1 Knowledge-Based Pharmacophore Hybridization

A distinct pharmacophore featuring a highly privileged biaryl moiety connected to the chelating core by a one atom linker seems to be common to many RNase H inhibitors and could be the key for tight binding to the RNase H active site. Based on this pharmacophore model for RNase H inhibition, HPD chemotype (Fig. 9) was redesigned to obtain selective RNase H inhibitors (Tang et al. 2016). The biaryl group with different substituents was introduced at the C6 position of the HPD using different linkers. Meanwhile, structural simplification was performed, namely, by removing the crucial C5 isopropyl group and substituting the N-1 position with a methyl group (**26** and **27**) or a hydrogen atom (**28**).

The hydroxypyridone carboxylic acid pharmacophore, found in the influenza endonuclease inhibitor **29** and in the approved HIV integrase (IN) strand transfer inhibitor dolutegravir **30**, has been exploited as a chelating core to design new selective RNase H inhibitors (Ju et al. 2017). In 2016, Kankanala et al. reported novel derivatives of the hydroxypyridone carboxylic acid scaffold with critical biarylmethyl or N-1 benzyl moieties (Kankanala et al. 2016). Thus, compound **31** was identified as the most effective HIV-1 RNase H inhibitor (IC_{50} = 0.65 μM).

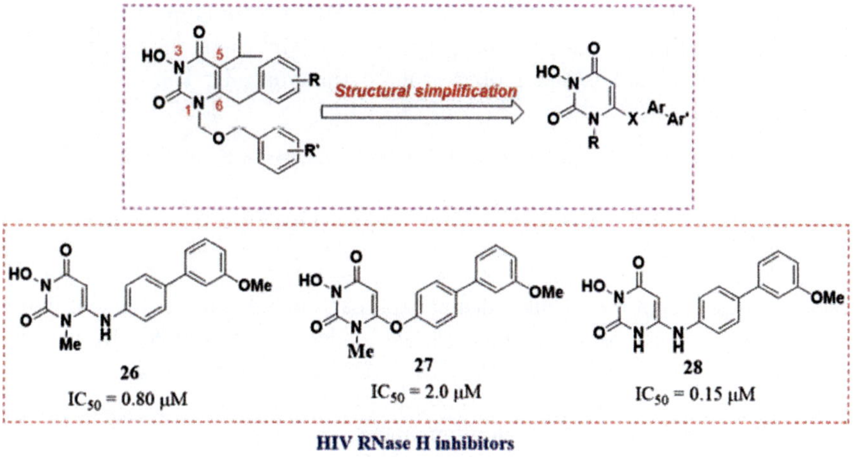

Fig. 9 Chemical structures of HIV RNase H inhibitors **26–28**

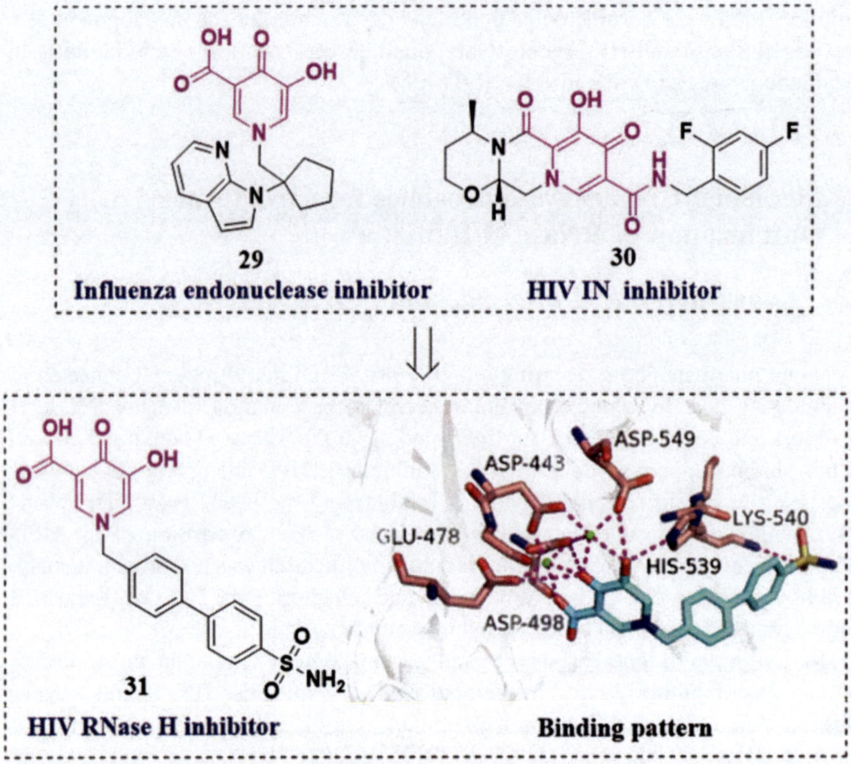

Fig. 10 Structures of compounds **29–31** and the binding mode of compound **31** in the HIV-1 RNase H active site (PDB code: 5J1E)

The recently reported X-ray structure of the HIV-1 RNase H domain bound to **31** represents a significant advance for understanding HIV RNase H inhibition. In the obtained crystal structure, compound **31** helates two Mg^{2+} ions through the carbonyl, hydroxyl and carboxylate groups of the pyrimidine (Fig. 10).

3.2 Structure-Based Design of Double-Winged RNase H Inhibitors

Structure-based design represents a distinct approach that could yield potent inhibitors to cut into the substrate dominance and guides synthesis molecules that fit complementarily to a given binding pocket. The specific conformation of the inhibitors at the active site once be recognized, the inhibitors exhibit antiviral activity and thus play key roles in bio-functions.

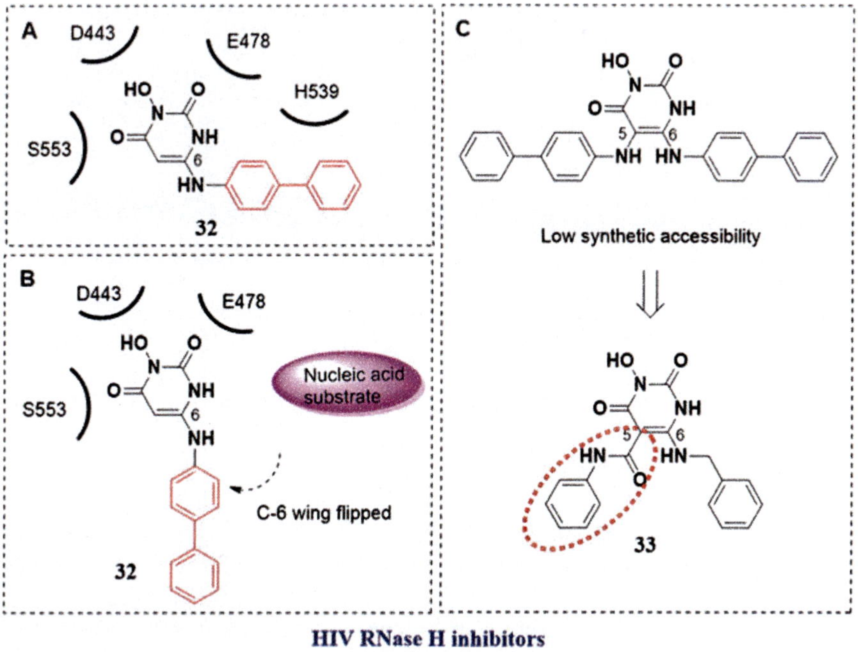

Fig. 11 Chemical structures of HIV-1 RNase H inhibitors **32–33**

The dominant RNA/DNA substrate at the RT RNase H active site may decrease antiviral activity of inhibitors. Like **32**, which inhibits RT RNase H in the sub micromolar range but does not exert antiviral activity, adopts two divergent binding modes in the absence or presence of the nucleic acid substrate respectively when docked into the active site of RNase H. Molecular docking indicates that C6 biarylamino moiety (the wing) of **32** makes a key contact with H539 without the substrate (Fig. 11a). In contrast, the wing of **32** is forced to flip in the presence of the competing nucleic acid substrate, resulting in a possible loss of key interaction with H539 (Fig. 11b). Given the feasibility of synthesis, **33** was obtained with the addition of a C5 wing, which inhibited RNase H with excellent potency (IC_{50} = 27 nM) and increased affinity as well as conferred significant antiviral activity (EC_{50} = 6.9 μM) due to the conformational restriction of C6 wing at the active site (Fig. 11c) (Vernekar et al. 2017).

3.3 RNase H Inhibitors Obtained from Privileged Fragment-Based Libraries

The privileged substructure-based diversity-oriented synthesis (pDOS) strategy, has proven to be another fruitful tool to rapidly discover biologically active lead

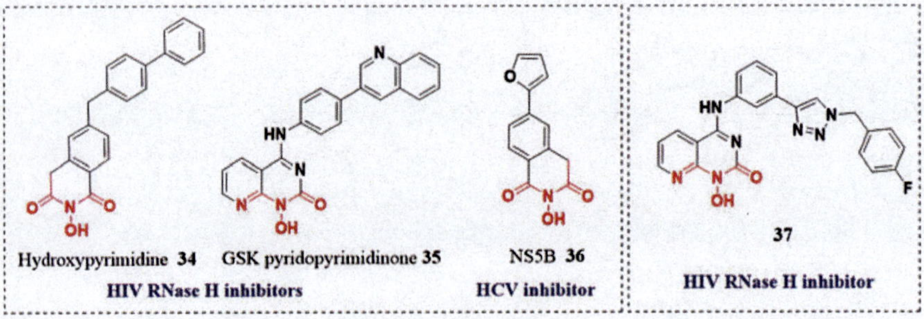

Fig. 12 Chemical structures of representative antiviral agents (**34–37**)

compounds by exploring the uncharted chemical space and constructing high-quality compound libraries (Oh and Park 2011; Kim et al. 2014).

For the construction of pDOS libraries, it is crucial to select a privileged substructure with potential for scaffold refining. It is known that hydroxy-(iso) quinazoline-2,4(1,3)-dione and its analogues are found in a number of molecules active against a broad spectrum of biological activities. Among them, 2-hydroxy-isoquinoline-1,3(2*H*,4*H*)-dione and N-hydroxypyridinediones(HPD) derivatives are considered the core structures in many antiviral agents. Examples are HIV-1 RNase H active-site inhibitors (**34–35**) (Vernekar et al. 2015) and the hepatitis C virus (HCV) NS5B polymerase inhibitor **36** (Fig. 12) (Chen et al. 2012). The presence of a terminal biaryl substituent seems to be important to increase the efficiency of those compounds.

Apart from the above inhibitors, a small library of pyridopyrimidinone derivatives was obtained by modification of the 1-hydroxypyrido [2,3-d]pyrimidin-2(1H)-one scaffold with copper(I)-catalyzed azide-alkyne cycloaddition (CuAAC) reactions (Gao et al. 2017). The most potent compound of this series, **37** (Fig. 12), also showed remarkable and selective potency against HIV-1 IN.

4 Dual anti- HIV-1 Integrase (iN) and RNase H Agents

Dual-acting agents could target two different viral proteins at the same time, and in consequence decrease the possibility for selection of drug resistant variants (Esposito et al. 2014). HIV-1 RNase H and IN are metalloenzymes with important functions in the viral life cycle and both of them are mechanistically related, belonging to the polynucleotidyl transferase superfamily (Nowotny 2009). Retroviral INs catalyze the integration of the viral DNA into the genome of infected host cells by the coordinated action of two enzymatic activities (3′-processing and strand transfer).

In 2008, a series of madurahydroxylactone derivatives was serendipitous discovered for dual inhibition of HIV IN and RNase H with micromolar IC_{50} values.

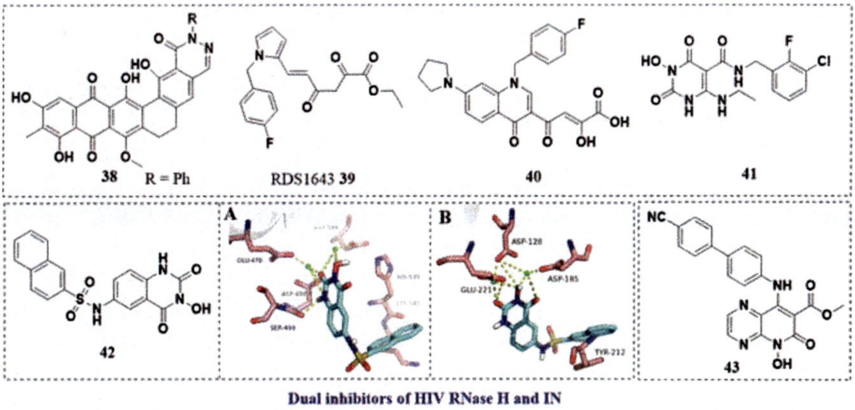

Fig. 13 Chemical structures of dual inhibitors (**38–43**) of HIV RNase H and IN as well as binding modes of **43** in the HIV-1 RNase H active site (**a**, PDB code: 5J1E) and the IN active site (**b**, PDB code: 3OYA)

Among them, **38** was the most potent dual inhibitor of RNase H and IN with IC_{50} values of 2.58 and 0.41, respectively (Fig. 13) (Marchand et al. 2008).

Diketo acid (DKA) inhibitors are active-site Mg^{2+}-binding inhibitors that were initially identified as inhibitors of the strand transfer activity of HIV-1 IN. As expected from the structural similarity between the catalytic sites of HIV-1 RNase H and IN, compounds bearing diketo acid moieties were frequently identified as dual inhibitors of both enzymes. Thus, the ester derivative RDS1643 (**39**) has inhibitory activity against both HIV-1 IN and RNase H in the low micromolar range (Corona et al. 2014). In addition, several pyrrolyl DKA derivatives, having a quinolinonyl-or a pyrroyl-based scaffold, such as compound **40**, have been reported as RNase H inhibitors with dual activity against HIV-1 IN and RT-associated RNase H (Fig. 13) (Costi et al. 2014; Cuzzucoli Crucitti et al. 2015).

Related to DKA derivatives, compound **41** was obtained after adding a HPDcore. The molecule was found to be a strong inhibitor of the HIV-1 IN strand transfer and RT-associated RNase H activities with IC_{50} values of 21 nM and 29 nM, respectively (Wu et al. 2016). Related molecules containing 3-hydroxyquinazoline-2,4(1H,3H)-diones were also effective inhibitors of HIV-1 RNaseH and IN strand transfer activities at sub to low micromolar concentrations. Among them, compound **42** (Fig. 13) was the most promising one with potent inhibitory activity against HIV-1 RNase H (IC_{50} = 0.41 μM) (Gao et al. 2019). This molecule was also a potent IN inhibitor, showing an IC_{50} value of 0.85 μM. These results reveal the potential of the 3-hydroxyquinazoline-2,4(1H,3H)-dione chemotype as a scaffold to develop potent HIV-1 RNase H and IN inhibitors.

In 2018, Sun et al. identified 5-hydroxypyrido [2,3-b]pyrazin-6(5*H*)-one scaffold as potent dual HIV-1 RNase H and IN inhibitors. Further studies revealed that compound **43** conferred significant antiviral effect against HIV-1 RNase H and IN strand transfer activities with IC_{50} values of 1.77 μM and 1.18 μM, respectively (Sun et al. 2018).

5 Conclusion

The effectiveness of current antiviral treatments is threatened by the emergence of drug resistance, and development of new drugs is important to treat HIV/HBV infections caused by transmitted drug-resistant strains. In this context, drugs acting on novel targets or having stronger inhibitory effects will be helpful to develop more effective treatments. In this chapter, we have covered progress recently made to inhibit the activity of HIV/HBV RNase H using small molecules, while focusing on drug discovery strategies and mechanisms of action of the identified inhibitors.

As described above, the vast majority of HIV/HBV RNase H inhibitors act by coordinating to the catalytic center metal ions. The remarkably potent RNase H inhibitors identified for both HIV-1 and HBV are active site chelating molecules containing a ring with fixed angles, which proved to be the best option to achieve inhibition of the enzymatic function. From a pure speculative point of view, given the promising data on HBV RNase H inhibitors, the development of compounds able to block the two viruses at once seems also a really appealing possibility to treat HIV-1/HBV co-infections.

Although several effective RNase H active site inhibitors has been described in this section, neither of those molecules is currently moving into advanced pre-clinical development. Despite their remarkable efficiency in vitro, those compounds show modest effects when tested ex vivo or in animal models. This may be due to poor membrane permeability and cell uptake (originating from the presence of several hydroxyl groups), poor solubility, or other physicochemical limitations. Nevertheless, the in vitro activity of those compounds indicates that they are good candidates for development into prodrugs by introducing lipophilic groups to improve cell permeability. In the case of HBVRNase H inhibitors, a more classical ligand-based approachshould be applied to better delineate the pharmacophore requirements needed for optimization. Nevertheless, potent, selective, orally active HIV and HBV RNase H metal chelating inhibitors to treat those viral infections still have good perspectives for future development.

Funding We gratefully acknowledge financial support from the National Natural Science Foundation of China (NSFC No. 81973181), National Science and Technology Major Projects for "Major New Drugs Innovation and Development' (2019ZX09301126), Science Foundation for Outstanding Young Scholars of Shandong Province (ZR2020JQ31), Foreign cultural and educational experts Project (GXL20200015001), Qilu Young Scholars Program of Shandong University, the Taishan Scholar Program at Shandong Province, the Young Scholars Program of Shandong University (YSPSDU No. 2016WLJH32), Key Project of NSFC for International Cooperation (No. 81420108027), and the Key research and development project of Shandong Province (No. 2017CXGC1401, 2019JZZY021011).

References

Billamboz M, Bailly F, Lion C, Touati N, Vezin H, Calmels C, Cotelle P. Magnesium chelating 2-hydroxyisoquinoline-1,3(2H,4H)-diones, as inhibitors of HIV-1 integrase and/or the HIV-1 reverse transcriptase ribonuclease H domain: discovery of a novel selective inhibitor of the ribonuclease H function. J Med Chem. 2011;54(6):1812–24.

Billamboz M, Suchaud V, Bailly F, Lion C, Andréola ML, Christ F, Cotelle P. 2-hydroxyisoquinoline-1,3(2H,4H)-diones (HIDs) as human immunodeficiency virus type 1 integrase inhibitors: Influence of the alkylcarboxamide substitution of position 4. Eur J Med Chem. 2019;117:256–68.

Budihas SR, Gorshkova I, Gaidamakov S, Wamiru A, Bona MK, Parniak MA, Crouch RJ, Le Grice SF. Selective inhibition of HIV-1 reverse transcriptase-associated ribonuclease H activity by hydroxylated tropolones. Nucleic Acids Res. 2005;33(4):1249–56.

Cai CW, Lomonosova E, Moran EA, Cheng X, Patel KB, Bailly F, Tavis JE. Hepatitis B virus replication is blocked by a 2-hydroxyisoquinoline-1,3(2H,4H)-dione (HID) inhibitor of the viral ribonuclease H activity. Antiviral Res. 2014;108:48–55.

Cerritelli SM, Crouch RJ. Ribonuclease H: the enzymes in eukaryotes. FEBS J. 2009;276 (6):1494–505.

Chen Y, Marion PL. Amino acids essential for RNase H activity of hepadnaviruses are also required for efficient elongation of minus-strand viral DNA. J Virol. 1996;70(9):6151–6.

Chen YL, Tang J, Kesler MJ, Sham YY, Vince R, Geraghty RJ, Wang Z. The design, synthesis and biological evaluations of C-6 or C-7 substituted 2-hydroxyisoquinoline-1,3-diones as inhibitors of hepatitis C virus. Bioorg Med Chem. 2012;20:467–79.

Corona A, Di Leva FS, Thierry S, Pescatori L, CuzzucoliCrucitti G, Subra F, Tramontano E. Identification of highly conserved residues involved in inhibition of HIV-1 RNase H function by Diketo acid derivatives. Antimicrob Agents Chemother. 2014;58(10):6101–10.

Costi R, Métifiot M, Chung S, CuzzucoliCrucitti G, Maddali K, Pescatori L, Messore A, Di Santo R. Basic quinolinonyldiketo acid derivatives as inhibitors of HIV integrase and their activity against RNase H function of reverse transcriptase. J Med Chem. 2014;57(8):3223–34.

Cuzzucoli Crucitti G, Métifiot M, Pescatori L, Messore A, Noemi Madia V, Pupo G, Saccoliti F, Di Santo R. Structure-activity relationship of pyrrolyldiketo acid derivatives as dual inhibitors of HIV-1 integrase and reverse transcriptase ribonuclease H domain. J Med Chem. 2015;58 (4):1915–28.

Desantis J, Massari S, Corona A, Astolfi A, Sabatini S, Manfroni G, Tabarrini O. 1,2,4-Triazolo [1,5-a]pyrimidines as a Novel Class of Inhibitors of the HIV-1 Reverse Transcriptase-Associated Ribonuclease H Activity. Molecules. 2020;25(5):E1183.

Edwards TC, Lomonosova E, Patel JA, Li Q, Villa JA, Gupta AK, Tavis JE. Inhibition of hepatitis B virus replication by N-hydroxyisoquinolinediones and related polyoxygenated heterocycles. Antiviral Res. 2017;143:205–17.

Edwards TC, Ponzar NL, Tavis JE. Shedding light on RNaseH: a promising target for hepatitis B virus (HBV). Expert Opin Ther Targets. 2019;23(7):559–63.

Esposito F, Tramontano E. Past and future. Current drugs targeting HIV-1 integrase and reverse transcriptase-associated ribonuclease H activity: single and dual active site inhibitors. Antivir Chem Chemother. 2014;23(4):129–44.

Gao P, Zhang L, Sun L, Huang T, Tan J, Zhang J, Liu X. 1-Hydroxypyrido[2,3-d]pyrimidin-2 (1H)-ones as novel selective HIV integrase inhibitors obtained via privileged substructure-based compound libraries. Bioorg Med Chem. 2017;25:5779–89.

Gao P, Cheng X, Sun L, Song S, Álvarez M, Luczkowiak J, Liu X. Design, synthesis and biological evaluation of 3-hydroxyquinazoline-2,4(1H,3H)-diones as dual inhibitors of HIV-1 reverse transcriptase-associated RNase H and integrase. Bioorg Med Chem. 2019;27 (17):3836–45.

Greig SL, Deeks ED. Abacavir/dolutegravir/lamivudine single-tablet regimen: a review of its use in HIV-1 infection. Drugs. 2015;75(5):503–14.

Hawkins T. Understanding and managing the adverse effects of antiretroviral therapy. Antiviral Res. 2010;85(1):201–9.

Herman BD, Sluis-Cremer N. Transient kinetic analyses of the ribonuclease H cleavage activity of HIV-1 reverse transcriptase in complex with efavirenz and/or a β-thujaplicinol analogue. Biochem J. 2013;455(2):179–84.

Himmel DM, Myshakina NS, Ilina T, Van Ry A, Ho WC, Parniak MA, Arnold E. Structure of a dihydroxycoumarin active-site inhibitor in complex with the RNase H domain of HIV-1 reverse transcriptase and structure-activity analysis of inhibitor analogs. J Mol Biol. 2014;426 (14):2617–31.

Hu Y, Cheng X, Cao F, Huang A, Tavis JE. β-Thujaplicinol inhibits hepatitis B virus replication by blocking the viral ribonuclease H activity. Antiviral Res. 2013;99(3):221–9.

Huber AD, Michailidis E, Tang J, Puray-Chavez MN, Boftsi M, Wolf JJ, Sarafianos SG. 3-Hydroxypyrimidine-2,4-diones as novel hepatitis B virus antivirals targeting the viral ribonuclease H. Antimicrob Agents Chemother. 2017;61(6):e00245-e217.

Jadhav MP. High-throughput screening (HTS) for the identification of novel antiviral scaffolds. Clin Pharmacol Drug Dev. 2014;3(2):79–83.

Jones LH, Bunnage ME. Applications of chemogenomic library screening in drug discovery. Nat Rev Drug Discov. 2017;16(4):285–96.

Ju H, Zhang J, Huang B, Kang D, Huang B, Liu X, Zhan P. Inhibitors of influenza virus polymerase acidic (pa) endonuclease: contemporary developments and perspectives. J Med Chem. 2017;60(9):3533–51.

Kankanala J, Kirby KA, Liu F, Miller L, Nagy E, Wilson DJ, Wang Z. Design, synthesis, and biological evaluations of hydroxypyridonecarboxylic acids as inhibitors of HIV reverse transcriptase associated RNase H. J Med Chem. 2016;59(10):5051–62.

Kim J, Kim H, Park SB. Privileged structures: efficient chemical "navigators" toward unexplored biologically relevant chemical spaces. J Am Chem Soc. 2014;136:14629–38.

Kolb HC, Finn MG, Sharpless KB. Click chemistry: diverse chemical function from a few good reactions. Angew Chem Int Ed. 2001;40(11):2004–21.

Lu G, Lomonosova E, Cheng X, Moran EA, Meyers MJ, Le Grice SF, Tavis JE. Hydroxylated tropolones inhibit hepatitis B virus replication by blocking viral ribonuclease H activity. Antimicrob Agents Chemother. 2015;59(2):1070–9.

Marchand C, Beutler JA, Wamiru A, Budihas S, Möllmann U, Heinisch L, Mellors JW, Pommier Y. Madurahydroxylactone derivatives as dual inhibitors of human immunodeficiency virus type 1 integrase and RNase H. Antimicrob Agents Chemother. 2008;52(1):361–4.

Massari S, Nannetti G, Desantis J, Muratore G, Sabatini S, Manfroni G, Tabarrini O. A Broad Anti-influenza Hybrid Small Molecule That Potently Disrupts the Interaction of Polymerase Acidic Protein-Basic Protein 1 (PA-PB1) Subunits. J Med Chem. 2015;58(9):3830–42.

Menéndez-Arias L, Sebastián-Martín A, Álvarez M. Viral reverse transcriptases. Virus Res. 2017;234:153–76.

Nowotny M. Retroviral integrase superfamily: the structural perspective. EMBO Rep. 2009;10 (2):144–51.

Nowotny M, Gaidamakov SA, Ghirlando R, Cerritelli SM, Crouch RJ, Yang W. Structure of human RNase H1 complexed with an RNA/DNA hybrid: insight into HIV reverse transcription. Mol Cell. 2007;28(2):264–76.

Oh S, Park SB. A design strategy for drug-like polyheterocycles with privileged substructures for discovery of specific small-molecule modulators. Chem Commun (Camb). 2011;47:12754–61.

Pala N, Esposito F, Rogolino D, Carcelli M, Sanna V, Palomba M, Naesens L, Corona A, Grandi N, Tramontano E, Sechi M. Inhibitory effect of 2,3,5,6-Tetrafluoro-4-[4-(aryl)-1H-1,2,3-triazol-1-yl]benzenesulfonamide derivatives on HIV reverse transcriptase associated RNase H activities. Int J Mol Sci. 2016;17:e1371.

Pei Y, Wang C, Yan SF, Liu G. Past, current, and future developments of therapeutic agents for treatment of chronic hepatitis B virus infection. J Med Chem. 2017;60(15):6461–79.

Sanna C, Scognamiglio M, Fiorentino A, Corona A, Graziani V, Caredda A, Esposito F. Prenylatedphloroglucinols from Hypericumscruglii, an endemic species of Sardinia (Italy), as new dual HIV-1 inhibitors effective on HIV-1 replication. PLoS One. 2018;13(3):e0195168.

Sarafianos SG, Marchand B, Das K, Himmel DM, Parniak MA, Hughes SH, Arnold E. Structure and function of HIV-1 reverse transcriptase: molecular mechanisms of polymerization and inhibition. J Mol Biol. 2009;385(3):693–713.

Schneider A, Corona A, Spöring I, Jordan M, Buchholz B, Maccioni E, Wöhrl BM. Biochemical characterization of a multi-drug resistant HIV-1 subtype AG reverse transcriptase: antagonism of AZT discrimination and excision pathways and sensitivity to RNase H inhibitors. Nucleic Acids Res. 2016;44(5):2310–22.

Schultz SJ, Champoux JJ. RNase H activity: structure, specificity, and function in reverse transcription. Virus Res. 2008;134(1–2):86–103.

Semenova EA, Johnson AA, Marchand C, Davis DA, Yarchoan R, Pommier Y. Preferential inhibition of the magnesium-dependent strand transfer reaction of HIV-1 integrase by alpha-hydroxytropolones. Mol Pharmacol. 2006;69(4):1454–60.

Singh KP, Crane M, Audsley J, Avihingsanon A, Sasadeusz J, Lewin SR. HIV-hepatitis B virus coinfection: epidemiology, pathogenesis, and treatment. AIDS. 2017;31(15):2035–52.

Sonar VP, Corona A, Distinto S, Maccioni E, Meleddu R, Fois B, Cottiglia F. Natural product-inspired esters and amides of ferulic and caffeic acid as dual inhibitors of HIV-1 reverse transcriptase. Eur J Med Chem. 2017;130:248–60.

Strittmatter SM. Overcoming drug development bottlenecks with repurposing: old drugs learn new tricks. Nat Med. 2014;20(6):590–1.

Suchaud V, Bailly F, Lion C, Calmels C, Andréola M-L, Christ F, Debyser Z, Cotelle P. Investigation of a novel series of 2-hydroxyisoquinoline-1,3(2H,4H)-diones as human immunodeficiency virus type 1 integrase inhibitors. J Med Chem. 2014;57(11):4640–60.

Sun J, Groppi VE, Gui H, Chen L, Xie Q, Liu L, Omary MB. High-throughput screening for drugs that modulate intermediate filament proteins. Methods Enzymol. 2016;568:163–85.

Sun L, Gao P, Dong G, Zhang X, Cheng X, Ding X, Wang X, Daelemans D, De Clercq E, Pannecouque C, Menéndez-Arias L, Zhan P, Liu X. 5-Hydroxypyrido[2,3-b]pyrazin-6(5H)-one derivatives as novel dual inhibitors of HIV-1 reverse transcriptase-associated ribonuclease H and integrase. Eur J Med Chem. 2018;155:714–24.

Tachikawa N. Reverse transcriptase inhibitors, protease inhibitors. Nihon Rinsho. 2012;70 (4):579–95.

Takayama Y, Kusamori K, Nishikawa M. Click chemistry as a tool for cell engineering and drug delivery. Molecules. 2019;24:e172.

Tang J, Liu F, Nagy E, Miller L, Kirby KA, Wilson DJ, Wang Z. 3-Hydroxypyrimidine-2,4-diones as selective active Site Inhibitors of HIV reverse transcriptase-associated RNase H: design, synthesis, and biochemical evaluations. J Med Chem. 2016;59(6):2648–59.

Tatyana K, Francesca E, Luca Z, Giovanni F, Cheng YC, Ginger ED, Enzo T. Inhibition of HIV-1 ribonuclease H activity by novel frangula-emodine derivatives. Med Chem. 2009;5 (5):398–410.

Tavis JE, Lomonosova E. The hepatitis B virus ribonuclease H as a drug target. Antiviral Res. 2015;118:132–8.

Tavis JE, Zoidis G, Meyers MJ, Murelli RP. Chemical approaches to inhibiting the hepatitis B virus ribonuclease H. ACS Infect Dis. 2019;5(5):655–8.

Tramontano E, Corona A, Menéndez-Arias L. Ribonuclease H, an unexploited target for antiviral intervention against HIV and hepatitis B virus. Antiviral Res. 2019;171:104613.

Vernekar SK, Liu Z, Nagy E, Miller L, Kirby KA, Wilson DJ, Kankanala J, Sarafianos SG, Parniak MA, Wang Z. Design, synthesis, biochemical, and antiviral evaluations of C6 benzyl and C6 biarylmethyl substituted 2-hydroxylisoquinoline-1,3-diones: dual inhibition against HIV reverse transcriptase-associated RNase H and polymerase with antiviral activities. J Med Chem. 2015;58:651–64.

Vernekar SKV, Tang J, Wu B, Huber AD, Casey MC, Myshakina N, Wang Z. Double-winged 3-hydroxypyrimidine-2,4-diones: potent and selective inhibition against HIV-1 RNase H with significant antiviral activity. J Med Chem. 2017;60(12):5045–56.

Villa JA, Pike DP, Patel KB, Lomonosova E, Lu G, Abdulqader R, Tavis JE. Purification and enzymatic characterization of the hepatitis B virus ribonuclease H, a new target for antiviral inhibitors. Antiviral Res. 2016;132:186–95.

Wu B, Tang J, Wilson DJ, Huber AD, Casey MC, Ji J, Wang Z. 3-Hydroxypyrimidine-2,4-dione-5-N-benzylcarboxamides Potently Inhibit HIV-1 Integrase and RNase H. J Med Chem. 2016;59(13):6136–48.

Wu G, Zhao T, Kang D, Zhang J, Song Y, Namasivayam V, Zhan P. Overview of Recent Strategic Advances in Medicinal Chemistry. J Med Chem. 2019;62(21), 9375–414.

Viral Vaccines

Vaccines for Emerging Viruses—A Comprehensive Update

Roger Hewson

Abstract Over the past decade several notable viruses have suddenly emerged from obscurity to become serious public health problems and global threats, provoking concern about sustained epidemic and pandemic transmission in immunologically naïve human populations. With each newly-emerged virus hazard comes the call for rapid vaccine development. Vaccines are an important component of disease prevention and control for emerging viral infections since in many cases other medical options are limited or non-existent and vaccines are also highly cost effective. Some infections for example may result in such a rapid clinical deterioration that the effectiveness of therapeutics is limited. Other new human virus infections maybe symptomless or cryptic and by the time a diagnosis is possible, a therapeutic maybe of limited use. While many classic approaches to vaccine development are amenable to emerging viruses, the application of molecular techniques in virology has profoundly influenced our understanding of virus biology and virus-host interactions, and so facilitated the development of new vaccine strategies and vaccine vector platforms. Vaccination approaches based on replicating, attenuated and non-replicating virus vector schemes have become useful vaccine tools. If our growing understanding of viral disease emergence and increasing range of vaccine strategies are coupled with an international commitment to underpin vaccine development, vaccine interventions for new and emerging viruses become a possibility.

1 Introduction

Immunisation is one of the most appropriate ways of preventing infectious disease, and the control of many important viral pathogens by vaccination is arguably one of the most outstanding achievements of medical science.

R. Hewson (✉)
Department of Infection Biology, London School of Hygiene and Tropical Medicine, London WC1E 7HT, UK
e-mail: Roger.Hewson1@lshtm.ac.uk

S. I. Ahmad (ed.), *Human Viruses: Diseases, Treatments and Vaccines*, https://doi.org/10.1007/978-3-030-71165-8_29

Vaccine-induced immunity that is established in advance of an otherwise hazardous virus infection relies primarily on the induction of adaptive immune responses to establish protective efficacy. Critically, vaccination depends on the properties of antigen recognition, activation, expansion, memory, trafficking, and the multitude of specialist functions of lymphocytes. The extent to which vaccine-induced immunity is successful also determines the spread and maintenance of a viral pathogen within a population. Viral vaccines have had profound and enduring consequences for human and animal health; the world-wide eradication of smallpox and rinderpest are testament to their outstanding contribution to modern society.

Infectious diseases still pose one of the greatest threats to public health, and the past three decades have coincided with a barrage of emerging new human pathogens. Significantly over 70% of these infections are zoonotic (Jones et al. 2008; Woolhouse et al. 2005), entering either directly from wildlife reservoirs, or indirectly via an intermediate domestic animal host (Field 2009). Avian influenza, Coronavirus Disease (COVID-19), Crimean-Congo haemorrhagic fever (CCHF), Ebola virus disease (EVD), Lassa fever, Middle East Respiratory Syndrome (MERS), Severe Acute Respiratory Syndrome (SARS), Rift Valley fever (RVF) and Tick-borne encephalitis are all examples of zoonoses that have recently emerged from wildlife. Such emerging zoonoses present a serious and increasing threat to human health, biosecurity and economies worldwide as made abundantly clear by the COVID-19 pandemic. Due to the serious threat to global health posed by the increasing emergence of pathogens from wildlife reservoirs, the mechanisms underlying disease transmission from animals to humans have received increasing attention (Jones et al. 2008; Woolhouse et al. 2005; Field 2009; Kreuder Johnson et al. 2015; Olival et al. 2017). Current evidence suggests that transmission occurs in a non-uniform pattern and is localised to distinct geographic 'hotspots' in Africa, Asia and South America. Each high threat pathogen is weighted towards a key wildlife species such as bats, rodents or non-human primates (Olival et al. 2017), with bats harbouring a significantly higher proportion of zoonotic viruses than all other mammalian orders. Such diseases place a substantial burden on global public health, especially in dense human populations where the pressures on environmental and economic resources are greatest. Over one billion cases of human zoonotic disease are estimated to occur annually, and emerging zoonoses result in enormous economic losses (Kreuder Johnson et al. 2015; Grace et al. 2012; Karesh et al. 2012). Increased urbanisation, international travel, commerce and climate change increase the likelihood that emerging zoonosis will continue, if not worsen, in the future.

When a zoonotic virus spills over into a susceptible new species, the new host has no pre-existing immunity, enabling attachment, entry and replication of the virus in receptive cells. The amplified virus may then evade clearance by host defences for long enough to be transmitted to another susceptible host. The lack of any herd immunity can result in the rapid dissemination of the virus, if it is sufficiently infectious, leading to an outbreak of new disease. However, each of the steps in the new virus's infection process represents an opportunity for vaccine intervention. Through such interventions, transmitters and susceptible hosts are removed from a population by the pre-emptive development of protective immunity, so that the spread of infection becomes less likely.

Vaccination is therefore a powerful strategy for preventing and controlling emerging zoonotic infectious disease. The development of vaccines for emerging infections however, needs to contend with several key challenges associated with such viruses. An emerging infection may be a recently discovered virus and the result of a rare outbreak for which basic biological information such as correlates of protection, antigenic variability, and immunodominance are unknown. There may be a lack of time to develop an appropriate animal model of disease in which to study viral immunology and evaluate vaccine candidates for pre-clinical assessment of protective efficacy and safety. Additionally, many emerging viruses have high case fatality rates, are transmitted and spread easily, and are not known to be treatable by existing therapeutics. These characteristics mandate that all experimental investigations with such infectious material be carried out at high levels of bio-safety such as containment level 3, or containment level 4 (BSL4) for the most hazardous. Unfortunately, the availability of such resource-heavy laboratory infrastructure is a serious bottleneck to basic research on these pathogens. Moreover, the standard vaccine approach of using attenuated virus strains or inactivated viruses as vaccines is not always feasible because of the possibility of reversion to virulence or the requirement for large scale culture and production in high containment facilities. In addition to these significant hurdles, the economic costs of novel human vaccine development for rare pathogens which are unlikely to provide an effective payback on investment, has been a major impediment to progress. Thus, basic research into many emerging pathogens has been neglected for years.

In 2014 the unpredicted size, speed and reach of the Ebola virus outbreak from West Africa (Agua-Agum et al. 2016) acted as a wake-up call for researchers, pharmaceutical companies and governments, emphasising the importance of investment into the study of emerging pathogens. Spurred on by this development and at the request of its 194 member states, in May 2015, the World Health Organization (WHO) convened a broad coalition of experts to develop a Research and Development (R&D) Blueprint for Action to Prevent Epidemics. The R&D Blueprint focuses on severe emerging diseases with the potential to generate serious public health emergencies or international epidemics, and for which no, or insufficient, preventive and curative solutions exist. It specifies innovative requirements including vaccine research and through international governance, the programme aims to define R&D roadmaps for prioritised pathogens and to catalyse funding strategies (WHO 2020a).

Given the zoonotic emergence of pathogenic coronaviruses (CoVs) in the first two decades of the 21st century, the WHO listed SARS-CoV, MERS-CoV and other (unknown at the time) highly pathogenic coronaviral diseases as part of the R&D Blueprint campaign. However, despite this awareness, the sudden emergence of a SARS-CoV-2 from bats into humans, possibly via an intermediary host, in Wuhan, China, in December 2019 still caught the world by surprise.

2 Emergence of SARS-CoV-2/COVID-19

The zoonotic emergence of this new SARS-CoV-2 was followed by rapid and sustained human to human transmission in January 2020, resulting in widespread respiratory illness in Wuhan and other urban areas of Hubei Province. The new Coronavirus continued to spread across China and at least 20 other countries during January (Li et al. 2020), prompting the WHO to declare the outbreak a Public Health Emergency of International Concern on January 30th 2020. On February 11th, the WHO and the International Committee on Taxonomy of Viruses named the virus SARS-CoV-2 and the disease syndrome COVID-19, or Coronavirus Disease 2019 (World Health Organization 2020b). It was formally declared a pandemic on March 11th and as of July 14th, 2020, more than 13.1 million cases of COVID-19 have been reported across 188 countries and territories, resulting in more than 573,000 deaths. While SARS-CoV-2 is not as lethal as SARS-CoV (which was responsible for the SARS outbreak in 2003) which had an estimated case fatality rate of 11%, COVID-19 is still characterized by severe respiratory illness and causes significant mortality, especially among individuals over the age of 60 years and those with underlying chronic illnesses such as diabetes and hypertension (Li et al. 2020). Moreover, SARS-CoV-2 is highly transmissible with an estimated reproductive number (R_0) of between 2.76 and 3.25, and an incubation period of 5–6 days (ranging from 1 to 14 days) (Eurosurveillance Editorial Team 2020). While notable research on the highly pathogenic human CoVs, SARS-CoV and MERS-CoVs, has been underway since these viruses emerged in 2003 and 2012 respectively, there are no clinically approved vaccines or antiviral drugs available for either. Nevertheless, the basic research on these viruses is contributing to the current global efforts to develop effective therapeutic and preventive strategies for SARS-CoV-2/COVID-19. Currently many countries are controlling their epidemics by relying on the behavioural changes of their populations including, amongst others, self-isolation of symptomatic individuals; increased hand hygiene; physical distancing; increased home-working; and school and business closures. Tracking, testing and isolation approaches are also reducing the cases of COVID-19. However, while these strategies are effective in reducing the circulation of SARS-CoV-2, they are not sustainable, given their devastating economic and social impacts.

Importantly, SARS-CoV-2 can be transmitted from infected individuals without symptoms (Rothe et al. 2020) and this, together with SARS-CoV-2's ability to cause pandemic disease within weeks of its emergence, indicates that control of this viral infection will be challenging without the availability of a vaccine.

3 Vaccine Platform Strategies

Key to the development of successful and effective vaccines is the design of an antigen delivery system that optimizes antigen presentation and induces broad protective immune responses. Recent advances in vector delivery technologies, immunology and basic virology have led to a deeper understanding of the molecular and cellular mechanisms by which vaccines should stimulate the adaptive immune response. This section discusses some current vaccine strategies that are being developed for safe and effective vaccination, based on recombinant technology involving the manipulation of virus biology for vector delivery of antigens, nucleic acid vaccines and self-disseminating vaccines (Table 1). Other types of vaccine platform such as inactivated whole viruses and subunit vaccines are also discussed. These include some of the classic vaccine methodologies, many of which are the basis of human vaccines used commercially for some time.

4 Viral Vector Technology

Advances in recombinant virology and virus reverse genetics have provided key insights into the replication and pathogenesis of a wide range of viruses. Notably these have facilitated the development of viral vectors for protein expression and vaccination. To date, several virus families have been exploited as vectors (Hewson 2000; Liu 2010; Small and Ertl 2011; Rollier et al. 2011) including many for vaccination (Ljungberg and Liljeström 2015; Gilbert and Warimwe 2017; Sánchez-Sampedro et al. 2015; Clarke et al. 2016) (Fig. 1). A basic advantage of viral vectored vaccines is that the choice antigen is expressed in the context of an active heterologous viral infection strategy, which stimulates the full gamut of innate immune responses required for the development of adaptive humoral and T cell-mediated immunity (Liu 2010).

An important aspect of a virus-vectored vaccine for emerging viruses is that the characteristics, type and intensity of the immune response, as well as safety considerations and manufacturing techniques are determined predominantly by the vector rather than the pathogen. Therefore, developing and testing a vaccine against a newly discovered virus can be significantly shortened by the use of a viral vector platform which has an extensive record of safety and efficacy.

4.1 Replication Competent (Live—Attenuated) Viral Vectors

Vesicular stomatitis virus (VSV): VSV which is a negative sense RNA virus of the Rhabdoviridae family, has become a prominent replication competent vaccine vector platform (Rose and Clarke 2015). VSV is non-pathogenic in humans and has

Table 1 Advantages/disadvantages of listed vaccine platform technologies

Platform	Benefits	Disadvantages
Viral vector • Replication competent	Development of good humoral and T cell responses Several platforms available. Simple production via virus culture	Selective pressures may result in reduced expression or loss of heterologous antigens or in some case reversion to virulence
Viral vector • Replication defective	Safe. Development of good humoral and T cell responses	Production may require additional steps and cell lines
Virus-like particles	Safe, amenable to regulatory rules and good potential to be quickly licensed	Production costs may be high and require multiple steps. Limited immunogenicity. Not possible for all viruses
Naked nucleic acid • DNA	Rapid synthesis, construction and production	Low levels of induced immune responses. Some safety concerns over integration
• RNA	Rapid synthesis, construction and production. Safe	Lipid vesicular encapsulation has been cumbersome for routine manufacture and requires development
Synthetic peptides	Safe. Ease of production and storage	Uncertain immunogenicity linked to target populations
Inactivated viruses	Safe. Well defined regulatory framework and good potential for rapid licensure	Requirement for high containment to culture before inactivation. Questionable (but sometimes effective) immunogenicity

an inherent ability to elicit strong cellular and humoral immune responses. One of the most useful aspects of this virus vector platform is its almost promiscuous ability to assemble recombinant VSV (rVSV) with many different types of heterologous viral glycoproteins. The platform is designed such that the VSV glycoprotein spike (G) protein is replaced with a heterologous envelope glycoprotein from, for example, a virus such as Zaire ebolavirus (ZEBOV), while this arrangement renders the rVSV replication competent, the recombinant viruses are generally highly attenuated (Roberts et al. 1999). The rVSV-ebolavirus glycoprotein recombinant (rVSV-ZEBOV-GP) was initially developed in 2005 (Jones et al. 2005) and became a useful vaccine candidate in the light of the EVD outbreak in 2014–15. Marketed under the brand name "Ervebo" this vaccine was approved for medical use in the European Union and United States in 2019 and has since been used in EVD outbreaks in the Congo to break chains of infection (FDA 2019).

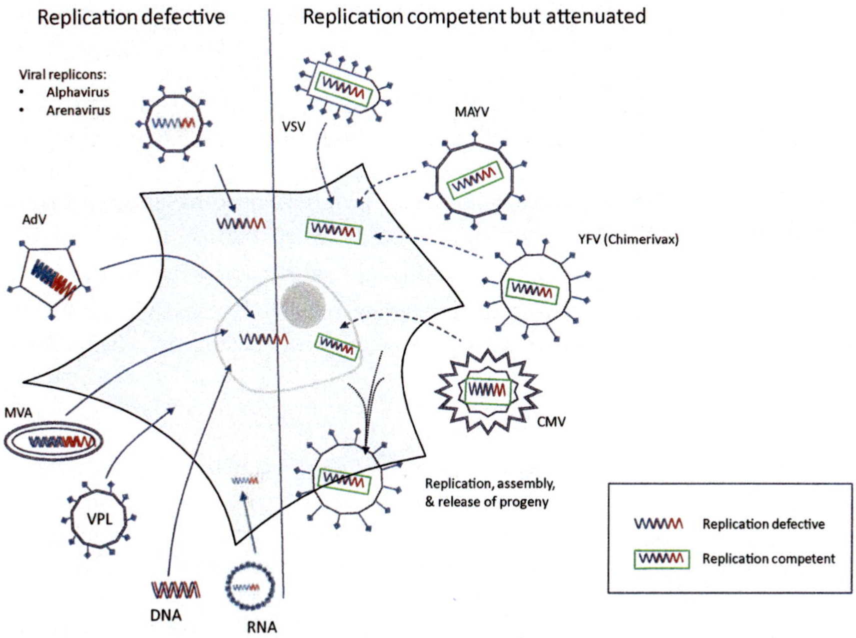

Fig. 1 Virus vector platform technologies. Replication competent but attenuated virus vectors (RHS) deliver heterologous antigen targets resulting in the induction of cellular and humoral responses. Their complement of genes enables a full round of replication and assembly of progeny virus which can amplify and spread the vaccine effect to new susceptible cells. While these platforms are based on highly attenuated viral backgrounds, replication competence may lead to the development of mutations and reversion to virulence. Replication defective vectors (LHS) support effective cellular entry and a single round of expression of the target gene/ antigen; they result in effective induction of cellular and humoral responses. They are unable to generate new infectious progeny and are considered safer than replication competent vectors. AdV: Adenovirus; MVA: Modified vaccinia virus Ankara; VPL: Virus particle; CMV: Cytomegalovirus; YFV: Yellow fever virus; MAYV: Mayaro virus; VSV: Vesicular stomatitis virus

Yellow fever (YF) virus: In 1937, the attenuation of a YF virus via successive rounds of serial passage in tissue culture led to the development of the YF vaccine termed 17D. The impact of this successful vaccine was recognised by the award of a Nobel Prize in 1956 (Norrby 2007) and it has been widely adopted as a live attenuated vaccine for human immunization for over 70 years (Monath and Barrett 2003). Based on the utility of this vaccine, an infectious cDNA clone of the YF17D virus (Rice et al. 1989) has enabled the development of a 17D platform that can be used to drive antigen delivery of pathogens of interest (Bonaldo et al. 2014). The technology (licensed as ChimeriVax™) is well suited to related flaviviruses, and successful recombinants have been constructed through a simple swap of the PrM/M-E genes of YF17D for the same membrane envelope antigens of other emerging flaviviruses such as Japanese encephalitis, Dengue and West Nile. This strategy results in live-attenuated recombinant virus vaccines which can be efficiently delivered in the context of the safety profile afforded by the 17D non-structural genes (Guy et al. 2011). A licensed vaccine for Japanese encephalitis (ChimeriVax™-JE) using this technology has been developed by Sanofi Pasteur, Lyon, France.

Other replication competent platforms: The ease of direct manipulation of viral genomes together with a growing understanding of their biology has led to the development of more precise ways of attenuating virus pathogens. Manipulated viral genomes can be generated as vaccine candidates with the potential to induce strong and specific immunogenicity, without causing disease. Many of these approaches are at the research development stage where features of induced immunity need to be balanced with safety. Of note is a recently developed vaccine candidate for Rift Valley fever virus (RVFV), in which a viral virulence factor was deleted, resulting in a highly attenuated but immunogenic replicating virus (Brennan et al. 2011; Volkova et al. 2008). A similar recombinant approach has been used to attenuate the emerging and neglected pathogen, Mayaro virus (MAYV), by swapping the sub-genomic promoter of this alphavirus for an internal ribosome entry site (Volkova et al. 2008; Rossi et al. 2013), which reduces the expression of MAYV structural proteins. While this arrangement makes the virus unable to infect mosquito cells, replication in mammalian cells is still possible, resulting in high immunogenicity. Similar studies on infectious clones of other viruses (Burns et al. 2006) have demonstrated that the genome-wide de-optimization of codon usage dramatically reduces gene expression and can be used to attenuate otherwise pathogenic traits. This strategy has been used for the prototype arenavirus lymphocytic choriomeningitis virus (LCMV) (Cheng et al. 2015); the study showed attenuation in an otherwise fatal mouse model of LCM disease and the ability to induce protective immunity, together with a maintained and robust ability to multiply in cell culture. Work to underpin confidence in the genetic stability of such attenuated viruses is critical before further consideration can be given to using this approach for clinical interventions. For example, the arenavirus polymerase has an inherent high error rate; reversion to virulence is a distinct possibility after multiple rounds of replication and safeguards would be essential to reduce this risk.

Live attenuated vaccines remain a focus of research because they have a track record of inducing protective immunity and a good safety profile. However, in rare circumstance some live attenuated vaccines (e.g. YF vaccine) have caused disease and this is always a concern for live replication competent vaccines.

4.2 Replication Defective Approaches

These have attracted considerable interest because they introduce increased levels of safety. They avoid possibilities of the virus vector reverting to wild type or virulence and focus on the delivery and expression of choice antigens.

Adenovirus vectors: Recombinant adenoviruses have been adopted as promising tools for antigen delivery and vaccine efficacy for several years (Tatsis and Ertl 2004). Deleting the E1 gene from the adenoviral genome and supplying it in *trans* from a packaging cell line allows replication-deficient, recombinant adenovirus to be produced, with the novel heterologous antigen gene of interest inserted at the E1 locus of the manipulated adenovirus genome. It is not expressed as a protein on the surface of the viral vector. Once the recombinant adenovirus vector enters cells of the vaccinated individual the heterologous gene is expressed to high levels. The first adenoviral vaccine vectors to be developed were based on human adenovirus serotype 5 (HAd5). However, it was found that pre-existing anti-HAd5 antibodies which are present in a large proportion of the human population could significantly reduce immune responses (Fausther-Bovendo and Kobinger 2014). Early setbacks were resolved by adopting simian adenoviruses as vaccine vectors and several different replication-deficient vaccine vectors (Dicks et al. 2012) have more recently been developed from chimpanzee adenoviral vectors. The platform has been applied to emerging pathogens such as Ebola virus, RVFV and MERS-CoV; some of these vaccine vectors are in clinical trials, including accelerated clinical trials with a vaccine originally based on MERS-CoV rapidly developed for SARS CoV-2/COVID-19 (van Doremalen et al. 2020).

Modified Vaccinia Virus Vaccine Ankara: Modified Vaccinia virus vaccine Ankara (MVA) is licensed as a third-generation vaccinia type vaccine against smallpox and serves as a potent vector system for development of new candidate vaccines against a range of infectious diseases, including those caused by emerging pathogens. Historically, MVA was developed by serial tissue culture passage in primary chicken cells of vaccinia virus strain Ankara, and was used clinically to avoid the undesirable side effects of conventional vaccinia vaccines (Brown et al. 1986; Stickl et al. 1974). Adapted to growth in avian cells, MVA does not replicate in mammalian hosts and lacks many of the pathogenic viral immune evasion genes (Goossens et al. 2013). The features of MVA, such as its capacity to accommodate large gene inserts (Smith and Moss 1983), thermostability for application in remote regions without established cold chain facilities (Alcock et al. 2010), ease of inexpensive manufacture to GMP standards, and established regulatory package for development as an Investigational New Drug, make the recombinant MVA platform (Volz and Sutter 2017) highly versatile. In the context of emerging infections, the recombinant MVA platform has shown encouraging preclinical efficacy against Ebola, Zika, Chikungunya (Nagata et al. 2018) and CCHF (Buttigieg et al. 2014) viruses. Additionally, MVA elicits a strong immunological response against a range of other Orthopoxviruses (OPXVs) including Variola, and vaccines based on this platform can be considered as providing added value, simultaneously protecting against possible OPXV emergence, since human immunity to OPXVs is low (after the cessation of the Smallpox vaccination campaign).

Virus replicon systems: RNA replicons derived from either positive or negative-strand RNA virus genomes embody manipulated and disabled virus genomes packed inside virus particles that are non-pathogenic and unable to revert to virulence. Once introduced into a susceptible cell, by way of the entry features of the virus particle, autonomous RNA replication of the replicon results in the high level, cytosolic expression of recombinant heterologous antigens, stimulating both the humoral and cellular arms of the immune system. Replicon vaccine approaches have closely followed technical developments to genetically manipulate viral genomes. For RNA viruses, replicons based on positive stranded picornaviruses were some of the first (Andino et al. 1994), these were followed by those based on alphaviruses (Frolov et al. 1999) and then negative strand RNA viruses (Lundstrom 2016).

While a series of different vaccine replicon systems are available, new capabilities with negative strand viruses have opened up opportunities to protect against a wider range of emerging viruses. Recent developments (Kainulainen et al. 2018) include the construction of LASV replicons packaged into LASV-like particles which allow a single round of replication and are able to confer protection against an otherwise lethal challenge of LASV in a guinea pig model of disease. This system enables the scalable propagation of replicon particles in a way that aims to combine the safety of replicon-delivered LASV antigens with the convenience of simply and rapidly producing an attenuated virus. Similar replicon approaches have been used to develop promising replicon-based vaccine candidates for Ebola virus (Halfmann et al. 2009) and RVFV (Dodd et al. 2012), although these have not yet entered in clinical trials. Replicon approaches have the potency of live attenuated vaccines but are inherently safer since their design ensures a single cycle of replication; this contrasts with fully replicating, live attenuated vaccine viruses.

5 Virus-Like Particle Antigens

Interest in the use of virus-like particles (VLPs) as vaccine candidates stem from the inherent ability of virus protein assemblies to present ordered and highly antigenic structures to the immune system. At the same time, unlike live attenuated viruses, they lack a viral genome, potentially yielding safer vaccines as there is no viral sequence that can cause disease. They induce strong B-cell responses in the absence of adjuvant by efficiently cross-linking specific receptors on B cells (Roldão et al. 2010), they can also trigger T cell-mediated responses (Buonaguro et al. 2010). The basis of ordered viral self-assembly from protein subunits that is noted in many different virus families provides the foundation for work with VLPs, with more than 30 different viruses that infect humans or other animals being identified as able to produce VLPs. They are structurally diverse, having single or multiple capsid proteins, or a lipid envelope. VLPs can elicit a protective response without requiring multiple booster shots; and so, can significantly reduce vaccine costs. To date, VLP-based vaccines for human papilloma virus (HPV), hepatitis B virus (HBV), and hepatitis E virus (HEV) have been licensed and are commercially available worldwide (Jain et al. 2015). Several VLP-based vaccine candidates for human diseases are under clinical development including those directed against Norwalk virus, Ebola and Marburg viruses and hepatitis C virus. VLP vaccines combine many of the immunogenic advantages of inactivated vaccines with the safety advantages of recombinant protein/antigen subunit vaccines.

6 Synthetic Peptides

Synthetic peptide-based epitope-vaccines (EVs) make use of short antigen-derived peptide fragments that can be presented either to T cells or B cells (Purcell et al. 2007). EVs offer several advantages over other forms of vaccines, particularly with regards to safety, ease of production, storage and distribution without cold chain issues. They also offer the opportunity to vaccinate against several pathogens, or multiple epitopes from the same pathogen. However, drawbacks include poor immunogenicity and the restriction of the approach to patients of a given tissue type (human leukocyte antigen (HLA) haplotype) (Paris et al. 2012); as such, they need to be tailored to accommodate the natural variation in HLA genes. Although initially this was thought to be a major impediment, new technologies have made this personalized-medicine approach feasible (Singh-Jasuja et al. 2004; Sirskyj et al. 2011). Recently, bioinformatics tools have been developed to identify putative CD4 + T-cell epitopes, mapped to the surface glycoproteins of the emerging viruses LASV, NipV and Hendra (Oyarzun et al. 2015). While these vaccine candidates still need to be experimentally tested, the approach represents an interesting and novel strategy that shows promise for vaccination and which could also address immunity in particular target populations.

7 Inactivated Viruses

The induction of immune responses by the delivery of inactivated pathogens has been a standard and successful vaccination approach for many years, and licensed, inactivated vaccines for diseases such as poliomyelitis (Polio vaccines: WHO position paper 2016) and rabies (Denis et al. 2019) are commercially available. The long history of this approach is underpinned by a well-defined regulatory framework that can be readily applied to new disease targets (Elmgren et al. 2013). The major challenge for the inactivated virus approach is that infection is not established and therefore a full adaptive immune response is generally not achieved. However, because of the absence of living pathogens, these types of vaccines are safe, and a basic capability to prepare such vaccines especially for emergency use, might be worthwhile as a stop-gap whilst alternative longer-term approaches are developed. In this regard, studies of virus inactivation with X-ray radiation (as a simple and cheap alternative to gamma irradiation by the use of radioactive isotopes), which maintain the tertiary antigenic structures of virus particles while destroying infectivity have shown useful promise for a range of applications including vaccination (Afough 2019).

8 Nucleic Acid Vaccines (Synthetic Vaccinology)

8.1 Nucleic acid vaccines have emerged as a safer alternative to standard live and inactivated vaccines for treating human and animal infections (Delany et al. 2014). They exhibit several advantages over the strategies mentioned above in terms of safety, stability, ease of manufacturing, and immunogenicity (Liu et al. 2006). They offer potential advantages for vaccination against emerging viruses in that nucleic acids expressing a viral antigen can be produced rapidly. Furthermore, once effectively delivered the antigen is expressed intra cellularly and induces both humoral and cell-mediated immune responses. Large quantities of DNA can be produced in a short time at low cost, and plasmid DNA preparations are more stable than other types of vaccines, which is desirable for a vaccine that may be used in remote areas. However, the main limitation in the development of DNA vaccines is their low intrinsic immunogenicity. Work to improve this has focused on optimising delivery approaches with the use of gene guns or electroporation to target immune effector cells, and the use of potent adjuvants. DNA vaccines are also frequently used in combination with other vaccine platforms in heterologous prime-boost strategies. A DNA vaccine against West Nile virus is currently licensed to immunize horses (Powell 2004) and has undergone Phase I clinical trials in humans (Ledgerwood et al. 2011). DNA vaccines have also been evaluated as candidates against many emerging viruses, including Ebola virus (Martin et al. 2006), RVFV (Boshra et al. 2011), Dengue virus (Porter et al. 2012), and Chikungunya virus (Mallilankaraman et al. 2011).

8.2 Vaccination with mRNA has various advantages over other nucleic acid-based approaches; RNA is directly taken up into the cytosol resulting in rapid antigen expression, simple synthetic production and the lack of chromosomal integration make mRNA a safe and fully controllable delivery platform. Vaccines based on mRNA have also been shown to induce a robust innate immune response resulting in chemokine and cytokine production at the inoculation site (Edwards et al. 2017), which may play an important role in successful immunization. Strategies for RNA vaccine delivery centre on the protection and stability of mRNA and can encompass the injection of naked mRNA, carrier-mediated delivery, electroporation, gene gun complexes, nanoparticle carriers and formulations with liposomes and protamine

(Iavarone et al. 2017). Certain issues relating to potency and tolerability in humans will be governed by the defined nature of the desired mRNA vaccine and the delivery system utilized. Nevertheless, recent mRNA vaccine designs have shown improved stability and protein translation efficiency, demonstrating good induction of robust immune responses (Pardi et al. 2018). The potential step change in vaccinology that mRNA vaccines can offer has led to their description as a "disruptive technology", with flexible production being based only on nucleic acid sequence information about the desired antigen. They are an attractive vaccine platform technology offering useful advantages over other types of approach.

9 Self-disseminating Wildlife Reservoir Vaccines

Self-disseminating vaccines provide an unconventional vaccine approach in which an emerging (or potentially emerging) virus is immunologically contained within its non-human animal reservoir. This approach is designed to exploit the ability of the viral vector technology discussed previously, to spread through animal host populations, so avoiding the need for direct inoculation of every animal. At the same time this approach should reduce the transmission opportunities of the potential human pathogen in the reservoir population. In this way, vaccinations of a limited number of initiator animals can be used to introduce the vaccine into a target population. Such a vaccine can be engineered to express target antigens from the emerging pathogen of interest, and so its transmission from vaccinated to non-vaccinated animals results in the coordinated spread of specific immunity against the emerging pathogen throughout the targeted animal population.

A proof of principle, for a disseminating vaccine against an animal pathogen, was first demonstrated with an attenuated myxoma virus (Bárcena et al. 2000; Spiesschaert et al. 2011) which spread in a rabbit population and protected against myxomatosis. Extensions of this work were developed in relation to rodent-born Hantaviruses in the mid-2000s. Hantaviruses are medically important zoonotic agents responsible for bouts of emerging and re-emerging disease in many parts of the world. This approach focused on *Sin Nombreorthohantavirus* (SNV) in its rodent reservoir, the deer mouse (*Peromyscusmaniculatus*), and used an engineered recombinant vaccine vector based on cytomegalovirus (CMV) (which causes a benign but transmissible infection in the host) expressing the SNV envelope glycoprotein G1 recombinantly (Rizvanov et al. 2006). A related approach has been developed to interrupt the zoonotic transmission of Ebola virus (Tsuda et al. 2015), in this case disseminating CMV vaccines specific to great apes and expressing EBOV antigens (Ghai 2014). While the principle goal of this work is to protect great apes themselves from Ebola virus disease, the strategy also protects humans, since approximately 30% of human Ebola virus outbreaks result from the direct handling of infected ape carcasses (Murphy et al. 2016). The applicability of this strategy to the wide range of emerging viruses that bats can reservoir has not been investigated. However, this could be an important area of research for limiting the spill-over into other animals and ultimately to humans of pathogens such as such as Nipah virus and the closely related SARS/MERS-CoV type viruses, now we know of their devastating pandemic capability. Assessment of the risks associated with introducing live transmissible vaccines into wildlife populations is a crucial area of study, and concerns that such live vaccines could evolve into pathogens need to be alleviated for the approach to gain traction. Such work is underway with studies focused on engineering a wildlife reservoir vaccine that is weakly transmitting, such that its reproduction number is below 1 ($R_0 < 1$). In this case, transmission chains would be kept short, so that the vaccine could not be ultimately maintained within the population. Indeed, mathematical models (Nuismer et al. 2016) have been developed to illustrate the value of this approach, which could offer exciting new ways of controlling disease.

10 General Perspectives and Outlook in the Context of COVID-19

Vaccines play a pivotal role in host protection against infectious diseases and have significantly reduced mortality worldwide. However, many vaccine candidates for emerging diseases have failed to make it into clinical development. This is perhaps surprising given the breadth of vaccine technology available and the nature of many of the diseases in question. A case in point is Lassa fever, a viral haemorrhagic illness endemic to many parts of West Africa, responsible for more than 300,000 cases of serious disease and approximately 3000 deaths each year since its discovery over 50 years ago. It is often in the headlines, being the most commonly exported viral haemorrhagic fever to other territories, including the UK which has received a disproportionate share of traveller-related cases (each incident placing substantial burden on public health resources).

In 2018, Lassa fever virus (LASV) caused an unusually large increase in cases in Nigeria, which led the WHO to classify it as a grade 2 public health emergency. However, despite the high burden of Lassa fever, its common exportation to other counties and public attention, no vaccines against this disease have so far been approved. Interestingly however, a range of effective pre-clinical candidates have been developed using multiple approaches over the last three decades and include attenuated vaccines (Lukashevich et al. 2005), replication competent vaccines, (Geisbert et al. 2005; Jiang et al. 2011; Clegg and Lloyd 1987), non-replicating LASV vaccines (Kainulainen et al. 2018), a rationally designed live attenuated vaccine (Cheng et al. 2017), and DNA vaccines (Cashman et al. 2013). Additionally, many of these platforms have shown efficacy in animal models including non-human primates (NHPs) (Geisbert et al. 2005; Lukashevich et al. 2008; Cashman et al. 2017). These data illustrate that a plethora of vaccine platforms have the potential to yield effective Lassa fever vaccines and so the lack of a vaccine must be due to other factors, such as economics and safety issues—

conceivably connected with the growing burden of regulatory thresholds required for human medical interventions. Thus, bringing a LASV vaccine and, by analogy, other potential vaccines for emerging diseases from the bench to the clinic is very difficult, for non-technical reasons. With these issues in mind, there has been renewed interest from multiple international agencies such as the WHO and the Coalition for Epidemic Preparedness Innovations (CEPI), to better progress human vaccines for key emerging pathogens (http://www.who.int/medicines/ebola-treatment/WHO-listof-top-emerging-diseases/en/; CEPI 2017). Key investment and focused attention from a broad range of experts on the many aspects of vaccine development is now supporting a step change in the way vaccines are developed and manufactured.

Currently, the world must develop appropriate controls and interventions to stem the COVID-19 pandemic, and vaccine research has become a priority. Stimulated by international alliances and government efforts to urgently organise resources, many programmes are underway and are utilizing a range of platform technologies (Le Thanh et al. 2020; Gates 2020; Fauci et al. 2020), with the most advanced candidates having already moved into clinical trials (Table 2).

This current vaccine effort is unique in terms of its scale and speed. Moreover, given the urgent need, there is an indication that human vaccines could be available under emergency use or similar regulations by autumn 2020 (H.H.S. USA 2020). This would represent a fundamental step change from the traditional vaccine development pathway, which generally takes over 10 years (Gouglas et al. 2018). Importantly this new accelerated approach to vaccine development will require novel concepts, including parallel and adaptive development phases, pioneering regulatory processes, and scaling of manufacturing capacity (Berkley 2020).

Traditional vaccine development standards cite attrition rates for vaccine candidates at more than 60% (Fauci et al. 2020), which explains the need to try out several different approaches for any given disease. The approaches being applied to COVID-19 vaccine development (which involve a new virus target, a range of vaccine platforms and novel development pathways), are likely to increase the risks associated with delivering a licensed vaccine. Careful evaluation of the effectiveness and safety of each vaccine candidate is required at each step. Additionally, such assessments of efficacy require the ready availability of standardised COVID-19-specific animal models, which are being developed (W.R.D. Blueprint 2020), including ACE2-transgenic mice, hamsters, ferrets and NHPs. Furthermore, experimental containment level 3 facilities are needed for animal studies involving live virus challenges, and the demand for these facilities is likely to require international coordination to ensure that sufficient laboratory capacity is available.

Table 2 Selection of vaccine candidates for COVID-19 in clinical trials in May 2020. Bold text indicates 1st date of approval for emergency use and country list of subsequent approval as of 23rd January 2021. A comprehensive list of COVID-19 vaccine candidates, trails and approved vaccines is available at https://covid19.trackvaccines.org/vaccines/

Vaccine candidate	Vaccine characteristics	Developer	Study ID and location	Status (May 2020) estimated completion date
mRNA-1273	LNP-encapsulated mRNA vaccine encoding SARS-CoV-2 spike (S) protein	Moderna. NIAID	5 Trials in 1 country NCT04283461 NCT04405076 NCT04649151 NCT04470427	Phase I FDA approval for Phase II Estimated completion: August 2021 Phase III planned July 2020 **FDA approval for emergency use in individuals >18 years. December 18 2020** **Approval in 37 countries (AUT, BEL, BGR, CAN, HRV, CYP, CZE, DNK, EST, FIN, FRA, DEU, GRC, HUN, ISL, IRL, ISR, ISR, ITA, LVA, LIE, LTU, LUX, MLT, MNG, NLD, NOR, POL, PRT, ROU, SEY, SVK, SVN, ESP, SWE, CHE, GBR, USA)**
DNA(INO-4800)	DNA plasmid vaccine encoding SARS-CoV-2 S protein	Inovio	NCT04336410 Missouri and Pennsylvania	Phase I Estimated completion: July 2021 Phase II/III planned Summer 2020
Rec-Adnovirus ChAdOx-1nCoV-19 AZD1222	Replication defective vector encoding SARS-CoV-2 S	Oxford University/ AsrtaZeneca	16 Trails in 12 countries CTRI/2020/08/027170 EudraCT 2020-001228-32, NCT04400838 ISRCTN89951424, NCT04536051 NCT04516746	Phase I complete Phase II/IIIb May 2020 Phase III Estimated completion: July 2021 **MHRA approval for emergency use in individuals >18 years. December 30 2020** **Approval in 11 countries (ARG, BRA, DOM, SLV, IND, IRQ, MEX, PAK, GBR)**

(continued)

Table 2 (continued)

Vaccine candidate	Vaccine characteristics	Developer	Study ID and location	Status $^{(May\ 2020)}$ estimated completion date
LV-SMENP-DC	Replication defective lentiviral vector encoding SARS-CoV-2	Shenzhen Geno-Immune Medical Institute	NCT04276896 China, Guangdong	Phase I/II Estimated completion: July 2023
Formalin-inactivated & alum-adjuvanted CoronaVac	Inactivated whole virus	Sinovac Research &Development Co., Ltd	11 Trails in 5 countries NCT04651790 NCT04456595 NCT04582344 NCT04617483 NCT04352608	Phase I/II Estimated completion August 2020 Phase III Estimated completion October 2021 **Chinese approval for emergency use: 31 August 2020** **Approval in 5 countries (BRA, CHN, CHL, IDN, TUR)**
BioNTech mRNA (BNT162b2)	Self-amplifying mRNA encoding SARS-CoV-2 S COL,	Pharmaceuticals GmbH + Pfizer Inc.	7 Trails in 8 countries EudraCT N0 2020-001038-36 NCT04368728 NCT04649021	Phase I/II Estimated completion August 2021 **MHRA^{UK} approval for emergency use in individuals >16 years: 2nd December 2020** **Approval in 54 countries (ARG, AUT, BHR, BEL, BGR, CAN, CHL, COL, CRI, HRV, CYP, CZE, DEK, ECU, EST, FIN, FRA, DEU, GRC, HUN, ISL, IRQ, IRL, ISR, ITA, JOR, KWT, LVA, LIE, LTU, LUX, MLT, MEX, MNG, NLD, NOR, OMN, PAN, SVK, SVN, ESP, SWE, CHE, ARE, GBR, USA)**
LNP-nCoVsaRNA	Self-amplifying mRNA encoding SARS-CoV-2 S	Imperial College London	ISRCTN17072692	Phase I Estimated completion July 2021

11 Conclusions

Disease emergence of viral pathogens into humans from wildlife reservoirs is a clear danger (Jones et al. 2008; Woolhouse et al. 2005; Field 2009; Kreuder Johnson et al. 2015; Olival et al. 2017), and perpetual challenge to global health. Unfortunately however, even recent technical developments in vaccinology have not adequately been able to meet this challenge. This is because of the development costs, required time and regulatory burdens associated with progressing vaccine candidates from the bench to the clinic.

Many novel vaccination strategies have the potential to specifically address the growing threat of new and emerging disease. The use of well-defined vaccine vector platforms, which have an extensive record of safety and efficacy against similar pathogens, can speed up development, validation and production (Table 1). Accordingly, the design and licensure for platform vaccine technologies will help to accelerate the development of new vaccines, since only the simple substitution of a new antigen gene into a vector platform is required. This allows manufacturers to move to a new target disease with minimal changes in chemistry, manufacturing, and controls. Thus, new vaccine development can focus on the safety and efficacy of the inserted antigen gene. In addition, the capability of platforms to target multiple pathogens helps to justify the investment required to build and maintain specialist manufacturing infrastructure for a given platform, because a single manufacturing facility can be ready to produce multiple vaccines at any time. Hence, while the antigen may change, the basic materials and methods such as cell substrates, production approaches, analytical quality control assays used as release criteria for products and purification processes, can all be defined under regulatory standards of Good Manufacturing Procedures. For these reasons, technological approaches that employ "plug and play" vaccine platforms such as viral vector technology and nucleic acid vaccines seem well suited to large-scale manufacture and clinical implementation and are likely to become increasingly important.

Accumulated evidence about zoonotic spill-over and the pandemic properties of emerging viruses (Jones et al. 2008; Woolhouse et al. 2005; Field 2009; Kreuder Johnson et al. 2015; Olival et al. 2017) demonstrate that future zoonotic pathogens will most likely be transmitted from a few key animal species in resource-poor areas of the world. In the past, effective containment of an emerging pathogen, before an epizootic has the opportunity to spill-over into the human population, has best been achieved by large-scale culling or mass vaccination (Jones et al. 2013; Swayne 2012) of animals. Nevertheless, the ability to contain even known emerging viruses such as Ebola virus in wildlife is currently not possible. Furthermore, the management of diseases that involve livestock, such as RVF and CCHF, pose additional problems (Indran and Ikegami 2012; Bukbuk et al. 2016) in that conventional vaccines are not suited for use in these situations. A major limitation being that conventional vaccination is the requirement for individual inoculation of each animal—a costly and impractical strategy for the target/reservoir species of those animals frequently involved in the emergence of high-risk pathogens (Jones et al.

2008). However, the development of self-disseminating wildlife vaccines could be useful in such circumstances.

Based on recent history, it is likely that new emerging pathogens will never have been seen before, which makes the task of identifying, controlling and preventing zoonoses a difficult and daunting goal. Surveillance work on known viruses, focusing on epizootics, that are or may become human pathogens to effectively prepare for and respond to escalating emerging threats, is useful and achievable. Nevertheless, predicting which animal pathogens will become established as globally significant causes of new and emerging human disease is currently guess work. In the future, a combination of surveillance work and artificial intelligence may provide early warning of potential pandemic events (Bogoch 2020) and give vaccine developers a head start. The exploitation of modern-day technological advances in virus research is essential to pre-emptively establish detailed information and comprehensive understanding of each family of viral pathogens. At the same time focused attention and investment in more infrastructure for surveillance and zoonotic spill-over events in developing countries and other identified hotspots of disease emergence is needed to expedite pathogen identification and catalyse the process of vaccine development using new platform technologies.

Emerging pathogens are described as representing one of the greatest risks to global health. However, as recent circumstances illustrate, they are potentially more far-reaching, impacting economies, society and conceivably civilisation. The rapid zoonotic emergence of SARS-CoV-2 and its ability to cause pandemic disease and economic chaos within weeks of its emergence is a working example. Indeed, the true cost of COVID-19 will be far greater than the direct health costs of treating cases; indirect costs of the disease will far outstrip the costs of testing, treating and hospitalising patients. The full extent of the economic damage will depend on how protracted the pandemic is, the steps that governments take to contain it, the impact of (and public adherence to) social distancing measures imposed by authorities, and the levels of economic support that governments and development agencies are willing to deploy during the pandemic's immediate impact and aftermath. However, easing the world out of lock down, living with the virus and ultimately controlling it through vaccination requires a deeper understanding of COVID-19. While we have learned a lot about SARS-CoV-2 in the months since it was first identified, knowledge about its interactions with the human immune system remains limited. What are the specifics of the immune system's response? How does this response affect the severity of illness? How long does immunity last? The answers to these questions will determine how effectively vaccination can control the disease and help the world to recover. Our response to SARS-CoV-2 will also highlight the amount of progress we have made, globally, in vaccine technologies.

References

Agua-Agum J, et al. After Ebola in West Africa—unpredictable risks, preventable epidemics. N Engl J Med. 2016;375(6):587–96.

Alcock R, et al. Long-term thermostabilization of live poxviral and adenoviral vaccine vectors at supraphysiological temperatures in carbohydrate glass. Sci Transl Med. 2010;2(19):19ra12.

Andino R, et al. Engineering poliovirus as a vaccine vector for the expression of diverse antigens. Science. 1994;265(5177):1448–51.

Afough, B. Inactivation with X-Ray radiation. 2019. Personal Communication.

Bárcena J, et al. Isolation of an attenuated myxoma virus field strain that can confer protection against myxomatosis on contacts of vaccinates. Arch Virol. 2000;145(4):759–71.

Berkley S. COVID-19 needs a big science approach. Science. 2020.

Bogoch II, et al. Pneumonia of unknown aetiology in Wuhan, China: potential for international spread via commercial air travel. J Travel Med. 2020;27(2).

Bonaldo MC, Sequeira PC, Galler R. The yellow fever 17D virus as a platform for new live attenuated vaccines. Hum Vaccin Immunother. 2014;10(5):1256–65.

Boshra H, et al. A DNA vaccine encoding ubiquitinated Rift Valley fever virus nucleoprotein provides consistent immunity and protects IFNAR(-/-) mice upon lethal virus challenge. Vaccine. 2011;29(27):4469–75.

Brennan B, et al. Creation of a recombinant Rift Valley fever virus with a two-segmented genome. J Virol. 2011;85(19):10310–8.

Brown F, Schild GC, Ada GL. Recombinant vaccinia viruses as vaccines. Nature. 1986;319 (6054):549–50.

Bukbuk DN, et al. Serological and virological evidence of Crimean-Congo Haemorrhagic Fever virus circulation in the human population of Borno state, Northeastern Nigeria. PLoS Negl Trop Dis. 2016;10(12):e0005126.

Buonaguro L, Tornesello ML, Buonaguro FM. Virus-like particles as particulate vaccines. Curr HIV Res. 2010;8(4):299–309.

Burns CC, et al. Modulation of poliovirus replicative fitness in HeLa cells by deoptimization of synonymous codon usage in the capsid region. J Virol. 2006;80(7):3259–72.

Buttigieg KR, et al. A novel vaccine against Crimean-Congo Haemorrhagic Fever protects 100% of animals against lethal challenge in a mouse model. PLoS One. 2014;9(3):e91516.

Cashman KA, et al. Enhanced efficacy of a codon-optimized DNA vaccine encoding the glycoprotein precursor gene of Lassa virus in a Guinea Pig disease model when delivered by dermal electroporation. Vaccines (Basel). 2013;1(3):262–77.

Cashman KA, et al. A DNA vaccine delivered by dermal electroporation fully protects cynomolgus macaques against Lassa fever. Hum Vaccin Immunother. 2017;13(12):2902–11.

CEPI. A global coalition to create new vaccines for emerging infectious diseases. 2017.

Cheng BY, et al. Development of live-attenuated arenavirus vaccines based on codon deoptimization. J Virol. 2015;89(7):3523–33.

Cheng BYH, et al. Development of live-attenuated arenavirus vaccines based on codon deoptimization of the viral glycoprotein. Virology. 2017;501:35–46.

Clarke DK, et al. Live virus vaccines based on a vesicular stomatitis virus (VSV) backbone: standardized template with key considerations for a risk/benefit assessment. Vaccine. 2016;34 (51):6597–609.

Clegg JC, Lloyd G. Vaccinia recombinant expressing Lassa-virus internal nucleocapsid protein protects guineapigs against Lassa fever. Lancet. 1987;2(8552):186–8.

Delany I, Rappuoli R, De Gregorio E. Vaccines for the 21st century. EMBO Mol Med. 2014;6 (6):708–20.

Denis M, et al. An overview of the immunogenicity and effectiveness of current human rabies vaccines administered by intradermal route. Vaccine. 2019;37(Suppl 1):A99–106.

Dicks MD, et al. A novel chimpanzee adenovirus vector with low human seroprevalence: improved systems for vector derivation and comparative immunogenicity. PLoS One. 2012;7(7):e40385.

Dodd KA, et al. Single-dose immunization with virus replicon particles confers rapid robust protection against Rift Valley fever virus challenge. J Virol. 2012;86(8):4204–12.

Edwards DK, et al. Adjuvant effects of a sequence-engineered mRNA vaccine: translational profiling demonstrates similar human and murine innate response. J Transl Med. 2017;15(1):1.

Elmgren L, et al. A global regulatory science agenda for vaccines. Vaccine. 2013;31(Suppl 2):B163–75.

Eurosurveillance Editorial Team. Updated rapid risk assessment from ECDC on coronavirus disease 2019 (COVID-19) pandemic: increased transmission in the EU/EEA and the UK. Euro Surveill. 2020;25(12).

Fauci AS, Lane HC, Redfield RR. Covid-19—navigating the uncharted. N Engl J Med. 2020;382(13):1268–9.

Fausther-Bovendo H, Kobinger GP. Pre-existing immunity against Ad vectors: humoral, cellular, and innate response, what's important? Hum Vaccin Immunother. 2014;10(10):2875–84.

FDA. First FDA-approved vaccine for the prevention of Ebola virus disease, marking a critical milestone in public health preparedness and response. 2019. https://www.fda.gov/news-events/press-announcements/first-fda-approved-vaccine-prevention-ebola-virus-disease-marking-critical-milestone-public-health. Accessed 28 May 2020; Press release.

Field HE. Bats and emerging zoonoses: henipaviruses and SARS. Zoonoses Public Health. 2009;56(6–7):278–84.

Frolov I, et al. Selection of RNA replicons capable of persistent noncytopathic replication in mammalian cells. J Virol. 1999;73(5):3854–65.

Gates B. Responding to Covid-19—a once-in-a-century pandemic? N Engl J Med. 2020;382(18):1677–9.

Geisbert TW, et al. Development of a new vaccine for the prevention of Lassa fever. PLoS Med. 2005;2(6):e183.

Ghai R. Ebola: outbreaks cause crisis for great apes and humans. Toronto: The Jane Goodall Institute of Canada; 2014.

Gilbert SC, Warimwe GM. Rapid development of vaccines against emerging pathogens: the replication-deficient simian adenovirus platform technology. Vaccine. 2017;35(35 Pt A):4461–4.

Goossens M, et al. Environmental risk assessment of clinical trials involving modified vaccinia virus Ankara (MVA)-based vectors. Curr Gene Ther. 2013;13(6):413–20.

Gouglas D, et al. Estimating the cost of vaccine development against epidemic infectious diseases: a cost minimisation study. Lancet Glob Health. 2018;6(12):e1386–96.

Grace D, Mutua F, Ochungo P, Kruska R, Jones K, Brierlry L, Lapar L, Said M, Herrero M, Duc-Phuc P, Thao N, Akuku I, Ogutu F. Mapping of poverty and likely zoonoses hotspots—Zoonoses Project 4 Report to Department for International Development, UK. 2012. International Livestock Research Institute, Kenya.

Guy B, et al. From research to phase III: preclinical, industrial and clinical development of the Sanofi Pasteur tetravalent dengue vaccine. Vaccine. 2011;29(42):7229–41.

Halfmann P, et al. Replication-deficient ebolavirus as a vaccine candidate. J Virol. 2009;83(8):3810–5.

Hewson R. RNA viruses: emerging vectors for vaccination and gene therapy. Mol Med Today. 2000;6(1):28–35.

H.H.S. USA. HHS Accelerates Clinical Trials, Prepares for Manufacturing of COVID-19 Vaccines. 2020. https://www.hhs.gov/about/news/2020/03/30/hhs-accelerates-clinical-trials-prepares-manufacturing-covid-19-vaccines.html.

Iavarone C, et al. Mechanism of action of mRNA-based vaccines. Expert Rev Vaccines. 2017;16(9):871–81.

Indran SV, Ikegami T. Novel approaches to develop Rift Valley fever vaccines. Front Cell Infect Microbiol. 2012;2:131.

Jain NK, et al. Formulation and stabilization of recombinant protein based virus-like particle vaccines. Adv Drug Deliv Rev. 2015;93:42–55.
Jiang X, et al. Yellow fever 17D-vectored vaccines expressing Lassa virus GP1 and GP2 glycoproteins provide protection against fatal disease in guinea pigs. Vaccine. 2011;29 (6):1248–57.
Jones SM, et al. Live attenuated recombinant vaccine protects nonhuman primates against Ebola and Marburg viruses. Nat Med. 2005;11(7):786–90.
Jones KE, et al. Global trends in emerging infectious diseases. Nature. 2008;451(7181):990–3.
Jones BA, et al. Zoonosis emergence linked to agricultural intensification and environmental change. Proc Natl Acad Sci U S A. 2013;110(21):8399–404.
Kainulainen MH, et al. Use of a scalable replicon-particle vaccine to protect against lethal Lassa virus infection in the Guinea Pig model. J Infect Dis. 2018;217(12):1957–66.
Karesh WB, et al. Ecology of zoonoses: natural and unnatural histories. Lancet. 2012;380 (9857):1936–45.
Kreuder Johnson C, et al. Spillover and pandemic properties of zoonotic viruses with high host plasticity. Sci Rep. 2015;5:14830.
Le Thanh T, et al. The COVID-19 vaccine development landscape. Nat Rev Drug Discov. 2020;19 (5):305–6.
Ledgerwood JE, et al. A West Nile virus DNA vaccine utilizing a modified promoter induces neutralizing antibody in younger and older healthy adults in a phase I clinical trial. J Infect Dis. 2011;203(10):1396–404.
Li Q, et al. Early transmission dynamics in Wuhan, China, of novel coronavirus-infected pneumonia. N Engl J Med. 2020;382(13):1199–207.
Liu MA. Immunologic basis of vaccine vectors. Immunity. 2010;33(4):504–15.
Liu MA, Wahren B, Karlsson Hedestam GB. DNA vaccines: recent developments and future possibilities. Hum Gene Ther. 2006;17(11):1051–61.
Ljungberg K, Liljeström P. Self-replicating alphavirus RNA vaccines. Expert Rev Vaccines. 2015;14(2):177–94.
Lukashevich IS, et al. A live attenuated vaccine for Lassa fever made by reassortment of Lassa and Mopeia viruses. J Virol. 2005;79(22):13934–42.
Lukashevich IS, et al. Safety, immunogenicity, and efficacy of the ML29 reassortant vaccine for Lassa fever in small non-human primates. Vaccine. 2008;26(41):5246–54.
Lundstrom K. Replicon RNA viral vectors as vaccines. Vaccines (Basel). 2016;4(4).
Mallilankaraman K, et al. A DNA vaccine against chikungunya virus is protective in mice and induces neutralizing antibodies in mice and nonhuman primates. PLoS Negl Trop Dis. 2011;5 (1):e928.
Martin JE, et al. A DNA vaccine for Ebola virus is safe and immunogenic in a phase I clinical trial. Clin Vaccine Immunol. 2006;13(11):1267–77.
Monath TP, Barrett AD. Pathogenesis and pathophysiology of yellow fever. Adv Virus Res. 2003;60:343–95.
Murphy AA, Redwood AJ, Jarvis MA. Self-disseminating vaccines for emerging infectious diseases. Expert Rev Vaccines. 2016;15(1):31–9.
Nagata LP, et al. Vaccinia-based vaccines to biothreat and emerging viruses. Biotechnol Genet Eng Rev. 2018;34(1):107–21.
Norrby E. Yellow fever and Max Theiler: the only Nobel Prize for a virus vaccine. J Exp Med. 2007;204(12):2779–84.
Nuismer SL, et al. Eradicating infectious disease using weakly transmissible vaccines. Proc Biol Sci. 2016;283(1841).
Olival KJ, et al. Host and viral traits predict zoonotic spillover from mammals. Nature. 2017;546 (7660):646–50.
Oyarzun P, et al. A bioinformatics tool for epitope-based vaccine design that accounts for human ethnic diversity: application to emerging infectious diseases. Vaccine. 2015;33(10):1267–73.
Pardi N, et al. mRNA vaccines—a new era in vaccinology. Nat Rev Drug Discov. 2018;17 (4):261–79.

Paris R, et al. HLA class II restriction of HIV-1 clade-specific neutralizing antibody responses in ethnic Thai recipients of the RV144 prime-boost vaccine combination of ALVAC-HIV and AIDSVAX(®) B/E. Vaccine. 2012;30(5):832–6.

Polio vaccines: WHO position paper—Mar 2016. Wkly Epidemiol Rec. 2016;91(12):145–68.

Porter KR, et al. Immunogenicity and protective efficacy of a vaxfectin-adjuvanted tetravalent dengue DNA vaccine. Vaccine. 2012;30(2):336–41.

Powell K. DNA vaccines—back in the saddle again? Nat Biotechnol. 2004;22(7):799–801.

Purcell AW, McCluskey J, Rossjohn J. More than one reason to rethink the use of peptides in vaccine design. Nat Rev Drug Discov. 2007;6(5):404–14.

Rice CM, et al. Transcription of infectious yellow fever RNA from full-length cDNA templates produced by in vitro ligation. New Biol. 1989;1(3):285–96.

Rizvanov AA, et al. Replication and immunoactivity of the recombinant Peromyscus maniculatus cytomegalovirus expressing hantavirus G1 glycoprotein in vivo and in vitro. Vaccine. 2006;24 (3):327–34.

Roberts A, et al. Attenuated vesicular stomatitis viruses as vaccine vectors. J Virol. 1999;73 (5):3723–32.

Roldão A, et al. Virus-like particles in vaccine development. Expert Rev Vaccines. 2010;9 (10):1149–76.

Rollier CS, et al. Viral vectors as vaccine platforms: deployment in sight. Curr Opin Immunol. 2011;23(3):377–82.

Rose JK, Clarke DK. Rhabdoviruses as vaccine vectors: from initial development to clinical trials. In: Biology and pathogenesis of Rhabdo- and Filoviruses. 2015. p. 199–230.

Rossi SL, et al. IRES-based Venezuelan equine encephalitis vaccine candidate elicits protective immunity in mice. Virology. 2013;437(2):81–8.

Rothe C, et al. Transmission of 2019-nCoV infection from an asymptomatic contact in Germany. N Engl J Med. 2020;382(10):970–1.

Sánchez-Sampedro L, et al. The evolution of poxvirus vaccines. Viruses. 2015;7(4):1726–803.

Singh-Jasuja H, Emmerich NP, Rammensee HG. The Tübingen approach: identification, selection, and validation of tumor-associated HLA peptides for cancer therapy. Cancer Immunol Immunother. 2004;53(3):187–95.

Sirskyj D, et al. Innovative bioinformatic approaches for developing peptide-based vaccines against hypervariable viruses. Immunol Cell Biol. 2011;89(1):81–9.

Small JC, Ertl HC. Viruses—from pathogens to vaccine carriers. Curr Opin Virol. 2011;1(4):241–5.

Smith GL, Moss B. Infectious poxvirus vectors have capacity for at least 25 000 base pairs of foreign DNA. Gene. 1983;25(1):21–8.

Spiesschaert B, et al. The current status and future directions of myxoma virus, a master in immune evasion. Vet Res. 2011;42(1):76.

Stickl H, et al. MVA vaccination against smallpox: clinical tests with an attenuated live vaccinia virus strain (MVA) (author's transl). Dtsch Med Wochenschr. 1974;99(47):2386–92.

Swayne DE. Impact of vaccines and vaccination on global control of avian influenza. Avian Dis. 2012;56(4 Suppl):818–28.

Tatsis N, Ertl HC. Adenoviruses as vaccine vectors. Mol Ther. 2004;10(4):616–29.

Tsuda Y, et al. A cytomegalovirus-based vaccine provides long-lasting protection against lethal Ebola virus challenge after a single dose. Vaccine. 2015;33(19):2261–6.

van Doremalen N, et al. ChAdOx1 nCoV-19 vaccination prevents SARS-CoV-2 pneumonia in rhesus macaques. bioRxiv. 2020. https://doi.org/10.1101/2020.05.13.093195.

Volkova E, et al. IRES-dependent replication of Venezuelan equine encephalitis virus makes it highly attenuated and incapable of replicating in mosquito cells. Virology. 2008;377(1):160–9.

Volz A, Sutter G. Modified vaccinia virus Ankara: history, value in basic research, and current perspectives for vaccine development. Adv Virus Res. 2017;97:187–243.

WHO. Prioritizing diseases for research and development in emergency contexts. 2020a. https://www.who.int/blueprint/priority-diseases/en/. Accessed 2020.

World Health Organization. WHO Director-General's remarks at the media briefing on 2019-nCoV on 11 Feb 2020. 2020b. https://www.who.int/dg/speeches/detail/who-director-general-sremarks-at-the-media-briefing-on-2019-ncov-on-11-february2020. Accessed 13 Feb 2020.

WHO publishes list of top emerging diseases likely to cause major epidemics. 2015. http://www.who.int/medicines/ebola-treatment/WHO-listof-top-emerging-diseases/en/.

Woolhouse ME, Haydon DT, Antia R. Emerging pathogens: the epidemiology and evolution of species jumps. Trends Ecol Evol. 2005;20(5):238–44.

W.R.D. Blueprint. COVID-19 animal models—summary of the progress made by the WHO COVID-19 modelling ad hoc expert working group. 2020. https://www.who.int/blueprint/priority-diseases/key-action/WHO-ad-hoc-Animal-Model-Working-Group_Summary.pdf.

Global Polio Eradication: Progress and Challenges

Ananda Sankar Bandyopadhyay and Grace Ruth Macklin

Abstract In 1988, the World Health Assembly passed a resolution to eradicate poliomyelitis, a disease which was endemic in 125 countries and paralyzed 350,000 children per year. Primarily through the large-scale use of oral poliovirus vaccine (OPV), as of 2020, two of the three wild poliovirus (WPV) serotypes (types 2 and 3) have been eradicated and transmission of type 1 remains uninterrupted only in Afghanistan and Pakistan. Unknown at the time of declaring eradication, vaccine-derived poliovirus (VDPVs)—rare strains that have genetically mutated from the poliovirus contained in OPV, pose a major challenge to eradication. In this chapter, we discuss the basic principles for eradication of poliomyelitis, the evolving epidemiology, with a focus on the recent period (2017–2020) of increasing cases of both WPV and VDPVs, and prospective scientific and strategic developments to achieve eradication in the near future.

Keywords Polio · Poliovirus · Poliomyelitis · Vaccination · Eradication · Endgame · Vaccine-derived poliovirus · Oral poliovirus vaccine · Inactivated poliovirus vaccine

1 Introduction

Eradication of human diseases continues to be a major area of interest for global health practice (Dowdle 1998). The idea of permanent reduction of global incidence of a disease to zero has triggered several attempts to identify pathogens suitable for eradication over the past several decades. However, with smallpox being the only

A. S. Bandyopadhyay (✉)
Bill and Melinda Gates Foundation, 1432 Elliot Ave W., Seattle, WA 98119, USA
e-mail: ananda.bandyopadhyay@gatesfoundation.org

G. R. Macklin
World Health Organization, 20 Avenue Appia, 1202 Genève, Switzerland

G. R. Macklin
London School of Hygiene and Tropical Medicine, Keppel Street, London WC17HT, UK

S. I. Ahmad (ed.), *Human Viruses: Diseases, Treatments and Vaccines*, https://doi.org/10.1007/978-3-030-71165-8_30

human disease to be ever eradicated, it is evident that even though the benefits of disease eradication are well acknowledged, the likelihood of success is complicated by a range of social, political and economic factors that go beyond the virologic or immunologic dimensions of disease control (Henderson 1987).

Galvanized by the success of smallpox eradication, understanding of biologic feasibility of poliovirus eradication and the initial success in controlling polio transmission in the World Health Organization (WHO) Region of Americas, the World Health Assembly passed a resolution in 1988 to eradicate polio by the year 2000 (World Health Assembly 1993). Twenty years from the target of eradication, several major milestones have been achieved: there has been a dramatic, more than 99.99% reduction in global incidence of poliomyelitis; five of the six WHO Regions have been certified free of wild poliovirus (WPV) transmission; and two of the three WPV serotypes have been declared as eradicated (type 2 and 3) (Chard et al. 2020).

Despite such progress, the aim to stop all transmission everywhere remains elusive. Circulation of type 1 WPV (WPV1) within the two endemic countries of Pakistan and Afghanistan has expanded in the recent years with cases of paralytic polio on a steady rise—marking a reversal of consistent trend of decline over the decades (Chard et al. 2020). Moreover, the expanding nature of outbreaks of circulating vaccine-derived polioviruses (cVDPVs) has become a major concern. Such outbreaks of cVDPVs, mostly caused by the type 2 strain of Sabin oral polio vaccine (OPV) have become particularly difficult to interrupt in the recent past with the elective, globally synchronized cessation of all routine use of type 2 containing OPV (Macklin et al. 2020).

Innovations on vaccine, operations and diagnostic fronts spiked in the past decade to address the evolving need of the endgame. Among several promising vaccine-related initiatives, a novel OPV type 2 (nOPV2) that is genetically more stable with less risk of reversion to neurovirulence compared to the current Sabin OPV type 2 is at the fore-front of new tools to be introduced in the program in the near-term (Van Damme et al. 2019a, b, c; World Health Organisation 2020a, b). Direct detection and novel sequencing methodologies hold the promise of making the outbreak response faster and more efficient. Understanding the social, political and economic dimensions of disease control in remaining areas of circulation remains a key priority to enable the existing and new tools to reach the last reservoirs of transmission.

2 Polio: Basic Characteristics

Poliomyelitis is a viral disease caused by infection with any of the three poliovirus serotypes. Infection results in one of four clinical outcomes: inapparent infection without symptoms (72%), minor illness (24%), aseptic meningitis (nonparalytic poliomyelitis) (4%), or paralytic poliomyelitis (<1%) (Sutter et al. 2018).

Humans are the only known natural host of poliovirus that are able to sustain transmission which typically occurs by person-to-person spread, through faecal-oral or oral-oral routes, with infants and young children driving virus transmission in most settings (Sutter et al. 2018). Polioviruses establish initial infection in the gastrointestinal tract and replicate in the oropharyngeal and intestinal mucosa. From these primary sites of replication, the virus is excreted in the faeces and saliva and can drain into lymph nodes and to the blood, causing a temporary viremia. Most human infections end at this stage and are asymptomatic, with some having minor disease comprising nonspecific symptoms such as sore throat, fever, and malaise. Rarely, however, the virus can spread to the central nervous system. Viral replication in motor neuron cells in the spinal cord cause cell destruction and flaccid paralysis in the muscles the neurons innervate. Infection of the brainstem in rare cases results in paralysis of respiratory muscles, which can be fatal (bulbar paralysis).

People remain most infectious immediately before and two weeks after infection, although virus is typically excreted in the feces for around 3–6 weeks and 2 weeks in saliva (Sutter et al. 2018). The incubation period between infection and mild illness is 3–6 days, and from infection to onset of paralytic disease is usually 7–21 days (Sutter et al. 2018). In immunodeficient individuals, an inability to mount an immune response can lead to prolonged viral replication and shedding in faeces (Sutter et al. 2018).

Polioviruses are classified as enteroviruses and belong to the family Picornaviridae. The viral genome, a single-stranded plus-strand RNA is enclosed in a non-enveloped capsid composed of four viral capsid proteins: VP1, VP2, VP3, and VP4 (Kew et al. 2005). There are three individual and immunologically-distinct serotypes of polioviruses: serotype 1, serotype 2 and serotype 3, which differ in their capsid proteins.For each of the three poliovirus serotypes, are two broad categories of polioviruses: (1) wild polioviruses, or the naturally occurring strains not linked with the live attenuated vaccine viruses, and (2) vaccine-related polioviruses (Burns et al. 2014).

Vaccine-related viruses for each of the three serotypes are categorized based on the divergence of the genetic sequence in the VP1 gene from the original OPV strain: (1) Sabin-like, which have limited divergence from their parental OPV strains and are ubiquitous wherever OPV is used, and (2) vaccine-derived polioviruses (VDPV), whose higher level of divergence from their parental OPV strains (>1% [types 1 and 3] or >0.6% [type 2]) indicates prolonged replication (or transmission) of the vaccine virus (Kew et al. 2005; Burns et al. 2014).

VDPVs resemble wild polioviruses phenotypically, can cause paralytic polio in humans and have the potential for sustained circulation (Kew et al. 2005; Burns et al. 2014). The clinical signs and severity of paralysis associated with VDPV and wild poliovirus infections are indistinguishable. VDPVs are categorized as (1) circulating VDPVs (cVDPVs), when there is evidence of person-to-person transmission in the community; (2) immunodeficiency-associated VDPVs (iVDPVs), which are isolated from persons with primary immune deficiencies (PIDs)who have prolonged VDPV infections; and (3) ambiguous VDPVs (aVDPVs), which do not fit into the previous two types, and are typically either clinical isolates from persons with no known immunodeficiency or sewage isolates where the primary source may be unknown (Kew et al. 2005; Burns et al. 2014).

3 Polio: Control and Eradication

The key biological characteristics that indicated poliovirus could be eradicated were (a) absence of a persistent carrier state (b) virus spread is by person-to-person transmission, (c) active immunization interrupts virus transmission, (d) absence of any nonhuman reservoir hosts capable of sustaining virus transmission, and (e) finite virus survival time in the environment (Kew et al. 2005). The strategy to achieve polio eradication has been focused on high vaccination coverage of infants and young children through routine immunisation and supplementary immunization activities (SIAs) and sensitive surveillance to detect poliovirus.

4 Vaccines

Over the past six decades, two types of vaccines have been used to protect against polio. The inactivated poliovirus vaccine (IPV), developed by Salk, is an injectable vaccine consisting of all three poliovirus serotypes inactivated with formalin (Vidor 2018). The OPV, developed by Sabin, is an oral vaccine composed of live attenuated polioviruses (Sabin strains), and can be monovalent (mOPV, type-specific), bivalent (bOPV, types 1 and 3), or trivalent OPV (tOPV, all serotypes) (Sutter et al. 2018). The Sabin strains were developed by serial passage of the three poliovirus serotypes in nonhuman primates and in cultured primate cells until strains with reduced neurovirulence were obtained (Sutter et al. 2018). OPV has been extensively used in routine immunization and SIAs due to the ease of administration, ability to induce intestinal immunity (which is critical to limit faecal-oral transmission), low cost and ability to provide immunity through secondary exposure (Kew et al. 2005).

Despite its many advantages, the live attenuated vaccine strains in OPV (Sabin strains) may re-acquire neurovirulence leading to vaccine-associated paralytic poliomyelitis (VAPP) in the vaccine recipient or close contacts, or generation of VDPVs (Sutter et al. 2018; Kew et al. 2005). In settings of persistently low immunization coverage, VDPVs can circulate in the community and cause paralytic outbreaks of cVDPV (Sutter et al. 2018). The estimated average rate of VAPP is about 4.7 per million births globally (Platt et al. 2014). Immunocompromised individuals with B cell deficiencies are at the highest risk of VAPP with >3,200 times the risk of VAPP compared to the general population. Type 3 virus is the most commonly isolated virus in people with VAPP who do not have immunodeficiencies and type-2 virus is the most commonly isolated virus in those with immunodeficiencies (Platt et al. 2014).

The pattern of immunogenicity and factors affecting it vary between IPV and OPV (Sutter et al. 2018; Vidor 2018). Humoral immunity, assessed by serological responses and considered predictive of protection from paralytic poliomyelitis, is consistent across geographies and populations with IPV in contrast to OPV where per dose immunogenicityhas been low in developing countries, attributed to several factors including malnutrition and enteropathy (Sutter et al. 2018). On the other hand, maternally derived antibodies in early weeks of life are known to be the biggest risk factors for lack of vaccine take with IPV (Dayan et al. 2007). For induction of mucosal immunity, impact of IPV remains less clear compared to the proven effect of OPV (Vidor 2018; Onorato et al. 1991; Hird and Grassly 2012). Based primarily on OPV challenge studies, IPV is considered comparable to OPV in reducing oro-pharyngeal excretion. However, it is inferior to OPV in inducing primary intestinal mucosal immunity (Macklin et al. 2019; Bandyopadhyay et al. 2018).

Given the understanding that fecal—route drives the transmission in areas with persistent poliovirus circulation, the lack of a meaningful impact on primary intestinal mucosal immunity is considered a critical limitation of IPV for the purposes of its use for polio eradication. Increasing antigen content or dose alone does not appear to have any favorable impact on IPV induced intestinal immunity although higher doses do translate into a rise in titers of serum neutralizing antibodies (Macklin et al. 2019; Saez-Llorens et al. 2016; Saleem et al. 2018). In contrast, several recent studies demonstrated a measurable reduction in virus excretion when IPV was administered to OPV-primed children where the impact on intestinal immunity is comparable to or no less than what was observed with an additional dose of OPV (John et al. 2014; Jafari et al. 2014; Gamage et al. 2018). This is encouraging for wider use of IPV beyond routine immunization and as an extension, analyses of acute flaccid paralysis and environmental surveillance data from Nigeria and Pakistan showed that compared to the traditional approach of OPV-only SIAs, a combined IPV—OPV campaigns had a bigger impact on interrupting VDPV and WPV transmission respectively (Grassly et al. 2018; Shirreff et al. 2017).

Given the near ten-fold cost differential per dose between IPV and OPV, several initiatives have focused on feasibility of dose sparing options such as administering a fractional dose (1/5th of the normal dose) through intradermal (ID) or more recently, intramuscular (IM) route (Resik et al. 2020; Okayasu et al. 2017). Overall, based primarily on the age at first dose and number of doses administered, fractional dose administration of IPV has generally shown encouraging results for seroconversion (Macklin et al. 2019). The difficulties in ID administration could be overcome with jet injector devices or with the option of IM fractional dose IPV. A recent randomized controlled clinical trial conducted in Cuba reported non-inferior seroconversion rates for all three serotypes for fractional IPV administered via IM compared with fractional IPV administered via the ID route after two doses, given at 4 and 8 months old infants (Resik et al. 2020). Further exploration of immunogenicity of fractional dose IPV administered intramuscularly in different age groups and settings will be important for possible policy impact of this option.

5 Surveillance

The primary means of detecting poliovirus is through identifying cases of paralytic poliomyelitis via surveillance of acute flaccid paralysis (AFP). An AFP case (for the purposes of polio surveillance) is defined as a child under 15 years of age presenting with recent, sudden onset of floppy paralysis or muscle weakness due to any cause, or any person of any age with paralytic illness if poliomyelitis is suspected by a clinician (World Health Organization 1999). For identified AFP cases, collection of adequate stool specimens and viral isolation is required to distinguish the case as poliomyelitis or non-polio AFP (World Health Organization 1999).

In many locations, AFP surveillance is supplemented by environmental surveillance (ES), the regular collection and testing of sewage to detect polioviruses (Asghar et al. 2014). Functioning environmental surveillance is a mechanism to detect poliovirus circulation in the absence of paralytic cases and provide information on the extent of circulation.

Laboratory testing of stool specimens and environmental samples is conducted by the Global Polio Laboratory Network (GPLN), which includes 146 WHO-accredited poliovirus laboratories in all WHO regions. The GPLN member laboratories follow standardized protocols to (1) isolate poliovirus, (2) conduct intratypic differentiation and (3) conduct genomic sequencing of the VP1 region to distinguish Sabin-like, VDPV and WPV types (Lickness et al. 2020).

6 Epidemiology: Wild Poliovirus

6.1 Progress Towards Eradication: 1988–2016

In 1988, when the World Health Assembly passed a resolution to eradicate poliomyelitis, poliovirus was endemic in over 100 countries and paralyzed an estimated number of 350,000 children per year. Between 1988 and 2001, polio incidence declined by 99%, primarily due to administration of tOPV through large-scale vaccination campaigns. The reported number of cases reduced from 35,000 to 3,000, and the number of polio endemic countries from 125 to 10 (Fig. 1) (Centers for Disease C, Prevention 2002). The WHO regions of the Americas and the Western Pacific were certified polio-free in 1994 and 2000 respectively, and the European Region in 2002 (Centers for Disease C, Prevention 2002). In addition, the last indigenous case of WPV2 occurred in 1999 in Aligarh, India (Centers for Disease C, Prevention 2001).

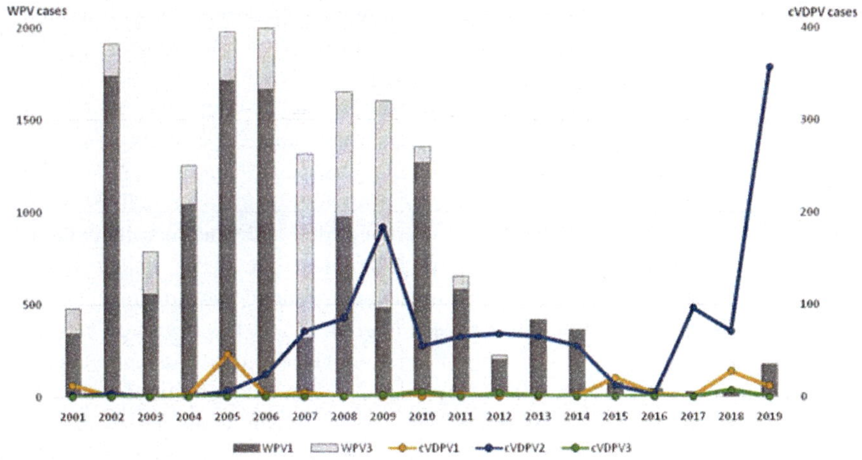

Fig. 1 Annual incidence of reported wild poliovirus (WPV) and circulating vaccine-derived poliovirus (cVDPV) cases, by year and serotype isolated, 2001–2019. Data as of August 2020 (World Health Organisation 2020a, b; Global Polio Eradication Initiative 2020)

In the early millennium, there was an expectation that global eradication of WPV was imminent. However, in 2006, WPV1 and WPV3 continued to persist in four endemic countries: India, Nigeria, Pakistan, and Afghanistan (Wassilak et al. 2014). The co-circulation of WPV1 and WPV3 was difficult to control due to requiring frequent interchange of tOPV, mOPV1, and mOPV3 during immunization activities. As the type 2 component of Sabin vaccine is more immunogenic than the other types, the immune response to type 1 and type 3 from tOPV was suboptimal. To overcome this challenge, development of bOPV that had superior immunogenicity compared with tOPV and non-inferiority compared with mOPV1 and mOPV3, was critical (Sutter et al. 2010).

After the introduction of bOPV in vaccination campaigns in 2009–2010, it was possible to simultaneously sustain reduction in WPV1 and WPV3 cases (John and Vashishtha 2013). India achieved a level of population immunity adequate to interrupt transmission, with the last confirmed WPV3 and WPV1 cases in October 2010 and January 2011, respectively (John and Vashishtha 2013). India, which was once considered the most difficult setting to achieve polio elimination, was removed from the list of endemic countries and the South East Asian Region was declared polio-free (John and Vashishtha 2013). The last case of WPV3 was detected on November 10, 2012, in Yobe, Nigeria (Moturi et al. 2014) and has subsequently been declared as eradicated by the Global Commission for the Certification of the Eradication of Poliomyelitis.

In 2013, WPV1 was the only wild-type poliovirus serotype circulating, and was endemic in three countries: Nigeria, Pakistan and Afghanistan (Wassilak et al. 2014; Moturi et al. 2014). However, importation of WPV1 lead to polio cases reported from previously polio-free countries: from Nigeria into the Horn of Africa resulted in 217 cases (9 in Ethiopia, 14 in Kenya, and 194 in Somalia); from Pakistan into Syria resulted in 35 cases; and from Nigeria to Cameroon resulted in 4 cases (Wassilak et al. 2014; Moturi et al. 2014; World Health Organisation 2020b).

The number of WPV1 cases in each of the three endemic countries continued to decline between 2013 and 2016: from 37 to 13 in Afghanistan, 93 to 20 in Pakistan and 53 to 4 in Nigeria (Fig. 1) (World Health Organisation 2020a, b). In Nigeria, no cases of WPV1 were reported for two years (from July of 2014 until June 2016). However, in July 2016, four WPV1 cases were reported and genetic sequencing of the excreted virus indicated that it might have been circulating undetected in Nigeria for 5 years (Nnadi et al. 2017). Subsequently, there were major efforts to improve access and surveillance in Northern Nigeria. There has been no detection of WPV1 since 2016 and Nigeria has been removed from the list of endemic countries. In August 2020, the Africa Regional Certification Commission certified the WHO African Region as wild polio-free after four years without a case.

6.2 *The Last Reservoirs, Afghanistan and Pakistan: 2017–2020*

There have been no WPV cases outside of Afghanistan and Pakistan since 2016. From a programmatic perspective, these two countries constitute a single epidemiologic block, with ongoing cross-border transmission accompanying large population movements (Chard et al. 2020). Whilst endemic transmission has been confined to these two countries, the risk of exportation remains: in 2019, WPV1 of Pakistan origin was detected in three environmental surveillance samples in Iran (Macklin et al. 2020).

Interruption of transmission seemed increasingly achievable with 22 WPV1 cases in 2017 (14 in Afghanistan and 8 in Pakistan) and 33 cases in 2018 (21 in Afghanistan and 12 in Pakistan) (Fig. 1) (World Health Organisation 2020a, b). However, a sharp increase in transmission followed in 2019 with 176 cases (29 in Afghanistan and 147 in Pakistan), widespread detection in environmental surveillance, and multiple genetic lineages circulating (Fig. 1) (Chard et al. 2020; World Health Organisation 2020a, b).

6.2.1 *Pakistan*

Polio in Pakistan began its resurgence in late 2018 and intensified, with 147 cases reported in 2019: a 12-fold increase over the cases reported in 2018 (Chard et al. 2020). Transmission of several viral genetic lineages persisted in the traditional three core reservoirs (Karachi, Peshawar and Quetta), with expansion among previously polio-free districts and provinces (Fig. 2) (Chard et al. 2020). In 2019, cases were reported from 43 districts, compared to only six districts in 2018 (Chard et al.

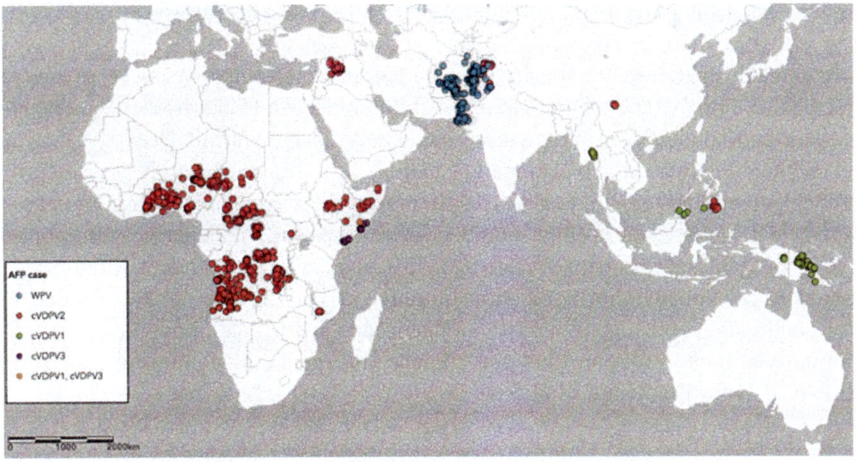

Fig. 2 Location of reported wild poliovirus (WPV) and circulating vaccine-derived poliovirus (cVDPV) cases, 01 January 2017 to 01 January 2020. Colour of circle indicates the poliovirus serotype. Data as of August 2020 (Global Polio Eradication Initiative 2020)

2020). Environmental surveillance detected WPV1 from 37 of 58 (64%) sites in 2018, which increased to 56 of 60 (93%) sites in 2019 (Chard et al. 2020).

The deteriorating epidemiology in Pakistan has primarily been attributed to a failure to vaccinate children at the doorstep. Independently assessed technical performance (LQAS) of vaccine rounds in Islamabad, Punjab, and Sindh in 2019 scored 44%, 64%, 56% against a benchmark of 90% (Independent Monitoring Board of the Global Polio Eradication Initiative 2019). The spread of misinformation and propaganda, fueled by social media, has resulted in mistrust in the polio vaccine and community resistance to vaccination (Independent Monitoring Board of the Global Polio Eradication Initiative 2019; Yusufzai 2020). Community fatigue towards repeated polio vaccination is present in those communities that are deprived of other basic services (Independent Monitoring Board of the Global Polio Eradication Initiative 2019; Yusufzai 2020). In addition, a lack of sustained political will or country-ownership of the program has been identified as another reason for this resurgence (Independent Monitoring Board of the Global Polio Eradication Initiative 2019).

6.2.2 *Afghanistan*

Among 176 WPV1 cases reported during 2019, 29 (16%) were reported by Afghanistan, representing a 38% increase over the 21 cases reported in 2018 (Chard et al. 2020). Cases were reported from 20 districts, a 43% increase from the 14 districts reporting cases during 2018 (Chard et al. 2020). In Afghanistan, WPV1 was detected in 83 (25%) of 336 sewage samples collected from 15 of 20 (75%)

sites at regular intervals in 2018 and 56 (22%) of 259 samples from 12 of 21 (57%) sites in 2019 (Chard et al. 2020).

In Afghanistan, insecurity and bans on vaccination campaigns have severely affected the ability of the program to reach children. House-to-house vaccinations in southern and southeastern provinces were halted during May to December 2018 and eventually all vaccination campaigns nationally were interrupted between April and September 2019 (Chard et al. 2020). In high-risk areas, there are on-going gaps in campaign quality due to high numbers of refusals: in Kandahar City, the proportion of refusals was between 30 and 43% in 2019 (Independent Monitoring Board of the Global Polio Eradication Initiative 2019; Yusufzai 2020). Kandahar City continues to be main source of transmission, with continuous detection of poliovirus through environmental surveillance (Independent Monitoring Board of the Global Polio Eradication Initiative 2019).

7 Epidemiology: circulating vaccine-derived poliovirus (cVDPV)

The first documented outbreak of cVDPV occurred in the Dominican Republic and Haiti, between 2000 and 2001 and was associated with the type 1 Sabin strain, likely originating from OPV dose given in 1998–1999 (Kew et al. 2002). The 21 confirmed cases occurred in communities with very low (7–40%) rates of coverage with OPV, with 20/21 individuals either unvaccinated or incompletely vaccinated (Kew et al. 2002).

Following their discovery, cVDPVs have been acknowledged as a cause of paralytic poliomyelitis outbreaks in certain settings and a barrier to achieving eradication (Kew et al. 2005; Fine et al. 2004). Between January 2000 and December 2016 there were a total of 810 paralytic cVDPV cases reported: 86% (697/810) cVDPV2, 13% (103/810) cVDPV1, and 1% (10/810) cVDPV3 (Fig. 1). These cases were from 42 genetically unique cVDPV outbreaks: 20 occurring in the African Region (Nigeria, Democratic Republic of Congo, Madagascar, Chad and Guinea); 11 in the Eastern Mediterranean Region (Ethiopia, Somalia, Afghanistan, Yemen, Pakistan and South Sudan); 5 in the Western Pacific Region (Philippines, China, Cambodia, Lao People's Democratic Republic); 4 in South East Asian Region (Indonesia, Myanmar and India); 1 in the European Region (Ukraine) and 1 in the American Region (Dominican Republic and Haiti) (O'Reilly et al. 2017).

Since the removal of type 2 OPV from routine immunization use in April 2016, there has been a substantial increase in cVDPV2 outbreaks from 2017 to 2019, discussed in the following Sect. 5. For cVDPV1 and cVDPV3, there have been 38 and 7 cases reported, respectively, between 2017 and 2019 (Fig. 2). cVDPV1 outbreaks have been reported in Indonesia (2018–2019), Malaysia (2019), Myanmar (2019), Papua New Guinea (2018), and Philippines (2019) and a cVDPV3 outbreak in Somalia (2018) (Fig. 2) (Alleman et al. 2020; Jorba et al. 2019).

The most significant risk factor for cVDPV outbreaks is insufficient population immunity and inadequate OPV vaccination coverage over time: using data from 2003 to 2016, significant risk factors associated with an increased probability of cVDPV outbreaks were the percentage of unimmunized children, the percentage of the population displaced, and the numbers of children born per year (O'Reilly et al. 2017). The prior elimination of indigenous poliovirus circulation also increases the risk, as nonimmune individuals accumulate rapidly in the absence of highvaccination coverage or naturally acquired immunity (Kew et al. 2005).

7.1 The Switch and cVDPV2 Resurgence

In April 2016, the first phase of OPV cessation occurred, with the removal of type 2 OPV from routine immunisation and a synchronized switch from tOPV (containing types 1, 2 and 3) to bOPV (containing types 1 and 3)—termed 'the Switch'(Hampton et al. 2016). The type 2 component was the first to be removed as it was responsible for most cVDPV outbreaks, whilst the WPV2 had been certified as eradicated in 2015, with the last indigenous WPV2 case in 1999 (Centers for Disease C, Prevention 2001). As a risk mitigation strategy, at least one dose of IPV was introduced into routine immunization in all OPV-only using countries, to protect against paralysis from serotype 2 in case of re-introduction of WPV2 or persistence of VDPV2 (Hampton et al. 2016). However, due to IPV supply shortages, many countries had to delay IPV introduction or faced stock-outs of IPV (Hampton et al. 2016).

Prior to the Switch, large-scale campaigns were conducted with tOPV to increase population immunity against type 2 poliovirus (Pons-Salort et al. 2016). Some cVDPV2 outbreaks were expected and a global stockpile of monovalent OPV2 (mOPV2) was created for outbreak response. However, any use of mOPV2 retains a risk of creating new VDPV2 emergences, which increases with time since the switch due to declining population immunity (McCarthy et al. 2017).

Between 01 January 2017 and 01 January 2020, there have been 50 genetically distinct outbreaks of cVDPV2 reported from 22 countries across four WHO regions (Eastern-Mediterranean, African, South-East Asian and Western Pacific) (Fig. 2) (Macklin et al. 2020; Alleman et al. 2020). These cVDPV2 outbreaks have resulted in >1,000 virus isolations and 525 paralytic poliomyelitis cases (Fig. 1).

The incidence and geographic scope of cVDPV2 outbreaks have been notably increasing over time since 2016, when there were 3 cVDPV2 emergences in 2 countries (Nigeria, Pakistan). In 2017 there were 4 new cVDPV2 emergences in 3 countries (Democratic Republic of Congo (DRC), Somalia, Syria); in 2018 there were 8 new emergences in 7 countries (China, DRC, Kenya, Mozambique, Niger, Nigeria and Somalia); and, in 2019, there were 44 new emergences in 19 countries (Angola, Benin, Burkina Faso, Cameroon, Central African Republic, Chad, China, Cote d'Ivoire, DRC, Ethiopia, Ghana, Malaysia, Niger, Nigeria, Pakistan, Philippines and Somalia) (Fig. 2) (Macklin et al. 2020; Alleman et al. 2020).

The two largest cVDPV2 outbreaks emerged in Somalia in 2017 (SOM-BAN-1 emergence) and Nigeria in 2018 (NIE-JIS-1). Failure to control the virus has resulted in spread across national borders to establish transmission in neighboring countries: from Somalia to Kenya and Ethiopia and from Nigeria to Benin, Burkina Faso, Cameroon, Chad, Côte d'Ivoire, Ghana, Niger and Togo. These two outbreaks, which have not yet been controlled, are the longest in duration, circulating for periods of 26 and 24 months, respectively, as of January 2020 (Macklin et al. 2020; Alleman et al. 2020).

Analysis based on the mutation rate of polioviruses and the genetic sequencing of cVDPV2 isolates, predicted that 65.5% (548/837) of VDPV2 viruses detected between May 2016 and November 2019 had a $\geq$ 90% probability of being seeded after the Switch (Macklin et al. 2020). Further, the source of 71.5% (392/548) of these isolates was consistent with mOPV2 outbreak response campaigns conducted within the country of emergence and that the source of 24.6% (135/548) was consistent with mOPV2 campaigns conducted within a neighbouring country (Macklin et al. 2020).

8 Looking Forward

8.1 Policy and Endgame Strategy

In accordance with the complexity of evolving epidemiology and timelines, the GPEI have produced several strategic plans, which outline the key goals to achieve and sustain a world free from polioviruses and are summarised below (Polio Endgame Strategy 2019; Global Polio Eradication Initiative 2013, 2018):

***Goal 1: Interrupt* transmission**: The first objective is to stop all wild poliovirus transmission and control new outbreaks of cVDPV within 120 days of confirmation of the index case.

Goal 2: Containment of polioviruses: Following interruption of WPV transmission globally, the safe handling and containment of infectious materials in laboratory and vaccine production facilities will be essential to minimize the risk of reintroducing WPV into the population. A reintroduction of WPV from a poliovirus facility would risk the potentially serious consequences of re-establishing WPV circulation. Additionally, after the cessation of all OPV use, the reintroduction of an OPV/Sabin virus strain from a poliovirus facility would risk the emergence of a cVDPV, and again the potentially serious consequences of re-establishing its circulation.

Most facility-associated poliovirus risks can be eliminated through the destruction of WPV and OPV/Sabin infectious and potentially infectious materials. However, poliovirus facilities will be necessary in several countries to continue essential international functions, including IPV production, OPV stockpile management, vaccine quality assurance, diagnostic reagent production, virus reference functions and research.

Goal 3: Certification of eradication of WPV: The Global Certification Committee *(GCC)* will be responsible for global certification of WPV, following the successful certification of all six WHO regions. The primary requirements for certifying a WHO region as free of WPV are:

- the absence of any WPV for a minimum of three years in all countries of the region;
- the presence of certification standard surveillance in all countries during that three-year period;
- the completion of Phase I biocontainment activities for all facility-based WPV stocks.

Four WHO regions have been certified as having interrupted transmission of WPVs: Region of the Americas (1994), the Western Pacific Region (2000), the European Region (2002), and the South-East Asia Region (2014). The GCC have also certified the eradication of WPV2 in 2015 and WPV3 in 2019.

Goal 4: Complete OPV withdrawal: The complete cessation of OPV use from routine immunisation and SIAs is essential to decisively eliminate the risk of VDPVs and VAPP. Global OPV cessation is planned approximately one year following certification of WPV eradication and would be followed by period of IPV-only vaccination. The first phase of OPV cessation was conducted in April 2016, with the globally-synchronised withdrawal of the type 2 OPV component from routine immunisation and switch from trivalent OPV use to bivalent OPV use. In addition, the introduction of at least one dose of IPV into the routine immunization schedule of all OPV-using countries has been completed. However, the lessons learned from the Switch will be key in future policy: the rapidly increasing outbreaks of cVDPV2 and the increasing evidence of seeding new emergence with mOPV2 use are key indicators that additional tools and knowledge will be necessary to plan the withdrawal of OPV and eliminate risks of cVDPVs.

Goal 5: Validation of absence of cVDPVs. Following complete OPV cessation, there should be no new cVDPV outbreaks seeded. After a period without of detection and high quality surveillance, the absence of cVDPV outbreaks can be verified. Any iVDPV excretors that were infected prior to OPV cessation and continue to excrete will require monitoring and treatment.

8.2 Research and Development

We identified the following research and development areas that could have major policy implications to facilitate both interruption of transmission of all polioviruses and maintain polio-free status for long-term.

OPV with less risk of VAPP/VDPVs: A central priority of polio vaccine development is to develop OPV strains that are significantly more genetically stable compared to Sabin strains. Enhanced genetic stability as a result of stabilizing key areas of the vaccine virus genome would translate into less risk of losing the attenuations that are known to be linked with reversion to neurovirulence. Two such novel OPV type 2 (nOPV2) candidates have been under pre-clinical and clinical development with heightened focus in accelerated clinical development over the recent past given the rapidly deteriorating cVDPV2 situation. The modifications in nOPV2 candidates include changes ribonucleic acid (RNA) sequence in the 5′ untranslated region (5′ UTR), the non-structural protein 2C, the capsid protein coding region (P1), and the polymerases (Yeh et al. 2020; Konopka-Anstadt et al. 2020). Pre-clinical development that begun early in the past decade has been successfully completed, and the first-in-human study was implemented under contained conditions in Belgiumin 2017 with promising results confirming safety and immunogenicity with enhanced genetic and phenotypic stability of the novel strains (Van Damme et al. 2019a, b, c). Following this, phase II studies and manufacturing processes have been accelerated with the WHO Executive Board in February 2020 urging for review and assessment of nOPV2 through the Emergency Use Listing (EUL) procedure of WHO—a process to expedite the availability of unlicensed medical products for PHEICs (World Health Organisation 2020a, b; Bandyopadhyay 2020).

IPV with mucosal immunogenicity: Although the novel OPV strains have shown significant promise to reduce the risk of generating VAPP and VDPV cases, a live attenuated vaccine strain would still carry some risk of reversion, recombination and potentially neurovirulence on prolonged excretion or circulation. A new or modified IPV with intestinal immunogenicity has therefore been considered an "ideal" solution to the issue of vaccine-related poliovirus disease and circulation. In recent times, IPV adjuvanted with enterotoxin-based mucosal vaccine antigens such as the double-mutant heat-labile enterotoxin (dmLT) has demonstrated rise in fecal IgA secretion and upregulation of expression of the intestinal homing receptor $\alpha 4\beta 7$, indicating potential induction of mucosal immunity in pre-clinical studies (Clements and Norton 2018; Norton et al. 2015). Human clinical data are awaited and would be key in informing the next phases of development.

IPV with easier delivery tools: A major issue restricting the broader use of IPV in SIAs or house-to-house campaigns has been the requirement of trained personnel for administering injections and managing associated logistics. Alternative modes and options for IPV administration such as needle-free, jet injector devices and microarray patches (MAP) hold the promise of making IPV more usable in the peripheral settings, outside of immunization clinics (Daly et al. 2020). Several studies have reported the user dynamics; safety and immunogenicity with such devices, and tools such as the MAPs are under further evaluation for technological and scientific merit and the potential to enhance equitable vaccine access for in low- and middle-income countries (Peyraud et al. 2019; Muller et al. 2017; Resik et al. 2015; Yousafzai et al. 2017).

Minimizing risk of containment failure: As successive types of WPVs are eradicated, any potentially infectious poliovirus material stored in laboratories, and vaccine production facilities will continue to have a risk of reintroduction of the homotypic poliovirus into communities. Several recent examples of containment failures from manufacturing sites in emphasize the importance of timely implementation of poliovirus containment measures to prevent potential long-lasting, damaging fall out of re-establishment of poliovirus transmission in the post-eradication era, especially when population immunity against poliovirus is expected to be on the decline (Bandyopadhyay et al. 2019). Major new initiatives in developing IPV and novel laboratory assays from non- or less- infectious materials or constructs have been reported in the recent past, including attenuated Sabin strains or S19 strains, that could significantly reduce the public health impact of any accidental containment failure of polio-essential facilities (Bandyopadhyay et al. 2019; Modlin and Chumakov 2020; Fox et al. 2019; Jiang et al. 2019).

Treating immunodeficient excretors: A unique challenge for long-term maintenance and completeness of polio eradication comes from individuals with the rare, inherited immunodeficiency disorders who are at risk of prolonged excretion of polioviruses and thus could give rise to polio outbreaks in communities (Macklin et al. 2017; Kalkowska et al. 2019). At least two drugs are under advanced stages of development to mitigate this risk in individuals with specific immunodeficiency disorders. Pocapavir, a capsid inhibitor with proven efficacy demonstrated in human OPV challenge study and V-7404, a 3C protease inhibitor, are in clinical development with initial trends suggesting the need of a combination product for effective virus clearance and to reduce development of resistance (McKinlay et al. 2014; Collett et al. 2017; Copelyn et al. 2020). Building a robust surveillance system to identify and track such individuals at risk of prolonged shedding will be a key contributory step to the success of these anti-viral agents.

Faster, easier detection of polioviruses : In addition to considerations for expanded and risk-based deployment of environmental surveillance sites for poliovirus detection, use of newer tools such as water-quality probes to evaluate physical attributes of sewage collection sites and tools with potential advantages of sample shipment and higher sensitivity of detection such as the Bag-mediated filtration system (BMFS) could be important for the final phases of eradication (Zhou et al. 2020; Estivariz et al. 2019; Hamisu et al. 2020). Wider, more targeted use of principles of molecular epidemiology to interpret sequencing results of poliovirus isolates could play a pivotal role in planning effective scale of outbreak response. The importance of using advanced molecular technologies to expedite the process of poliovirus detection and final classification of the isolates has been emphasized in recent reviews (Jorgensen et al. 2020). For example, methodologies with nested PCR with nanopore sequencing protocols have demonstrated promising sensitivity for detection of WPVs, VDPV2 and Sabin-like viruses in a pilot conducted in Pakistan, generating sequencing information in less than 3 days from the time of initiation of sample, compared to 2–4 weeks with the current, culture-based method (Shaw et al. 2020).

9 Conclusion

In 1988, when the World Health Assembly declared its commitment to eradication and the Global Polio Eradication Initiative (GPEI) was formed in pursuit of this goal, there were 350 000 annual cases of WPV in 125 countries. By 2020, only two countries remain endemic (Afghanistan and Pakistan) and two of the three WPV serotypes have been certified as eradicated. Unknown at the time of declaring eradication, VDPVs—rare poliovirus strains that have genetically mutated from the poliovirus contained in OPV—have been discovered and pose a major challenge. Now the GPEI must confront a dual emergency: interrupting WPV1 in the two remaining endemic countries and stopping outbreaks of cVDPVs.

The barriers to reaching WPV eradication Afghanistan and Pakistan are not an exclusive matter of science or virology anymore; they are instead social and political realities that impede the delivery of polio vaccine in these settings. The programme has not been able to vaccinate every child for several reasons, including inaccessibility of some areas due to geographical isolation, insecurity or bans on vaccination activities by political or religious leaders. Even when the programme does have access, pockets of vaccine refusals are growing due to misinformation, mistrust, cultural beliefs, fatigue or other, urgent, health priorities (such as access to water and basic healthcare).

The problems of VDPV, on the other hand, are more biological and require scientific innovation and strategies for improved program implementation. It remains clear that cessation of Sabin OPV use is essential to stop all cases of paralytic poliomyelitis. However, the epidemiology that has evolved since Sabin OPV2 removal has implications for existing strategies outlined for total OPV cessation, which need urgent attention.

In 2019, we have observed the largest numbers of outbreaks and countries experiencing cVDPV2 transmission to date primarilydue to new emergences of cVDPV2 outbreaks seeded through inadequate mOPV2 response to outbreaks. It is not currently possible to control cVDPV2 outbreaks without inducing intestinal mucosal immunity through mOPV2 use; however, any sub-optimal use of mOPV2 risks generating cVDPV2. The spread of cVDPV2 is increasing over time as the immunity of the global population against type 2 poliovirus rapidly decreases.

The GPEI began the introduction of nOPV2 in March 2021, developed to be more genetically stable and less likely to revert to a neurovirulent genotype than Sabin OPV2. However, uncertainty remains about when nOPV2 will be rolled out for widespread use in outbreak response and how effective they will be in interrupting outbreaks in high-risk areas of polio transmission.

Several modelling groups have developed prospective mathematical models to inform and evaluate the endgame strategic plans including a recent global model update that emphasised that the GPEI is not on track to achieve WPV1 eradication prior to 2024 without improved implementation, or to successfully stop the transmission of VDPV2 viruses using current tools (Thompson et al. 2000). This indicates the current trajectory of the polio endgame will continue to have significant challenges.

In the final phases of the global eradication program, it would be important to maintain high population immunity against all three WPV serotypes through strengthening of routine and supplementary immunization delivery systems (Bandyopadhyay et al. 2015, 2020). Alongside, effective and urgent incorporation of technological advances such as use of novel, more genetically stable vaccine options and innovative program strategies to have better and faster outbreak response will be necessary to complete and sustain eradication of all types of polioviruses.

References

Alleman MM, Jorba J, Greene SA, Diop OM, Iber J, Tallis G, et al. Update on vaccine-derived poliovirus outbreaks—worldwide, July 2019-February 2020. MMWR Morb Mortal Wkly Rep. 2020;69(16):489–95.

Asghar H, Diop OM, Weldegebriel G, Malik F, Shetty S, El Bassioni L, et al. Environmental surveillance for polioviruses in the Global Polio Eradication Initiative. J Infect Dis. 2014;210 (Suppl 1):S294-303.

Bandyopadhyay AS. Clinical data from novel type-2 oral polio vaccine trials and plan for emergency use listing. Meeting of the Strategic Advisory Group of Experts (SAGE) on Immunization, March–April 2020. 2020. https://www.who.int/immunization/sage/meetings/2019/october/Bandyopadhyay_polio_sage_october_2019.pdf?ua=1.

Bandyopadhyay, Ananda S., Macklin, Grace R. Final frontiers of the polio eradication endgame, current opinion in infectious diseases: August 06, 2020—Volume Publish Ahead of Print – Issue. https://doi.org/10.1097/QCO.0000000000000667.

Bandyopadhyay AS, Modlin JF, Wenger J, Gast C. Immunogenicity of new primary immunization schedules with inactivated poliovirus vaccine and bivalent oral polio vaccine for the polio endgame: a review. Clin Infect Dis. 2018;67(suppl_1):S35-41.

Bandyopadhyay AS, Singh H, Fournier-Caruana J, Modlin JF, Wenger J, Partridge J, et al. Facility-associated release of polioviruses into communities—risks for the posteradication era. Emerg Infect Dis. 2019;25(7):1363.

Bandyopadhyay AS, Garon J, Seib K, Orenstein WA. Polio vaccination: past, present and future. Future Microbiol. 2015;10(5):791–808.

Burns CC, Diop OM, Sutter RW, Kew OM. Vaccine-derived polioviruses. J Infect Dis. 2014;210 (Suppl 1):S283–93.

Centers for Disease C, Prevention. Apparent global interruption of wild poliovirus type 2 transmission. MMWR Morb Mortal Wkly Rep. 2001;50(12):222–4.

Centers for Disease C, Prevention. Progress toward global eradication of poliomyelitis, 2001. MMWR Morb Mortal Wkly Rep. 2002;51(12):253–6.

Chard AN, Datta SD, Tallis G, Burns CC, Wassilak SGF, Vertefeuille JF, et al. Progress toward polio eradication—worldwide, January 2018-March 2020. MMWR Morb Mortal Wkly Rep. 2020;69(25):784–9.

Clements JD, Norton EB. The Mucosal Vaccine Adjuvant LT(R192G/L211A) or dmLT. mSphere. 2018;3(4).

Collett MS, Hincks JR, Benschop K, Duizer E, van der Avoort H, Rhoden E, et al. Antiviral activity of Pocapavir in a randomized, blinded, placebo-controlled human oral poliovirus vaccine challenge model. J Infect Dis. 2017;215(3):335–43.

Copelyn J, Hincks JR, Wilmshurst JM, Petersen W, Howard W, Jallow S, et al. Clearance of immunodeficiency-associated vaccine-derived poliovirus infection with Pocapavir. Pediatr Infect Dis J. 2020;39(5):435–7.

Daly C, Molodecky NA, Sreevatsava M, Belayneh AD, Chandio SA, Partridge J, et al. Needle-free injectors for mass administration of fractional dose inactivated poliovirus vaccine in Karachi, Pakistan: a survey of caregiver and vaccinator acceptability. Vaccine. 2020;38(8):1893–8.

Dayan GH, Thorley M, Yamamura Y, Rodriguez N, McLaughlin S, Torres LM, et al. Serologic response to inactivated poliovirus vaccine: a randomized clinical trial comparing 2 vaccination schedules in Puerto Rico. J Infect Dis. 2007;195(1):12–20.

Dowdle WR. The principles of disease elimination and eradication. Bull World Health Organ. 1998;76(Suppl 2):22–5.

Estivariz CF, Perez-Sanchez EE, Bahena A, Burns CC, Gary HE Jr, Garcia-Lozano H, et al. Field performance of two methods for detection of poliovirus in wastewater samples, Mexico 2016–2017. Food Environ Virol. 2019;11(4):364–73.

Fine PE, Oblapenko G, Sutter RW. Polio control after certification: major issues outstanding. Bull World Health Organ. 2004;82(1):47–52.

Fox H, Carlyle S, Minor P, Macadam A. Use of hyperattenuated poliovirus as a replacement for Sabin or wild-type strains for laboratory assays in a post eradication world. Access Microbiol. 2019;1(1A).

Gamage D, Mach O, Palihawadana P, Zhang Y, Weldon WC, Oberste MS, et al. Boosting of mucosal immunity after fractional-dose inactivated poliovirus vaccine. J Infect Dis. 2018;218 (12):1876–82.

Global Polio Eradication Initiative. Polio Post-Certification Strategy Geneva, Switerland: World Health Organisation;2018. https://polioeradication.org/wp-content/uploads/2018/04/polio-post-certification-strategy-20180424-2.pdf.

Global Polio Eradication Initiative. Polio Information Systems (PolIS) Geneva, Swizerland: World Health Organisation;2020. https://extranet.who.int/polis.

Grassly NC, Wadood MZ, Safdar RM, Mahamud AS, Sutter RW. Effect of inactivated poliovirus vaccine campaigns, Pakistan, 2014–2017. Emerg Infect Dis. 2018;24(11):2113–5.

Hamisu AW, Blake IM, Sume G, Braka F, Jimoh A, Dahiru H, et al. Characterizing environmental surveillance sites in nigeria and their sensitivity to detect poliovirus and other enteroviruses. J Infect Dis. 2020.

Hampton LM, Farrell M, Ramirez-Gonzalez A, Menning L, Shendale S, Lewis I, et al. Cessation of trivalent oral poliovirus vaccine and introduction of inactivated poliovirus vaccine—worldwide, 2016. MMWR Morb Mortal Wkly Rep. 2016;65(35):934–8.

Henderson DA. Principles and lessons from the smallpox eradication programme. Bull World Health Organ. 1987;65(4):535–46.

Hird TR, Grassly NC. Systematic review of mucosal immunity induced by oral and inactivated poliovirus vaccines against virus shedding following oral poliovirus challenge. PLoS Pathog. 2012;8(4):e1002599.

Independent Monitoring Board of the Global Polio Eradication Initiative. The Art of Survival: The Polio Virus Continues To Exploit Human Frailties. Seventeenth Report, November 2019: Global Polio Eradication Initiative;2019.

Initiative GPE. Polio eradication and endgame strategic plan 2013–2018. World Health Organisation: World Health Organisation; 2013.

Jafari H, Deshpande JM, Sutter RW, Bahl S, Verma H, Ahmad M, et al. Polio eradication. Efficacy of inactivated poliovirus vaccine in India. Science. 2014;345(6199):922–5.

Jiang Z, Liu G, Guo-yang L, Sun M, Xu K, Ying Z, et al. A simple and safe antibody neutralization assay based on polio pseudoviruses. Human Vaccines & Immunotherapeutics. 2019;15(2):349–57.

John TJ, Vashishtha VM. Eradicating poliomyelitis: India's journey from hyperendemic to polio-free status. Indian J Med Res. 2013;137(5):881–94.

John J, Giri S, Karthikeyan AS, Iturriza-Gomara M, Muliyil J, Abraham A, et al. Effect of a single inactivated poliovirus vaccine dose on intestinal immunity against poliovirus in children previously given oral vaccine: an open-label, randomised controlled trial. Lancet. 2014;384 (9953):1505–12.

Jorba J, Diop OM, Iber J, Henderson E, Zhao K, Quddus A, et al. Update on vaccine-derived poliovirus outbreaks—worldwide, January 2018-June 2019. MMWR Morb Mortal Wkly Rep. 2019;68(45):1024–8.

Jorgensen D, Pons-Salort M, Shaw AG, Grassly NC. The role of genetic sequencing and analysis in the polio eradication program. Virus Evol. 2020.

Kalkowska D, Pallansch M, Thompson K. Updated modelling of the prevalence of immunodeficiency-associated long-term vaccine-derived poliovirus (iVDPV) excreters. Epidemiol Infect. 2019;147.

Kew O, Morris-Glasgow V, Landaverde M, Burns C, Shaw J, Garib Za, et al. Outbreak of poliomyelitis in Hispaniola associated with circulating type 1 vaccine-derived poliovirus. Science. 2002;296(5566):356–9.

Kew OM, Sutter RW, de Gourville EM, Dowdle WR, Pallansch MA. Vaccine-derived polioviruses and the endgame strategy for global polio eradication. Annu Rev Microbiol. 2005;59:587–635.

Konopka-Anstadt JL, Campagnoli R, Vincent A, Shaw J, Wei L, Wynn NT, et al. Development of a new oral poliovirus vaccine for the eradication end game using codon deoptimization. NPJ Vaccines. 2020;5:26.

Lickness JS, Gardner T, Diop OM, Chavan S, Jorba J, Ahmed J, et al. Surveillance to track progress toward polio eradication—worldwide, 2018–2019. MMWR Morb Mortal Wkly Rep. 2020;69(20):623–9.

Macklin G, Liao Y, Takane M, Dooling K, Gilmour S, Mach O, et al. Prolonged excretion of poliovirus among individuals with primary immunodeficiency disorder: an analysis of the world health organization registry. Front Immunol. 2017;8:1103.

Macklin GR, Grassly NC, Sutter RW, Mach O, Bandyopadhyay AS, Edmunds WJ, et al. Vaccine schedules and the effect on humoral and intestinal immunity against poliovirus: a systematic review and network meta-analysis. Lancet Infect Dis. 2019;19(10):1121–8.

Macklin GR, O'Reilly KM, Grassly NC, Edmunds WJ, Mach O, Santhana Gopala Krishnan R, et al. Evolving epidemiology of poliovirus serotype 2 following withdrawal of the serotype 2 oral poliovirus vaccine. Science. 2020;368(6489):401–5.

McCarthy KA, Chabot-Couture G, Famulare M, Lyons HM, Mercer LD. The risk of type 2 oral polio vaccine use in post-cessation outbreak response. BMC Med. 2017;15(1):175.

McKinlay MA, Collett MS, Hincks JR, Oberste MS, Pallansch MA, Okayasu H, et al. Progress in the development of poliovirus antiviral agents and their essential role in reducing risks that threaten eradication. The J Infect Dis. 2014;210(suppl_1):S447-53.

Modlin JF, Chumakov K. Sabin Strain Inactivated Polio Vaccine for the Polio Endgame. Oxford University Press US;2020.

Moturi EK, Porter KA, Wassilak SG, Tangermann RH, Diop OM, Burns CC, et al. Progress toward polio eradication–worldwide, 2013–2014. MMWR Morb Mortal Wkly Rep. 2014;63 (21):468–72.

Muller DA, Fernando GJ, Owens NS, Agyei-Yeboah C, Wei JC, Depelsenaire AC, et al. High-density microprojection array delivery to rat skin of low doses of trivalent inactivated poliovirus vaccine elicits potent neutralising antibody responses. Sci Rep. 2017;7(1):1–10.

Nnadi C, Damisa E, Esapa L, Braka F, Waziri N, Siddique A, et al. Continued endemic wild poliovirus transmission in security-compromised areas—Nigeria, 2016. MMWR Morb Mortal Wkly Rep. 2017;66(7):190.

Norton EBBD, Weldon WC, Oberste MS, Lawson LB, Clements JD. The novel adjuvant dmLT promotes dose sparing, mucosal immunity and longevity of antibody responses to the inactivated polio vaccine in a murine model. Vaccine. 2015;33(16):1909–15.

Okayasu H, Sein C, Chang Blanc D, Gonzalez AR, Zehrung D, Jarrahian C, et al. Intradermal administration of fractional doses of inactivated poliovirus vaccine: a dose-sparing option for polio immunization. J Infect Dis. 2017;216(suppl_1):S161–S7.

Onorato IM, Modlin JF, McBean AM, Thoms ML, Losonsky GA, Bernier RH. Mucosal immunity induced by enhanced-potency inactivated and oral polio vaccines. J Infect Dis. 1991;163(1):1–6.

O'Reilly KM, Lamoureux C, Molodecky NA, Lyons H, Grassly NC, Tallis G. An assessment of the geographical risks of wild and vaccine-derived poliomyelitis outbreaks in Africa and Asia. BMC Infect Dis. 2017;17(1):367.

Peyraud N, Zehrung D, Jarrahian C, Frivold C, Orubu T, Giersing B. Potential use of microarray patches for vaccine delivery in low- and middle- income countries. Vaccine. 2019;37 (32):4427–34.

Polio endgame strategy 2019–2023: eradication, integration, certification and containment [press release]. Geneva, Switzerland: World Health Organisation;2019.

Platt LR, Estivariz CF, Sutter RW. Vaccine-associated paralytic poliomyelitis: a review of the epidemiology and estimation of the global burden. J Infect Dis. 2014;210(Suppl 1):S380–9.

Pons-Salort M, Burns CC, Lyons H, Blake IM, Jafari H, Oberste MS, et al. Preventing vaccine-derived poliovirus emergence during the polio endgame. PLoS Pathog. 2016;12(7): e1005728.

Resik S, Tejeda A, Mach O, Sein C, Molodecky N, Jarrahian C, et al. Needle-free jet injector intradermal delivery of fractional dose inactivated poliovirus vaccine: association between injection quality and immunogenicity. Vaccine. 2015;33(43):5873–7.

Resik S, Mach O, Tejeda A, Jeyaseelan V, Fonseca M, Diaz M, et al. Immunogenicity of intramuscular fractional dose of inactivated poliovirus vaccine. J Infect Dis. 2020;221(6):895–901.

Saez-Llorens X, Clemens R, Leroux-Roels G, Jimeno J, Clemens SA, Weldon WC, et al. Immunogenicity and safety of a novel monovalent high-dose inactivated poliovirus type 2 vaccine in infants: a comparative, observer-blind, randomised, controlled trial. Lancet Infect Dis. 2016;16(3):321–30.

Saleem AF, Mach O, Yousafzai MT, Khan A, Weldon WC, Steven Oberste M, et al. Immunogenicity of different routine poliovirus vaccination schedules: a randomized, controlled trial in Karachi, Pakistan. J Infect Dis. 2018;217(3):443–50.

Shaw AG, Majumdar M, Troman C, O'Toole Á, Benny B, Abraham D, et al. Rapid and sensitive direct detection and identification of poliovirus from stool and environmental surveillance samples using nanopore sequencing. bioRxiv. 2020.

Shirreff G, Wadood MZ, Vaz RG, Sutter RW, Grassly NC. Estimated effect of inactivated poliovirus vaccine campaigns, Nigeria and Pakistan, January 2014–April 2016. Emerg Infect Dis. 2017;23(2):258–63.

Sutter RW, John TJ, Jain H, Agarkhedkar S, Ramanan PV, Verma H, et al. Immunogenicity of bivalent types 1 and 3 oral poliovirus vaccine: a randomised, double-blind, controlled trial. Lancet. 2010;376(9753):1682–8.

Sutter RW, Kew OM, Cochi SL, Aylward RB. Poliovirus Vaccine– Live. Plotkin's Vaccines: Elsevier;2018. p. 866–917. e16.

Thompson KM, Kalkowska DA. Review of poliovirus modeling performed from, 2000 to 2019 to support global polio eradication. Expert Rev Vaccines. 2020;1–25.

Van Damme P, De Coster I, Revets H, Bandyopadhyay AS. Poliopolis. The Lancet. 2019c;394 (10193):115.

Van Damme P, De Coster I, Bandyopadhyay AS, Revets H, Withanage K, De Smedt P, et al. The safety and immunogenicity of two novel live attenuated monovalent (serotype 2) oral poliovirus vaccines in healthy adults: a double-blind, single-centre phase 1 study. Lancet. 2019;394(10193):148–58.

Van Damme P, Coster ID, Bandyopadhyay AS, Suykens L, Rudelsheim P, Neels P, et al. Poliopolis: pushing boundaries of scientific innovations for disease eradication. Future Microbiol. 2019b;14(15):1321–30.

Vidor E. Poliovirus Vaccine–Inactivated. Plotkin's Vaccines: Elsevier;2018. p. 841–65.e10.

Wassilak SG, Oberste MS, Tangermann RH, Diop OM, Jafari HS, Armstrong GL. Progress toward global interruption of wild poliovirus transmission, 2010–2013, and tackling the challenges to complete eradication. The Journal of infectious diseases. 2014;210(suppl_1):S5–15.

World Health Assembly. Resolution WHA 41.28: Global eradication of poliomyelitis by the year 2000. Handbook of resolutions and decisions of the World Health Assembly and the Executive Board. III: (1985–92). Geneva, Switzerland: World Health Organization;1993.

World Health Organization. WHO-recommended standards for surveillance of selected vaccine preventable diseases. World Health Organization;1999.

World Health Organisation. Executive Board, 146th session, Agenda item 16.1—Polio Eradication World Health Organisation: World Health Organisation;2020a. https://apps.who.int/gb/ebwha/pdf_files/EB146/B146(11)-en.pdf.

World Health Organisation. Polio Case Count Geneva, Swizerland: World Health Organisation;2020b. https://extranet.who.int/polis/public/CaseCount.aspx.

Yeh MT, Bujaki E, Dolan PT, Smith M, Wahid R, Konz J, et al. Engineering the live-attenuated polio vaccine to prevent reversion to virulence. Cell Host Microbe. 2020;27(5):736–51 e8.

Yousafzai MT, Saleem AF, Mach O, Baig A, Sutter RW, Zaidi AKM. Feasibility of conducting intradermal vaccination campaign with inactivated poliovirus vaccine using Tropis intradermal needle free injection system, Karachi, Pakistan. Heliyon. 2017;3(8):e00395.

Yusufzai A. Efforts to eradicate polio virus in Pakistan and Afghanistan. The Lancet Child & Adolescent Health. 2020;4(1):17.

Zhou NA, Fagnant-Sperati CS, Komen E, Mwangi B, Mukubi J, Nyangao J, et al. Feasibility of the bag-mediated filtration system for environmental surveillance of poliovirus in Kenya. Food Environ Virol. 2020;12(1):35–47.

Passive Immunization to Pandemic Viruses Through Herd Parties

Nitish Aggarwal and Pushkar Aggarwal

Abstract The general population is always concerned about the devastating effects of a pandemic. Based upon the 1918 Spanish flu pandemic, scientific publications and theories, the idea that herd parties may achieve passive immunity against pandemic viruses has been proposed. In this chapter, we dynamically modelled the effect of swine flu parties on the immunity achieved and associated mortality for a period of two years. In our analysis on swine flu, most of the people who attended the herd party would become ill with the virus. The final impact of such strategies can only be fully understood several years after the pandemic ceases to exist.

Keywords Swine flu · Coronavirus · COVID-19 · Passive immunization · Herd parties · Stella software · Mathematical modelling · Herd immunity · Computer simulation · Public health · Pandemic

1 Introduction

The global pandemics from Swine Flu (H1N1) and Coronavirus (COVID-19) viruses in the last decade have led to increased public health awareness for airborne viruses. Other than prevention, the public generally desires treatment modalities including vaccinations. However, early in the pandemic, a safe and effective vaccination is unlikely to be available. This is where the idea of passive vaccinations through herd parties is intriguing. There has been interest in herd parties among health officials and the general public wary of deadly strains of the virus, both in developed as well as undeveloped countries. The concept of herd parties is based upon achieving passive immunity against the pandemic similar to the immunity achieved during the first wave of the Spanish Flu pandemic of 1918 and chicken

N. Aggarwal (✉)
Yale New Haven Hospital, 20 York St Tompkins Bldg - 3rd Fl, New Haven, CT 06510, USA
e-mail: Nitish.aggarwal@yale.edu

P. Aggarwal
University of Cincinnati College of Medicine, 3230 Eden Ave, Cincinnati, OH 45267, USA

S. I. Ahmad (ed.), *Human Viruses: Diseases, Treatments and Vaccines*, https://doi.org/10.1007/978-3-030-71165-8_31

pox parties (Barry et al. 2008). People exposed to a less virulent strain of the virus may achieve immunity from the highly virulent strain of the virus and cross-clade viruses. However, the translation of this concept clinically is usually met with skepticism. For example, during the swine flu pandemic, officials of the British Medical Association's public health committee and US Centers for Disease Control and Prevention had warned against deliberately exposing people to the H1N1 virus (http://www.cnn.com/2009/HEALTH/06/30/flu.party/). Furthermore, UK's Chief Medical Officer had also stated that such parties are the result of "seriously flawed thinking" (Time.com 2009). However, during the COVID-19 pandemic, Sweden instituted a nationwide healthcare model of herd immunity.

As the new pandemic arises, the scientific community tries to assess the infectivity and pathogenesis of the virus. Furthermore, scientists try to assess these parameters in animals including ferrets, which generally mimic the infection in humans (Maines et al. 2009). The symptoms for the viruses vary but can have a similar prodrome based on colonization. For instance, both the H1N1 2009 and COVID-19 colonize the upper respiratory tract, lower respiratory tract, and the gut (Maines et al. 2009). As such, both produce fever, cough, shortness of breath and general malaise. H1N1 was more virulent and caused more severe and extensive disease than seasonal influenza but was much less virulent and lethal than the 1918 Spanish flu virus and H5N1 avian flu (Itoh et al. 2009). It has been suggested that there is relative protection for persons who were exposed to H1N1 strains during childhood before the 1957 flu pandemic (Chowell et al. 2009). One expects a reward of lifelong "friendship—read immunity" to pandemic H1N1 2009 and COVID-19 after surviving an attack. This is not too dissimilar to the concept of transfusion of convalescent plasma from the recovered individual, with the main difference being that herd immunity is noninvasive. The general public is not fully aware of the reasoning for attending or not attending the herd parties and frustrated at not knowing what course of action to take if a highly lethal variant of the virus descends. Through simulation, we attempted to assess the consequences of swine flu parties in the absence of effective active vaccination. Similar universal models can be constructed for other viral pandemics. However, a model for COVID-19, would not be accurate enough due to highly variable infectivity and mortality rates that were observed due to virulence of various circulating strains and unequal access of the earth's inhabitants to healthcare facilities.

2 Materials and Methods

Simulation software Stella version 7.0.3 (ISEE systems) was used to develop a compartment model using epidemiological parameters SIR (Susceptibility, Infectivity, and Recovery) (Hethcote 1976). The model was developed and applied to predict pandemic H1N1 2009 progression in the US, Australia, and Mexico. The infection coefficients for the model were derived using linear regression from the data published up to July 2009 by the World Health Organization (WHO) and Centers for

Disease Control (CDC) (World Health Organisation 2009; Centers for Disease Control 2009). The starting date for the initiation of the model for a country was the first date when pandemic H1N1 2009 cases were reported by the World Health Organization for that country. For Mexico and the United States of America (US), the basic model data are from 10 weeks, while for Australia they are from 6 weeks. The model is simulated for a period of two years from the first reported date by the WHO. Euler's method of integration was used for the simulation with a time interval (dt) of one day. The initial population, birthrate and mortality rate of the three countries were taken from the CIA World Factbook (The World Factbook 2009).

3 Results

The line diagram for the model developed is shown in Fig. 1. The pandemic H1N1 2009 infection rate for Australia (6.238 per million population per day) was higher than that of the US (1.773 per million population per day) and Mexico (1.282 per million population per day). The US's higher rate over Mexico may be attributed to difference in weather between Mexico and US as Mexico lies within latitudes 16° N and 32° N while US lies between 25° N and 50° N. The thought is that warmer weather would inactivate or reduce the infectivity of the virus. As such, the winter season coefficient factor for US was calculated to be 3.5 ± 0.37 and is based upon the winter season infectivity rate of Australia. Similarly, the summer season coefficient factor for Australia (0.25 ± 0.04) was calculated based upon the summer season infectivity rate of the US and Mexico. The crude case fatality rates were 5.70, 11.96, and 1.88 per 1,000 cases for the US, Mexico, and Australia respectively over the reported period. In the last 15 days, the crude case fatality rates were 7.85, 2.27, and 3.14 per 1,000 cases for US, Mexico, and Australia respectively and these numbers were used in the model to predict future values. These rates show an upward trend in the US and Australia but a downward trend in Mexico. Mexico's downward trend may be due to fewer numbers of reported/diagnosed infections during the early stages of the spread of the disease. The model was simulated for two years to predict the number of pandemic H1N1 2009 infected persons and the associated deaths. The results are shown in Table 1.

A second model was built with a swine flu party construct added. Ninety percent of the population going to the parties was assumed to get exposure to the pandemic H1N1 2009 virus. It was assumed that 0.1% of the population per day attends the parties to receive passive immunization. The parties start after 10 weeks of reported cases in a country. The results of the population developing immunity due to parties and without parties are exhibited in Fig. 2 and Table 1. The case fatality rate of persons attending parties is assumed to be 10 times less than the mortality from natural pandemic H1N1 2009 transmission, as those exhibiting adverse symptoms are likely to get immediate medical care if needed. In the US, 22.8% of the population would achieve immunity from pandemic H1N1 2009 after Year 1 and 44.2% after Year 2. Similarly, in Australia, 22.8% of the population would achieve

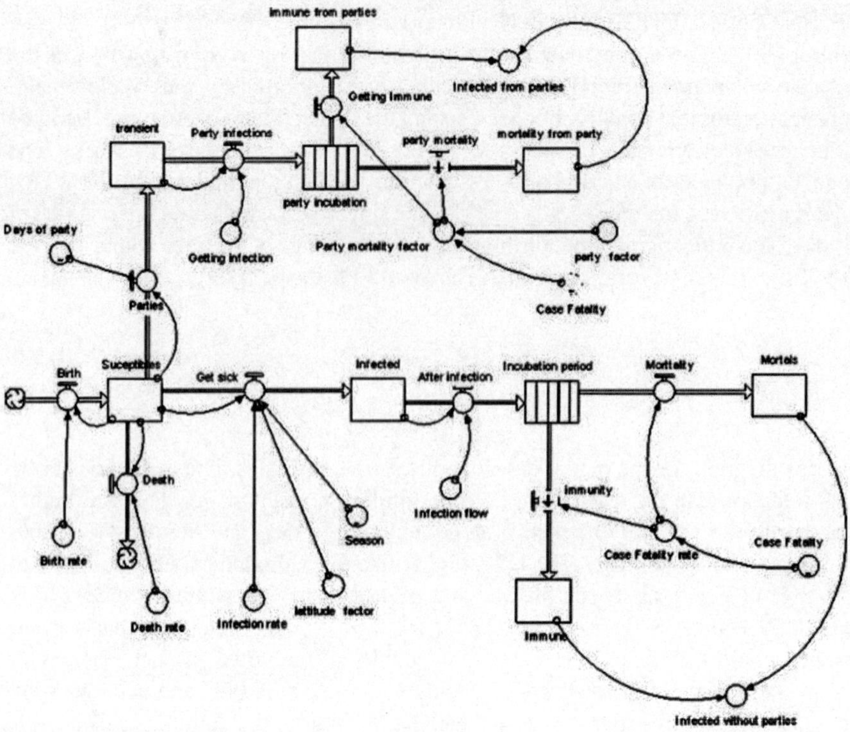

Fig. 1 Line diagram of the simulation model of immunity and mortality of pandemic H1N1 2009

Table 1 Immune population and mortality for pandemic H1N1 2009 with and without swine flu parties as predicted by simulation

Country	Days since start	Immune without swine party	Immune due to swine party	Deaths without swine party	Deaths with swine party
USA	70	37,329	–	213	–
	365	331,837	70,000,283	2,524	54,993
	730	670,920	136,261,232	5,185	107,049
Australia	42	5376	–	10	–
	365	21,873	4,818,028	57	1,513
	730	43,761	9,375,075	126	2,945
Mexico	70	9,746	–	117	–
	365	87,010	25,402,114	293	5,768
	730	176,777	49,653,853	497	11,274

immunity from pandemic H1N1 2009 after Year 1 and 44.2% after Year 2. The corresponding values for Mexico were predicted to be 22.7 and 43.8%. The mortality for the US, Australia, and Mexico are predicted to be 179.2 per million per year, 71.7 per million per year, and 51.64 per million per year, respectively. These values are compared to natural mortality without swine flu parties in Fig. 3. The Monte-Carlo simulations were conducted to simulate the effect of best-and worst-case assumptions based upon the confidence intervals of linear regression model coefficients and changes in mortality rates. The concept of attending herd parties were found to be unadvisable in all scenarios except when the mortality rate from H1N1 was less than or equal to the normal mortality rate of a country. The model has a few limitations. The model was based upon coefficients derived from data published by the WHO and CDC up to 10 July 2009. Patient-specific immune determinants such as age, gender, previous disease, or immunological characteristics have not been considered or adjusted for in the model. Crude case fatalities were used and were not adjusted for any confounders. The effect of any breakthrough vaccination—prophylactic or therapeutic—has not been considered in this model or the simulation thereof. Any mutations of the virus or the effects of different types of influenza infecting the same individual have not been considered. The contribution of swine flu parties to unintended further spread of pandemic H1N1 2009 or mutations of the virus has not been accounted for. Even with these limitations, the model was able to predict the immunity achieved and mortalities expected for guidance and policy analysis.

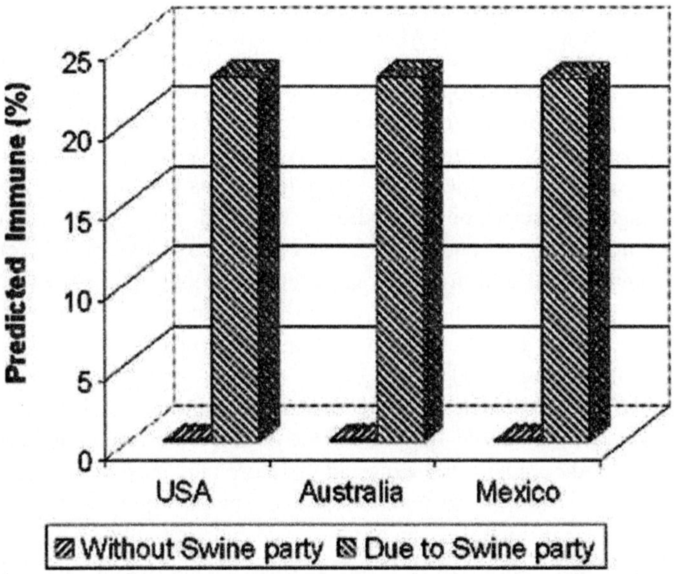

Fig. 2 Immune population (%) at the end of one year of pandemic H1N1 2009 with or without swine flu parties as predicted by simulation

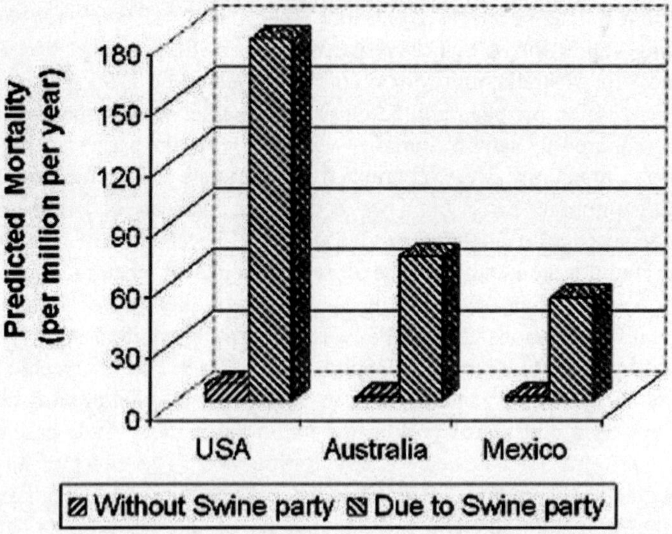

Fig. 3 Mortality per million per year of pandemic H1N1 2009 with or without swine flu parties as predicted by simulation

4 Discussion

Passive vaccination is not a novel idea. It has been repeatedly employed in pandemic conditions from Spanish Flu of 1918 to Measles outbreak of 1934 in US to Argentine hemorrhagic fever (Fischer et al. 2020). The difference in employment techniques between the latter and other two is the advancement of technology. Passive vaccinations can be broadly categorized as invasive and non-invasive. The invasive ones include convalescent plasma, which has the most convincing data for use so far in Argentine hemorrhagic fever (Abraham 2020). In it, plasma from patients already infected with the virus is given to patients without any symptoms or those exhibiting initial stages of the prodrome. The goal is for the body to develop antibodies to the 'mild' infectious agents as opposed to 'severe' ones exhibited during an infection. The non-invasive ones include the herd parties where people visit the person infected with the virus, in hopes of getting passive immunization. As such passive vaccinations can be a bridging therapy until the availability of vaccines or antiviral drugs.

In our analysis on swine flu, most of the people who attended the herd party would become ill with the virus. Based upon the person's immune system, he or she will develop antibodies against the virus and therefore may become immune to new infections from the virus in the future. However, based upon the virulence and concentration of the virus inhaled, the immune system may not be able to cope with this viral load of the disease and the person's health may deteriorate, leading to death. The high use of drugs to combat a surge in cases may possibly increase resistance against the drugs. Examples of herd parties include summer camps; the exception being that the students are in contact with probable pandemic cases for a longer period. This can foster contact with the virus for longer period than necessary, causing a full blown infectious course. Indeed, some camps/nursing homes had already been hit by flu cases before pandemic was officially declared in 2009 (New York Times 2009). The majority of these cases were assumed to be pandemic H1N1 2009 in the absence of proper and speedy tests. "This flu is not over " said Dr. Thomas R. Frieden, the head of the Centers for Disease Control and Prevention, describing its continuing spread in more than 50 summer camps and the initial detections of three cases resistant to the drug Tamiflu (New York Times 2009). Per World Health Organization (WHO) guidelines, the following public health measures are recommended to individuals and communities as the most feasible measures available to reduce or delay disease (morbidity) caused by pandemic influenza: social distancing, respiratory etiquette, hand hygiene, and household ventilation (World Health Organization 2009). In cases of mild illness, patients should be provided with supportive care at home by a designated caregiver and only referred to health care facilities if they deteriorate or develop danger signs. Separation of sick from well individuals as well as rigorous respiratory etiquette and hygiene measures, should be practiced. In health care settings, a system of triage, patient separation, prioritization of use of antiviral medicines and personal protective equipment (PPE) according to risk of exposure, and patient management should be in place to focus efforts on the most effective interventions to reduce mortality and any further morbidity.

The herd parties are not without severe consequences. Deliberate exposure of individuals to pandemic influenza can increase morbidity and mortality as observed in measles in Niger, Nigeria, and Chad (Grais et al. 2007). This escalation will add an additional burden on the public health systems and hospital facilities of developing countries and resource-limited economies. Individuals with other underlying health conditions as HIV, neurological diseases, respiratory complications and cardiovascular diseases would be at higher risk of complications and higher mortality by deliberate exposure as well as indirect contact with exposed individuals (Chowell et al. 2009). It is concluded based on our model that even though an appreciable amount of the population of a country can achieve passive immunization from herd parties in the absence of vaccination, this result comes at the cost of a high number of individuals who might have an adverse outcome by natural pandemic viral transmission. It is recommended that people do not resort to herd parties to achieve the low-cost passive immunity as mortality would increase, but each pandemic can have different virulence and mortality rates for assessment.

The concept of passive immunity is not cerebral only, as it has been clinically applied. During the COVID-19 pandemic, unlike its Nordic neighbors, Sweden was the only major industrialized country to institute a policy of herd immunity. Four months after the onset of cases, the preliminary results showed that Sweden had a total of 80,000 cases while the neighbors had between 7,000 and 15,000 cases (Kavaliunas et al. 2020). Similarly, the mortality rate for Sweden was 570 deaths per million, while for the neighbors, it was between 50 and 110 deaths per million.

This mortality rate of Sweden was similar to that of United States (503), which instituted a quarantine lockdown nationwide. There is also concern for people moving from herd immunity allowable areas to quarantined states, and thus, increasing the infection rate. Furthermore, there is much debate on what percentage of population needs to be infected or vaccinated before herd immunity is reached. The estimates vary from 10 to 20% to as high as 70% (Aguas et al. 2020). In addition, once a patient is infected with COVID-19, that person tends to make detectable antibodies 90% of the time (Long et al. 2020). This implies that the passive immunity can possibly be achieved by attending herd parties. However, the success of this strategy requires not only cooperation from states and governments, but also individual responsibilities to curtail the spread of the virus (Orlowski and Goldsmith 2020). The final impact of such strategies can only be fully understood several years after the pandemic ceases to exist.

5 Conclusion

The general population is always concerned about the devastating effects of a pandemic. Based upon the 1918 Spanish flu pandemic, scientific publications and theories, the idea that herd parties may achieve passive immunity against pandemic viruses has been proposed. Public health officials have generally asked the general public not to resort to these parties. In this chapter, we dynamically modeled the effect of swine flu parties on the immunity achieved and associated mortality for a period of two years. The verdict for passive vaccinations for other viruses will depend on the virulence and mortality rate.

References

Abraham J. Passive antibody therapy in COVID-19. Nat Rev Immunol. 2020;20(7):401–3.

Aguas R, et al. Herd immunity thresholds for SARS-CoV-2 estimated from unfolding epidemics. medRxiv. 2020. https://doi.org/10.1101/2020.07.23.20160762.

Barry JM, Viboud C, Simonsen L. Cross-protection between successive waves of the 1918–1919 influenza pandemic: epidemiological evidence from US Army camps and from Britain. J Infect Dis. 2008;198(10):1427–34.

Centers for Disease Control. CDC Novel H1N1 Flu Situation Update. 2009. http://www.cdc.gov/h1n1flu/update.htm.

Chowell G, et al. Severe respiratory disease concurrent with the circulation of H1N1 influenza. N Engl J Med. 2009;361(7):674–9.

Fischer JC, et al. The role of passive immunization in the age of SARS-CoV-2: an update. Eur J Med Res. 2020;25(1):16.

Grais RF, et al. Unacceptably high mortality related to measles epidemics in Niger, Nigeria, and Chad. PLoS Med. 2007;4(1):e16.

Hethcote HW. Qualitative analyses of communicable disease models. Math Biosci. 1976;28:335–56.

http://www.cnn.com/2009/HEALTH/06/30/flu.party/, C.c.D.w.a.s.f.p.A.

Itoh Y, et al. In vitro and in vivo characterization of new swine-origin H1N1 influenza viruses. Nature. 2009;460(7258):1021–5.

Kavaliunas A, et al. Swedish policy analysis for Covid-19. Health Policy Technol. 2020.

Long Q-X, et al. Antibody responses to SARS-CoV-2 in patients with COVID-19. Nat Med. 2020;26(6):845–8.

Maines TR, et al. Transmission and pathogenesis of swine-origin 2009 A(H1N1) influenza viruses in ferrets and mice. Science. 2009;325(5939):484–7.

New York Times. Obama Warns of Return of Swine Flu in the Fall. 2009. New York Times, 10 July 2009: A18.

Orlowski EJW, Goldsmith DJA. Four months into the COVID-19 pandemic, Sweden's prized herd immunity is nowhere in sight. J R Soc Med. 2020;113(8):292–8.

The World Factbook. CIA. 2009. https://www.cia.gov/library/publications/the-world-factbook/.

Time.com. Swine Flu in Britain: Nothing to Party About. 2009. http://www.time.com/time/health/article/0,8599,1908407,00.html. Accessed 13 July 2009.

World Health Organisation. Situation updates -Pandemic (H1N1) 2009-Update 58. 2009. http://www.who.int/csr/disease/swineflu/updates/en/index.html.

World Health Organization. Pandemic influenza prevention and mitigation in low resource communities. 2009. http://www.who.int/csr/resources/publications/swineflu/low_resource_measures/en/index.html.

General Virology

Phage Therapy of Human Bacterial Infections: A Systematic Review

Shamim I. Ahmad

Abstract The discovery of penicillin by Sir Alexander Fleming in 1928 and its publication in the *British Journal of Experimental Pathology* in 1929 could not be taken seriously until in 1931 when Howard Flory and Ernst Chain, the two researchers at the University of Oxford took it over and finished. Thus these two scientists were credited with the development of penicillin and use who experimented it on mice. Then the demand of this antibiotic increased during World War II and its serious commercialisation started by a US drug company in 1943 when the war ended. The production of this drug reached to 650 billion units per month. To be noted that although there were no proper treatment for most bacterial infections prior to the discovery of penicillin and hence the death rate was uncontrollably high, in Tbilisi, Georgia (then a Russian State) around 1921 and in Poland around 1946 a therapy was ongoing when people routinely were drinking mixture of crude bacteriophages to get rid of some kind of stomach problems most likely infected by bacteria. Then industrial production of penicillin spread globally and interest in phage cocktail was suppressed. More and more types of antibiotics were discovered and many of them were engineered for their increased action, stability and safety. Then came the problem of resistance to antibiotics and the reasons were the incidence of mutations in bacteria leading to antibiotic resistance which at times were of multiple drugs. The mutations were either chromosomal or by receiving antibiotic resistant plasmids from the other bacteria, including interspecies or intraspecies mating or by transformation of plasmids in the environment. When this problem started becoming out of control then the interest of scientific communities and the clinicians started growing up towards the phage therapy. This chapter is presenting a comprehensive review of the subject.

S. I. Ahmad (✉)
Former Senior Lecturer, School of Science and Technology, Nottingham Trent University, Clifton Lane, Nottingham NG8 2RF, England
e-mail: shamim.ahmad@ntu.ac.uk

S. I. Ahmad (ed.), *Human Viruses: Diseases, Treatments and Vaccines*, https://doi.org/10.1007/978-3-030-71165-8_32

Keywords Phage therapy · Pseudomonas aeruginosa · Escherichia coli · Staphylococcus aureus · Streptococcus agalactiae · Shigella flexneri · RNA phage MS2 · Double stranded DNA tailed phages · T2 · T4 · T7 · Phage lytic cycle · Phage lysogenic cycle · Multiple drug resistant (MDR) · Murein hydrolase · Holin · Corynebacterium diphtheriae · Vibrio cholerae · Extensively Drug Resistant (XDR) bacteria · Pan Drug Resistant (PDR) bacteria · Acinetobacter baumannii · Bacillus anthracis · Human vaccine · Nano particle encapsulation · Mycobacterium tuberculosis · Breast cancer · Xeroderma Pigmentosum

1 Introduction

Almost all living organisms, so far studied, are infected by one or more groups of viruses acting as pest, can invade and grow inside the cells of algae, fungi, protozoa, plant and animals. Viruses infecting and growing in bacteria are known as bacteriophages or just phages. Living cells are mostly the targets; phages adsorb, enter in the cell or just transfer their nucleic acid inside the cell, complete their life cycle and come out to attack the remaining population constituting a cyclic process in a steady state condition. Although a number of human viruses known as zoonotic viruses which can infect and complete their life cycle on animals as well as on human, most phages are highly host specific infecting selected species of bacteria. However few exceptions exist such as *Escherichia coli* phage *sfin-1* can also infect *Shigella flexneri, S. dysenteriae* and *S. sonnei* (Ahamed et al. 2019). Another interesting feature of phages is that for one species of bacteria there may be a number of different viruses infecting them. An example is the bacterium *Escherichia coli*, for which there are at least 19 different phages such as small single-stranded DNA phage ΦX174, single stranded RNA phage MS2, double stranded DNA tailed phages: λ, P1, M13, T1, T4, T6, T7, as well phages Φ6, Φ29, G4, P2, P4, T2 T12, 186, N4 and R17 phages. Their specificity and number adds further importance in employing bacteriophages in phage therapy (Bragg et al. 2014).

2 Structures and Compositions of Bacteriophages

Phages are of different shapes and sizes; a simple spherical, filamentous and the complex tadpole shaped. T phages are of the latter type which contains a capsid or head attached with a hollow tube surrounded by contractile sheath. In between them are present collar and with the sheath are attached tail fibres and short tail pins.

Most phages have an external coat known as a capsid which encloses their genetic materials. The capsids are made up of smaller protein subunits known as capsomers. These are identical building blocks assembled in simple geometrical principles. Basically there are two types of capsids, helical or isometric. Although rare but certain bacteriophages such as PM2 retains an external membranous envelop covering the capsid made of lipids bilayer.

Inside the capsid is enclosed viral nucleic acid which can be DNA or RNA. Also encapsidated are a number of proteins required for the early stage of phage metabolism. Present in some phages are endolysins which are the enzymes used by the organisms to hydrolyze bacterial cell wall to get entrance in the cell. Three forms of endolysins have been discovered including holin-endolysin, pinholin-SAR and spanins (Cahill and Young 2018).

3 Life Cycle of Bacteriophages

Two main types of phage life cycles are the lytic in which the phage nucleic acid enters the bacterial cell, complete its cycle, the cell is broken open by lysis and the progeny phages are released. The second type is known as lysogenic cycle in which the phage nucleic acid remains inside the host cell until such time it is induced to enter in lytic cycle and the phage progeny is released. In lysogenic cycle the phage nucleic acid in the host, known as prophage either integrates in bacterial nucleic acid and replicate along with host nucleic acid or as in phage P4 it can stay independently as a plasmid and replicates under the control (or in some cases uncontrolled) of the host cells nucleic acid replication (Briani et al. 2001). Phages able to enter in lytic cycle are known as virulent phages and those enter in lysogenic cycle are known as temperate phages.

4 Lytic and Lysogenic Cycles of Bacteriophages

Lytic cycle

Here phage T4 of *E. coli* is presented as an example. The first stage of the cycle is the attachment of phage at the surface of the susceptible host cell. Although the bacterial cells are protected by a layer of polysaccharides (Temple et al. 1986) preventing attack for example by antibiotics, phage attaching to its receptor on the cell surface which can be teichoic acid, a protein, flagella or even the polysaccharides. This process can differ for different phages and is known as adsorption. Two proteins, murein hydrolase (lysin) and holin, a membrane protein, help to get access of the phage nucleic acid into the host cell. The function of holin is to generate a lesion in the cell membrane and through this murein hydrolase can pass. The phage sheath contracts and the internal hollow tube inserted in the cell like a hypodermic needle. Phage head is squeezed and the nucleic acid along with some essential proteins is transferred in the lumen of the cell. Together with nucleic acid certain phage contents are also transferred and these then take over the control of the host metabolism through degrading certain cellular constituents. Bacterial strains can develop resistance to their specific phage(s) by mutation or by removing the receptor site.

Soon after invasion of the phage genome, host metabolism is paralysed and normal synthesis of its proteins and nucleic acid is disrupted. Ribosome of the host cell starts translating certain early viral mRNA into proteins. For RNA phages RNA replicase is synthesised and the proteins modify the bacterial RNA polymerase, and the viral RNA is transcribed. Soon after, a rapid synthesis of phage nucleic acid and its structural proteins occur followed by the assembly of progeny phages. Phage assembly is a well organised process in which the base plates are assembled first flowed by the tails being built upon them. The phage head is built separately and the nucleic acid is packed. Finally the phage head and phage tails are joined completing its progeny production. Depending upon the type of phage, the number of particles produced and the time taken from start to end inside the host can vary significantly.

Final stage of the phage lytic cycle is the release of progeny phages from the cell which occurs by lysing the cell or in certain case by budding. Endolysin is used which breaks down the cell wall. Certain filamentous phages such as of *Mycoplasma* release its progeny particle by continuous secretion in which case the cell is not destroyed (Maniloff 1988). The process, by which virus progeny is released to find new hosts, is called shedding. This is the final stage in the viral lytic cycle. The newly produced viruses find new cells to continue production. For phage therapy mostly virulent phages are employed (Bragg et al. 2014).

Lysogenic cycle

In lysogenic cycle the virus in some way "hide within the cell", either to evade the host cell defences or immune system, or simply because it is not in the best interest of the virus to continually replicate. Phage λ (lambda) is a good example which can undergo both, lytic and lysogenic cycles. For this, after invasion, most phage infections undergo lytic cycle but a small number of phage DNA attaches itself at certain specific region of the host genome. In that case the host survives and known as lysogenic cell. *E. coli* have been an important host which has been extensively used to study lysogenic phages and their cycles.

During lysogenisation the viral nucleic acid, after integration in the host genome becomes one replicon and in this form the phage cannot produce any progeny. The lysogenic cell on the other hand can keep on producing lysogenic cells in which the phage nucleic acid remains attached. The cycle continues until an external stimulus such as ultraviolet light or agent, able to damage DNA, or another environmental stress prompt prophage to become activated and in this case the viral genome is released from the cellular genome and enter in lytic phase (as described above). Many phage particles are produced, lyse the cell and release the progeny particles in the medium. Interestingly the lysogenic bacterial cells remain resistant to any super-infection by the same phage for which the host cell is lysogenic (Kasman and Porter 2020). A intriguing phenomenon known as Lysogenic Conversion noted with certain host cells, in that the harmless strains *Corynebacterium diphtheriae* (Mokrousov 2009) and *Vibrio cholerae* have been found to become highly virulent (Charles and Ryan 2011).

5 History of Bacteriophage Therapies

Bacteriophage therapy started as early as in 1920 in Tbilisi, Georgia when Richard Bruynoghe and his student Joseph Maisin published the first use of bacteriophages in phage therapy (Lavigne and Robben 1951) and in Poland 1946 (Zabezhinsky and Gorstkina-Shevandronova 1946). This idea remained suppressed until 1980, when the scientists all over the world started taking interest in phages to be as an alternative or combined with conventional antibiotics to become a promising agent in the treatment or prophylaxis of bacterial infectious diseases (Sula and Sulová 1984; Smith and Huggins 1982). Subsequent universal appearance of multiple drug resistant (MDR), Extensively Drug Resistant (XDR) and Pan Drug Resistant (PDR) bacteria (Moghadam et al. 2020) had induced considerable interest in phage therapy which were presented during nineties (Barrow and Soothill 1997; Alisky et al. 1998; Carlton 1999).

Now in Tbilisi there exists a Phage Therapy Centre, claiming to have "effective treatment solutions for patients who have bacterial infections that are difficult/ nonhealing, chronic, drug-resistant or do not otherwise respond to conventional antibiotic therapies and a number of different kind of diseases associated with bacterial infections are being treated and success rate being claimed to be 80%.

Why phage therapy remained silent from 1920s until 1950s? The answers to this question have been put forward by Carlton (1999) that: (i) there was no adequate scientific methodologies applied by the so called physicians such as placebo controlled studies, (ii) how endotoxins could be removed from the preparations and to confirm phage viability after going through sterilizing process, (iii) the host range and rapid clearance from the body was not known, (iv) the physiology of lysogenic and lytic cycles were not discovered.

Since its first publication in 1946, the number of publication to date (PubMed, October 25, 2020) remains a mere 4,227, reflecting still a lack of interest in this research field in compare to studies, such as in cancer therapy reaching to 766,019.

According to the report (Ahmad 2002), the success rates of phage therapy in Tbilisi and in Poland were having 80% to treat Enterococcus infections and up to 90%, against *S. aureus*, *P. aeruginosa, Escherichia coli* and *Klebsiella pneumoniae*.

According to Moelling et al. (2018) to date there have been no reports of any serious adverse events caused by phage therapy. In fact in Eastern Europe, for several decades' phage therapy has been successfully going on to cure certain human diseases. However, these authors emphasise that the regulatory and legal frameworks in the Western World are causing major hurdles to the introduction of phage therapy,these regulations may take a decade or longer to be fulfilled, albeit of an urgent need to speed up the availability to be utilised. In another study McCallin et al. (2013) investigated the composition of phage cocktail from Russian Pharmaceutical Company. The cocktail were to target *E. coli* and *Proteus* infections. EM study identified 6 phage types amongst them T7-like phages dominated over T4-like phages. Further detailed studies identified 18 distinct types including 7 genera of Podoviridae, 4 genera of Myoviridae and 2 genera of Siphoviridae. Small volunteer trial did not associate any adverse effects when given orally (McCallin et al. 2013). Further importance of phage therapy has been highlighted by Weber-Dabrowska et al. (2000) in which a large number of patients with suppurative bacterial infection, caused by multidrug-resistant bacteria were treated with specific bacteriophages. It was found that phage therapy was highly effective showing full recovery (85.9%). Transient improvement was observed in 10.9% and only 3.8% was non-effective. The conclusion was that bacteriophage treatment can be a highly valuable tool in combating bacterial infection specially the multiple resistant species of bacteria (Weber-Dabrowska et al. 2000).

The history of phage therapy may be concluded by referring the review compiled by Alisky et al. 1998. They pooled the data from Medline citations dating from 1966 to 1996 which included 27 papers from Poland, the Soviet Union, Britain and the USA. The study combined a variety of human diseases including dermatitis, empyemas, gastroenteritis, osteomyelitis, pneumonia, sepsis and suppurative wound infection. Pathogens responsible for the diseases included *Escherichia, Klebsiella, Proteus, Pseudomonas, Shigella, Staphylococcus, Streptococcus, Vibrio and Salmonella spp.* The success rate of therapy described by Polish and Soviets researchers was 80–95% with few cases of gastrointestinal and allergic reactions. About the dosage and clinical criteria no clear indications were presented. Demonstrable efficacy, however, could be shown when studies were done against *Escherichia, Acinetobacter, Pseudomonas and Staphylococcus spp*. The conclusion drawn was that bacteriophages may be used for treating antibiotic resistant pathogens. To be noted that this kind of treatment may be more valuable in the Third World countries where there exists lack of stringent safety regulations and proper hygiene, as well the cocktail of antibiotics may not be available or affordable.

6 Bacterial Species Resistant to Antibiotics and/or Phages

Antibiotic resistance bacterial species and their resistant genes have most likely been present since ancient time in response to naturally produced antibiotics by organisms in nature; naturally produced streptomycin by streptomyces and penicillin by Penicillium are good examples. A variety of antibiotics isolated from microbes, their chemical modifications and their heavily industrial production and their use have been ongoing for many years. For these reasons resistant bacteria for various antibiotics have also been emerging due to their overuse, misuse especially where the drugs are available at the counters without prescriptions by the physicians. So much so, that certain bacterial species have become resistant to a number of antibiotics known as MDR, PDR or FDR bacterium. This has been leading to human fatalities of those infections which were easily controllable by single antibiotic in the past. This situation has led clinicians and industries to find alternative solution to overpower these multiple drug resistant bacteria. This gave rise to a new surge to develop phages to treat these pathogens, now called Bacteriophage Therapy or just Phage Therapy.

Besides developing resistance against the conventional antibiotics, the infectious bacteria can also develop resistance against the phages used in phage therapy. It is likely that for certain phage therapy appropriate phages may not be available, in that, phages found are either lysogenic which can lysogenise bacterial species making it resistant to further attack or sustaining genomic mutation making them resistant. Experimental data are available that phage resistant variants occurred in up to 80% studies targeting the intestinal milieu and 50% of studies using sepsis models (Oechslin 2018).

7 Infectious Pathogens and Their Phages

Since the discovery of phage cocktails, used by the Georgians, a surge began to find how and where the bacteriophages can be used specially for those diseases infected by MDR XDR and PDR bacteria (Moghadam et al. 2020). MDR bacteria were noticed not only emerged in hospitals but in community settings and therefore requires extensive vigilance as community based resistance emergence which can stay unnoticed for a long time. The basic mechanism of antibiotic resistance is based on the Darwinian Theory of "Survival of the Fittest" including genetic plasticity of the pathogen which results in adaptation to mutation, acquisition of genetic material (plasmids and phage nucleic acid), alteration of gene expression leading a large number of antibiotics available (Munita and Arias 2016). Below is the list of bacterial species and their phages suitable to be used in phage therapy.

Acinetobacter baumannii: For this bacterium a number of bacteriophages have been isolated one of them was TAC1 of Myoviridae family and was able to infect a broad range of intraspecies hosts (Asif et al. 2002). Other bacteriophage, SH-Ab 15,497 isolated, characterised and found to be able of infecting *A. baumannii* and be used in phage therapy to treat infection by multidrug resistant bacteria (Hua et al. 2019). Due to the presence of extensive number of worldwide XDR stains of *A. baumannii,* attempts continued to find more bacteriophages able to infect this bacterium. The phage AB1801 of the order Caudovirales, family Siphoviridae could infect at least 70% of XDR strains. Also inhibit the biofilm formation of the host leading to reduce performed biofilm in dose-dependent manner (Wintachai et al. 2019). In another study two new strains of virulent, vB_AbaS_D0 and vB_AbaP_D2, were isolated belonging to the Siphoviridae family with an icosahedral head. Although this phage had a very weak bactericidal activity, it could infect a wide range of its host. When a cocktail of both bacteriophages were applied their synergistic therapeutic efficacy improved significantly (Yuan et al. 2019). Subsequently three new temperate new phages, vB_AbaS_fEg-Aba01, vB_AbaS_fLi-Aba02 and vB_AbaS_fLi-Aba03 were able to infect only one clinical strain of *A. baumannii* (Badawy et al. 2020). Yet in another study phage, vB_AbaM_PhT2 showed 28% host range against 150 MDR isolates of

A. baumannii. This phage was capable to produce an endotoxin which was non-toxic to human cell line in vitro. This bacteriophage was proposed to have a potential for phage therapy at least to treat surface microbial infection (Styles et al. 2020). In another study with *A. baumannii* therapeutic action of phage Abp1 was demonstrated on PDR strain. This phage too was proposed to be an excellent candidate for therapy against *A. baumannii* (Yin et al. 2017).

Bacillus anthracis: This bacterium is considered to be one of the most unwanted species in bacterial world. It has also been considered and may already have been used in Biological Warfare. Due to this reason and the availability of multidrug resistance species, an alternative to antibiotic treatment, phage therapy remains to be an imminent kind of treatment. And for this a myoviridae phage vB-BceM_HSE3 has been isolated and analysed. Besides *B. Anthracis*, this phage can also infect infects MDR *B. cereus* group strains, and *B. thuringiensis*. Further characterisations of this phage led to the conclusion that it can be used in phage therapy (Peng and Uan 2018). In another study three highly effective *B. anthracis* phages, Negev SA, CarmelSA and TavorSA have been isolated. Based on other features such as complete genome sequencing it was proposed to have potentials in phage therapy against anthrax (Alkalay et al. 2018).

Bacillus cereus: This bacterium is a food- born pathogen causing intestinal illness such as diarrhoea, vomiting as well infection of eyes, respiratory tract and wound through producing a number of toxins. MDR of this species has emerged hence an alternative treatment of antibiotic must be found. A myoviridae family phage vB_BceM-HSE3 has been isolated and analysed. This phage can produce the endolysin, PlyHSE3, and hence can infect and lyse *B. cereus, B. anthracis* and *B. thuringiensis*. It is proposed that this phage can play an important role in phage therapy (Peng and Uan 2018).

Bordetella bronchiseptica: Although the true role of this bacterium in human disease is not clear, hypervirulent strains of this bacterium may be responsible for respiratory infection in human. Due to the presence of antibiotic resistance gene in this bacterium, for phage therapy, a lytic bacteriophage vB_BbrP_BB8 of this virus of Podoviridae family has been isolated. This phage can cause complete lysis of the bacteria, including removal of bacterial biofilm with full efficiency (Alkalay et al. 2018) and hence can be used in phage therapy.

Clostridium perfringens is a gram-positive anaerobic bacterium present mostly in soil and the gastrointestinal tracts of human causing gas gangrene and food poisoning. The phage PlyCP41 engineered and used its glycosyl hydrolase endolysin which is produced by infecting plant tissue with potential for the use of crude plant sap as an effective agent against *C. perfringens* (Hammond et al. 2019).

Clostridium difficile: The new name of this bacterium is *Clostridioides difficile*, one of the intestinal pathogen, and its phage JD03 isolated, propagated and characterised. It has been analysed for the global transcriptomic changes during phage JD032 infection in the hypervirulent RT078 difficile strain, TW11. Phage was found to have a complex interaction and suggestion is that the phage undergoing the lytic cycle, can be different in host similar to prophage and inspire phage therapy (Li et al. 2020).

Elizabethkingia anopheles: This organism can cause meningitis in neonats and in immunosuppressed patients and lead to high mortality. For their MDR bacteria, it is only phage therapy as alternative for patient's survival. For this a phage TCUEAP1 was isolated from waste water in a hospital. This double stranded DNA phage belonged to Siphoviridae family was tested on the host bacteria which displayed narrow spectrum infectivity out of 16 strains tested of *E. anopheles* (Peng et al. 2020).

Enterococcus faecalis: This gram positive bacterium is frequently found in oral cavities and gastrointestinal tract causing a variety of diseases including endocarditis, urinary tract infection and bacteraemia. It is specifically been associated with persistence infections in root canal treatment failure. A novel phage has been isolated, vB_EfaS_HEf13 (phage HEf 13), which showed broad lytic activity against 12 different strains of S. *faecalis*. A detailed analysis suggested that this phage has the lytic cycle and can be a potential therapeutic agent to treat or prevent of *E. Faecalis*-associated infectious diseases (Lee et al. 2019). Its MDR variant due to its biofilm becomes more difficult to be treated. Next a phage EFDG1 belonging to Spounavirinae subfamily infecting as anti-*E. faecalis* and *E. faecium* has been isolated and analysed. Experiments have shown that this phage has effective lytic activity against various *E. faecalis* and *E. faecium* and can be especially suitable to treat root canal infection (Khalifa et al. 2015).

Enterococcus faecium In a recent study a phage EfV12-Φ of *E. faecium* was isolated and analysed but no data are available about its ability to be used in phage therapy (Wandro et al. 2019). However, in an earlier study another *E. faecium* phage IME-EFm5 was isolated and analysed and was found that this phage produces an endolysin LysEFm5, an amidase which has a bactericidal activity (Gong et al. 2016). The phage enzyme has been detailed below in section: "Phage enzymes and lytic proteins in phage therapies".

Escherichia coli: Although this commensal bacterium is present in most human colon, is considered commonly as a harmless organism. This bacterium which supplies vitamin K for human metabolism, when it receives a plasmid with toxin producing gene, can become pathogenic. Examples of pathogenic *E. coli* are: O157: H7, O104:H4, O121, O26, O103, O111, O145, DE192, DE205B and O104:H21 produce potentially lethal toxins and its pathogenic variants can cause serious food poisoning, including septic shock, meningitis, or urinary tract infections in humans. Recent studies include extra-intestinal pathogenic *E. coli* (ExPEC) adherent invasive *E. coli* (AIEC) and Shiga-toxin producing enteroaggregative *E. coli* (STEAEC). In fact during 2007–2017 an increase of septicaemia in Chinese children became a public concern. Analysis of an isolate from the patients was found to have become resistant to a number of first generation derivatives of cephalosporin; cephazolin, cephalothin, the second generation cephalosporin, cefoxitin, and cephuroxime, the third generation derivative of cephalosporin, cephotaxime, ceftazidime, cefoperazome, cefoperazome/sulbactam, ceftriaxone and cephepime and the fourth generation cephalosporin, cephepime (Wang et al. 2007). Due to this large number of pathogenic *E. coli* attempts have been ongoing to find phages which can be used in phage therapy to treat MDR *E. coli* strains.

For this, a number of bacteriophages have been isolated including the DNA double stranded phage vB_Eco4M-7 which effectively can infect a number of *E. coli* strains including O157 exhibited the properties of an early isolated phage ECML-117 (Necel et al. 2020; Montso et al. 2019). In another study two phages WG01 and QL01 were engineered to generate a phage WGqlae which showed significant effects on *E. coli* DE192 and DE205B strains. Interestingly this phage had an additional inhibitory effect on the biofilm and its clearance (Li et al. 2020). Another bacteriophage, VB_EcoS-Golestan has been isolated and analysed showing its potential to treat MDR *E. coli* strain (Yazdi et al. 2020). A recent alarming statement is in which was isolated a MDR *E. coli* strain having addionally developed resistance to the last effective antibiotic, colistin. Genetic study of this bacterial strain showed that it is carrying a mutation in mcr-1 gene encoding resistance the collistin (Johura et al. 2020). For further studies on *E. coli* EV36 bacterial infection, causing neonatal meningitis, a bacteriophage K1F was isolated and tested in vitro on human cerebral microvascular endothelial cells. The result was encouraging to adapt the phage therapy for the treatment of the disease (Moller-Olsen et al. 2020).

Edwardsiella tarda: This bacterium has been associated with gastrointestinal disease in human and an increasing number of cases have been observed as Entrada bacteraemia. It can also cause mycotic aneurism which although is rare disease but can be fatal. Thus phage therapy remains an alternative treatment and for this treatment Edwardsiella tarda phage, ETP-1, has been isolated and analysed; its experimental data suggest that this phage can become a potential bio-therapeutic for the MDR *Edwardsiella tarda* (Nikapitiya et al. 2020).

Helicobacter pylori: This bacterium is involved in causing stomach ulcer and may also induce stomach cancer. Although a number of phages for this bacterium have been isolated but most studies have focussed on genome structure and function. One isolated phage of *H. pylori*, KHP30, with spherical shape and containing lipid around the shell is highly unusual in that it is stable from pH 2.5 to pH 10. In bacterial cells it is present as an episome and is considered to be a completely new virus different from all others isolated. Information on its role in phage therapy is very limited (Uchiyama etal. 2013) and may be more effort is required in finding a suitable phage to be used in therapy.

7.1 Klebsiella Species

A number of *Klebsiella* spp. have been identified including *K. aerogenes, K. bacteraemia, K. Oxytoca, K. pneumoniae, K. variicola and K. pneumoniae var KN2*. They induce different human diseases including urinary tract infections (UTIs) which are the most common infections in renal transplant recipients, nosocomial infections in immunocompromised patients, nosocomial pneumonia and intraabdominal infections, bloodstream infections, wound or surgical site infections, and meningitis. For these reasons the phage selection to treat different *Klebsiella* spp. has been addressed separately.

Klebsiella aerogenes: A jumbo phage N1M2 has been isolated from maize silage which is capable of successfully removing a pre-formed biofilm from *K. aerogenes.* When this phage was tested combined with phage K, another phage of S. *aureus,* the combination worked significantly better than the phages employed individually (Lewis et al. 2020).

Klebsiella oxytoca: Genus Klebsiella is well known opportunistic pathogen linked in human suffering from diabetes and chronic pulmonary obstructions and in animals induce mastitis which is an inflammation of the mammary gland that occurs when pathogen enters the udder. A number of strains of this pathogen have been found which are resistant to multiple antibiotics. To control this pathogen by phage therapy a phage PKO11 infecting *K. oxytoca* has been isolated and studied. This phage appears to be stable at 4–60 °C and at pH 3–11. No human virulence related genes which is important from safety point of views has been found. Thus this phage may be used in phage therapy (Park et al. 2017). In another study two bacteriophages were isolated from sewage-contaminated river in Georgia (Russian State). These are vB_Klp_5 and vB_Klox_2 belonging to the family Podoviridae and Siphoviridae respectively were studied in full detail to be used in phage therapy. Good burst size, efficient lysis, efficient rates of adsorption and stability maintaining at different adverse conditions were noted. These properties reflected that these phages have potential utility in phage therapy cocktail (Karumidze et al. 2013).

Klebsiella pneumoniae: This bacterial species seems to have attracted most researcher because of its being opportunistic nosocomial pathogen. Thus, in a study by Gao et al. (2020) a novel virulent phage, vB_KpnP_IME337 was isolated which was capable of infecting carbapenem resistant *K. pneumoniae.* In another recent study a new phage is isolated called vB_KleS-HSE3 which displays high antibacterial activity and high physical stability towards four MDR strains of *K. pneumoniae*. The phage belonging to Siphoviridae and has potentials for controlling infections of MDR bacterial strain (Peng et al. 2020).

Due to the high diversity of this bacterial species it became difficult to isolate a phage to target all the species. *K. pneumoniae* is capable of producing a hypervariable extracellular matrix which makes it to produce a dozen or more of its variants. However, researchers were able to isolate four novel microbes infecting *K. Pneumoniae* capsular type K22. Analysis of these phages revealed that they are highly host specific and encoded putative depolymerases that could digest the capsule of K22 *K. pneumoniae* K22. It was proposed that these phages may prove suitable for combating MDR species of *K. pneumoniae* (Domingo-Calap et al. 2020). In other study Wu et al. isolated a novel bacteriophage SH-KP152226, encoding polysaccharide depolymerase. This phage was found to confer specific activity against multidrug-resistant *K. Pneumoniae* via biofilm (Wu et al. 2019). The most recent development for the phage associated therapy of *K. pneumoniae* is the isolation of a T7 like lytic phage kpssk3, its characterization and genomic analysis is postulated to lead to our understanding of phage biology and diversity and provide a potential strategy for controlling carbapenem-resistant *K. pneumoniae* (Shi et al. 2020).

Mycobacterium tuberculosis: To be noted that by June 2019, 10,427 mycobacteriophages had been isolated and 1670 of them had their genomes sequenced. This group of bacteriophages must be on the top of the list to be isolated and study their possible use in phage therapy to treat mycobacterial diseases, the most notorious disease tuberculosis which is caused by *Mycobacterium tuberculosis* and even their non-MDRA strain are difficult to be treated due to their unusual mycolic acid-containing cell wall through which most drugs cannot permeate inside the cell. Also MDR-TB and XDR-TB have been found. In mycobacteriophage SWU1 has been found a novel gene SWU1gp39 whose function, although is unknown but when this gene is expressed in *M. smegmatis* leads to susceptibility of multiple antibiotics, isoniazid, erythromycin, norfloxacin, ampicillin, ciprofloxacin, ofloxacin, rifamycin and vancomycin resistance (Li et al. 2016). It is surprising that despite such a valuable gene product found in bacteriophage, no further information about this gene is available (personal communication with the author).

Three other mycobacterial phages isolated, MTCC993, kansai MTCC3058, avium subsp., avium MTCC1723 and tuberculosis MTCC300, found to infect four different species of *Mycobacterium fortuitum*. It was proposed that this could be the first step of exploitation of phage therapy of this group of bacteria (Satish and Desouza 2019). In another study yet another mycobacteriophage Bo4 isolated and analysed which could infect and lyse *M. tuberculosis* and *M. smegmatis*. This can be achieved when the bacteria is present in blood and lysosomal macrophages. Furthermore this phage did not contain any such harmful gene whose product can increase mycobacterial virulence or decrease human immunity and possesses the potential to destroy *M. tuberculosis* (Gan et al. 2014). Yet in another study a virulent bacteriophage D29 was isolated and analysed. The phage was able to infect and kill several species of *Mycobacterium* including *M. tuberculosis*. This ability was achieved by producing three proteins: peptidoglycan hydrolase (LysA), micolylarabinogalactan ester (LysB) and holin. Combined together the lysis of the host occurred in a timely manner (Bavda and Jain 2020).

Mycobacterium leprae is the bacterial strain which causes leprosy. This disease prevails globally and often difficult to be treated. The disease has attracted a limited number of attentions as the number of publications on its therapies is limited to 2,157 (PubMed October, 2020). According to a recent report the bacterium causing leprosy has started developing resistance to a number of drugs, including rifampicin, dapsone and ofloxacin (Mahajan et al. 2020). Considering the demand to treat the multi-resistant *M. leprae,* Sula and Sulová (1984) have carried out a search for the phages infecting of *M. Leprae* murium “Douglas”. Their newly isolated phage, Al-1 showed lytic activity of this bacterium with the possibility of its use in the phage therapy of MDR bacteria (Sula and Sulová 1984).

Pseudomonas aeruginosa: It is a one of the most notorious species of bacterium, gram negative, the leading cause nosocomial infections, and inducing human cystic fibrosis (CF); the disease most commonly is a genetic conditions caused due to mutation in any one of the 1700 different human genes. The patients suffer for the thickening of the mucus due to the infection of airways leading to lung damage. Mostly present in immunocompromised patients, and the bacterial natural resistance to a number of antibiotics phage therapy has been considered to alternative action requires treating patients suffering from CF (Ahmad Shamim 2017). Their resistance is based on a number of factors including the chromosomally encoded efflux pumps, poor permeability of outer membrane partly based on the production of biofilm around the cell for which treatment of CF and other infections become very difficult to be treated. Due to poor permeability of membrane only a limited number of antibiotics can be used. Also those antibiotics which can be employed, sooner or later the bacterium develops resistant and most commonly it becomes impossible to be treated and leads to fatality. For this reasons persistent efforts are continuing to find phages which can be used to treat Pseudomonas infection.

A number of phages isolated and studied of *P. aeruginosa* include PA-YS35 whose genes are fully sequenced with a view to be used in phage therapy (Jiang et al. 2020). Then a lytic phage 352 was studied to understand its infectivity in different structured communities,it was found that in different growth media and different condition the same phage differ in their infectivity. The result makes aware of the use of this phage (and may be others) that even the lytic phage ought to be tested in different conditions and media when used as bactericidal agent (Testa et al. 2019). In another study, two phages ΦPA01 (66,220 bp) and ΦPA02 (279,095 bp) with different host spectra belong to the genus Pbunalikevirus and Phikzlikevirus respectively were isolated and analysed. Application of the cocktail of these phages plus ciprofloxacin and meropenem showed a remarkable effect of suppression of the growth of *P. aeruginosa* up to 96 h (Ong et al. 2020). Other drug resistant bacteria for which the phage MA-1 has been isolated and found to cause significant reduction in treatment compared to control (Adnan et al. 2020). In another study were isolated three phages, E79, F116 and one novel clinically derived isolate, P5. These phages were tested against 21 *P. Aeruginosa* strains isolated from children with CF. E79 phage was found to be best amongst all, which showed broad antibacterial activity (91% of tested strains sensitive) (Trends et al. 2018).

Another phages isolated with the view to use in phage therapy include are: phage, MIJ3(vB_PaeM_MIJ3) (Imam et al. 2019), phage K25055 (Bogovazova et al. 1991), Jumbo phage vB_PaeM_PA5oct (Lood et al. 2020), ACQ, UTI and BLB (Schrader et al. 1997), B86 (Kilbane and Miller 1988), template bacteriophage UNL1 (Shaffer et al. 1999), G-101 (Carrigan and Krishnapillai 1979), template E79, F8, pf20, B3 (Holloway et al. 1971), bacteriophages 16, 44, 109, F8, PBI, Phl 50, 73, 21, 68, D3 (Kropinski et al. 1977; Kropinsky 2000), Bacteriophage vB_Paep_PA01_PhiC725A (short name, Phi C725A), F116, PM105, and B3 (Pourcel et al. 2016; Pourcel et al. 2017).

***Salmonella enterica* Serovars**: is a human and animal pathogen and to control its MDR strains through phage therapy the phage isolated is ZCSE2 which shows broad-spectrum activity against this bacterium. Its potential to treat MDR *S. enterica* is highlighted (Mohamed et al. 2020).

Salmonella enteritidis: This bacterium is a common pathogen mostly infecting farm animals also is a major public health burden worldwide. A lytic phage vB_SenM-PA13076 (PA13076) is isolated to be used in phage therapy. Studies showed that mice infected with 13076^{AMP} were protected by the phage from a lethal dose of *S. enteritidis* (Bao et al. 2020).

Staphylococcus aureus: Prevalence of endovascular infection by methicillin resistant *Staphylococcus aureus* (MRSA) has becoming more common and life-threatening infection with limited options to be treated by antibiotics remains insufficient. Phage therapy therefore is considered as an alternative treatment. For this phages, VB_SauS_SH-St 15,644, belonging to Siphoviridae was isolated and a clinical trial on MRSA revealed that in vitro at least 32% of the cells were lysed by the phage, concluding that this phage may be used to treat MRSA skin infection (Ji etal. 2020). Subsequently, a *S. aureus* mutant strain isolated in Japan gave an insight about the MRSA strain which was resistant to vancomycin (Hiramatsu et al. 1997), and an insight remains about its prevalence and search for alternative treatment. This group subsequently presented a detailed analysis of the genetics of β-lactam resistance in *S. aureus* due to mec gene and ccr gene complexes were responsible for recombinase activity (Hiramatsu et al. 2001).

In order to find an alternative treatment of infection by *S. aureus*, three lytic bacteriophages viz vB_SauM-A, vB_SauM-C and vB_SauM-AD were isolated and chacterise to be able to grow and destroy the infective agent. Various phage activities against the host bacterium were characterised and found that the lytic activity worked on broad host-range MDR *S. aureus* (Lubowska et al. 2019). In other study was obtained a bacteriophage ΦMR11, which had proven to be highly effective in removing the infection by *S. aureus*. Mice were tested by infecting the bacterium and treated by intraperitoneal administration of purified phage. Rapid appearance of this phage observed in the circulation which remained at substantial levels until entire bacterial population was inactivated. It was concluded that the phage may be a potential prototype for gene-modified, advanced therapeutic *S. aureus* phages (Matsuzaki et al. 2003).

Shigella sonnei* and *S. flexneri: These bacterial species are responsible for causing acute human gartrointestinal infections and for their treatment through phage therapy two types of bacteriophages vB-SflS-ISF001 and vB-SsoS-ISF002 were isolated and tested against the two species of Shigella. The cocktail was effective against a wide range of extended spectrum, β lactemase-positive and negative isolates (Shahin et al. 2020).

Streptococcus agalactiae or Group B Streptococcus: This bacterium is a leading cause of sepsis in neonates. For the phage therapy four different phages from LF-1 to LF4 were isolated and analysed. In a therapeutic concept the broad host range activity of these phages observed in clinical in vitro trial to be a promising for future therapies of *Streptococcus agalactiae* including bioengineered phage or lysin applications (Furfaro et al. 2020).

Streptococcus pyogenes: Streptococcus pyogenes is a significant pathogen of humans, annually causing over 700,000,000 infections and 500,000 deaths. A less important report appeared in a publication of 2018 in which it has been shown that two isolated bacteriophages, Str01 and Str03 showed activity against *Streptococcus* species, especially against pyogens (Harhala et al. 2018). Their further studies are required to find if these phages can be employed in Phage therapy.

Vibrio parahaemolyticus: This gram-negative marine bacterium is capable of causing gastroenteritis when inappropriately prepared sea foods are consumed. Phage ΦVP-1, a lytic phage, has been isolated which can be used to treat infection by multiple drug resistant *V. parahaemolyticus* and *V. alginolyticus* strains of mangrove and seafood origin as an alternative to antibiotics (Matamp and Bhat 2020). In another study two types of vibriophage, KPV 40 and KVP241, with their 6 relatives were isolated from seawater employing *V. parahaemolyticus* as the host. This phage may be checked for its ability to infect and destroy this highly infective bacterial species (Matsazuki et al. 1992; Matsuzaki et al. 2000).

Vibrio alginolyticus is a human pathogen as well of marine animals, ubiquitously found in marine environments and causes soft tissue and bloodstream infection. For phage therapy a phage, ValSw3-3, with morphology of siphoviruses has been isolated which efficiently can infect four species of the *Vibrio* genus. Also has been shown that this phage can act as a potent candidate to eliminate infection caused by *V. Alginolyticus*. Due to its high pH and thermal stability as well stronger infectivity it is highly recommended in phage therapy. Other Vibrio phages are: P23 (MK097141.1), pYD8-B (NC_021561.1), 2E1 (KX507045.1) and 12G5 (HQ632860.1) (Chen et al. 2020) which may be considered for phage therapy.

Yersinia enterocolitica: Yersinia phage, vB_YenP_Φ80-18 was isolated using *Y. enterocolitica* serotype O:8 strain 8081 from sewage in Turku, Finland. Phage belongs to the Autographivirinae subfamily of the Podoviridae family. This phage can also infect strain of serotype O:4, 32, O20 and O:21 (Filik et al. 2020). Another lytic Yersinia phage X1 with double stranded DNA, belonging to Caudovirales and Myoviridae, isolated which showed its broad activities to lyse 27/51 *Y. enterocolitica* strain of various serotypes , O3 and 08, that cause yersiniosis in human and animal. This phage too can be a promising candidate to control infection by *Y. enterocolitica* (Xue et al. 2020).

8 Phage Plus Antibiotic Combination Therapies

Phage combination therapy is used when a bacterial infection cannot be treated by a single type of phage or that phage is lysogenic in which case it can harbour in the host cell and make it resistant to the same type of phage or for long-standing, persistent, or chronic bacterial infection or sustaining biofilm components (Abedon 2019). Therapy then requires considering the phage interaction with immune system. Also may include phage–antibiotic combinations, phage derived enzymes, exploitation of phage resistance mechanisms and phage bioengineering.

For such infections and treatment it is important that the phage(s) to be used are first tested for their lysogenic and lytic life cycles. As suggested for phage combination therapy in *P. aeruginosa* infection include: determination of the timing of application, order of exposure, the strength, cost and mutational basis of resistance evolution in the opportunistic pathogens (Wright et al. 2019). The following section highlights the selected results obtained using various combinations of phages and antibiotics therapy on bacteria resistance to antibiotic(s).

Acinetobacter baumannii: In a combination therapy of *A. baumannii*: was applied its phage vB_AbaP_AGC01 belonging to *Friunavirus* genus from a subfamily of *autographivirinae*. The phage adsorbed the cells efficiently and although showed certain amount of bacteriolytic activities significant reduction in cells was noted until the phage was combined with gentamycine, ciprofloxacin and meropenem when a synergistic action occurred (Grygorcewicz et al. 2020).

Escherichia coli: Phage EcSw isolated and treated *E. coli* SW1 and 0157:H7. The combined activity of the phage with kanamycine or chloramphenicol was more effective than either of them applied alone (Easwaran et al. 2020).

Pseudomonas aeruginosa: To treat *P. aeruginosa* infection a number of tests were carried out such as bacteriophage PEV20 plus the antibiotic ciprofloxacin in nebulised or in powder forms were used. For this the antibiotic resistant *P. aeruginosa* was isolated from a patient suffering from cystic fibrosis. From the tests the conclusion drawn was that stable and inhalable combination powder formulation of phage PEV20 plus ciprofloxacin may be a potential treatment for cystic fibrosis (Lin et al. 2019). In another intricate study a phage resistant mutant *P. aeruginosa* in which resistance developed through loss or changing of antigen structure, was infected by a RNA PhiYY phage that uses core Lipopolysaccharide as receptor and this led to efficiently killing the O-antigen deficient mutant. Also was included the phage PaoP5-m1, obtained from a phage PaoP5, creating a cocktail mixing of these two phages was found to efficiently killing a diverse selection of clinical isolate of *P. aeruginosa* (Yang et al. 2020).

Yet in another study was employed two strains of *P. Aeruginosa* one which was phage sensitive and ciprofloxacin resistant and the other resistant to phage but sensitive to the antibiotic. It was found that combination therapy of these agents outperformed the results obtained either by phage or antibiotic alone and that therapeutic effectiveness was enhanced given interaction with innate immune responses (Rodriguez-Gonzalez et al. 2020). In other study a phage cocktail of ΦPA01 and ΦPA02 with ciprofloxacin and meropenem was found to suppress the growth of this bacterial species up to 96 h, hence considered as potential combination for phage therapy (Testa et al. 2019).

Staphylococcus aureus: In this study a methicillin resistant *S. aureus* (MRSA) was tested in a combination therapy involving bacteriophage Sb-1 with daptomycin and vancomycin. The result showed that the combination therapy with the phage and the two antibiotics was significantly more effective than a single agent therapy (Kebriaei et al. 2020).

9 Phage Enzymes and Lytic Proteins in Phage Therapies

It is about the last one cntuary phage therapy has been going around mostly using live viruses but its recent demand relies not only the lytic phages but also purified phage lytic proteins. For the bacterial cell lysis is being used two proteins, murein hydrolase or lysin and the membrane protein, holin. The role of holin is to generate a lesion in the cytoplasmic membrane. The murein hydrolase then passes through this lesion to gain access to the murein layer. Young (1992) presented a review on these proteins basing information obtained from the work on phage phiX174 and MS2 of *E. coli* (Young 1992). In another study a speedy destruction of cell wall of *S. pneumoniae* in nasopharynx of mice was found. Subsequent presentation of lysozyme and an amidase protected the animal from bacteraemia and death through enhanced destruction of the bacterial cell wall (Jado et al. 2003). In the same year in another study, a novel phage lytic enzyme Cpl-1 was tested against *S. pneumoniae* causing pneumococcal bacteraemia. It was found very effective when employed as a topical nasal treatment against colonization by several serotypes of *S. pneumoniae* (Loeffler et al. 2003). Yet in another study a phage IME-EFm5 belonged to Siphoviridae family with double-stranded genome producing an endolysin was isolated and analysed. The phage showed bacteriolytic activity against vancomycin-resistant *E. faecium* and hence suggested to be an alternative treatment for this multidrug-resistant bacterium (Gong et al. 2016).

Another study carried out was on *Streptococcus pyogens,* a methicillin-resistant *S. aureus* (MRSA) and *S. pyogens* (group A streptococcus [GrAs] which could cause serious and often fatal illness in human. Most fatalities are due to its resistance to several antibiotics. Thus as an alternative, phage therapy was developed in which phage lysin (Ply Ss2), isolated from *Streptococcus suis* phage, was tested with broad lytic activity against a number of antibiotic resistant bacteria including MRSA, VISA (vancomycin-intermediate *S. aureus*) *S. epidermidis*, *Streptococcus equi*, *S. agalactiae*, *S. pyogenes*, *S. sanguinis* group G Streptococci GGS), group *E. streptococci (GES)* and, *S. pneumoniae.* According to the authors, no other lysin has shown such notable broad lytic activity, stability and efficacy against these MDR bacterial species (Gilmer et al. 2013). Another bacteriophage recently isolated is cpl-711 and its endolysin tested to find its synergistic effect with cephotaxime. The combination therapy showed synergistic effects on MDR clinical isolates of *S. pneumoniae* (Letrado et al. 2018).

Further studies on a recombinant endolysin (LysSS) containing a lysozyme-like domain displayed action against a number of bacterial species including 16 strains of *A. baumannii, E. coli, K. pneumoniae, P. Aeruginosa* and *Salmonella* which had developed multiple drug resistance against a number of antibiotics as methicillin resistant *S. aureus* (MRSA). LysSS had no cytotoxic effects on A549 human lung cells suggesting that this protein can be a safe and promising antimicrobial agent against MDR bacteria (Kim et al. 2020).

10 Phage-Based Human Vaccines

As most phages have simpler structural components, the nucleic acid and the proteins, based on this simplicity attempts have been made to produce and use the phage delivery vaccines (i) in the form of immunogenic peptides attached to modified phage coat proteins and (ii) employ phage as delivery vehicles for DNA vaccines by incorporating a eukaryotic promoter driven vaccine gene within their genome. Both approaches were considered promising creating a hybrid phage combining both components that are cheap, easy and rapid to produce.

Phage based vaccine can express multiple copies of an antigen on the surface of immunogenic phage particles and can elicit a powerful and effective immune response. This kind of vaccine display system has been found to be highly stable under all kinds of environmental conditions, can be simple and inexpensive, large scale production and contain potent adjuvant capacities. Future of this technology appears to have great potentials for future development of different phage-based vaccine platforms (Clark and March 2004; Aghebati-Maleki et al. 2016).

This programme was tested on mice and rabbits which were vaccinated by phage lambda containing a DNA vaccine expression cassette under the control of the Cytomegalo virus (CMV) promoter (enhanced green fluorescent protein [lambda-EGFP] or hepatitis B surface antigen [lambda-HBsAg]). The experiments were successful in that following two vaccinations carried out with rabbits. The result was that one of the four rabbits responded but following third vaccination all four rabbits showed similar high level responses and this lasted at least for six months (March et al. 2004).

Virus-like particle-based L2 vaccines have also been produced against Human Papilloma Viruses which is responsible to globally sexually transmitted infections. The prophylactic vaccine, Gardasil-9 has been found to protect against seven HPV types associated with a large number of cervical cancer and against two HPV types linked to genital warts. Production of vaccine is based on viral-like particles obtained from the major capsid protein, L1. The other minor capsid protein, L2 has been an alternative target for a broadly protective HPV vaccine (Yadav et al. 2019).

Another phage based anti-HER2 (Human epidermal growth factor receptor 2) vaccine was shown to circumvent immune tolerance against breast cancer. Δ16HER2, which is a splice variant of HER2 known as the transforming isoform in HER2-positive breast cancer was used. A DNA vaccine was developed against the Δ16HER2 isoform and this showed that the immunotherapies hampered carcinogenesis in a breast cancer transplantation model (Bartolacci et al. 2018). In another study a messenger RNA vaccine was developed against prostate cancer which was based on recombinant MS2 virus-like particles (Li et al. 2014) and in another study Inovirus-associated vector vaccine was developed for which was used phage display technologies. Inovirus-associated vectors (IAVS) were derived from filamentous bacteriophages and were responsible to displaying of specific antigenic peptides which induce antibody production to prevent infection (Stern et al. 2019).

Two bacterial species *Bacillus anthracis* and *Yersinia pestis* causing Anthrax and Plague respectively are considered to be the deadliest pathogens which have been and can be used as biological warfare weapons. Food and Drug Administration's approved vaccine Anthrax which is known as Biothrax but no vaccine exists for plague. Tao et al. (2018) has developed a dual anthrax-plague nanoparticle vaccine using T4 phage as platform. For this T4 heads (capsids) were arrayed with the two bacterial antigens fused to the small outer capsid protein Soc. The antigens included anthrax protective antigen, the mutated capsular antigen and the low-calcium-response V antigen of the type 3 secretion system from *Y. pestis* (F1mutV). It has been claimed that that this vaccine can be used as a strong candidate for biothreat agents (Tao et al. 2018).

Finally a lambda phage-based vaccine has been developed which induces anti-tumor immunity in hepatocellular carcinoma. This cancer is difficult to be treated with poor prognosis. Aspartate β-hydroxylase (ASPH) is the enzyme overexpressed on the cell surface of the human HCC cells. The therapeutic nanoparticle lambda phage vaccine was constructed against ASPH and found to be therapeutically active against murine liver tumors. Antitumor activities was assessed by generation of antigen specific cellular immune responses and suggested that this vaccine may serve as a highly antigenic target for immunotherapy (Iwagami et al. 2017).

11 Phage Therapies of Systemic Diseases

Skin burn: Skin burn especially of serious type, covering a large area of skin, most likely ends with fatality. Factors involved leading to death include destruction of vascular supply resulting in low supply of antibiotics, disruption of skin barrier and lack of and disturbance of immunosuppression. Besides infection by pathogens such as fungi and viruses most infections occur by bacterial species. In a detail study in India 920 different isolates were analysed and shown that besides *P. aeruginosa* and Klebsiella spp. (the two most common pathogens) were present *E. faecalis*, *E. coli*, Acinetobacter spp., *Salmonella senftenberg S. aureus* and Proteus spp. (Revathi et al. 1998). Also some of these infections could have developed resistant to a number of antibiotics, for alternative to antibiotics,in a study phage therapy was proposed. Special emphasis was presented for the treatment of ubiquitously present Pseudomonas spp. notoriously resistant to antibiotics. Also a table was presented for 24 bacterial spp. with the availability of their phages which could be used in therapy (Ahmad 2002). In another study for the burn wounds, infected primarily by *S. aureus*, leading to morbidity and mortality, was presented an ointment LAA developed by adding phage lysin LysGH15 and Apigenin to aquaphor. This ointment was found to be having antibacterial activities against *S. aureus* induced haemolysis. Also this ointment reduced the levels of proinflammatory cytokines, accelerating wound healing (Cheng et al. 2018). Yet in another recent studies carried out by a number of researchers, cocktails of viruses were used to treat wound infections (Duplessis and Biswas 2020; Patel et al. 2019;

Chadha et al. 2017; Basu et al. 2015; Merabishvili etal. 2009). However, in a unique study on the cocktail of phages, Kpn1, Kpn2, Kpn3, Kpn4 and Kpn5, were used to study the burn wound infection in BALB/c mice with Klebsiella infected burn wound and concluded that that the animal who did not respond to conventional antibiotic therapy had sustained effective treatment (Chadha et al. 2016).

Skin acnes: A number of different microbes have been associated with skin acnes; the most prominent is Propionibacterium which remains on top. *P. acne* phage transmission has shown to be a common transmission between individuals. A type II P phage has been tested on the bacterium and found either to lyse, leads to pseudolysogeny or even resistance to the further phage infection (Liu et al. 2015).

Buruli ulcer: It is a skin disease induced by *Mycobacterium ulcerans*. Mycolactone secreted by the bacterium affects the massive destruction of soft skin tissues. Mycobacteria have its outer membrane composed of micolylarabinogalactan and phage lipolytic enzyme Lysin B can overcome this problem by degrading this barrier and can prevent further bacterial proliferation (Fraga et al. 2019).

Diabetic foot infection: In this infection *S. aureus*, especially of MRSA type plays most prominent role contributing to 15–30% of infections. Reduction or even loss of immune function in diabetes causes additional wall in treating the infections; this in several cases can lead to foot amputation. To overcome this problem phage therapy has been considered for which a lytic bacteriophage MR-10 was tested to BALB/c mice infected with *S. aureus* ATCC43300. Combination therapy with phage and antibiotic linezolid showed much more satisfactory result than either of the agent employed on its own (Chhibber et al. 2013). In another in vitro and in vivo studies human skin infection by MDR *P. aeruginosa* and phage PA709 were used to treat and found this phage efficiently inactivates this bacterium (Vieira et al. 2012) which means that phage therapy can be used to treat a difficult infection such as MDR type bacteria specially in diabetic foot infection.

Xeroderma Pigmentosum: Patients suffering from Xeroderma Pigmentosum (XP) sustain defective DNA repair which can develop in a number of medical conditions including cutaneous malignancies (Ahmad Shamim and Hanaoka 2008). A technique involving phage therapy has been developed in which phage T4 endonuclease 5 (T4N5) in a liposomal lotion being used in clinical trials and this had received a Fast Track Designation from the FDA. The treatment is considered to be a potential tool for future treatment of skin cancers in patients suffering from XP (Zahid and Brownell 2008). The same treatment using T4 endonuclease V has been tried on actinic keratoses and non-melanoma skin cancer in XP patients, the results show that it could traverse the stratum corneum and become incorporated in the epidermal keratinocytes and Langerhans cells. The results also revealed that basal cell carcinomas were reduced by 30% and actinic keratoses by greater than 68% without giving any adverse effects (Cafardi and Elmets 2008).

Breast cancer immunotherapies: Applications of phage therapy especially in breast cancer is relatively a new idea and a number of research papers have been published on this subject. The role of phages in cancer treatment principally is associated with systematic anti-tumour immune response. Immunological peptides on phages are being considered as a new and strong cancer vaccine delivery strategy. In phage display immunotherapy a protein or peptide antigen is presented by genetic fusions to the phage coat-proteins and this construct acts as a protective vaccine against cancer. A few peptides, E75, AE37, and GP2 have already been tested on breast cancer xenograft tumors of BALB/c mice. This therapy is concluded to retain great promise in the future of cancer immunotherapy. Virus and phage display are being considered as one of the most powerful tool with potential applications in cancer prevention and treatment (Arab et al. 2019).

Periodontal disease: Phage treatment against bacterial infections in dentistry such as caries, periodontal disease, gingivitis, endodontic infections and peri-implantitis can be more beneficial than conventional antibiotics. This is because most antibiotics have a limited access and effects on biofilm in oral disease. Lytic phages with the ability to digest biofilm can offer new horizon to dentistry (Shlezinger et al. 2017).

Tuberculosis: For the treatment of tuberculosis by *M. tuberculosis*, a combination of two best drugs is isoniazid and rifampicin. These drugs are commonly given for 3–6 months. Due to this long term application (and some other reason such as indiscriminate use), fairly commonly the rifampicin resistant bacteria have been found. To overcome this hurdle to treat MDR and XDR *M. Tuberculosis* efforts have been continuing to find phages which could be used to treat tuberculosis (Azimi et al. 2019 and for more information see reference Gan et al. 2014).

Tissue re-generation: As phages almost never been found to infect human cells, they are considered for a number of unusual applications. Filamentous phages have specially selected to be more appropriate for this purpose due to their unique nanofiber-like morphology. They can self assemble into ordered scaffolds, can be produced in an error-free format, display multiple signalling peptides site, and serve as a platform for identifying novel signalling or homing peptides. Hence they can develop the ability to direct stem cell differentiation into specific cell type when they are organised into proper pattern or display suitable peptides (Cao et al. 2016).

Encapsulation in micro- and nanoparticle for effective phage delivery: Employment of phages in phage therapy requires their stability in solution during processing and storage, so that high concentration of the particles can be presented at the site of infection and their ability to eradicate high number of infecting agents. For this phage encapsulation is another driver of phage therapy.

Phage encapsulation in micro-and nanoparticle for effective delivery is a relatively new technique through which phages are allowed to sustain the adverse effect of critical environmental conditions as well the improvement in the stability and their controlled delivery. One method involves two compatible materials: a lipid cationic mixture and a combination of alginate with $CaCO_3$ and for its application, purified lysate of very high concentration, may be up to 10^{11} pfu/ml is used in this process (Cortes et al. 2018, Malik et al. 2017).

12 Conclusion

In recent years phage therapy has been receiving a good momentum and many scientists around the world are carrying a variety of experiments culminating in publication of 4180 research papers (PubMed, dated 19-9-2020). Reports started culminating from 1980 were those experiments carried on animal models and claimed that appropriate use of living phages can be used to treat lethal infectious diseases inflicted by a variety of bacteria including: *P. aeruginosa*, *K. pneumoniae*, *E. coli*, *Salmonella* species, *S. aureus*, *B. anthracis* and *Group B Streptococci.* Phages for these bacteria have been tested against the infective organisms and have resulted in strong bacteriolytic activities when applied exogenously.

Multiple drug resistance bacterial infections are arguably one of the biggest current global threats. The reason such infections are increasing with unexpectedly high rates and in certain cases becoming almost impossible to treat, leading to high morbidity, mortality and heavy cost are mainly due to abuse and misuse of the antibiotics. Although the momentum now is developing to find alternative solutions of treating those bacterial infections which have developed resistance to multiple antibiotics, the question arise as to why simple cocktail of phage therapies which were discovered and used over a century ago has been kept suppressed. A number of reasons can be put forward. One is the discovery of antibiotic penicillin by Alexander Fleming, considered as panacea. This agent overpowered the market and the old type phage therapies were almost forgotten. Subsequently more search and development of existing antibiotics to better suit treatments continued for a long time, which is still continuing.

A second reason for the silencing of phage therapy was the limited scientific information about the use of phages, especially there was no information on the innate and adaptive immunity existed for the clearance of phage from the body. Also there exists variation for the rout of administration which depends on the type of live phages used. It was when a number of pathogenic MDR bacterial species emerged that scientists started taking serious interest in this alternative treatment and obtain more knowledge about this treatment.

Other reason that incited scientific community taking deeper interest in phage therapies is that phages do not present much adverse safety concerns. However, the efficacy of therapy outcomes were considered to be of importance, should these factors be resolved and accepted at international levels: (i) the dose of the phage (ii) delivery site of the infection (iii) ability of phage to target and inactivate bacteria and (iv) clearance of infection. On the other hand for phage therapy a number of cares are also required including the requirement of relatively narrow host range of phages, the presence of toxins in phage lysate and the capacity of host defence systems to remove phage particles from the circulatory system.

Other issues that are required to reduce the development of MDR, to be addressed and agreed at international levels are: (a) that antibiotics, especially in the Third World countries must not be dispensed at the counter without prescription from the doctors. Also the medical practitioners and dispensers must assure the patients to complete the course of the drugs prescribed.

In conclusion it can be said that phage therapy is an important alternative tool which must be tested independently scientifically for its use under the control of relevant Drug Administrative Authorities.

References

Abedon, ST, Use of phage therapy to treat long-standing, persistent, or chronic bacterial infections. Adv Drug Deliv Rev. 2019;145:18–39. https://doi.org/10.1016/j.addr.2018.06.018. Epub 2018 Jul 3.

Adnan, M Alishah MR, Jamal M et al. Isolation and characterization of bacteriophage to control multidrug-resistant Pseudomonas aeruginosa planktonic cells and biofilm. Biologicals. 2020;63:89–96.

Aghebati-Maleki L, Bakhshinejad B, Bradaran B et al. Phage display as a promising approach for vaccine development. J Biomed Sci. 2016;23(1):66. https://doi.org/10.1186/s12929-016-0285-9

Ahmad SI. Treatment of post-burns bacterial infections by bacteriophages, specifically ubiquitous Pseudomonas spp. notoriously resistant to antibiotics. Medical Hypotheses. 2002. 58(4). 327–31. https://doi.org/10.1054/mehy.2001.152, www.idealibrary.com.

Ahamed Sk Tousif, Roy B, Basu U, et al. Genetic and proteomic characterization of sfni-1, a novel lytic phage infecting multidrug-resistant Shigella spp. and Escherichia C. Front Miccrobiol. 2019;10:1876. https://doi.org/10.3389/fmicb.2019.01876. e.Collection 2019.

Ahmad Shamim I, Hanaoka F (Eds.) Molecular mechanisms of Xeroderma Pigmentosum. Advances in Experimental Medicine and Biology, Volum 637. Springer Science + Business Media, LLC, Landes Bioscience; 2008.

Ahmad Shamim I. Novel treatments of microbial burn infections. JSM Burns and Trauma, Sci Med Central. 2017;2(2):1017; ISSN 2475-9406.

Alisky J, Iczkowski K, Rapoport A. et al. Bacteriophages show promise as antimicrobial agents. J. Infect 1998;36:5–15.

Alkalay S, Sternberg S, Coppenhegen-Glazer et al. Complete genome sequences of three Bacillus anthracis bacteriophages. Genome Announc. 2018;6(1):e01164–17. https://doi.org/10.1128/genomeA.01164-17.

Arab A, Behravan N, Razazn A et al. The viral approach to breast cancer immunotherapy. J Cell Physiol. 2019;234(2):1257–67. https://doi.org/10.1002/jcp.27150. Epub 2018 Aug 26.

Asif M, Alvi A, Tabassum R, et al. TAC1, an unclassified bacteriophage of the family myoviridae infecting Acinetobacter baumannii with a large burst size and a short latent period. Arch Virol. 2002;165:419–24.

Azimi T, Mosadegh M, Nasiri MJ, et al. Phage therapy as a renewed therapeutic approach to mycobacterial infections: a comparative review. Infect Drug Resist. 2019;12:2943–59.

Badawy S, Pajunen M, Haiko J, et al. Identification and functional analysis of temperate Siphoviridae bacteriophages of Acinetobacter baumannii. Viruses. 2020;12(6):E604. https://doi.org/10.3390/v12060604.

Bao H, Zhou Y, Shahin K et al. The Complete genome of lytic Salmonella phage vB_SenM-PA13076 and therapeutic potency in the treatment of lethal Salmonells enteritidis infections in mice. Microbiol Res. 2020237:126471. https://doi.org/10.1016/j.micres.2020.126471. Epub 2020 Mar 19.

Barrow PA, Soothill JS. Bacteriophage therapy and prophylaxis: recovery and renewal assessment of potential. Trends Microbiol. 1997;5(7):268–71. https://doi.org/10.1016/S0966-842X(97)01054.8.

Bartolacci C, Andreani C, Curcio C. et al. Phage-based anti-HER2 vaccination can circumvent immune tolerance against breast cancer. Cancer Immunol Res. 2018;6(12):1486–98. https://doi.org/10.1158/2326-6066.CIR-18-0179. Epub 2018 Oct 16.

Basu S, Agarwal M, Bhartiya et al. An in vivo wound model utilizing bacteriophage therapy of Pseudomonas aeruginosa biofilms. Ostomy Wound Manage. 2015;61(8):16–23.

Bavda VR, Jain V. Deciphering the role of holin in mycobacteriophage D29 physiology. Front Microbiol. 2020;8(11):883.

Bogovazova GG, Voroshilova NN, Bondarenko VM. The efficacy of Klebsiella pneuminiae bacteriophage in the therapy of experimental Klebsiella infection. Zh Mikrobiol Epidemiol Immunobiol. 1991;(4):5–8.

Bragg R, Westhuizen W. Van der, Lee Ji-Yun et al. Bacteriophages as potential treatment option for antibiotic resistant bacteria. Adv Exp Med Biol. 2014;807:97–110. https://doi.org/10.1007/978-81-322-1777-0_7

Briani F, Deho G, Forti F, et al. The plasmid status of satellite bacteriophage P4. Plasmid. 2001;45(1):1–17. https://doi.org/10.1006/plas.2000.1497.

Cafardi JA, Elmets CA. T4 endonuclease V: review and application to dermatology. Expert Opin Biol Ther. 2008;8(6):829–38. https://doi.org/10.1517/14712598.8.6.829.

Cahill J, Young Ry. Phage lysis: multiple genes for multiple barriers. Adv Virus Res. 2019;103:33–70. https://doi.org/10.1016/bs.aivir.2018.09.003. Epub 2018 Nov 28.

Cao B, Li Y, Bao Q. Bacteriophage based biomaterials for tissue regeneration. Adv Drug Deliv Res. 2016;Nov 16.pil`

Carlton RM. Phage therapy: past history and future prospects. Arch Immunol Ther Exp (Warsz). 1999;47(5):267–74.

Carrigan JM, Krishnapillai V. Identification of cistrons involved in conjugal transfer of narrow-host-range R plasmid R91–5 of pseudomonas aeruginosa. J Abcteriol. 1979;140(3):809–16.

Chadha P, Katare OP, Chibber S. In vivo efficacy of single phage versus phage cocktail in resolving burn wound infection in BALB/c mice. Microb Pathog. 2016;99:68–78. https://doi.org/10.1016/j.micpath.2016.08.001 Epub 2016 Aug 4.

Chadha P, Katare OP, Chibber S. Liposome loaded phage cocktail: enhanced therapeutic potential in resolving Klebsiella pneumoniae burn wound infection. Burns. 2017;43(7):1532–43. https://doi.org/10.1016/j.burns.2017.03.029. Epub 2017 May 11.

Charles RC, Ryan ET. Cholera in the 21st century. Curr Opin Infect Dis. 2011;24(5):472–7. https://doi.org/10.1097/QCO.0b013e32834a88af.

Chen L, Liu Q, Fan J, et al. Characterization and genomic analysis of ValSw3-3, a new Siphoviridae bacteriophage infecting Vibrio alginolyticus. J Virol. 2020;94(10):e00066–120. https://doi.org/10.1128/JVI.00066-20.Print2020May4.

Cheng M, Zhang L, Zhang H, et al. An ointment consisting of the phage lysin LysGH15 and apigenin for decolonization of methicillin-resistant Staphylococcus aureus from skin wound. Viruses. 2018;10(5):244. https://doi.org/10.3390/v10050244.

Chhibber S, Kaur T, Kaur S. Co-therapy using lytic bacteriophage and linezolid: effective treatment in eliminating methicillin resistant Staphylococcus aureus (MRSA) from diabetic foot infection. PLoS One. 2013;8(2);e56022. https://doi.org/10.1371/journal.pone.0056022. Epub 2013 Feb 13.

Clark JR, March JB. Bacterial viruses as human vaccines? Expert Rev Vaccines. 2004;3(4):463–76.

Cortes P, Cano-Sarabia M, Colom J, et al. Nano/Micro formulation of bacteriophage delivery. Methods Mol Biol. 2018;1693:271–83. https://doi.org/10.1007/978-14939-7395-8_20.

Domingo-Calap P, Beamud B, Vienne J. et al. Isolation of four lytic phages infecting Klebsiella pneumoniae K22 clinical isolates from Spain. Int J Mol Sci. 2020;21(2); Pil: E425. https://doi.org/10.3390/ijms21020425.

Duplessis CA, Biswas B. A review of topical phage therapy for chronically infected wounds and preparations for a randomized adaptive clinical trial evaluating topical phage therapy in chronically infected diabetic foot ulcers Antibiotics (Basel) 2020;9(7):377. https://doi.org/10.3390/antibiotics9070377

Easwaran M, De Zoysa M, Shin H. Application of phage therapy: synergistic effect of phage EcSw and antibiotic combination towards antibiotic-resistant Escherichia coli. Transbound Emerg Dis. 2020. https://doi.org/10.1111/tbed.13646.

Filik K, Szermer-Olearnik B, Wernecki M et al. The podovirus Φ80–18 targets the pathogenic American biotype 1B strain of Yersinia enterocolitica. Front Microbiol. 2020;11:1356. https://doi.org/10.3389/fmicb.2020.01356. ECollection 2020.

Fraga AG, Trigo G, Murthy RK et al. Antimicrobial activity of Mycobacteriophage D29 lysin B during mycobacterium ulcerans infection. PLoS Neg1 Trop Dis. 2019;13(8):e0007113. https://doi.org/10.1371/journal.pntd.0007113. ECollection 2019 Aug.

Furfaro LL, Payne MS, Chang BJ. Host range, morphological and genomic characterization of bacteriophages with activity against clinical Streptococcus agalactiae isolate. PLoS One. 2020;15(6):e0235002.

Gan Y, Wu T, Liu P, et al. (2014) Characterization and classification of Bo4 as a cluster G mycobacteriophage that can infect and lyse M tuberculosis. Arch Microbiol. 2014;196(3):209–18. https://doi.org/10.1007/s00203-014-0954-6 Epub 2014 Feb 7.

Gao M, Wang C, Qiang X et al. Isolation and characterisation of a novel bacteriophage infecting carbapenem-resistant Klebsiella pneumoniae. Curr Microbiol. 2020. https://doi.org/10.1007/s00284-019-01849-8 [Epub ahead of print].

Gilmer DB, Schmitz JE, Euler CW et al. Novel bacteriophage lysis with broad lytic protects against mixed infection by Streptococcus pyogens and methicillin-resistance. Antimicrob Agents Chemother. 2013;57(6):2743–50. https://doi.org/10.1128/AAC.02526-12. Epub 2013 Apr 9.

Gong P, Cheng M, Li X. Characterization of Enterococcus faecium bacteriophage IME-EFm5 and its endolysin lys EFm5. Virology. 2016;492:11–20. https://doi.org/10.1016/j.virol.2016.02.006 . Epub 2016 Feb 19.

Grygorcewicz B, Roszak M, Golec P, et al. Antibiotic act with vB_AbaP_AGC01 phage against Acinetobacter Baumannii in human heat-inactivated plasma blood and Galleria mellonella model. Int J Mol Sci. 2020;21(12):E4309.

Hammond RW, Swift SM, Foster-Frey JA et al. Optimised production of a biologically active clostridium perfringens glycol hydrolase phage endolysin PlyCP41 in plants using virus based systemic expression. Biotechnol. 2019;19(1):101.

Harhala M, Barylskki J, Huminska-Lisowska K et al. Two novel temperate bacteriophages infectingStreptococcus pyogens: their genomes, morphology and stability. PLoS One. 2018;13(10):e0205995. https://doi.org/10.1371/journal.pone.0205995. eCollection 2018.

Hiramatsu K, Aritaka N, Hanaki H, et al. Dissemination in Japanese hospitals of strains of Staphylococcus aureus heterogenously resistant to vancomycin. Lancet. 1997;350(9092):1670–3. https://doi.org/10.1016/SO140-6736(97)07324-8.

Hiramatsu K, Cui L, Kuroda M, et al. The emergence and evaluation of methicillin-resistant Staphylococcus aureus. Trends Microbiol. 2001;9(10):486–93.

Holloway BW, Krishnapillai V, Stanisich V. Pseudomonas genetics. Annu Rev Genet. 1971;5:425–46. https://doi.org/10.1146/annurev.ge.05.120171.002233.

Hua Y, Xu M, Wang R et al. Characterisation and whole genome analysis of a novel bacteriophage SH-Ab 15497 against multidrug resistant Acinetobacter baumannii. Acta biochimica et biophysica Sinica. 2019;51(10):1079–81.

Imam M, Alrashid B, Patel F et al. vB_PaeM_MIJ3, a novel jumbo phage infecting Pseudomonas aeruginosa, possess unusual genomic features. Front Microbiol 2019;2772. https://doi.org/10.3389/fmicb.2019.02772. eCollection 2019.

Iwagami Y, Casulli S, Nagaoka K et al. Lambda phage-based vaccine induces antitumor immunity in hepatocellular carcinoma. Heliyon 2017 Sep 26;3(9):e00407. https://doi.org/10.1016/j.heliyon.2017.e00407. eCollection 2017 Sep

Jado I, Lopez R, Garcia E et al. Phage lytic enzymes as therapy for antibiotic-resistant Streptococcus infection in murine sepsis model. J Antimicrob Chemother. 2003;52(6):967–73. https://doi.org/10.1093/jac/dkg485. Epub 2003 Nov 12

Ji J, Liu Q, Wang R et al. Identification of a novel phage targeting methicillin-resistant Staphylococcus aureus in vitro and in vivo. Microb Pathog. 2020;149:104317. https://doi.org/10.1016/j.micpath.2020.104317. Online ahead of print

Jiang YH, Liu JQ, Zhao CY, et al. Isolation and genome sequencing of a novel Pseudomonas aeruginosa phage PA-YS35. Curr Microbiol. 2020;77(1):123–8. https://doi.org/10.1007/s00284-019-01792-8 Epub 2019 Oct 29.

Johura FT, Tasnim J, Barman I. Colistin resistant Escherichia coli carrying mcr-1 in food, water, hand rinse, and healthy human gut in Bangladesh. Gut PathogJanuary. 2020;27(12):5. https://doi.org/10.1186/s13099-020-0345-2. eCollection 2020.

Karumidze N, Kusradze Ia, Rigvava S et al. Isolation and characterization of lytic bacteriophagesof Klebsiella pneumoniae and Klebsiella oxytoca. Curr Microbiol. 2013;66(3):251–8. https://doi.org/10.1007/s00284-012-0264-7. Epub 2012 Nov 11.

Kasman LM, Porter LD. Bacteriophages. In: Statpearls [Internet]. Treasure Island (FL): StatPerls Publishing; 2020.

Kebriaei R, Ley K, Morrisette T et al. Bacteriophage-antibiotic combination strategy: an alternative against methicillin-resistant phenotype of Staphylococcus aureus. Antimicrob Agents Chemother. 2020: https://doi.org/10.1128/AAC.00461-20. Print 2020 Jun 23.

Khalifa L, Brosh Y, Gelman D, et al. Targetting Enterococcus faecalis biofilms with phage therapy. Appl Environ Microbiol. 2015;81(8):2696–705. https://doi.org/10.1128/AEM.00096-15 Epub 2015 Feb 6.

Kilbane JJ, Miller RV. Molecular characterization of Pseudomonas aeruginosa bacteriophages: identification and characterization of the novel virus B86. Virology. 1988; 164(1):193–200.

Kim S, Lee DW, Jin JS et al. Antimicrobial activity of LysSS, a novel phage endolysin, against Acinetobacter baumannii and Pseudomonas aeruginosa. J Glob Antimicrobe Resist. 2020 Jan 29 pil.

Kropinski AM, Chan L, Jarrell K. The nature of Pseudomonas aeruginosa strain PAO bacteriophage receptors. Cn J Microbiol. 1977;23(6):653–8.

Kropinsky AM. Sequence of the genome of the temperate, serotype-converting, Pseudomonas aeruginosa bacteriophage D3. J Bacteriol. 2000;182(21):6066–74.

Lavigne R, Robben J. Professor Dr. Richard Bruynoghe: a 1951 overview of his bacteriophage research spanning three decades. Bacteriophage. 2012;2(1):1–4. https://doi.org/10.4161/bact.20024.

Lee D, Im J, Na H et al. The novel Enterococcus phage vB_EfaS_Hef13 has broad lytic activity against clinical isolates of Enterococcus faecalis. Front Microbiol. 2019;10:2877.

Letrado P, Corsini B, Diez-martinez R et al. Bacterial synergism between antibiotics and phage endolysin Cpl-711 to kill multidrug-resistant pneumococcus. 2018;13(11):1215–23. https://doi.org/10.2217/fmb-2018-0077. Epub 2018 Sep 21.

Lewis R, Clooney AG, Stockdale S et al. Isolation of a novel jumbo bacteriophage effective against Klebsiella aerogenes. Front Med (Lausanne). 2020;7:67.

Li J, Sun Y, Jia T et al. Messeger RNA vaccine based on recombinant MS2 virus-like particles against prostate cancer. Int J Cancer. 2014;134(7) :1683-1694. https://doi.org/10.1002/ijc.28482. Epub 2013 Oct 8

Li Q, Zhou M, Fan X. Mycobacteriophage SWU1 gp39 can potentiate multiple antibiotics against mycobacterium via altering the cell wall permeability. Sci Rep. 2016;28(6):28701. https://doi.org/10.1038/srep28701.

Li T, Zhang Y, Dong K. Isolation and Characterization of the novel phage JD032 and global transcriptomic response during JD032 infection of Clostridioides difficile ribotype 078. mSystems 2020;5(3):e00017–20.

Li M, Shi D, Li Y. Recombinant of T4-like phages and its activity against pathogenic Escherichia coli in plantonic and biofilm. Virol Sin. 2020. https://doi.org/10.1007/s12250-020-00233-2. Online ahead of print.

Lin Y, Chang RYK, Britton WJ et al. Inhalable combination powder formulation of phage and ciprofloxacin for P. Aeruginosa respiratory infection. Eur J Pharm Biopharm. 2019;142:543–52.

Liu J, Yan R, Zhong Q et al. The diversity and host interaction of Propionibacterium acnes bacteriophages on human skin. 2015;9(9):2078–93. https://doi.org/10.1038/ismej.2015.47. Epub 2015 Apr 7.

Loeffler JM, Djurkovic S, Fischetti VA. Phage lytic enzyme Cpl-1 as a novel antimicrobial for Pneumococcal bacteremia. Infect Immun. 2003;71(11):6199–204. https://doi.org/10.1128/iai.71.11.6199-6204.2003.

Lood C, Danis-Wlodarczyk K, Blasdel BG. Integrative omics analysis of Pseudomonas aeruginosa virus PA5oct highlights the molecular complexity of jumbo phages. Environ Microbiol. 2020. https://doi.org/10.1111/1462-2920.14979. [Epub ahead of print]

Lubowska N, Grygorcewicz B, Kosznik-Kwasnicka K et al. Characterization of the three new kayviruses and their lytic activity against multidrug-resistant Staphylococcus aurous. Microorganisms. 2019;7(10).pil:E471.doi

Mahajan NP, Lavania M, Singh I, et al. Evidence for Mycobacterium leprae drug resistance in a large cohort of leprous neuropathy patients from India. Am J Trop Med Hyg. 2020;102(3):547–52.

Malik DJ, Sokolov IJ, Vinner GK et al. Formulation, stabilization and encapsulation of bacteriophage for phage therapy. Adv Colloid Interface Sci. 2017;249:100–33.

Maniloff J. Mycoplasma viruses. Crit Rev Microbiol. 1988;15(4):339–89. https://doi.org/10.3109/10408418809104462.

March JB, Clark JR, Jepson CD. Genetic immunization against hepatitis B usuing whole bacteriophage lambda particles. Vaccines. 2004;22(13–14):1666–71.

Matamp N, Bhat SG. Genome characterization of novel lytic myoviridae bacteriophage ΦVP-1 enhances its applicability against MDR-biofilm-forming Vibrio parahaemolyticus. Arch Virol. 2020;165:387–96.

Matsazuki S, Tanaka S, Koga T, et al. A broad host range vibriophage, KVP40, isolated from sea water. Microbiol Immunol. 1992;36(1):93–7.

Matsuzaki S, Inoue T, Tanaka S. Characterization of a novel Vibrio parahaemolyticus phage, KVP241, and its relatives frequently isolated from seawater. Microbiol Immunol. 2000;44(11):953–6.

Matsuzaki S, Yasuda M, Nishikawa et al. Experimental protection of mice against lethal staphylococcus aureus infection by novel bacteriophage phi MR11. J Infect Dis. 2003;187(4):613–24. Epub 2003 Feb 7.

McCallin S, Alam Sarker S, Barrretto C, et al. Safety analysis of Russian phage cocktail: from metagenomic analysis to oral application in healthy human subjects. Virology. 2013;443(2):187–96.

Merabishvili M, Pirnay J, Verbeken G et al. Quality-controlled small-scale production of a well-defined bacteriophage cocktail for use in human clinical trial. PLoS Onc. 2009;4(3):e4944. https://doi.org/10.1371/journal.pone.0004944. Epub 2009 Mar 20.

Moelling K, Broecker F, Willy C. A wake-up call: we need phage therapy now. Viruses. 2018;10(12). Pili:E688.

Moghadam MT, Amirmozafari N, Shariati A et al. How phage overcome the challenges of drug resistance bacteria in clinical infections. Infect Drug Resist. 2020; 45–61.

Mohamed A, Taha O, El-Sherif et al. Bacteriophage ZCSE2 is a potent antibacterial against Salmonella enterica Serovar: ultrastructure, genomics and efficacy. Viruses. 2020;12(4):424. https://doi.org/10.3390/v1204024.

Mokrousov I. Corynebacterium diphtheriae: genome diversity, population structure and genotyping perspectives. Infect Genet Evol. 2009;9(1):1–15. https://doi.org/10.1016/j.meegid.2008.09.011. Epub 2008 Oct 19.

Moller-Olsen C, Ross T, Leppard KN. Bacteriophage K1F targets Escherichia coli K1 in cereberal endothelial cells and influences the barrier function. Sci Rep. 2020;10(1):8903. https://doi.org/10.1038/s441598-020-55867-4.

Montso PK, Mlambo V, Ateba CN. Characterisation of lytic bacteriophages infecting multidrug-resistant shiga toxigenic atypical Escherichia coli 0177 strain isolated from cattle faeces. 2019;7:355.

Munita JM, Arias CA. Mechanisms of antibiotic resistance. Microbiol Spectr. 2016;4(2).

Necel A, Bloch S, Nejman-Falenczyk. Charecterization of a bacteriophage, vB_Eco4M-7, that effectively infects many Escherichia coli 0157 strains. Sci Rep. 2020;10(1):3743.

Nikapitiya C, Chandrarathna HPSU, Dananjaya SHS et al. Isolation and characterization of phage (ETP-1) specific to multidrug resistant pathogenic Edwardsiella tarda and its in vivo biocontrol efficacy in zebrafish (Danio rerio). Biologicals 2020 Jan;63:14–23. https://doi.org/10.1016/j.biologicals.2019.12.006. Epub 2019 Dec 27.

Oechslin F. Resistance development to bacteriophages occurring during bacteriophage therapy. Viruses. 2018;10(7) pil.

Ong SP, Azam AH, Sasahara T. Characterization of Pseudomonas lytic phages and their application as a cocktail with antibiotics in controlling Pseudomonas aeruginosa. J Biol Bioeng. 2020. Pil: S1389–1723(19)30928–4. https://doi.org/10.1016/j.jbiosc.2020.02.001. [Epub ahead of print].

Park E-A, KimY-T, Chao J-H. Characterization and genome analysis of novel bacteriophages infecting the opportunistic human pathogens Klebsiella oxytoca and K. pneumoniae. Arch Virol. 2017;162(4):1129–39. https://doi.org/10.1007/s00705-016-3202-3. Epub 2016 Dec 27.

Patel DR, Bhartiya S, Kumar R. Use of customised bacteriophages in the treatment of chronic nonhealing wounds: a prspective study. Int J Low Extrem Wounds. 2019;1534734619881076. https://doi.org/10.1177/1534734619881076. Online ahead of print.

Peng Q, Uan Y. Characterization of a novel phage infecting the pathogenic multidrug-resistant Bacillus cereus and functional analysis of its endolysin. Appl Microbiol Biotechnol. 2018;102 (18):7901–12. https://doi.org/10.1007/s00253-018-9219-7 Epub 2018 Jul 14.

Peng O, Fang M, Liu X. Isolation and characterization of a novel phage for controlling multidrug-resistant Klebsiella pneumoniae. Microoganisms. 2020;8(4):542.

Peng S, Chen L, Wu W. Isolation and characterization of a new phage infecting Elizabethkingia anopheles and evaluation of its therapeutic efficiency in vitro and in vivo. Front Microbiol 2020 May 13;11:728. https://doi.org/10.3389/fmicb.2020.00728. eCollection 2020.

Pourcel C, Midoux C, Bourkaltseva M et al. Complete genome sequence of PM105, a new Pseudomonas aeruginosa B3-like transposable phage. Genome Announc. 2016;4(2). Pil: e0143–15. https://doi.org/10.1128/genomeA.01543-15.

Pourcel C, Midoux C, Hauck Y et al. Large preferred region for packing of bacterial DNA by phiC725A, a novel Pseudomonas aeruginosa F116-like bacteriophage. PLoS One. 2017;12(1): e0169684. https://doi.org/10.1371/journal.pone.0169684. eCollection 2017.

Revathi G, Puri J, Jain BK. Bacteriology of burns. Burns. 1998;24(4):347–9. https://doi.org/10.1016/s0305-4179(98)00009-6.

Rodriguez-Gonzalez RA, Leung CY, Chan BK et al. Quantitative models of phage-antibiotic combination therapy. mSystems. 2020:5(1). pil

Satish R Desouza A. Study of characteristics of mycobacteriophage—a novel tool to treat mycobacterium spp. Int J Mycobacteriol. 2019;8(2):170–4. https://doi.org/10.4103/ijmy.ijmy_42_19.

Schrader HS, Schrader JO, Walker JJ, et al. Bacteriophage infection and multiplication occur in Pseudomonas aeruginosa starcved for 5 years. CanJ Microbiol. 1997;43(12):1157–63.

Shaffer JJ, Jacobsen LM, Schrader JO, et al. Characterization of Pseudomonas aeruginosa bacteriophage UNL-1, a bacterial virus with novel UV-A-inducible DNA damage reactivation phenotypr. Appl Environ Microbiol. 1999;65(6):2606–13.

Shahin K, Bouzari M, Komijani M et al. A new phage cocktail against multidrug, ESBL-producer isolates of Shigella sonnei and Shigella flexneri with highly efficient bacteriolytic activity. Microb Drug Resist. 2020;26(7):831–841. https://doi.org/10.1089/mdr.2019/mdr.2019.0235. Epub 2020 Feb 7.

Shi Y, Chen Y, Zhang Y et al. Characterization and genome sequencing of a novel T7-like lytic phage, kpssk3, infecting carbapenem-resistant Klebsiella pneumoniae. Arch Virol. 2020;165 (1):97–104

Shlezinger M, Khalifa L, Houri-Haddad et al. Phage therapy: a new horizon in the antibacterial treatment of oral pathogens. CurrTop Med Chem. 2017;17(10):1199–211.

Smith HW, Huggins MB. Successful treatment of experimental Escherichia coli infections in mice using phage: its general superiority over antibiotics. J Gen Microbiol. 1982;128(2):307–18.

Stern Z, Stylianou DC, Kostrikis LG. The development of inovirus-associated vector vaccines using phage display technologies. Expert Rev Vaccines. 2019;18(9):913–20. https://doi.org/10.1080/14760584.2019.1651649 Epub 2019 Sep 8.

Styles KM, Thummeeepak R, Leungtongkam U et al. Investigating bacteriophages targeting the opportunistic pathogen Acinetobacter baumannii. Antibiotics (Basel). 2020;9(4);200. https://doi.org/10.3390/antibiotics9040200.

Sula L, Sulová J. Lytic potency against various mycobacterial strains of the phage isolated from Mycobacterial strains of the phage isolated from Mycobacterium leprae murium "Douglas." Czech Med. 1984;7(3):187–173.

Tao P, Mahalingam M, Zhu J. A bacteriophage T4 nanoparticle-based dual vaccine against anthrax and plague. mBio. 2018;9(5):01926–18. https://doi.org/10.1128/mBio.01926-18.

Temple GS, Ayling PD, Wilkinson SG. The role of polysaccharide as a receptor for some bacteriophages of Pseudomonas aeruginosa. Microbios. 1986;45(183):93–104.

Testa S, Berger S, Piccardi et al. Spatial structire affects phage efficacy in infecting dual-strain biofilms of Pseudomonas aerugonosa. Commun Biol. 2019;2:405. https://doi.org/10.1038/s42003-019-0633-x. eCollection 2019.

Trends S, Chang B, O'Dea M et al. Use of primary epithelial cells screening tool to investigate phage therapy in cystic fibrosis. Front Pharmacol. 2018;9:1330.

Uchiyama J, Takeuchi H, Kato S. Characterization of Helicobacter pylori bacteriophage KHP30 Appl Environ Microbiol. 2013;79(10):3176–3184. https://doi.org/10.1128/AEM.03530-12. Epub 2013 Mar 8.

Vieira A, Silva YJ, Cunha A, et al. Phage therapy to control multidrug-resistant Pseudomonas aeruginosa skin infection: in vitro and ex vivo experiments. Eur J Clin Microbiol Infect Dis. 2012;31(11):3241–9. https://doi.org/10.1007/s10096-012-1691-x Epub 2012 Jul 10.

Wandro S, Oliver A, Gallagher T et al. Predictable molecular adaptation of coevolving Enterococcus faecium and lytic phage. Front Microbiol. 2019;9:3192. https://doi.org/10.3389/fmicb.2018.03192. eCollection 2018.

Wang M, Zu X, Zhao Z et al. Cephalosporin resistance in Escherichia coli isolated from children with septicemia in mainland china from 2007 to 2017: a systematic review and meta analysis. Microb Drug Resist. 2020. https://doi.org/10.1089/mdr.2018.0368. [Epub ahead of print]

Weber-Dabrowska B, Mulczyk M, Gorski A. Bacteriophage therapy of bacterial infection: an update of our institute's experience. Arch Immunol Ther Exp (Warsz). 2000;48(6):547–51.

Wintachai P, Naknaen A, Pomwised R, et al. Isolation and characterization of Siphoviridae phage infecting extensively drug-resistant Acinetobacter baumannii. J Med Microbiol. 2019;68:1096–108.

Wright RCT, Friman VP, Smith MCM. Resistance evolution against phage combination depends on the timing and order of exposure. M Bio.2019;10(5) pil:

WuY, Wang R, Xu M et al. A novel polysaccharide depolymerase encoded by the phage SH_KP152226 confers specific activity against multidrug-resistant Klebsiella pneumoniae biofilm. 2019 10:2768.

Xue Y, Zhai S, Wang Z. The Yersinia phage XI administered orally efficiently protects a murine chronic enteritis model against Yersinia enterocolitica infection. Front Microbiol. 2020;11:331. https://doi.org/10.3389/fmicb.2020.00351. eCollection 2020.

Yadav R, Zhai L, Tumban E. Virus-like particle-based L2 vaccines against HPVs: where are we today? Viruses. 2019;12(1):18. https://doi.org/10.3390/v12010018.

Yang Y, Shen W, Zhong Q. Development of bacteriophage cocktail to constrain the emergence of phage-resistant Pseudomonas aeruginosa. Front Microbiol. 2020;11:327. https://doi.org/10.3389/fmicb.2020.00327. eCollection2020.

Yazdi M, Bouzari M, Ghaemi EA, et al. Isolation, characterization and genomic analysis of a novel bacteriophage VB_EcoS-Golastan infectiong multidrug-resistant Escherichia coli isolated from urinary tract infection. Sci Rep. 2020;10(1):7690. https://doi.org/10.1038/s41598-020-63048-x .

Yin S, Huang G, Zhang Y. Phage Abp1 rescues human cells and mice from infection by Pan-drug resistant Acinetobacter baumannii. Cell Physiol Biochem. 2017;44(6):2337–2345. https://doi.org/10.1159/000486117. Epub.2017 Dec 15.

Young R. Bacteriophage lysis: mechanism and regulation. Microbiol Rev. 1992;56(3):430–481.

Yuan Y, Wang L, Li X et al. Efficacy of a phage cocktail in controlling phage resistance development in multidrug resistant Acinetobacter baumannii. Virus Res. 2019; 272: 197734.

Zabezhinsky LM, Gorstkina-Shevandronova LA. Phage therapy of typhoid. Zh Mikrobiol Epidemiol Immunobiol. 1946;4:25–7.

Zahid S, Brownell I. Repairing DNA damage in Xeroderma Pigmentosum: T4N5 lotion and gene therapy. J Drug Dermatol. 2008;7(4):405–8.

Oligomerization of Fusion Proteins: A Common Symptom for Class I Viruses

Geetanjali Meher and Hirak Chakraborty

Abstract Protein oligomerization is an important regulatory process in biology. Several proteins maintain equilibrium with their oligomeric species to optimize their functions. In addition, oligomerization is an important event in generating larger force for biomechanical processes. Membrane fusion is a crucial process in the survival of eukaryotes. It is an integral part of several constitutive biological processes, such as cellular trafficking, sexual reproduction, compartmentalization, intercellular communication, endocytosis and exocytosis, and cell division. Apart from these useful events, lipid sheathed viruses exploit membrane fusion to enter into the host cell, thereby making this process as the early step of viral infection. The fusion between virus and the host cell is being orchestrated by fusion protein, which is also known as fusion machine. The X-ray crystallographic results in tandem with cryo-electron microscopic measurements clearly demonstrate that the viral fusion machine is trimer of a heterodimer. The heterodimers are formed by the non-covalent association of receptor binding and transmembrane subunits of the fusion proteins. This chapter describes the oligomeric structure of four important fusion machines from influenza virus, human immunodeficiency virus, severe acute respiratory syndrome coronavirus (SARS-CoV) and SARS-CoV-2, and their importance in the entry of virus into the host cell through membrane fusion. We have further discussed about the strategies to develop drugs and/or vaccines targeting the oligomerization of fusion machines.

Keywords Membrane fusion · Fusion protein · Protein oligomerization · Activation barrier · Influenza, hemagglutinin · HIV · gp41 · Severe acute respiratory syndrome · SARS-coronavirus · spike protein

G. Meher · H. Chakraborty (✉)
School of Chemistry, Sambalpur University, Jyoti Vihar, Burla, Odisha 768019, India
e-mail: hirak@suniv.ac.in

S. I. Ahmad (ed.), *Human Viruses: Diseases, Treatments and Vaccines*, https://doi.org/10.1007/978-3-030-71165-8_33

1 Protein Oligomerization

Oligomerization is a common process that proteins utilize to recognize their interacting partners for the execution of biological functions. The oligomerization process takes place either between two identical proteins (homo-) or different proteins (hetero-), and the degree of oligomerization ranges from dimer to any higher order oligomer. As the oligomerization is critical for the function of proteins, it is therefore tightly regulated by various factors. The oligomerization is mainly induced by the change in folding of polypeptide chain under thermal stress, binding with partner proteins, interaction with metal cofactors, and allosteric effects. In addition, conformational change upon binding to effector proteins may leads to structural rearrangement, that induces formation of oligomers (Gabizon and Friedler 2014).

The protein oligomerizes through either via covalent bonding (mainly disulfide bonding) or through weak-bond-network associations. Both types of oligomerization are generally reversible in nature, at least up to a low degree of polymerization. Protein requires the presence of specific amino acid for the formation of covalent bound oligomers, whereas there is no such requirement for oligomerization through non-bonding interactions (Marianayagam et al. 2004). Domain swapping in proteins involves the interactions of exposed interfaces of different protein molecules which lead to oligomerization (Bennett et al. 1995). The dynamic nature of the protein oligomerization is important for the biological function of proteins. Disulfide bonding and domain swapping can act together to form larger aggregates, which might be harmful in nature (Liu 2015).

The oligomerization properties of proteins can be applied as therapeutic strategy as oligomerization is crucial for functioning of several proteins. Therefore, significant efforts have been given in understanding the structure of the oligomeric proteins utilizing X-ray crystallography, cryo-electron microscopy, and nuclear magnetic resonance (NMR) and various fluorescence spectroscopy based methodologies. In addition of providing structural information, these methods offer the kinetic and thermodynamics information of the oligomerization process. The advent of cryo-electron microscopy and small angle X-ray scattering offers tremendous potential to determine structure of protein oligomers in solution. In addition to the oligomeric structure, the oligomeric state is also extremely important in biology.

2 Membrane Fusion and Fusion Proteins

Membrane fusion is essential for the life of eukaryotic cells. In this process, two closely apposed bilayers merge into a single bilayer, and their internal materials mix with each other (Jahn et al. 2003). Several events such as cellular trafficking, sexual reproduction, compartmentalization, intercellular communication, endo- and exocytosis, and cell division are dependent on this basic process (Verkleij and Post

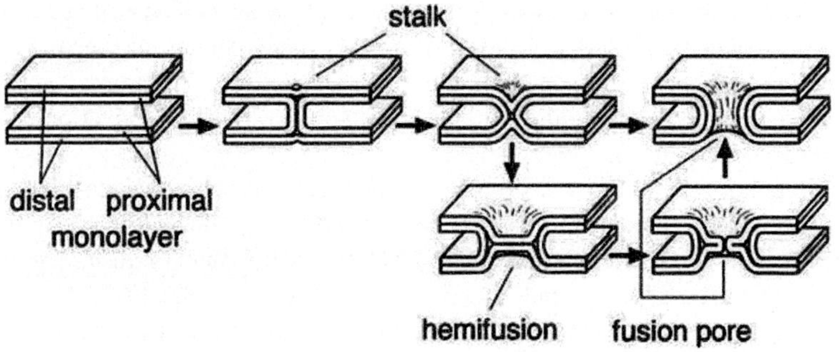

Fig. 1 Schematic representation of two closely apposed flat bilayers, intermediate states, and fusion pore during membrane fusion. The monolayers are described as flat but bendable sheets as described by the stalk hypothesis. The figure has been adapted from reference (Jahn et al. 2003) with appropriate permission

2000; Sollner and Rothman 1994; Primakoff and Myles 2002; Risselada et al. 2011). Apart from these constitutive processes, enveloped viruses exploit membrane fusion to enter into the host cell. Some of the viruses fuse with the plasma membrane (such as HIV, SARS, MERS), whereas other viruses enter in the cell through receptor mediated endocytosis, followed by fusing with the cell organelle like endosome (such as influenza). The cell surface receptors play a key role in the fusion of the viruses at the plasma membranes. Despite this diversity, all fusion reactions follow a basic process where the outer leaflets of two apposing membranes (docked) merge first, followed by the mixing of the inner leaflets during the opening of fusion pore (shown in Fig. 1). Interestingly, this intermediate stage, called hemifusion, is an important event similar in endocytosis, exocytosis, protein trafficking and viral entry.

The docking of two membranes and the intermixing of lipids to form the intermediate stages, and fusion pore are energy uphill process. Our knowledge derived from viral entry into the host cell and cellular secretory releases suggest that a protein machine (fusion protein) facilitates this kinetically unfavored fusion process toward completion (Lentz et al. 2000). Interestingly, each virus envelope contains a protein that is basically responsible for docking and fusion between the virus and host cell membranes (Pecheur et al. 1999). Several studies have shown that the isolated fusion proteins are capable of inducing fusion of model membranes (White et al. 2008; Wickner and Schekman 2008). The study of membrane fusion in model systems provides detailed understanding on the molecular details of membrane fusion. Based on these studies, three mechanisms of membrane fusion have been hypothesized. The stalk-pore hypothesis of membrane fusion deals with the formation and expansion of stalk that occurs between two interacting membranes. The gradual expansion of the stalk structure increases the mechanical tension on the separating bilayer, which eventually ruptures to form the pore (Kozlov et al. 1989; Chernomordik et al. 1995a, b). The second mechanism proposes pore formation at the very beginning of bilayer merging (Siegel 1993; Kuzmin et al. 2001). The third hypothesis describes the Brownian dynamics of membrane bilayer fusion (Muller et al. 2003; Noguchi and Takasu 2001). In spite of many studies still we have little understanding on the formation of initial intermediate and final pore. These two steps involve lipid/water transient rearrangement to relieve activation barrier (Chakraborty et al. 2012).

In biological systems, fusion proteins act as machines which trigger the fusion process. The entire fusion process is being carried out by three different types of proteins, namely docking proteins that select the membrane to fuse, regulatory proteins control the process, and fusion proteins that make the fusion to happen (Lentz et al. 2000). The protein systems that carry out these works are called protein machines, and all the enveloped (lipid-sheathed) viruses are hosting protein machines on their surfaces. Generally, viral fusion glycoproteins undergo enzymatic cleavage to acquire fusogenic property. Among the few well-characterized fusion proteins, the viral glycoproteins responsible for fusion of influenza, human immunodeficiency virus (HIV), severe acute respiratory syndrome coronavirus (SARS-CoV), and severe acute respiratory syndrome coronavirus-2 (SARS-CoV-2) have been discussed in this chapter.

3 Oligomerization: A Key Aspect of the Fusion Proteins

This has been found that more than 35% proteins are oligomeric in any organism (Goodsell and Olson 2000). The oligomerization behaves as a regulatory mechanism of protein activation, reducing genome load, and minimizing transcriptional error. In addition, oligomerization diversifies protein function, introduces allosteric regulation, and provides resistance against degradation or denaturation (Ali and Imperiali 2005). Interestingly, it has been observed that the tetramer is the most common oligomeric state of proteins (Goodsell 1991). The oligomers are being stabilized either due to hydrophobic effect (Miller 1989) or having significant electrostatic interaction. The geometrical shape complementarity provides the specificity for homo- or hetero-oligomerization (Jones and Thornton 1995).

Generally, the viral fusion proteins oligomerize to form a trimeric structure (pre-fusion structure), which undergoes series of conformational changes leading to the formation of a higher order hairpin-like structure through coil-coil interaction at the later stage of the fusion process (post-fusion) (Colman and Lawrence 2003). Membrane fusion is an energetically unfavorable event due to the strong repulsive hydration, electrostatic repulsion and steric barriers (Helm, Israelachvili, and McGuiggan 1992; Chernomordik et al. 1995a, b). The energy barrier can be overcome by fusion proteins by facilitating local dehydration, inducing local perturbation in the lipid bilayer, and providing energy through their conformational transition (Monck and Fernandez 1994; Kanaseki et al. 1997; Chernomordik et al. 1997). Taken together, viral fusion proteins are directly responsible for membrane merging (Hoekstra 1990). In spite of dissimilarities among viruses in terms of genomic type, entry pathways, and viral fusion glycoproteins, the fusion process

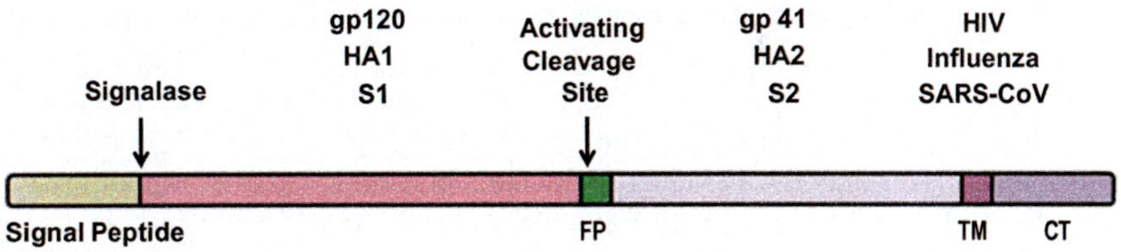

Fig. 2 The schematic representation of general organization of the class I fusion protein

share several common features for the fusion between virus and host cell. The general organization of class I fusion protein is shown in Fig. 2. All the class I fusion proteins contain a receptor binding domain and a fusion domain. These two domains get separated during the process of membrane fusion. A signal peptide is present at the N-terminal of the receptor binding domain, which recognizes the location, where cell surface receptors are populated. The fusion domain hosts fusion peptide, heptad region, transmembrane domain and the cytoplasmic tail.

All class I viral fusion proteins possess about 20–25 amino acid hydrophobic stretch that plays a crucial role in perturbing host membrane and inducing membrane fusion. The other similarity is that all viral fusion protein forms trimeric pre-fusion structure, which undergoes conformational change to form a six-helix bundle at the later stage of the fusion process (White 1990; Stegmann, Doms, and Helenius 1989). The oligomerization might be crucial for generating sufficient mechanical force to bring two membranes together. Moreover, conformational change in the trimeric structure provides more energy than the monomer of the corresponding protein. This extra energy might be critical to overcome the activation barrier for the membrane fusion. In this particular chapter, we have focused our discussion on four different class-I fusion proteins, i.e., hemagglutinin of influenza virus, gp41 of HIV, S2 protein from SARS coronavirus, and S2 protein of SARS coronavirus-2. The functional stringency of these four proteins is lying in their oligomeric status.

4 Hemagglutinin Fusion Protein from Influenza Virus

Influenza, commonly known as ‘flu’ is an extremely infectious disease that affects lungs and airways. The viruses spread easily through respiratory droplets, contaminated surfaces, saliva and skin-to-skin contact. Approximately, 3–5 million severe cases were observed per year worldwide, responsible for the cause of highest number of deaths. Influenza virus contains a lipid membrane, obtains from the plasma membrane of an infected cell during maturation. The membrane contains cellular lipids, but the membrane proteins are coded by the virus. Influenza virus has two membrane glycoproteins, the hemagglutinin (HA) and the neuraminidase (NA) (Wiley and Skehel 1987). The hemagglutinin (HA) fusion protein from influenza virus is responsible for fusion of virus with the host cell. HA is a trimeric transmembrane protein having both receptor binding and fusion activities domains of influenza virus. HA is synthesized as a single chain fusion-inactive precursor, HA0 which is cleaved by host-cell proteases by removing arginine 329 to produce two subunits, HA1 and HA2 (Meher and Chakraborty 2017, 2019; Benton et al. 2018). HA1 (328 amino acids) is mainly responsible for receptor binding, whereas HA2 (221 amino acids), the transmembrane subunit is crucial for fusion. These chains are covalently connected by a disulfide bond between HA1 position 14 and HA2 position 137. The HA1 binds with the sialic acid (N-acetyl neuraminic acid) present on the target cell surface, which promotes the sticking of the virus particle to the host cell. The C-terminal of HA2 contains a hydrophobic region (transmembrane domain) that anchors to the viral membrane and N-terminal possesses a conserved amphipathic region known as fusion peptide which is being exposed due to the conformational changes induced by the endosomal pH around 5.3. The transmembrane domain and the fusion peptide are connected via heptad region and a flexible linker region.

This low pH triggered viral fusion destabilizes the HA1 trimer contacts at the head region leading to dissociation of globular head domains. This movement changes the polypeptide segment of HA2 from loop to helix transition. This extended structure has been proposed to form pre-hairpin intermediate in the mechanism of fusion which is anchored both in target (through fusion peptide) and virus membrane (through transmembrane domain) (Wilson et al. 1981; Bullough et al. 1994; Eckert and Kim 2001). The crystal structure of HA trimer is shown in Fig. 3 (DuBois et al. 2011).

The fusion activity of HA has been measured both in vivo and vitro and by different techniques like hemolysis (Wilson et al. 1981; Durell et al. 1997), polykaryon formation, resonance energy transfer (Poyry and Vattulainen 2016; Wilson et al. 1981), liposome cell fusion, spin labeled phospholipid transfer and electron microscopy (Ali and Imperiali 2005). The determination of pH dependence fusion by these processes corresponds to the pH at which the characteristics structural changes of HA occurs. However, HA mediated membrane fusion has also been reported to occur at neutral pH (Ali and Imperiali 2005), however the efficiency is much higher in acidic pH. Sialic acids are found in cell-surface glycoproteins and glycolipids, and neuraminidase treatment removes the sialic acid by destroying receptor binding and preventing infection. Further, sialic acid containing glycolipids serve as cellular receptors restoring infectivity to neuraminidase-treated cells (Bergelson et al. 1982; Suzuki et al. 1985).

In 1968, the structures of three conformations of the ectodomain of the Hong Kong influenza virus HA have been determined by X-ray crystallography (Chen et al. 1999). Though the three-dimensional structure of fusion active conformation of the HA is not yet known. However, the results of several experiments provide a partial picture of the active conformation of HA2. The hydrophobic N-terminal peptide of HA2 is exposed as indicated by the hydrophobic properties and thermolytic sensitivity, and the loss of the hydrophobic properties after thermolytic treatment removes HA2. The antigenic changes in the HA1-HA1 interfacial region

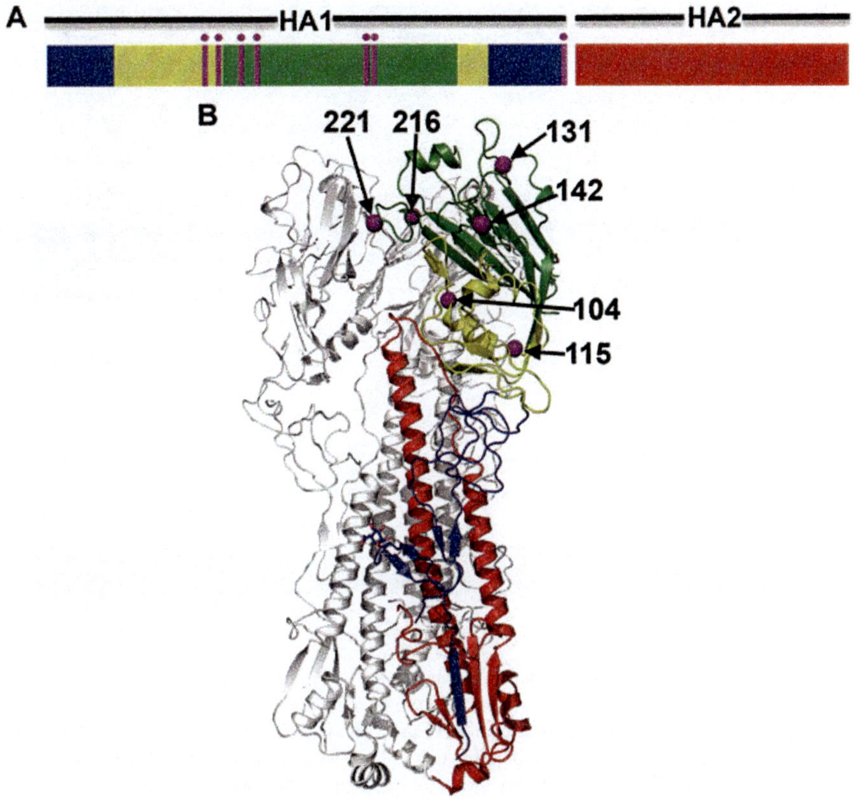

Fig. 3 **a** Schematic of the HA protein after proteolytic cleavage into HA1 and HA2 subunits. The receptor-binding domain (RBD) consists of the receptor-binding sub-domain (green) and the vestigial esterase sub-domain (yellow). The F' fusion sub-domain in HA1 is colored blue and the HA2 stalk domain is colored red. **b** Crystal structure of HA trimer determined at 2.50Å. One protomer is colored as in panel A (with HA1 in blue, HA2 in red). The remaining 2 HA protomers are colored grey. The amino acids those are different in highly pathogenic and moderately pathogenic strains. The figure has been adapted from reference (DuBois et al. 2011) with appropriate permission

destabilize the nature of mutants and affect the pH of the membrane fusion. The HA2-HA2 trimer interface along the whole length, from virus membrane to the top of long helix suggests location of amino acid substitutions in fusion mutants. Thermolysin-solubilized BHA2 aggregate indicates that the stem region of the molecule remains trimeric, although the rearranged interface appears weakened since low-pH BHA2 can dissociate at sufficiently low protein concentrations (Wiley and Skehel 1987).

The crystal structure of HA2 and its fusion activity results indicate that the protein undergoes significant conformational changes at acidic pH, where N-terminal and C-terminal regions undergo extensive refolding. Also, the

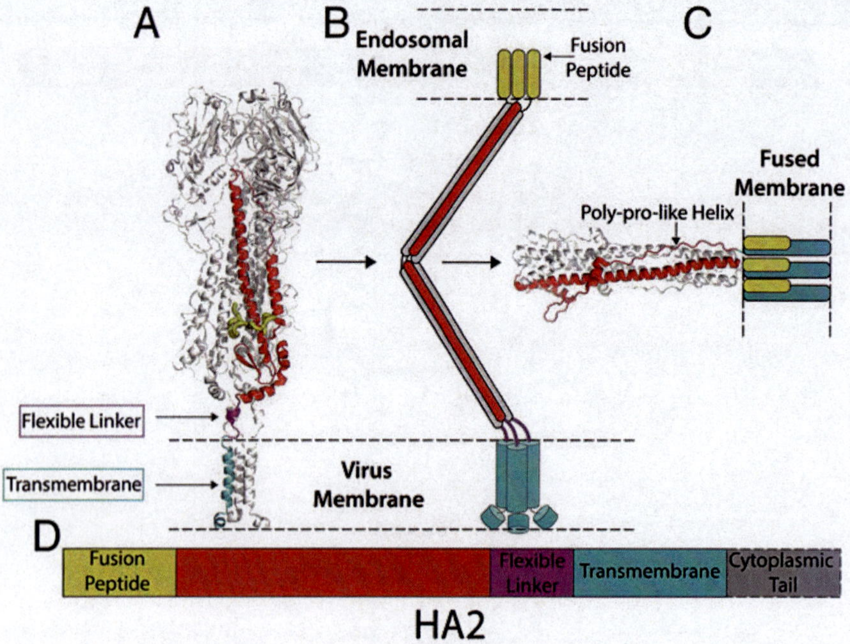

Fig. 4 Structural snapshots toward membrane fusion. **a** The structure of straight full-length HA from cryo-EM (this study). **b** Schematic representation of possible fusion pH intermediate (Calder and Rosenthal 2016). **c** The structure of fusion pH HA2 (18), with modeled membrane anchor and fusion peptide colocalized in a fused membrane. **d** A color-coded diagram showing the location of the different regions of HA2 in its primary structure. The figure has been adapted from (Benton et al. 2018) with permission

conformation of HA2 assumes different pre- and post-fusion conformation. The C-terminal chain adopts an extended polyproline type II helix that is packed into the outward-facing grooves between the α-helices of the central trimeric coiled coil (Chen et al. 1999) and forms a six helix bundle (Fig. 4). The N-terminal (attached to the host cell) and transmembrane domain (in the viral membrane) approach to each other, which in turn bring two interacting membranes closely apposed. The flexibility of the linker region may also be important to facilitate the substantial refolding event at low pH. An understanding of this final part of the process will require determination of the structure of HA2, containing both the fusion peptide and the membrane anchor, in the fusion pH conformation.

The influenza viruses are relatively simple, RNA-containing viruses with strongly immunogenic surface proteins, especially the HA. However, their segmented genomes undergo antigenic shift and drift, and results an adaptive immune response in a range of mammalian and avian species, including humans. Because of their adaptive ability, influenza viruses make dilemma in producing long-lasting vaccines against the disease (Weis et al. 1990).

5 Human Immunodeficiency Virus Envelope Glycoprotein

The human immunodeficiency virus (HIV) infection causes AIDS and interferes with body's ability to fight. This virus transmits through infected blood, semen or vaginal fluids. In current scenario, almost 40 million people are living with HIV worldwide. The virus enters into target cells through series of receptor triggered conformational changes of the fusion machine that elevate virus-cell fusion (Dey et al. 2008). HIV expresses a glycoprotein, gp160 on its cell surface which proteolytically cleaved into two subunits. The larger subunit, gp120 contains the binding site for the host cell surface receptor, CD4, whereas the smaller subunit gp41 contains a transmembrane anchoring highly hydrophobic C-terminal sequence and an N-terminal hydrophobic fusion peptide (Blumenthal, Durell, and Viard 2012; Skehel and Wiley 2000; Chen, Skehel, and Wiley 1999). The first step in HIV entry is docking, which involves a stepwise recognition of gp120 by CD4 and co-receptor on the target cells. Binding of gp120 to CD4 produces conformational changes in gp120 that lead to recognition of co-receptor, CXCR4 or CCR5. The conformational change in gp120 results the exposure of N-terminal fusion peptide of gp41, and insertion of fusion peptide into the lipid bilayer of the target cell membrane (Blumenthal et al. 2012). The heterodimer of gp120 and gp41 forms a mushroom shaped trimeric structure, which is incorporated into the viral envelope through the transmembrane region of gp41, while virus particles bud from the cell surface (White et al. 2010). Julien et al., solved the crystal structure of an antigenically near-native, cleaved, stabilized and soluble trimer of gp160 (gp120/gp41) in complex with a broadly neutralizing PGT122 antibody at 4.7 Å resolution (Julien et al. 2013). The trimeric structure demonstrates the pre-fusion state of gp41, and the interaction between gp120 and gp41 subunits. It has been shown that the gp120/gp41 trimer is unstable when separated from its membrane-anchored environment. The gp120 is highly glycosylated and interacts with CD4 receptor, and the co-receptor proteins of the target cells.

In terms of membrane fusion, the structure of gp41 is important as this subunit is responsible for inducing fusion between virus and host cell membranes. HIV gp41 consists of three major domains, e.g., the extracellular domain or ectodomain, the transmembrane domain, and the cytoplasmic domain or cytoplasmic tail (CT). The extracellular domain of gp41 carries out multiple functions and considers as the most important part for the fusion of virus with the host cell. The extracellular domain can be further subdivided into the following five functional regions: the fusion peptide (FP) followed by the N-terminal heptad repeat (NHR), the loop region (LR), the C-terminal heptad repeat (CHR), and finally, the membrane-proximal external region (MPER) (Garg et al. 2011). It has been shown that three NHR helices form a coiled-coil core, and the three CHR helices dock in the three hydrophobic grooves around the core, forming the iconic six-helix bundle (6HB) structure (Caffrey et al. 1998; Weissenhorn et al. 1997; Chan et al. 1997). This six-helix bundle formation is the unique trait of class I fusion proteins. The crystallographic study has expanded our understanding by including the fusion

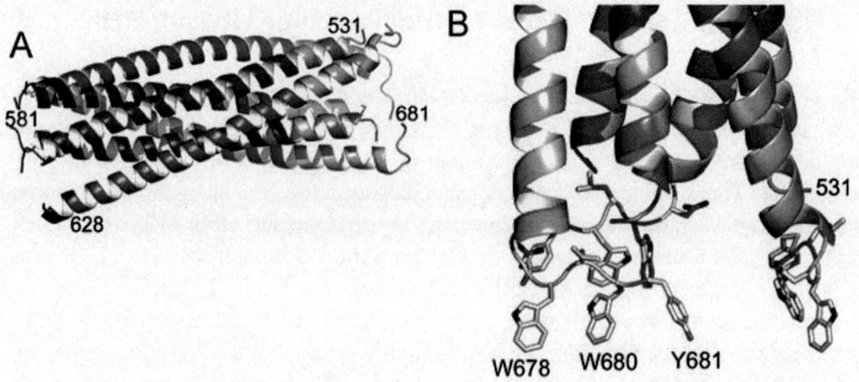

Fig. 5 X-ray crystallographic structure of the gp41 ectodomain in the six-helix bundle conformation. The advantage of this structure is the inclusion of the FP proximal region and MPER segments. The figure has been adapted from (Buzon et al. 2010) with permission

peptide proximal region (FPPR) and membrane proximal external region (MPER) segments with the classic 6HB structure of gp41 (Fig. 5) (Buzon et al. 2010). These extra segments occur as helical extensions from the core coiled-coil structure but still interact to enhance the thermodynamic stability. An interesting bent at the C terminus of the MPER segment exposes conserved hydrophobic residues that may insert into the outer layer of the viral membrane to increase the curvature.

6 Spike Protein (S) from SARS-CoV

Severe acute respiratory syndrome (SARS) is a viral respiratory disease caused by the SARS coronavirus (SARS-CoV). The SARS-CoV transmits through droplets when someone with the disease coughs or sneezes to someone else. Since 2002, SARS spread virtually infecting thousands of people and killed 35% of infected ones. SARS-CoV contains spike protein in the viral envelope which is responsible for the fusion of virus and T-cell. The coronavirus spike protein (S) is a protein machine that plays the key role for viral entry to the host cells. The protein S undergoes proteolytic cleavage resulting two non-covalently associated subunits, S1 and S2, however the cleavage is not an absolute requirement for inducing membrane fusion (Masters 2006). The coronavirus docks to the host cell through the interaction between S1 spike protein and the cell surface receptor protein, angiotensin converting enzyme 2 (ACE2). The S2 glycoprotein promotes the fusion between SARS coronavirus and the host cell (Jeffers et al. 2004; Li et al. 2003). The S2 subunit contains two highly conserved heptad regions namely HR1 and HR2, (Xu et al. 2004b; Ingallinella et al. 2004; Howard et al. 2006), N-terminal fusion peptide, and C-terminal transmembrane domain like other class 1 virus fusion proteins such as gp41, HA, Ebola virus glycoprotein and paramyxovirus F protein

(Chambers, Pringle, and Easton 1990; Gallaher 1996). The N-terminal signal peptide in S-protein facilitates transport of viral particle into endoplasmic reticulum where proteins are extensively glycosylated.

Generally, the ACE2 receptor activates the fusion proteins in two ways, by binding to receptor and/or engaging the receptor to trigger the viral molecules into endosomes where protonation activates membrane fusion (Hofmann and Pohlmann 2004). The receptor binding of S1 leads to conformational change in S2 subunit that exposes the fusion peptide, which interacts with the target host cell, similar to other class 1 fusion proteins (Epand 2003). The cryo-electron microscopy images showed that the S glycoprotein of SARS-CoV (without the transmembrane domain and cytoplasmic tail) forms a homotrimer (Gui et al. 2017). In the SARS-CoV S glycoprotein, the S1 subunit is rich in β-strand, and contains N-terminal domain (NTD, residues 14-294) and three C-terminal domains (CTD1, residues 320-516; CTD2, residues 517-578 and CTD3, residues 579-663). The S1 protein binds with the cellular receptor ACE2 through the CTD1 domain of S1 protein (Li et al. 2003; 2005).

The S2 subunit, which begins after the cleavage of S1 at the residue 667, is majorly α-helical. The S2 subunit houses the functionally important fusion peptide (residues 798-815), HR1 (residues 880-967), HR2 (residues 1154-1183, and a transmembrane domain. Refolding of HR1 into a long α-helix, exposes the fusion peptide so that it can interact with the host-cell membranes. Thereafter, HR1 and HR2 interact to form a coiled coil, which brings the host and viral membranes close to each other. Like other class I virus fusion protein, S2 protein of SARS-CoV forms a fusion-active conformation (Xu et al. 2004a). This hairpin structure brings the viral and cellular membranes, which assists the fusion between two membranes and facilitates viral entry. The fusion core protein of SARS-CoV spike protein was crystallized by Xu et al., and structure was determined at 2.8 Å resolutions (Xu 2004a). It has been found that the three HR2 helices packed against in the hydrophobic grooves on the coiled coil formed by three parallel HR1 helices (Fig. 6).

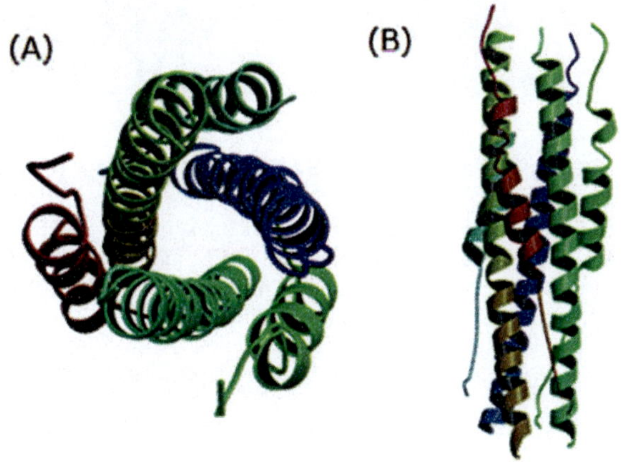

Fig. 6 Overall views of the fusion core structure. **a** top view of the SARS-CoV spike protein fusion core structure showing the 3-fold axis of the trimer, **b** side view of the SARS-CoV spike protein fusion core structure showing the six-helix bundle. The figure has been adapted from (Xu et al. 2004a) with permission

7 Spike Protein (S) from SARS-CoV-2

Three coronaviruses have crossed the species boundary to cause deadly pneumonia in human are SARS-CoV, middle-east respiratory syndrome coronavirus (MERS-CoV) and SARS-CoV-2. SARS-CoV-2 is associated to the recent epidemic of atypical pneumonia that has already infected more than three million people and killed more than 2,00,000 worldwide. Like SARS-CoV, SARS-CoV-2 hosts a transmembrane spike glycoprotein that induces the fusion between virus and host cell membranes. The SARS-CoV-2 S protein is almost 76% amino acid sequence identity to its counterpart SARS-CoV. It has been found that the equilibrium constants of binding of SARS-CoV and SARS-CoV-2 with ACE-2 receptor are 5.0 nM and 1.2 nM, respectively with similar kinetic rate constants. The difference that SARS-CoV-2 S protein harbors from S protein of SARS-CoV is the presence of a furin cleavage site at the boundary of S1/S2 subunits. This might have been incorporated during biogenesis, and sets this virus apart from SARS-CoV and SARS-related CoVs (Walls et al. 2020).

The cryo-electron microscopic structure of the ectodomain of S protein has been solved in 2.8 Å resolutions (Fig. 7). The structure showed that it is a trimer of heterotrimer of S1 and S2 proteins as observed for SARS-CoV.

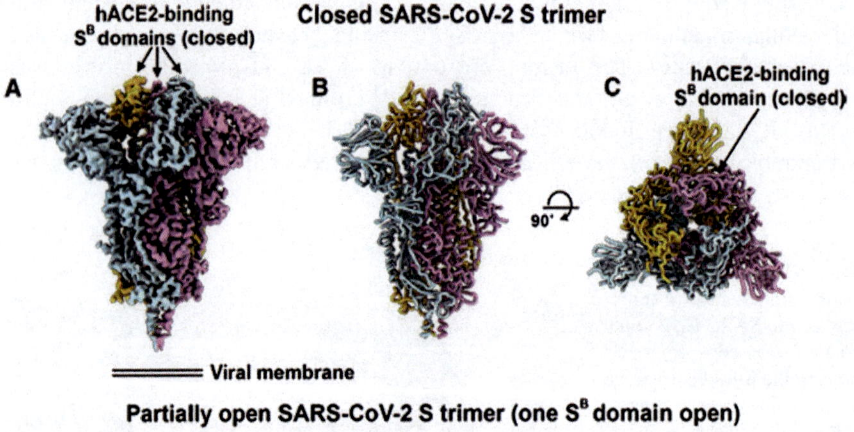

Fig. 7 Cryo-electron microscopic structures of the SARS-CoV-2 S glycoprotein **a** Closed SARS-CoV-2 S trimer unsharpened cryo-EM map. **b** and **c** Two orthogonal views from the side (**b**) and top (**c**) of the atomic model of the closed SARS-CoV-2 S trimer. The figure has been adapted from (Walls et al. 2020) with permission

8 Fusion Peptides and Their Oligomeric Status

Interestingly, all class I fusion proteins include a highly conserved amphipathic peptide of 20–25 amino acid residues at the N-terminus of the fusion domain (HA2, gp41, S2), which is crucial for fusion activity. This unique stretch of amino acid residues is called 'fusion peptide' and even a single mutation in the fusion peptide region alters the fusogenic activity of the protein (Gething et al. 1986). The fusion peptide gets exposed either due to receptor binding of the fusion protein (HIV, SARS-CoV) or exposed to low pH (influenza). The exposed fusion peptide interacts with the host cell membranes, and induces perturbation to the organization and dynamics of the host membranes. The amphipathic nature of the fusion peptide reduces the change of crystallization of the fusion protein. Therefore, the crystal structures of all the fusion proteins are devoid of the fusion peptide residues. This catalyzed to evaluate the secondary structure of fusion peptides and their oligomeric status in membrane milieu.

The N-terminal fusion peptide of hemagglutinin is being exposed due to the conformational changes brought out by the endosomal low pH ($\sim$5.3) (White 1990; Durell et al. 1997). The function of fusion peptide has been studied extensively in vitro and in vivo (Chakraborty et al. 2013; Epand et al. 1994; Haque et al. 2011; Kozlovsky et al. 2002), and suggested that the fusion peptide induces membrane fusion by altering bending modulus (Tristram-Nagle and Nagle 2007), filing void volume (Malinin and Lentz 2004), promoting positive curvature (Chakraborty et al. 2013) or negative intrinsic curvature (Ge and Freed 2009) to apposed bilayer and reducing Gaussian energy (Chang et al. 2000). The NMR structure of HA fusion peptide was solved in dodecylphosphocholine (DPC) micelles, and ESR measurements were carried out in utilizing several depth-dependent spin probes in the vesicular system, and the structure of the fusion peptide has been proposed to be α-helical with a bent between Glu11 and Asn12 (Han et al. 2001; Lai et al. 2006) followed by short 3_{10} helix. The bent conformation in between two helices make an 'inverted V-shaped' structure of hemagglutinin fusion peptide. The 23-residue fusion peptide of gp41 was reported to have α-helical conformation in phosphatidylglycerol–phosphatidylcholine membranes (Lai et al. 2012). Interestingly, the fusion peptide converts from α- helix to β-sheet depending on the concentration of cholesterol on the membrane. However, both α-helical and β-sheet conformations are capable of inducing membrane fusion. The NMR of S2 fusion peptide from SARS-CoV in DPC micelles suggest an 'inverted V-shaped' α-helical conformation (Mahajan and Bhattacharjya 2015).

Weliky group synthesized chemically cross-linked oligomers of gp41 fusion peptide and evaluated the fusogenic properties of different oligomers. It has been found from their work that trimer is the smallest catalytically efficient oligomer (Qiang and Weliky 2009). This strongly supports the trimer formation of the fusion peptide region of gp41, which is generally absent in the X-ray crystallographic structures. We have recently showed that SARS-CoV fusion peptide oligomerizes in the membrane containing higher concentration of cholesterol (Meher et al. 2019).

De Grado and his coworkers showed that a synthesized triple-stranded coiled-coil X31 peptide that mimics influenza virus hemagglutinin demonstrated higher lipid mixing and leakage of liposomal content (Lau et al. 2004). These results support the oligomerization, especially formation of trimer of fusion peptides of envelope proteins from HIV, influenza and SARS-CoV.

9 Fusion Proteins Are Potential Targets for Vaccines

The fusion proteins are potential target for vaccine development because of their lead role in virus entry by promoting fusion between virus and host cell membranes, and their location on the outer surface of the viral membranes. There are three broad ways to block virus entry, (i) preventing the virus from reaching to host cell by immobilizing the virus, (ii) using antibody to cover the receptor binding domain of the fusion protein to block virus attachment, and (iii) preventing entry through fusion inhibition (Graham 2013). In addition, the six-helix bundle formation by the heptad regions of the fusion domain has also been successfully utilized to inhibit virus entry. T20 (also called as DP178, enfuvirtide, pentafuside, fuzeon$^{(TM)}$) is one of the well characterized members of the entry inhibitors (Baldwin, Sanders, and Berkhout 2003). T20 is a 36 amino acid homologous peptide of HR2 of gp41 (Wild et al. 1994a, b), which blocks six helix bundle formation, that is necessary for viral entry by competitively binding to the HR1 (Lawless et al. 1996; Kliger and Shai 2000; Wild et al. 1994b). C34 and T649 are, two other C-peptides similar to T20 and have been developed based on the gp41 HR2 region, known to inhibit virus entry following a similar mechanism as T20 (Malashkevich et al. 1998; Lu et al. 2004). Two compounds, MBX2329 and MBX2546, with aminoalkyl phenol ether and sulphonamide scaffolds, respectively are known to inhibit the hemagglutinin-mediated entry of influenza virus (Basu et al. 2014). FI6v3 and CR9114, two peptides based on complementarity of loop region of hemagglutinin, exhibit nanomolar affinity in neutralizing H1N1 and H5N1 influenza strains. These inhibitors bind to the highly conserved stem epitope and inhibit pH-induced conformational changes of hemagglutinin, thereby inhibiting fusion between virus and host cells (Kadam et al. 2017). Recent studies showed that the SARS-CoV S protein mediated fusion can be inhibited by the HR2 peptide, however the HR1 peptide did not show any inhibitory effect (Bosch et al. 2004; Liu et al. 2004; Zhu et al. 2004), as we observed in HIV. Taken together, it is important to know the structures of fusion proteins and their oligomers, as they provide crucial information on the mechanism of membrane fusion. The proper understanding of the mechanism opens up avenues to develop inhibitors against the entry of the viruses.

10 Concluding Remarks

Although we have discussed the only few rèpresentative examples of fusion protein, it is actually a general feature of all class I viruses. Presence of large amount of hydrophobic amino acid residues in fusion proteins, especially in the fusion peptide and transmembrane regions, makes them difficult to crystallize. Therefore, most of the crystal structures have been solved for the ectodomains of the fusion proteins. However, with the advent of the cryo-electron microscopy the oligomeric structures of the fusion proteins are being solved. The appropriate surface mapping of the fusion protein oligomers provides an edge in designing entry inhibitor-based vaccines. The fusion studies in model membranes demonstrate the role of different regions (such as fusion peptide, transmembrane domain, MPER, etc.) in the fusion process, and allows an opportunity to evaluate the potential strategies for fusion inhibition. Taken together, it is clear that the detailed structural knowledge in tandem with the functional data offers an opportunity to control the viral infection.

Acknowledgements This work was supported by research grant from Science and Technology Department, Government of Odisha. H.C. and G.M. thank the University Grants Commission (UGC) for the UGC Assistant Professor position and UGC-BSR fellowship, respectively. We acknowledge Department of Science and Technology, New Delhi and UGC for providing instrument facility to the School of Chemistry, Sambalpur University under the FIST and DRS programs, respectively. We thank members of Chakraborty laboratory for their comments and discussions.

References

Ali MH, Imperiali B. Protein oligomerization: how and why. Bioorg Med Chem. 2005;13:5013–20.

Baldwin CE, Sanders RW, Berkhout B. Inhibiting HIV-1 entry with fusion inhibitors. Curr Med Chem. 2003;10:1633–42.

Basu A, Antanasijevic A, Wang M, et al. New small molecule entry inhibitors targeting hemagglutinin-mediated influenza a virus fusion. J Virol. 2014;88:1447–60.

Bennett MJ, Schlunegger MP, Eisenberg D. 3D domain swapping: a mechanism for oligomer assembly. Protein Sci. 1995;4:2455–68.

Benton DJ, Nans A, Calder LJ, et al. Influenza hemagglutinin membrane anchor. Proc Natl Acad Sci U S A. 2018;115:10112–7.

Bergelson LD, Bukrinskaya AG, Prokazova NV, et al. Role of gangliosides in reception of influenza virus. Eur J Biochem. 1982;128:467–74.

Blumenthal R, Durell S, Viard M. HIV entry and envelope glycoprotein-mediated fusion. J Biol Chem. 2012;287:40841–9.

Bosch BJ, Martina BE, Van Der Zee R, et al. Severe acute respiratory syndrome coronavirus (SARS-CoV) infection inhibition using spike protein heptad repeat-derived peptides. Proc Natl Acad Sci U S A. 2004;101:8455–60.

Bullough PA, Hughson FM, Skehel JJ, Wiley DC. Structure of influenza haemagglutinin at the pH of membrane fusion. Nature. 1994;371:37–43.

Buzon V, Natrajan G, Schibli D, Campelo F, Kozlov MM, Weissenhorn W. Crystal structure of HIV-1 gp41 including both fusion peptide and membrane proximal external regions. PLoS Pathog. 2010;6:

Caffrey M, Cai M, Kaufman J, et al. Three-dimensional solution structure of the 44 kDa ectodomain of SIV gp41. EMBO J. 1998;17:4572–84.

Calder LJ, Rosenthal PB. Cryomicroscopy provides structural snapshots of influenza virus membrane fusion. Nat Struct Mol Biol. 2016;23:853–8.

Chakraborty H, Tarafdar PK, Bruno MJ, Sengupta T, Lentz BR. Activation thermodynamics of poly(ethylene glycol)-mediated model membrane fusion support mechanistic models of stalk and pore formation. Biophys J. 2012;102:2751–60.

Chakraborty H, Tarafdar PK, Klapper DG, Lentz BR. Wild-type and mutant hemagglutinin fusion peptides alter bilayer structure as well as kinetics and activation thermodynamics of stalk and pore formation differently: mechanistic implications. Biophys J. 2013;105:2495–506.

Chambers P, Pringle CR, Easton AJ. Heptad repeat sequences are located adjacent to hydrophobic regions in several types of virus fusion glycoproteins. J Gen Virol. 1990;71:3075–80.

Chan DC, Fass D, Berger JM, Kim PS. Core structure of gp41 from the HIV envelope glycoprotein. Cell. 1997;89:263–73.

Chang DK, Cheng SF, Deo Trivedi V, Yang SH. The amino-terminal region of the fusion peptide of influenza virus hemagglutinin HA2 inserts into sodium dodecyl sulfate micelle with residues 16-18 at the aqueous boundary at acidic pH. Oligomerization and the conformational flexibility. J Biol Chem. 2000;275:19150–8.

Chen J, Skehel JJ, Wiley DC. N- and C-terminal residues combine in the fusion-pH influenza hemagglutinin HA(2) subunit to form an N cap that terminates the triple-stranded coiled coil. Proc Natl Acad Sci U S A. 1999;96:8967–72.

Chernomordik L, Kozlov MM, Zimmerberg J. Lipids in biological membrane fusion. J Membr Biol. 1995a;146:1–14.

Chernomordik L, Leikina E, Cho MS, Zimmerberg J. Control of baculovirus gp64-induced syncytium formation by membrane lipid composition. J Virol. 1995b;69:3049–58.

Chernomordik LV, Leikina E, Frolov V, Bronk P, Zimmerberg J. An early stage of membrane fusion mediated by the low pH conformation of influenza hemagglutinin depends upon membrane lipids. J Cell Biol. 1997;136:81–93.

Colman PM, Lawrence MC. The structural biology of type I viral membrane fusion. Nat Rev Mol Cell Biol. 2003;4:309–19.

Dey AK, David KB, Ray N, et al. N-terminal substitutions in HIV-1 gp41 reduce the expression of non-trimeric envelope glycoproteins on the virus. Virology. 2008;372:187–200.

DuBois RM, Zaraket H, Reddivari M, Heath RJ, White SW, Russell CJ. Acid stability of the hemagglutinin protein regulates H5N1 influenza virus pathogenicity. PLoS Pathog. 2011;7:

Durell SR, Martin I, Ruysschaert JM, Shai Y, Blumenthal R. What studies of fusion peptides tell us about viral envelope glycoprotein-mediated membrane fusion. Mol Membr Biol. 1997;14:97–112.

Eckert DM, Kim PS. Mechanisms of viral membrane fusion and its inhibition. Annu Rev Biochem. 2001;70:777–810.

Epand RF, Martin I, Ruysschaert JM, Epand RM. Membrane orientation of the SIV fusion peptide determines its effect on bilayer stability and ability to promote membrane fusion. Biochem Biophys Res Commun. 1994;205:1938–43.

Epand RM. Fusion peptides and the mechanism of viral fusion. Biochim Biophys Acta. 2003;1614:116–21.

Gabizon R, Friedler A. Allosteric modulation of protein oligomerization: an emerging approach to drug design. Front Chem. 2014;2:9.

Gallaher WR. Similar structural models of the transmembrane proteins of Ebola and avian sarcoma viruses. Cell. 1996;85:477–8.

Garg H, Viard M, Jacobs A, Blumenthal R. Targeting HIV-1 gp41-induced fusion and pathogenesis for anti-viral therapy. Curr Top Med Chem. 2011;11:2947–58.

Ge M, Freed JH. Fusion peptide from influenza hemagglutinin increases membrane surface order: an electron-spin resonance study. Biophys J. 2009;96:4925–34.

Gething MJ, Doms RW, York D, White J. Studies on the mechanism of membrane fusion: site-specific mutagenesis of the hemagglutinin of influenza virus. J Cell Biol. 1986;102:11–23.

Goodsell DS. Inside a living cell. Trends Biochem Sci. 1991;16:203–6.
Goodsell DS, Olson AJ. Structural symmetry and protein function. Annu Rev Biophys Biomol Struct. 2000;29:105–53.
Graham BS. Advances in antiviral vaccine development. Immunol Rev. 2013;255(1):230–42.
Gui M, Song W, Zhou H, et al. Cryo-electron microscopy structures of the SARS-CoV spike glycoprotein reveal a prerequisite conformational state for receptor binding. Cell Res. 2017;27:119–29.
Han X, Bushweller JH, Cafiso DS, Tamm LK. Membrane structure and fusion-triggering conformational change of the fusion domain from influenza hemagglutinin. Nat Struct Biol. 2001;8:715–20.
Haque ME, Chakraborty H, Koklic T, Komatsu H, Axelsen PH, Lentz BR. Hemagglutinin fusion peptide mutants in model membranes: structural properties, membrane physical properties, and PEG-mediated fusion. Biophys J. 2011;101:1095–104.
Helm Christiane A, Israelachvili Jacob N, McGuiggan Patty M. Role of hydrophobic forces in bilayer adhesion and fusion. Biochemistry. 1992;31:1794–805.
Hoekstra D. Membrane fusion of enveloped viruses: especially a matter of proteins. J Bioenerg Biomembr. 1990;22:121–55.
Hofmann H, Pohlmann S. Cellular entry of the SARS coronavirus. Trends Microbiol. 2004;12:466–72.
Howard MW, Tripet B, Jobling MG, Holmes RK, Holmes KV, Hodges RS. Dissection of the fusion machine of SARS-coronavirus. Adv Exp Med Biol. 2006;581:319–22.
Ingallinella P, Bianchi E, Finotto M, et al. Structural characterization of the fusion-active complex of severe acute respiratory syndrome (SARS) coronavirus. Proc Natl Acad Sci U S A. 2004;101:8709–14.
Jahn R, Lang T, Sudhof TC. Membrane fusion. Cell. 2003;112:519–33.
Jeffers SA, Tusell SM, Gillim-Ross L, et al. CD209L (L-SIGN) is a receptor for severe acute respiratory syndrome coronavirus. Proc Natl Acad Sci U S A. 2004;101:15748–53.
Jones S, Thornton JM. Protein-protein interactions: a review of protein dimer structures. Prog Biophys Mol Biol. 1995;63:31–65.
Julien JP, Cupo A, Sok D, et al. Crystal structure of a soluble cleaved HIV-1 envelope trimer. Science. 2013;342:1477–83.
Kadam RU, Juraszek J, Brandenburg B, et al. Potent peptidic fusion inhibitors of influenza virus. Science. 2017;358:496–502.
Kanaseki T, Kawasaki K, Murata M, Ikeuchi Y, Ohnishi S. Structural features of membrane fusion between influenza virus and liposome as revealed by quick-freezing electron microscopy. J Cell Biol. 1997;137:1041–56.
Kliger Y, Shai Y. Inhibition of HIV-1 entry before gp41 folds into its fusion-active conformation. J Mol Biol. 2000;295:163–8.
Kozlov MM, Leikin SL, Chernomordik LV, Markin VS, Chizmadzhev YA. Stalk mechanism of vesicle fusion. Intermixing of aqueous contents. Eur Biophys J. 1989;17:121–9.
Kozlovsky Y, Chernomordik LV, Kozlov MM. Lipid intermediates in membrane fusion: formation, structure, and decay of hemifusion diaphragm. Biophys J. 2002;83:2634–51.
Kuzmin PI, Zimmerberg J, Chizmadzhev YA, Cohen FS. A quantitative model for membrane fusion based on low-energy intermediates. Proc Natl Acad Sci U S A. 2001;98:7235–40.
Lai AL, Moorthy AE, Li Y, Tamm LK. Fusion activity of HIV gp41 fusion domain is related to its secondary structure and depth of membrane insertion in a cholesterol-dependent fashion. J Mol Biol. 2012;418:3–15.
Lai AL, Park H, White JM, Tamm LK. Fusion peptide of influenza hemagglutinin requires a fixed angle boomerang structure for activity. J Biol Chem. 2006;281:5760–70.
Lau WL, Ege DS, Lear JD, Hammer DA, DeGrado WF. Oligomerization of fusogenic peptides promotes membrane fusion by enhancing membrane destabilization. Biophys J. 2004;86:272–84.
Lawless MK, Barney S, Guthrie KI, Bucy TB, Petteway SR Jr, Merutka G. HIV-1 membrane fusion mechanism: structural studies of the interactions between biologically-active peptides from gp41. Biochemistry. 1996;35:13697–708.

Lentz BR, Malinin V, Haque ME, Evans K. Protein machines and lipid assemblies: current views of cell membrane fusion. Curr Opin Struct Biol. 2000;10:607–15.
Li F, Li W, Farzan M, Harrison SC. Structure of SARS coronavirus spike receptor-binding domain complexed with receptor. Science. 2005;309:1864–8.
Li W, Moore MJ, Vasilieva N, et al. Angiotensin-converting enzyme 2 is a functional receptor for the SARS coronavirus. Nature. 2003;426:450–4.
Liu S. A Review on Protein Oligomerization Process. Int J Pr Eng Man. 2015;16:2731–60.
Liu S, Xiao G, Chen Y, et al. Interaction between heptad repeat 1 and 2 regions in spike protein of SARS-associated coronavirus: implications for virus fusogenic mechanism and identification of fusion inhibitors. Lancet. 2004;363:938–47.
Lu J, Sista P, Giguel F, Greenberg M, Kuritzkes DR. Relative replicative fitness of human immunodeficiency virus type 1 mutants resistant to enfuvirtide (T-20). J Virol. 2004;78:4628–37.
Mahajan M, Bhattacharjya S. NMR structures and localization of the potential fusion peptides and the pre-transmembrane region of SARS-CoV: implications in membrane fusion. Biochim Biophys Acta. 2015;1848:721–30.
Malashkevich VN, Chan DC, Chutkowski CT, Kim PS. Crystal structure of the simian immunodeficiency virus (SIV) gp41 core: conserved helical interactions underlie the broad inhibitory activity of gp41 peptides. Proc Natl Acad Sci U S A. 1998;95:9134–9.
Malinin VS, Lentz BR. Energetics of vesicle fusion intermediates: comparison of calculations with observed effects of osmotic and curvature stresses. Biophys J. 2004;86:2951–64.
Marianayagam NJ, Sunde M, Matthews JM. The power of two: protein dimerization in biology. Trends Biochem Sci. 2004;29:618–25.
Masters PS. The molecular biology of coronaviruses. Adv Virus Res. 2006;66:193–292.
Meher G, Bhattacharjya S, Chakraborty H. Membrane cholesterol modulates oligomeric status and peptide-membrane interaction of severe acute respiratory syndrome coronavirus fusion peptide. J Phys Chem B. 2019;123:10654–62.
Meher G, Chakraborty H. Organization and dynamics of Trp14 of hemagglutinin fusion peptide in membrane mimetic environment. Chem Phys Lipids. 2017;205:48–54.
Meher G, Chakraborty H. Membrane composition modulates fusion by altering membrane properties and fusion peptide structure. J Membr Biol. 2019;252:261–72.
Miller S. The structure of interfaces between subunits of dimeric and tetrameric proteins. Protein Eng. 1989;3:77–83.
Monck JR, Fernandez JM. The exocytotic fusion pore and neurotransmitter release. Neuron. 1994;12:707–16.
Muller M, Katsov K, Schick M. A new mechanism of model membrane fusion determined from Monte Carlo simulation. Biophys J. 2003;85:1611–23.
Noguchi H, Takasu M. Self-assembly of amphiphiles into vesicles: a Brownian dynamics simulation. Phys Rev E Stat Nonlin Soft Matter Phys. 2001;64:
Pecheur, E. I., J. Sainte-Marie, A. Bienven e, and D. Hoekstra. 1999. Peptides and membrane fusion: towards an understanding of the molecular mechanism of protein-induced fusion. J Membr Biol. 167:1–17.
Poyry S, Vattulainen I. Role of charged lipids in membrane structures—insight given by simulations. Biochim Biophys Acta. 2016;1858:2322–33.
Primakoff P, Myles DG. Penetration, adhesion, and fusion in mammalian sperm-egg interaction. Science. 2002;296:2183–5.
Qiang W, Weliky DP. HIV fusion peptide and its cross-linked oligomers: efficient syntheses, significance of the trimer in fusion activity, correlation of beta strand conformation with membrane cholesterol, and proximity to lipid headgroups. Biochemistry. 2009;48:289–301.
Risselada HJ, Kutzner C, Grubmuller H. Caught in the act: visualization of SNARE-mediated fusion events in molecular detail. ChemBioChem. 2011;12:1049–55.
Siegel DP. Energetics of intermediates in membrane fusion: comparison of stalk and inverted micellar intermediate mechanisms. Biophys J. 1993;65:2124–40.
Skehel JJ, Wiley DC. Receptor binding and membrane fusion in virus entry: the influenza hemagglutinin. Annu Rev Biochem. 2000;69:531–69.

Sollner T, Rothman JE. Neurotransmission: harnessing fusion machinery at the synapse. Trends Neurosci. 1994;17:344–8.

Stegmann T, Doms RW, Helenius A. Protein-mediated membrane fusion. Annu Rev Biophys Biophys Chem. 1989;18:187–211.

Suzuki Y, Matsunaga M, Matsumoto M. N-Acetylneuraminyllactosylceramide, GM3-NeuAc, a new influenza A virus receptor which mediates the adsorption-fusion process of viral infection. Binding specificity of influenza virus A/Aichi/2/68 (H3N2) to membrane-associated GM3 with different molecular species of sialic acid. J Biol Chem. 1985;260:1362–5.

Tristram-Nagle S, Nagle JF. HIV-1 fusion peptide decreases bending energy and promotes curved fusion intermediates. Biophys J. 2007;93:2048–55.

Verkleij AJ, Post JA. Membrane phospholipid asymmetry and signal transduction. J Membr Biol. 2000;178:1–10.

Walls AC, Park YJ, Tortorici MA, Wall A, McGuire AT, Veesler D. Structure, function, and antigenicity of the SARS-CoV-2 spike glycoprotein. Cell. 2020;181(281–292):

Weis WI, Brunger AT, Skehel JJ, Wiley DC. Refinement of the influenza virus hemagglutinin by simulated annealing. J Mol Biol. 1990;212:737–61.

Weissenhorn W, Dessen A, Harrison SC, Skehel JJ, Wiley DC. Atomic structure of the ectodomain from HIV-1 gp41. Nature. 1997;387:426–30.

White JM. Viral and cellular membrane fusion proteins. Annu Rev Physiol. 1990;52:675–97.

White JM, Delos SE, Brecher M, Schornberg K. Structures and mechanisms of viral membrane fusion proteins: multiple variations on a common theme. Crit Rev Biochem Mol Biol. 2008;43:189–219.

White TA, Bartesaghi A, Borgnia MJ, et al. Molecular architectures of trimeric SIV and HIV-1 envelope glycoproteins on intact viruses: strain-dependent variation in quaternary structure. PLoS Pathog. 2010;6:

Wickner W, Schekman R. Membrane fusion. Nat Struct Mol Biol. 2008;15:658–64.

Wild C, Dubay JW, Greenwell T, et al. Propensity for a leucine zipper-like domain of human immunodeficiency virus type 1 gp41 to form oligomers correlates with a role in virus-induced fusion rather than assembly of the glycoprotein complex. Proc Natl Acad Sci U S A. 1994a;91:12676–80.

Wild CT, Shugars DC, Greenwell TK, McDanal CB, Matthews TJ. Peptides corresponding to a predictive alpha-helical domain of human immunodeficiency virus type 1 gp41 are potent inhibitors of virus infection. Proc Natl Acad Sci U S A. 1994b;91:9770–4.

Wiley DC, Skehel JJ. The structure and function of the hemagglutinin membrane glycoprotein of influenza virus. Annu Rev Biochem. 1987;56:365–94.

Wilson IA, Skehel JJ, Wiley DC. Structure of the haemagglutinin membrane glycoprotein of influenza virus at 3 A resolution. Nature. 1981;289:366–73.

Xu Y, Lou Z, Liu Y, et al. Crystal structure of severe acute respiratory syndrome coronavirus spike protein fusion core. J Biol Chem. 2004a;279:49414–9.

Xu Yanhui, Zhu Jieqing, Liu Yiwei, et al. Characterization of the heptad repeat regions, HR1 and HR2, and design of a fusion core structure model of the spike protein from Severe Acute Respiratory Syndrome (SARS) coronavirus. Biochemistry. 2004b;43:14064–71.

Zhu J, Xiao G, Xu Y, et al. Following the rule: formation of the 6-helix bundle of the fusion core from severe acute respiratory syndrome coronavirus spike protein and identification of potent peptide inhibitors. Biochem Biophys Res Commun. 2004;319:283–8.

Geetanjali Meher received her M.Sc. and M.Phil. degrees in Chemistry from School of Chemistry, Sambalpur University (India) in 2012 and 2013, respectively. In 2016, she joined School of Chemistry, Sambalpur University for her Ph.D. degree under the supervision of Dr. Hirak Chakraborty, and currently she is working to understand the relation among peptide structure and dynamics and membrane fusion.

Hirak Chakraborty received his M.Sc. degree in Chemistry (2000) from Jadavpur University (India). He obtained his Ph.D. from Jadavpur University while working at Saha Institute of Nuclear Physics in 2006 and was a Postdoctoral Research Associate at the University of North Carolina at Chapel Hill, USA. Before joining Sambalpur University as Assistant Professor, he was a Pool Scientist at the Centre for Cellular and Molecular Biology, India. Dr. Chakraborty's work is focused to understand lipid-peptide/protein interaction in the context of membrane fusion and developing peptide-based membrane fusion inhibitor. His other research interest is to understand protein structure and function.

Fluorescence Imaging Approaches in Flavivirus Research

Jorge L. Arias-Arias and Rodrigo Mora-Rodríguez

Abstract The genus *Flavivirus* within the family *Flaviviridae* contains many arthropod-borne infectious agents of medical relevance like dengue virus (DENV), Zika virus (ZIKV), and West Nile virus (WNV), among others, that cause epidemics of hemorrhagic fevers and encephalitis for which there are no antiviral treatments and effective vaccines available. Fluorescence imaging is a powerful and versatile research tool for the study of flaviviral diseases. This tool can be complemented with biochemical and molecular methods to gain insight into the mechanisms of flavivirus infection and immunity, in order to develop feasible prophylactic and therapeutic interventions to lower their impact on public health. Here we discuss the basics of the fluorescence imaging techniques currently employed in flavivirus research, including immunofluorescence assay (IFA), fluorescence in situ hybridization (FISH), fluorescence-labeled viral particles, fluorescent labeling of cytopathic effect (CPE), subgenomic reporter replicons (SRRs)/reporter virus particles (RVPs), and cell-based molecular reporters (CBMRs). We also address the advantages of each application based on our own experience and included some of our protocols to facilitate its applications.

Keywords Dengue virus (DENV) · Zika virus (ZIKV) · Yellow fever virus (YFV) · West Nile virus (WNV) · Flavivirus · Immunofluorescence assay (IFA) · In situ hybridization (FISH) · Fluorescence-labeled viral particles · Cytopathic effect (CPE) · Subgenomic replicons · Reporter virus particles (RVPs) · Cell-based molecular reporters · Viral plaques · Live-cell imaging

J. L. Arias-Arias · R. Mora-Rodríguez (✉)
Facultad de Microbiología, Centro de Investigación en Enfermedades Tropicales, Universidad de Costa Rica, San José 11501-2060, Costa Rica
e-mail: rodrigo.morarodriguez@ucr.ac.cr

S. I. Ahmad (ed.), *Human Viruses: Diseases, Treatments and Vaccines*, https://doi.org/10.1007/978-3-030-71165-8_34

1 Introduction

The genus *Flavivirus* within the family *Flaviviridae* contains more than 70 species of arthropod-borne viruses, transmitted to animals and humans by the bite of infected mosquitoes or ticks, including the medically relevant species dengue virus (DENV), Zika virus (ZIKV), yellow fever virus (YFV), West Nile virus (WNV), Japanese encephalitis virus (JEV), Tick-borne encephalitis virus (TBEV), and St. Louis encephalitis virus (SLEV), among others (Gould and Solomon 2008).

Flaviviruses possess small enveloped icosahedral particles of about 50 nm in diameter, which harbor a positive sense RNA genome of approximately 11 kb in length. This genome encodes three structural proteins: capsid (C), membrane precursor (prM), and envelope (E), and seven nonstructural proteins: NS1, NS2A, NS2B, NS3, NS4A, NS4B, and NS5, that form a precursor polyprotein (Fig. 3a), which must be cleaved by both cellular and viral proteases in order to generate the individual viral proteins (Lindenbach et al. 2007).

During the last seven decades, flaviviruses have continuously emerged and re-emerged, constituting a global threat that cause epidemics of hemorrhagic fevers and encephalitis for which there are no specific treatments more than life support upon hospitalization. This highlights the critical need for a detailed understanding of the biology of flavivirus, the interplay between the virus and the host cell, and the immunological responses elicited in order to develop feasible prophylactic and therapeutic approaches to lower their impact on public health (Lindenbach et al. 2007; Pierson and Diamond 2020).

In this scenario, fluorescence imaging techniques represent powerful and versatile research tools for visualization and study of flavivirus infected cells, that can be complemented with biochemical and molecular methods to gain knowledge about the mechanisms of infection and to test candidate antiviral drugs and vaccines to combat flaviviral diseases (Chong et al. 2014). Among them, immunofluorescence assay (IFA), fluorescence *in situ* hybridization (FISH), fluorescent labeling of cytopathic effect (CPE), fluorescence-labeled viral particles, subgenomic reporter replicons (SRRs)/reporter virus particles (RVPs), and cell-based molecular reporters (CBMRs), have been applied in flavivirus research and will be individually discussed in the present chapter.

2 Immunofluorescence Assays (IFA)

Immunofluorescence assays (IFA) is the most popular and widely applied fluorescence imaging approach for flaviviruses, since this method have been employed in diagnostics and research of flaviviral infections for more than 50 years (Atchison et al. 1966). Indeed, their presence and cellular localization can be visualized inside infected cells by means of fluorophore-tagged antibodies directed against either structural or non-structural flavivirus antigens (Chong et al. 2014).

Zuza et al. (2016) observed a perinuclear localization of the SLEV proteins upon immunolabeling of infected astrocytes with a polyclonal ascitic fluid from immunized mic. Likewise, Miorin et al. reported that the TBEV E, prM, and NS1 proteins were localized at the perinuclear region and within irregularly shaped foci of infected BHK-21 cells (Miorin et al. 2013). This is a common fluorescence pattern observed among other members of the genus, including DENV 1-4, ZIKV, YFV, JEV, and WNV (Ledizet et al. 2007; Ricciardi-Jorge et al. 2017; Slon Campos et al. 2017), since flaviviruses replication and assembly occurs on the cytoplasmic side of the endoplasmic reticulum (ER) membrane (Rothan and Kumar 2019).

IFA also remains as an excellent method to evaluate and visualize the permissiveness of cell lineages to flavivirus infection. Růžek et al. (2009) tested the susceptibility of different human neural cell lines and the TBEV-induced cytopathic effect using an anti-E antibody.. Besides, we applied immunostaining of E and NS3 proteins as part of our experiments to demonstrate the permissiveness of primary human umbilical artery smooth muscle cells (HUASMC) to clinical isolates of both DENV-2 and DENV-3 (Fig. 1c) (Arias-Arias et al. 2018). Others also employed immunolabeling to evaluate the susceptibility of JEG-3 and hCMEC/D3 cell lines, as well as Sertoli cells in muse testis to ZIKV infection (Chiu et al. 2020; Sheng et al. 2017).

IFA also enables the assessment of colocalization as a first screening for studies of protein–protein interactions by the combination of different antibodies directed against both host and viral antigens, as these interactions could play important roles during flavivirus infection. Hung et al. employed IFA colocalization of host secreted heat-shock protein 90 beta (Hsp90β) and viral E protein during JEV infection. They performed a subsequent validation with sucrose-density fractionation and Western blot analysis to demonstrate that this interaction is required for JEV infectivity in BHK-21 cells (Hung et al. 2011). In addition, using IFA and cryoimmunoelectron microscopy colocalization with monospecific antibodies, NS3 and NS2B proteins were found to be present in WNV-induced membrane structures in Vero cells (Westaway et al. 1997).

One of the greatest methodological advantages of applying IFA in flavivirus research is the wide commercial availability of group cross-reactive and type-specific antibodies. This is based on the fact that some flaviviral proteins possess both conserved and variable antigens, e.g., the fusion loop at the extremity of domain II of protein E carries the flavivirus group conserved epitopes, whereas domains I and III contain the variable antigens (Lai et al. 2008). This observation guided the in vitro production of the hybridoma clone D1–4G2–4–15, which produces the most popular flavivirus group monoclonal antibody (4G2), used in diagnostics for the screening of viral isolates from clinical samples and in research for the monitoring of the infection by agents like DENV, WNV, JEV, YFV, and ZIKV (Garg et al. 2020; Göertz et al. 2017; Lai et al. 2008; Martins et al. 2019), among other flaviviruses. Here we described in detail a fast and very effective IFA protocol applied in our laboratory for the immunostaining of DENV and ZIKV antigens using commercial antibodies (Fig. 1).

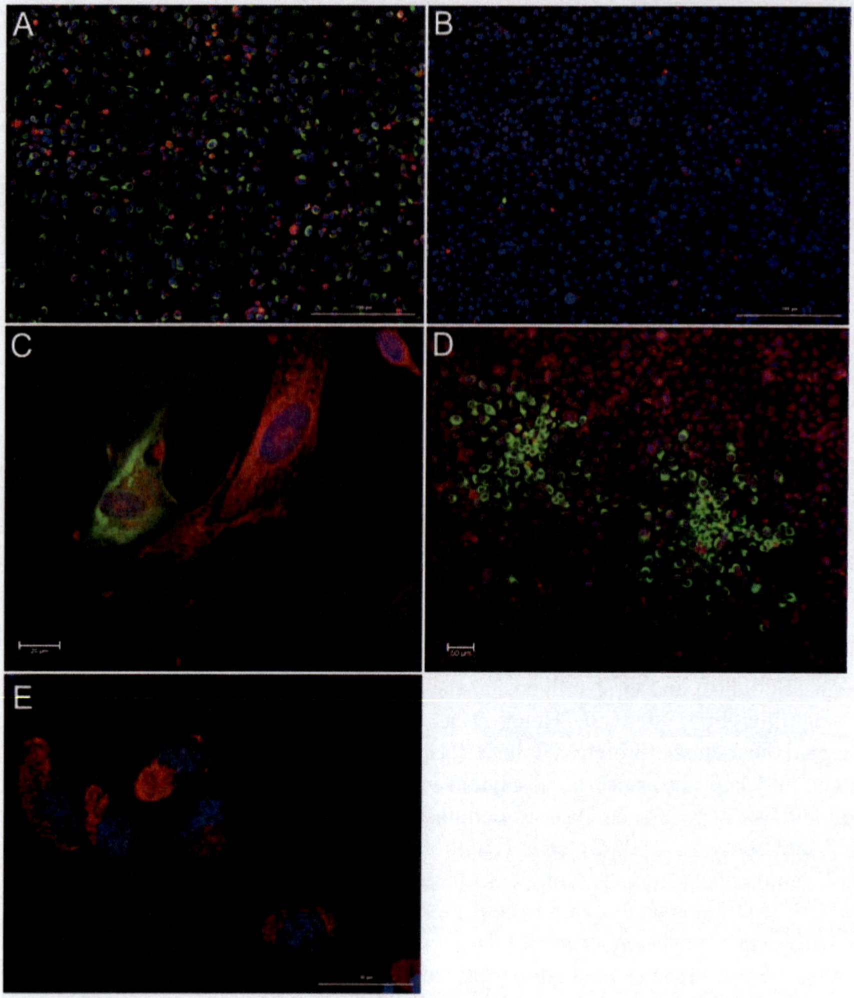

Fig. 1 Immunofluorescence assay (IFA) in DENV/ZIKV-infected cell cultures. Epifluorescence images of ZIKV-infected (**A**) and mock-infected (**B**) Vero cells after immunostaining with an anti-ZIKV E protein monoclonal antibody (green) and cytoplasmic (Evans blue, red)—nuclear (Hoechst 33,342, blue) counterstains (total magnification of 100 X; scale bar = 200 μm). DENV-infected HUASMC (**C**) and LLC-MK2 (**B**) cells immunostained with an anti-DENV 1–4 E protein monoclonal antibody (green) and counterstained with cytoplasmic (Evans blue, red) and nuclear (Hoechst 33,342, blue) fluorescent dyes (total magnification of 400 X and 100 X; scale bars = 20 μm and 50 μm, respectively). Immunolabeling of DENV in BHK-21 cell with an anti-DENV NS3 protein polyclonal antibody (orange) and a nuclear (Hoechst 33,342, blue) counterstain (total magnification of 600 X; scale bar = 30 μm). Images by Jorge L. Arias-Arias, Universidad de Costa Rica

3 Protocol: Rapid IFA for Labeling DENV/ZIKV Structural and Nonstructural Proteins

Seed and infect the model cells with the DENV/ZIKV strain of interest at the desired multiplicity of infection (MOI), on a μClear black 96-well plate (Greiner Bio-One 655,090).

Remove the culture media and wash once with 100 μL/well of phosphate-buffered saline (PBS, Gibco 10,010,023) to remove detached cells and cellular debris.

Fix the cell monolayers with 50 μL/well of a 3,5% paraformaldehyde (Sigma 158,127) solution in PBS for 15 min at room temperature. Remove the fixative and wash once with 100 μL/well of PBS.

Permeabilize cells with 50 μL/well of 70% ethanol in water for 15 min at room temperature. Remove the ethanol and wash once with 100 μL/well of PBS.

Incubate for 1 h at 37 °C with 50 μL/well of one of the following primary antibodies diluted in a 0,001% Triton X-100 (Sigma 10,789,704,001) solution in PBS:

- 1:400 dilution of mouse anti-DENV 1–4 E protein monoclonal antibody (GeneTex GTX29202).
- 1:400 dilution of mouse anti-ZIKV E protein monoclonal antibody (GeneTex GTX634157).
- 1:800 dilution of rabbit anti-DENV NS3 protein polyclonal antibody (GeneTex GTX124252).
- 1:400 dilution of rabbit anti-ZIKV NS3 protein polyclonal antibody (GeneTex GTX133309).

Wash twice with 100 μL/well of PBS and once with 100 μL/well of 0,001% Triton X-100 solution in PBS.

Incubate for 30 min at 37 °C with 50 μL/well of one of the following secondary antibodies (as appropriate) diluted in a solution composed by 0,001% Triton X-100, 0.02% Evans blue (Sigma E2129), and 1 μg/mL Hoechst 33,342 (Invitrogen H3570, 1:10 000 dilution) in PBS:

- 1:400 dilution of Alexa Fluor 488 goat anti-mouse IgG, IgM (Invitrogen A-10684), Alexa Fluor 568 goat anti-mouse IgG (Invitrogen A-11031), or Alexa Fluor 647 goat anti-mouse IgG (Invitrogen A-21237).
- 1:400 dilution of Alexa Fluor 488 goat anti-rabbit IgG (Invitrogen A-11034), Alexa Fluor 594 goat anti-rabbit IgG (Invitrogen A-11037), or Alexa Fluor 647 goat anti-rabbit IgG (Invitrogen A-21245).

Wash three times with 100 μL/well of PBS, add 100 μL/well of FluoroBrite Dulbecco's modified Eagle's medium (DMEM, Gibco A1896701), and acquire images at the desired magnification with an automated fluorescence microscope (e.g., BioTek Lionheart FX).

Analyze the images using an image analysis software (e.g., CellProfiler 3.0, Broad Institute https://www.cellprofiler.org/) to quantify the percentages of infected cells.

Notes: Cells could be also seeded and infected on round glass coverslips into a 24-well plate, stained and mounted in slides to analyze the results with a conventional fluorescence microscope. We have also labeled all the above-mentioned primary antibodies using commercial labeling kits for Alexa Fluor 488 (Invitrogen A20181), Alexa Fluor 568 (Invitrogen A20184), and Alexa Fluor 647 (Invitrogen A20186), which allows the performing of a much faster direct IFA protocol applying dilutions in the range 1:75–1:150, as well as flow cytometry assays with dilutions 1:300–1:400.

4 Fluorescence in Situ Hybridization (FISH)

FISH is a classical cytogenetic technique used to detect both RNA and DNA within tissues and cells by the application of fluorochrome-abelled probes that are complementary to the sequence of interest, with their subsequent visualization by fluorescence microscopy (Rudkin and Stollar 1977). One of the major goals in RNA viruses research is the understanding of the coordination of the intracellular trafficking of viral RNA and proteins during the assembly of virions, as well as the deciphering of the involved interactions between the viral genome and other components of the host and the virus itself (Vyboh et al. 2012). Combinations of FISH with other cellular and molecular techniques have been applied in recent years to tackle the above-mentioned challenges in flavivirus infection research.

FISH is the method of choice to visualize the localization, transcription, and replication of flavivirus RNA within infected cells. Raquin et al. (2012) developed a set of highly specific oligonucleotide probes that hybridize to the viral RNA from a broad range of DENV isolates including all the four serotypes, but not to the closely related YFV and WNV genomes, when used in vitro on infected C6/36 cells and in vivo with dissected salivary glands from infected *Aedes albopictus* specimens. Similar FISH approaches have been used for the observation of ZIKV (Hou et al. 2017; Liu et al. 2019; Martinez-Lopez et al. 2019) and YFV genome replications (Sinigaglia et al. 2018), and for the visualization of WNV noncoding RNAs (Roby et al. 2014), among other flaviviruses.

When combined with IF, FISH serves as a useful starting point to study protein-viral RNA interactions. Hirano et al. (2017) applied FISH/IF and immunoprecipitation/RT-PCR to demonstrate that neuronal granules, involved in the transportation and local translation of dendritic mRNAs, also transport the TBEV genomic RNA. Viral RNA interacts with a RNA-binding protein present in the neuronal granules and impairs the transport of dendritic mRNAs, which seems to be involved in the neuropathogenesis of the TBEV infection. Besides, Hou et al. 2017) observed co-localization of ZIKV E protein with its own viral RNA by FISH/IF and demonstrated its interaction by an RNA chromatin immunoprecipitation (RNA-ChIP) assay, which implies that E protein may have a role in ZIKV replication.

The availability of online bioinformatic tools and databases assisting probe design (e.g., https://www.arb-silva.de/fish-probes/probe-design/), together with the custom probe synthesis services offered by many biotech companies, is increasing the application of RNA FISH in the field of virology, including flavivirus research. For a detailed and versatile FISH protocol specially standardized for RNA viruses, please refer to the work of Lindquist and Schmaljohn (Lindquist and Schmaljohn 2018).

5 Fluorescence-Labeled Viral Particles

In recent years, with the improvements in confocal and super-resolution fluorescence microscopy, several researchers have exploited the direct labeling of virions for the visualization of the early events of virus-cell interactions, even at a single particle level (Sakin et al. 2016). To assess viral membrane fusion, the use of lipophilic dyes that get inserted into the lipid bilayer membrane of virions and are released in the endosomal membrane after the fusion event, have been reported for some flaviviruses (Hoffmann et al. 2018). Nour et al. (2013) applied octadecyl rhodamine B chloride (R18) labeling of JEV and YFV particles to study the kinetics of viral membrane fusion and nucleocapsid delivery into the cytoplasm. Moreover, labeling of virions with 1,1′-dioctadecyl-3,3,3′,3′-tetramethylindodicarbocyanine, 4-chlorobenzenesulfonate salt (DiD) was used to analyze the cellular entry of WNV (Makino et al. 2014) and to demonstrate the involvement of autophagy machinery during the early stages of DENV infection (Chu 2013).

Furthermore, Zhang et al. (2010) developed a simple and efficient method to covalently tag the amino free groups of DENV E protein with the dye Alexa Fluor 594 succinimidyl ester, a useful labeling approach employed to monitor virus binding, uptake, and intracellular trafficking. A similar procedure was used to label ZIKV particles with the dye Atto647N-NHS ester, which enabled the visualization of viral transcytosis through both the placental and the blood brain barriers (Chiu et al. 2020).

Labeling of virions is an advantageous technique since the above-mentioned dyes show bright fluorescence, high photostability, are suitable to be applied in both fixed and live-cell imaging protocols, and exist in a broad range of colors over the ultraviolet–visible spectrum (Hoffmann et al. 2018). The work of Zhang et al. (2011) describes a detailed protocol for the fluorescence-labeling of flavivirus particles. A comprehensive description of labeling procedures applied to a broader repertory of viruses is discussed by Hoffmann and collaborators (Hoffmann et al. 2018).

6 Fluorescent Labeling of Cytopathic Effect (CPE)

Detection of CPE as an indirect way to monitor viral infections have been largely exploited since the early days of virology and constitutes a fundamental part of the principle behind classical virological methods like Dulbecco plaque assay (Dulbecco 1952). Monitoring the morphological changes on flavivirus infected cells is generally accomplished by bright-field light microscopy, directly on living cells or after fixation and staining with conventional dyes as crystal violet or hematoxylin–eosin (Bakonyi et al. 2005; Chong et al. 2014).

Taking into account that in many cases the CPE is a result of virus–cell interactions leading to cell demise by host-encoded programs like programmed necrosis and apoptosis (Agol 2012), we envisaged a simple but very effective way to label and monitor the CPE in real-time by fluorescence live-cell imaging. Using fluorescent DNA dyes commonly employed in cell biology and cancer research such as nuclei stains (Hoechst 33,342) and cell dead markers (propidium iodide, SYTOX green, TO-PRO-3 iodide), we were able to visualize early (chromatin condensation) and late (membrane permeabilization) events of the CPE correlating with cell damage and cell death, respectively.

Our research group successfully applied the above-mentioned approach to perform kinetics of CPE detection and real-time plaque assays on living cells infected with DENV, ZIKV, and YFV (Fig. 2). This allowed us to analyze the viral plaques growth over time at a single-cell level using an image analysis software (Arias-Arias et al. 2020).

7 Protocol: Real-time CPE Labeling Kinetics in Flavivirus Infected BHK-21 / Vero Cells

Seed 15 000 BHK-21 (ATCC CCL-10) or Vero (ATCC CCL-81) cells per well on a µClear black 96-well plate (Greiner Bio-One 655,090) using 100 µL/well of minimum essential medium (MEM, Gibco 41,090,101) supplemented with 2% fetal bovine serum (FBS, Gibco 10,100,147) and incubate for 18–24 h at 37 °C –5% CO_2. *Do not use the wells at the plate's periphery as those must be reserved for the addition of PBS in long-term experiments.*

Remove the medium and immediately add 50 µL/well of flavivirus inoculum at the desired MOI in MEM –2% FBS. Likewise, load a mock-infected control and allow to adsorb for 2 h at 37 °C –5% CO_2. *Work fast and do not handle too many wells at the same time in order to avoid desiccation that could kill and label cells with the dead markers.*

Label cell nuclei by adding over the inoculum 50 µL/well of a 2 µg/mL Hoechst 33,342 (Invitrogen H3570, 1:5000 dilution) solution in MEM –2% FBS and incubate for 10 min at 37 °C –5% CO_2.

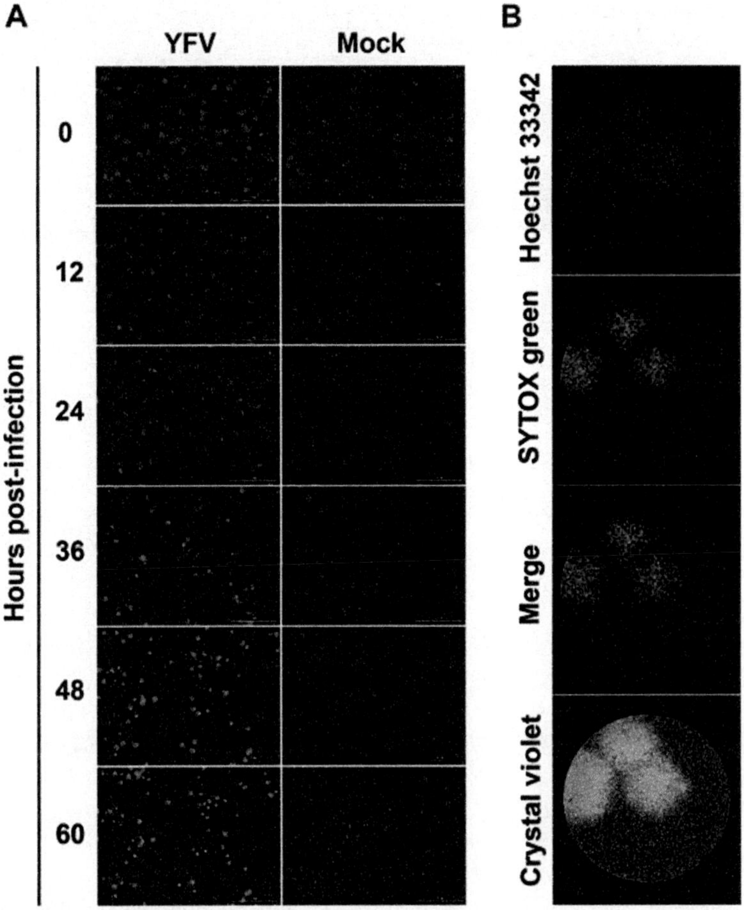

Fig. 2 Fluorescent labeling of cytopathic effect (CPE) in flavivirus-infected cell cultures. **A** CPE imaging kinetics in YFV-infected BHK-21 cells with the DNA stainings Hoechst 33,342 (cells with condensed chromatin, saturated blue) and TO-PRO-3 iodide (dead cells, red) at a total magnification of 200 X (scale bar = 100 μm). **B** Dulbecco's plaque assay on unfixed Vero cells by CPE labeling with the nucleic acid dyes Hoechst 33,342 (DNA/chromatin condensation, blue) and SYTOX green (cell death, green), at 96 hours post-infection with ZIKV (total magnification of 40 X; scale bar = 1000 μm). Images by Jorge L. Arias-Arias, Universidad de Costa Rica

Remove the inoculum and immediately add 100 μL/well PBS-1% FBS to make a wash for removing the unbound viruses and the excess of Hoechst 33,342 solution as it results toxic over long incubation periods. *Work fast and do not handle too many wells at the same time in order to avoid desiccation that could kill and label cells with the dead markers.*

Remove the washing solution and immediately add 100–300 μL/well of FluoroBrite DMEM (Gibco A1896701) supplemented with 2% FBS and containing one of the following cell dead dyes: 2,5 μg/mL propidium iodide (Invitrogen

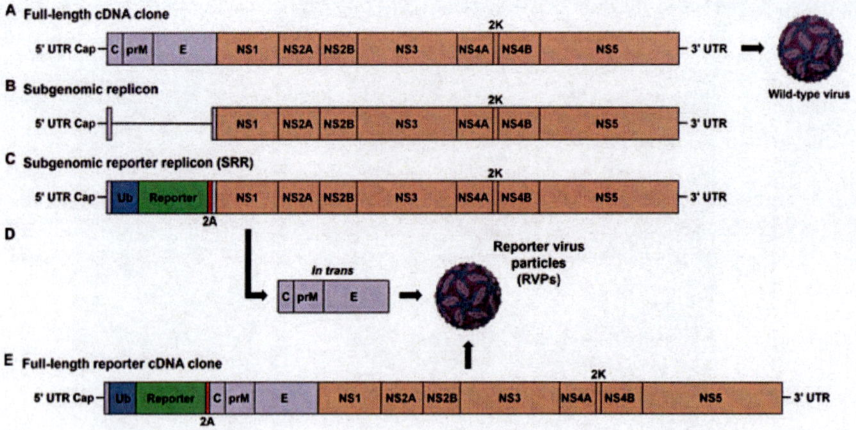

Fig. 3 Schematic presentation of genetic constructs encoding flavivirus full-length cDNA clones (**A**), subgenomic replicons (**B**), subgenomic reporter replicons (SRRs, **C**), and full-length reporter cDNA clones (**E**), which enable the production of reporter virus particles (RVPs, **D**). UTR: untranslated region; Ub: ubiquitin coding sequence; 2A: ribosomal skipping 2A peptide. Modified and adapted from Kümmerer (2018)

P3566, 1:400 dilution), 500 nM SYTOX green (Invitrogen S7020, 1:10 000 dilution), or 200 nM TO-PRO-3 iodide (Invitrogen T3605, 1:5000 dilution). *Work fast and do not handle too many wells at the same time in order to avoid desiccation that could kill and label cells with the dead markers. The volume of medium to use per well is based on the duration of the experiment: 24 h-100 μL, 48 h-150 μL, 72 h-200 μL, 96 h-250 μL, and ≥ 120 h-300 μL. Fill the wells of the periphery with 300 μL/well of PBS in order to reduce the evaporation of media in the wells with experimental conditions during long-term assays.*

Incubate and read out at 37 °C –5% CO_2 into an automated fluorescence microscope (e.g., BioTek Lionheart FX) for the desired incubation period at the preferred magnification and time-lapse imaging acquisition. *To avoid cell toxicity produced by the exposure to the excitation light wavelengths, it is recommended to acquire images at time-lapses ≥ than 30 min.*

Analyze the images using an image analysis software (e.g., CellProfiler 3.0, Broad Institute https://www.cellprofiler.org/) to quantified the percentages of dead cells and cells with condensed chromatin over time, as read-outs of the CPE.

Note: A similar protocol can be used for real-time kinetic plaque assays by adding the cell dead dyes to a MEM AutoMod (Sigma M0769) supplemented with 2% FBS and 1% carboxymethylcellulose (Sigma C4888).

8 Subgenomic Reporter Replicons (SRRs) and Reporter Virus Particles (RVPs)

Flaviviruses harbor positive strand RNA genomes that are per se infectious. Thus, transfection of RNA produced by in vitro transcription from a cDNA clone containing the reverse transcribed full-length flavivirus genome results in the production of infectious recombinant viral particles (Fig. 3a). In contrast, flavivirus subgenomic replicons possess all the essential genetic elements for self- replication and production of nonstructural proteins, but lack the complete coding sequences of the structural C-prM-E proteins (Fig. 3b) and consequently do not allow the generation of virions. Using such replicons as templates, SRRs are established by the introduction of reporter genes that code for bioluminescent or fluorescent proteins in place of the deleted structural genes (Fig. 3c) (Kümmerer 2018). This enables the easy tracking of the replication/translation of subgenomic replicons and the screening of antiviral compounds by the direct visualization and measurement of the signal produced by the reporter proteins (Kato and Hishiki 2016). Flavivirus SRRs have been developed and validated for YFV (Jones et al. 2005), JEV (Li et al. 2013), WNV (Shi et al. 2002), ZIKV (Mutso et al. 2017), and DENV (Pang et al. 2001; Usme-Ciro et al. 2017), among others.

SRRs are also used for the generation of single-round infectious RVPs, by providing the deleted C-prM-E genes in trans with another genetic construct (Fig. 3d). Such single-round RVPs are extremely useful as surrogate pseudoviruses in studies of BSL3 flaviviruses like JEV (Lu et al. 2017) and WNV (Li et al. 2017; Velado Fernández et al. 2014). Single-round infectious RVPs have also been used on the development of easy fluorescent neutralization assays for the detection of flavivirus-specific antibodies in the serum of individuals and the assessment of the humoral immune response elicited by candidate vaccines, as shown for DENV (Mattia et al. 2011), TBEV (Yoshii et al. 2009), ZIKV (Garg et al. 2017), and WNV (Pierson et al. 2006).

However, SRRs and single-round infectious RVPs cannot be used to study the pathogenesis, transmission, and dynamics of the complete virus life cycle, as well as for the screening of antivirals targeting the structural proteins (Kato and Hishiki 2016). For such applications, whole genome RVPs have been engineered by the insertion of reporter genes into full-length flavivirus cDNA clones (full-length reporter cDNA clone, Fig. 3e), as described for JEV (Jia et al. 2016), DENV (Schmid et al. 2015; Schoggins et al. 2012; Suphatrakul et al. 2018), WNV (Pierson et al. 2005), and ZIKV (Gadea et al. 2016), among others. As example, Schmid and collaborators developed and characterized a far-red DENV-2 reporter virion that allows the monitoring of the viral infection kinetics in animal cells by live imaging (Schmid et al. 2015).

For further details, the work by Kümmerer (2018) describes in detail the molecular genetics, development and applications of flavivirus subgenomic and full-length replicons.

9 Cell-Based Molecular Reporters (CBMRs)

SRRs and RVPs are valuable tools to perform kinetic studies by live-cell imaging, but its development is expensive, time consuming, and limited to the chosen molecular clones derived from specific flavivirus strains, which precludes the direct work with clinical isolates and native virus strains. To overcome this limitation, in recent years a few manuscripts have outlined the use of CBMRs as an alternative to carry out kinetics of infection with raw flaviviruses in living cells (Arias-Arias et al. 2020).

So far, all the published flavivirus CBMRs are based on the monitoring of the proteolytic activity of the flaviviral NS2B–NS3 serine protease. Medin et al. (2015) devised a DENV 1–4 plasmid-based reporter system containing the cleavage site between the NS4B and NS5 proteins attached to an EGFP by a nuclear localization sequence (NLS). Upon cleavage, the EGFP relocalizes from the cytoplasm to the nucleus of infected cells. The same principle was adapted by McFadden and collaborators for imaging ZIKV infection in living Huh7 cells (McFadden et al. 2018). In addition, a modification of this approach by Hsieh and collaborators exploited the DENV NS3 cleavage between NS4B/NS5 to activate a Cre recombinase-based nuclear reporter. This system showed superior performance compared to traditional methods for DENV 1–4 titration (Hsieh et al. 2017).

Moreover, we developed a fluorescence-activatable reporter of flavivirus infection by the modification of previously published caspase reporters (Wu et al. 2013). Instead of the caspase cleavage sequence, we inserted the internal NS3 cleavage site (conserved among many members of the *Flavivirus* genus) between a fluorescent protein and a quenching peptide (QP, Fig. 4a). This system was used to generate a BHK-21 reporter cell line suitable for monitoring the kinetics of infection by DENV, ZIKV, and YFV both in viral plaques and at a single-cell level using live-cell imaging (Fig. 4b) (Arias-Arias et al. 2020). We are currently working on the improvement of such systems for the high-throughput screening of antiviral compounds targeting the flavivirus NS2B–NS3 protease.

10 Conclusions

With the current advances in super-resolution microscopy, biological sciences are facing a change in the methodological paradigm, evolving from techniques that rely on indirect inferences to those that allow the visualization of the phenomenon of interest. Research on emerging diseases, as those caused by flaviviruses, is playing an active role in that revolution, taking advantage of the already discussed fluorescence imaging approaches. Such methodologies are pivotal for the deciphering and understanding of the mechanisms involved in virus-host cell interactions, a crucial knowledge for the development of the urgently needed flavivirus antivirals and vaccines in the close future.

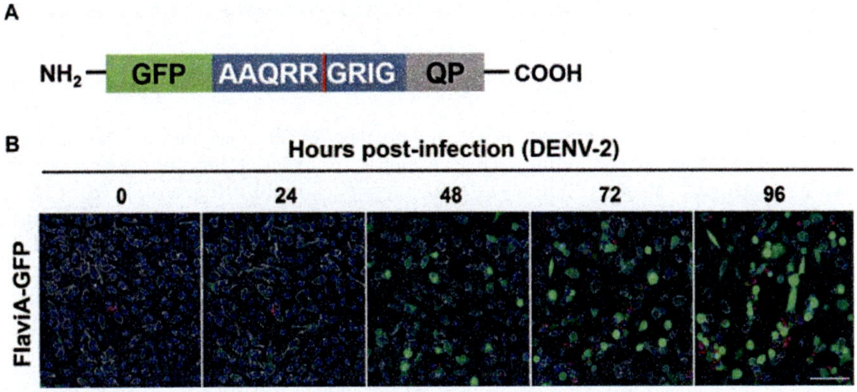

Fig. 4 The flavivirus cell-based molecular reporter. **A** The flavivirus-activatable GFP reporter (FlaviA-GFP) contains a GFP with a C-terminal quenching peptide (QP) joined by a linker composed by a cleavage site of the flaviviral NS3 protease. When the protease cuts the linker, the quenching peptide is removed, and the GFP adopts the fluorescent conformation. **B** DENV-2 infection kinetics in BHK-21 reporter cells. An automated image analysis protocol was programmed in CellProfiler 2.0 for the quantification of activated FlaviA-GFP fluorescent cells (green), live cells (white outline) and dead cells (red outline). Total magnification of 200 X; scale bar = 100 µm. Adapted from Arias-Arias et al. (2020)

Acknowledgements We want to thank Universidad de Costa Rica (Project VI-803-B9–505) and International Centre for Genetic Engineering and Biotechnology (Grant CRP/CRI18-02) for supporting our research on flaviviruses, an endemic Problem that affects thousands of people in our region. This work is dedicated to our students.

References

Agol VI. Cytopathic effects: virus-modulated manifestations of innate immunity? Trends Microbiol. 2012;20:570–6. https://doi.org/10.1016/j.tim.2012.09.003.

Arias-Arias JL, MacPherson DJ, Hill ME, Hardy JA, Mora-Rodríguez R. A fluorescence-activatable reporter of flavivirus NS2B–NS3 protease activity enables live imaging of infection in single cells and viral plaques. J Biol Chem. 2020;295:2212–26. https://doi.org/10.1074/jbc.RA119.011319.

Arias-Arias JL, Vega-Aguilar F, Corrales-Aguilar E, Hun L, Loría GD, Mora-Rodríguez R. Dengue virus infection of primary human smooth muscle cells. Am J Trop Med Hyg. 2018;99:1451–7. https://doi.org/10.4269/ajtmh.18-0175.

Atchison RW, Ordonez JV, Sather GE, Hammon WM. Fluorescent antibody, complement fixation method for detection of dengue viruses in mice. J Immunol. 1966;96:936–43.

Bakonyi T, Lussy H, Weissenböck H, Hornyák Á, Nowotny N. In vitro host-cell susceptibility to usutu virus. Emerg Infect Dis. 2005;11:298–301. https://doi.org/10.3201/eid1102.041016.

Chiu CF, Chu LW, Liao IC, Simanjuntak Y, Lin YL, Juan CC, Ping YH. The mechanism of the Zika virus crossing the placental barrier and the blood-brain barrier. Front Microbiol. 2020;11:1–15. https://doi.org/10.3389/fmicb.2020.00214.

Chong MK, Chua AJS, Tan TTT, Tan SH, Ng ML. Microscopy techniques in flavivirus research. Micron. 2014;59:33–43. https://doi.org/10.1016/j.micron.2013.12.006.

Chu L-W. Single-virus tracking approach to reveal the interaction of Dengue virus with autophagy during the early stage of infection. J Biomed Opt. 2013;19:011018. https://doi.org/10.1117/1.jbo.19.1.011018.

Dulbecco R. Production of plaques in monolayer tissue cultures by single particles of an animal virus. Proc Natl Acad Sci. 1952;38:747–52.

Gadea G, Bos S, Krejbich-Trotot P, Clain E, Viranaicken W, El-Kalamouni C, Mavingui P, Desprès P. A robust method for the rapid generation of recombinant Zika virus expressing the GFP reporter gene. Virology. 2016;497:57–162. https://doi.org/10.1016/j.virol.2016.07.015.

Garg H, Mehmetoglu-Gurbuz T, Joshi A. Virus Like Particles (VLP) as multivalent vaccine candidate against Chikungunya, Japanese Encephalitis, Yellow Fever and Zika Virus. Sci Rep. 2020;10:1–13. https://doi.org/10.1038/s41598-020-61103-1.

Garg H, Sedano M, Plata G, Punke EB, Joshi A. Development of virus-like-particle vaccine and reporter assay for Zika virus. J Virol. 2017;91:1–16. https://doi.org/10.1128/jvi.00834-17.

Göertz GP, Vogels CBF, Geertsema C, Koenraadt CJM, Pijlman GP. Mosquito co-infection with Zika and chikungunya virus allows simultaneous transmission without affecting vector competence of Aedes aegypti. PLoS Negl Trop Dis. 2017;11:1–22. https://doi.org/10.1371/journal.pntd.0005654.

Gould E, Solomon T. Pathogenic flaviviruses. Lancet 2008;371 :500–509. https://doi.org/https://doi.org/10.1016/S0140-6736(08)60238-X.

Hirano M, Muto M, Sakai M, Kondo H, Kobayashi S, Kariwa H, Yoshii K. Dendritic transport of tick-borne flavivirus RNA by neuronal granules affects development of neurological disease. Proc Natl Acad Sci USA. 2017;114:9960–5. https://doi.org/10.1073/pnas.1704454114.

Hoffmann AB, Mazelier M, Léger P, Lozach PY. Deciphering virus entry with fluorescently labeled viral particles. Methods Mol Biol. 2018;1836:159–83. https://doi.org/10.1007/978-1-4939-8678-1_8.

Hou W, Armstrong N, Obwolo LA, Thomas M, Pang X, Jones KS, Tang Q. Determination of the cell permissiveness spectrum, mode of RNA replication, and RNA-protein interaction of Zika virus. BMC Infect Dis. 2017;17:1–12. https://doi.org/10.1186/s12879-017-2338-4.

Hsieh M-S, Chen M-Y, Hsieh C-H, Pan C-H, Yu G-Y, Chen H-W. Detection and quantification of dengue virus using a novel biosensor system based on dengue NS3 protease activity. PLoS One. 2017;12:1–19. https://doi.org/10.1371/journal.pone.0188170.

Hung CY, Tsai MC, Wu YP, Wang RYL. Identification of heat-shock protein 90 beta in Japanese encephalitis virus-induced secretion proteins. J Gen Virol. 2011;92:2803–9. https://doi.org/10.1099/vir.0.033993-0.

Jia F, Zhu X, Xu F. A single adaptive point mutation in Japanese encephalitis virus capsid is sufficient to render the virus as a stable vector for gene delivery. Virology. 2016;490:109–18. https://doi.org/10.1016/j.virol.2016.01.001.

Jones CT, Patkar CG, Kuhn RJ. Construction and applications of yellow fever virus replicons. Virology. 2005;331:247–59. https://doi.org/10.1016/j.virol.2004.10.034.

Kato F, Hishiki T. Dengue virus reporter replicon is a valuable tool for antiviral drug discovery and analysis of virus replication mechanisms. Viruses. 2016;8:1–11. https://doi.org/10.3390/v8050122.

Kümmerer BM. Establishment and application of flavivirus replicons. In: Caputo AT, Alonzi DS, Kiappes JL, Struwe WB, Cross A, Basu S, Darlot B, Roversi P (Eds.), Dengue and Zika: Control and Antiviral Treatment Strategies. Springer, pp. 165–173. https://doi.org/10.1007/978-981-10-8727-1_12.

Lai C-Y, Tsai W-Y, Lin S-R, Kao C-L, Hu H-P, King C-C, Wu H-C, Chang G-J, Wang W-K. Antibodies to envelope glycoprotein of dengue virus during the natural course of infection are predominantly cross-reactive and recognize epitopes containing highly conserved residues at the fusion loop of domain II. J Virol. 2008;82:6631–43. https://doi.org/10.1128/jvi.00316-08.

Ledizet M, Kar K, Foellmer HG, Bonafé N, Anthony KG, Gould LH, Bushmich SL, Fikrig E, Koski RA. Antibodies targeting linear determinants of the envelope protein protect mice against West Nile virus. J Infect Dis. 2007;196:1741–8. https://doi.org/10.1086/523654.

Li SH, Li XF, Zhao H, Deng YQ, Yu XD, Zhu SY, Jiang T, Ye Q, Qin ED, Qin CF. Development and characterization of the replicon system of Japanese encephalitis live vaccine virus SA14-14-2. Virol J. 2013;10:1–6. https://doi.org/10.1186/1743-422X-10-64.

Li W, Ma L, Guo LP, Wang XL, Zhang JW, Bu ZG, Hua RH. West Nile virus infectious replicon particles generated using a packaging-restricted cell line is a safe reporter system. Sci Rep. 2017;7:1–10. https://doi.org/10.1038/s41598-017-03670-4.

Lindenbach, B., Thiel, H.-J., Rice, C.M., 2007. Flaviviridae: The viruses and their replication, in: Fields Virology. pp. 1101–1152.

Lindquist, M.E., Schmaljohn, C.S., 2018. Intracellular detection of viral transcription and replication using RNA FISH, in: Salvato, M.S. (Ed.), Hemorrhagic Fever Viruses: Methods and Protocols. Springer, pp. 201–207. https://doi.org/https://doi.org/10.1007/978-1-4939-6981-4

Liu D, Tedbury PR, Lan S, Huber AD, Puray-Chavez MN, Ji J, Michailidis E, Saeed M, Ndongwe TP, Bassit LC, Schinazi RF, Ralston R, Rice CM, Sarafianos SG. Visualization of positive and negative sense viral RNA for probing the mechanism of direct-acting antivirals against Hepatitis C virus. Viruses. 2019;11. https://doi.org/10.3390/v11111039.

Lu CY, Hour MJ, Wang CY, Huang SH, Mu WX, Chang YC, Lin CW. Single-round infectious particle antiviral screening assays for the Japanese encephalitis virus. Viruses. 2017;9:1–16. https://doi.org/10.3390/v9040076.

Makino Y, Suzuki T, Hasebe R, Kimura T, Maeda A, Takahashi H, Sawa H. Establishment of tracking system for West Nile virus entry and evidence of microtubule involvement in particle transport. J Virol Methods. 2014;195:250–7. https://doi.org/10.1016/j.jviromet.2013.10.002.

Martinez-Lopez A, Persaud M, Chavez MP, Zhang H, Rong L, Liu S, Wang TT, Sarafianos SG, Diaz-Griffero F. Glycosylated diphyllin as a broad-spectrum antiviral agent against Zika virus. EBioMedicine. 2019;47:269–83. https://doi.org/10.1016/j.ebiom.2019.08.060.

Martins LC, Da Silva EVP, Casseb LMN, Da Silva SP, Cruz ACR, De Sousa Pantoja JA, De Almeida Medeiros DB, Filho AJM, Da Cruz EDRM, De Araújo MTF, Cardoso JF, Da Cunha MACR, Almada GL, Romano APM, Santos MGDP, Rodrigues GAP, Chiang JO, Quaresma JAS, Carvalho VL, Da Costa Vasconcelos PF. First isolation of west nile virus in brazil. Mem Inst Oswaldo Cruz. 2019;114:1–7. https://doi.org/10.1590/0074-02760180332.

Mattia K, Puffer BA, Williams KL, Gonzalez R, Murray M, Sluzas E, Pagano D, Ajith S, Bower M, Berdougo E, Harris E, Doranz BJ. Dengue reporter virus particles for measuring neutralizing antibodies against each of the four dengue serotypes. PLoS ONE. 2011;6. https://doi.org/10.1371/journal.pone.0027252.

McFadden MJ, Mitchell-Dick A, Vazquez C, Roder AE, Labagnara KF, McMahon JJ, Silver DL, Horner SM. A fluorescent cell-based system for imaging zika virus infection in real-time. Viruses. 2018;10:13–8. https://doi.org/10.3390/v10020095.

Medin CL, Valois S, Patkar CG, Rothman AL. A plasmid-based reporter system for live cell imaging of dengue virus infected cells. J Virol Methods. 2015;211:55–62. https://doi.org/10.1016/j.jviromet.2014.10.010.

Miorin L, Romero-Brey I, Maiuri P, Hoppe S, Krijnse-Locker J, Bartenschlager R, Marcello A. Three-dimensional architecture of Tick-borne encephalitis virus replication sites and trafficking of the replicated RNA. J Virol. 2013;87:6469–81. https://doi.org/10.1128/jvi.03456-12.

Mutso M, Saul S, Rausalu K, Susova O, Žusinaite E, Mahalingam S, Merits A. Reverse genetic system, genetically stable reporter viruses and packaged subgenomic replicon based on a Brazilian zika virus isolate. J Gen Virol. 2017;98:2712–24. https://doi.org/10.1099/jgv.0.000938.

Nour AM, Li Y, Wolenski J, Modis Y. Viral membrane fusion and nucleocapsid delivery into the cytoplasm are distinct events in some flaviviruses. PLoS Pathog. 2013;9. https://doi.org/10.1371/journal.ppat.1003585.

Pang X, Zhang M, Dayton AI. Development of Dengue virus type 2 replicons capable of prolonged expression in host cells. BMC Microbiol. 2001;1:1–7. https://doi.org/10.1186/1471-2180-1-18.

Pierson TC, Diamond MS. The continued threat of emerging flaviviruses. Nat Microbiol. 2020;5:736–812. https://doi.org/10.1038/s41564-020-0714-0.

Pierson TC, Diamond MS, Ahmed AA, Valentine LE, Davis CW, Samuel MA, Hanna SL, Puffer BA, Doms RW. An infectious West Nile virus that expresses a GFP reporter gene. Virology. 2005;334:28–40. https://doi.org/10.1016/j.virol.2005.01.021.

Pierson TC, Sánchez MD, Puffer BA, Ahmed AA, Geiss BJ, Valentine LE, Altamura LA, Diamond MS, Doms RW. A rapid and quantitative assay for measuring antibody-mediated neutralization of West Nile virus infection. Virology. 2006;346:53–65. https://doi.org/10.1016/j.virol.2005.10.030.

Raquin V, Wannagat M, Zouache K, Legras-Lachuer C, Moro CV, Mavingui P. Detection of dengue group viruses by fluorescence in situ hybridization. Parasites and Vectors. 2012;5:1. https://doi.org/10.1186/1756-3305-5-243.

Ricciardi-Jorge T, Bordignon J, Koishi A, Zanluca C, Mosimann AL, Duarte Dos Santos CN. Development of a quantitative NS1-capture enzyme-linked immunosorbent assay for early detection of yellow fever virus infection. Sci Rep. 2017;7:1–9. https://doi.org/10.1038/s41598-017-16231-6.

Roby JA, Pijlman GP, Wilusz J, Khromykh AA. Noncoding subgenomic flavivirus RNA: Multiple functions in West Nile virus pathogenesis and modulation of host responses. Viruses. 2014;6:404–27. https://doi.org/10.3390/v6020404.

Rothan HA, Kumar M. Role of endoplasmic reticulum-associated proteins in flavivirus replication and assembly complexes. Pathogens. 2019;8. https://doi.org/10.3390/pathogens8030148.

Rudkin GT, Stollar BD. High resolution detection of DNA-RNA hybrids in situ by indirect immunofluorescence. Nature. 1977;265:472–3. https://doi.org/10.1038/265472a0.

Růžek D, Vancová M, Tesařová M, Ahantarig A, Kopecký J, Grubhoffer L. Morphological changes in human neural cells following Tick-borne encephalitis virus infection. J Gen Virol. 2009;90:1649–58. https://doi.org/10.1099/vir.0.010058-0.

Sakin V, Paci G, Lemke EA, Müller B. Labeling of virus components for advanced, quantitative imaging analyses. FEBS Lett. 2016;590:1896–914. https://doi.org/10.1002/1873-3468.12131.

Schmid B, Rinas M, Ruggieri A, Acosta EG, Bartenschlager M, Reuter A, Fischl W, Harder N, Bergeest JP, Flossdorf M, Rohr K, Höfer T, Bartenschlager R. Live cell analysis and mathematical modeling identify determinants of attenuation of dengue virus 2'-O-methylation mutant. PLoS Pathog. 2015;11:1–36.

Schoggins JW, Dorner M, Feulner M, Imanaka N, Murphy MY, Ploss A, Rice CM. Dengue reporter viruses reveal viral dynamics in interferon receptor-deficient mice and sensitivity to interferon effectors in vitro. Proc Natl Acad Sci USA. 2012;109:14610–5. https://doi.org/10.1073/pnas.1212379109.

Sheng ZY, Gao N, Wang ZY, Cui XY, Zhou DS, Fan DY, Chen H, Wang PG, An J. Sertoli cells are susceptible to ZIKV infection in mouse testis. Front Cell Infect Microbiol. 2017;7:1–13. https://doi.org/10.3389/fcimb.2017.00272.

Shi PY, Tilgner M, Lo MK. Construction and characterization of subgenomic replicons of New York strain of West Nile virus. Virology. 2002;296:219–33. https://doi.org/10.1006/viro.2002.1453.

Sinigaglia L, Gracias S, Décembre E, Fritz M, Bruni D, Smith N, Herbeuval JP, Martin A, Dreux M, Tangy F, Jouvenet N. Immature particles and capsid-free viral RNA produced by Yellow fever virus-infected cells stimulate plasmacytoid dendritic cells to secrete interferons. Sci Rep. 2018;8:1–15. https://doi.org/10.1038/s41598-018-29235-7.

Slon Campos JL, Poggianella M, Marchese S, Mossenta M, Rana J, Arnoldi F, Bestagno M, Burrone OR. DNA-immunisation with dengue virus E protein domains I/II, but not domain III, enhances Zika, West Nile and yellow fever virus infection. PLoS ONE. 2017;12:1–19. https://doi.org/10.1371/journal.pone.0181734.

Suphatrakul A, Duangchinda T, Jupatanakul N, Prasittisa K, Onnome S, Pengon J, Siridechadilok B. Multi-color fluorescent reporter dengue viruses with improved stability for analysis of a multi-virus infection. PLoS ONE. 2018;13:1–19. https://doi.org/10.1371/journal.pone.0194399.

Usme-Ciro JA, Lopera JA, Alvarez DA, Enjuanes L, Almazán F. Generation of a DNA-launched reporter replicon based on Dengue virus type 2 as a multipurpose platform. Intervirology. 2017;59:275–82. https://doi.org/10.1159/000476066.

Velado Fernández I, Okamoto N, Ito A, Fukuda M, Someya A, Nishino Y, Sasaki N, Maeda A. Development of a novel protocol for generating flavivirus reporter particles. J Virol Methods. 2014;208:96–101. https://doi.org/10.1016/j.jviromet.2014.08.002.

Vyboh K, Ajamian L, Mouland AJ. Detection of viral RNA by fluorescence in situ hybridization (FISH). J Vis Exp. 2012;2–6. https://doi.org/10.3791/4002.

Westaway EG, Mackenzie JM, Kenney MT, Jones MK, Khromykh AA. Ultrastructure of Kunjin virus-infected cells: colocalization of NS1 and NS3 with double-stranded RNA, and of NS2B with NS3, in virus-induced membrane structures. J Virol. 1997;71:6650–61. https://doi.org/10.1128/jvi.71.9.6650-6661.1997.

Wu P, Nicholls SB, Hardy JA. A Tunable, modular approach to fluorescent protease-activated reporters. Biophys J. 2013;104:1605–14. https://doi.org/10.1016/j.bpj.2013.01.058.

Yoshii K, Ikawa A, Chiba Y, Omori Y, Maeda J, Murata R, Kariwa H, Takashima I. Establishment of a neutralization test involving reporter gene-expressing virus-like particles of tick-borne encephalitis virus. J Virol Methods. 2009;161:173–6. https://doi.org/10.1016/j.jviromet.2009.05.016.

Zhang S, Tan C, Ooi E. Visualizing dengue virus through Alexa fluor labeling. J Vis Exp. 2011;8–11. https://doi.org/10.3791/3168.

Zhang SLX, Tan HC, Hanson BJ, Ooi EE. A simple method for Alexa Fluor dye labelling of dengue virus. J Virol Methods. 2010;167:172–7. https://doi.org/10.1016/j.jviromet.2010.04.001.

Zuza AL, Barros HLS, de Mattos Silva Oliveira TF, Chávez-Pavoni JH, Zanon RG. Astrocyte response to St. Louis encephalitis virus. Virus Res. 2016;217:92–100. https://doi.org/https://doi.org/10.1016/j.virusres.2016.03.005.

The Genetics of Post-polio Syndrome—An Overview

Radha Saraswathy

Abstract Post-polio syndrome (PPS) was introduced by Halstead in 1986 for a collection of symptoms in polio affected individuals several years after infection by polio virus. Now, despite considerable advancements in medical sciences, major global health problems induced by human viral infections causing morbidity and mortality in humans remain. Although genetics has metamorphosed and millions of PPS survivors are globally living, a genotype–phenotype correlation study has to be carried out for the PPS management in the patients to lead a healthy and meaningful life. This also can lead to eradication of polio completely. In this chapter an overview of the genetics of PPS is presented.

Keywords Post-polio syndrome · Alanine (Ala) 67 Threonine (Thr) polymorphism · Cytogenetic · Chromosome damage · Polio virus · Muscular atrophy · Genomic medicine · Poliovirus receptor · PCR RFLP · *CD155/PVR* gene · Cytokinesis blocked micronuclei · Poliomyelitis

1 Introduction

Poliomyelitis, now commonly known as Polio, appears to be one of the oldest human infectious diseases caused by the polio virus—a member of the genus Enterovirus. Started from gut it moves to the central nervous system and leads to muscle weakness. Globally, as many as 20 million people continue to live with the disabling consequences of the disease (Groce et al. 2014), and which is mainly prevailing in the sub-continents of Asian countries. Polio virus is known to infect humans, although other primates expressing CD155 polio virus receptors can also be infected.

R. Saraswathy (✉)
120TT Biomedical Genetics Research Lab (BMGRL), Department of Biomedical Sciences, School of Biosciences & Technology, VIT University, Vellore 632014, Tamil Nadu, India
e-mail: radhasaraswathy60@gmail.com; radhasaraswathy@vit.ac.in

S. I. Ahmad (ed.), *Human Viruses: Diseases, Treatments and Vaccines*, https://doi.org/10.1007/978-3-030-71165-8_35

Polio is a public health problem, for the survivors it remains a concern at least in the developing countries (Gonzalez et al. 2010), and good care of these polio survivors is required so that they can spend a normal life as much as possible. As a consequence of polio, limb paralysis results due to an extensive damage to the anterior horn cells of the spinal cord, including joint deterioration, malformation of the bones, accelerating muscle atrophy due to motor neuron degeneration, fatigue and hence, decreased physical activity leading to minimal active life.

Post-polio syndrome (PPS, 2020 ICD-10-CM Diagnosis Code G14) is a collection of neurological and muscular malfunctioning observed in polio affected patients several years after of infection (Dalakas 1995; Lo and Robinson 2018; Kay et al. 2018; Oluwasanmi et al. 2019). It is also referred as 'late effects of polio' (LEoP) and "post-polio muscular atrophy".

1.1 Genomic Structure of the Virus and Its Pathogenesis

Among the RNA and DNA viruses, RNA viruses and especially retroviruses are considered to be most dangerous. The infectivity and infection of RNA viruses are enhanced due to its recombinational properties which were discovered by Hirst (1962) and Ledinko (1963) followed by the biological properties and molecular basis which were studied in other picornaviruses (Pringle 1965), (+) RNA viruses (Lai 1992; Simon and Bujarski 1994) and retroviruses (Coffin 1979; Zhang and Temin 1993) to understand certain molecular mechanisms of Polio virus.

Despite the fact that enormous advances in medical sciences have occurred, increase in major global health problems due to viral infections and the prevalence of morbidity and mortality are continuing (Whitton and Oldstone 1996). The suggestion is that there is a dire need to understand the mechanisms of genome evolution of the RNA viruses for better understanding of the viral diseases (Figlerowicz et al. 2003). We are aware that genetics and genomic medicine has revolutionized globally; yet much still remains to be unravelled in polio survivors (Saraswathy 2016). In this chapter the genetics of PPS is presented.

2 Molecular Genetics of Polio Virus

2.1 Poliovirus Receptor (PVR)/CD155

Poliovirus is an enterovirus belongs to family Picornaviridae that includes human and animal pathogens. Its RNA genome, a single-stranded (+)-strand contains approximately of 7,500 nucleotides, enclosed in a non-enveloped capsid comprised of 60 copies of four different polypeptides arranged in icosahedral symmetry. The PVR/CD155 is a cell surface receptor for all three serotypes. CD155 consists of three extracellular immunoglobulin-like domains: (1) membrane-distal V-type domain that binds poliovirus, (2) two C2-type domains. CD155 is a recognition molecule for natural killer (NK) cells and it interacts with CD226 and CD96 on NK cells to stimulate their cytotoxicity (Bottino et al. 2003; Fuchs et al. 2004). The type I transmembrane glycoprotein is a member in the immunoglobulin super family of proteins present as receptor on the cell surface (Mendelsohn et al. 1989) and located on chromosome 19q13.2 (https://www.ncbi.nlm.nih.gov/gene/5817).

The gene locations from one end to another are as follows: in region 1 includes the domain 1 which is involved in the synthesis of positive and negative strands during genome replication. The region 2 includes domain 2–7 as VP4, VP2, VP3 and VP1 and is responsible for internal ribosomal entry site. The region 2A includes 2A 2B, 2C 3A/3B 3C and 3D, responsible for the synthesis of non-structural proteins, protease, polymerase, helicase and host shut off proteins. Finally domain 3UTR is poll (A) tail to terminate the genome's round of functions (Baj et al. 2015).

2.2 *Alanine (Ala) 67 Threonine (Thr) Polymorphism*

DNA polymorphisms in the PVR are associated with poliovirus infection in cell culture. Earlier studies had reported immunodeficiency and HLA alleles as factors for poliomyelitis. However, Picard et al. (2006) reported that mutations in specific genes in the virus receptors are predisposed to infection with specific microorganisms and variations in certain genes related to resistance to particular pathogens. It was observed that certain individuals developed paralytic poliomyelitis whereas others were asymptomatic which lead to a hypothesis that some genetic factors might contribute to the disease (Kindberg et al. 2009). Thus, two nucleotides (G/A) have been found at position 199 in the mRNA of CD155 resulting in an Ala or a Thr amino acid found at position 67 in human epithelial cells and thus leading to an increased susceptibility to poliomyelitis with the Ala67Thr mutation established. Ala67Thr mutation also has been reported in the human neuroblastoma cells infected with PV and an increased resistance exhibited against polio virus inducing cell lysis and apoptosis (Pavio et al. 2000).

Alanine (Ala) 67 Threonine is a neuropeptide that stimulates food intake through inhibition of central melanocortin receptors. There have been reports on genetic susceptibility associated with this neuropeptide and infectious diseases such as the progressive muscular atrophy which leads to a significant increased frequency of heterozygous Ala67Thr/*PVR gene polymorphism as compared to* controls (Chapman and Hill 2012; Zhang et al. 2014; Saunderson et al. 2004). This report encouraged more researchers to work in this field. Thus, in Italian subjects with neuromuscular weakness, using PCR RFLP and sequencing techniques, Kindberg et al. (2009) reported that the heterozygous poliovirus receptor Ala67Thr genotype was present in 13.3% of them tested and 8.5% of the controls (Odds Ratio = 1.667) This polymorphism in PVR gene was concluded to be a risk factor for poliomyelitis patients. Furthermore, Nandi et al. (2016) developed a SNP assay to screen population for this polymorphism and the single nucleotide substitution (GCG → ACG) in *CD155/PVR* gene was recorded. This assay showed the carrier rate of heterozygous polymorphism (Ala67Thr) in 150 DNA samples and the frequency was 27% higher than that of the controls reported. Polymorphism in PVR gene has also been reported in several motor neuron diseases (Saunderson et al. 2004) such as progressive muscular atrophy (PMA) and amyotrophic lateral sclerosis (ALS) and also considered as a tumour marker (KučanBrlić et al. 2019; Molfetta et al. 2020) and in anti tumour therapy. This SNP assay approach could be used for population screening to analyse the heterozygous carrier as an important tool in Polio disease as well the others shown above.

3 Cytogenetic Studies of Polio Virus

The cytogenetic techniques in human biology were developed in 1960s but there are no reports of such study carried out on PPS. It is known that DNA and RNA viruses can cause human chromosome damage after infecting their cells. Bartsch et al. (1967) reported chromosome breaks in human cells lines by polio virus infection. Other in vitro studies showed considerable chromosome damage in leucocytes of patients with measles (Nichols et al. 1965), with mumps (Aula 1965; Gripenberg 1965), chromosome pulverization in HeLa cells infected with mumps and Sendai viruses (Cantell et al. 1966), patients with yellow fever virus (Harnden 1964), hepatitis patients (Aya and Makino 1966; El Alfi and Smith 1965; Emerit and Emerit 1972), patients with congenital rubella (Kuroki et al. 1966; Makino and Aya 1968; Nusbacher et al. 1967), in vitro human cells infected with rubella virus (Boué and Boué 1969; Chang et al. 1966) and in PPS (Bhattacharya et al. 2011). However, no chromosome rearrangements was observed in the human leukocyte chromosomes infected by rubella virus (Chang et al. 1966).

In PPS patients, the frequency of chromosomal damage was measured by analysing the chromosomal aberrations, sister chromatid exchanges (SCE) and cytokinesis blocked micronuclei (CBMN) assays (Bhattacharya et al. 2011). A significant increase in the frequency of chromosomal anomalies was reported in the PPS than in the controls. However a larger sample size has to be screened and analysed to come to any conclusion. A specific chromosome rearrangement if established could also serve as a marker for screening for PVR gene in individuals in a large population.

4 Treatment

Ironically there is no cure for polio disease. Patients, however, are provided relief of symptoms to reduce the rate of complications and wherever possible speedy recovery. The patients who sustain weak muscle ought to be cared by providing updated latest technologically gadgets to evade the disability. Also moderate exercise and better nutritional diets is recommended. Other cares to be provided when available include rehabilitation, physical and occupational therapies, appropriate special shoes, braces and in rare cases orthopedic surgery.

5 Conclusion

Based on the limited molecular and cytogenetic studies on PPS, a large scale screening in various populations is recommended using the SNP approach to identify the polymorphism and to analyse for any specific chromosome anomalies which could serve as a marker. Also microarray approach is recommended for identifying specific variant in the population. Also great care should be taken of the PPS survivors so that they lead a normal life. The genetic mechanisms of this viral disease and molecular cross talk between host and the viral infection have been established to some extent. Further microarray analysis and role of epigenetic might throw light on the predisposition of individuals to specific variant in a particular population.

Acknowledgements The author is grateful to VIT University, Vellore for providing all facilities to contribute this chapter and to carry out genetics analysis in PPS patients shared here.

References

Aula P. Virus associated chromosome breakage. A cytogenetic study of chicken pox, measles and mumps patients and of cell cultures infected with measles virus. AM Acad Sci Fenn, Ser A, IV, Biol. 1965;89:1–78.

Aya T, Makino S. Notes on chromosome abnormalities in cultured leucocytes from serum hepatitis patients. Proc Jap Acad. 1966;42:648–53.

Baj A, Colombo M, Headley JL, et al. Post-poliomyelitis syndrome as a possible viral disease. Int J Infect Dis. 2015;35:107–16.

Bartsch HD, Habermehl KO, Diefentahl W. Chromosomal damage after infection with poliomyelitis virus. Exp Cell Res. 1967;48:671–5.

Bhattacharya SK, Saraswathy R, Sivakumar E. Genotoxic assessment in peripheral blood lymphocytes of post-polio individuals using sister chromatid exchange analysis & micronucleus assay. Hum Exp Toxicol. 2011;30:636–48.

Bottino C, Castriconi R, Pende D, et al. Identification of PVR (CD155) and Nectin-2 (CD112) as cell surface ligands for the human DNAM-1 (CD226) activating molecule. J Exp Med. 2003;198(4):557–67.

Boué A, Boué JG. Effects of Rubella virus infection on the division of human cells. Am J Dis Child. 1969;118(1):45–8.

Cantell K, Saksela E, Aula P. Virological studies on chromosome damage of HeLa cells induced by myxoviruses. Ann Med Exp Biol Fenn. 1966;44:255–259.

Chang TH, Moorhead PS, Boué JG, et al. Chromosome studies of human cells infected in utero and in vitro with rubella virus. Proc Soc Exp Biol Med. 1966;122(1):236–43.

Chapman SJ, Hill AV. Human genetic susceptibility to infectious disease. Nat Rev Genet. 2012;13:175–88.
Coffin JM. Structure, replication, and recombination of retrovirus genomes: some unifying hypotheses. J Gen Virol. 1979;199:47–59.
Dalakas MC. The post-polio syndrome as an evolved clinical entity: definition and clinical description. Ann NY Acad Sci. 1995;753(1):68–80.
El Alfi OS, Smith PM, Biesele JJ. Chromosomal breaks in human leukocyte cultures induced by an agent in the plasma of infectious hepatitis patients. Hereditas. 1965;52:285–94.
Emerit I, Emerit J. Virus induced chromosome breakage. Lancet. 1972;2:494.
Figlerowicz M, Alejska M, Kurzyńska-Kokorniak A, et al. Genetic variability: the key problem in the prevention and therapy of RNA-based virus infections. Med. Res. Rev. 2003;23 (4):488-518.
Fuchs A, Cella M, Giurisato E, et al. Cutting edge: CD96 (tactile) promotes NK cell-target cell adhesion by interacting with the poliovirus receptor (CD155). J Immunol. 2004;172(7):3994–8.
Gonzalez H, Olsson T, Borg K. Management of postpolio syndrome. Lancet Neurol. 2010;9 (6):634–42.
Gripenberg U. Chromosome studies in some virus infections. Hereditas. 1965;54(1):1–18.
Groce NE, Banks LM, Stein MA. Surviving polio in a post-polio world. Soc Sci Med. 2014;107:171–8.
Harnden DG. Cytogenetic studies on patients with virus infections and subjects vaccinated against yellow fever. Am J Hum Genet. 1964;16:204–13.
Hirst GK. Genetic recombination with Newcastle disease virus, poliovirus, and influenza. Cold Spring Harbor Symp Quant Biol. 1962;27:303–9.
Kay L, Nielsen NM, Wanscher B, et al. Neurological symptoms in Danes with a history of poliomyelitis: lifelong follow-up of late symptoms, their association with initial symptoms of polio, and presence of postpolio syndrome. Eur Neurol. 2018;80(5–6):295–303.
Kindberg E, Ax C, Fiore L, Svensson L. Ala67Thr mutation in the poliovirus receptor CD155 is a potential risk factor for vaccine and wild-type paralytic poliomyelitis. J Med Virol. 2009;81 (5):933–6.
KučanBrlić P, LenacRoviš T, Cinamon G, et al. Targeting PVR (CD155) and its receptors in anti-tumor therapy. Cell MolImmunol. 2019;16(1):40–52.
Kuroki Y, Makino S, Aya T, Nagayama T. Chromosome abnormalities in cultured leucocytes from rubella patients Jap. J Hum Genet. 1966;11:17–23.
Lai MMC. RNA recombination in animal and plant viruses. Microbiol Rev. 1992;56:61–79.
Ledinko N. Genetic recombination with poliovirus type 1 studies of crosses between a normal horse serumresistant mutant and several guanidine-resistant mutants of the same strain. Virology. 1963;20:107–19.
Lo JK, Robinson LR. Post-polio syndrome and the late effects of poliomyelitis: Part 2. Treatment, management, and prognosis, Muscle Nerve. 2018;58(6):760–9.
Makino S, Aya T. Cytogenetic studies in leukocyte cultures from patients with some viral diseases and in those infected with HSV. Cytologia (Tokyo). 1968;33(3):370–96.
Mendelsohn CL, Wimmer E, Racaniello VR. Cellular receptor for poliovirus: molecular cloning, nucleotide sequence, and expression of a new member of the immunoglobulin superfamily. Cell. 1989;56(5):855–65.
Molfetta R, Zitti B, Lecce M, Milito ND, et al. CD155: a multi-functional molecule in tumor progression. Int J Mol Sci. 2020;21(3):922.
Nandi SS, Sharma DK, Deshpande JM. Assay for identification of heterozygous single-nucleotide polymorphism (Ala67Thr) in human poliovirus receptor gene. Indian J Med Res. 2016;144 (1):38–45.
Nichols WW, Levan, Aula P, Norrby E. Chromosome damage associated with the measles virus in vitro. Hereditas. 1965;54:101–18.
Nusbacher J, Hirschhorn K, Cooper LZ. Chromosomal abnormalities in congenital rubella. New Eng J Med. 1967;276:1409–13.

Oluwasanmi OJ, Mckenzie DA, Adewole IO, et al. Postpolio syndrome: a review of lived experiences of patients. Int J Appl Basic Med Res. 2019;9(3):129–34.

Pavio N, Couderc T, Girard S, et al. Expression of mutated poliovirus receptors in human neuroblastoma cells persistently infected with poliovirus. Virology. 2000;274:331–42.

Picard C, Casanova JL, Abel L. Mendelian traits that confer predisposition or resistance to specific infections in humans. Curr Opin Immunol. 2006;18(4):383–90.

Pringle CR. Evidence of genetic recombination in foot-and-mouth disease virus. Virology. 1965;25:48–54.

Saraswathy R. The genetics of post-polio syndrome - much to be unravelled! Indian J Med Res. 2016;144(1):9–10.

Saunderson R, Yu B, Trent RJ, et al. A polymorphism in the poliovirus receptor gene differs in motor neuron disease. Neuro Rep. 2004;15:383–6.

Simon AE, Bujarski JJ. RNA–RNA recombination and evolution in virus-infected plants. Annu Rev Phytopathol. 1994;32:337–62.

Whitton JL, Oldstone MBA, et al. Immune response to virus. In: Fields BN, Knipe DM, Howley PM, et al., editors. Fields virology. 3rd ed. Philadelphia: Lippincott-Raven Publishers; 1996. p. 345–74.

Zhang J, Temin HM. Rate and mechanism of nonhomologous recombination during a single cycle of retroviral replication. Science. 1993;259:234–8.

Zhang X, Xu H, Chen X, et al. Association of functional polymorphisms in the MxA gene with susceptibility to enterovirus 71 infection. Hum Genet. 2014;133:187–97.

Index

S. I. Ahmad (ed.), *Human Viruses: Diseases, Treatments and Vaccines*, https://doi.org/10.1007/978-3-030-71165-8